2019 International Conference on Simulation of Semiconductor Processes and Devices (SISPAD 2019)

Udine, Italy
4 – 6 September 2019

IEEE Catalog Number: CFP19SSD-POD
ISBN: 978-1-7281-0941-1

**Copyright © 2019 by the Institute of Electrical and Electronics Engineers, Inc.
All Rights Reserved**

Copyright and Reprint Permissions: Abstracting is permitted with credit to the source. Libraries are permitted to photocopy beyond the limit of U.S. copyright law for private use of patrons those articles in this volume that carry a code at the bottom of the first page, provided the per-copy fee indicated in the code is paid through Copyright Clearance Center, 222 Rosewood Drive, Danvers, MA 01923.

For other copying, reprint or republication permission, write to IEEE Copyrights Manager, IEEE Service Center, 445 Hoes Lane, Piscataway, NJ 08854. All rights reserved.

*** *This is a print representation of what appears in the IEEE Digital Library. Some format issues inherent in the e-media version may also appear in this print version.*

IEEE Catalog Number:	CFP19SSD-POD
ISBN (Print-On-Demand):	978-1-7281-0941-1
ISBN (Online):	978-1-7281-0940-4
ISSN:	1946-1569

Additional Copies of This Publication Are Available From:

Curran Associates, Inc
57 Morehouse Lane
Red Hook, NY 12571 USA
Phone: (845) 758-0400
Fax: (845) 758-2633
E-mail: curran@proceedings.com
Web: www.proceedings.com

Wednesday September 4

09.00-09.10	**Conference opening**

09.10-10.00
Room T9   **Plenary invited**
Neuro-Inspired Computing with Nanoelectronic Devices: Experimental Progressesand Modeling Opportunities
Shimeng Yu,
Georgia Institute of Technology

10.00-10.20	**Coffee break**

Room T9   **Session 1**
Reliability of devices and interconnects
Chairpersons: L. Filipovic, Technische Universität Wien,Austria, F. Driussi, University of Udine

10.20-10.40   1.1 - 3D Kinetic Monte Carlo Simulation of Electromigration in Multi-layer Interconnects   1
Linlin Cai, Wangyong Chen, Xing Zhang, Yudi Zhao, and XiaoyanLiu
Institute of Microelectronics, Peking University, Beijing, China

10.40-11.00   1.2 - Metallic ions drift in hybrid bonding integration modeling, towards the evolution of failure criterion   5
Manzanarez Hervé, Moreau Stéphane, Cueto Olga
Univ. Grenoble Alpes, CEA, LETI, Grenoble France.

11.00-11.20   1.3 - TCAD Framework to Estimate the NBTI Degradation in FinFET and GAANSFET Under Mechanical Strain   9
Ravi Tiwari[1], Narendra Parihar[1], Karansingh Thakor[1], Hiu-Yung Wong[2], andSouvik Mahapatra[1]
[1]Department of Electrical Engineering, Indian Institute of Technology Bombay, Mumbai, India
[2]San Jose State University, San Jose, CA, USA

11.20-11.40   1.4 - A Stochastic Hole Trapping-Detrapping Framework for NBTI, TDDS and RTN   13
Sharang Bhagdikar and Souvik Mahapatra
Department of Electrical Engineering, Indian Institute of Technology Bombay, Mumbai, India

11.40-12.00   1.5 - TCAD-Enabled Machine Learning Defect Prediction to Accelerate Advanced Semiconductor Device Failure Analysis   17
Chea-Wei Teo[1 2], Kain Lu Low[1], Vinod Narang[2], and Aaron Voon-YewThean[1],
[1]Department of Electrical and Computer Engineering, National University of Singapore, Singapore
[2]Device Analysis Lab, Advanced Micro Devices Pte Ltd,Singapore

12.00-12.20   1.6 - TCAD Augmented Machine Learning for Semiconductor Device FailureTrouble shooting and Reverse Engineering   21
Y.S. Bankapalli and H. Y. Wong
Electrical Engineering, San Jose State University, San Jose, CA, USA

Room T4   **Session 2**
Advanced methods for numerical calculations
Chairpersons: A. Godoy, University of Granada, Spain, D. Connelly, Atomera,USA

10.20-10.40   2.1 - On the Simulation of Plasma Waves in HEMTs and the Dyakonov-Shur Instability   25
Christoph Jungemann, Tobias Linn, Zeinab Kargar
Chair of Electromagnetic Theory, RWTH Aachen University, Aachen,Germany

10.40-11.00   2.2 - Exact Correction of the Self-Force Problem in Monte Carlo DeviceSimulation   29
Andrea Ghetti
Technology Development, Micron Technology Inc., Vimercate, Italy

11.00-11.20   2.3 - A Robust Simulation Method for Breakdown with Voltage BoundaryCondition Utilizing Negative Time Constant Information   33
Shigetaka Kumashiro, Tatsuya Kamei, Akira Hiroki, Kazutosi Kobayashi
Kyoto Institute of Technology, Matsugasaki, Sakyo-ku, Kyoto, Japan

SISPAD 2019
Wednesday-Friday, 4-6 September

| 11.20-11.40 | 2.4 - Extending the Numerov Process to the Semiconductor Transport Equations | 37 |

11.20-11.40   2.4 - Extending the Numerov Process to the Semiconductor Transport Equations   37
Nicolò Speciale[1], Rossella Brunetti[2], Massimo Rudan[1]
[1]"E. De Castro" Advanced Research Center on Electronic Systems (ARCES) and Department DEI, University of Bologna, [2]FIM Department, University of Modena and Reggio Emilia, Modena

11.40-12.00   2.5 - Implementation of Automatic Differentiation to Python-based Semiconductor Device Simulator   41
Tsutomu Ikegami, Koichi Fukuda, Junichi Hattori
National Institute of Advanced Industrial, Science and Technology (AIST), Tsukuba, Japan

12.00-12.20   2.6 - Deep Neural Network for Generation of the Initial Electrostatic Potential Profile   45
Seung-Cheol Han and Sung-Min Hong
School of Electrical Engineering and Computer Science, Gwangju Institute of Science and Technology, Gwangju, Republic of Korea

12.20-13.30   **Lunch**

Room T9   **Session 3**
Memories
Chairperson: Seong-dong Kim, SK Hynix, South Korea

13.30-14.00   Invited: The evolution of TCAD as virtual design solutions: Fully atomistic TCAD, TCAD assisted AI, TCAD mediated DTCO and real-time TCAD
Dae Sin Kim
Samsung Electronics, Hwasung-si, Republic of Korea

14.00-14.20   3.1 - Multiscale Modeling of Charge Trapping in Molecule Based Flash Memories   49
Oves Badami[1], Toufik Sadi[2], Vihar Georgiev[1], Fikru Adamu-Lema[1], Vasanthan Thiruna vukkarasu[1], Jie Ding[3], Asen Asenov[1]
[1]School of Engineering, University of Glasgow, Glasgow, United Kingdom, [2]School of Science, Aalto University, Aalto, Finland, [3]College of Electrical and Power Engineering, Taiyuan University of Technology, China

14.20-14.40   3.2 - TCAD Model for Ag-GeSe -Ni CBRAM Devices   53
Kiraneswar Muthuseenu[1], E. Carl Hylin[2], Hugh J. Barnaby[1], Priyanka Apsangi[1], Michael N. Kozicki[1], Garrett Schlenvogt[2], Mark Townsend[2]
[1]School of Electrical, Computer and Energy, Engineering, Arizona State University, Tempe, AZ, USA
[2]Silvaco Inc., North Chelmsford, MA, USA

14.40-15.00   3.3 - Comprehensive Comparison of Switching Models for Perpendicular Spin-Transfer Torque MRAM Cells   57
Simone Fiorentini[1], Roberto Orio[1], Wolfgang Goes[2], Johannes Ender[1], Viktor Sverdlov[1]
[1]Christian Doppler Laboratory for Magnetoresistive Nonvolatile Memory and Logic Institute for Microelectronics, TU Wien Vienna, Austria, [2]Silvaco Europe Ltd Cambridge, United Kingdom

15.00-15.20   3.4 - 3D TCAD Model for Poly-Si Channel Current and Variability in Vertical NAND Flash Memory   61
D. Verreck[1], A. Arreghini[1], F. Schanovsky[2], Z. Stanojevic[2], K. Steiner[2], F. Mitterbauer[2], M. Karner[2], G. Van den bosch[1], A. Furnemont[1],
[1]imec, Leuven, Belgium, [2]Global TCAD Solutions GmbH., Vienna, Austria

15.20-15.40   3.5 - Optimization of select gate transistor in advanced 3D NAND memory cell   65
Jin Cho, Derek Kimpton, Eric Guichard
Silvaco, Inc, Santa Clara, CA, USA

Room T4   **Session 4**
Quantum transport in nanoscale devices
Chairperson: W. Vandenberghe, University of Texas at Dallas, USA

13.40-14.00   4.1 - A Hybrid Mode-Space/Real-Space Scheme for DFT+NEGF Device Simulations   69
F. Ducry, M. H. Bani-Hashemian, and M. Luisier
Integrated Systems Laboratory (ETH Zurich)

14.00-14.20   4.2 - Full band quantum transport modelling with EP and NEGF methods: application to nanowire transistors   73
M. Pala[1], D. Esseni[2]
[1]C2N, Universitè Paris-Saclay, Palaiseau, France, [2]DPIA, University of Udine, Udine, Italy

14.20-14.40   4.3 - A Quantum Element Reduced Order Model   77
Ming-C. Cheng
Dept. of ECE, Clarkson University, Potsdam, NY, USA

14.40-15.00   4.4 - Quantum Mechanical Simulations of the Impact of Surface Roughness on Nanowire TFET performance   81
Yunhe Guan[1], ZunChao Li[1], Hamilton Carrillo-Nuñez[2], Vihar P.Georgiev[2], Asen Asenov[2]
[1]School of Microelectronics, Xi'an Jiaotong University, Xi'an, China
[2]School of Engineering, University of Glasgow, Glasgow, UnitedKingdom

15.00-15.20   4.5 - Efficient Coupled-mode space based Non-Equilibrium Green's Function Approach for Modeling Quantum   85
Transport and Variability in Vertically Stacked SiNW FETs
V.Thirunavukkarasu[1], H. Carrillo-Nunez[1], F.D. Alema[1], S. Berrada[1], O. Badami[1], C. Medina-Bailón[1],
T.Datta[1], J. Lee[1], Y.Guen[2], V.Georgiev[1] and A. Asenov[1]
[1]School of Engineering, University of Glasgow, Glasgow, Scotland,UK.
[2]Department of Electrical Engineering, Xian Chiao Tung University, People Republic of China.

15.20-15.40   4.6 - Surface Roughness Scattering in NEGF using self-energy formulation   89
Oves Badami, Salim Berrada, Hamilton Carrillo-Nunez, Cristina Medina-Bailon, Vihar Georgiev, A. Asenov
School of Engineering, University of Glasgow, Glasgow, United Kingdom

15.40-16.00   **Coffee break**

16.00-17.30   **Poster Session**
Velario

P01 - Impact of BEOL Design on Self-heating and Reliability in Highly-scaled FinFETs   93
Jaehee Choi, Udit Monga, Yonghee Park, Hyewon Shim[1], Uihui Kwon, Sangwoo Pae[1], Dae Sin Kim
Semiconductor R&D Center, [1]Foundry Business, Samsung Electronics Co., Hwasung-si, Korea

P02 - Modeling 1/f and Lorenzian noise in III-V MOSFETs   97
E. Caruso[1], F.Bettetti[2], L. Del Linz[2], D. Pin[2], M. Segatto[2], P. Palestri[2]
[1]Tyndall National Institute, University College Cork, Cork,Ireland, [2] DPIA, University of Udine,Italy.

P03 - Investigation of TCAD Calibration for Saturation and Tail Current of 6.5kV IGBTs   101
Takeshi Suwa and Shigeaki Hayase
Toshiba Electronic Devices & Storage Corporation, Kawasaki, Japan

P04 - Simulation of Statistical NBTI Degradation in 10nm Doped Channel pFinFETs   105
F. Adamu-Lema, V.Georgiev, and A. Asenov,
Device Modeling Group, University of Glasgow, Glasgow,UK

P05 - A SPICE Compatible Compact Model for Process and Bias Dependence of HCD in HKMG FDSOI   109
MOSFETs
Uma Sharma and Souvik Mahapatra
Department of Electrical Engineering, Indian Institute of Technology Bombay, Mumbai, India

P06 - Quantum Transport Simulations of the Zero Temperature Coefficient in Gate-all-around Nanowire   113
pFETs
Hyeongu Lee, Junbeom Seo, Mincheol Shin
School of Electrical Engineering Korea Advanced Institute of Science and Technology Daejeon, Korea

P07 - Electro-Thermal Analysis and Edge Termination Techniques of High Current β-Ga2O3 Schottky   117
Rectifiers
Ribhu Sharma[1], Erin Patrick[2], Jiancheng Yang[3], Fan Ren[3], Mark Law[2], Stephen Pearton[1]
[1]Material Science and Engineering, University of Florida, Gainesville, USA, [2]Electrical and Computer
Engineering, University of Florida, Gainesville, Fl, USA, [3]Chemical Engineering) University of Florida,
Gainesville, Fl, USA

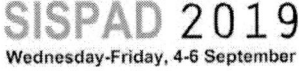

P08 - Thermal Conductivity of Silicon Nanowire Using Landauer Approach for Thermoelectric Applications 121
Ming-Yi Lee[1], Min-Hui Chuang[1], Yiming Li[1,3], Seiji Samukawa[2,3]
[1]Institute of Communications Engineering, National Chiao Tung University, Hsinchu, Taiwan
[2]Institute of Fluid Science and WPI-AIMR, Tohoku University, Aoba-ku, Sendai, Japan
[3]Center for mmWave Smart Radar Systems and Technologies, National Chiao Tung University, Taiwan

P09 - Simulation of Chemically Reacting Flow in Plasma Native Oxide Cleaning Process 125
Seung-Min Ryu[1], Yunho Kim[1], Dylan Pederson[1], Jonghyun Lee[2], Youngkwon Kim[2], Laxminarayan L. Raja[1], Jiho Uh[2], Sang-Jin Choi[2]
[1]Dept. of Aerospace Engineering and Engineering Mechanics, University of Texas at Austin, U.S.
[2]Memory Technology Innovation Team Samsung Electronics Hwaseong-si, South Korea

P10 - Simulation of deep level transient spectroscopy using circuit simulator with deep level trap model 129
implemented by Verilog-A language
Koichi Fukuda[1], Junichi Hattori[1], Hidehiro Asai[1], Mitsuaki Shimizu[1], Tamotsu Hashizume[2]
[1]National Institute Of Adivanced Industrial Science and Technology (AIST), Tsukuba, Japan
[2]Research Center for Integrated, Quantum Electronics (RCIQE), Hokkaido University, Hokkaido, Japan

P11 - A New Computer-Aided Calibration Technique of Physics Based IGBT & Power-Diode Compact 133
Models with Verilog-A Implementation
Arnab Biswas, Daniel Ludwig, Maria Cotorogea
Infineon Technologies AG, Neubiberg, Germany

P12 - First-principles investigation of paramagnetic centers in P2O5 based glasses Luigi Giacomazzi[1], Layla 137
Martin-Samos[2], Nicolas Richard[3], Matjaz Valant[1], Nadege Ollier[4]
[1]Materials Research Laboratory, (University of Nova Gorica), Nova Gorica, Slovenia, [2]CNR-IOM Democritos, Trieste, Italy, [3]DAM-DIF CEA, Arpajon, France, [4]LSI-IRAMIS, CEA, Palaiseau, France

P13 - Theoretical Study of the Edge Effect of Dumbbell-shape Graphene Nanoribbon with a Dual Electronic 141
Properties by First-principle Calculations
Qinqiang Zhang[1], Takuya Kudo[2], Jowesh Gounder[2], Ying Chen[3], Ken Suzuki[3], Hideo Miura[3]
[1]Dept. of Finemechanics, Tohoku University, Sendai, Japan, [2]Dept. of Finemechanics, Graduate School of Engineering, Sendai, Japan, [3]Fracture and Reliability Research Institute, Tohoku University, Sendai, Japan

P14 - A new wet etching method for black phosphorus layer number engineering: experiment, modeling and 145
DFT simulations
Teren Liu[1], Tao Fang[1,2,3], Karen Kavanagh[4], Hongyu Yu[5], Guangrui (Maggie) Xia[1]
[1]Department of Materials Engineering, and [2]Department of Physics and Astronomy, University of British Columbia, Vancouver, Canada, [3]Stewart Blusson Quantum Matter Institute, University of British Columbia, Vancouver, Canada, [4]Department of Physics, Simon Fraser University, Vancouver, Canada, [5]School of Microelectronics, line 3: Southern University of Science of Technology, Shenzhen, China

P15 - OFF Current Suppression by Gate-gontrolled Strain in The N-type GaAs Piezoelectric FinFETs 149
Yuxiong Long[1], Jun Z. Huang[2], Zhongming Wei[1], Jun-Wei Luo[1], Xiangwei Jiang[1],
[1]Institute of Semiconductors, Chinese Academy of Sciences, Beijing, China, [2]MaxLinear Inc., Carlsbad, CA, USA

P16 - Numerical Investigation of the Photogating Effect in MoTe2 Photodetectors 153
J.M. Gonzalez-Medina[1,2], E.G. Marin[1,3], A. Toral-Lopez[1,2], F.G. Ruiz[1,2], A. Godoy[1,2]
[1]Dpto. Electrónica, Universidad de Granada, Spain, [2]Pervasive Electronics Advanced Research Laboratory, CITIC, Universidad de Granada, Spain, [3]Dipartimento di Ingegneria dell'Informazione, Università di Pisa, Italy.

P17 - Physical Insights into the Transport Properties of RRAMs Based on Transition Metal Oxides 155
Toufik Sadi[1], Oves Badami[2], Vihar Georgiev[2], Jie Ding[3], Asen Asenov[2]
[1]Engineered Nanosystems Group, School of Science, Aalto University, AALTO, Finland
[2]School of Engineering, Electronic and Nanoscale Engineering, University of Glasgow, Glasgow, Scotland, UK
[3]College of Electrical and Power Engineering, Taiyuan University of Technology, China.

P18 - Molecular Dyanamics Simulation of Thermal Chemical Vapor Deposition for Hydrogenated   159
Amorphous Silicon on Si (100) Substrate by Reactive Force-Field
Naoya Uene[1], Takuya Mabuchi[2], Masaru Zaitsu[3], Shigeo Yasuhara[3], TakashiTokumasu[4]
[1]Graduate school of engineering Tohoku university Sendai, Japan, [2]Frontier research institute for
interdisciplinary sciences Tohoku university Sendai, Japan, [3]Research & development Japan advanced
chemicals ltd. Sagamihara,Japan, [4]Institute of fluid science Tohoku university Sendai,Japan

P19 - Device-to-circuit modeling approach to Metal – Insulator – 2D material FETs targeting the design of   163
linear RF applications
Alejandro Toral-López[1], Francisco Pasadas[2], Enrique G. Marín[1], Alberto Medina-Rull[1], Francisco J. G. Ruiz[1],
David Jiménez[2], Andrés Godoy[1]
[1]Departamento de Electrónica y Tecnología de Computadores, Universidad de Grana da, Granada, Spain
[2]Departament d'Enginyeria Electrònica, Universitat Autònoma de Barcelona, Barcelona,Spain

P20 - Simulation and Investigation of Electrothermal Effects in Heterojunction Bipolar Transistors   167
Xujiao Gao, Gary Hennigan, Lawrence Musson, Andy Huang, Mihai Negoita
Electrical models and simulation, Sandia National Laboratories, Albuquerque,USA

P21 - Variability of Threshold Voltage Induced by Work-Function Fluctuation and Random Dopant   171
Fluctuation on Gate-All-Around Nanowire nMOSFETs
Wen-Li Sung, Min-Hui Chuang, Yiming Li
Department of Electrical and Computer Engineering, Center for mmWave SmartRadar System and
Technologies, National Chiao Tung University, Hsinchu, Taiwan.

P22 - A First Principle Insight into Defect Assisted Contact Engineering at the Metal-Graphene and Metal-   175
Phosphorene Interfaces
Jeevesh Kumar, Adil Meersha, Ansh and Mayank Shrivastava
Department of Electronic Systems Engineering, Indian Institute of Science, Bengaluru, India

P23 - Transient Simulation of Field-Effect Biosensors: How to Avoid Charge Screening Effect   179
Kyoung Yeon Kim, Byung-Gook Park
Department of Electrical and Computer Engineering, Seoul National University, Seoul, South Korea

P24 - Electronic and structural properties of interstitial titanium in crystalline silicon from first-principles   183
simulations
Gabriela Herrero-Saboya[1][2], Layla Martin-Samos[3], Anne Hemeryck[2], Denis Rideau[4], Nicolas Richard[1]
[1]CEA, DAM, DIF Arpajon, France, [2]LAAS-CNRS, Université de Toulouse, Toulouse, France, [3]CNR-IOM /
Democritos National Simulation Center, c/o SISSA, Trieste, Italy, [4]STMicroelectronics, Crolles,France

P25 - Hybrid method for electromagnetic modelling of coherent radiation in semiconductor lasers.   187
Mateusz Marek Krysicki, Bartlomiej Salski, Pawel Kopyt
Warsaw University of Technology, Institute of Radioelectronics and Multimedia Technology, Warsaw, Poland

P26 - NEGF simulations of stacked silicon nanosheet FETs for performance optimization   191
Hong-Hyun Park[1], Woosung Choi[1], Mohammad Ali Pourghaderi[2], Jongchol Kim[2],Uihui Kwon[2], Dae Sin
Kim[2]
[1]Device Laboratory Samsung Semiconductor Inc. San Jose, California, USA
[2]Semiconductor R&D Center Samsung Electronics, Hwasung-si, Gyeonggi-do, Korea

18.00-19.00 **Cocktail Reception** – Palazzo di Toppo Wassermann

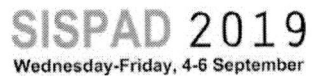

Thursday September 5

09.00-09.10	**Presentation of SISPAD 2020**

09.10-10.00 **Plenary invited**
Room T9   Modeling Silicon CMOS devices for quantum computing   195
Venitucci Benjamin, Li Jing, Bourdet Léo, Niquet Yann-Michel CEA/IRIG/MEM/L_Sim, Grenoble, France

10.00-10.20 **Coffee break**

Room T9 **Session 5**
Atomistic and ab-initio modeling
Chairperson: L. Martin-Samos, SISSA, Trieste, Italy

10.20-10.40 5.1 - ATOMOS: An ATOmistic MOdelling Solver for dissipative DFT transport in ultra-scaled HfS_2 and Black   199
phosphorus MOSFETs
Aryan Afzalian, Geoffrey Pourtois
Imec, Leuven, Belgium

10.40-11.00 5.2 - Atomistic modeling of nanoscale ferroelectric capacitors using a density functional theory and non-   203
equilibrium Green's-function method
Daniele Stradi, Ulrik G. Vej-Hansen, Petr A. Khomyakov, Maeng-Eun Lee, Gabriele Penazzi, Anders Blom,
Jess Wellendorff, Søren Smidstrup, Kurt Stokbro
Synopsys Denmark ApS, Copenhagen, Denmark

11.00-11.20 5.3 - Trigonal Tellurium Nanostructure Formation Energy and Band gap   207
Aaron Kramer[1], Maarten L. Van de Put[2], Christopher L. Hinkle[3], and William G. Vandenberghe[2]
[1]Department of Physics and [2]Department of Materials Science and Engineering, University of Texas at Dallas,
Richardson, USA, [3]Department of Electrical Engineering, University of Notre Dame, Notre Dame, USA

11.20-11.40 5.4 - Defect creation and Diffusion under electric fields from first-principles: the prototypical case of silicon   211
dioxide
N. Salles[1,3], L. Martin-Samos[1], S. de Gironcoli[2], L. Giacomazzi[1,3], M. Valant[3], A. Hemeryck[4], P. Blaise[5],
B. Sklenard[5], N. Richard[6]
[1]CNR-IOM/Democritos National Simulation Center, Istituto Officina dei Materiali, c/o SISSA, Italy, [2]SISSA, Italy
[3]Materials Research Laboratory, University of Nova Gorica, Slovenija, [4]LAAS-CNRS, Université de
Toulouse, CNRS, France, [5]Univ. Grenoble Alpes, CEA, LETI, France, [6]CEA, DAM, DIF, Arpajon, France

11.40-12.00 5.5 - Effective work-function tuning of $TiN/HfO_2/SiO_2$ gate-stack; a density functional tight binding study   215
Hesameddin Ilatikhameneh[1], Hong-Hyun Park[1], Zhengping Jiang[1], Woosung Choi[1], Mohammad Ali
Pourghaderi[2], Jongchol Kim[2], Uihui Kwon[2], Dae Sin Kim[2]
[1]Device lab, Samsung Semiconductor Inc., San Jose, California, USA
[2]Semiconductor Research and Development Center, Samsung Electronics, Hwasung-si, Korea

Room T4 **Session 6**
Compact and circuit oriented modelling
Chairperson: S. Martinie, CEA-LETI, Grenoble, France

10.20-10.40 6.1 - Single Event Transient Compact Model for FDSOI MOSFETs Taking Bipolar Amplification and Circuit   219
Level Arbitrary Generation Into Account
Neil Rostand[1], Sébastien Martinie[2], Joris Lacord[2], Olivier Rozeau[2], Thierry Poiroux[2], Guillaume Hubert[1]
[1]DPHY ONERA, Toulouse, France, [2]DCOS CEA-LETI, Grenoble, France

10.40-11.00 6.2 - From devices to circuits: modelling the performance of 5nm nanosheets   223
Andrew R. Brown[1], Liping Wang[1], Plamen Asenov[1], Fabian J. Klüpfel[2], Binjie Cheng[1], Sébastien Martinie[3],
Olivier Rozeau[3], Sylvain Barraud[3], Jean-Charles Barbé[3], Campbell Millar[1], Jürgen K. Lorenz[2]
[1]Synopsys Northern Europe Ltd. Glasgow, UK, [2]Fraunhofer IISB, Erlangen, Germany,
[3]CEA Leti, 38054 Grenoble, France

11.00-11.20 6.3 - Compact Modelling of Resistive Switching Devices based on the Valence Change Mechanism   227
Camilla La Torre[1], Alexander F. Zurhelle[1], Stephan Menzel[2]
[1]Instutitut für Werkstoffe der Elektrotechnik II, RWTH Aachen University & JARA-FIT, Aachen, Germany
[2]Peter Grünberg Institut (PGI-7), Forschungszentrum Juelich GmbH & JARA-FIT, Juelich, Germany

Room T4 **Session 7**
Temperature related effects
Chairperson: Y.Li, National Chiao Tung University, Taiwan

11.20-11.40 7.1 - Effect of Stacking Faults on the Thermoelectric Figure of Merit of Si Nanowires 231
Kantawong Vuttivorakulchai, Mathieu Luisier, and Andreas Schenk
Integrated Systems Laboratory ETH Zürich, Zürich, Switzerland

11.40-12.00 7.2 - Modeling of Temperature-Dependent MOSFET Aging 235
Fernando Ávila Herrera[1], Mitiko Miura-Mattausch[1], Hideyuki Kikuchihara[1],Takahiro Iizuka[1], Hans Jürgen
Mattausch[1], Hirotaka Takatsuka[2]
[1]HiSIM Research Center, Hiroshima University, Higashihiroshima,Japan
[2]Technology Development Division, Mie Fujitsu Semiconductor Limited, Yokoyama,Japan

12.00-12.20 7.3 - TCAD analysis of FinFET temperature-dependent variability for analog applications 239
S. Donati Guerrieri, F.Bonani, G. Ghione
Dipartimento di Elettronica e Telecomunicazioni, Politecnico di Torino, ITALY

12.20-13.30 **Lunch**

Room T9 **Session 8**
Power devices and wide band-gap semiconductors
Chairperson: S. Reggiani, University of Bologna, Italy

13.30-14.00 Invited: Nitride electronics exploiting ultrawide bandgapAlN
Debdeep Jena
Cornell University, USA

14.00-14.20 8.1 - Numerical Investigation of the Leakage Current and Blocking Capabilities of High-Power Diodes with 243
Doped DLC Passivation Layers
Luigi Balestra[1], Susanna Reggiani[1], Antonio Gnudi[1], E. Gnani[1], G.Baccarani[1], J. Dobrzynska[2], J. Vobecký[2]
[1]ARCES and DEI, University of Bologna, Bologna, Italy.
[2]ABB Switzerland Ltd.Semiconductors, Lenzburg, Switzerland.

14.20-14.40 8.2 - Influence of Accurate Electron Drift Velocity Modelling on the Electrical Characteristics in GaN-on-Si 247
HEMTs
Korbinian Reiser[1,2], John Twynam[1], Christian Eckl[1], Helmut Brech[1], Robert Weigel[2]
[1]Infineon Technologies AG, Regensburg, Germany, [2]Institute for Electronics Engineering, Friedrich-
Alexander-Universitaet Erlangen-Nuernberg, Erlangen-Nuernberg, Germany

14.40-15.00 8.3 - TCAD Simulations Combined with Free Carrier Absorption Experiments Revealing the Physical Nature of 251
Hydrogen-Related Donors in IGBTs
Andreas Korzenietz[1], Frank Hille[2], Franz-Josef Niedernostheide[2], Christian Sandow[2], Gerhard Wachutka[1],
Gabriele Schrag[1]
[1]Chair for Physics of Electrotechnology, Technical University of MunichMunich, Germany
[2]Infineon Technologies AG Neubiberg, Germany

15.00-15.20 8.4 - TCAD investigation of zero-cost high voltage transistor architectures forlogic memory circuits 255
Jordan Locati[1], Christian Rivero[1], Julien Delalleau[1], Vincenzo Della Marca[2],Karine Coulié[2], Jordan
Innocenti[1], Olivier Paulet[1], Arnaud Regnier[1], StephanNiel[1]
[1]STMicroelectronics, Rousset, France, [2]Aix-Marseille University, CNRS, Marseille, France

15.20-15.40 8.5 - Barrier Engineering of Lattice Matched AlInGaN/ GaN Heterostructure Toward High Performance E-mode 259
Operation
Niraj Man Shrestha[1], Chao-Hsuan Chen[2], Zuo-Min Tsai[3], Yiming Li[3], Jenn-Hawn Tarng[3], Seiji Samukawa[4]
[1]Department of Electrical and Computer Engineering and Center for mmWave Smart Radar System and
Technologies, National Chiao Tung University, Hsinchu, Taiwan, [2]Institute of Communications Engineering,
National Chiao Tung University,Hsinchu, Taiwan, [3]Institute of Communications Engineering, Department of
Electrical and Computer Engineering, and Center for mmWave Smart Radar System and Technologies,
National Chiao Tung University, Hsinchu, Taiwan, [4]Center for mmWave Smart Radar System and
Technologies, National Chiao Tung University, Hsinchu, Taiwan and Institute of Fluid Science, Tohoku
University,Sendai, Japan

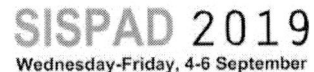

Room T4   **Session 9**
Technology Optimization
Chairperson: V. Moroz, Synopsys, USA

14.00-14.20   9.1 - RF performance improvement on 22FDX® platform and beyond   263
Tom Herrmann, Alban Zaka, Nandha Kumar Subramani, Zhixing Zhao, Steffen Lehmann, Yogadissen Andee
GLOBALFOUNDRIES Dresden Dresden, Germany

14.20-14.40   9.2 - Leakage Performance Improvement in Multi-Bridge-Channel Field Effect Transistor (MBCFET) by   267
Adding Core Insulator Layer
Saehoon Joung[1,2], SoYoung Kim[2]
[1]Samsung Electronics Co. Foundry Division, Yield Enhancement, Process Integration Engineering Group,
Ltd Kiheung, Republic of Korea, [2]College of Information and Communication Engineering, Sungkyunkwan
University, Suwon, Gyeounggi-do, Republic of Korea

14.40-15.00   9.3 - Impact of MOL/BEOL Air-Spacer on Parasitic Capacitance and Circuit Performance at 3 nm Node   271
Ashish Pal, Sushant Mittal, El Mehdi Bazizi, Angada Sachid, Mehdi Saremi, Benjamin Colombeau, Gaurav
Thareja, Samuel Lin, Blessy Alexander, Sanjay Natarajan and Buvna Ayyagari
Applied Materials, Santa Clara, USA

15.00-15.20   9.4 - Scaling-aware TCAD Parameter Extraction Methodology for Mobility Prediction in Tri-gate Nanowire   275
Transistors
Cristina Medina-Bailon[1], Tapas Dutta[1], Fabian Klupfel[2], S. Barraud[3], V. Georgiev[1], J. Lorenz[2], A.Asenov[1]
[1]School of Engineering, University of Glasgow, Glasgow, United Kingdom
[2]Fraunhofer Institut für Integrierte, Systeme und Bauelementetechnologie, Erlangen, Germany
[3]CEA, LETI, MINATEC campus and University Grenoble Alpes, Grenoble, France

15.40-16.00   **Coffee break**

Room T9   **Session 10**
Beyond CMOS materials and devices
Chairperson: A. Schenk, ETH Zurich, Switzerland

16.00-16.20   10.1 - DFT study of graphene doping due to metal contacts   279
P. Khakbaz[1], F. Driussi[1], A. Gambi[1], P. Giannozzi[2], S. Venica[1], D. Esseni[1], A. Gahoi[3], S. Kataria[3], M. C.
Lemme[3,4]
[1]DPIA, University of Udine, Udine, Italy, [2]DMIF, University of Udine, Udine, Italy, [3]RWTH Aachen
University, Aachen, Germany, [4]AMO GmbH, Advanced Microelectronic Center, Aachen, Germany

16.20-16.40   10.2 - Device and Circuit Level Gate Configuration Optimization for 2D Material Field-Effect Transistors   283
Devin Verreck[1], Goutham Arutchelvan[1,2], Marc M. Heyns[1,2], Iuliana P.Radu[1]
[1]imec, Leuven, Belgium, [2]Department of Materials Engineering, KU Leuven, Leuven, Belgium

16.40-17.00   10.3 - Accurate and Efficient Dynamic Simulations of Ferroelectric Based Electron Devices   287
T.Rollo[1], L. Daniel[2], D. Esseni[1]
[1]DPIA, University of Udine, Italy, [2]Electrical Engineering and Computer Science Depart., MIT, USA

17.00-17.20   10.4 - Precise Transient Mechanism of Steep Subthreshold Slope PN-Body-Tied   291
SOI-FET and Proposal of a New Structure for Reducing Leakage Current upon Turn-off Takayuki Mori[1], Jiro
Ida[1], Hiroki Endo[1], Yasuo Arai[2]
[1]Kanazawa Inst. of Tech. Nonoichi, Japan, [2]High Energy Accelerator, Research Org., KEK, Tsukuba, Japan

17.20-17.40   10.5 - Negative Capacitance Field-Effect Transistor Based on a Two-Dimensional Ferroelectric   295
M. Soleimani, N. Asoudegi, P.Khakbaz, M. Pourfath
School of Electrical and Computer Eng., University College of Engineering, University of Tehran, Iran

Room T4   **Session 11**
Interfaces, traps and defects
Chairperson: R. Rideau, STMicroelectronics, France

16.00-16.20   11.1 - Progress in dislocation stress field model and its appications   299
Uihui Kwon[1], Jeong-Guk Min[1], Seon-Young Lee[1], Alexander Schmidt[1], Dae Sin Kim[1], Yasuyuki Kayama[2],
Yutaka Nishizawa[2], Kiyoshi Ishikawa[2]
[1]Semiconductor R&D Center, Samsung Electronics Corp. Ltd., Hwasung-si, Gyeonggi-do, Korea
[2]Device Solution Center, Samsung R&D Institute Japan, Tshurumi-ku, Yokohama, Japan

16.20-16.40   11.2 - Effect of Trap on Carrier Transport in InAs FET with Al_2O_3 Oxide: DFT-based NEGF simulations   303
Mincheol Shin, Yucheol Cho, Seonghyeok Jeon
School of Electrical Engineering, Korea Advanced Institute of Science and Technology, Daejeon, Korea

16.40-17.00   11.3 - Trap Dynamics based 3D Kinetic Monte Carlo Simulation for Reliability Evaluation of UTBB MOSFETs   307
Wangyong Chen, Linlin Cai, Xiaoyan Liu, Gang Du
Institute of Microelectronics, Peking University, Beijing, China

17.00-17.20   11.4 - Polarization Effect Induced by Discrete Impurity at Semiconductor/Oxide Interface in Si-FinFET   311
Katsuhisa Yoshida, Kohei Tsukahara, Nobuyuki Sano
Institute of Applied Physics, University of Tsukuba, Ibaraki, Japan

17.20-17.40   11.5 - Relationship between capacitance and conductance in MOS capacitors   315
E. Caruso[1], J. Lin[1], S. Monaghan[1], K. Cherkaoui[1], L. Floyd[1], F. Gity[1], P. Palestri[2], D. Esseni[2], L. Selmi[3], P.K. Hurley[1]
[1]Tyndall National Institute, University College Cork, Ireland, [2]DPIA, University of Udine, Udine, Italy, [3]DIEF, University of Modena and Reggio Emilia, Modena, Italy

19.30-22.30 **Gala Dinner** at Palazzo Kechler

Friday September 6

09.10-10.00   **Plenary invited**
Room T9
Compact Modeling Perspective – Bridge to Industrial Applications   319
Mitiko Miura-Mattausch
Hiroshima University, Higashi-Hiroshima, Japan

10.00-10.20   **Coffee break**

Room T9   **Session 12**
Process modelling
Chairperson: J. Lorenz, Fraunhofer Institute for Integrated Systems and Device Technology IISB, Germany

10.20-10.50   Invited: Interaction of oxygen and dopants in oxygen-inserted silicon
Daniel Connelly, Nyles Cody, Marek Hytha, Robert Stephenson, Hideki Takeuchi, Keith Doran Weeks, R. Mears
Atomera

10.50-11.10   12.1 - Modeling and Simulation of Atomic Layer Deposition   323
Lado Filipovic
Institute for Microelectronics, Technische Universität Wien, Vienna, Austria

11.10-11.30   12.2 - Novel Numerical Dissipation Scheme for Level-Set Based Anisotropic Etching Simulations   327
Alexander Toifl[1], Michael Quell[1], Andreas Hössinger[2], Artem Babayan[2], Siegfried Selberherr[3], Josef Weinbub[1]
[1]Christian Doppler Laboratory for High Performance TCAD, Institute for Microelectronics, TU Wien, Austria
[2]Silvaco Europe Ltd. Cambridge, United Kingdom, [3]Institute for Microelectronics, TU Wien, Austria

11.30-11.50   12.3 - Numerical simulations of nanosecond laser annealing of Si nanoparticles for plasmonic structures   331
A-S. Royet[1], S. Kerdilès[1], P. Acosta Alba[1], C. Bonafos[2], V. Paillard[2], F. Cristiano[3], B. Curvers[4], K. Huet[4]
[1]CEA-LETI, MINATEC Campus, Université Grenoble Alpes, Grenoble, France, [2]CEMES-CNRS, Université de Toulouse, Toulouse, France, [3]LAAS, CNRS, Université de Toulouse, Toulouse, France, [4]LASSE, SCREEN SPE, Gennevilliers, France

11.50-12.10   12.4 - Parallelized Level-Set Velocity Extension Algorithm for Nanopatterning Applications   335
Michael Quell[1], Alexander Toifl[1], Andreas Hössinger[2], Siegfried Selberherr[3], Josef Weinbub[1]
[1]Christian Doppler Laboratory for High Performance TCAD, Institute for Microelectronics, TU Wien, Austria
[2]Silvaco Europe Ltd. Cambridge, United Kingdom, [3]Institute for Microelectronics, TU Wien, Austria

12.10-12.30   12.5 - Process Simulation in the Browser: Porting ViennaTS using WebAssembly   339
Xaver Klemenschits, Paul Manstetten, Lado Filipovic, Siegfried Selberherr
Institute for Microelectronics, TU Wien, Wien, Austria

Room T4   **Session 13**
                   Sensors and optoelectronic devices
                   Chairperson: C. Jungemann, University of Aachen, Germany

10.30-10.50   13.1 - A model of the interface charge and chemical noise due to surface reactions in Ion Sensitive FETs   343
              Leandro Julian Mele[1], Pierpaolo Palestri[1], Luca Selmi[2]
              [1]DPIA, University of Udine, Udine, Italy, [2]DIEF, University of Modena and Reggio Emilia, Modena, Italy

10.50-11.10   13.2 - Investigation and Modelling of Single-Molecule Organic Transistors   347
              Fabrizio Torricelli[1], Eleonora Macchia[2][3], Paolo Romele[1], Kyriaki Manoli[2], Cinzia Di Franco[4], Zsolt M. Kovacs-
              Vajna[1], Gerardo Palazzo[2][5][6], Gaetano Scamarcio[4][5], Luisa Torsi[2][3][6]
              [1]Department of Information Engineering, University of Brescia, Brescia, Italy, [2]Dipartimento di Chimica,
              Università degli Studi di Bari, Italy, [3]The Faculty of Science and Engineering, Åbo Akademi University,
              Turku, Finland, [4]CNR, Istituto di Fotonica e Nanotecnologie, Bari, Italy, [5]Dipartimento InterAteneo di Fisica
              "M. Merlin", Università degli Studi di Bari, Italy, [6]CSGI (Centre for Colloid and Surface Science), Bari, Italy

11.10-11.30   13.3 - Advances in 3D CMOS image sensors optical modeling: combining realistic morphologies with FDTD   351
              Benjamin Vianne[1], Axel Crocherie[1], Sofiane Guissi[2], Daniel Sieger[3], Stéphane Calderon[3], D. Rideau[1], Hélène
              Wehbe-Alause[1]
              [1]STMicroelectronics, Crolles, France, [2]Lam Research, Meylan, France, [3]Coventor a Lam Research Company,
              Villebon sur Yvette, France

11.30-11.50   13.4 - Simulation of quantum dot based single-photon sources using the Schrödinger-Poisson-Drift-Diffusion-   355
              Lindblad system
              Markus Kantner, Thomas Koprucki, Hans-Jürgen Wünsche and Uwe Bandelow Weierstrass Institute for
              Applied Analysis and Stochastics Mohrenstr. 39, Berlin, Germany

11.50-12.10   13.5 - A generalized multi-particle drift-diffusion simulator for optoelectronic devices   359
              Daniele Rossi, Matthias Auf der Maur, Aldo DiCarlo
              Dept. of Electronic Engineering, Università degli Studi di Roma "Tor Vergata", Rome, Italy

12.10-12.30   13.6 - An Efficient Method for Modeling Parasitic Light Sensitivity in Global Shutter CMOS Image Sensors   363
              Federico Pace[1][2], Olivier Marcelot[1], Philippe Martin-Gonthier[1], Olivier Saint-Pé[2], Michel Breart de
              Boisanger[2], Rose-Marie Sauvage[3], Pierre Magnan[1]
              [1]ISAE-SUPAERO, Université de Toulouse, Toulouse, France, [2]Airbus Defence & Space, ZI du Palais,
              Toulouse, France, [3]Direction Générale de l'Armement, Paris, France

12.30-13.30   **Lunch**

Proceedings of

2019 International Conference on Simulation of

Semiconductor Processes and Devices (SISPAD)

SISPAD 2019

September 4-6, 2019

Udine, Italy

Edited by Francesco Driussi

WELCOME

We would like to welcome you to the 2019 International Conference on Simulation of Semiconductor Processes and Devices (SISPAD 2019), to Udine and to Palazzo di Toppo Wassermann, a premises of the Università degli Studi di Udine.

The Università degli Studi di Udine is proud to host the 24th edition of SISPAD, whose longevity stems from its ability to stay at the forefront of the hot topics in the field of simulation of semiconductor processes and devices. This is the reason why, for almost a quarter of a century, leading scientists, researchers, and students have chosen the SISPAD as the forum where to share the latest developments in advanced modelling of novel semiconductor devices, processes and equipment for integrated circuits and nanoelectronics.

This year's conference program consists of 3 plenary invited presentations, 3 invited talks, 64 contributed papers and 26 posters, that were selected out of 136 submitted abstracts. These presentations have been arranged in 13 sessions and one poster session, and the conference program covers two and a half days, namely from the opening on Wednesday September 4th to the closing right after lunch on Friday September 6th.

The program of the conference is augmented and complemented by a tutorial entitled "Atomistic simulations for nanoelectronic and optoelectronic devices" and a workshop entitled "Leti seminar on Simulation and Modeling for Emerging Non-Volatile Memories", both held on Tuesday September 3rd in the afternoon.

This conference has been made possible thanks to the support, expertise and work of many people.

We are very thankful to the members of the Steering Committee for their support and useful suggestions, to the members of the Technical Program Committee for their commitment and expertise to critically analyse many submitted abstracts and then select those that now form the technical program of this conference. We are particularly thankful to the Chairman of the previous European SISPAD edition, Jürgen Lorenz, and to the Chairman of last year edition, Leonard Franklin Register, for sharing their expertise in the conference organization and for many useful words of advice. We would like to thank our invited speakers for sharing their renowned expertise with the attendees of the conference.

We are also thankful to the Università degli Studi di Udine, that is hosting the conference in Palazzo di Toppo Wassermann, to the Conference Secretariat (Centro Congressi Internazionale, Torino) and particularly to Giulia Datta and Rosaria Petrolo for their professionalism and tireless commitment, to the offices and people of the IEEE Electron Devices Society and IEEE Meetings, Conferences & Events Department, who helped us in several respects in the organization of the conference.

We extend our sincere thanks to our sponsors for their generous contributions.

We are finally deeply thankful to all authors and presenters for having chosen SISPAD to share their scientific work, and to each and every attendee of the conference, who literally made this conference possible.

David Esseni	Conference Chair
Pierpaolo Palestri Denis Rideau	Technical Program Co-Chairs
Francesco Driussi	Publication Chair

Committees

Conference Chair

David Esseni, University of Udine, Italy

Technical Program Co-Chairs

Pierpaolo Palestri, University of Udine, Italy
Denis Rideau, STMicroelectronics, France

Honorary Committee:

R. Dutton, Stanford University, USA
S. Selberherr, TU Wien, Austria
K. Taniguchi, Osaka University, Japan

International Steering Committee:

A. Asenov, Univ. of Glasgow, UK
V. Axelrad, Sequoia Design Systems, USA
D. Esseni, University of Udine, Italy
N. Goldsman, University of Maryland, USA
J. Lorenz, Fraunhofer Institute for Integrated Systems and Device Technology IISB, Germany
Y. Kamakura Osaka University, Japan
N. Mori, Osaka University, Japan
L. F. Register, University of Texas at Austin, USA
K. Sonoda, Renesas, Japan

Technical Program Committee:

C. Millar, Synopsys Inc, UK
C. Jungemann, University of Aachen, Germany
G. Eneman, IMEC, Belgium
S. Reggiani, University of Bologna, Italy
L. Filipovic, Technische Universität Wien, Austria
J. Lorenz, Fraunhofer Institute for Integrated Systems and Device Technology IISB, Germany
L. Martin-Samos Colomer, SISSA, Trieste, Italy
P. Palestri, University of Udine, Udine, Italy
S. Martinie, CEA-LETI, Grenoble, France
D. Rideau, STMicroelectronics, France
A. Schenk, ETH Zurich, Switzerland
A. Godoy, University of Granada, Spain
V. Moroz, Synopsys, USA
D. Connelly, Atomera, USA
S. Hasan, Intel, USA
G. Xia, The University of British Columbia, Canada
S. Gupta, Purdue University, USA
W. Vandenberghe, University of Texas at Dallas, USA
S.-D. Kim, SK Hynix, South Korea
K. Tatsuya, Renesas Electronics, Japan
U. Kwon, Samsung, South Korea
Y. Li, National Chiao Tung University, Taiwan
S. Souma, Kobe University, Japan
J. Wu, TSMC, Taiwan

Special Thanks to Our Sponsors:

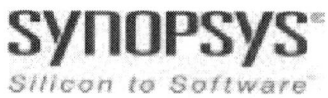

3D Kinetic Monte Carlo Simulation of Electromigration in Multi-layer Interconnects

Linlin Cai, Wangyong Chen, Xing Zhang, Yudi Zhao+, and Xiaoyan Liu*

Institute of Microelectronics
Peking University
Beijing 100871, China
Email: +zhaoyd@pku.edu.cn; *xyliu@ime.pku.edu.cn;

Abstract—A 3D kinetic Monte Carlo simulator is developed to describe the electromigration (EM) behaviors in multi-layer interconnects based on the proposed physical mechanism including the metal ions activation, hopping and aggregation processes. The effects of e-wind, hydrostatic stress and Joule heat on EM are implemented in the simulator. The void locations in two directions of upstream and downstream current flow are well reproduced by the simulator, consistent with the experimental observations. The microscopic void morphology and EM degradation are investigated with different operation schemes.

Keywords—kinetic Monte Carlo simulator, electromigration, interconnect, void

I. INTRODUCTION

The electromigration (EM) reliability of back-end-of-line (BEOL) becomes more challenging in high-density integration due to the contradiction between the scaled cross-sectional area and excessive current density [1-3]. The EM simulation by atomic kinetic Monte Carlo (KMC) method provides a better physical understanding of vacancies motion and void formation, also gives the guidelines of interconnects design and reliability prediction [4-5].

In this work, a 3D KMC simulator is developed to capture the void evolution and EM degradation in multi-layer interconnects. The microscopic physical mechanism coupled with the impacts of electron wind (e-wind), stress and heat is proposed and applied to the simulator. The void locations and EM failure with different current operations are investigated by using the simulation tool.

II. SIMULATION METHOD AND VALIDATION

The microscopic mechanism of EM contains the physical processes of metal ions activation, hopping and aggregation, as shown in Fig. 1. The hopping direction is dominated by the e-wind which can be hindered by the stress- and electric field-induced backflow. The interface barrier $E_{h,i}$ is lower than the bulk barrier $E_{h,b}$ according to the material characteristic [6]. Due to the fact that the metal ions migration trends to the minimum energy locations, the ions with the least nearest ion numbers (n_{ij}) have the smallest activation energy E_a. The metal ions with more n_{ij} are easier to aggregate, which contributes to the void formation. The probability equations of ions motion are described as (1-3) by considering the coupling effects of e-wind, stress and temperature in (4-10) [7-8].

$$P_a = f \cdot \exp\left(-(E_a(n_{ij}) - \Delta E)/k_B T\right) \quad (1)$$

$$P_h = f \cdot \exp\left(-(E_h - \gamma E_{ew} - \lambda E_{st})/k_B T\right) \quad (2)$$

$$P_{ag} = f \cdot \exp\left(-(E_{ag}(n_{ij}) - \Delta E)/k_B T\right) \quad (3)$$

$$E_{ew} = -Z^* eU \quad (4)$$

$$I = U / R_1 \quad (5)$$

$$I = \sinh(\alpha U) / R_2 \quad (6)$$

$$E_{st} = -\Omega \cdot \Delta\sigma_i \quad (7)$$

$$\Delta\varepsilon_i = -f_i \Omega \quad (8)$$

$$\varepsilon_{i,k} = -v\sigma_{i,klm}\delta_{ik} / E + (1+v) \cdot \left(\sigma_{i,klm}\delta_{il} + \sigma_{i,klm}\delta_{im}\right) / 2E \quad (9)$$

$$C \cdot \partial T / \partial t = \nabla(k \cdot \nabla T) + Q \quad (10)$$

Where ΔE is the electric field modulated barrier. I is the node current in the resistor network, U is the potential difference between two adjacent nodes. γ, λ, α are the fitting coefficients. $\Delta\sigma_i$ is the hydrostatic stress gradient, and $\Delta\varepsilon_i$ is the lattice strain deformation. T is the local temperature and Q is the Joule heat power density. The simulated parameters and related physical quantities are listed in Table I.

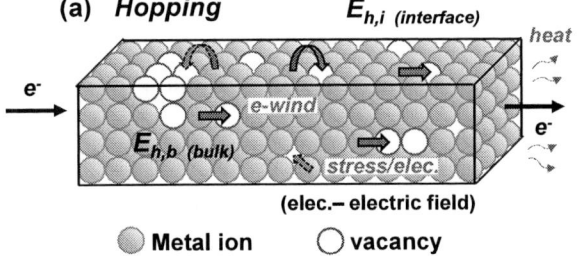

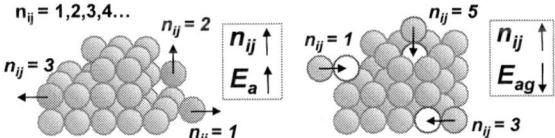

Fig. 1. Microscopic physical behaviors of metal ions during the EM degradation: (a) hopping (b) activation (c) aggregation.

Fig. 2 shows the EM simulation flowchart. The initial parameter input depends on the simulated metal material and size. The grain boundaries are randomly generated by KMC method. The electrical properties of interconnects are calculated by the developed 3D-resistor network with cross nodes [9]. The lattice strain is introduced by vacancy relaxation factor f_i, which stands for the deformation between vacancy volume and atom volume. The relations of lattice strain and stress conform to Hooke's law. The Joule heat is calculated by Fourier equation of heat conduction. Each metal ion motion is determined by comparing the probabilities of different events. Once the ions distribution is changed, the

978-1-7281-0941-1/19 $31.00 © 2019 IEEE

node potential, current, stress gradient and local temperature would be updated. The simulator visualizes the void microscopic evolution and investigates the resistance-time degradation during EM as well as the prediction of time to failure (TTF).

TABLE I. SIMULATION PARAMETERS IN THIS WORK

Symbol	Value	Description
$E_{h,i}$	0.9 eV	Hopping barrier at interfaces
$E_{h,b}$	1.1 eV	Hopping barrier in bulk
$E_{a,nij=1}$	0.85 eV	Activation energy
$E_{ag,nij=1}$	0.8 eV	Aggregation energy
f	10^{13} Hz	Vibration frequency
R_1	$10^2\,\Omega$	Resistance between two ions
R_2	$10^6\,\Omega$	Resistance between ion and vacancy or two vacancies
Z^*	4	Effective charge
Ω	1.6×10^{-23} cm^3	Atom volume
f_i	0.9	Vacancy relaxation factor
E	130 GPa	Young's modulus
v	0.34	Poisson's ratio
C	3×10^6 J/K·m^3	Thermal capacity
k	380 W/K·m	Thermal conductivity

Interconnects	Width	Height
M1-M2	10 nm	20 nm
Via	8 nm	24 nm

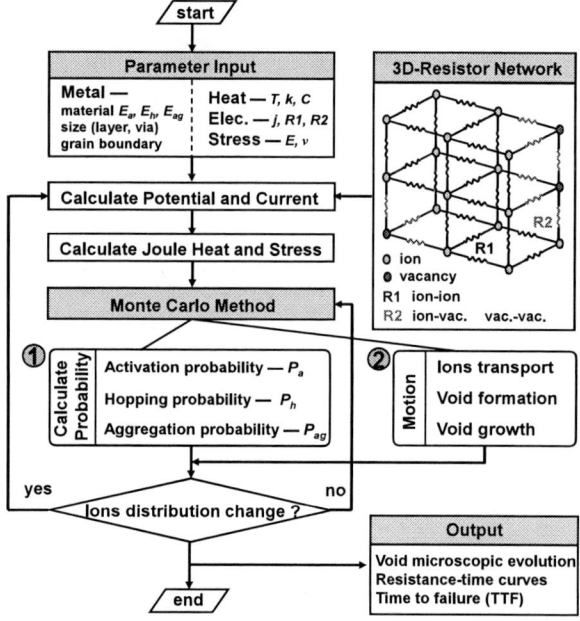

Fig. 2. The EM simulation flowchart. The potential and current distribution are calculated by 3D resistor network.

The simulator is further validated with the experimental measurements [10]. Fig. 3-4 indicate that the simulated results show the excellent agreements with the measured resistance-time curves and cumulative failure distribution respectively. The hopping barrier affected by e-wind and stress plays the dominant role in controlling the TTF of EM. The changes of relative resistance before and after the EM failure are determined together by R_1 and R_2. The gradual increase of resistance after the failure is due to the void lateral growth along the metal wires. The statistic failure distribution is simulated by considering the differences of grain boundaries and the variations of metal ions movements. The resistivity of copper is calibrated with the different metal line widths for the following EM evaluation at 7nm technology node, as shown in Fig. 5 [11].

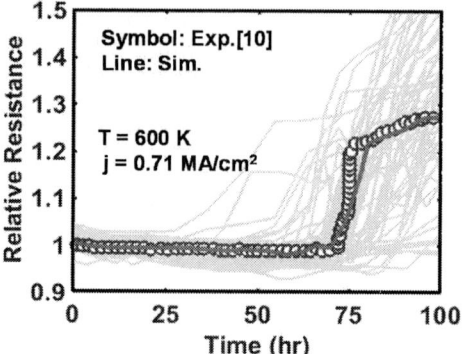

Fig. 3. The simulated resistance-time curves compared with experiment [10] during the EM degradation.

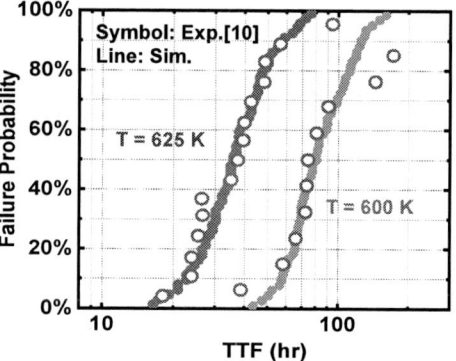

Fig. 4. Cumulative failure distribution between simulation and experiment [10] at different temperatures.

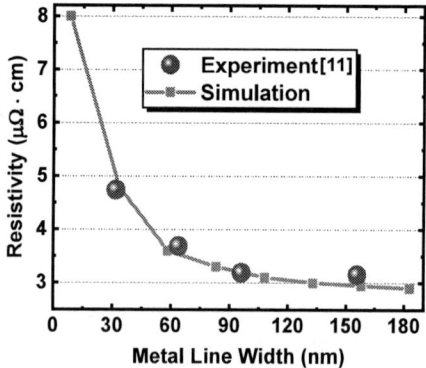

Fig. 5. Calibration of simulated copper resistivity at scaled nodes with experiment [11].

III. RESULTS AND DISCUSSION

Fig. 6 (a) shows the void formation in M1-via-M2 Cu interconnects with upstream current flow. It can be seen that the interconnects are fragile at the via regions near M1 as well

978-1-7281-0941-1/19 $31.00 © 2019 IEEE

as the corner of via to M2 due to the shift of metal ions driven by the upstream current flow, which is consistent with the experimental FIB-SEM image in [10]. The concentrated vacancies in the via increase the local current density of the wire, leading to the large Joule heat as shown in Fig. 6 (b)-(c). The increased temperature accelerates the void growth, further aggravates the EM degradation. The void evolution over time including the void accumulation, full void formation and growth is shown in Fig. 6 (d). The EM of interconnects undergoes failure after the full void formation in the via. And the void size becomes shrinking over time only when the current flow is cut off. From Fig. 6 (e), the slight recovery of resistance can be found at 600K without the power input of interconnects. Here the EM failure is defined as the 20% resistance increase of initial state.

Fig. 7 shows the simulated EM degradation with downstream current flow from M2 to M1. From Fig. 7 (a), the void location is visualized at the M1 regions under the via corresponding to the experimental observation in [12]. The related current and temperature distribution at 25 hours are demonstrated in Fig. 7 (b)-(c). The results show that the void formation with the downstream current flow is slower than that with the upstream current flow, as shown in Fig. 7 (d). And the resistance in Fig. 7 (e) indicates the less degradation and more recovery compared to Fig. 6 (e) due to the incomplete void formation.

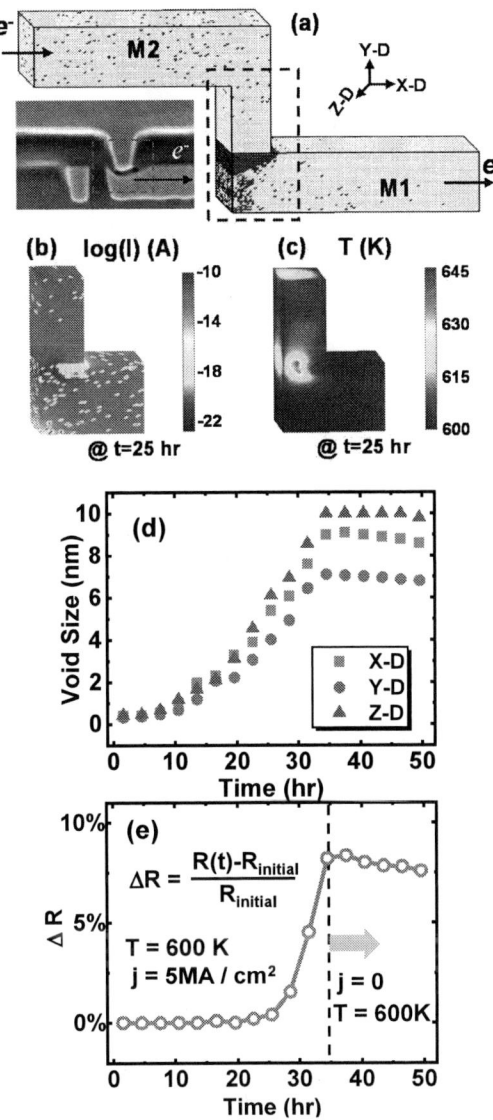

Fig. 7. Downstream current flow from M2 to M1 in EM simulation. (a) void formation in M1 under the via compared with experimental observation [12] (b)-(c) local current and temperature distribution of via-M1 (d) void size evolution over time (e) R-t curve.

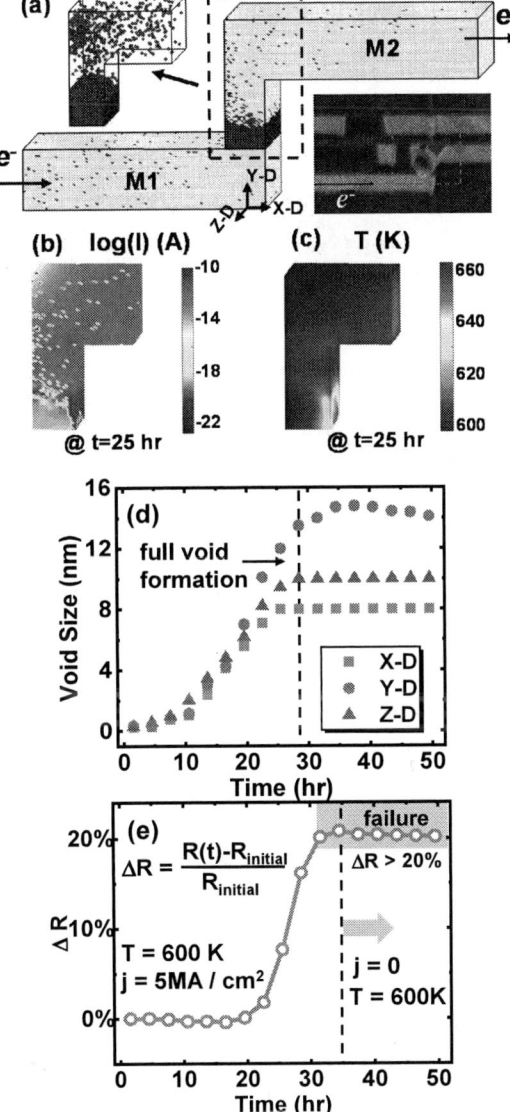

Fig. 6. Upstream current flow from M1 to M2 in EM simulation. (a) void formation in the via compared with FIB-SEM image [10] (b)-(c) local current and temperature distribution of via-M2 (d) void accumulation, full void formation and growth over time (e) R-t curve.

Three cases of different current operations of interconnects are investigated in Table II. The results in Fig. 8 present that the void forms at the middle of interconnects in Case I with the unidirectional current. Adding the current branch from the left via, M2 shows the better EM in Case II in comparison with Case I because the metal ions migration driven by the left current fills up the vacancies in the middle of line. However, the interconnects with the bidirectional current operation in

Case III suffer the worst resistance degradation due to the superposed influence of vacancies. The void locations and EM degradation in three cases at 600K are summarized in Table II.

TABLE II. THREE CASES OF CURRENT OPERATIONS OF INTERCONNECTS

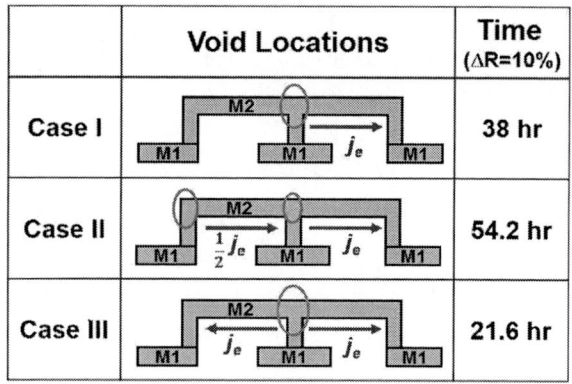

Fig. 8. The void morphology in M2 during EM degradation in three cases of current operations.

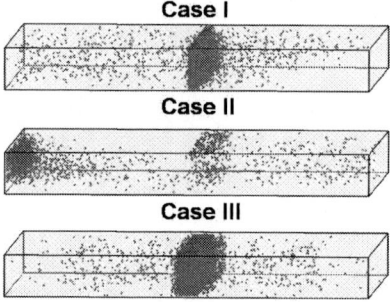

IV. CONCLUSIONS

The 3D KMC simulator including the physical behaviors of metal ions is developed to investigate the EM reliability with the coupling effects of e-wind, stress and Joule heat. The simulator validated by the experimental results can well visualize the void locations in interconnects with both upstream and downstream current flow. The void morphology and resistance degradation during the EM simulation are successfully captured with different current operations in three cases. The simulator provides an effective tool for EM prediction of multi-layer interconnects, covering the advanced material, flexible current operation, varying temperature and different interconnect architecture at scaled technology node.

ACKNOWLEDGMENT

This work is supported by National Natural Science Foundation of China 2016YFA0202101, No. 61674008, No. 61421005.

REFERENCES

[1] D. Prasad, and A. Naeemi, "Interconnect Design and Technology Optimization for Conventional and Emerging Nanoscale Devices: A Physical Design Perspective," in IEEE International Electron Devices Meeting, pp. 103–106, Dec. 2018.

[2] International Technology Roadmap for Semiconductors, 2015, http://www.itrs2.net/itrs-reports.html.

[3] Szu-Tung Hu, Linjun Cao, Laura Spinella, and Paul S. Ho, "Microstructure Evolution and Effect on Resistivity for Cu Nanointerconnects and Beyond," in IEEE International Electron Devices Meeting, pp. 115–117, Dec. 2018.

[4] S. V. Kolesnikov, A. M. Saletsky, S. A. Dokukin, and A. L. Klavsyuk, "Kinetic Monte Carlo Method: Mathematical Foundations and Applications for Physics of Low-Dimensional Nanostructures," Mathematical Models and Computer Simulations, vol. 10, pp. 564-587, 2018.

[5] Cher Ming Tan, Zhenghao Gan, Wei Li, and Yuejin Hou, "Applications of Finite Element Methods for Reliability Studies on ULSI Interconnections," Springer, pp. 73-110, 2011.

[6] J. R. Lloyd, J. Clemnes, and R. Snede, "Copper metallization reliability," Microelectronics Reliability, vol. 39, pp. 1595–1602, Nov. 1999.

[7] Y. D. Zhao, P. Huang, Z. H. Guo, Z. Y. Lun, B. Gao, X. Y. Liu, and J. F. Kang, "Atomic Monte-Carlo Simulation for CBRAM with Various Filament Geometries," in International Conference on Simulation of Semiconductor Processes and Devices, pp. 153-156, Sept. 2016.

[8] Shidong Li, Michael S. Sellers, Cemal Basaran, Andrew J. Schultz, and David A. Kofke, "Lattice Strain Due to an Atomic Vacancy," International Journal of Molecular Sciences, vol. 10, pp. 2798-2808, June 2009.

[9] Yudi Zhao, Peng Huang, Zhe Chen, Chen Liu, Haitong Li, Bing Chen, Wenjia Ma, Feifei Zhang, Bin Gao, Xiaoyan Liu, and Jinfeng Kang, "Modeling and Optimization of Bilayered TaOx RRAM Based on Defect Evolution and Phase Transition Effects," IEEE Trans. Electron Devices, vol. 63, pp. 1524-1532, April 2016.

[10] Hui Zheng, Binfeng Yin, Ke Zhou, Leigang Chen, and Chinte Kuo, "Temperature-dependent activation energy of electromigration in Cu/porous low-k interconnects," Journal of Applied Physics, vol. 122, pp. 074501-1-7, August 2017.

[11] Seungman Choi, Cathryn Christiansen, Linjun Cao, James Zhang, Ronald Filippi, Tian Shen, Kong Boon Yeap, Sean Ogden, Haojun Zhang, Bianzhu Fu, and Patrick Justison, "Effect of Metal Line Width on Electromigration of BEOL Cu Interconnects", in International Reliability Physics Symposium, pp. 4F.4-1-6, March 2018.

[12] A. S. Oates, and M. H. Lin, "Analysis and modeling of critical current density effects on electromigration failure distributions of Cu dual-damascene vias," in International Reliability Physics Symposium, pp. 385-391, April 2008.

978-1-7281-0941-1/19 $31.00 © 2019 IEEE

Metallic ions drift in hybrid bonding integration modeling, towards the evolution of failure criterion

MANZANAREZ Hervé   MOREAU Stéphane   CUETO Olga

Univ. Grenoble Alpes, CEA, LETI, F-38000 Grenoble France.

stephane-nico.moreau@cea.fr

Abstract—Copper ions drift is modeled in the case of hybrid bonding integration. The continuity equation is coupled to the Poisson's equation and a copper ion concentration saturation is assumed. A 1D geometry simulation is initially realized to validate the model and 2D geometry simulations of hybrid bonding are analyzed by looking the time to percolate (TTP).

Index Terms—Hybrid bonding, copper ions drift, failure criterion, time to percolate.

I. INTRODUCTION

The continuous race to higher performance and compactness has led to the introduction of 3D integration in CMOS ICs. Among the different schemes to obtain 3D devices, the last decade has shown that direct bonding and more recently hybrid bonding (HB) was a key enabler for 3D high density integration [1]. Roughly, it consists in putting into contact two surfaces (wafer-to-wafer or die-to-wafer) to create interconnects by direct bonding (Fig. 1). This process requires to solve technological challenges, in particular a high level of cleanliness interface, an ultra smooth surface, and an ultra-precise aligment accuracy between the two surfaces. The misalignment cause copper to be directly in contact with the surrounding dielectric material. This could lead to reliability issues, dielectric breakdown [2]–[4] mainly, and must be investigated to evaluate the risk level.

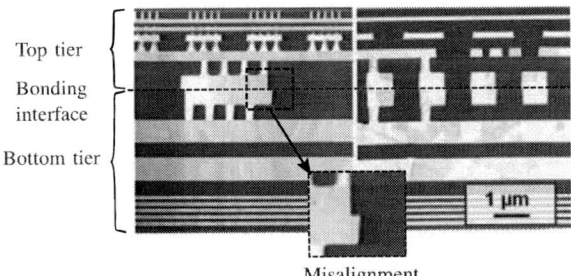

Fig. 1. TEM cross sections of 3D stacked image sensors with 8.8 (left) and 1.44 μm-pitch (right) [1]. Dashed box illustrates the top/bottom misalignement.

This paper aims at investigating the copper diffusion/drift phenomena in the context of HB technology by means of numerical approach. The "reference" diffusion/drift model provides good results for time dependant dielectric breakdown (TDDB) in 1D condition [5]. However, the HB geometry

This work was funded thanks to the French National programme "Programme d'Investissement d'Avenir IRT Nanoelec" ANR-10-AIRT-05.

intrinsically requires a 2D or 3D representation and the "reference" model fails to take into account the concentration variation due to the complexity of the boundary conditions. Consequently, a second model is proposed to overcome this problem. Both models are compared and the influence of voltage, temperature, misalignment, saturation concentration and failure criteria is presented and discussed.

II. MODELS

The time evolution of the copper ions concentration C is described by the usual variational principle of the continuity equation coupled with the Poisson's equation for electrostatics:

$$\frac{\partial C}{\partial t} = \nabla \cdot (\Lambda \nabla \mu) \tag{1}$$

$$\nabla \cdot (\epsilon_r \epsilon_0 \nabla V) = -qC \tag{2}$$

where the electrochemical potential μ and the mobility Λ can be described in terms of derivatives of Gibbs free energy [6]:

$$\mu = \frac{\partial G}{\partial C} + qV \text{ and } \Lambda = D \left(\frac{\partial^2 G}{\partial C^2} \right)^{-1}$$

In a first approach, the diffusion coefficient D is assumed to be independent of the copper ions concentration $D = D_0 exp(-\beta E_a)$ with $\beta = 1/k_B T$ and E_a is the activation energy.

The dimensionless forms of the equations (1) and (2) are obtained by defining the scaling parameters $t_0 = L_0^2/D$ and $V_0 = (\beta q)^{-1}$:

$$\frac{d\phi}{dt} = \tilde{\nabla} \cdot \left[\tilde{\nabla}\phi + \left(\frac{\partial^2 G}{\partial \phi^2} \right)^{-1} \tilde{\nabla}\tilde{V} \right] \tag{3}$$

$$\tilde{\nabla}^2 \tilde{V} = -\Gamma \phi \; ; \; \Gamma = \frac{\beta C_0 (qL_0)^2}{\epsilon_r \epsilon_0} \tag{4}$$

where $\phi = C/C_0$, $\tilde{V} = V/V_0$, $\tilde{\nabla} = L_0 \nabla$ and Γ are dimensionless quantities.

Some studies [5], [7], [8] assume an approximation of the Gibbs free energy such as:

$$G = \frac{C_0}{\beta} \phi ln\phi + \Delta H \tag{5}$$

Equation (5) is a good approximation for low ions copper concentration value, in other words $\phi \ll 1$.

The second derivative Gibbs free energy:

$$\frac{\partial^2 G_1}{\partial \phi^2} = \frac{1}{\phi}\left(1 + \beta\phi\frac{\partial^2 \Delta H}{\partial \phi^2}\right) \approx \frac{1}{\phi} \quad (6)$$

In this work, a second model is introduced. We assume that the copper ions Cu^{i+} drift is done via a number of limited free sites available in the dielectric material. In these conditions, the divergence concentration at the cathode is not possible and a solution is found by the assumption that an equilibrium concentration saturation occurs at $C = C_0$, which results in the modification of the following Gibbs free energy like:

$$G = \frac{C_0}{\beta}\left[\phi ln\phi + (1 - \phi)\, ln\,(1 - \phi)\right] + \Delta H \quad (7)$$

This condition can be imposed easily with the thermodynamical condition to the second derivative Gibbs free energy of mixing:

$$\frac{\partial^2 G_2}{\partial \phi^2} = \frac{1}{\phi(1-\phi)}\left[1 + \beta\phi(1-\phi)\frac{\partial^2 \Delta H}{\partial \phi^2}\right] \approx \frac{1}{\phi(1-\phi)} \quad (8)$$

Such an approach allows to stop the contribution of the electric potential on the concentration evolution when the saturation equilibrium is obtained. Both models (equations (6) and (8)) postulate that the entropy contribution is greater than the enthalpy contribution ΔH.

This paper will initially present a comparison of the "reference" model (equation (3), (4) and (6)) and our "saturation" model (equation (3), (4) and (8)) in one dimension under the boundary conditions shown in Fig. 2. The scaling geometric parameter is taken to be equal to the dielectric domain length $L_0 = L$.

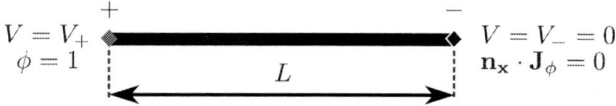

Fig. 2. 1D model for metal ions diffusion/drift study (dielectric illustrated only); bondary conditions.

Secondly, a 2D structure is used to investigate the HB technology from the diffusion/drift point of view (Fig. 3). Misalignement induces a source term of copper ions in the dielectric media (Fig. 3.a2) for both anode and cathode boundaries. Parameters used are listed in TABLE I.

TABLE I
PHYSICAL AND DESIGN PARAMETERS

Description	Symbol [unit]	Value
Electric potential	$V_+ [V]$	0 - 30
Homogeneous temperature	$T\ [^{\circ}C]$	100 - 350
Saturation concentration	$C_0\ [at.m^{-3}]$	10^{23} - 10^{26}
Dielectric domain length	$L\ [nm]$	200 and 600*
Misalignment *	$L^*_{shift}\ [nm]$	20* - 150*
Energy activation	$E_a\ [eV]$	0.653
Pre-exponential factor	$D_0\ [m^2.s^{-1}]$	$1.68.10^{-14}$
Dielectric constant	ϵ_r	3.9
Dielectric breakdown	$E_{bd}\ [V.m^{-1}]$	10^8

*only in 2D

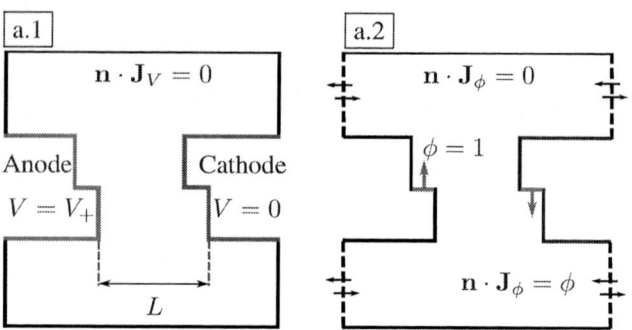

Fig. 3. 2D model for hybrid bonding-based interconnect (dielectric illustrated only), (a.1) electric potential boundary conditions, and (a.2) copper ions concentration boundary conditions.

III. SIMULATION RESULTS AND DISCUSSION

A. "Reference" model vs. "saturation" model: a 1D geometry

COMSOL Multiphysics® is used to solve the previous dimensionless set of equations. Fig. 4 shows the time evolution of the ions concentration profile and the electrical potential V for the 1D geometry (Fig. 2). The "reference" model results (Fig. 4.a.1) shows a divergence of the concentration at the cathode (x=L). This divergence induced a strong electric potential variation localised mainly at the cathode [9]. Achanta et al. [9] defined the time to failure (TTF) as the instance in time when the electric field at the cathode exceeds the breakdown strength. First, we have chosen this criterion to compare our model to the "reference" one.

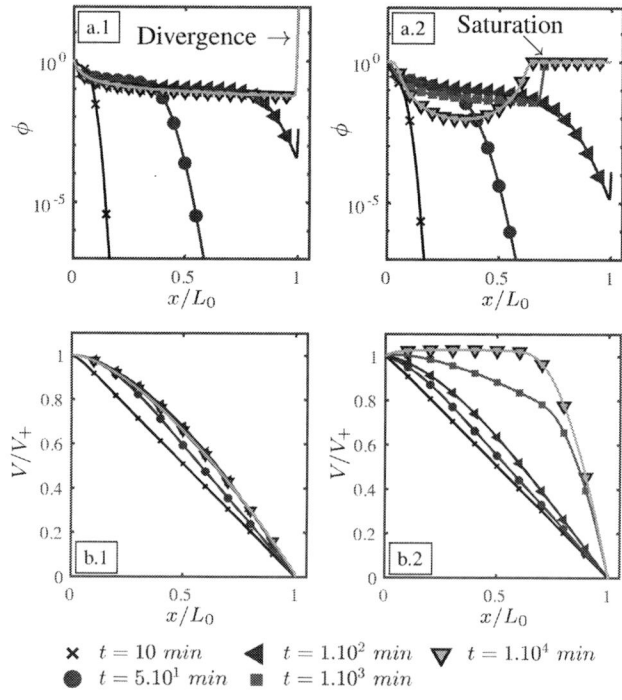

Fig. 4. ϕ vs. $\frac{x}{L_0}$ (a.1 and a.2) and $\frac{V}{V_+}$ vs. $\frac{x}{L_0}$ (b.1 and b.2) for different times in 1D geometry. The results of the "reference" model and the "saturation" model are respectively represented by the couple of plots (a.1, b.1) and (a.2, b.2). $C_0 = 10^{24}\ at.m^{-3}$, $V_+ = 20\ V$.

Our model (Fig. 4.a.2) induces the accumulation of copper ions in free sites until a maximum saturation value is reached.

Due to the saturation mechanism, an electric potential variation inside the dielectric domain can lead, near the cathode, to electric field values larger than the dielectric strength. In addition, for long periods ($t > 10^4\,s$) the maximum of the electric potential slightly exceeds the maximum initial value imposed on the anode, and this for a wide spatial area ($0 < x < 0.6$). This electric potential field significantly deforms the near-cathode field.

The influence of C_0 is shown in Fig. 5. While the dielectric breakdown value is reached in all cases for the "reference" model (Fig. 5.a.), our model shows that there is a saturation concentration limit value for the internal electric field to exceed the failure criterion (Fig. 5.b.). When the concentration saturation value is less than $5.10^{24}\,at.m^{-3}$, the copper ions drift does not modify the electric field sufficiently to induce a dielectric breakdown (i.e. $E < E_{db}$). Thus, the estimation of this concentration becomes a crucial issue.

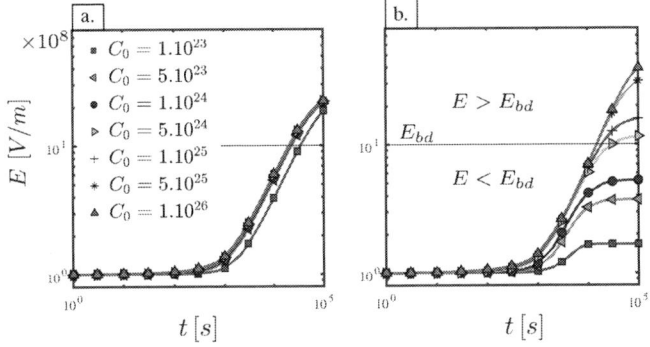

Fig. 5. E vs. t for different concentration saturation C_0. The results of the "reference" model and the "saturation" model are respectively represented by the plot (a.) and (b.).

B. "Saturation" model and hybrid bonding-based geometry

Currently, wafer-to-wafer HB technology provides interconnect pitches in the range of 1 to $10\,\mu m$ and an average misalignment between tiers of $200\,nm$ [1], [10]. Pitch shrinkage requires by future applications potentially increases reliability issues. In that context, this study focused on extreme case with a pitch of $1.2\,\mu m$, which corresponds to a typical dielectric domain length between the anode and the cathode of $0.6\,\mu m$. The saturation concentration limit was set to $C_0 = 10^{26}\,at.m^{-3}$ to calibrate our model with previous studies [11].

Fig. 6 shows the copper ions concentration evolution only produced by the thermal diffusion mechanism. In order to understand the internal electric field induced by the presence of copper ions, two cases were compared: the diffusion equation (3) is solved (i) without coupling with the Poisson's equation $\tilde{\nabla}\tilde{V} = 0 \,\forall t$ (usual Fick law) (Fig. 6.a.1-3), or (ii) by coupling with the Poisson's equation but the anode electric potential is fixed to zero (Fig. 6.b.1-3).

For the first case (i), the two source terms create two areas that grow symmetrically until they gradually merge in the middle of the longer path between the anode and the cathode.

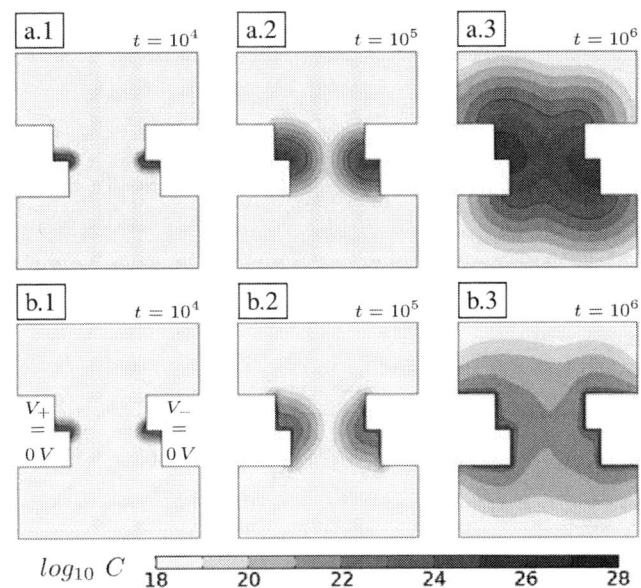

Fig. 6. 2D model for hybrid bonding-based interconnect with the "saturation" model: a- copper ions concentration evolution without electric physics, b- copper ions concentration evolution with electric physics and the anode electric potential $V = 0\,V$.

For the second case (ii), the inclusion of ionized particles creates an internal electric field that opposes to the conventional thermal diffusion. When the copper ions concentration is very low ($\phi \ll 1$), the electric driving force is weaker than the thermal diffusion driving force ($\tilde{\nabla}\tilde{V} \ll \tilde{\nabla}\phi$). However, when the copper ions concentration is very high ($\phi \approx 1$), the electric driving force is stronger than thermal diffusion driving force ($\tilde{\nabla}\tilde{V} \gg \tilde{\nabla}\phi$) changing significantly the diffusion mechanism. Thus, the internal electric field induced by the copper ions inclusion under the dielectric material seems to act as a diffusion barrier for high concentrations.

Fig. 7 shows the effect of the external electric field on the diffusion dynamics. As can be seen, the presence of an external electric field breaks the symmetry between the two copper ions sources. If we only look at the diffusion along the x-axis, the copper ions displacement is accelerated in the direction of the cathode. This implies that the ions from the anode will penetrate more quickly in the dielectric medium and in greater number. At the opposite, the ions from the cathode will be more easily prevented to penetrate the dielectric region.

Furthermore, the intensity of the external electric field changes the topology of copper ions concentration between anode and cathode. As an example, in Fig. 7a.1-3, a copper ions concentration evolves until there is a connection between anode and cathode (percolation mechanism) for a threshold value of $C_{th} = 10^{21}\,at.m^{-3}$ (light orange/deep orange on log scale of Fig. 7) at $t = 10^6\,s$ (arbitrary time). For $C_{th} = 10^{22}\,at.m^{-3}$, the domains never connect. In Fig. 7b.1-3, the connected space for $C_{th} = 10^{21}\,at.m^{-3}$ appears earlier ($t = 10^5\,s$) and there is no connected space for $C_{th} = 10^{22}\,at.m^{-3}$. Finally, when the anode electric potential is sufficiently high, a connected space

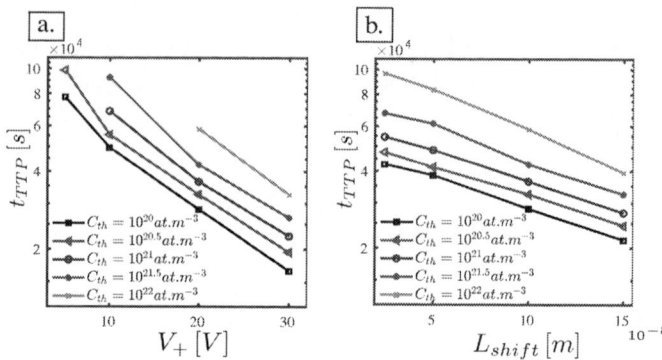

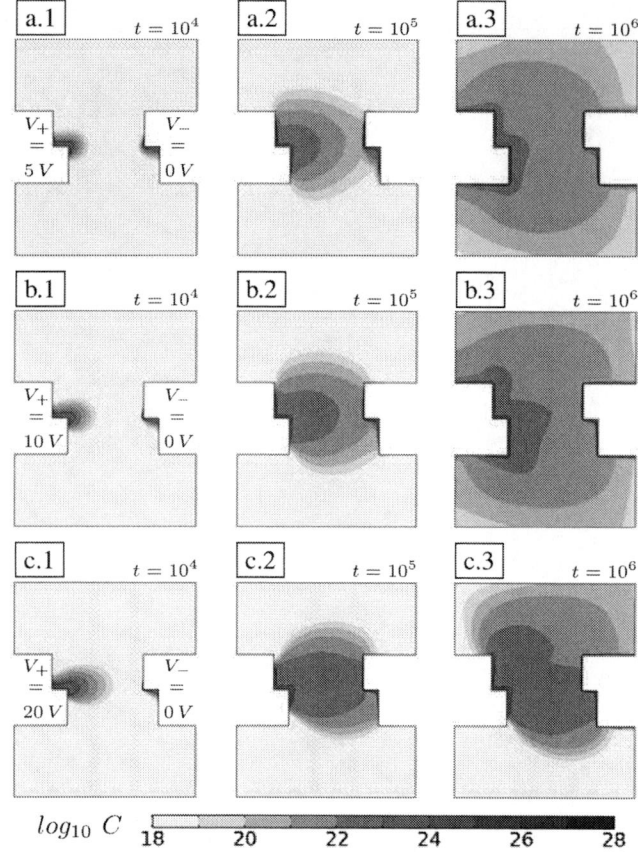

Fig. 7. Effect of the external electric potential for different times $t\,[s]$ and different electric potential: a- $V_+ = 5\,V$; b- $V_+ = 10\,V$ and c-$V_+ = 20\,V$.

for $C_{th} = 10^{22}\,at.m^{-3}$ exists (Fig. 7.c.3).

Regarding the standard TTF estimation, a direct calculation of the electric field at the cathode is inappropriate because peak effect occurs with consequently mesh-dependance. Hwang et al. [12] defined the TTF as the time at which the copper ions concentration at the cathode is $0.1\,\%$ of the solubility (saturation concentration in our case). Although interesting this failure criterion is not satisfactory in our case because of the term source at the cathode. Finally, we defined the percolation time when, for a chosen copper ions concentration value, a topological connected space exists between anode and cathode. A range of values between 10^{20} and $10^{24}\,at.m^{-3}$ is analysed at each time step and a time to percolate (TTP) is extracted when a connection is encountered.

Fig. 8 shows the TTP for different external electric potentials (8.a) and different misalignments (8.b). As it was seen in previous studies in 1D [12], the TTP seems to follow exponential laws:

$$t_{TTP}(V) \propto exp\left(\alpha_1 V_+\right)$$
$$t_{TTP}(L_{shift}) \propto exp\left(\alpha_2 L_{shift}\right)$$

were $\alpha_1 \approx -0.6\,V^{-1}$ and $\alpha_2 \approx -5.10^6\,m^{-1}$. To the extent that misalignment is an experimental stochastic variable

Fig. 8. TTP variation for different conditions: a- electric potential V_+ ranging from 5 to 30 V for a misalignment value $L_{shift} = 100\,nm$; b- misalignment L_{shift} ranging from 25 to 150 nm for $V_+ = 20\,V$.

ranging from 0 to $200\,nm$ [1], the dependence of the TTP on misalignment cannot be overlooked.

IV. CONCLUSIONS AND PERSPECTIVES

To the best of our knowledge, this paper presents the first model to investigate copper ions diffusion/drift in dielectric for hybrid bonding-based IC. This model introduces a copper concentration saturation mechanism that modifies the electrical field until the dielectric breakdown. A new failure criterion, time to percolate, is defined to overcome limitations of current ones dedicate to 1D geometry. In a near future, the TTP must be linked to the TTF using experiments and more complete topological analysis, in particular on the variation of the electric field.

REFERENCES

[1] J. Jourdon et al., in *2018 IEEE International Electron Devices Meeting (IEDM)*, Dec 2018, pp. 7.3.1–7.3.4. https://doi.org/10.1109/IEDM.2018.8614570

[2] J. P. Borja et al., *Dielectric Breakdown in Gigascale Electronics: Time Dependent Failure Mechanisms*. Cham: Springer International Publishing, 2016, ch. General Theories, pp. 11–19.

[3] J. W. McPherson, *Microelectronics Reliability*, vol. 52, no. 9, pp. 1753–1760, 2012. [Online]. Available: https://doi.org/10.1016/j.microrel.2012.06.007.

[4] T. K. S. Wong, *Materials,*. vol. 5, pp. 1602–1625, 2012.https://doi.org/10.3390/ma5091602

[5] R. S. Achanta et al.,*Journal of Applied Physics*, vol. 103, no. 1, p. 014907, 2008. https://doi.org/10.1063/1.2828048

[6] N. Pottier, Physique statistique hors d'équilibre. EDP Sciences, 2007, ch. 2, p. 42.

[7] J. D. McBrayer et al., *J. Electrochem. Soc*, vol. 133, no. 6, pp. 1242–1246, 2002. https://doi.org/10.1149/1.2108827

[8] K.-S. Kim et al., *Jpn. J. Appl. Phys*, vol. 41, no. L99, pp. L99–L101, 2002. https://doi.org/10.1143/JJAP.41.L99

[9] R. S. Achanta et al., *Applied Physics Letters*, vol. 91, no. 23, p. 234106, 2007. https://doi.org/10.1063/1.2823576

[10] E. Beyne et al., 2017 *IEEE International Electron Devices Meeting (IEDM)*, pp. 32.4.1-32.4.4, 2017. https://doi.org/10.1109/IEDM.2017.8268486

[11] J. L. Plawsky et al., *Journal of Materials Science: Materials in Electronics*, vol. 23, no. 1, pp. 48–55, Jan 2012. https://doi.org/10.1007/s10854-011-0406-x

[12] S.-S. Hwang et al., *Journal of Applied Physics*, vol. 101, no. 7, p. 074501, 2007. https://doi.org/10.1063/1.2714668

978-1-7281-0941-1/19 $31.00 © 2019 IEEE

TCAD Framework to Estimate the NBTI Degradation in FinFET and GAA NSFET Under Mechanical Strain

Ravi Tiwari[1], Narendra Parihar[1], Karansingh Thakor[1], Hiu-Yung Wong[2], and Souvik Mahapatra[1]

[1]Department of Electrical Engineering, Indian Institute of Technology Bombay, Mumbai 400076, India
[2]San Jose State University, San Jose, CA 95192, USA
*Phone: +91-222-572-0408, Email: souvik@ee.iitb.ac.in

Abstract--A physics-based TCAD framework is used to estimate the interface trap generation (ΔN_{IT}) during Negative Bias Temperature Instability (NBTI) stress in P-channel FinFET and Gate All Around (GAA) Nano-Sheet (NS) FET. The impact of mechanical strain due to channel length scaling (L_{CH}) on ΔN_{IT} generation is estimated. The bandstructure calculations are used to explain the impact of mechanical strain on ΔN_{IT} generation.

Keywords—NBTI, RD model, GAA-NSFET, FinFETs, mechanical strain, RD model, L_{CH} scaling.

I. INTRODUCTION AND BACKGROUND

The NBTI continues to remain a critical reliability issue in P-channel FinFETs [1,2] and GAA NSFETs [3]. It results in buildup of positive charges in the gate insulator which shift various device parameters over time and hampers the longtime operation of devices and circuits. Therefore, the accurate modeling of NBTI degradation is necessary for reliable operation of devices and circuits at advanced technology nodes. Although the physical mechanism of NBTI is debated [4], a framework explained in [5] has been used to predict the DC and AC stress and recovery kinetics in various technologies such as Si and Si capped SiGe planar, Si and SiGe FinFET and FDSOI, and Si SOI FinFET having HKMG gate stacks [5]-[11]. The framework uses uncorrelated contributions from interface (ΔV_{IT}) and bulk (ΔV_{OT}) trap generation and trapping of holes (ΔV_{HT}) in pre-existing traps.

The framework of [5] uses Reaction-Diffusion Model (RD Model) to calculate trap density (ΔN_{IT}) at channel/interlayer (IL) and IL /High-K interfaces, Transient Trap Occupancy Model (TTOM) to calculate their charge occupancy and contribution to ΔV_{IT}, and analytical models to model the ΔV_{HT} and ΔV_{OT} time kinetics. Analyses of multiple technologies [5]-[11] have established that although the overall ΔV_T at shorter stress time is due to the multiple subcomponents, ΔV_{IT} dominates end of life (EOL) ΔV_T at lower bias use conditions.

Recently, RD model with proper physics has been implemented in Sentaurus Device using the Multi-State-Configuration (MSC) framework and Capture-Emission Depassivation (CED) model for inversion hole and IL field assisted breaking of interfacial Si-H bonds [12]-[14]. Figure 1 shows the schematic of MSC hydrogen transport model. In this model, hydrogen (H) passivated bonds (X–H) at the channel/gate insulator interface get broken, which create defects (X–) and H atoms during the stress phase. These H atoms diffuse in the gate insulator bulk and react with other H passivated bonds (Y–H), which result in additional defects (Y–) and H_2 molecules, which diffuse away in the gate stack layers. Note that, all the bulk insulator defects are assigned to IL/High-K interface for analysis. During recovery, the diffused H_2 reach

out and react with Y- defects and create Y-H bonds and H atoms. These H atoms further react (re-passivation) with X-defects and form X-H bonds. With proper capacitance ratio, trap contribution from both the interfaces is used to compute ΔN_{IT} CED model is used for inversion hole (hole density p_H) and IL field (E_{ox}) assisted breaking of interfacial Si-H bonds as shown in Fig.2 [12]-[14]. In the presence of oxide electric field (E_{ox}), inversion layer hole tunnel to X-H polarized bonds, get captured and broken due to thermal dissociation (activation E_{AKF1}). The framework calculates reactions at channel/IL and IL/High-K interfaces and diffusion of H and H_2.

The RDM framework has been successfully validated against the measured ΔN_{IT} time kinetics in Si and SiGe channel FinFETs [12]-[13]. The model can accurately predict the stress and recovery time kinetics for different V_{GSTR} and T using only four process dependent parameters: Pre-factor for Si-H bond dissociation (K_{FIT}), Activation energy (E_{AKF}), Tunneling parameter (Γ_0), and bond polarization factor (α).

II. SCOPE OF THIS WORK AND EXPERIMENTAL DETAILS

Figure 3 shows the schematic of the simulation framework used to calculate the ΔN_{IT} time kinetics in FinFET and NS-FET devices. The Sentaurus Process [15] is used to generate FinFET and NSFET structures with a process flow that is consistent with actual practice including the epitaxial SiGe Source-Drain, whose volume decides the stress distribution in the fin and the three nano-sheets [16-17], as shown in Fig. 4. The structure uses a metal backend for H_2 diffusion. Note that unlike Sentaurus Structure Editor, the Sentaurus Process rightly consider the process induced strain in the channel due to variation in L_{CH}. The Bandstructure calculations [18] by the tight binding method are used to determine the impact of mechanical strain on the tunneling effective mass (m_T) and valence band offset (φ_B) which in turn affect the parameter K_{F10} and Γ_0 (Fig.2 of reference [14]). Table-1 shows the dimensional description of the devices used in this work. The RDM is used to calculate generation and passivation of ΔN_{IT} at channel/ IL and IL/High-K interfaces in FinFET and GAA NSFET for different L_{CH}.

III. EVALUATION OF STRAIN IMPACT

Figure 5 shows the 3D isometric view of Source/Drain SiGe epi induced stress distribution along the channel direction ([110]) in FinFET and NSFET for a fixed channel length of 14nm. Figure 7 shows the 1D cut of the stress profile along the channel direction (Stress-ZZ) for FinFET and NSFET. Note that SiGe epi dominantly impacts the top part of the Fin in FinFET (Fig.7(a)) and top sheet in case of NSFET (Fig.7(b)).

978-1-7281-0941-1/19 $31.00 © 2019 IEEE

The 1D cut of the stress distribution in all the three directions of FinFET and NSFET is shown in Fig. 6. The stress magnitude in the other two directions (Stress-XX and Stress-YY) is negligible in both FinFET and NSFET. Figure 8 shows the L_{CH} dependence of integrated stress along the channel direction for FinFET and NSFET. The integrated stress in the FinFET is higher than the average of the integrated stress in all sheets of NSFET for different L_{CH}, and it increases with a decrease in L_{CH}.

The impact of uniaxial compressive stress (UCS) on the valence bandstructure of (110) and (100) oriented Si surface is shown n Fig. 9. The impact of channel stress on m_T and φ_B is different in FinFET and NSFET due to different conducting surface domination; (110) in FinFET and (100) in the case of NSFET. The relative increase in m_T is higher for FinFET compared to NSFET (Fig.10) while relative change in φ_B is smaller for both FinFET and NSFET (Fig. 11). Figure 12 shows the impact of increase in strain (or decrease in L_{CH}) on the K_{F10} and Γ_0. With increases in strain (or decreases in L_{CH}), the K_{F10} reduces and Γ_0 increases. The reduction in K_{F10} and increase in Γ_0 are higher for FinFET compared to NSFET (Fig.11).

IV. MSC (DEVICE) SIMULATION

For device simulations, both FinFET and NSFET are calibrated for the same I_{OFF} current by changing the metal work function. The MSC simulations are then performed for different V_{GSTR} and for different L_{CH} in both the architectures. Both the FinFET and NSFET show similar time kinetics and longtime power-law time dependence, as shown in Fig.13. Figure 14 compares the fixed time ΔN_{IT} with and without mechanical strain for different L_{CH} at operating conditions for both the architectures. For both FinFET and NSFET, the ΔN_{IT} reduces with a reduction in L_{CH}. For a fixed L_{CH}, the NSFET shows smaller degradation compared to FinFET when strain is not considered. This ascribed to the lower field in the NSFET because of the fully depleted sheet and lower precursor bond density for (100) surface compared to (110) surface. These results are consistent with the measurement data shown in [3] for long channel FinFET and NSFET where strain due to SiGe epi will be minimal. However, when the effect of strain is considered, the actual ΔN_{IT} becomes lower for FinFET compared to NSFET due to a significant increase in m_T for (110) surface compared to (100) surface (Fig. 13(a)). The reduction in ΔN_{IT} is higher for shorter L_{CH} due to higher mechanical strain. The Voltage Acceleration Factor (VAF), an important parameter to extrapolate the degradation measured at higher stress voltages to operating voltage, of NSFET is smaller compared to FinFET, as shown in Fig. 15. The lower VAF for NSFET suggests higher NBTI degradation at the operating voltage. Figure 16 compares the VAF for different L_{CH} in FinFET and NSFET. Although the unstrained VAF is smaller for FinFET compared to NSFET, the strained VAF is higher for FinFET compared to NSFET due to higher m_T variation which increases the Γ_0 (Fig. 11).

V. CONCLUSION

Sentaurus Device framework with proper physical models and Sentaurus Process are used to study the NBTI in FinFET and NSFET devices considering the impact of mechanical strain for different L_{CH}. The NSFET shows lower degradation compared to FinFET when the effect of strain is not

considered. This is the case for long channel devices where the SiGe epi induced strain is small. However, the NSFET shows higher degradation than FinFET when effect of strain is considered. This is primarily due to (110) dominated surface in FinFET which shows relatively higher increase in m_T compared to (100) dominated surface in NSFET.

REFERENCES

[1] S. Ramey, et al, "Intrinsic transistor reliability improvements from 22nm tri-gate technologyv," in *Proc. Int. Rel. Phys. Symp.*, pp. 4C.5.1-4C.5.5, Apr. 2013.

[2] Anisur Rahman et al., "Reliability Studies of a 10nm High-performance and Low-power CMOS Technology Featuring 3rd Generation FinFET and 5th Generation HK/MG", in *Proc. Int. Rel. Phys. Symp. 2018*, pp. 6F.4-1-6F.4-6.

[3] Miaomiao Wang et al, "Bias Temperature Instability Reliability in Stacked Gate-All-Around Nanosheet Transistor", in *Proc. Int. Rel. Phys. Symp. 2019*.

[4] J. H. Stathis, Souvik Mahapatra, Tibor Grasser, "Controversial issues in negative bias temperature instability", *Microelectronics Reliability*, Volume 81, 2018, Pages 244-251.

[5] N. Parihar, N. Goel, S. Mukhopadhyay and S. Mahapatra, "BTI Analysis Tool—Modeling of NBTI DC, AC Stress and Recovery Time Kinetics, Nitrogen Impact, and EOL Estimation," *in IEEE Trans. Electron Devices*, vol. 65, no. 2, pp. 392-403, Feb. 2018.

[6] N. Parihar and S. Mahapatra, "Prediction of NBTI stress and recovery time kinetics in Si capped SiGe p-MOSFETs," *2018 IEEE IRPS*, Burlingame, CA, 2018, pp. P-TX.5-1-P-TX.5-7,

[7] N. Parihar, R. Southwick, M. Wang, J. Stathis, and S. Mahapatra, "Modeling of NBTI time kinetics and T dependence of VAF in SiGe p-FinFETs," in *IEDM Tech. Dig.*, Dec. 2017, pp. 167–170.

[8] N. Parihar, R. G. Southwick, M. Wang, J. H. Stathis and S. Mahapatra, "Modeling of NBTI Kinetics in RMG Si and SiGe FinFETs, Part-I: DC Stress and Recovery," in *IEEE Transactions on Electron Devices*, vol. 65, no. 5, pp. 1699-1706, May 2018

[9] N. Parihar, R. G. Southwick, M. Wang, J. H. Stathis and S. Mahapatra, "Modeling of NBTI Kinetics in RMG Si and SiGe FinFETs, Part-II: AC Stress and Recovery," in *IEEE Transactions on Electron Devices*, vol. 65, no. 5, pp. 1699-1706, May 2018.

[10] V. Huard, C. Ndiaye, M. Arabi, N. Parihar X. Federspiel, S. Mhira, S. Mahapatra and A. Bravaix,, "Key parameters driving transistor degradation in advanced strained SiGe channels," in *Proc. Int. Rel. Phys. Symp.*, 2018, pp. P-TX.4-1-P-TX.4-6.

[11] N. Parihar, U. Sharma, R. G. Southwick, M. Wang, J. H. Stathis and S. Mahapatra, "Ultrafast Measurements and Physical Modeling of NBTI Stress and Recovery in RMG FinFETs Under Diverse DC–AC Experimental Conditions," *IEEE Trans. Electron Devices*, vol. 65, no. 1, pp. 23-30, Jan. 2018.

[12] Ravi Tiwari, et al, "A 3-D TCAD Framework for NBTI, Part-I: Implementation Details and FinFET Channel Material Impact" *in IEEE Trans. Electron Devices*, vol. 66, issue: 5, pp. 2086 - 2092, May 2019.

[13] Ravi Tiwari, et al, "A 3-D TCAD Framework for NBTI, Part-II: Impact of Mechanical Strain, Quantum Effects and FinFET Dimension Scaling " *in IEEE Trans. Electron Devices*, vol. 66, issue: 5, pp. 2093 - 2099, May 2019.

[14] N. Parihar et al, "Modeling of Process (Ge, N) Dependence and Mechanical Strain Impact on NBTI in HKMG SiGe GF FDSOI p-MOSFETs and RMG p-FinFETs," in proc. *Simulation of Semiconductor Processes and Devices (SISPAD)*, 2018, pp. 167-171.

[15] Sentaurus™ Process user guide, N-2017.09.

[16] P. Hashemi et al, "Replacement High-K/Metal-Gate High-Ge-Content Strained SiGe FinFETs with High Hole Mobility and Excellent SS and Reliability at Aggressive EOT ~7Å and Scaled Dimensions Down to Sub-4nm Fin Widths", in *Symp. on VLSI Tech.*, 2017.

[17] N. Loubet et al, "Stacked Nanosheet Gate-All-Around Transistor to Enable Scaling Beyond FinFET", in *Symp. on VLSI Tech.*, 2017.

[18] https://nanohub.org/resources/bandstrlab

Acknowledgement: Victor Moroz, Steve Motzny, and Munkang Choi from Synopsys Inc. for useful discussion.

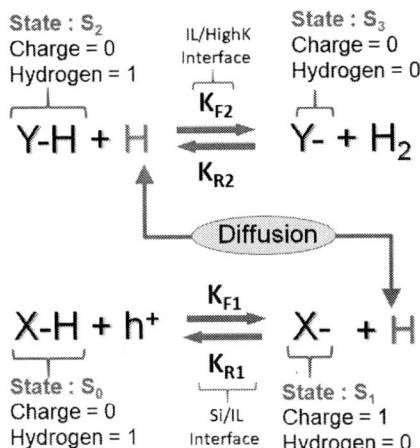

State : S_2
Charge = 0
Hydrogen = 1

IL/HighK Interface

State : S_3
Charge = 0
Hydrogen = 0

K_{F2}
K_{R2}

$Y{-}H + H \rightleftarrows Y^{-} + H_2$

Diffusion

$X{-}H + h^{+} \rightleftarrows X^{-} + H$

K_{F1}
K_{R1}

State : S_0
Charge = 0
Hydrogen = 1

Si/IL Interface

State : S_1
Charge = 1
Hydrogen = 0

Fig.1. Schematic of the Multi-State-Configuration (MSC) Hydrogen transport degradation model and state diagram of Hydrogen depassivation (by Capture Emission Depassivation or CED model, Fig.2). MSC is used to model reactions between the mobile hydrogen elements and localized hydrogen states such as silicon-hydrogen bonds at the Si–SiO2 interface. Electrically active defects at Si–SiO2 interface are denoted as X-H bonds because of its unknown nature. Y-H bonds are located at IL/HighK interface. S_0, S_1, S_2 and S_3 are the state occupation probabilities used in MSC model. K_{F1}, K_{R1}, K_{F2} and K_{R2} are the forward and reverse reaction rates at Si/IL and IL/HighK interface respectively.

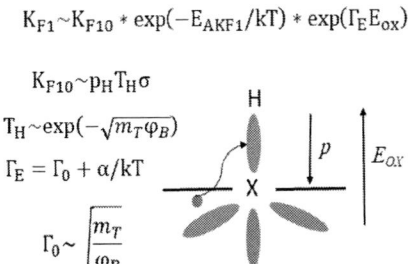

$$K_{F1} \sim K_{F10} * \exp(-E_{AKF1}/kT) * \exp(\Gamma_E E_{ox})$$

$$K_{F10} \sim p_H T_H \sigma$$

$$T_H \sim \exp(-\sqrt{m_T \varphi_B})$$

$$\Gamma_E = \Gamma_0 + \alpha/kT$$

$$\Gamma_0 \sim \sqrt{\frac{m_T}{\varphi_B}}$$

Fig.2. Schematic of H passivated bond dissociation process at the channel/IL interface used in Capture Emission Depassivation (CED) model. Inversion layer holes tunnel into polarized interfacial X-H bonds, aided by oxide electric field.

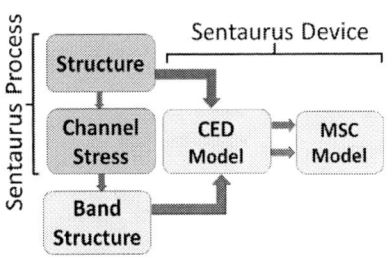

Fig.3. TCAD simulation framework: structure, material and strain are calculated from process simulation, band structure is calculated using tight binding method, CED and MSC models are used to calculate ΔN_{IT} kinetics.

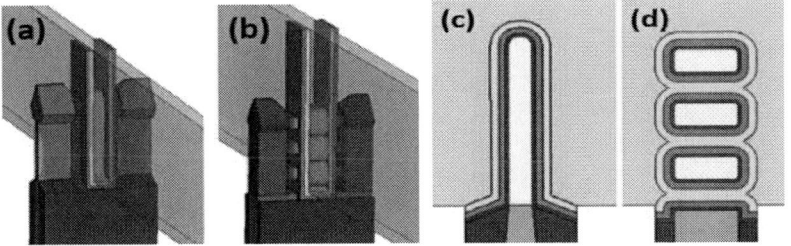

Fig.4. Isometric view of 3D (a) p-FinFET, and (b) p-NSFET structure with raised Source drain and lateral backend for hydrogen diffusion. 2D cross section of the channel in (c) p-FinFET, and (d) p-NSFET showing IL, High-K, TiN-Cap and Tungsten layers. Equivalent Oxide Thickness (EOT) of 1nm is used.

FinFET	NSFET
Fin Height = 50nm	Nano-Sheet thickness = 6nm
Fin Width = 6nm	Nano-sheet width = 15nm
	Inter-Sheet Gap = 5nm

Table I. Dimensions details of the device used in this work.

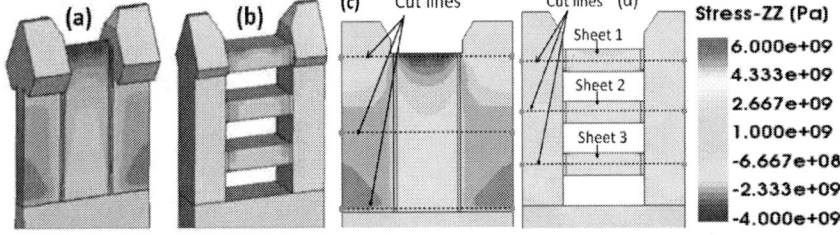

Fig.5. Isometric view of 3D (a) p-FinFET, and (b) p-NSFET showing stress distribution along the channel. Corresponding 2D view of (c) p-FinFET and (d) p-NSFET from source to drain with cut-lines for next 1D plot.

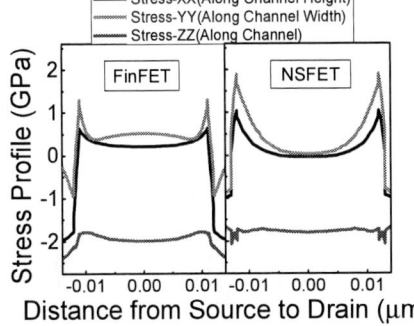

Fig.6. Stress profile from Source to Drain in (a) p-FinFET in the middle of the fin, and (b) p-NSFET in the middle of the sheet-2.

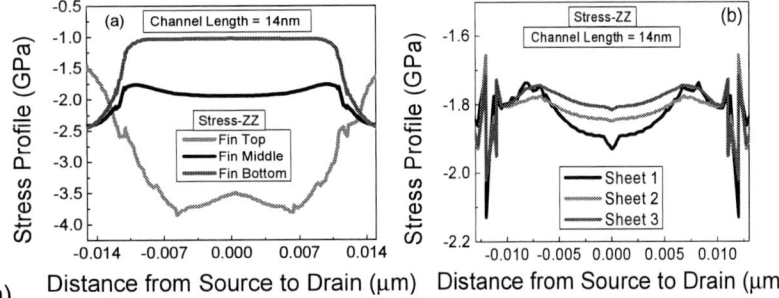

Fig.7. 1D stress profile (Stress-ZZ) in (a) p-FinFET at three positions (cut lines are shown in Fig. 5) and (b) all three sheets of p-NSFET. Fin top in FinFET and Sheet1 in NSFET are showing higher stress because of SiGe- epitaxial SD volume. TCAD simulation.

978-1-7281-0941-1/19 $31.00 © 2019 IEEE

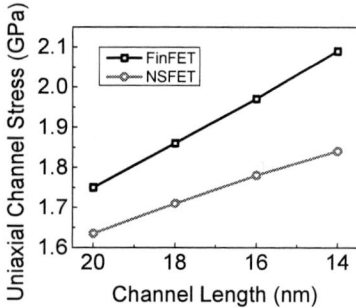

Fig.8. Uniaxial compressive stress along the channel for different channel length for p-FinFET and p-NSFET. TCAD simulation.

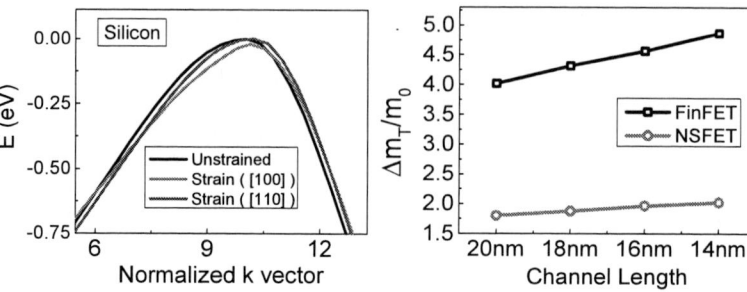

Fig.9. E-K diagram of the top most valence band with and without Uniaxial Compressive Stress (UCS) for [110] (dominating surface in FinFET) and [100] (dominating surface in NSFET).

Fig.10 Impact of UCS on barrier height φ_B for p-FinFET and p-NSFET.

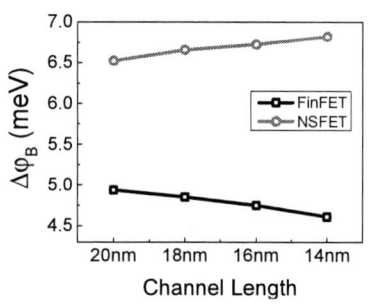

Fig.11. Impact of UCS on barrier height φ_B for p-FinFET and p-NSFET.

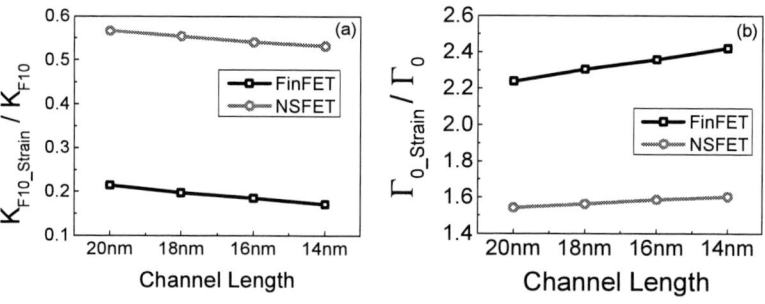

Fig.12. Mechanical strain impact on bond dissociation pre-factor (a) $K_{F10,}$ and (b) field acceleration parameter (Γ_0) for different channel length. Parameters are normalized to unstrained values. TCAD simulation.

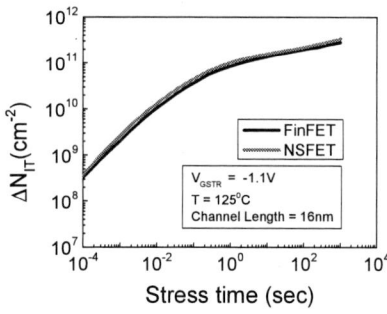

Fig.13. Consistency in ΔN_{IT} time kinetics in p-FinFET and p-NSFET for same V_{GSTR}-T and channel length. TCAD simulation.

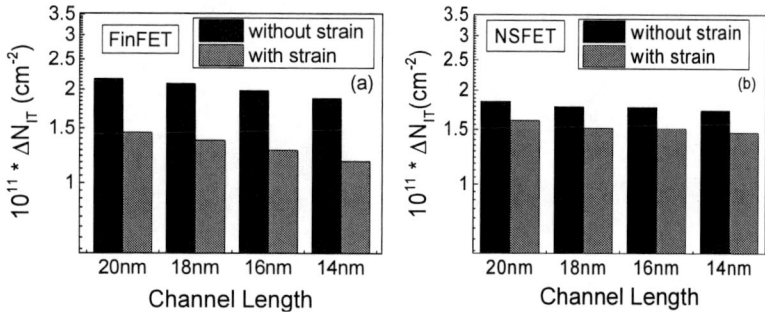

Fig.14. Mechanical strain impact on fixed time ΔN_{IT} of 1Ks for (a) p-FinFET, and (b) p-NSFET, showing higher degradation in NSFET as compared to p-FinFET when proper strain calculation is incorporated in TCAD simulation, at fixed V_{GSTR} = -0.8V and T= 125°C.

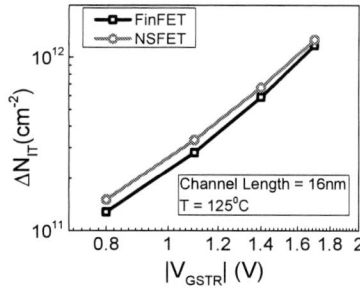

Fig.15. Fixed time ΔN_{IT} comparison between p-FinFET and p-NSFET at 1Ks vs V_{GSTR} at fixed channel length. TCAD simulation.

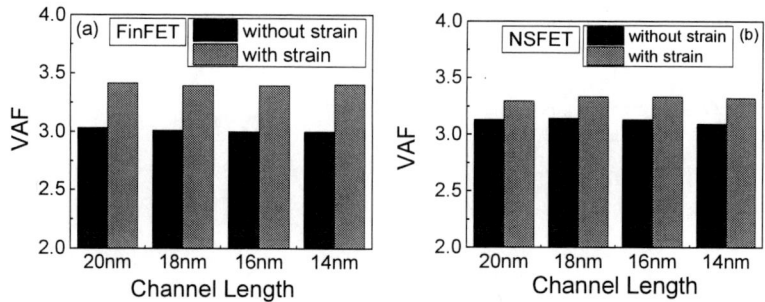

Fig.16. Mechanical impact on voltage acceleration factor (VAF $\propto \Gamma_0$) for both (a) p-FinFET, and (b) p-NSFET for different channel lengths. T=125°C. TCAD simulation.

978-1-7281-0941-1/19 $31.00 © 2019 IEEE

A Stochastic Hole Trapping-Detrapping Framework for NBTI, TDDS and RTN

Sharang Bhagdikar and Souvik Mahapatra*
Department of Electrical Engineering
Indian Institute of Technology Bombay
Mumbai 400076, India
*Phone: +91-222-572-0408, Email: souvik@ee.iitb.ac.in

Abstract--A stochastic framework is presented to model hole trapping and detrapping into and out of individual defects that are present in the gate dielectric of a p-channel MOS transistor. The model calculates thermionic reactions between uncharged and charged states of a defect that are separated by an energy barrier, by using the Gillespie Stochastic Simulation Algorithm (GSSA). The model is validated using experimental data from small area devices under Negative Bias Temperature Instability (NBTI), Random Telegraph Noise (RTN) and Time Dependent Defect Spectroscopy (TDDS) studies.

Keywords—GSSA, NBTI, RTN, TDDS.

I. INTRODUCTION

NBTI is a key reliability concern in small and large area pMOSFETs [1]. The degradation and recovery of ΔV_T during and after NBTI stress is due to the cumulative contributions from interface trap generation (ΔV_{IT}), hole trapping (ΔV_{HT}) and bulk trap generation (ΔV_{OT}) subcomponents. Accurate modelling of the same requires understanding the physical mechanisms that govern these processes.

Hole trapping and detrapping results in drain current and threshold voltage instabilities during NBTI, RTN and TDDS studies [2]-[4]. Historically, the Shockley-Read-Hall (SRH) mechanism is used to calculate the time kinetics [5], usually with temperature (T) activated capture cross section to account for trap relaxation and phonon coupling effects [6]. Alternatively, trap relaxation effect is handled by using thermionic [7] or non-radiative multi-phonon (NMP) [8] processes, in models that treat the uncharged and charged states of a defect as two energy levels; either separated by an energy barrier [7] or represented as intersecting parabolic potentials of a quantum harmonic oscillator [8]. Deterministic implementation of these models are used for NBTI kinetics in large area devices [9], [10], and the stochastic implementation of the multi-phonon model are used for NBTI and TDDS kinetics in small area devices [11]. Note, the original thermionic model of [7] is modified in [9] using a thermally activated barrier for explanation of NBTI stress-recovery kinetics under different T. This deterministic Activated Barrier Double Well Thermionic (ABDWT) model [9] is further validated in [12] by using NBTI stress-recovery data from diverse experimental conditions (wide stress V_G and T ranges) and different technologies. Moreover, the dependence of capture (τ_c) and emission (τ_e) times, associated with hole trapping and detrapping during TDDS and RTN on gate voltage (V_G) is not shown in [11]. This particular aspect is addressed by the extended multi-phonon model of [4]. However large number of parameters (mean + spread = 22) makes it hard to implement in practical situations. Moreover, explanation of NBTI kinetics (like in [12]) is not shown (yet) even using [4].

II. SCOPE OF THIS WORK

The success of deterministic ABDWT model in explaining measured NBTI stress-recovery kinetics over wide T range (– 40°C to +150°C) and for multiple technologies, such as Silicon Oxynitride (SiON), Gate First (GF) High-K Metal Gate (HKMG) and Replacement Metal Gate (RMG) HKMG based planar and FinFET devices [9], [12] has motivated the stochastic version of the same in this work. The mean of multiple stress-recovery simulations matches mean NBTI kinetics measured in multiple small area devices under different stress V_G and T. The V_G and T dependence of τ_c and τ_e for RTN and TDDS studies can be explained. The time kinetics of individual defects during TDDS can also be explained.

III. MODEL FRAMEWORK

The ABDWT model provides transition rates for charge (hole in p-FET) capture and emission within a trap, Fig.1. The hole capture reaction is modeled as a transition from a reference neutral state (E_1) to the charged state (E_2) via a barrier (E_B). The barrier E_B and state E_2 reduces when a gate bias (V_G) is applied [12]. The barrier energy E_B is distributed in energy to model the spatial and energetic distribution of traps. For setting up the stochastic simulation, a finite number of defects are randomly distributed at the Si/oxide interface. Each defect is assigned random barrier energy E_B obtained from a normal distribution. Simulations are performed on multiple devices by invoking GSSA [13] to generate individual trapping and detrapping transients. A hole trapping (or detrapping) event manifests onto the stochastic ΔV_{HT} transient as a positive (or negative) discrete jump. Macroscopic simulation with identical ABDWT model parameters [12] is shown to match mean of multiple stochastic simulations.

IV. NBTI VALIDATION

Fig.2 depicts measured ΔV_T stress kinetics from multiple small area devices of type D2 [12] at a reference stress V_{GSTR} and temperature (T) along with their mean. Mean of measured ΔV_T kinetics is modeled by the macroscopic BAT framework [2], which isolates the subcomponents ΔV_{IT} (generation of interface traps) and ΔV_{HT}. It is seen that the impact of ΔV_{HT} is relatively higher at shorter stress time and lower T whereas

978-1-7281-0941-1/19 $31.00 © 2019 IEEE

ΔV_{IT} dominates at long stress time and high T. Mean of individual stochastic ΔV_{HT} traces is shown to converge with the macroscopic model ΔV_{HT}, Fig.3. Comparison of stochastic mean with experimental ΔV_{HT} (mean) is performed for a range of V_{GSTR} (Fig.4) and with macroscopic ΔV_{HT} curves for a range of T (Fig.5) to affirm the veracity of the model in a rigorous fashion. In Figs.6-9, similar treatment is accorded to the ΔV_{HT} transients during recovery: matching of mean and macroscopic (Fig.6), isolating the mean ΔV_{HT} from measured data (Fig.7) using BAT, and model experimental data for various V_G (Fig.8) and T (Fig.9). Fig.10 and 12 depict the measured ΔV_{HT} time kinetics during stress and recovery respectively over an extended temperature range (-40°C to +150°C) for device type D3 [12]. The model calculated time transients are reproduced in Fig.11 and 13 and are shown to be concurrent with experimental data over the large temperature range.

V. RTN AND TDDS VALIDATION

Fig.14 shows dependence of τ_c and τ_e on V_G obtained from RTN measurements at various T [3]. This is reproduced by model (Fig.15) that reveals similar T activation trends. Correlation of τ_c and τ_e with change in temperature [3] is depicted in Fig.16. A slope greater than 1 indicates τ_e is more strongly coupled with T whereas a slope less than 1 suggests a stronger coupling for τ_c. For slopes ~1, τ_c and τ_e show equal T acceleration. These trends are modeled in Fig.17 by simulating individual traps with appropriate ABDWT model parameters to generate different T accelerations for each.

Recovery step heights versus emission time plot (Fig.18) from TDDS measurement [11] for ΔV_{HT} dominated (shorter time of stress and lower T) and ΔV_{IT} dominated (longer time of stress and higher T) kinetics is reproduced (Fig.19). ΔV_{HT} dominated kinetics show shorter emission times and hence recover faster. Step heights generated from the stochastic model (Fig.19) are found to be consistent with above.

Fig.20 models the V_G dependence of τ_c and τ_e acquired from TDDS measurements for a non-switching trap [4]. The bias dependence is reproduced across two different T using suitable model parameters for the trap. The vanilla ABDWT model is unable to reproduce the distinct tapering off of capture time constants towards saturation at higher biases, as illustrated in Fig.20. This is addressed by modifying the field activated barrier lowering by introducing a weakly quadratic dependent term in addition to the linearly dependent term which is present in the original model. The same is depicted schematically in Fig.21. The modified ABDWT model yields the correct capture time bias dependence, Fig.22. Prediction of switching trap [4] time constants is also shown using the modified ABDWT model, Fig.23. Prediction of experimental TDDS data of trap 'A1' is performed using the classical NMP transition rates in [14]. These require a correction factor to accurately model the data. In Fig.24, ABDWT is invoked to model the TDDS data of trap 'A1'.

Fig.25 enumerates the distinct types of bias couplings observed in RTN data [15]. The different V_G dependencies can be reproduced by the model upon selection of appropriate parameters.

V. CONCLUSION

GSSA is used to implement stochastic version of the ABDWT model for hole trapping-detrapping. Experimental time kinetics for NBTI stress-recovery are accurately predicted over a range of biases, temperatures and technologies. RTN measured capture-emission time constants are reproduced and their T activation trends are captured Switching and non-switching trap characteristics are modelled using TDDS data. Different V_G dependence of τ_e and τ_c of different RTN traps can be explained.

REFERENCES

[1] S. Ramey, et al, "Intrinsic transistor reliability improvements from 22nm tri-gate technologyv," in *Proc. Int. Rel. Phys. Symp.*, pp. 4C.5.1-4C.5.5, Apr. 2013.

[2] N. Parihar, et al, "BTI Analysis Tool-Modeling of NBTI DC, AC Stress and Recovery Time Kinetics, Nitrogen Impact, and EOL Estimation," in IEEE *Trans. Electron Devices*, vol. 65, no. 2, pp. 392-403, Feb.2018.

[3] H. Miki, et al, "Voltage and temperature dependence of random telegraph noise in highly scaled HKMG ETSOI nFETs and its impact on logic delay uncertainty", in *Symposium on VLSI Technology* (VLSIT), pp. 138, June 2012.

[4] T. Grasser, et al, "On the Microscopic Origin of the Frequency Dependence of Hole Trapping in pMOSFETs", IEDM, Dec. 2012, pp. 19.6.4.

[5] W. Shockley and W. T. Read, Jr., "Statistics of the Recombinations of Holes and Electrons", Phys. Rev. 87, 835, Sep. 1952.

[6] F. Schanovsky, et al, "Advanced modeling of charge trapping at oxide defects", *Simulation of Semiconductor Processes and Devices* (SISPAD), Oct. 2013, pp. 451.

[7] D. Ielmini, et al, "A unified model for permanent and recoverable NBTI based on hole trapping and structure relaxation", 2009 IEEE *International Reliability Physics Symposium*, Apr. 2009, pp. 26.

[8] C. H. Henry and D. V. Lang, "Nonradiative capture and recombination by multiphonon emission in GaAs and GaP", Phys. Rev. B 15, pp. 989, Jan. 1977.

[9] S. Desai, S. Mukhopadhyay, N. Goel, N. Nanaware, B. Jose, K. Joshi and S. Mahapatra, "A comprehensive AC / DC NBTI model: Stress, recovery, frequency, duty cycle and process dependence," 2013 IEEE *International Reliability Physics Symposium* (IRPS), Anaheim, CA, 2013, pp. XT.2.1-XT.2.11. doi: 10.1109/IRPS.2013.6532117.

[10] G Rzepa, et al, "Efficient physical defect model applied to PBTI in high-κ stacks", 2017 IEEE *International Reliability Physics Symposium* (IRPS), pp. XT.11.1-XT.11.6.

[11] Anandkrishnan R, et al, "A Stochastic Modeling Framework for NBTI and TDDS in Small Area p-MOSFETs", *Simulation of Semiconductor Processes and Devices* (SISPAD) , Austin TX, Sep. 2018, pp. 181.

[12] Nilotpal Choudhury, Thirunavukkarasu A, Narendra Parihar, Nilesh Goel and Souvik Mahapatra, "A Semi-Physical Model for Hole Trapping-Detrapping Kinetics During NBTI", IEEE *Trans. Electron Devices* 2019, under review.

[13] Gillespie, Daniel T., "A General Method for Numerically Simulating the Stochastic Time Evolution of Coupled Chemical Reactions", Journal of Computational Physics. 22 (4): 403–434, 1976.

[14] Goes, et al, "Identification of oxide defects in semiconductor devices: A systematic approach linking DFT to rate equations and experimental evidence" in *Microelectronics Reliability* 87, 2018 pp. 286-320.

[15] H. Miki, et al, "Understanding short-term BTI behavior through comprehensive observation of gate-voltage dependence of RTN in highly scaled high-κ / metal-gate pFETs", *Symposium on VLSI Technology* - Digest of Technical Papers, June 2011, pp. 149.

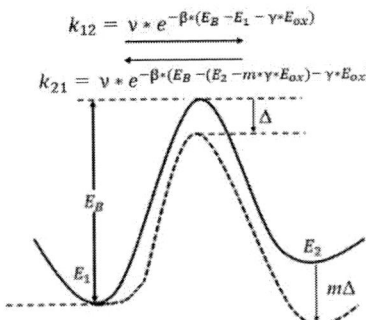

$$k_{12} = \nu * e^{-\beta*(E_B - E_1 - \gamma*E_{ox})}$$

$$k_{21} = \nu * e^{-\beta*(E_B - (E_2 - m*\gamma*E_{ox}) - \gamma*E_{ox})}$$

Fig.1. Schematic of ABDWT model. E_1, E_2 and E_b define the energetic configuration of the trap whereas m and γ control the bias coupling.

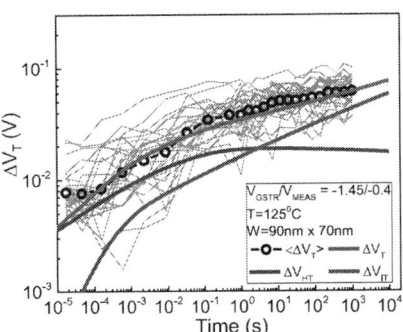

Fig.2. Individual (gray) and mean (black) measured ΔV_T traces during stress along with model calculated mean (red) and decomposition into subcomponents

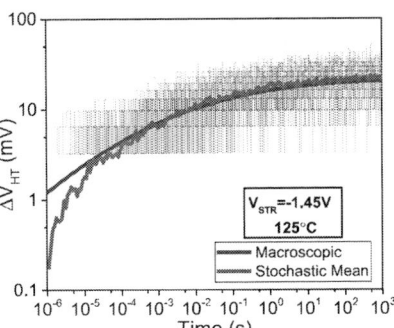

Fig.3. Individual stochastic ΔV_{HT} stress traces (gray) and their mean alongside macroscopic ΔV_{HT} curve.

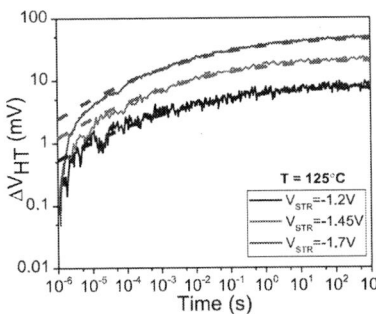

Fig.4. Matching of mean stochastic ΔV_{HT} stress curves (solid lines) with corresponding experimental data (dashed lines) for different stress biases.

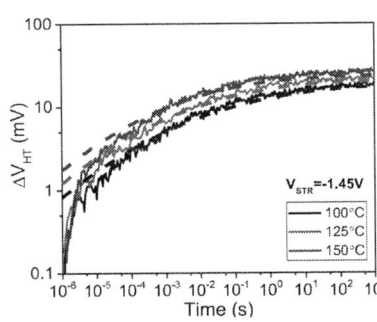

Fig.5. Matching of mean stochastic ΔV_{HT} stress curves (solid lines) with corresponding macroscopic curves (dashed lines) at different temperatures.

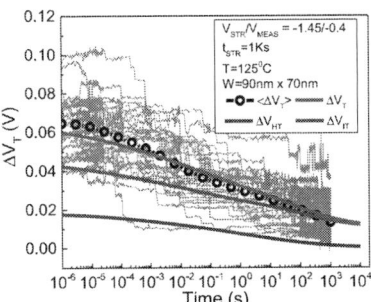

Fig.6. Individual (gray) and mean (black) measured ΔV_T traces during recovery along with model calculated mean (red) and decomposition into subcomponents

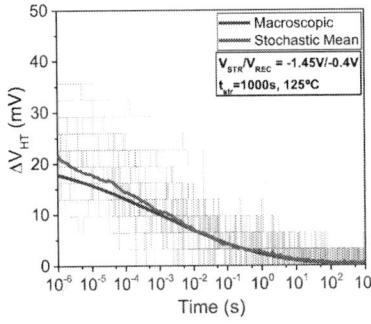

Fig.7. Individual stochastic ΔV_{HT} recovery traces (gray) and their mean alongside macroscopic ΔV_{HT} curve.

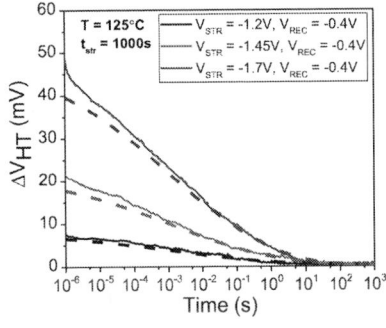

Fig.8. Matching of mean stochastic ΔV_{HT} recovery curves (solid lines) with corresponding experimental data (dashed lines) for different stress biases.

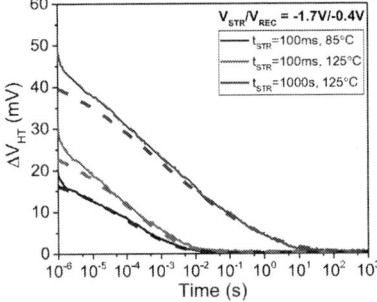

Fig.9. Matching of mean stochastic ΔV_{HT} recovery curves (solid lines) with corresponding experimental data (dashed lines) for different stress time and temperature.

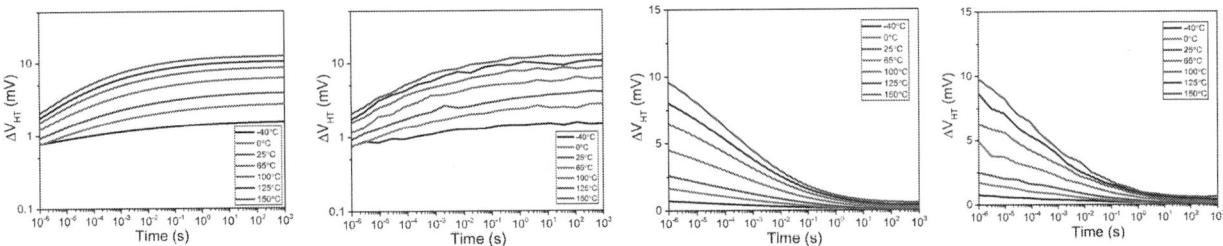

Figs.10-13. Comparison of mean stochastic ΔV_{HT} degradation with experimental data over extended temperature range. Stress and recovery curves are generated and shown to be consistent with measured data. Device is type D3 (RMG HKMG SOI FinFET).

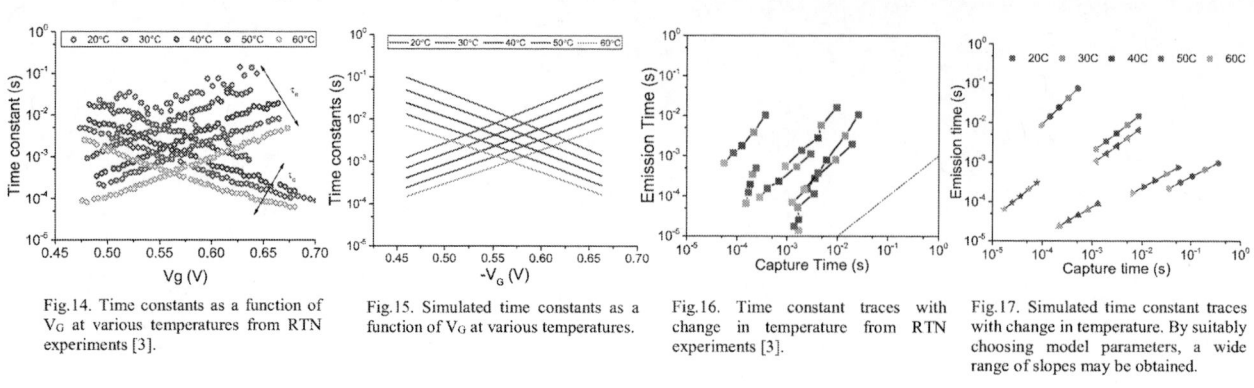

Fig.14. Time constants as a function of V_G at various temperatures from RTN experiments [3].

Fig.15. Simulated time constants as a function of V_G at various temperatures.

Fig.16. Time constant traces with change in temperature from RTN experiments [3].

Fig.17. Simulated time constant traces with change in temperature. By suitably choosing model parameters, a wide range of slopes may be obtained.

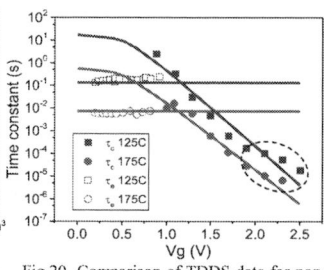

Fig.18. Measured ΔV_T step height versus emission time from TDDS recovery traces.

Fig.19. Simulated step height versus emission time generated from stochastic ΔV_{HT} recovery traces.

Fig.20. Comparison of TDDS data for non-switching trap [4] (symbols) with model calculated time constants (lines). Model calculation predicts linearly decreasing capture time constant whereas experimental results show tapering off towards saturation at higher biases.

Fig.21. Modified ABDWT schematic diagram. At low biases barrier lowering is linear in Eox while at higher biases it becomes quadratic in Eox.

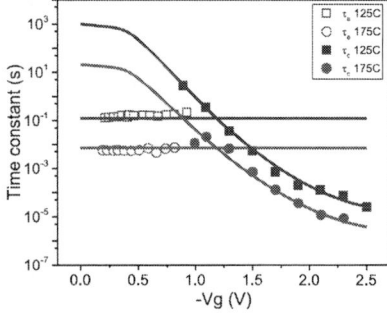

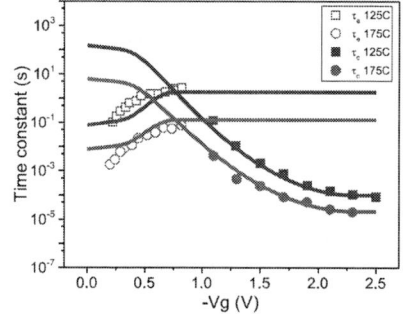

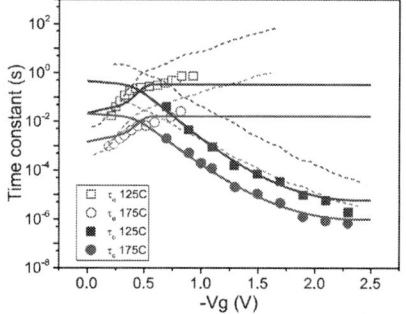

Fig.22. Simulated capture times using the modified ABDWT correctly predicts the tapering off of time constants at higher biases.

Fig.23. Prediction of TDDS capture and emission time constants (symbols) for switching trap [4] using modified ABDWT (lines).

Fig.24. Comparison of TDDS data for trap 'A1' [14] (symbols) with ABDWT model calculated time constants (solid lines) as well as NMP calculated time constants (dashed lines).

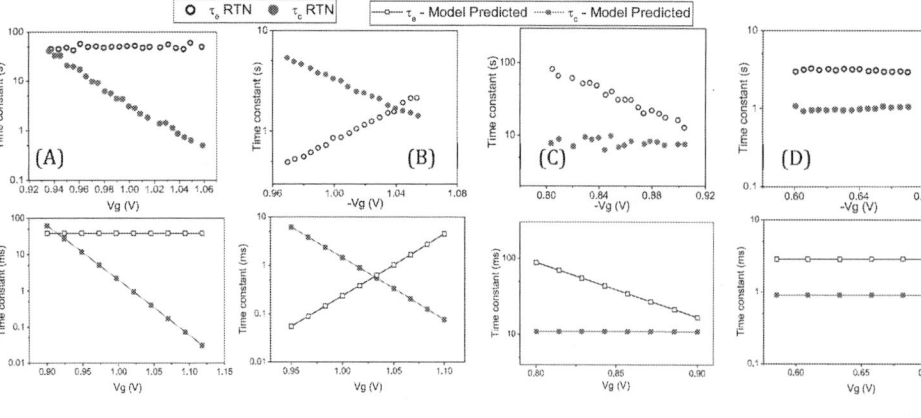

Fig.25. Upper panel depicts various couplings of time constants to V_G extracted from RTN data [15].
(A) $\tau_c < 0$, $\tau_e \sim 0$, (B) $\tau_c < 0$, $\tau_e > 0$, (C) $\tau_c \sim 0$, $\tau_e < 0$, (D) $\tau_c \sim 0$, $\tau_e \sim 0$.
Lower panel shows time constants extracted from ABDWT model simulation. Suitable selection of ABDWT model parameters m and γ yields the corresponding bias couplings: (A) $m \sim 1$, (B) $m > 1$, (C) $m < 0$ with small γ, (D) $m \sim 1$ with small γ.

978-1-7281-0941-1/19 $31.00 © 2019 IEEE

TCAD-Enabled Machine Learning Defect Prediction to Accelerate Advanced Semiconductor Device Failure Analysis

Chea-Wei Teo[1, 2], Kain Lu Low[1], Vinod Narang[2], and Aaron Voon-Yew Thean[1, *]

[1]Department of Electrical and Computer Engineering, National University of Singapore, Singapore

[2]Device Analysis Lab, Advanced Micro Devices Pte Ltd, Singapore

*Tel: +65 6516 6471, *email: aaron.thean@nus.edu.sg

Abstract- **In this work, we present a unique approach of combining TCAD modelling and machine learning to detect the defect locations of a bridging defect in a single-fin FinFET. The prediction of the defect location is guided by the predictive model consisting of Random Forest algorithm which is trained with the measureable electrical attributes from the I-V. High accuracy in predicting the defect location is achieved by the proposed scheme which can further enhance the FA success rate, expediting the cycle of design to product.**

Keywords– Defect Location Prediction, FinFET, Machine Learning, TCAD.

I. Introduction

Failure Analysis (FA) has been a critical process for driving semiconductor yield enhancement, reliability, and accelerating product development cycle. In today complementary metal-oxide semiconductor (CMOS) technology, the number of transistors in an integrated circuits (IC) approximately doubles every two years as predicted by the Moore's law [1]. Continued scaling of CMOS devices results in an exponential growth in the number of transistors on IC chips. Higher packing density allows more logic circuits to be fabricated on a given IC chip area which in turn reduces the cost per function.

However, as transistor dimensions are aggressively scaled to the deep sub-micrometer regime and beyond, several serious challenges arise, including the short-channel effects (SCEs) [2]. Short channel effects, such as drain induced barrier lowering (DIBL), V_{TH} roll-off, and punch-through, significantly increase the OFF-state current (I_{OFF}) of highly scaled MOSFETs. To overcome the issues associated with SCEs, FinFET devices [3] have been introduced in recent years to meet the high performance and low power requirement for state-of-the-art electronic products. However, the complexity for fault identification has significantly increased [4] due to the use of FinFET in the electronic industry. Firstly, nanoscale non-planar device structures have led to more occurrences of Non-Visual Defects (NVD), and this lowers the chance of getting to the root cause of failure. Secondly, the complex multi-layer structure of FinFET devices leads to more complex Transmission Electron Microscopy (TEM) analysis, which is both time consuming and difficult to prepare. As such, defect identification workflows are becoming increasingly reliant on electrical nanoprobing. However, the electrical interactions between the defect, the transistors, the complex interconnects can be difficult to partition. In the light of the above, we propose a new approach of defect prediction to improve the defect identification and location success rates.

The organization of this paper is as follows. Section II details the methodology, covering the setup of TCAD model, dataset generation and predictive model for machine learning. In Sections III-A and III-B, the analysis of current-voltage (I-V) for a single FIN with different defect configurations is presented. It is followed by the discussion on the performance of random forest model on the prediction of the defect location in III-C. Finally, the conclusions are drawn in Section IV.

II. Approach

Due to rare occurrence of defects in chips, it is extremely difficult to collect enough statistically-significant failing samples. In this context, the defect modeling and simulation using calibrated TCAD models can serve as a forward prediction model to generate electrical responses for many defect-device configurations. In this work, we focus on bridging defects that lead to leakage and electrical shorts. Guided by actual defect and nanoprobing results, we built a database of bridging defects in different location, size, and the electrical responses. We curated key electrical features that may identify the defects. A 3D TCAD model of the transistor and local interconnect is used to generate labeled data set for the machine training. We tested the machine learning model against TCAD-generated defect-device configurations to evaluate its prediction accuracy.

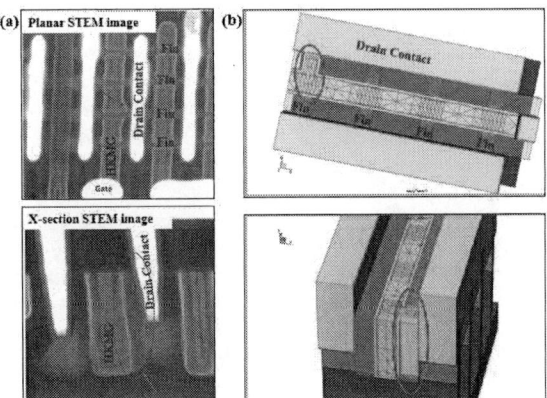

Figure 2: Gate pattern defect: Planar view (top) and Cross-sectional view (bot): (a) STEM Image. (b) TCAD Defect model.

Parameters	Dimensions
Fin height	50 nm
Fin width	10 nm
Gate length	20 nm
Source / Drain doping	1×10^{19} cm^{-3}
Body doping	5×10^{17} cm^{-3}
GOX HfO$_2$ thickness	2 nm
GOX SiO$_2$ thickness	1 nm

Figure 1: FinFET Structure and paramaters for TCAD.

978-1-7281-0941-1/19 $31.00 © 2019 IEEE

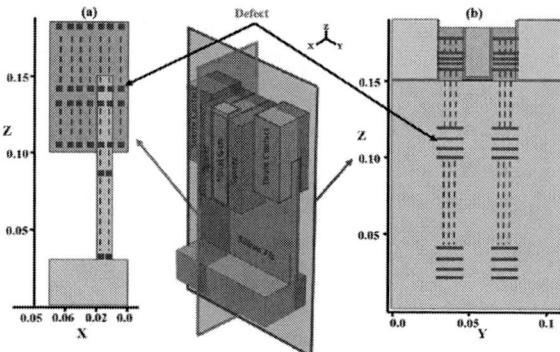

Figure 3: Placement of defect in a single-FIN FinFET: (a) Z-X view. (b) Z-Yview.

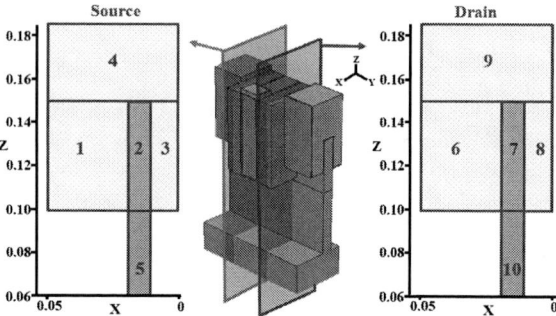

Figure 4: Classification of the regions serving as the classes for the dataset.

A. TCAD Model Setup

A single-fin FinFET structure is considered in this work. It is constructed using Synopsys Sentaurus Structure Editor Process Emulation method [5], as shown in Fig. 1 with the parameters detailed. Correlation between the simulated result and actual device failure was validated with the existing FA cases which involve a gate-pattern defect on a multi-fin FinFET, as described in [4]. The gate pattern defect was introduced into the TCAD model as shown in Fig. 2. The simulated drain/source current vs gate voltage characteristics (I_D/I_S-V_G) give a qualitative resemblance to the IV behavior of actual device.

B. Dataset Generation for Machine Learning

A single-fin FinFET with a fixed dimension of bridging defect is considered for this work. The bridging defect consists of Titanium Nitride as material. It has a fixed X, Y, and Z dimension of 5 nm, 18 nm, 3 nm, respectively.

As shown in Fig. 3, the defects are distributed at various X and Z positions and Y-position is fixed at 30 nm and 62 nm, respectively. The region is further broken down into 10 sub-regions which are the classes for the dataset, as captured in Fig. 4. By employing the models used in [4], the electrical characteristics of FinFET with introduced defects are simulated using the Synopsys Sentaurus device simulator.

Measureable electrical attributes are subsequently extracted from the I-V which serve as the feature set in the database for the machine learning. Once the dataset is setup, the supervised learning algorithm based on Random Forest (RF) [6] is adopted for training and predicting the defect location based on the electrical attributes provided. A total number of 273 samples were consolidated which

constitute the dataset for training and validating the predictive model.

C. Predictive Model Setup for Machine Learning

The machine learning component starts with data preprocessing step to ensure the integrity of the inputs. Prior to training the predictive model, the dataset is split into training and testing sets in a random manner [7]. To minimize the situation of underfitting or overfitting, cross-validation approach is used to ensure that the model is generalized to an independent or unseen data set. The optimal parameters for Random Forest algorithm are obtained using the grid search technique via exhaustive searching from the range of parameters specified for the best cross validation score.

III. Results and Discussions

A. I-V: Defect within the FIN (S/D – Channel)

Firstly, the *I-V* of defect located between the drain and channel inside the FIN [Fig. 5(a)] is investigated. From the transfer characteristic shown in Fig. 5(b), the leakage current of the defective FIN is higher than that of the control device. In order to comprehend the trend observed, the band diagram at the OFF-state ($V_G = 0$ V, $V_D = 1$ V) is extracted in Fig. 6. It is found that the metallic defect (TiN) alters the band diagram around the channel and drain substantially. The 1-dimensional (1-D) band diagram along the defect reveals that the source barrier significantly reduced by the defect compared to the one of control device, resulting high leakage current.

On the contrary, the magnitude of the current level appears to be lower than that of the control device for the case where the defect is located in between the source and the channel [Fig. 7(a)], as demonstrated in Fig. 7(b).

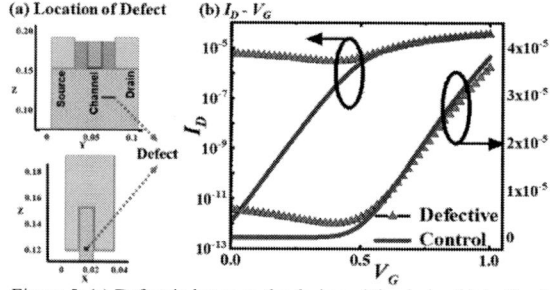

Figure 5: (a) Defect in between the drain and the drain. (b) I_D-V_G of defective FIN and the control device in linear and logarithmic scale.

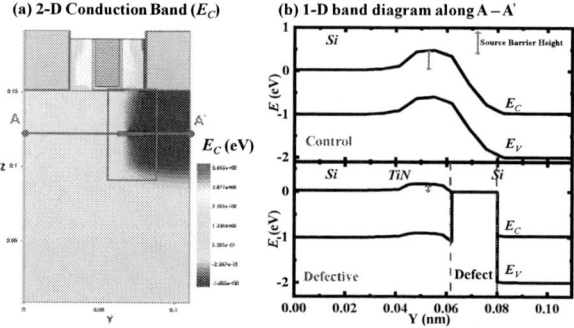

Figure 6: (a) 2-D conduction band (E_C). (b) 1-D band diagram along the cut line where the defect is located for control and defective FIN.

978-1-7281-0941-1/19 $31.00 © 2019 IEEE

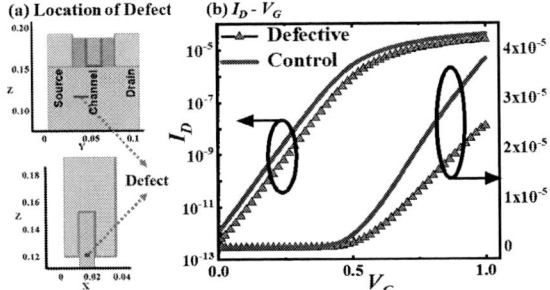

Figure 7: (a) Defect in between the source and the channel. (b) I_D-V_G of defective FIN and control device in linear and algorithmic scale.

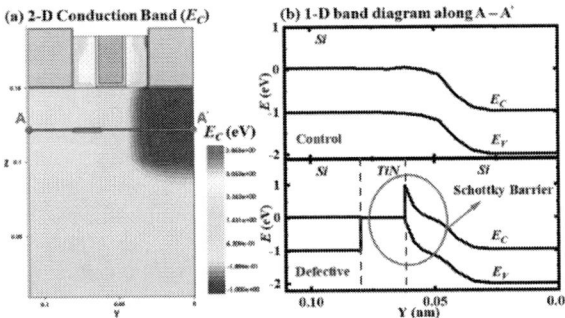

Figure 8: (a) 2-D conduction band (E_C). (b) 1-D band diagram along the cut line (A to A').

Similarly, the 2-D and 1-D band diagrams are examined to understand the underlying physical insights leading to lower current level observed in the defective FIN. The 1-D band diagram along the defect [extracted at the condition of $V_G = 0.6$ V and $V_D = 1$ V] in Fig. 8(b) shows that the degree of reduction in the source barrier height with V_G is smaller for the defective FIN. The Schottky barrier formed in between the defect in the source and the channel increases the source barrier. Consequently, smaller amount of carriers can surpass the potential barrier, resulting in lower current level in the defective FIN.

B. I-V: Defect outside the FIN (S/D – Channel)

The schematic diagram in Fig. 9(a) shows the defect located outside of the FIN electrically connect the drain and the gate electrode. The defect forms a resistive con-

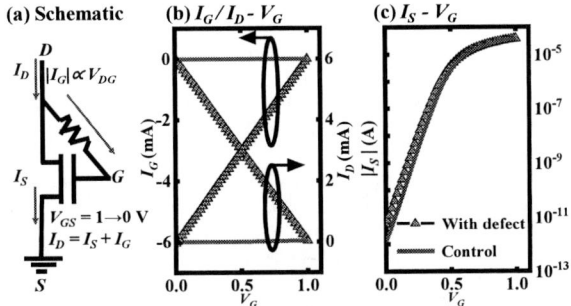

Figure 9: (a) Transistor Schematic showing that the gate and drain electrode are electrically connected via the defect, as represented by a resistive path ($I_{D \to G}$). (b) I_D / I_G versus V_G of FinFET exhibiting linear dependence on the V_G. (c) I_S versus V_G follows the conventional current characteristic of FinFET.

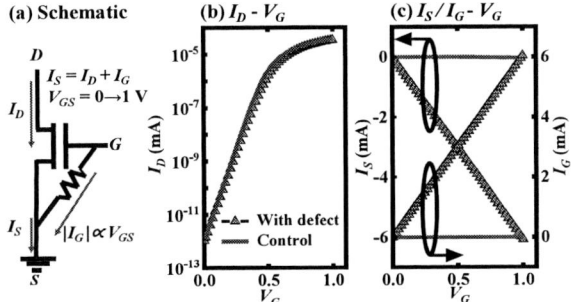

Figure 10: (a) Transistor Schematic: the gate and source electrode shorted electrically by the defect, as represented by a resistive path ($I_{S \to G}$). (b) I_D versus V_G follows the conventional current characteristic of Fin-FET. (c) I_S / I_G versus V_G showing the magnitude of I_S and I_G increases linearly with V_G.

ducting path between the drain and the gate electrode. By the Kirchhoff's current law (KCL), the current of drain is contributed by the gate and the source current. From the current characteristics of the drain (I_D) and gate (I_G) versus the V_G in Fig. 9(b), both I_D and I_G exhibit a linear dependence on the potential difference between the drain and gate electrode due to the resistive path formed by the defect. It is also noted that the magnitude of I_D and I_G is much larger than the source current (I_S). The I_S follows the current characteristic of a transistor [Fig. 9(c)].

Similar analysis and justifications could be applied in the scenario where the defect forms a resistive conducting path in between the source and the gate electrode [Fig. 10(a)]. As depicted in Fig. 10(b), the characteristic of I_D is similar to that of the control device. On the other hand, the magnitude of both I_S and I_G increases linearly with V_G due to the flow of current through the resistive conducting path between the source and the gate electrode, shown in Fig. 10(c).

C. Performance of the Random Forest

Based on the Random Forest model with optimal parameters, high accuracy is achieved with an average accuracy score of 0.9612 which is obtained by running 1000 randomly-split training and testing sets on the model. From the evaluation of the confusion matrix (Fig. 11), all samples, except for those from region 5, are classified

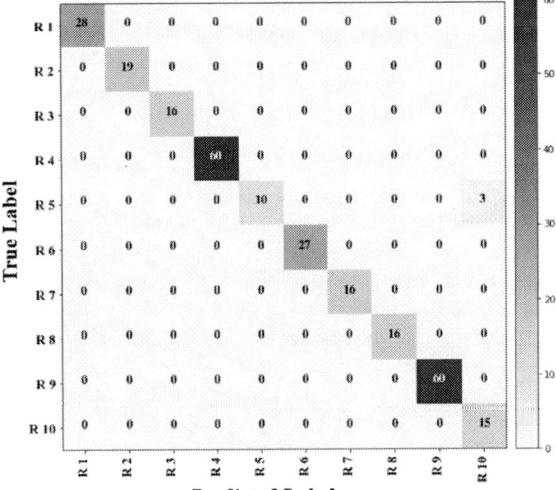

Figure 11: Confusion matrix with all features considered. R stands for Region.

978-1-7281-0941-1/19 $31.00 © 2019 IEEE

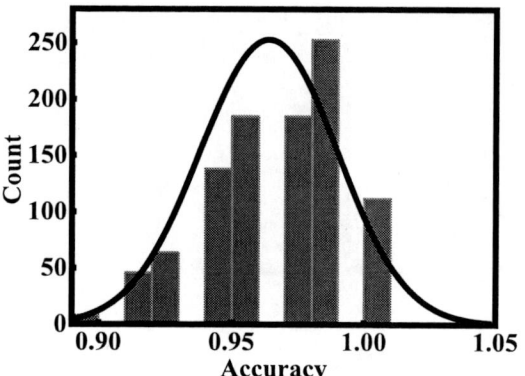

Figure 12: Distribution of the accuracies of Random Forest Model considering only the important features

Table I. Average accuracy score

Model	Average Accuracy
Random Forest (All features)	0.9612
Random Forest (Important features only)	0.9629

Table II. Features Importance of Random Forest

Features	Importance
S_{min_S} (minimum subthreshold swing of I_S)	0.122769
S_{min_D} (minimum subthreshold swing of I_D)	0.113200
V_{TH_D} (threshold voltage at I_D = 1e-8 A)	0.077937
S_{avg_D} (average subthreshold swing of I_D)	0.072414
S_{avg_S} (average subthreshold swing of I_S)	0.071190
S_{max_D} (maximum subthreshold swing of I_D)	0.071001
S_{max_S} (maximum subthreshold swing of I_S)	0.060934
V_{TH_S} (threshold voltage at I_S = 1e-8 A)	0.055311
I_{S_slope} (average slope I_S-V_G)	0.050797
I_{Gsat} (Drain current at V_G = 1V, V_D = 1 V)	0.047725
I_{Dsat}/I_{Ssat} (ratio of I_{Dsat} and I_{Ssat})	0.044750
I_{Dsat} (Drain current at V_G = 1V, V_D = 1 V)	0.038916

correctly. It also reveals the issue of imbalanced dataset with more samples from region 4 and region 9. To circumvent this issue, the "class_weight" argument in the predictive model is set to "balanced" in order to achieve a balanced mix of each class in the dataset.

Among the important features, as depicted in Table II, are the features related to the subthreshold swing (S_{min}, S_{avg}, and S_{max}), threshold voltage (V_{TH}), I_{S_slope}, I_{Gsat}, and I_{D_slope}. From the perspective of device physics, the differentiation for the defect in region 1, 3, 4, 6, 8, and 9 is related to the subthreshold swing (S_{min}). Due to the closer proximity of the defects residing in region 1, 3, 6, and 8 to the semiconductor area (region 2 and 7), the influence of defect in region 1, 3, 6, and 8 on the electrostatic of the semiconductor is more pronounced than that of defect in region 4 and 9. This leads to different characteristic in the subthreshold swing (SS), subsequently affecting the threshold voltage (V_{TH}) which is extracted based on constant-current method. This justifies the importance of SS and V_{TH} in classifying the regions of defect.

As discussed in the section III (A) and III (B), the defect located outside of the FIN results in high gate leakage current where $I_S/I_D/I_G$ show linear dependence on the V_G. This explains I_{S_slope} and I_{Gsat} being the important features that distinguish the defect located outside the FIN from the one within the FIN.

Based upon the importance score of the features, the Random Forest model are retrained using the reduced dataset which only considers features with an importance score higher than 0.05. The distribution of the model accuracy obtained from 1000 runs presented in Fig. 12 illustrates that the accuracy is at least 0.86. The average accuracy is further improved to 0.9629 relative to the one with all features considered. This implies that the selected features are sufficient for accurate prediction, reducing the noise in the dataset as well as enabling the model pick up the relevant features.

IV. Conclusion

We successfully demonstrated a systematic approach for predicting the locations of bridge defect in a single-fin FinFET using a combination of TCAD-generated defect database and machine learning. The Random Forest algorithm as predictive model is trained with the electrical attributes from the simulated I-V. The proposed scheme showcases high accuracy in predicting the defect location. It can be easily extended to predict other type of defects and more complex circuits, such as multiple-fin FinFET transistors and SRAM bitcell structure. Once a properly calibrated TCAD transistor model is set up, it can be employed for predicting real failing device failures. Finally, this machine-learning-aided guidance defect detection system will further enhance the FA success rate for advanced nanoscale devices.

Acknowledgements

This work is supported in part by A*Star Accelerated Materials Development for Manufacturing Grant no: A1898b0043.

© 2019 Advanced Micro Devices, Inc. All rights reserved. AMD, the AMD Arrow logo, and combinations thereof are trademarks of Advanced Micro Devices, Inc. Other product names used in this publication are for identification purposes only and may be trademarks of their respective companies.

References

[1] G. E. Moore, "Cramming more components onto integrated circuits," Proceedings of the IEEE, vol. 86, no. 1, pp. 82-85, Jan. 1998.

[2] D. J. Frank, R. H. Dennard, E. Nowak, P. M. Solomon, Y. Taur, and H. S. P. Wong, "Device scaling limits of Si MOSFETs and their application dependencies," Proc. IEEE, vol. 89, issue 3, pp. 259–288, 2001.

[3] C. Auth et al., "A 22nm high performance and low-power CMOS technology featuring fully-depleted tri-gate transistors, self-aligned contacts and high density MIM capacitors," 2012 Symposium on VLSI Technology, Honolulu, HI, 2012, pp. 131-132.

[4] C.W. Teo, V. Narang, and A. Thean, "Electrical Characterization of FEOL Bridge Defects in Advanced Nanoscale Devices Using TCAD Simulations," 2018 IEEE International Symposium on the Physical and Failure Analysis of Integrated Circuits (IPFA), Singapore, 2018, pp. 1-4.

[5] Synopsys Sentaurus Structure Editor user guide.

[6] L.Breiman, "Random Forests", Machine Learning, 45(1), 5-32, 2001.

[7] Scikit-learn: Machine Learning in Python, Pedregosa et al., JMLR 12, pp. 2825-2830, 2001.

TCAD Augmented Machine Learning for Semiconductor Device Failure Troubleshooting and Reverse Engineering

Y. S. Bankapalli and H. Y. Wong*
Electrical Engineering
San Jose State University
San Jose, CA, USA
*hiuyung.wong@sjsu.edu

Abstract— **In this paper, we show the possibility of using Technology Computer Aided Design (TCAD) to assist machine learning for semiconductor device failure trouble shooting and device reverse engineering. When TCAD simulation models and parameters are properly chosen and calibrated, large number of devices with random defects and structural characteristics can be generated and simulated. The results can then be used to train machine learning algorithms to predict the defect and structural characteristics of a device with given electrical characteristics (such as IV's and CV's). 1D PIN diode with various layer thicknesses and doping concentrations are used in this study. It is showed that with less than 2000 training samples, by using simple linear regression, one can achieve good prediction of layer thickness and doping of a given IV curve.**

Keywords—Machine Learning, Reverse Engineering, TCAD, Semiconductor Defects

I. Introduction

Semiconductor device failure troubleshooting and device reverse engineering require expensive analyses such as SEM and TEM [1]. Machine learning (ML) has been used widely in the manufacturing process to enable early discovery of defects [2]. However, the authors are not aware of any extensive application of ML to analyze defects based on fabricated device electrical characteristics, such as Current-Voltage (IV) and Capacitance-Voltage (CV) curves, where defects include epitaxial layer thickness and doping level variations. This is probably because, for a matured process with high yield, the number of defective dies is limited, while for nascent process with low yield, the number of dies produced are limited. As result, it is difficult to obtain enough defective IV curves for accurate machine learning.

Using TCAD, in principle, a large number of IV's and CV's can be generated by changing the layer thicknesses (to model epitaxial layer variation) and doping levels (to model doping variation), and by including various defective models (such as trap assisted tunneling at various spatial location). ML can then be used to generate model to accurately correlate IV and CV curves to defect characteristics. Using the trained model, one can rapidly narrow down the possible cause of an abnormal IV or CV curve and, if necessary, perform further failure analysis (e.g. cutting TEM at the most probable failure spot predicted by ML). The same reasoning applies well in device reverse engineering.

In this paper, we demonstrate this idea by studying the relationship between 1-D PIN diode structural defects (epitaxial layer thickness and doping concentration variations) and its forward and reverse IV curves. Various machine learning models are tested. To reduce the number of TCAD simulations, epitaxial layer thickness and doping concentration studies are performed separately.

II. TCAD Simulations

Figure 1 inset shows the structure simulated in which only layer thickness variations are studied. About 2000 1D PIN diode structures are created using SProcess [3] with n+/i/p+ thicknesses being varied independently and uniformly within the range given in Figure 1. Figure 2 shows the scattering plots of n+/i/p+ thicknesses, which are uniform and independent. Sdevice [4] is then used to simulate the IV characteristics. Essential physics models are turned on, including Fermi-Dirac statistic, doping dependent and high field saturation models for carrier mobilities, Schottky-Reed-Hall Recombination (SRH) and non-local Band to Band tunneling (BTBT). 80-bit ExtendedPrecision is used to avoid noisy reversed curves. Poisson, electron and hole continuity equations are solved self-consistently to produce the curves in Fig. 1.

Figure 5 shows the IV's of another 2000 1D PIN diodes simulated with layer concentrations varied independently and uniformly in their logarithmic values. The corresponding inset shows the structure simulated and the variation range. Figure 6 shows that the layer concentrations are independent and their logarithmic values are uniformly distributed.

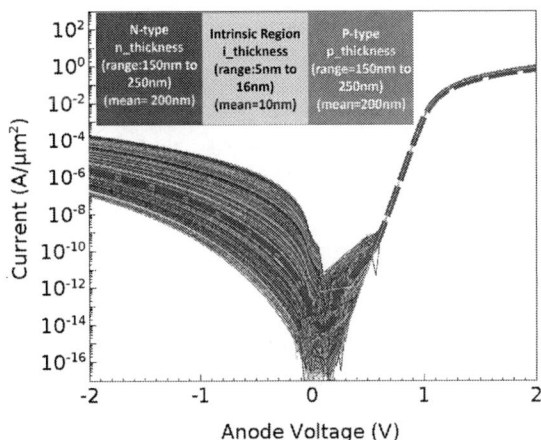

Figure 1: IV's of the 2000 devices (thickness variations only) simulated. The thick pink dash line is the IV of nominal device (200nm/10nm/200nm). Both n+ and p+ concentrations are 10^{20}cm^{-3}. i-layer concentration is 10^{17}cm^{-3}.

978-1-7281-0941-1/19 $31.00 © 2019 IEEE

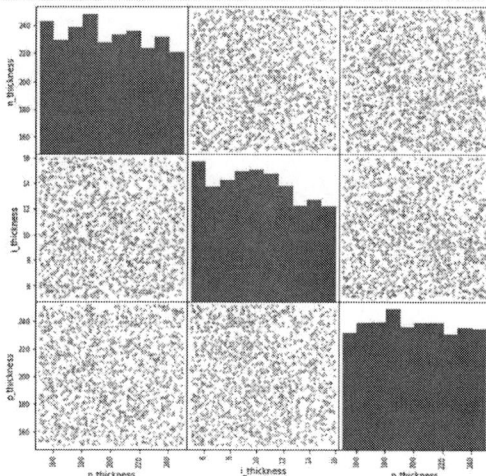

Figure 2: Scattering plot of n+/i/p+ thicknesses showing their frequencies and correlations.

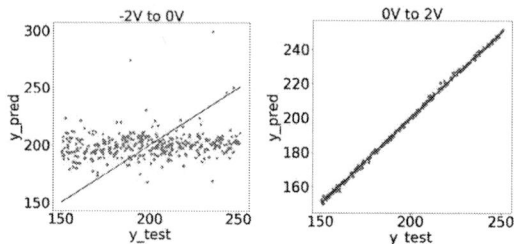

Figure 4: n_thickness prediction by LR machines trained by data from -2V to 0V (left) and data from 0V to 2V (right).

III. MACHINE LEARNING FOR THICKNESSES PREDICTION

Scikit-learn library is used for ML [5]. Four types of algorithms were tested on the layer-thickness data, namely, linear regression (LR), decision tress (DT), random forest (RF) and Multi-Layer Perceptron (MLP) Regressor. 80% of the data (~1600) are used for training and 20% of the data (~400) are used for validation. The input is the IV curve (102 current values for V = -2V to 2V) and the output are i_thickness and n_thickness. Various parts of the IV curve are used for training, namely, I(V=-2V), I(V=-2V to 0V), I(V=0V to 2V) and I(V=-2V to 2V).

The first attempt to train the machine with the raw data was not successful. This is because the current changes orders of magnitude for various thicknesses in reverse bias. As showed in Figure 3, the model fails to predict large i_thickness (prediction is capped at about 12nm) because reverse current (I) is indistinguishable numerically in the raw form for large i_thickness. Moreover, at sufficiently low current level (i.e. when i-thickness is sufficiently large), SRH will dominate and has very weak dependent on the layer thicknesses. If log(I) is used, it gives much better prediction. Therefore, in all trainings for layer thicknesses, log(I) is used.

Table 1 shows the i_thickness and n_thickness prediction Mean Squared Error (MSE) of various machines trained by

different data ranges and algorithms. The learning can be summarized as:

1) DT is a not a suitable algorithm as it often overfits (training MSE = 0, with large prediction MSE)

2) LR performs the best with low training and prediction MSE for both i_thickness and n_thickness

3) MLP performs similar to LR for i_thickness but fails with n_thickness

4) Wide voltage range (-2V to 2V) gives the most accurate results. However, depending on the problem of interest, reduced voltage range gives similar results and simulation time can be substantially reduced. For example, by using the current at -2V, high accuracy of i_thickness can be obtained already because i_thickness influences the BTBT current strongly.

5) It is important to perform the simulation in a regime where the relevant physics is captured. For example, reverse current is insensitive to n_thickness. Therefore, bad result is obtained if data is only available between -2V to 0V. Positive bias simulation is required for n_thickness as forward neutral region potential drop correlates strongly to n_thickness. (Figure 4)

Training of p+ layer thickness gives similar results as the n+ layer thickness and are not shown.

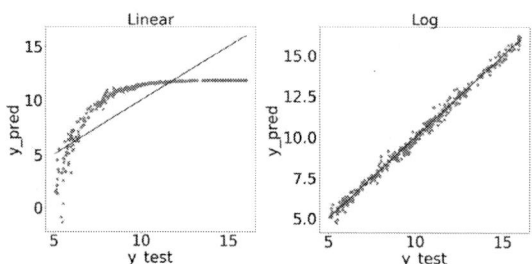

Figure 3: Prediction of i_thickness using linear regression model trained with raw current data (I, left) and processed current data (log(I), right) at V = -2V.

	Data Range Used	MLP	LR	DT	RF
	"-2V"	0.06/0.06	0.06/0.06	0.03/0.08	0.01/0.08
	"-2V to 0V"	0.26/0.22	0.04/0.05	0.02/0.06	0.01/0.05
	"0V to 2V"	0.86/0.88	0.09/0.09	0.00/0.29	0.03/0.18
i_thickness	"-2V to 2V"	0.05/0.05	0.03/0.03	0.01/0.04	0.01/0.04
	"-2V"	847/796	846/795	0.00/1741	163/1248
	"-2V to 0V"	1043/984	761/811	0.00/1456	114/751
	"0V to 2V"	514/434	0.96/0.89	0.00/294	22/162
n-thickness	"-2V to 2V"	473/407	0.80/0.86	86.91/236	22/163

Table 1: Training and prediction Mean Squared Errors (MSE) of i_thickness and n_thickness by machines trained by various data range and algorithms. The numbers are format in "training MSE/ prediction MSE".

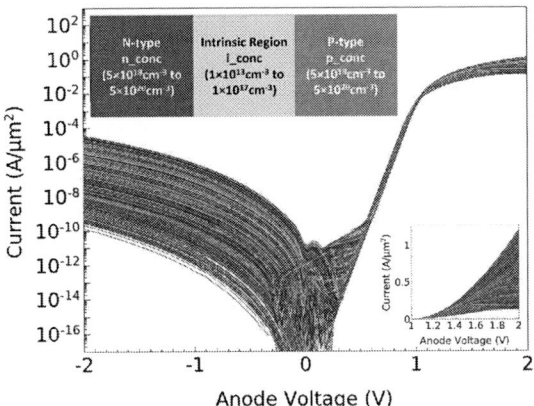

Figure 5: IV's of the 2000 devices simulated with layer concentrations varied. The thicknesses of n+/i/p+ are 200nm/10nm/200nm. The upper left inset shows the structure and the range of variation. The lower inset shows the forward IV's in linear scale.

IV. ML FOR CONCENTRATIONS PREDICTION

Since linear regression shows excellent results in predicting the layer thicknesses of an 1D PIN diode, it is also used to train the model to predict the layer concentrations as the non-linearity is expected to be the similar or less. The structures, IV's and variations are showed in Figures 5 and 6.

The IV distribution of concentration varying diodes is showed in Figure 5 and is very different from that of layer thickness varying one in Figure 1 in the forward region. In Figure 1, the thicknesses, thus the neutral region resistance, vary less than 2 times. But in Figure 5, the concentrations vary by 100 times, which results in large variation of resistance and, thus, forward current. Moreover, from the forward current traces in the inset of Figure 5, one can see that in additional to the magnitude, the shape and curvature vary for different doping concentrations. For example, when the concentration is high, the curvature is positive in the whole region. But when the concentration is low, the curvature changes from positive to negative as voltage increases.

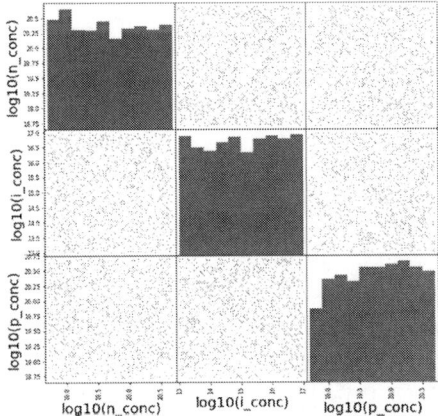

Figure 6: Scattering plot of n+/i/p+ layers concentrations showing their frequencies and correlations used in the study. Note that the logarithmic values of the concentrations are uniformly distributed.

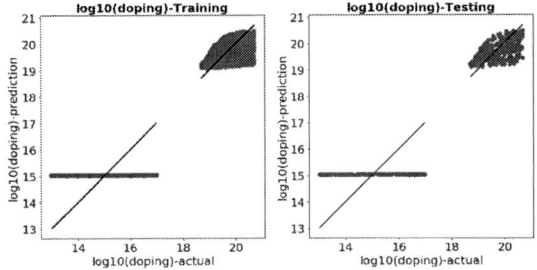

Figure 7: Training and validation of doping concentrations based on current from V = 1.8V to V = 2V.

Therefore, it is expected that the forward region IV will contain sufficient information of the n+ and p+ layer concentrations. Moreover, since it is in forward region, the current values are used directly in the training instead of their logarithmic values being used.

Since n+ and p+ concentrations both have big impact to the IV, multi-variate linear regression is used instead of multi-variation linear regression (which is equivalent to multiple independent linear regression).

Firstly, currents from V = 1.8V to V = 2V are used for training. As shown in Figure 7, although it is expected that doping should have strong impact on the diode current at high forward bias, the training result is very bad. There are two groups of data. The lower concentration one is of the i-layer while the higher concentration one is of the n+ and p+ regions. Since i-layer doping has very small effect on the forward current (due to its small thickness), the model cannot predict any variation in i-conc.

Next, the whole forward current curve is used for training (V = 0V to V = 2V) as showed in Figure 8. The n+ and p+ concentrations can be predicted accurately. It is worth to mention that heavily doped n+ and lightly doped p+ diode (e.g. 5×10^{20}cm^{-3} n+ / 5×10^{18}cm^{-3} p+) is expected to give similar current at V = 2V as the lightly doped n+ and heavily doped p+ diode (e.g. 5×10^{18}cm^{-3} n+ / 5×10^{20}cm^{-3} p+). However, the trained algorithm still can distinguish them clearly. This implies that the asymmetric of n+/p+ doping is captured in the forward IV curves. Indeed, Figure 9 shows that n/p = 5×10^{20}cm^{-3} / 5×10^{18}cm^{-3} and n/p = 5×10^{18} cm^{-3} / 5×10^{20}cm^{-3} give different forward IV shapes. This is probably

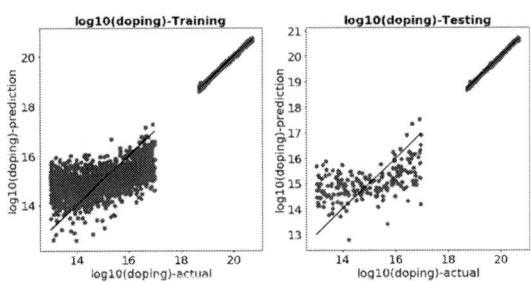

Figure 8: Training and validation of doping concentrations based on current from V = 0V to V = 2V.

978-1-7281-0941-1/19 $31.00 © 2019 IEEE   23

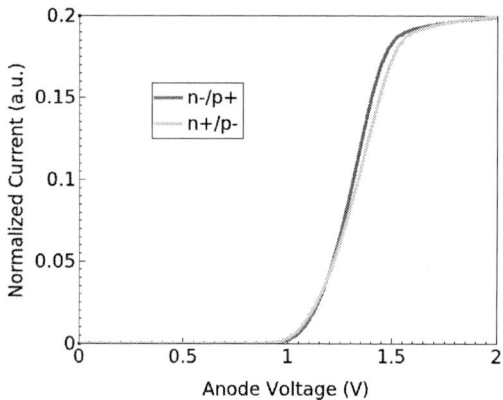

Figure 9: Forward IV of diode with n-/p+ = $(5\times10^{20}\text{cm}^{-3}$ $/5\times10^{18}\text{cm}^{-3})$ and n+/p- = $(5\times10^{18}\text{cm}^{-3}/5\times10^{20}\text{cm}^{-3})$. Doping of i-layer is p-type = 10^{17}cm^{-3}. The current at V = 2V are scaled to be the same.

the reason why the machine can distinguish n and p concentration from each other.

As shown in Figure 8, i-layer concentration still cannot be modeled well even the full forward IV is used in the training. This is because it does not have strong influence on the forward IV due to its small thickness.

Instead of using linear regression with the original 50 input features (i.e. currents at voltage 0V to 2V), second order linear regression is used in which the number of features is expanded to 1326. This gives better fitting in both training and validation. However, it is still not good enough (Figure 10).

Third order linear regression was also tried but it results in overfitting in which it gives perfect fitting to the training model but bad prediction in validation. Therefore, in order to capture the i-layer concentration, more data points are needed.

V. PROSPECT OF 3D TCAD SIMULATION WITH ML

The 1D PIN diode has about 300 mesh points. The simulation was performed in Intel Xeon E5-2603 with 1 core used. The total simulation time of each simulation (process and device) is about 90 seconds. As a result, it takes about 2 days to complete the data generation. A typical realistic 3D FinFET IV simulation is between 1 hour to 6 hours (process +

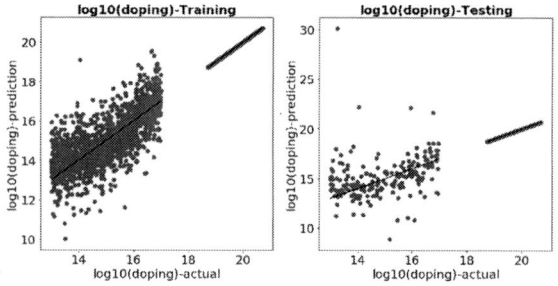

Figure 10: Training and validation of doping concentrations based on current from V = 0V to V = 2V using second order linear regression.

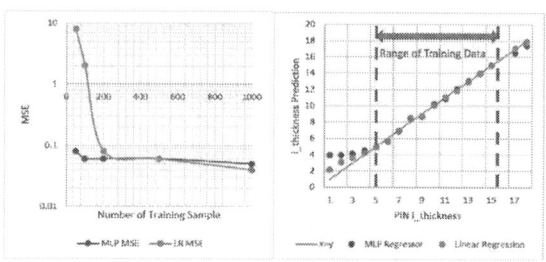

Figure 11: Prediction MSE as a function of the number of training samples (Left). Prediction of i_thickness as a function of PIN i_thickness (Right).

device) [6]. If computing farm with thousands of CPU are available, we anticipate similar study can be completed in <1 day for a realistic 3D FinFET structure. To reduce simulation time, one can reduce the number of training data point or/and the range of defect (e.g. i_thickness) variation. Figure 11 shows that even with 50 (or 200) data points, instead of 1600, MLP (or LR) is still very accurate. Moreover, LR can predict accurately 50% wider range of i_thickness than the training data. These make 3D TCAD augmented ML for defect trouble-shooting more feasible.

VI. CONCLUSIONS

Using PIN diode with various layer thicknesses and concentrations, we demonstrated that TCAD can be used to generate sufficient data to train machine to identify the "defect value" (variation of layer thickness and concentration) rapidly based on IV curves. It is found that 1) data processing before ML is critical to obtain accurate results but the type of preprocessing depends strongly on domain knowledge (e.g. forward and reverse currents require different treatments), 2) linear regression gives the best prediction and is better than Multi-Layer Perceptron (MLP), 3) the model is able to predict structure with thickness out of the range of training data set and 4) the full process (TCAD simulation and ML) can be completed in less than 2 days with 1 cpu core. We anticipate that by using computing farm with thousands of cores, such scheme can be implemented for more realistic 3D simulations. In certain algorithm, only number of training data as low as 50 is needed. Such TCAD augmented ML can expedite defect trouble-shooting and reverse engineering of semiconductor devices.

REFERENCES

[1] R. Torrance and D. James, "Reverse Engineering in the Semiconductor Industry," 2007 IEEE Custom Integrated Circuits Conference, San Jose, CA, 2007, pp. 429-436. doi: 10.1109/CICC.2007.4405767

[2] G. A. Susto, M. T. and A. Beghi, "Anomaly Detection Approaches for Semiconductor Manufacturing", Procedia Manufacturing 11 (2017) 2018 – 2024.

[3] Sentaurus™ Process User Guide Version O-2018.06, June 2018.

[4] Sentaurus™ Device User Guide Version O-2018.06, June 2018.

[5] https://scikit-learn.org/stable/

[6] H. Y. Wong, D. Dolgos, L. Smith and R. V. Mickevicius, "Modified Hurkx Band-to-Band-Tunneling Model for Accurate and Robust TCAD Simulations", submitted to Mircoelectronics Reliabily.

On the Simulation of Plasma Waves in HEMTs and the Dyakonov-Shur Instability

1st Christoph Jungemann
Chair of Electromagnetic Theory
RWTH Aachen University
Aachen, Germany
cj@ithe.rwth-aachen.de

2nd Tobias Linn
Chair of Electromagnetic Theory
RWTH Aachen University
Aachen, Germany
tl@ithe.rwth-aachen.de

3rd Zeinab Kargar
Chair of Electromagnetic Theory
RWTH Aachen University
Aachen, Germany
zk@ithe.rwth-aachen.de

Abstract—Modeling of plasma waves in HEMTs by moments-based transport models is investigated. The balance equations are derived from the Boltzmann transport equation by projection onto Hermitian polynomials. The discretized equations are stabilized by an approach based on matrix exponentials, which in the case of the drift-diffusion model reproduces the Scharfetter-Gummel stabilization. Simulations of a realistic HEMT show that plasma instabilities are rather unlikely to occur and that effects not considered by Dyakonov and Shur (e.g. real ohmic contacts) strongly damp the THz waves. Furthermore, quasi-ballistic transport can not be captured by higher-order models.

Index Terms—HEMT, plasma waves, drift-diffusion, Boltzmann transport equation, Scharfetter-Gummel stabilization

I. INTRODUCTION

Many promising high-frequency applications fall into the so-called THz gap (0.3-3THz), for which only very power-inefficient sources exist [1]. Dyakonov and Shur proposed THz wave sources based on plasma instabilities in high electron mobility transistors (HEMT), which might fill this gap [2], [3]. Due to an applied dc drain/source bias a current flows in the channel and the plasma waves in the direction of the electron flow (downstream) behave differently from the plasma waves in the opposite direction (upstream). They assumed that the ac electron density at the source-side of the channel is zero and that the ac electron current at the drain-side vanishes. These boundary conditions can lead to a plasma instability and the generation of THz waves. For a HEMT in the common source configuration these boundary conditions correspond to an ac-shorted gate/source (input) port and open drain/source (output) port. Since HEMTs with a negative differential output resistance are in general not unconditionally stable, such a configuration almost always leads to oscillations at rather low frequencies [4]. Experimental verification of THz wave generation by plasma instabilities is therefore difficult and the experimental results are rather inconclusive. Furthermore, the measured THz emissions are often barely above the black body radiation (e.g. [5]). It is thus not clear whether plasma instabilities can be used to generate THz waves or not and in this paper we investigate the accuracy of the underlying theory by device simulation.

Financial support by the German Science Foundation (DFG) under contract number JU406/14-1 is gratefully acknowledged.

II. MODEL

The hydrodynamic modeling approach used by Dyakonov and Shur is questionable for various reasons. Their model based on the Euler and continuity equations is similar to a drift-diffusion model which includes the convective derivative and a time derivative in the constitutive equation for the electron current density [6]. It can be derived from the Boltzmann transport equation (BTE) by taking the first two (velocity) moments together with a closure relation based on a drifted Maxwellian [7], [8]. This assumption holds only in the case of strong scattering, whereas THz wave generation requires extremely high mobilities and thus quasi-ballistic transport [2], [3]. A quasi-ballistic distribution function in a device is very different from a Maxwellian in the case of nonequilibrium and it is not clear whether electron-electron scattering is sufficiently strong to drive the distribution function towards a Maxwellian on the required time and length scales. In addition, the Maxwellian will be heated, a fact that was neglected by Dyakonov and Shur and leads to a strong increase of the channel resistance and damping of the plasma instability, especially at high mobilities [9]. Moreover, the restriction to two moments leads to a plasma dispersion relation with only two branches, whereas the BTE yields in addition a continuum of modes [10]. In order to capture the impact of the additional modes, the differential equations should be solved in the real space, which also allows to account for inhomogeneous channels and parasitics. In addition, it is possible to apply more realistic boundary conditions. The assumption of a drift-diffusion model together with Dirichlet boundary conditions at the source and drain terminals leads for high mobilities to unrealistically large conductivities, which exceed the ballistic limit for thermal bath boundary conditions. This can be at least partially avoided by assuming a finite surface recombination velocity (real ohmic contacts) [11].

In order to avoid some of the above mentioned problems, we solve moments-based models of arbitrary order l for a realistic 2D device structure (Fig. 1). While the Poisson equation for the quasi-stationary potential is solved in 2D, the electron transport for the electron gas in the channel is assumed to be 1D (charge sheet approximation) [12]. The Poisson equation is discretized by the finite volume method in conjunction with finite differences for the electric flux density [11]. On the contacts Dirichlet boundary conditions are applied to the potential, otherwise homogeneous Neumann boundary conditions. The 1D transport models are derived from the BTE by projection onto Hermitian polynomials assuming a macroscopic relaxation time approximation for the scattering integral and a parabolic band structure:

$$\frac{\partial g_n}{\partial t} + \frac{2neE_x}{\sqrt{2mk_BT_0}}g_{n-1} + \sqrt{\frac{2k_BT_0}{m}}\left(\frac{1}{2}\frac{\partial g_{n+1}}{\partial x} + n\frac{\partial g_{n-1}}{\partial x}\right) = -\frac{g_n - g_0\delta_{n,0}}{\tau} \tag{1}$$

e is the elementary charge, $E_x(x,t)$ is the x component of the electric field in the channel, m the effective mass, k_BT_0 the thermal energy and τ the macroscopic relaxation time with $\mu = e\tau/m$. For the sake of simplicity τ is assumed to be position independent. $g_n(x,t)$ is the nth moment of the distribution function,

$$f(x,\vec{k},t) = \sum_{n=0}^{\infty} \frac{g_n(x,t)H_n\left(\frac{\hbar k_x}{\sqrt{2mk_BT_0}}\right)}{2^n n!\sqrt{\pi}}e^{-\frac{\hbar^2\vec{k}^2}{2mk_BT_0}} \tag{2}$$

where $H_n(u)$ is the nth order Hermitian polynomial [13]. This special type of distribution function is due to the assumption of 1D transport in the real space and the macroscopic relaxation time approximation for the scattering integral. The electron density is proportional to the zeroth order component: $n(x,t) = N_C/\sqrt{\pi}\, g_0(x,t)$, and the electron current density is given by: $j_x(x,t) = N_C\sqrt{k_BT_0}/\sqrt{2m\pi}\, g_1(x,t)$, where N_C is the effective density of states of the conduction band. The terminal currents are evaluated by the extended Ramo-Shockley theorem [14].

In order to obtain a system of equations of finite size, the expansion in (2) is truncated at maximum order l and all components for $n > l$ are assumed to be zero ($g_{n>l}(x,t) = 0$). This closure relation has the advantage that the equations remain linear, but it can lead to stability problems for large electric fields. For $l = 1$ the drift-diffusion model is obtained without the convective derivative. To account for such transport effects in a more rigorous manner, transport models with a much larger l are solved (e.g. $l = 9$), where the convergence of the expansion must be checked.

For a finite l the balance equations (1) for different n can be aggregated into:

$$\frac{\partial\vec{g}}{\partial t} + \hat{A}\frac{\partial\vec{g}}{\partial x} + \hat{B}\vec{g} = \vec{0} \tag{3}$$

$\vec{g}(x,t)$ is the vector containing the l components. $\hat{A}$ is a constant $l \times l$ matrix and $\hat{B}$ depends on the electric field. A grid with N nodes x_i is used for the channel. First, the dc case is considered, for which (3) can be solved under the assumption of a position independent $\hat{C}_{i+\frac{1}{2}} = \hat{A}^{-1}\hat{B}_{i+\frac{1}{2}}$ in the interval $[x_i, x_{i+1}]$ with the matrix exponential [15], where $\vec{G}_i = \vec{G}(x_i)$ is the dc solution, $x_{i+\frac{1}{2}} = (x_{i+1} + x_i)/2$ and $\hat{D}_{i+\frac{1}{2}}(x) = -\hat{C}_{i+\frac{1}{2}}(x - x_{i+\frac{1}{2}})$:

$$\vec{G}(x) = e^{\hat{D}_{i+\frac{1}{2}}(x)}\vec{G}_{i+\frac{1}{2}} \tag{4}$$

The vector $\vec{G}$ is split into even and odd components with $\vec{G} = (\vec{G}_e^T, \vec{G}_o^T)^T$ and the even components on the grid points are given by:

$$\begin{pmatrix}\vec{G}_{e,i} \\ \vec{G}_{e,i+1}\end{pmatrix} = \begin{pmatrix}\hat{P}e^{\hat{D}_{i+\frac{1}{2}}(x_i)} \\ \hat{P}e^{\hat{D}_{i+\frac{1}{2}}(x_{i+1})}\end{pmatrix}\vec{G}_{i+\frac{1}{2}} = \hat{H}_{i+\frac{1}{2}}\vec{G}_{i+\frac{1}{2}} \tag{5}$$

$\hat{P}$ is a nonsquare matrix selecting the even components: $\vec{G}_e = \hat{P}\vec{G}$. On the grid nodes the fluxes must be continuous:

$$\hat{P}\hat{A}e^{\hat{D}_{i-\frac{1}{2}}(x_i)}\hat{H}_{i-\frac{1}{2}}^{-1}\begin{pmatrix}\vec{G}_{e,i-1} \\ \vec{G}_{e,i}\end{pmatrix} = \hat{P}\hat{A}e^{\hat{D}_{i+\frac{1}{2}}(x_i)}\hat{H}_{i+\frac{1}{2}}^{-1}\begin{pmatrix}\vec{G}_{e,i} \\ \vec{G}_{e,i+1}\end{pmatrix} \tag{6}$$

This equation links the even components on the nodes $i-1$, i and $i+1$. For $l = 1$ the well known Scharfetter-Gummel stabilization is obtained [16]. On the contacts a constant surface recombination velocity is used. At the source this yields:

$$-\left(\vec{G}_{e,1} - \vec{G}_{e,eq}\right)v_s = \hat{P}\hat{A}e^{\hat{D}_{\frac{3}{2}}(x_1)}\hat{H}_{\frac{3}{2}}^{-1}\begin{pmatrix}\vec{G}_{e,1} \\ \vec{G}_{e,2}\end{pmatrix} \tag{7}$$

$\vec{G}_{e,eq}$ is the equilibrium solution and v_s the surface recombination velocity. With a corresponding boundary condition for the drain a closed system of equations is obtained for the even components, which can be solved self-consistently with the Poisson equation by a Newton-Raphson method, where the electron density for the Poisson equation is evaluated on the grid nodes.

Small-signal analysis for the sinusoidal steady state condition is straightforward. Equation (3) is linearized with $\vec{g}(x,t) = \vec{G}(x) + \Re\{\vec{g}(x)e^{i\omega t}\}$ and an expression for the complex phasors $\vec{g}$, $\underline{E}_x$ is obtained:

$$i\omega\vec{g} + \hat{A}\frac{\partial\vec{g}}{\partial x} + \hat{B}\vec{g} = -\frac{\partial\hat{B}}{\partial E_x}\vec{G}\underline{E}_x \tag{8}$$

With $\hat{C}'_{i+\frac{1}{2}} = \hat{C}_{i+\frac{1}{2}} + i\omega\hat{A}^{-1}$ and $\hat{D}'_{i+\frac{1}{2}}(x)$, $\hat{H}'_{i+\frac{1}{2}}$ calculated with $\hat{C}'_{i+\frac{1}{2}}$ instead of $\hat{C}_{i+\frac{1}{2}}$ the solution is in the interval $[x_i, x_{i+1}]$:

$$\begin{aligned}\vec{g}(x) = {}&e^{\hat{D}'_{i+\frac{1}{2}}(x)}\left(\hat{H}'_{i+\frac{1}{2}}\right)^{-1}\begin{pmatrix}\vec{g}_{e,i} \\ \vec{g}_{e,i+1}\end{pmatrix} \\ &+ \left[\hat{M}_{i+\frac{1}{2}}(x) - e^{\hat{D}'_{i+\frac{1}{2}}(x)}\left(\hat{H}'_{i+\frac{1}{2}}\right)^{-1}\begin{pmatrix}\hat{P}\hat{M}_{i+\frac{1}{2}}(x_i) \\ \hat{P}\hat{M}_{i+\frac{1}{2}}(x_{i+1})\end{pmatrix}\right] \\ &\hat{H}_{i+\frac{1}{2}}^{-1}\begin{pmatrix}\vec{G}_{e,i} \\ \vec{G}_{e,i+1}\end{pmatrix}\underline{E}_{x,i+\frac{1}{2}} \end{aligned} \tag{9}$$

The matrix $\hat{M}$ is given by an integral:

$$\hat{M}_{i+\frac{1}{2}}(x) = \int_0^1 e^{\hat{D}'_{i+\frac{1}{2}}(x)(1-\alpha)}\frac{\partial\hat{D}_{i+\frac{1}{2}}(x)}{\partial E_x}e^{\hat{D}_{i+\frac{1}{2}}(x)\alpha}d\alpha \tag{10}$$

The integral is evaluated by numerical means. Equation (9) can be used to build a linear system of equations for the even components of the small-signal solution similar to the dc case.

III. SIMULATION RESULTS

A GaN HEMT similar to the one in Ref. [5] is investigated, where instead of 2000 gates only three are considered. The effective mass of the conduction band is $0.13m_0$, where m_0 is the free electron mass. The relative permittivity of GaN is 9.7, the temperature 300K and the surface recombination velocity $1.49 \cdot 10^7$cm/s. A constant grid spacing of 10nm is used in transport direction ($N = 311$). The device is

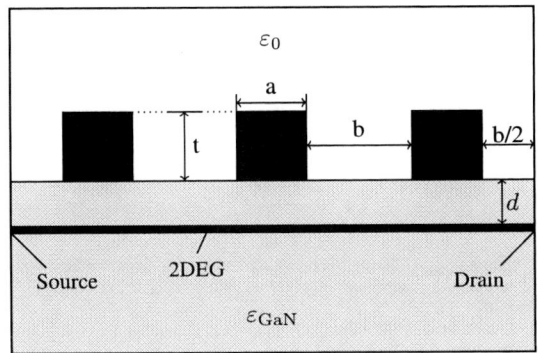

Fig. 1. HEMT with three gates and a 3μm long channel ($a = 0.34\mu$m, $b = 0.66\mu$m, $d = 0.026\mu$m, $t = 0.04\mu$m) [5].

simulated for the highest mobility mentioned in the paper (4170cm^2/Vs and $\tau = 0.308$ps) (Fig. 2). The lowest order

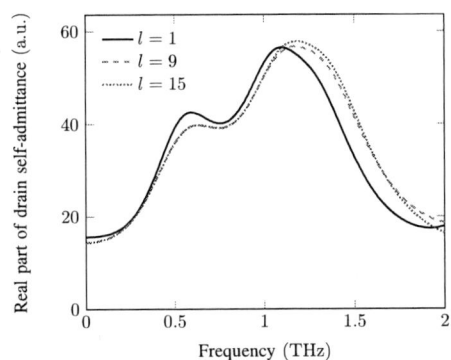

Fig. 2. Real part of the drain self-admittance ($\Re\{Y_{\mathrm{DD}}(j2\pi f)\}$) for $\mu = 4170$cm^2/Vs and $V_{\mathrm{DS}} = 0$V.

model ($l = 1$) corresponds at zero drain/source bias to the model used by Dyakonov and Shur, since the (linearized) convective derivative is zero in this case. For this low mobility the expansion with Hermitian polynomials converges and reproduces the results of the BTE, which are not shown. In Fig. 3 the absolute value of the drain self-admittance is shown in the complex plane ($j\omega = \sigma + j2\pi f$) for $l = 1$. As expected, at a real part of $\sigma = -1/2\tau = -1.62$/ps a series of zeros and poles is found in accordance with the theory of Dyakonov and Shur [2]. The zeros (poles of the drain self-impedance) correspond to the plasma instabilities and they are strongly damped due to the low mobility and zero drain/source bias.

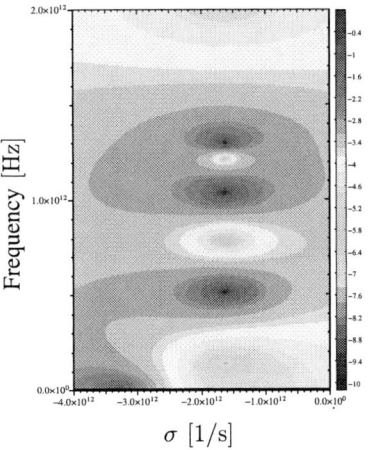

Fig. 3. Absolute value of the drain self-admittance ($|Y_{\mathrm{DD}}(\sigma + j2\pi f)|$) for $l = 1$, $\mu = 4170$cm^2/Vs and $V_{\mathrm{DS}} = 0$V.

In the case of a model with 10 moments additional poles and zeros occur at about $\sigma = -1/\tau = -3.25$/ps (Fig. 4), which have negligible impact on the admittance at $\sigma = 0$ (Fig. 2). In

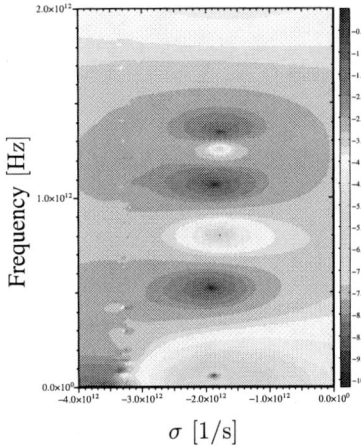

Fig. 4. Absolute value of the drain self-admittance for $l = 9$, $\mu = 4170$cm^2/Vs and $V_{\mathrm{DS}} = 0$V.

Ref. [5] an electric field of 450V/cm was applied resulting in the case of three gates in a drain/source bias of $V_{\mathrm{DS}} = 0.135$V. As shown in Fig. 5 the results barely change and no negative real part of the drain self-admittance occurs. At such low mobilities THz waves cannot be generated and a much higher mobility with $2\pi f\tau \gg 1$ is required. In Fig. 6 results are shown for a 100 times higher mobility. The expansion with Hermitian polynomials no longer converges and the additional poles due to the higher moments (Fig. 7) have a much stronger impact on the drain self-admittance than in the case of the lower mobility (Fig. 4). Thus, the hydrodynamic model used

978-1-7281-0941-1/19 $31.00 © 2019 IEEE

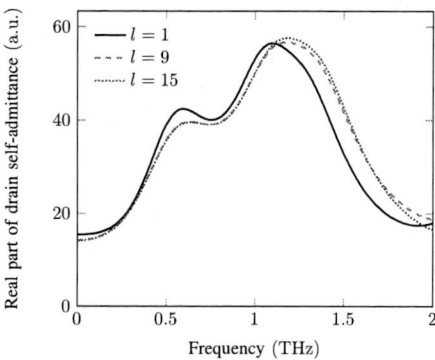

Fig. 5. Real part of the drain self-admittance for $\mu = 4170 \mathrm{cm}^2/\mathrm{Vs}$ and $V_{\mathrm{DS}} = 0.135\mathrm{V}$.

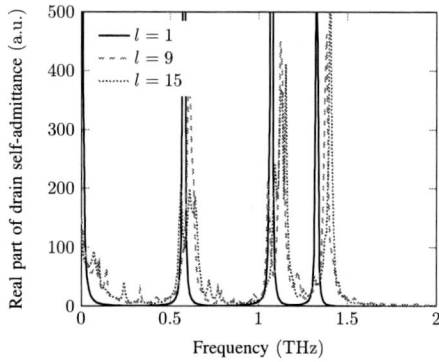

Fig. 6. Real part of the drain self-admittance for $\mu = 4.17 \cdot 10^5 \mathrm{cm}^2/\mathrm{Vs}$ and $V_{\mathrm{DS}} = 0\mathrm{V}$.

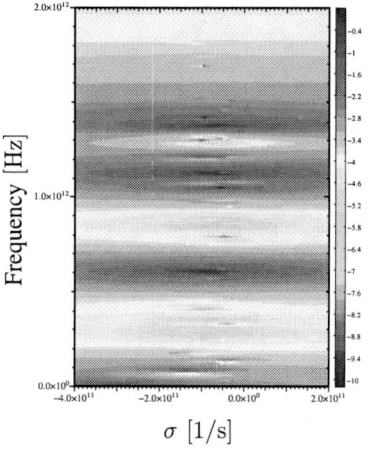

Fig. 7. Absolute value of the drain self-admittance for $l = 9$, $\mu = 4.17 \cdot 10^5 \mathrm{cm}^2/\mathrm{Vs}$ and $V_{\mathrm{DS}} = 0\mathrm{V}$.

by Dyakonov and Shur is not able to describe transport in devices with mobilities necessary for plasma oscillations even at zero drain/source bias. In the case of a nonzero bias the situation gets worse.

IV. CONCLUSIONS

Under quasi-ballistic conditions, the expansion with Hermitian polynomials no longer converges and plasma wave modeling requires a much more sophisticated transport model than the hydrodynamic model. Furthermore, it is not clear whether plasma instabilities can be used to generate THz waves with HEMTs.

REFERENCES

[1] T. Otsuji and M. Shur, "Terahertz plasmonics: Good results and great expectations," *Microwave Magazine, IEEE*, vol. 15, no. 7, pp. 43–50, Nov 2014.

[2] M. Dyakonov and M. Shur, "Shallow water analogy for a ballistic field effect transistor: New mechanism of plasma wave generation by dc current," *Phys. Rev. Lett.*, vol. 71, pp. 2465–2468, Oct 1993. [Online]. Available: http://link.aps.org/doi/10.1103/PhysRevLett.71.2465

[3] ——, "Plasma wave electronics: Novel terahertz devices using two dimensional electron fluid," *Electron Devices, IEEE Transactions on*, vol. 43, no. 10, pp. 1640–1645, Oct 1996.

[4] G. Gonzalez, *Microwave Transistor Amplifiers: Analysis and Design*, 2nd ed. Upper Saddle River, New Jersey: Prentice-Hall, 1997.

[5] V. Jakstas, I. Grigelionis, V. Janonis, G. Valusis, I. Kasalynas, G. Seniutinas, S. Juodkazis, P. Prystawko, and M. Leszczynski, "Electrically driven terahertz radiation of 2DEG plasmons in AlGaN/GaN structures at 110K temperature," *Applied Physics Letters*, vol. 110, no. 20, p. 202101, 2017. [Online]. Available: https://doi.org/10.1063/1.4983286

[6] Z. Kargar, T. Linn, D. Ruić, and C. Jungemann, "Investigation of transport modeling for plasma waves in THz devices," *IEEE Transactions on Electron Devices*, vol. 63, no. 11, pp. 4402–4408, Nov 2016.

[7] T. Grasser, T.-W. Tang, H. Kosina, and S. Selberherr, "A review of hydrodynamic and energy-transport models for semiconductor device simulation," *Proc. IEEE*, vol. 91, no. 2, pp. 251–274, 2003.

[8] T. Linn, Z. Kargar, and C. Jungemann, "Investigation of moments-based transport models applied to plasma waves and the Dyakonov–Shur instability," *Semiconductor Science and Technology*, vol. 34, no. 1, p. 014002, 2018.

[9] C. Jungemann, T. Grasser, B. Neinhüs, and B. Meinerzhagen, "Failure of moments-based transport models in nanoscale devices near equilibrium," *IEEE Trans. Electron Devices*, vol. 52, no. 11, pp. 2404–2408, 2005.

[10] N. Van Kampen, "The dispersion equation for plasma waves," *Physica*, vol. 23, no. 6-11, pp. 641–650, 1957.

[11] S. Selberherr, *Analysis and Simulation of Semiconductor Devices*. Wien: Springer, 1984.

[12] S.-M. Hong and J.-H. Jang, "Numerical simulation of plasma oscillation in 2-d electron gas using a periodic steady-state solver," *Electron Devices, IEEE Transactions on*, vol. 62, no. 12, pp. 4192–4198, Dec 2015.

[13] M. Abramowitz and I. A. Stegun, *Handbook of Mathematical Functions*. New York: Dover Publications, INC., 1972.

[14] H. Kim, H. S. Min, T. W. Tang, and Y. J. Park, "An extended proof of the Ramo-Shockley theorem," *Solid–State Electron.*, vol. 34, pp. 1251–1253, 1991.

[15] C. Moler and C. Van Loan, "Nineteen dubious ways to compute the exponential of a matrix, twenty-five years later," *SIAM review*, vol. 45, no. 1, pp. 3–49, 2003.

[16] D. L. Scharfetter and H. K. Gummel, "Large-signal analysis of a silicon read diode oscillator," *IEEE Trans. Electron Devices*, vol. ED-16, no. 1, pp. 64–77, 1969.

Exact Correction of the Self-Force Problem in Monte Carlo Device Simulation

Andrea Ghetti

Technology Development, Micron Technology Inc., Vimercate, Italy

Abstract—The self-force is a specific problem of self-consistent Monte Carlo-Poisson simulation resulting in an un-physical field component acting on a particle coming from the particle itself (the self-force). Several approaches have been proposed in literature to mitigate this problem, but all of them suffer to some extent of approximations and/or limitations. In this paper we propose a new and mathematically exact correction of the self-force problem based on a numerical approach. Although computationally expensive, it has no restriction and can be always applied. The new method has been tested on the difficult problem of plasma oscillation simulation providing the expected plasma energy from theory. Moreover, the same mathematical framework introduced here for the self-force correction can be readily applied also for the exact calculation of the reference force in the Particle-Particle-Particle-Mesh (P3M) method. The accuracy of such approach to P3M method is demonstrated by simulating the bulk low field mobility dependence on doping concentration.

I. INTRODUCTION

Since the seminal work of Hockney and Eastwood [1] the self-force (SF) problem has been recognized as a critical problem for self-consistent Monte Carlo-Poisson simulation (SC-MC-P). It arises from mixing point-like particles (charges) with finite element solution of the Poisson Eq. for the potential/field profile. Typically, SC-MC-P includes the following steps: 1) charge assignment to the mesh; 2) solution of the Poisson Eq. for the potential; 3) interpolation of the resulting electric field back to particle position. The problem stems from the unavoidable displacement of the charge implicit in step 1) and the approximation involved in step 3).

In order to understand the origin of the SF problem, we can consider the simple case of a single particle with zero initial velocity/energy and no external field (the so called 'lone' particle case). Obviously, the particle should stay fixed where it is, but this is not the case for SC-MC-P, as explained in Fig. 1. Indeed, let's consider for convenience the simplest charge assignment method scheme at step 1), that is assigning the particle charge to the nearest grid point (NGP). Then the computed potential (solid line) will feature a finite deep at that grid point (whose amplitude depends on the grid spacing), and the corresponding field would be piece-wise constant (dotted line). The key point to notice is that the computed field is not zero at the particle position as it should be. Instead there is a fictitious field acting on the particle due to the particle itself that should not be there. This is caused by the displacement of the computed potential profile on the mesh with respect to the real one (Coulomb potential, dashed line) due to space

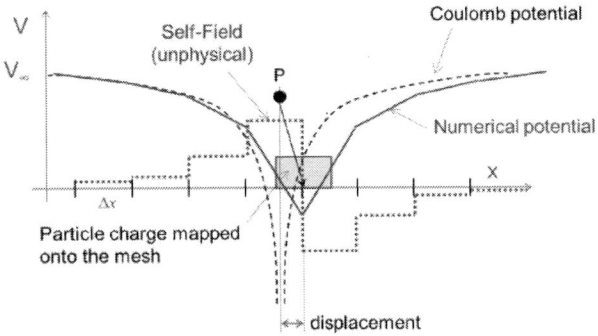

Fig. 1. Schematic description of the SF origin for a single particle P. Correct (Coulomb) potential: dashed line. Finite element solution of the Poisson Eq. (solid line) and corresponding field profile (dotted line).

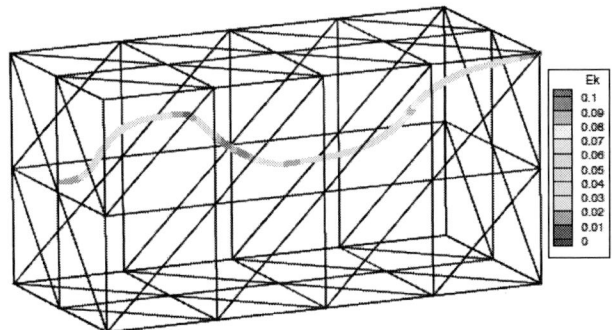

Fig. 2. Example of SC-MC-P simulation of the 'lone' particle case when the SF problem is not corrected. Color level represents the particle kinetic energy E_k in eV.

discretization, and the interpolation of the electric field at the particle position that is not able to resolve the symmetry of the field around the particle position.

An example of such un-physical results is reported in Fig. 2. A particle with no kinetic energy (E_k) has been placed on a mesh node. Then a SC-MC-P simulation is carried out with only this unique particle. Because of the SF problem, the particle starts to move around, attaining also a significant kinetic energy, whereas it should stay fixed at the initial position. Followed trajectory and gained energy are un-predictable as they depend on the mesh topology, initial position, how often the potential profile is updated and so on.

Clearly this problem is present only in SC-MC-P simulation, it worsens for decreasing mesh spacing, i.e mesh node distance (Δx), and it is more impacting where the external field is low.

Several methods have been proposed in the literature to mitigate the SF problem. Many of them are based on smoother charge assignment schemes such as Cloud-in-Cell (CIC) [1], nearest-element-center (NEC) [2], etc. Others, instead, are based on more complex field interpolation [3]. However, all of them suffer of some limitations, either imposing some constraint on the mesh (e.g. uniform spacing, or no interfaces, or equilateral triangles [2]), either not guaranteeing the consistency of potential and field, possibly leading to the not conservation of both momentum and energy [1].

In this paper we introduce a new approach to solve the SF problem in SC-MC-P simulation. It is based on a numerical correction of the potential profile. Although it might be computational expensive, it is mathematically exact, and does not suffer of any of the above-mentioned limitations, and thus can be generally applied.

Sect. II introduces the mathematical framework behind the new correction, whose results are then discussed in Sect. III. The same approach is leveraged in Sect. IV for an exact implementation of the P3M method. Finally Sect. V draws some conclusions.

II. EXACT SELF-FORCE CORRECTION

In principle, the SF can be avoided if the potential caused by the *i-th* particle p_i (φ_i) is removed from the total potential ψ. In practice, this can be done by applying to p_i a corrected potential $\psi_i^{cor} = \psi - \varphi_i$. Here ψ is the solution of the Poisson Eq. that, in case of point-like charges, can be written as

$$\nabla \cdot (-\epsilon \nabla \psi) = q \Big(\sum_i w_i \delta(\vec{r} - \vec{r}_{NGP(p_i)}) + N_c(\vec{r}) \Big) \quad (1)$$

with boundary conditions

$$\psi = \psi_a \quad \text{on electrodes} \quad (2)$$
$$\frac{d\psi}{d\vec{n}} = 0 \quad \text{on remaining boundary,} \quad (3)$$

where q is the elementary charge, ϵ the dielectric constant, N_c is any charge described by a continuous profile (e.g. doping), $\vec{r}_{NGP(p_i)}$ is the nearest grid point to p_i, and w_i is its statistical weight (with sign, i.e. positive for holes, negative for electrons). The difficulty lies in identifying φ_i.

The new method we are proposing calculates φ_i numerically, and, therefore, can be always applied, and it is mathematically exact within the framework of finite element calculation of ψ. To this purpose we follow the main idea in [4]. We start by noticing that ψ can be written as

$$\psi = \psi_c + \sum_i w_i \Phi_{NGP(p_i)}, \quad (4)$$

where ψ_c is the solution of

$$\nabla \cdot (-\epsilon \nabla \psi_c) = q N_c(\vec{r}) \quad (5)$$

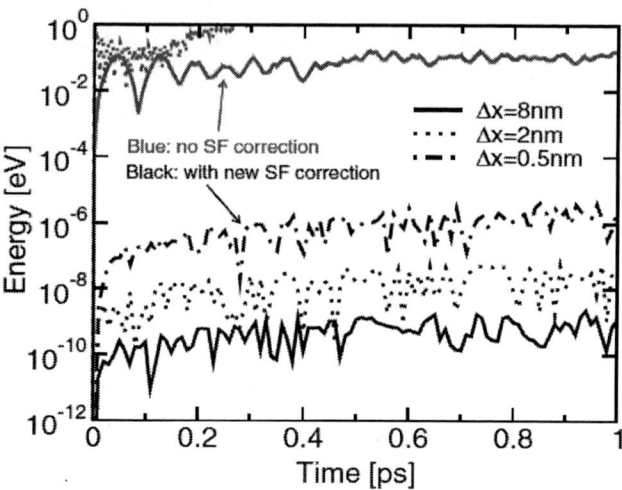

Fig. 3. 'Lone' particle energy SC-MC-P 3D simulation results for different mesh spacing Δx. Blue lines: no SF correction. Black lines: with exact SF correction. The correct value should be 0.

with boundary conditions (2) and (3), and $\Phi_{NGP(p_i)}$ is the solution of

$$\nabla \cdot (-\epsilon \nabla \Phi_{NGP(p_i)}) = q \delta(\vec{r} - \vec{r}_{NGP(p_i)}) \quad (6)$$

with boundary conditions (3) and

$$\Phi_{NGP(p_i)} = 0 \quad \text{on electrodes.} \quad (7)$$

Notice that: *i)* this method is somehow different from [4] because of the different form of (6) and the different boundary condition represented by (7) (compare them to (4) and (5) in [4] respectively); *ii)* (7) is mandatory in order for (4) to be true; *iii)* (7) sets a clear and unambiguous boundary condition regardless of the device structure.

From (4) it is clear that

$$\varphi_i = w_i \Phi_{NGP(p_i)}. \quad (8)$$

The new method then requires to compute and store the solution of (6) for all mesh nodes to be later used to correct the potential on a particle basis. This can be computationally expensive, but it can be speeded-up parallelizing the solution of (6), that are all independent on each other. Moreover, it also requires a lot of memory, but this is not a problem anymore for modern computers with hundreds of GB of memory.

III. RESULTS AND DISCUSSION

First, we verified the new method on the case of the single particle with zero initial energy/velocity. In this particular case, ψ and φ_i coincide, except for the numerical error in their calculation. Thus $\psi_i^{cor} = 0$ always, and, consequently, the particle keeps its state. SC-MC-P 3D simulation results with and without the new correction are reported in Fig. 3 for different mesh spacing. Clearly, without any SF correction, the particle energy rapidly diverges. Only for the larger mesh spacing, the particle energy attains a finite value in the range of hundreds of *meV*, which is anyway not acceptable (the correct one should be 0). On the contrary, the new SF correction always provides a particle energy below $10^{-6}eV$, i.e. comparable with the numerical precision with which ψ and φ_i have been computed.

Next, we tested the new SF correction on the 3D simulation of plasma oscillations. Plasma oscillations are a collective motion of a gas of free electrical particles. Let's consider the simple case of a gas of free electrons of density n on a uniform background of positive fixed charge. If an electron is displaced from its equilibrium position even by a small amount the Coulomb interaction with the other charges will tend to pull it back, resulting in an oscillation around the equilibrium configuration. These oscillations take place at a characteristics frequency

$$\omega_p = \sqrt{\frac{nq^2}{m^*\epsilon}}$$

where m^* is the carrier effective mass. In this process, the free electrons exchange energy with the Coulomb potential resulting in an increased average kinetic energy (ΔE_C) with respect to the thermal energy. Theory in [5] showed this extra energy to be

$$\Delta E_C = 1.451 \frac{q^2 n^{1/3}}{4\pi\epsilon_\infty} \left[1 + \left(\frac{1}{\omega_p \tau_m}\right)^2\right]^{-1/2} \quad (9)$$

where τ_m is the momentum relaxation time due to the other scattering mechanisms.

Plasma simulation is a very difficult task for SC-MC-P since it requires very short time step between Poisson Eq. solution ($\Delta t \ll 1/\omega_p$) to capture the potential oscillation, and very small grid spacing to accurately resolve the Coulomb force among particles according to the Particle-Mesh (PM) scheme [1]. We found that $\Delta x = 1/4\sqrt[3]{n}$ is as good trade-off between accuracy and computational time. Under these conditions SF becomes very important and thus must be avoided. In the simulation we adjust the size of the 3D cubic domain to hold approximately 1000 electrons at all densities, resulting in a mesh with $\approx 70K$ nodes. In this case, calculation of all $\Phi_{NGP(p_i)}$ takes approximately half of the simulation time, which is anyway dominated by the time spent solving (1), and $\approx 25GB$ of RAM.

Fig. 4 compares simulation results with and without the new SF correction against the theoretical value. Without SF correction the simulated average energy is much higher than expected, while the new and exact SF correction provides a good agreement with theory, demonstrating the validity of the approach.

IV. EXTENSION TO P3M METHOD

The calculation of φ_i (contribution of the particle p_i to the total potential ψ) is also required in the implementation of the Coulomb interaction among charges according to the Particle-Particle-Particle-Mesh (P3M) method [1]. In order to allow the use of coarser grid than needed by the PM method as discussed in the previous section, in the P3M method the Coulomb force

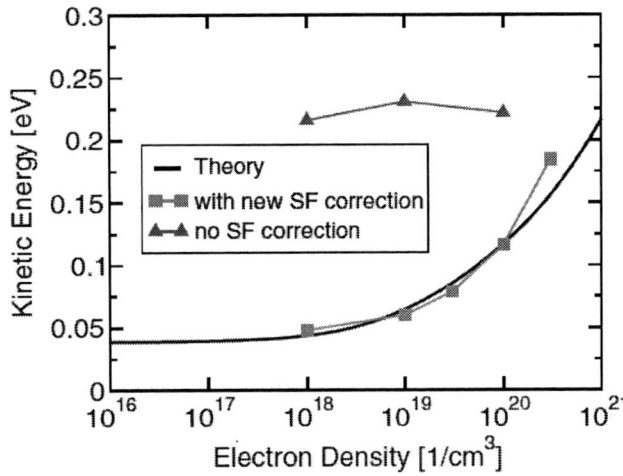

Fig. 4. Plasma energy simulation. Black line: theoretical value $1.5k_B T + \Delta E_C$. Blue triangles: no SF correction. Red squares: exact SF correction.

exchanged with surrounding charges is explicitly added to the electric field acting on a particle. In this way, Coulomb scattering with fixed dopants (impurity scattering) and with other mobile carriers (electron-electron scattering) are not treated anymore as explicit localized scattering events with their own rates, but they are substituted by the continuous interaction exchanged via the Coulomb force during particle motion (a sort of molecular dynamics approach). However, in order not to double count the interaction with surrounding particles, since also included in the mesh potential, their contribution to ψ, that in the framework of P3M is called reference potential(force), must be removed [1], [6]–[8]. This contribution is, by definition, the same as derived in Sect. II and it is given again by (8). Therefore, in the P3M method, the potential profile 'seen' by the *i-th* particle, including also the SF correction introduced in this paper, is

$$\psi_i^{P3M} = \psi \;+\; \sum_{j\neq i} \frac{q w_j}{4\pi\epsilon |\vec{r_j} - \vec{r_i}|} \qquad \text{Coulomb potential}$$
$$-\; \sum_{j\neq i} w_j \Phi_{NGP(p_j)} \qquad \text{Reference potential}$$
$$-\; w_i \Phi_{NGP(p_i)} \qquad \text{SF correction.} \quad (10)$$

The accuracy of such approach is demonstrated by simulating the bulk low field mobility dependence on doping (i.e. impurity) concentration. Usually, in the framework of MC device simulation, impurity scattering is treated as any other scattering mechanisms with its own scattering rate and after scattering state selection rules, provided by a number of models (Brook and Herring [9], Ridley [10], Conwell-Weisskopf [11], Kosina [12]), that are all depending on the doping concentration considered as a continuous profile. Often, these models have fitting parameters [13], as reproducing low field mobility doping dependence is a difficult task for MC.

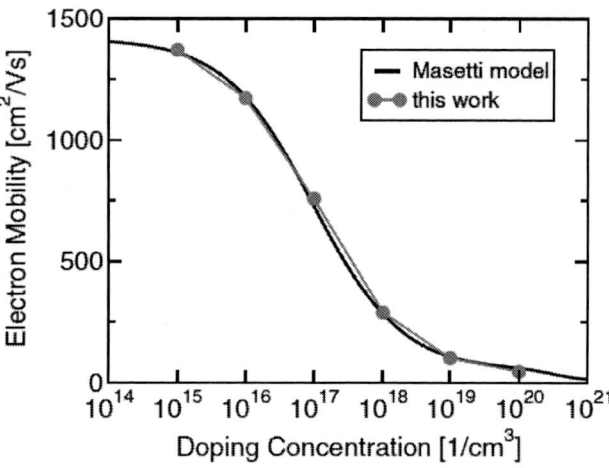

Fig. 5. Comparison of simulated low field electron bulk mobility in Silicon (symbols) with the analytical Masetti model [15].

However, with the restless shrinking of device physical dimension, nowadays only a few dopants are present in the channel of MOS transistors, and even in the source/drain extensions, mandating to treat them as discrete, and randomly placed, fixed particles [14]. Therefore, the aforementioned models cannot be applied anymore, whereas the P3M method is particularly suited for this as it was designed to handle Coulomb interaction with point-like particles.

SF correction and P3M method have been implemented in our MC code (MC++), which was then used to simulate the bulk low field mobility for several doping levels. These simulations are very similar to the ones in the previous section with the following differences: 1) background doping is described here by discrete particles placed randomly (no need to be located at mesh nodes though); 2) an external low field ($1 KV/cm$) is applied to induce a detectable average drift velocity; 3) mesh spacing has been relaxed as allowed by P3M and adapted to doping level to include in the 3D simulation domain a large number of impurities in order to reduce the statistical variability associated with the random placement while keeping to a manageable level the number of mesh nodes.

Results of such simulations are compared in Fig. 5 to the reference Masetti model that is a good fit to experimental data [15]. The agreement is quite good demonstrating the soundness of P3M for this kind of problem, and the effectiveness of the proposed implementation. It is worth noticing that: 1) in the simulation of Fig. 5 there is no free parameter as the only phenomenon at play is the Coulomb force; 2) SF correction is again mandatory to get the correct kinetic energy and, hence, velocity.

V. CONCLUSION

In this paper we have introduced a new method for the correction of the self-force problem in SC-MC-P simulation. This method is based on the identification of the contribution of each particle to the potential/field profile, so that it can be easily removed from the driving forces of particle motion. This operation is done numerically on the simulation mesh and, although computationally expensive, is mathematically exact within the framework of finite element calculation of the potential profile, and it has no restriction. We demonstrated its effectiveness by showing that this new method allows to reproduce the average kinetic energy in plasma oscillation simulation.

Moreover we have proposed to use the same approach also to compute the reference force needed for the implementation of the Coulomb interaction among point-like charges (either fixed or mobile) via the P3M method. We verified the correctness of this new approach to P3M by reproducing the electron low field mobility dependence on doping concentration with no fitting parameter.

REFERENCES

[1] R. W. Hockney and J. P. Eastwood, *Computer simulation using particles.* New York: McGraw Hill, 1981.

[2] S. Laux, "On particle-mesh coupling in Monte Carlo semiconductor device simulation," *IEEE Trans. Computer-Aided Design*, vol. 15, no. 10, pp. 1266–1277, 1996.

[3] M. Aldegunde, N. Seoane, A. Garca-Loureiro, and K. Kalna, "Reduction of the self-forces in monte carlo simulations of semiconductor devices on unstructured meshes," *Computer Physics Communications*, vol. 181, no. 1, pp. 24 – 34, 2010.

[4] M. Aldegunde and K. Kalna, "Energy conserving, self-force free monte carlo simulations of semiconductor devices on unstructured meshes," *Computer Physics Communications*, vol. 189, pp. 31–36, 2015.

[5] M. Fischetti and S. Laux, "Long-range Coulomb interactions in small Si devices. Part I: Performance and reliability," *Journal of Applied Physics*, vol. 89, no. 2, pp. 1205–1231, 2001.

[6] W. Gross, D. Vasileska, and D. Ferry, "A novel approach for introducing the electron-electron and electron-impurity interactions in particle-based simulations," *IEEE Electron Device Letters*, vol. 20, no. 9, pp. 463–465, 1999.

[7] C. Wordelman and U. Ravaioli, "Integration of a Particle-Particle-Particle-Mesh Algorithm with the Ensemble Monte Carlo Method for the Simulation of Ultra-Small Semiconductor Devices," *IEEE Trans. Electron Devices*, vol. 47, no. 2, pp. 410–416, 2000.

[8] W. Lee and U. Ravaioli, "A simple and efficient method for the calculation of carrier-carrier scattering in Monte-Carlo simulations," in *Proc. SISPAD Conference*, 2010, pp. 131–134.

[9] H. Brooks and C. Herring, "Scattering by ionized impurities in semiconductors," *Phys. Rev.*, vol. 83, p. 879, 1951.

[10] B. Ridley, "Reconciliation of the Conwell-Weisskopf and Brooks-Herring formulae for charged-impurity scattering in semiconductors: Third-body interference," *Journal of Physics C: Solid State Physics*, vol. 10, no. 10, pp. 1589–1593, 1977.

[11] E. Conwell and V. Weisskopf, "Scattering by ionized impurities in semiconductors," *Phys. Rev.*, vol. 73, no. 3, pp. 388–390, 1950.

[12] H. Kosina, "A method to reduce small-angle scattering in Monte Carlo device analysis," *IEEE Trans. Electron Devices*, vol. 46, no. 6, pp. 1196–1200, 1999.

[13] F. Bufler, A. Schenk, and W. Fichtner, "Efficient Monte Carlo Device Modeling," *IEEE Trans. Electron Devices*, vol. 47, no. 10, pp. 1891–1897, 2000.

[14] C. Alexander, G. Roy, and A. Asenov, "Random-Dopant-Induced Drain Current Variation in Nano-MOSFETs: A Three-Dimensional Self-Consistent Monte Carlo Simulation Study Using "Ab Initio" Ionized Impurity Scattering," *IEEE Trans. Electron Devices*, vol. 55, no. 11, pp. 3251–3258, 2008.

[15] G. Masetti, G. Severi, and M. Solmi, "Modeling of carrier mobility against carrier concentration in Arsenic-, Phosphorous- and Boron-doped silicon," *IEEE Trans. Electron Devices*, vol. ED-30, pp. 764–769, 1983.

A Robust Simulation Method for Breakdown with Voltage Boundary Condition Utilizing Negative Time Constant Information

Shigetaka Kumashiro
Kyoto Institute of Technology
Matsugasaki, Sakyo-ku,
Kyoto 606-8585, Japan
kumasiro@vlsi.es.kit.ac.jp

Tatsuya Kamei
Kyoto Intitute of Technology
Matsugasaki, Sakyo-ku,
Kyoto 606-8585, Japan

Akira Hiroki
Kyoto Institute of Technology
Matsugasaki, Sakyo-ku,
Kyoto 606-8585, Japan
hiroki@kit.ac.jp

Kazutosi Kobayashi
Kyoto Institute of Technology
Matsugasaki, Sakyo-ku,
Kyoto 606-8585, Japan
kazutoshi.kobayashi@kit.ac.jp

Abstract—**Dominant time constant analysis reveals that the semiconductor equations at hard breakdown turn into a positive feedback state where the convergence of steady state (DC) Newton iteration is substantially difficult even if continuation method is used. A robust simulation method for hard breakdown which detects the appearance of negative time constant during DC Newton iteration and then switches to transient (TR) simulation is proposed. The negative time constant value during the TR simulation is used for the time step restriction and the maximum time constant value is used for the determination of the final time of the TR simulation. By using the proposed method, a trace of the stable operation points in the snapback I-V trajectory corresponding to each DC bias can be obtained robustly with a simple voltage sweep at the voltage boundary.**

Keywords—*hard breakdown, negative time constant, positive feedback, DC Newton iteration, continuation method, voltage boundary condition*

I. Introduction

Breakdown voltage is a key design parameter for power devices, ESD protection devices and device isolations. Current capacity after the onset of breakdown is also important for ESD protection devices. To optimize the breakdown voltage, TCAD simulation is intensively used. Continuation method [1] which adopts a variable external resistance connected to a voltage boundary is usually used for the breakdown TCAD simulation because simple voltage sweep at the voltage boundary produces a convergence problem. Although the continuation method works well for variety of problems, the authors found that it sometimes suffers from the convergence problem if hard breakdown of PN junction is simulated. In this paper, the convergence problem is investigated and a robust simulation method for hard breakdown with voltage boundary condition utilizing negative time constant information is presented.

II. Problems with Conventional Breakdown Simulation Methods

The discretization scheme shown in Fig. 1 which assigns the carrier generation term by impact ionization to the downstream side of the carrier drift [2] is adopted. This scheme can avoid the instability due to the local self-feedback in a control volume under large electric field. The diode

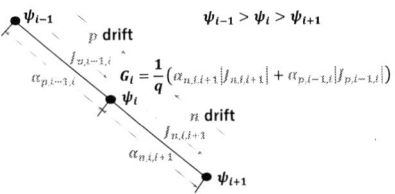

Fig. 1. Carrier generation term by impact ionization is assigned to the downstream side of the carrier drift.

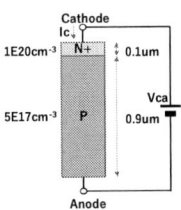

Fig. 2. Diode structure used for hard breakdown simulation.

structure used for hard breakdown simulation is shown in Fig. 2. Ic-Vca characteristics obtained by using conventional voltage sweep is shown in Fig. 3 (a). At Vca = 9.2 V, Newton iteration did not converge as shown by the blue line in Fig. 3 (b), where *dx_max* is the maximum variation of the variables at each Newton iteration. Fig. 4 shows the variation of the electrostatic potentials, the electron densities and the hole densities during the first 10 Newton iterations at Vca = 9.2 V. As shown by the thick red arrow, the carrier densities continue to decrease. This is due to the damper implemented in the program which suppresses the appearance of negative carrier density. In other words, the Newton method tries to solve the equations by introducing negative carrier density, which is an extraneous root. The grey line in Fig. 3 (b) shows the convergence behavior when another linear-search damper which tries to minimize the equation residual is used. In this case, the solution is rarely improved, which means the solution has been trapped in a local minimum. For sufficiently smooth and homeomorphic problems, even if they are highly nonlinear, Newton iteration converges if the variation of the variables is suppressed by a proper damper. [3] However, the hard breakdown simulation does not seem to be the case. To analyze the situation, 4 largest magnitude time constants of the

---

This work was supported by JSPS KAKENHI Grant-in-Aid for Scientific Research (C) Number 17K05142.

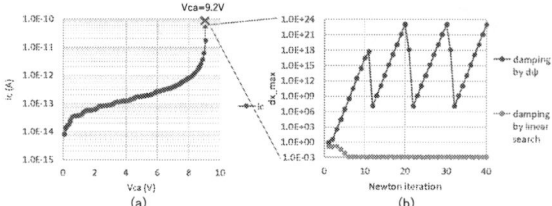

Fig. 3. (a) Ic-Vca characteristics obtained by using conventional DC voltage sweep. (b) Convergence behaavior at the non-converged bias point Vca = 9.2 V.

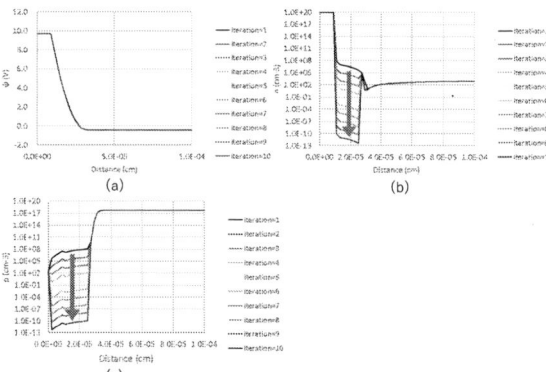

(a)   (b)

(c)

Fig. 4. Variation of (a) ψ, (b) n, and (c) p during first 10 Newton iterations at Vca = 9.2 V

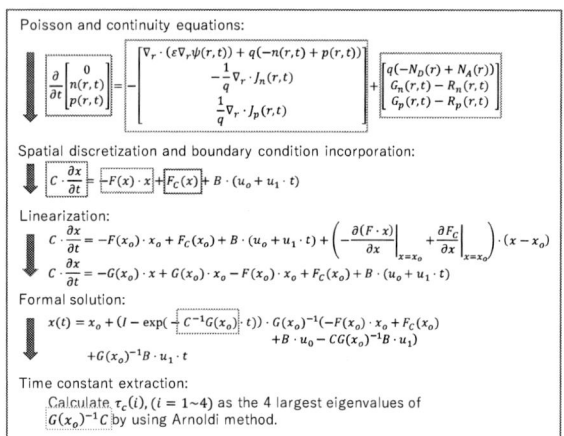

Fig. 5. Concept of time constant extraction of semiconductor device state equation.

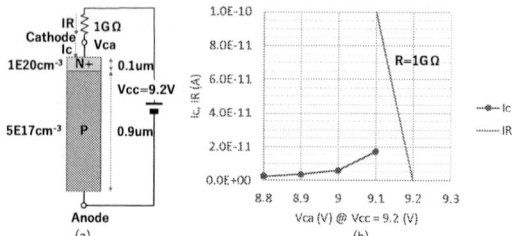

Fig. 7. (a) Diode + external resistance structure to mimic continuation method for DC breakdown simulation. (b) Diode characteristics before breakdown and 1 GΩ external resistance load line.

diode structure at each Newton iteration are extracted by applying Arnoldi method to the exponential decay factor in the formal solution of the device state equation shown in Fig. 5. [4][5] Fig. 6 shows the variation of the time constants during the Newton iterations in the hard breakdown simulation. The filled squares are the time constants when the Newton iteration converges and the open circles are the negative time constants appeared during the Newton iterations. At Vca = 9.2 V, negative time constants appear at almost all iterations, which means the system is in a positive feedback state from a view point of transfer function (i.e. positive poles). This is due to the negative differential resistance induced by the increasing carrier densities by impact ionization.

To improve the convergence, 1 GΩ external resistance is connected to the cathode to mimic the continuation method as shown in Fig. 7. By adding high resistance, it is expected that the negative differential resistance would be compensated and the system would become stable. However, the convergence behaviors are quite similar to the ones in Figs. 3 and 4. Fig. 8 shows the variation of the time constants. Although the largest time constants become positive, negative time constants still remain at Vca = 9.2 V. Therefore, adding external resistance can limit the voltage variation of the cathode terminal but it cannot turn the positive feedback system into a stable one.

III. PROPOSED BREAKDOEN SIMULATION METHOD

Since the negative time constants do not disappear by connecting the external resistance, the positive feedback loop seems to be closed within the high field impact ionization region as shown in Fig. 9, where $-\tau_c > 0$ is the positive feedback loop time constant. By using TR simulation with the time step width $\Delta t < |\tau_c|$, the feedback is suppressed and stable trace of the time development is expected. Suppose that the positive feedback system with a state variable x is described by the following equation.

$$\frac{dx}{dt} = -\frac{x}{\tau_c} \qquad (1)$$

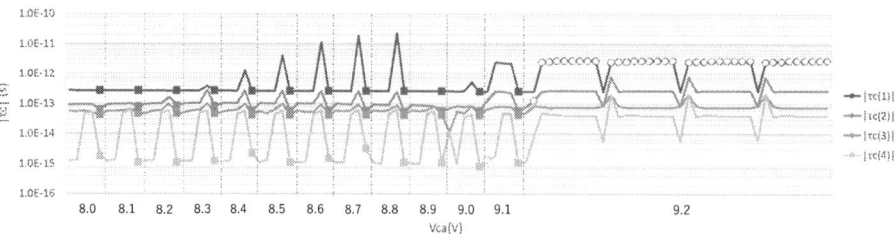

Fig. 6. Variation of time constants during Newton iteration in hard breakdown simulation.

978-1-7281-0941-1/19 $31.00 © 2019 IEEE

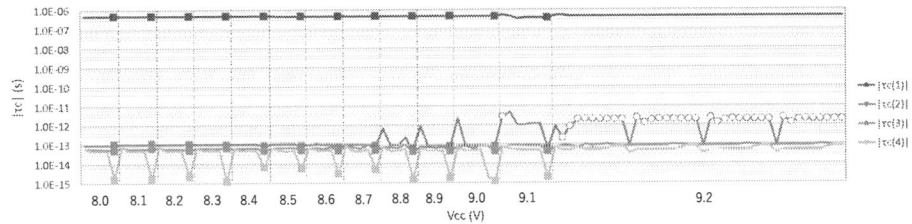

Fig. 8. Variation of the time constants during Newton iteration in DC breakdown simulation of diode + external resistance.

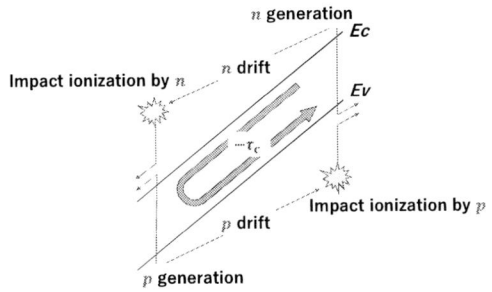

Fig. 9. Positive feedback loop closed within the high field impact ionization region. Here, $-\tau_c$ is the positive feedback loop time constant.

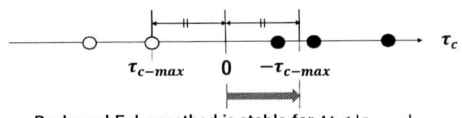

Backward Euler method is stable for $\Delta t < |\tau_{c-max}|$

Fig. 10. Backward Euler method becomes conditionally stable if both positive (filled circle) and negative (open circle) time constants coexist.

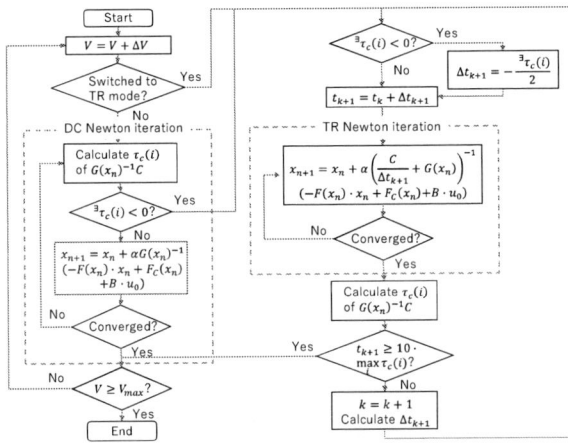

Fig. 11. Proposed breakdown simulation flow switching from DC to TR anaysis upon detecting a negative time constant.

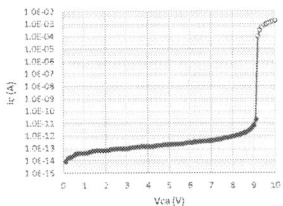

Fig. 12. Ic-Vca characteristics calculated by the proposed simulation method.

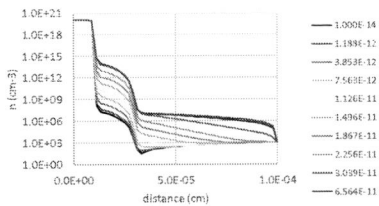

Fig. 13. Variation of electron concentration during TR simulation at Vca = 9.2 V.

By using Backward Euler method, (1) can be discretized as follows.

$$\frac{x_{k+1} - x_k}{\Delta t_{k+1}} = -\frac{x_{k+1}}{\tau_c} \tag{2}$$

where k is a time step number. By rearranging (2), the following relation is obtained.

$$x_{k+1} = \frac{\tau_c}{\Delta t_{k+1} + \tau_c} x_k \tag{3}$$

As $\tau_c < 0$ in this case, $\Delta t_{k+1} > |\tau_c|$ produces unphysical oscillatory solution with sign change. Therefore, $\Delta t_{k+1} < |\tau_{c-max}|$ where τ_{c-max} is the maximum negative time constant must be kept in Backward Euler method as shown in Fig. 10.

Since TR simulation is costly for TCAD, an algorithm shown in Fig. 11 is proposed. The algorithm switches DC simulation to TR simulation when negative time constant is detected during the DC Newton iteration. Fig. 12 shows Ic-

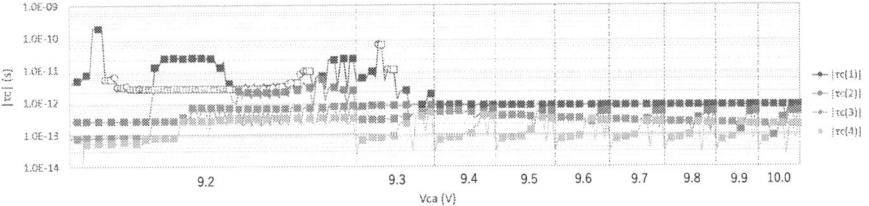

Fig. 14. Variation of the time constants at each time step during TR simulation for Vca = 9.2 − 10 V.

978-1-7281-0941-1/19 $31.00 © 2019 IEEE   35

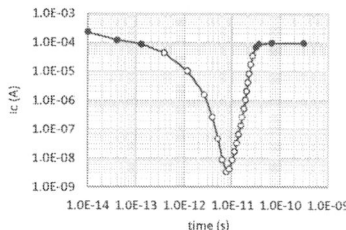

Fig. 15. Ic-t characteristic during TR simulation at Vca = 9.2 V.

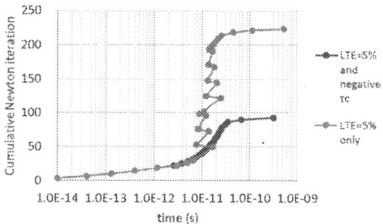

Fig. 16. Cumulative Newton iteration with / without time step limitation by negative time constant.

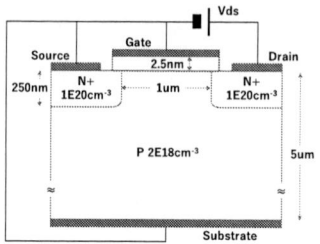

Fig. 17. GG-NMOS structure used for hard breakdown simulation.

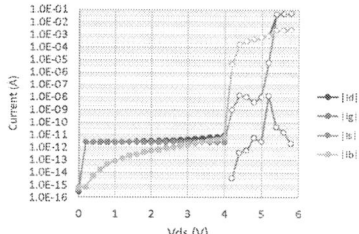

Fig. 18. Breakdown characteristics of GG-NMOS calculated by the proposed simulation method.

Vca characteristic calculated by using the algorithm. The open circles correspond to the solution by the TR simulation. Fig. 13 shows the variation of the electron density during the TR simulation at Vca = 9.2 V, where the electron density gradually increases as opposed to Fig. 4 (b). Variation of the time constants at each time step during the TR simulation for Vca = 9.2 V – 10 V is shown in Fig. 14, where the filled squares mean the positive cones and the open squares mean the negative ones. There are a lot of negative time constants at Vcs = 9.2 V and a few at Vca = 9.3 V. Since there is no negative time constant after Vca = 9.4 V, convergence may be obtained by using the conventional dampers in DC simulation. This stabilization comes from the increased Shockley-Red-Hall recombination. Ic-t characteristics during the TR simulation at Vca = 9.2 V is shown in Fig. 15, where the open circles stand for the time steps with negative time constants. The final time of the TR simulation is determined as 10 times of the largest magnitude time constant. Fig. 16 shows the cumulative Newton iteration when only Local Truncation Error (LTE) = 5 % is applied to the time step control (orange) [4][5] and when the time step width is limited to the half of the negative time constant magnitude in addition to LTE = 5 % (blue). In the case of LTE = 5 % only, a lot of retries occur due to non-convergence since the carrier density generated by impact ionization is still too small to affect the LTE. Therefore, it is very important to limit the time steps by taking the negative time constant information into account. The most CPU-time consuming part in TCAD is LU-factorization of Jacobian matrix. Since the time constant calculation in Arnoldi method in Fig. 5 uses the same Jacobian $G(x_n)$ as that of Newton iteration in Fig. 11, CPU-time overhead is small for the negative time constant detection in the DC simulation. On the other hand, in the TR simulation, as the Jacobian is $C/\Delta t_{k+1} + G(x_n)$, an extra LU-factorization is required for the time constant calculation. This cost is almost equivalent to one more extra Newton iteration at each time step [5] as shown in Fig. 11.

Fig. 17 shows a Grounded Gate (GG) NMOS structure which mimics an ESD protection device. The breakdown

simulation results are shown in Fig. 18, where the filled circles are with the original DC simulation and open circles are with the switched TR simulation. Hard breakdown of the drain-substrate junction occurs at Vds = 4.2 V and the second abrupt drain current increase due to parasitic bipolar effect occurs at Vds = 5.4 V. Since the simulation is performed with voltage boundary condition, no snapback I-V curve is obtained here. However, without using the continuation method, a trace of the stable operating points in the snapback I-V trajectory corresponding to each DC bias can be obtained robustly with a simple voltage sweep at the voltage boundary.

IV. CONCLUSION

A robust simulation method for hard breakdown is proposed. It detects the appearance of negative time constant during DC Newton iteration and then switches to TR simulation. The negative time constant value during the TR simulation is used for the time step restriction and the maximum time constant value is used for the determination of the final time of the TR simulation. By using the proposed method, a trace of the stable operating points in the snapback I-V trajectory corresponding to each DC bias can be obtained robustly with a simple voltage sweep at the voltage boundary.

REFERENCES

[1] R. Goosens, S. Beebe, Z. Yu, and R. Dutton, "An Automatic Biasing Scheme for Tracing Arbitrarily Shaped I-V Curves," IEEE Trans. CAD, vol. 13, no. 3, pp. 310-317, Mar. 1994.

[2] S. Kumashiro, "Method of Simulating Impact Ionization Phenomenon in Semiconductor Device," U. S. Patent 6,144,929, Nov. 2000.

[3] R. Bank and D. Rose, "Global Approximate Newton Method," Numeriche Mathematik, vol. 37, no. 2, pp. 279–295, Jun. 1981.

[4] S. Kumashiro, T. Kamei, A. Hiroki, and K. Kobayashi, "An Accurate Metric to Control Time Step of Transient Device Simulation by Matrix Exponential Method," in Proc. SISPAD, Kamakura, Japan, 2017, pp. 37-40.

[5] S. Kumashiro, T. Kamei, A. Hiroki, and K. Kobayashi, "An Efficient and Accurate Time Step Control Method for Power Device Transient Simulation Utilizing Dominant Time Constant Approximation," IEEE Trans. CAD Early Access, DOI:10.1109/TCAD.2018.2889673, Dec. 2018.

Extending the Numerov Process to the Semiconductor Transport Equations

Nicolò Speciale, Rossella Brunetti[1], <u>Massimo Rudan</u>

"E. De Castro" Advanced Research Center on Electronic Systems (ARCES) and Department DEI
University of Bologna, Viale Risorgimento 2, I-40136 Bologna, Italy
nicolo.speciale(massimo.rudan)@unibo.it

[1]*FIM Department, University of Modena and Reggio Emilia, Via Campi 213/A, I-41125 Modena, Italy*
rossella.brunetti@unimore.it

Abstract—Some classes of differential equations are amenable to a numerical solution based on the Numerov process (NP), whose accuracy can be up to two orders of magnitude superior with respect to the standard finite-difference or box-integration methods, with a negligible increase in the computational cost. The paper shows that the equations describing charge transport in solid-state devices can suitably be manipulated to make the application of NP possible. Also, thanks to a specifically-tailored algebraic solver, the 1D Poisson equation is fully decoupled from the transport equation, this reducing the procedure to the solution of a single non-linear equation. The example of an Ovonic device is considered, used as selector in phase-change memory applications.

Index Terms—Numerov Process, Transport equations

I. INTRODUCTION

A differential equation of the second order where the first derivative of the unknown function $z(x)$ is missing: $-z'' = Q(x)\, z + P(x)$, is amenable to a discretization scheme based on the Numerov process (NP), whose accuracy is $O(h^4)$, with h the grid size, in contrast to $O(h^2)$ of the standard methods (details in [1] and references therein). Although the original version of NP applies to a uniform grid, this constraint can be relieved as shown below. Letting $Q = 0$, the above equation yields the Poisson equation, with $z = u$ the normalized electric potential and $P = q/(\varepsilon\, k_B\, T)\, \varrho$, where ϱ is the charge density; if, instead, one lets $P = 0$, the time-independent Schrödinger equation is found, with $z = w$ the spatial part of the wave function and $Q = 2\, m\, (E - V)/\hbar^2$, where E and $V(x)$ are the total and potential energy, respectively. Due to the superior performance of NP with respect to the standard finite-difference method, it is of interest to seek for extensions of NP to other classes of equations; among these, those that model charge transport in solid-state devices.

II. THE MODEL EQUATIONS

Referring to the semiclassical model for charge transport in solids, a one-dimensional case with only one type of carriers, e.g., electrons, is considered. This situation is typical, among others, of devices like phase-change memories (e.g., [2] and references therein); the form of the Poisson equation thus becomes $u'' = u''(n, n_T)$, with n, n_T the concentrations of electrons and empty traps, respectively. Combining the steady-state continuity equation for the electrons, $J_n' = q\, U$, with the transport equation of the drift-diffusion form, $J_n = q\, D_n\, (n' - u'\, n)$, yields $n'' - u'\, n' - u''\, n = U/D_n$, where $U = U(n, n_T)$ is the net-recombination rate and $D_n = $ const the diffusion coefficient of the electrons. In a decoupled solution scheme, $-u'' = q/(\varepsilon\, k_B\, T)\, \varrho = P(n, n_T)$ plays in the latter equation the role of a coefficient known from the previous iteration. Letting $s = -(U/D_n)\, \exp(-u/2)$, $g = n\, \exp(-u/2)$, the equation is given a form suitable for the application of NP:

$$-g'' = c\, g + s, \qquad c = P/2 - (u')^2/4. \qquad (1)$$

Note that the transformation leading to (1) is not a reduction to the self-adjoint form, which would in fact read $[n'\, \exp(-u)]' = (U/D_n - P\, n)\, \exp(-u)$, nor an exponential fitting like the one typically adopted for solving the semiconductor equations, which would read $q\, D_n\, [n\, \exp(-u)]' = J_n\, \exp(-u)$ with $J_n = $ const over the segment connecting two nodes (the exponential-fitting scheme is also known as *Scharfetter-Gummel method* [3]).

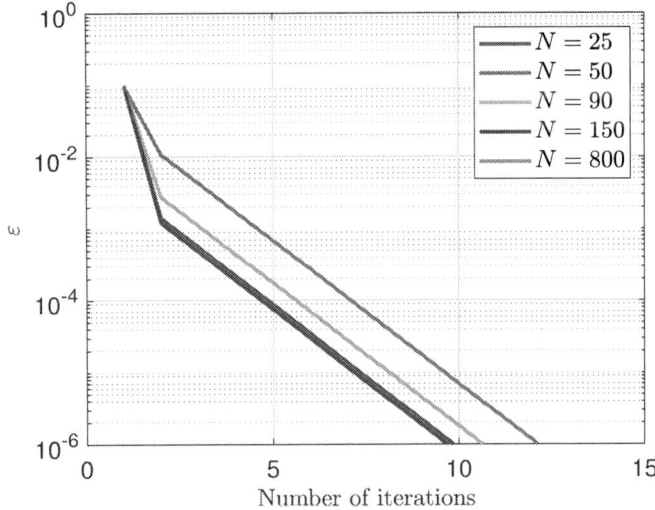

Fig. 1. Error ϵ vs. the number of iterations, with $V = 0.1$ V applied bias. The legend shows the number of nodes of each run.

III. THE NUMEROV PROCESS

Given a uniform grid having N internal nodes with an element size equal to h, application of NP to the Poisson equation yields the algebraic system

$$-u_{i-1} + 2\,u_i - u_{i+1} = \frac{h^2}{12}\left(P_{i-1} + 10\,P_i + P_{i+1}\right),\quad (2)$$

while its application to (1) yields the algebraic system

$$-\left(1 + \frac{h^2}{12}\,c_{i-1}\right) g_{i-1} + \left(2 - 10\,\frac{h^2}{12}\,c_i\right) g_i \quad (3)$$

$$-\left(1 + \frac{h^2}{12}\,c_{i+1}\right) g_{i+1} = \frac{h^2}{12}\left(s_{i-1} + 10\,s_i + s_{i+1}\right).$$

The unknowns of (3) are the nodal values $g_i = n_i \exp(-u_i/2)$, whereas c_i and s_i depend on the original unknown n through P and U. If the nodal values of $\exp(-u/2)$ are incorporated as they are into the coefficients of (3), the matrix may become stiff; it is preferable to restore the original unknown by multiplying both sides of (3) by $\exp(u_i/2)$. The new form of (3) thus obtained is more convenient because the potential differences between neighboring nodes appear, instead of the nodal potential alone. For instance, g_{i-1} is replaced with $n_{i-1} \exp[(u_i - u_{i-1})/2]$, and so on. Finally, one must express u' in the second relation of (1); using again NP, one easily finds

$$u'_i = \frac{u_{i+1} - u_{i-1}}{2\,h} + h\,\frac{P_{i+1} - P_{i-1}}{12}. \quad (4)$$

When $i = 2, \ldots N-1$, (4) involves only the internal nodes; when $i = 1$ or $i = N$, (4) involves one or the other boundary condition (u_0 and $P_0 = 0$ or, respectively, u_{N+1} and $P_{N+1} = 0$); finally, when $i = 0$ or $i = N+1$, (4) involves the potential and charge inside the contacts; as the latter are equipotential and neutral, one finds $u_{-1} = u_0$, $P_{-1} = 0$ and $u_{N+2} = u_{N+1}$, $P_{N+2} = 0$. By the same token one finds

$$g'_i = \frac{(6 + h^2\,c_{i+1})\,g_{i+1} - (6 + h^2\,c_{i-1})\,g_{i-1}}{12\,h} +$$

$$+ h\,\frac{s_{i+1} - s_{i-1}}{12}. \quad (5)$$

IV. FULLY DECOUPLING THE MODEL EQUATIONS

As the coefficients of (3) depend on the nodal values u_i of the electric potential, it is necessary to iterate between the solution of (3) and that of the algebraic system deriving from the Poisson equation. In general such solutions are obtained by algebraic solvers that provide the nodal values in a prescribed sequence; considering for instance the $A = LU$ decomposition, one obtains the ith nodal value of the potential only after calculating it at nodes 1 through $i-1$ (or at nodes N through $i+1$). A neater approach would be that in which u_i is obtained as soon as necessary, without the need of calculating the rest of the sequence; such a result is indeed achieved for the 1D Poisson equation of the form (2) by the method shown in [4, p. 769], having the advantage that each nodal value u_i can be calculated independently of the others; in fact, using

the short-hand notation C_i for the right hand side of (2), and letting

$$Z_j = h^2 \sum_{k=1}^{j} C_k,\quad Y_i = \sum_{j=1}^{i} Z_j,\quad i = 1 \ldots N, \quad (6)$$

and $R = (u_{N+1} - u_0 + Y_N)/(N+1)$, one finds $u_1 = u_0 + R$,

$$u_i = u_0 + i\,R - Y_{i-1},\quad i = 2 \ldots N. \quad (7)$$

The potential differences that appear in (3) are easily evaluated from (6–7):

$$u_i - u_{i-1} = R - Z_{i-1},\quad u_{i+1} - u_{i-1} = 2\,R - Z_{i-1} - Z_i.$$

If the discretized form of the Poisson equation is such that the method based on (6–7) is applicable, the calculation is cheaper than the $A = LU$ decomposition and completely decouples the transport equation from the Poisson equation.

It is also worth remarking other differences with respect to the standard discretization schemes: here, after discretization, all functions and derivatives belong to the nodes, whereas in the standard schemes the functions and the even derivatives belong to the nodes, while the odd derivatives belong (in one dimension) to the elements. Also, no hypothesis is necessary here about the behavior of the discretized functions along each element.

V. STABILITY

Iterations are in general necessary due to the non-linearity of the equations; for instance, in a semiconductor the normalized charge concentrations P_{i-1}, P_i, P_{i+1} that appear at the right hand side of (2) depend on the unknown u either exponentially or through a Fermi integral; in both cases, the derivatives $\mathrm{d}P_i/\mathrm{d}u$ are negative irrespective of the fact that electrons or holes are considered: in fact, hole concentration contributes positively to the charge density, and decreases with increasing u; electron concentration contributes negatively, and increases with increasing u. It follows that the extra terms obtained from the linearization with respect to u add weight to the main diagonal of the algebraic system (2), thereby improving convergence.

Coming now to (3), when the expressions of c_{i-1}, c_i, c_{i+1} that appear in the second equation of (1) are inserted into (3), the right hand side of the ith row of the resulting algebraic system is the sum of three terms: the first one, $A_i = -g_{i-1} + 2\,g_i - g_{i+1}$, has the same structure as that of the discretized Poisson equations (2). The other two terms read

$$B_i = \frac{(u'_{i-1})^2\,g_{i-1} + 10\,(u'_i)^2\,g_i + (u'_{i+1})^2\,g_{i+1}}{48/h^2}, \quad (8)$$

$$C_i = -\frac{P_{i-1}\,g_{i-1} + 10\,P_i\,g_i + P_{i+1}\,g_{i+1}}{24/h^2}. \quad (9)$$

As the coefficients in (8) are non negative, they add weight to all diagonals of the algebraic system (3); due to factor 10, the added weight of the main diagonal is dominant unless the ith node is in an extremum of u. As for (9), the analysis is

978-1-7281-0941-1/19 $31.00 © 2019 IEEE

complicated by the presence of the normalized charge density P, which may have either sign. On the other hand, (9) is of order 2 in h, whereas, due to a cancellation (compare with (4)), B_i is of the same order as A_i, namely, order 0.

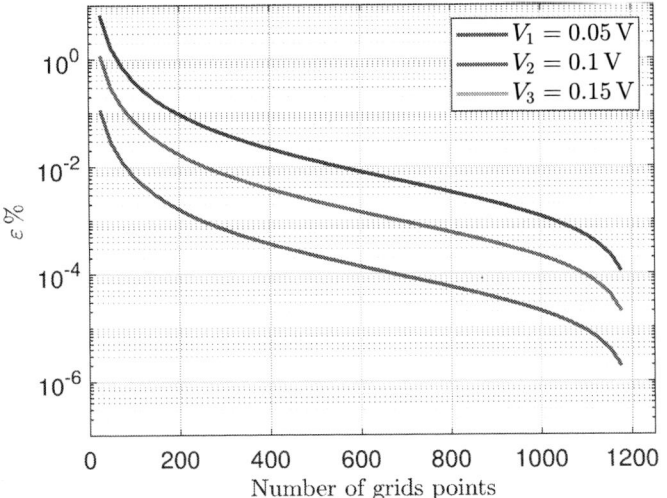

Fig. 2. Maximum error (in percentage) of the node values between the reference solution ($N = 2000$) and different runs for different biases ($V_1 = 0.05$ V, $V_2 = 0.1$ V, and $V_3 = 0.15$ V).

VI. RESULTS

The scheme for solving the transport model was applied here to a chalcogenide layer of thickness $L = 30$ nm and a Gaussian trap distribution in the direction of the external field. Uniformity was assumed in the direction normal to the field, and the applied bias V was kept below threshold. The analysis was carried out on different grids: we started with a large number of nodes (reference solution: $N = 2000$) and decreased it down to $N = 25$. Fig. 1 shows $\epsilon = \max_i(|n_i^{\text{new}} - n_i^{\text{old}}|/n_i^{\text{old}})$ vs. the number of iterations at different values of N, with $V = 0.1$ V applied bias. The convergence criterion was $\epsilon \leq 10^{-6}$. Letting $C(N)$ be the computational cost needed to fulfill this convergence criterion, the analysis showed that $C(1500) \simeq 26 \times C(25)$. To measure the accuracy of different solutions as function of N, the relative error was calculated in corrispondence of all nodes of the coarser grid ($N = 25$) by considering the difference between the values of the solution for all N with respect to the reference solution. Fig. 2 compares the maximum difference (in percentage) for different biases.

Next, the error of the present method has been compared with that of the exponential-fitting one; the results are shown in the Tables. As before, the starting point was a reference solution obtained with a dense grid ($N = 2000$); then, other solutions were run with a progressively-decreasing number of nodes, and the maximum difference with respect to the reference solution was extracted. More precisely, Tab. I lists $\eta(\varphi) = \max_i|\varphi_i - \varphi_i^{\text{ref}}|$ for different values of the grid nodes and applied bias; the errors of the exponential-fitting method are listed in column "SG". In turn, Tab. II lists $\eta(n) =$

$\max_i|n_i - n_i^{\text{ref}}|/n_M^{\text{ref}}$, with n_M^{ref} the maximum concentration of the reference solution. Despite the simplicity of the problem in hand, the improvement of the present method with respect to the standard one is about one order of magnitude in all cases.

A final check refers to the constancy of the current density; in fact, in a one-dimensional, steady-state case where the mobility of the trapped electrons is set to zero, the continuity equation yields $J_n' = qU = 0$. As remarked above, with the present method all functions and derivatives belong to the nodes: it follows that J_n and $J_n' = q D_n (n'' - u'n' - u''n)$ belong to the nodes as well; the latter quantity is shown in Fig. 3 as a function of position.

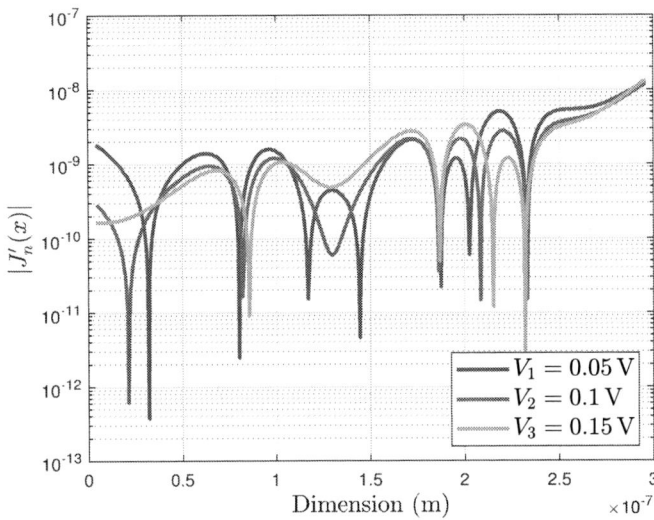

Fig. 3. Modulus $|J_n'|$ as a function of position, with different biases ($V_1 = 0.05$ V, $V_2 = 0.1$ V, and $V_3 = 0.15$ V). The outcome is similar for all grids examined.

To summarize, like in the case of the Schrödinger equation [1], the NP-based approach represents a fairly simple way to improve the solution of the semiconductor equations with a limited increase in the computational burden. Indeed, the drift-diffusion and current-continuity equations have been considered here by way of example; in fact, all pairs of moments of order $2k$ and $2k+1$, $k = 0, 1, \ldots$ of the Boltzmann transport equation have the same structure: those of even order have the form $-\operatorname{div} \boldsymbol{S} = C$, those of odd order have the form $\boldsymbol{S} = a \operatorname{grad} \sigma + \sigma \nabla b$, with a suitable meaning of symbols; it follows that the present method applies to any order of transport model in one dimension; also, considering that in the dynamic case the term C above embeds the time derivative $\partial \sigma / \partial t$, the method is not limited to the steady state.

It may be argued that the approach depicted here is applicable only in the one-dimensional case when a uniform grid is used; in fact this is not true: NP has been extended to the variable stepsize, still in one dimension [5]. The applicability to 1D cases is not too severe a constraint when the solution of transport problems in nano-sized cylindrical structures (e.g., nanowires or carbon-nanotube transistors) is sought, where the typical approach decouples the longitudinal coordinate from the transversal ones [6].

Conversely, the extension of NP to the two- and three-dimensional cases using tensor-product, uniformly-spaced grids has also been achieved (in [7], with reference to the Schrödinger equation), whereas that to non-uniform, multi-dimensional grids is still missing. The multi-dimensional form of (1) is readily obtained with the replacements $g'' \leftarrow \nabla^2 g$ and $u' \leftarrow |\nabla u|$, to find

$$-\nabla^2 g = c\,g + s, \qquad c = P/2 - |\nabla u|^2/4. \qquad (10)$$

Considering the case of a tensor-product, uniformly-spaced grid, the notation becomes awkward even in two dimensions; following [7] we adopt a matrix notation of the form

$$a_{11}\,g_{i+1}^{j-1} + a_{12}\,g_{i+1}^{j} + a_{13}\,g_{i+1}^{j+1} + a_{21}\,g_{i}^{j-1} + \cdots \qquad (11)$$

$$+ a_{23}\,g_{i}^{j+1} + \cdots + a_{33}\,g_{i-1}^{j+1} = \begin{bmatrix} a_{11} & a_{12} & a_{13} \\ a_{21} & a_{22} & a_{23} \\ a_{31} & a_{32} & a_{33} \end{bmatrix}_g,$$

where the lower (upper) index of $g_i^j, \ldots$ refers to the x (y) axis. Application of NP in two dimensions yields, first,

$$\begin{bmatrix} 1 & 0 & 1 \\ 0 & -4 & 0 \\ 1 & 0 & 1 \end{bmatrix}_g = 2\,h^2\,(\nabla^2 g)_i^j + \qquad (12)$$

$$+ \frac{h^4}{6}\,[\nabla^2(\nabla^2 g)]_i^j + \frac{2}{3} \begin{bmatrix} 1 & -2 & 1 \\ -2 & 4 & -2 \\ 1 & -2 & 1 \end{bmatrix}_g.$$

Taking the Laplacian of both sides of (12), and neglecting the 6th-order derivatives, allows one to eliminate $[\nabla^2(\nabla^2 g)]_i^j$. Finally, replacing $(\nabla^2 g)_i^j$ from (10) and letting

$$\boldsymbol{M} = -\begin{bmatrix} 1 & 4 & 1 \\ 4 & -20 & 4 \\ 1 & 4 & 1 \end{bmatrix}, \quad \boldsymbol{J} = \begin{bmatrix} 0 & 0 & 0 \\ 0 & 1 & 0 \\ 0 & 0 & 0 \end{bmatrix}, \quad (13)$$

eventually yields the two-dimensional generalization of (3):

$$\boldsymbol{M}_g - h^2 \left(6\,\boldsymbol{J}_{c\,g} - \frac{\boldsymbol{M}_{c\,g}}{12} \right) = h^2 \left(6\,\boldsymbol{J}_s - \frac{\boldsymbol{M}_s}{12} \right). \quad (14)$$

The result expressed by (14) extends that of [7] to the second-order equation of the general form; its derivation also shows that, like in the one-dimensional case, no hypothesis is necessary about the behavior of the discretized functions inside each elements or along its edges.

REFERENCES

[1] F. Buscemi, E. Piccinini, R. Brunetti, and M. Rudan, "High-Order Solution Scheme for Transport in Low-D Devices," in *IEEE International Conference on Simulation of Semiconductor Processes and Devices (SISPAD)*, Yokohama, September 2014, pp. 161–164.

[2] E. Piccinini, R. Brunetti, and M. Rudan, "Self-Heating Phase-Change Memory-Array Demonstrator for True Random Number Generation," *IEEE Transactions on Electron Devices*, vol. 64, no. 5, pp. 2185–2192, 2017.

[3] D. L. Scharfetter and H. Gummel, "Large-signal analysis of a silicon Read diode oscillator," *IEEE Trans. Electron Dev.*, vol. ED-16(1), pp. 64–77, 1969.

[4] M. Rudan, *Physics of Semiconductor Devices*. Springer, 2018.

[5] J. Vigo-Aguia and H. Ramos, "A variable-step Numerov method for the numerical solution of the Schrödinger equation," *Journal of Mathematical Chemistry*, vol. 37, no. 3, pp. 255–262, 2005.

[6] M. Rudan, A. Gnudi, E. Gnani, S. Reggiani, and G. Baccarani, "Improving the Accuracy of the Schrödinger-Poisson Solution in CNWs and CNTs," in *IEEE International Conference on Simulation of Semiconductor Processes and Devices (SISPAD)*, Bologna, September 2010, pp. 307–310.

[7] T. Graen and H. Grubmüller, "NuSol–Numerical solver for the 3D stationary nuclear Schrödinger equation," *Computer Physics Communications*, vol. 198, pp. 169–178, 2016.

TABLE I
MAXIMUM ERROR ON THE POTENTIAL $\varphi(x)$

Bias	# of nodes	SG	This work
0.05 V	50	$2.08 \cdot 10^{-05}$	$1.32 \cdot 10^{-06}$
	100	$2.79 \cdot 10^{-06}$	$3.26 \cdot 10^{-07}$
	150	$9.81 \cdot 10^{-06}$	$1.33 \cdot 10^{-07}$
	300	$6.72 \cdot 10^{-08}$	$6.57 \cdot 10^{-09}$
	500	$6.57 \cdot 10^{-08}$	$5.38 \cdot 10^{-09}$
0.1 V	50	$4.08 \cdot 10^{-05}$	$1.25 \cdot 10^{-05}$
	100	$3.74 \cdot 10^{-05}$	$3.00 \cdot 10^{-06}$
	150	$6.01 \cdot 10^{-05}$	$1.22 \cdot 10^{-06}$
	300	$4.97 \cdot 10^{-07}$	$6.13 \cdot 10^{-08}$
	500	$8.14 \cdot 10^{-07}$	$4.92 \cdot 10^{-08}$
0.15 V	50	$2.75 \cdot 10^{-04}$	$5.72 \cdot 10^{-05}$
	100	$3.44 \cdot 10^{-04}$	$1.36 \cdot 10^{-05}$
	150	$4.19 \cdot 10^{-05}$	$5.56 \cdot 10^{-06}$
	300	$2.38 \cdot 10^{-06}$	$2.78 \cdot 10^{-07}$
	500	$2.19 \cdot 10^{-06}$	$2.23 \cdot 10^{-07}$

TABLE II
MAXIMUM ERROR ON THE CONCENTRATION $n(x)$

Bias	# of nodes	SG	This work
0.05 V	50	$3.47 \cdot 10^{-02}$	$5.17 \cdot 10^{-03}$
	100	$2.71 \cdot 10^{-02}$	$1.08 \cdot 10^{-04}$
	150	$2.46 \cdot 10^{-03}$	$1.14 \cdot 10^{-04}$
	300	$3.11 \cdot 10^{-05}$	$5.72 \cdot 10^{-06}$
	500	$2.68 \cdot 10^{-05}$	$1.76 \cdot 10^{-06}$
0.10 V	50	$5.21 \cdot 10^{-02}$	$1.03 \cdot 10^{-02}$
	100	$4.23 \cdot 10^{-02}$	$1.04 \cdot 10^{-03}$
	150	$3.91 \cdot 10^{-03}$	$1.02 \cdot 10^{-03}$
	300	$4.89 \cdot 10^{-04}$	$5.02 \cdot 10^{-05}$
	500	$4.11 \cdot 10^{-04}$	$1.50 \cdot 10^{-05}$
0.15 V	50	$6.34 \cdot 10^{-01}$	$5.48 \cdot 10^{-02}$
	100	$5.17 \cdot 10^{-01}$	$1.45 \cdot 10^{-02}$
	150	$4.77 \cdot 10^{-02}$	$6.03 \cdot 10^{-03}$
	300	$6.02 \cdot 10^{-02}$	$3.05 \cdot 10^{-04}$
	500	$5.17 \cdot 10^{-03}$	$2.15 \cdot 10^{-04}$

Implementation of Automatic Differentiation to Python-based Semiconductor Device Simulator

Tsutomu Ikegami
National Institute of Advanced Industrial Science and Technology (AIST)
Tsukuba, Japan
t-ikegami@aist.go.jp

Koichi Fukuda
National Institute of Advanced Industrial Science and Technology (AIST)
Tsukuba, Japan
fukuda.koichi@aist.go.jp

Junichi Hattori
National Institute of Advanced Industrial Science and Technology (AIST)
Tsukuba, Japan
j.hattori@aist.go.jp

Abstract—**A Python-based device simulator named Impulse TCAD was developed. The simulator is built on top of a nonlinear finite volume method (FVM) solver. To describe physical behavior of non-standard materials, both device properties and their dominant equations can be customized. The given FVM equations are solved by the Newton method, where required derivatives of the equations are derived automatically by using an automatic differentiation technique. As a demonstration, a steady state analysis of the negative capacitance field effect transistors with ferroelectric materials is selected, where the coupled Poisson and Devonshire equations are implemented in several different ways.**

Keywords—**TCAD, device simulation, automatic differentiation, Python, negative capacitance**

I. Introduction

Reducing the power consumption of semiconductor devices is an urgent issue in both edge devices of the internet of things and server machines in data centers. To overcome the physical limitation inherent in conventional silicon devices, many researchers are trying to utilize singular physical phenomena like quantum tunneling and novel materials like ferroelectrics. On the other hand, semiconductor device simulation has long been a standard tool to realize the rigorous silicon roadmap [1]–[3]. Therefore, for the development of the next-generation semiconductor devices, non-standard physical models have to be implemented in the device simulators. A new device simulator named Impulse TCAD [4], [5] was thus developed to support such studies on the new physical models. Impulse TCAD is written mainly in the Python language, which is also used to describe the run script to control the simulation pathway. In Impulse TCAD, users can declare device properties that characterize target materials, and define dominant equations that describe their physical behaviors. It is even possible to intervene in the iterative solver of the equations, which helps several difficult cases to converge. Because all these customizations are managed in the Python domain, a special customization for a specific project can even be possible from the run script, without modifying the main body of the simulator. Modifications are also reflected instantly without the complicated compile and link process, which accelerates the development process of new models. In the following, the

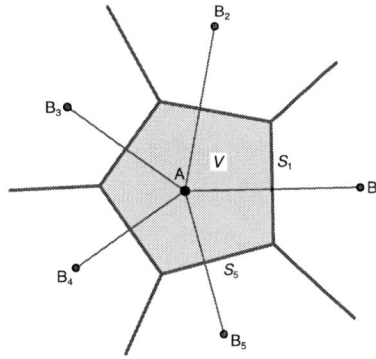

Fig. 1. Control volumes in FVM. Each material region is divided into control volumes, which are designated by node points (A and B_i).

basic equations used in Impulse TCAD are introduced along with the background mechanism, followed by a representative example. Applying this to the negative capacitance field effect transistors is also shown as a demonstration.

II. Impulse TCAD

A. Finite Volume Method

Impulse TCAD is a semiconductor device simulator based on the finite volume method (FVM). Device structures are represented in 3D unstructured meshes, on which the vertex-centered non-linear FVM is implemented. To setup FVM, each material region is divided into tiny control volumes (CV) as shown in Fig. 1. Device properties in CV, such as electric potential and electron density, are represented on "node" points (A and B_i). Similarly, properties on the boundary surface (or on the "edge") between neighbouring CVs A and B_i are dependent only on the properties of nodes A and B_i. In Impulse TCAD, the FVM equations are set up on the node A as

$$g = \sum_i f(A, B_i)S_i + s(A)V = 0, \qquad (1)$$

Where $f(A, B_i)$ is a flux term directing from B_i to A, $s(A)$ is a source term, S_i is an area of the boundary surface, V is a volume of the CV, and the summation is taken over all neighbouring nodes. The conservation rule Eq. (1) is satisfied

---

978-1-7281-0941-1/19 $31.00 © 2019 IEEE

simultaneously over all CVs, by adjusting variable properties v on each node. The Newton method is employed to calculate the adjustment δv:

$$
\begin{aligned}
\delta g &\simeq \sum_i \left(\frac{\partial f}{\partial v_A} \delta v_A + \frac{\partial f}{\partial v_{B_i}} \delta v_{B_i} \right) S_i + \frac{\partial s}{\partial v_A} \delta v_A V \\
&= -g, \qquad\qquad\qquad\qquad\qquad\qquad (2)
\end{aligned}
$$

which is written collectively as $(\partial \boldsymbol{G}/\partial \boldsymbol{V})\delta \boldsymbol{V} = -\boldsymbol{G}$. Here, v and g are collected from all nodes to form vectors $\boldsymbol{V}$ and $\boldsymbol{G}$, respectively. The Jacobian $\partial \boldsymbol{G}/\partial \boldsymbol{V}$ is a large and sparse matrix, which is constructed from derivatives $\partial f/\partial v$ and $\partial s/\partial v$. The update $\delta \boldsymbol{V}$ at each Newton iteration is obtained by solving the large linear equation.

In Impulse TCAD, users set up the FVM equations by providing the flux term $f(A, B_i)$ and the source term $s(A)$ symbolically. To illustrate, a setup of coupled Poisson and drift-and-diffusion equations is shown below:

```
# Poisson
f_Psi = eps * (Psi.j − Psi.i) / dx_d
s_Psi = q * (ND.i − NA.i + Hole.i − Elec.i)

# Drift & Diffusion
E = abs((Psi.j − Psi.i) / dx_d)
mu_n = CaugheyThomas(E, T, mu_n, beta_n, Vsat0_n, A_n, T0_n)
mu_p = CaugheyThomas(E, T, mu_p, beta_p, Vsat0_p, A_p, T0_p)
x = q / (kB*T) * (Psi.j − Psi.i)
f_Elec = −mu_n * kB*T / dx_d * C_Bernoulli( x, Elec.j, Elec.i)
f_Hole =  mu_p * kB*T / dx_d * C_Bernoulli(−x, Hole.j, Hole.i)
s_Elec = −q * ShockleyReadHall(Elec.i,Hole.i,ni,tau_n,tau_p)
s_Hole = −s_Elec

Eqn = ( [(f_Psi, s_Psi), (f_Elec, s_Elec), (f_Hole, s_Hole)],
        [Psi, Elec, Hole] )
```

Here, Psi, Elec, Hole, NA, and ND are the device properties for the electric potential, electron density, hole density, acceptor density, and donor density, respectively, where suffixes ".i" and ".j" denote the values on the primary (A) and adjacent (B_i) nodes, respectively. The property dx_d gives the edge length between A and B_i. Other symbols are defined elsewhere as numerical constants. At the last line, the FVM equations are defined as a list of (flux, src) terms, along with a list of variable properties for which the equations have to be solved. As shown in the listing, the flux and source terms are constructed by using the device properties in a straightforward manner.

B. Automatic Differentiation

The given FVM equations are solved by the Newton method, where the automatic differentiation facility of Theano [6] is used. The device properties listed above are implemented as Theano symbols, and equations written with them compose Theano expressions. Indeed, the functions CaugheyThomas, C_Bernoulli, and ShockleyReadHall return Theano expressions for the mobility [7], the interpolated electric current [8], and the generation-recombination rate [9], respectively.

Under the hood, differential expressions of the FVM equations are derived with respect to the given variable properties.

C source codes are then generated to calculate both the equations and their derivatives numerically, which are compiled and linked to build a module accessible from Python. The equations are also analyzed to extract a list of incorporated device properties. For example, [Psi.i, Psi.j, dx_d] and [Elec.i, Hole.i, ND.i, NA.i] are extracted from f_Psi and s_Psi, respectively. Note that eps and q in these equations are constants, which are embedded in the source codes directly. The module is called repetitively to build $\boldsymbol{G}$ and its Jacobian, where the required arguments are assembled according to the property list. This Jacobian construction is parallelized by MPI [10]–[14], as well as the direct solver for the linear equation [15]–[19].

C. Comparison with other simulators

In our survey, PRISM [20], [21] and Genius [22] employ an automatic differentiation scheme in the semiconductor device simulation. PRISM is based on a FORTRAN package IVPACK, where derivatives are automatically generated for functions programmed with IVPACK function calls. Mobility models and generation-recombination terms can be extended easily with the IVPACK scheme, though details are not available. Genius implements a C++ framework of automatic differentiation, with which users can customize materials. In Genius, dominant equations and associating physical models are predefined in the simulator, and users can modify physical parameters (such as electric permittivity), as well as the functional form of the physical models (such as the band gap narrowing model). To customize Genius, building a dynamic library with a set of prescribed APIs is mandatory, though rebuilding the whole simulator can be avoided.

In contrast to these simulators, the automatic differentiation scheme is hidden from users in Impulse TCAD. Instead of composing individual derivatives of physical models, the automatic differentiation is used to derive the Jacobian of the FVM equations (1) directly. This approach allows users to customize dominant equations themselves, without the hassles of defining auxiliary functions. The customizations are also reflected instantly, without the complicated compile and link process. These features are prerequisite to develop non-standard physical models of new materials, as shown in the next section.

III. APPLICATION TO FERROELECTRIC MATERIAL

This section demonstrates how Impulse TCAD helps the development of new physical models by taking the negative capacitance field effect transistors (NC FET) as an example [4], [23]–[26]. In the NC FET, ferroelectric materials like HfO_2 are embedded in the gate stack. The device structure used in the present study is shown in Fig. 2. To solve the steady states of the device, the standard Poisson and drift-and-diffusion equations are applied on materials other than HfO_2.

The electric potential ψ of the HfO_2 regions is solved by the Poisson equation coupled with the Devonshire equation [27].

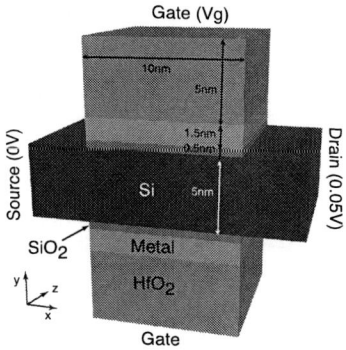

Fig. 2. Device structure of the negative capacitance FET.

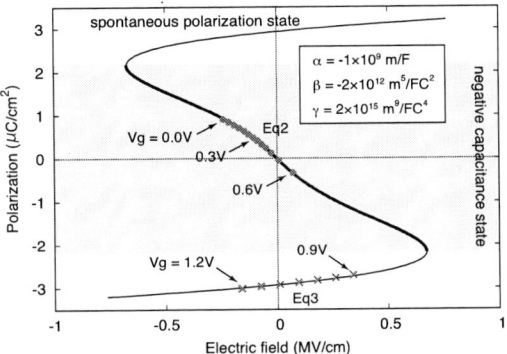

Fig. 3. The polarization of a ferroelectric material as a function of the electric field. The employed material parameters are shown in the inset. The calculated steady states are plotted by red dots and crosses for Eq2 and Eq3, respectively.

The polarization of the ferroelectrics is described by the Devonshire equation,

$$E = 2\alpha P + 4\beta P^3 + 6\gamma P^5, \quad (3)$$

where $E = -\nabla\psi$ is an electric field, P is a polarization, and α, β, γ are constant parameters. The E-P relationship of Eq. (3) is depicted in Fig. 3. The negative capacitance (NC) state is plotted with a thick line, which is of our major interest. The ferroelectric axis of HfO_2 is taken to be perpendicular to the channel (Y axis), and the polarizations parallel to the channel are treated as normal dielectrics.

Assuming that there is no charge in the HfO_2 regions, the Poisson equation is written in the integrated form as

$$\int_S (\epsilon_0 E + P) dS = 0. \quad (4)$$

Several approaches are possible to implement Eq. (4) in Impulse TCAD. To start with, the device properties are declared as follows:

```
1   Psi   = NodeProp("Psi")
2   Pol   = NodeProp("Pol")
3   Elf   = NodeProp("Elf")
4   Pos   = NodeVec("Pos")
5   dx_d  = EdgeProp("dx_d")
```

Psi, Pol, and Elf are the scalar properties giving electric potential, polarization, and electric field, respectively, while Pos is the 3D vector property giving spatial location. An edge property dx_d gives an edge length between A and B_i. Constant parameters eps0, eps, alpha, beta, and gamma are also prepared elsewhere.

In the first approach, the polarization Pol is supplied as a parameter:

```
6    E    = - (Psi.i - Psi.j) / dx_d
7    Dir  = (Pos.i - Pos.j) / dx_d
8    Evec = E * Dir
9    Px   = eps * Evec[0]
10   Py   = (Pol.i + Pol.j) / 2
11   Pz   = eps * Evec[2]
12   P    = Px*Dir[0] + Py*Dir[1] + Pz*Dir[2]
13   Eq1  = ([(eps0 * E + P, 0)], [Psi])
```

Before each Newton iteration, Pol is calculated from the gradient of Psi, by solving Eq. (3) for the NC state. The polarization vector (Px, Py, Pz) is calculated at the edge center, which is projected on the edge to define the electric flux density along the edge. In this approach, the dependence of Pol on Psi is not reflected directly in the Newton method, which causes severe difficulty in convergence. Note that the HfO_2 regions of Fig. 2 consist from 1500 node points, and Eq. (4) has to be satisfied on them simultaneously.

In the second approach, the ferroelectric polarization is symbolically calculated in the equation:

```
14   Py2 = P_of_E(Evec[1])
15   P2  = Px*Dir[0] + Py2*Dir[1] + Pz*Dir[2]
16   Eq2 = ([(eps0 * E + P2, 0)], [Psi])
```

Here, P_of_E() is a Theano operator defined in Impulse TCAD, which solves Eq. (3) to give P from E for the NC state. As a result, $\partial P / \partial \psi_A$ and $\partial P / \partial \psi_{B_i}$ are incorporated in the Jacobian, by way of the chain rule. This approach successfully calculates the steady states of the NC devices [24], [25], with the gate potential V_g varied at the step of 0.05 V. For the converged states, the electric field and the polarization are averaged over the upper HfO_2 region, and are plotted in Fig. 3 using red dot symbols. When V_g is increased beyond 0.6 V, some of the node points start to escape from the NC state, where the simulation stops.

In the previous two approaches, the spontaneous polarization (SP) states of the ferroelectrics (thin lines in Fig. 3) is ignored. In the third approach, the SP state is incorporated by solving Eq. (3) coupled together with the Poisson equation:

```
17   Devonshire = 2*alpha*Pol.i + 4*beta*Pol.i**3 \
18                + 6*gamma*Pol.i**5 - Elf.i
19   Eq3 = ([(eps0*E + P, 0), (0, Devonshire)], [Psi, Pol])
```

Both Psi and Pol are treated as variable node properties, which are solved in a coupled manner. Similarly to Pol in the first approach, the electric field Elf is prepared before each Newton iteration and supplied as a parameter, so that the dependence of Elf on Psi is ignored in the Newton method. This approach is still convergent, however, because Elf is less sensitive to V_g than Pol. Note that the full Newton approach is also possible for Elf [28], by constructing the electric field from those E defined on the edges. Starting from an initial condition tailored for the SP state, the simulation converges with some intervention in the iterative process, which are plotted in Fig. 3 using red cross symbols. The Newton method did not converge for V_g lower than 0.9 V, probably because the NC state becomes possible for some node points, causing oscillation in the Newton iteration. Because both of the SP and NC states can be described, this approach is employed in the transient analysis [4], [26], where the transition between the SP and NC states are simulated.

IV. CONCLUSION

A new semiconductor device simulator named Impulse TCAD is introduced, which allows easy handling of non-standard physical models. Both device properties and their dominant equations can be customized dynamically, which accelerates the development of simulation schemes for novel devices. Auxiliary functions required to solve the equations are generated automatically by using the automatic differentiation technique. Usability of the simulator is demonstrated by taking NC FET as an example, where the physical model of ferroelectric materials is implemented in several different ways on the same platform.

Impulse TCAD has several other features not mentioned in the manuscript that makes the simulator more accessible. A set of materials are predefined, where typical dominant equations and boundary conditions are packaged. They can be used as a starting point of the customization. Several utility functions are also provided, with which users can set up simulations according to prescriptions, export snapshots to files, import device properties from the snapshot files, check consistency of equations, and so on. The exported snapshots can be examined by using the state-of-the-art viewers [29] directly without conversion, which allows ready steering of simulations. With these features, we hope Impulse TCAD will contribute to the development of the next generation semiconductor devices.

REFERENCES

[1] S. Selberherr, A. Schutz, and H. W. Potzl, "MINIMOS - a two-dimensional mos transistor analyzer," *IEEE Journal of Solid-State Circuits*, vol. 15, no. 4, pp. 605–615, Aug 1980.

[2] C. S. Rafferty, M. R. Pinto, and R. W. Dutton, "Iterative methods in semiconductor device simulation," *IEEE Transactions on Computer-Aided Design of Integrated Circuits and Systems*, vol. 4, no. 4, pp. 462–471, October 1985.

[3] W. L. Engl, H. K. Dirks, and B. Meinerzhagen, "Device modeling," *Proceedings of the IEEE*, vol. 71, no. 1, pp. 10–33, Jan 1983.

[4] T. Ikegami, K. Fukuda, J. Hattori, H. Asai, and H. Ota, "A tcad device simulator for exotic materials and its application to a negative-capacitance fet," *Journal of Computational Electronics*, vol. 18, no. 2, pp. 534–542, Jun 2019.

[5] "Impulse TCAD." [Online]. Available: https://unit.aist.go.jp/neri/en/ImpulseTCAD/index.html

[6] Theano Development Team, "Theano: A Python framework for fast computation of mathematical expressions," *arXiv e-prints*, vol. abs/1605.02688, May 2016. [Online]. Available: http://arxiv.org/abs/1605.02688

[7] D. M. Caughey and R. E. Thomas, "Carrier mobilities in silicon empirically related to doping and field," *Proceedings of the IEEE*, vol. 55, no. 12, pp. 2192–2193, Dec 1967.

[8] D. L. Scharfetter and H. K. Gummel, "Large-signal analysis of a silicon read diode oscillator," *IEEE Transactions on Electron Devices*, vol. 16, no. 1, pp. 64–77, Jan 1969.

[9] W. Shockley and W. T. Read, "Statistics of the recombinations of holes and electrons," *Phys. Rev.*, vol. 87, pp. 835–842, Sep 1952.

[10] L. D. Dalcin, R. R. Paz, P. A. Kler, and A. Cosimo, "Parallel distributed computing using Python," *Advances in Water Resources*, vol. 34, no. 9, pp. 1124 – 1139, 2011, new Computational Methods and Software Tools.

[11] L. Dalcín, R. Paz, M. Storti, and J. D'Elía, "MPI for Python: Performance improvements and MPI-2 extensions," *Journal of Parallel and Distributed Computing*, vol. 68, no. 5, pp. 655 – 662, 2008.

[12] L. Dalcín, R. Paz, and M. Storti, "MPI for Python," *Journal of Parallel and Distributed Computing*, vol. 65, no. 9, pp. 1108 – 1115, 2005.

[13] MPI Forum, "MPI: a message passing interface standard," *Int. J. Supercomput. Appl.*, vol. 8, no. 3-4, pp. 159–416, 1994.

[14] ——, "MPI2: a message passing interface standard," *Int. J. High Perform. Comput. Appl.*, vol. 12, no. 1-2, pp. 1–299, 1998.

[15] P. R. Amestoy, I. S. Duff, J. Koster, and J.-Y. L'Excellent, "A fully asynchronous multifrontal solver using distributed dynamic scheduling," *SIAM Journal on Matrix Analysis and Applications*, vol. 23, no. 1, pp. 15–41, 2001.

[16] P. R. Amestoy, A. Guermouche, J.-Y. L'Excellent, and S. Pralet, "Hybrid scheduling for the parallel solution of linear systems," *Parallel Computing*, vol. 32, no. 2, pp. 136–156, 2006.

[17] X. S. Li, "An overview of SuperLU: Algorithms, implementation, and user interface," *ACM Trans. Math. Softw.*, vol. 31, no. 3, pp. 302–325, Sep. 2005.

[18] X. S. Li and J. W. Demmel, "SuperLU_DIST: A scalable distributed-memory sparse direct solver for unsymmetric linear systems," *ACM Trans. Mathematical Software*, vol. 29, no. 2, pp. 110–140, June 2003.

[19] X. Li, J. Demmel, J. Gilbert, iL. Grigori, M. Shao, and I. Yamazaki, "SuperLU Users' Guide," Lawrence Berkeley National Laboratory, Tech. Rep. LBNL-44289, September 1999, http://crd.lbl.gov/~xiaoye/SuperLU/. Last update: October 2014.

[20] W. Schoenmaker, R. Vankemmel, R. Cartuyvels, W. Magnus, and B. Tijskens, "Status of the device simulator PRISM," *COMPEL - The international journal for computation and mathematics in electrical and electronic engineering*, vol. 10, no. 4, pp. 631–640, 1991.

[21] E. Tijskens, W. Schoenmaker, and K. De Meyer, "Automatic numerical evaluation of derivatives and its use in device simulators," in *NUPAD IV. Workshop on Numerical Modeling of Processes and Devices for Integrated Circuits,*, May 1992, pp. 251–255.

[22] Cogenda, "Genius: 3D parallel device simulator." [Online]. Available: http://www.cogenda.com/article/Genius

[23] S. Salahuddin and S. Datta, "Use of negative capacitance to provide voltage amplification for low power nanoscale devices," *Nano Letters*, vol. 8, no. 2, pp. 405–410, 2008, pMID: 18052402.

[24] H. Ota, T. Ikegami, J. Hattori, K. Fukuda, S. Migita, and A. Toriumi, "Fully coupled 3-D device simulation of negative capacitance FinFETs for sub 10 nm integration," in *2016 IEEE International Electron Devices Meeting (IEDM)*, Dec 2016, pp. 12.4.1–12.4.4.

[25] H. Ota, K. Fukuda, T. Ikegami, J. Hattori, H. Asai, S. Migita, and A. Toriumi, "Perspective of negative capacitance FinFETs investigated by transient TCAD simulation," in *2017 IEEE International Electron Devices Meeting (IEDM)*, Dec 2017, pp. 15.2.1–15.2.4.

[26] H. Ota, T. Ikegami, K. Fukuda, J. Hattori, H. Asai, K. Endo, S. Migita, and A. Toriumi, "Multi-domain dynamics of ferroelectric polarization and its coherency-breaking in negative capacitance field-effect transistors," in *2018 IEEE International Electron Devices Meeting (IEDM)*, Dec 2018, pp. 9.1.1–9.1.4.

[27] A. F. Devonshire, "Xcvi. theory of barium titanate," *The London, Edinburgh, and Dublin Philosophical Magazine and Journal of Science*, vol. 40, no. 309, pp. 1040–1063, 1949.

[28] J. Hattori, T. Ikegami, K. Fukuda, H. Ota, S. Migita, and H. Asai, "Device simulation of negative-capacitance field-effect transistors with a ferroelectric gate insulator," in *2018 International Conference on Simulation of Semiconductor Processes and Devices (SISPAD)*, Sep 2018, pp. 214–219.

[29] U. Ayachit, *The ParaView Guide: A Parallel Visualization Application*. Kitware, 2015.

Deep Neural Network for Generation of the Initial Electrostatic Potential Profile

Seung-Cheol Han and Sung-Min Hong
School of Electrical Engineering and Computer Science
Gwangju Institute of Science and Technology
Gwangju, Republic of Korea
Email: smhong@gist.ac.kr

Abstract—A deep neural network is trained to learn the electrostatic potential of the semiconductor device. In order to demonstrate its feasibility, pn diodes are considered. Various pn diodes with different doping densities are generated and the numerical solutions are calculated. The resultant electrostatic potential profiles are used in the training phase. Our numerical results clearly demonstrate that the trained neural network can provide the initial electrostatic potential reasonably well. Since the initial electrostatic potential is improved, the Newton-Raphson loop for the nonlinear Poisson equation can be converged within a smaller number of iterations.

Keywords—Deep neural network; Initial electrostatic potential; Newton-Raphson method

I. INTRODUCTION

Recently, a deep neural network has been widely applied to many application areas such as image analysis, natural language processing, and expert systems. For example, in the ImageNet Large Scale Visual Recognition Challenge (ILSVRC) 2017, 29 of 38 competing teams had accuracy greater than 95 %. Inspired by its superior performance demonstrated in those applications, great research efforts have been made to apply the deep neural network to other unexplored fields.

As far as the semiconductor industry is involved, a deep neural network has been mainly used as a tool for the design-technology co-optimization (DTCO) [1], [2]. In [2], the neural network is applied to the yield estimation. Basically, the neural network is considered as just a component of an optimizer for predicting a better technology option, based upon the pre-existing results from the semiconductor device simulator. Therefore, in those works, the semiconductor device simulator itself is not modified. It is treated as a given building block of the technology development.

Except for the DTCO application discussed so far, it is difficult to find a report on application of the deep neural network to the semiconductor device simulation. In [3], the machine learning technique has been used in predicting the Hamiltonian operators for the density-functional theory. However, the density-function theory is quite different from the conventional device simulation based on the drift-diffusion equation. Also, the adopted machine learning technique is not the deep neural network.

In the present authors' opinion, the deep neural network can be employed in the semiconductor device simulation field in an alternative manner. To be specific, it can be used to solve a set of nonlinear equations more efficiently. For example, it is expected that a trained deep neural network can be used to provide a good initial guess for the electrostatic potential. With a better initial solution, the number of the Newton-Raphson iterations required to solve the nonlinear Poisson equation can be reduced.

In this preliminary report, a deep neural network which can predict the electrostatic potential profiles of pn diodes is introduced. The structure of this extended abstract is as follows: In Section II, the neural network proposed in this work is briefly introduced. The numerical results for pn diodes at equilibrium are shown in Section III. It is clearly demonstrated that the number of the Newton-Raphson iterations can be considerably reduced by adopting the neural network. Moreover, a brief discussion on the future research direction is provided. Finally, the conclusion is made in Section IV.

II. NEURAL NETWORK

The conceptual diagram for the proposed neural network is shown in Fig. 1. First, the neural network is trained as shown in Fig. 1(a). The training data set contains a list of the device specifications and the resultant electrostatic potential profiles. In this preliminary work, the device specifications are simplified as two scalar numbers, which represent the doping densities of a pn diode. It is prepared before starting the training phase. The supervised learning [4] is performed with the backpropagation algorithm. In addition to the training data set, the validation data set is prepared. Since the neural network is not trained with the validation data set, it is used to check the validity of the trained neural network. After trained with the predefined training data set, the neural network can be used to generate the initial potential profile in the semiconductor device as shown in Fig. 1(b).

The configuration of the proposed neural network is shown in Fig. 2. The doping densities of two regions (p-type and n-type regions) are used as the input parameters. Our goal is to generate the electrostatic potential profile as results of the last, output layer. For that purpose, two hidden layers have been introduced. Each hidden layer is fully connected

978-1-7281-0941-1/19 $31.00 © 2019 IEEE

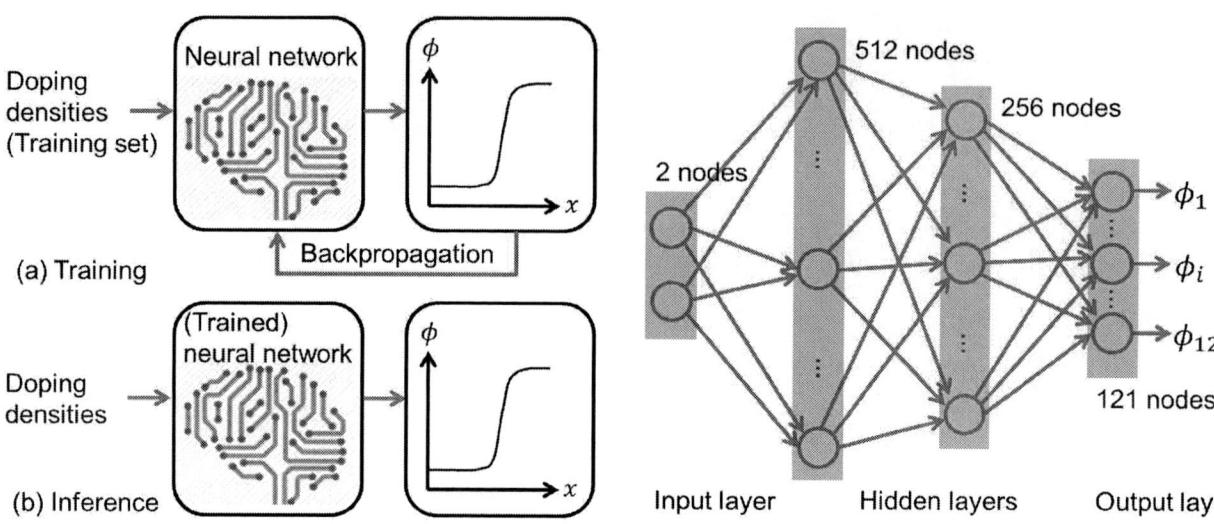

Fig. 1. Conceptual diagram for the proposed neural network. (a) Training and (b) inference phases.

Fig. 2. Layer structure of the proposed neural network. Two hidden layers are introduced. Each hidden layer is fully connected with its neighboring layers.

with its neighboring layers. The first layer consists of 512 artificial neurons and uses the rectified linear unit (ReLU) as the activation function,

$$ReLU(x) = max(0, x). \qquad (1)$$

The dropout parameter is set to be 0.2. The second layer has the same structure with the first one, except for the reduced number of artificial neurons, 256. The last, output layer has 121 artificial neurons. No activation function is used in the output layer.

It is noted that the layer structure shown in Fig. 2 has been obtained by performing several numerical experiments with various layer structures. Depending on the problem, other layer structures may exhibit better results. Unfortunately, we have not found a general rule to build the best layer structure without intensive numerical experiments. It would be an interesting research topic to establish an efficient method to find a sufficiently good layer structure.

As far as the actual implementation is concerned, we have used the sequential model among the Keras layer models [5]. We use the RMSprop (with the learning rate of 0.00005) as the optimizer and the mean square error as the loss function. The number of epochs is 1000.

III. NUMERICAL RESULTS

One-dimensional, abrupt pn diodes at equilibrium are considered. The material is silicon and the temperature is 300 K. The doping density of each region varies from 1×10^{15} cm^{-3} to 1.3×10^{17} cm^{-3}. The p-type and n-type doping densities vary independently. By adopting the depletion approximation, the depletion width varies from 0.13 μm (both regions with 1.3×10^{17} cm^{-3}) to 1.24 μm (both regions with 1×10^{15} cm^{-3}). A sufficiently long structure, whose length is 3 μm, is used for all the cases to ensure that the entire depletion region is included in the simulation domain. The real space is discretized with a uniform grid, whose spacing is 25 nm. The number of nodes in the output layer is matched to that of the grid points.

The difference between the predicted electrostatic potential and the numerical simulation result is regarded as an error. The numerical solution of the nonlinear Poisson equation is obtained by using our in-house tool and the intrinsic carrier density of Si at 300 K is set to be 1×10^{10} cm^{-3}. In Fig. 3, the mean absolute error of the electrostatic potential is shown as a function of the learning epoch. It is estimated after each learning epoch is finished. Throughout the training phase, the error is reduced for both the training data set and the validation data set. At the initial phase, the mean absolute error is rapidly decreasing with the epoch number. However, after a few hundred epochs, the error does not decrease any more. With the given layer structure shown in Fig. 2, it seems that the neural network is sufficiently trained after a few hundred epochs.

After the training phase is finished, the neural network can be used to predict the electrostatic potential profile. Fig. 4 shows the electrostatic potential profile generated by the trained neural network. Various structures with different doping densities are tested. The test structures shown in Fig. 4 are not included in the original training set. The doping densities are randomly selected in a range of (1×10^{15} cm^{-3}, 1.3×10^{17} cm^{-3}). Nevertheless, the generated profile agrees with the numerical simulation result reasonably well for each test structure. Boundary values at both ends are well reproduced and the depletion layers are accurately predicted.

Of course, it should be noted that the profile generated by the deep neural network cannot be perfectly matched to the numerical simulation result. In Fig. 4, slight differences between the symbols (the predicted profiles) and the solid lines

978-1-7281-0941-1/19 $31.00 © 2019 IEEE

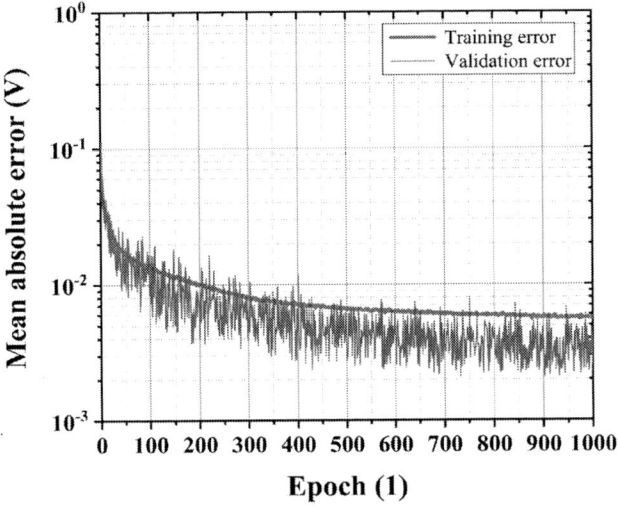

Fig. 3. Training and validation errors as functions of the learning epoch. The errors are rapidly decreasing with the epoch number.

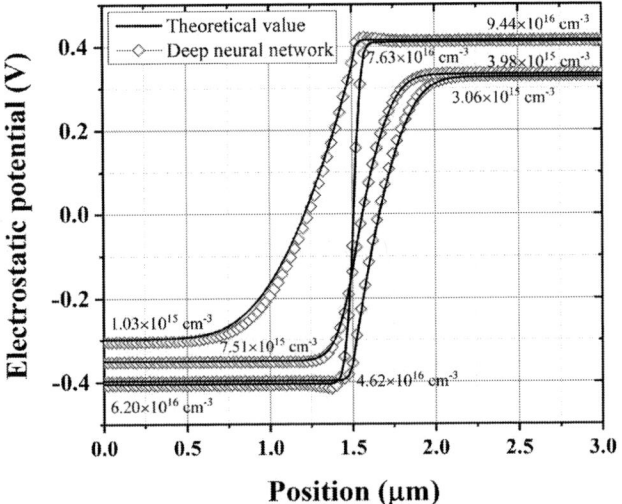

Fig. 4. Electrostatic potential profile generated by the trained neural network. Various structures with different doping densities are tested. The n-type region is located at the right side. For comparison, the numerical solutions are also drawn as solid lines.

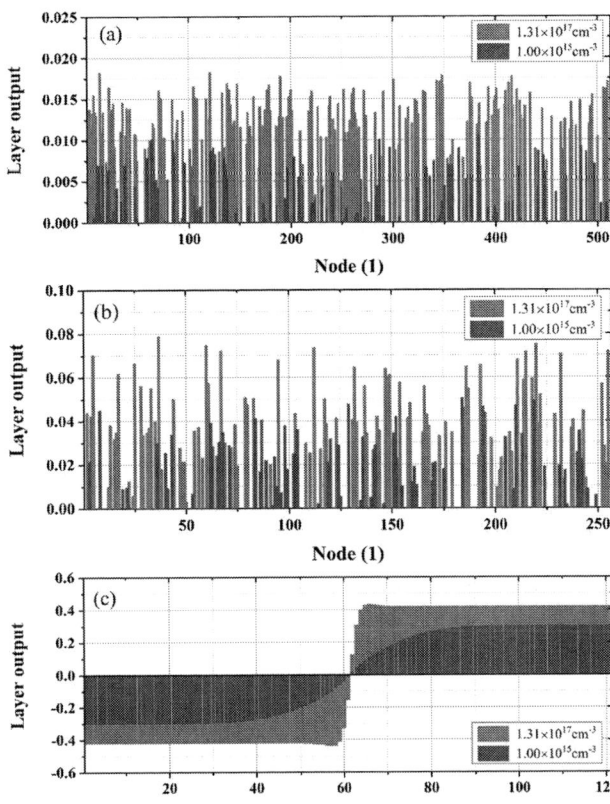

Fig. 5. Results of three layers. (a) The first hidden layer, (b) the second hidden layer, and (c) the output layer for two different symmetric pn diodes.

(the numerical solutions) are visible. Therefore, the predicted electrostatic profile cannot be treated as the converged solution.

In order to understand the internal procedure for the neural network to predict the electrostatic potential profile, results of three layers are shown in Fig. 5. Two symmetric pn diodes are selected as representative examples. One has a relatively high doping density of 1.3×10^{17} cm^{-3} and the other has a low doping density of 1×10^{15} cm^{-3}. It can be observed that the high doping concentration excites the nodes strongly. Although it is difficult to interpret the physical meaning of

node outputs in the first and second layers (Figs. 5(a) and 5(b)), the output layer gives appropriate potential profiles for both structures as shown in Fig. 5(c). The sensitivity analysis on the trained neural network would be an interesting research topic.

Up to now, it has been demonstrated that the trained neural network can be used to guess the electrostatic potential profile at equilibrium. The predicted electrostatic potential can be used as the initial solution for the nonlinear Poisson equation. Since the initial guess for the electrostatic potential is quite similar to the numerical solution, it is expected that the convergence behavior can be improved significantly.

Fig. 6 shows the convergence behavior of the nonlinear Poisson equation. Several structures with randomly selected doping densities are simulated. The maximum potential correction in the first Newton-Raphson iteration is much smaller than 0.1 V. Since the initial error is small, the converged solution can be obtained quickly. For example, in every case simulated in Fig. 6, the maximum potential correction in the fourth Newton-Raphson iteration is smaller than 10^{-5} V. Compared with the conventional method assuming the local charge balance, the convergence acceleration by the deep neural network is obvious. In order to obtain the maximum potential correction

978-1-7281-0941-1/19 $31.00 © 2019 IEEE

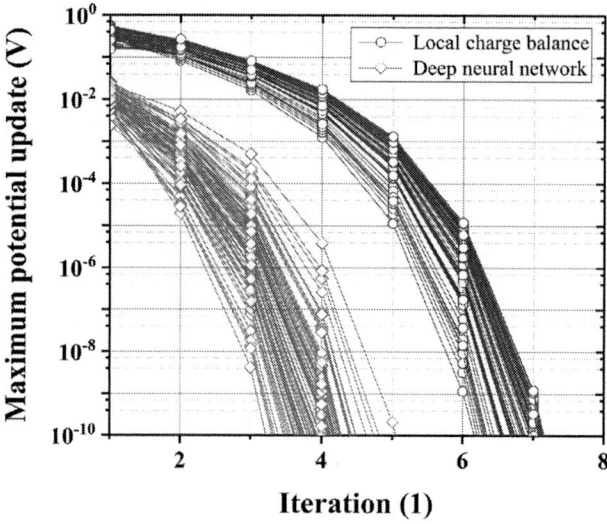

Fig. 6. Convergence behaviour of the equilibrium nonlinear Poisson equation. Two methods for the initial potential profiles (the local charge balance and the generated value by the neural network) are compared for several test structures.

smaller than 10^{-5} V, six or seven Newton-Raphson iterations are needed when the local charge balance approximation is adopted. Therefore, at least two or three iterations can be skipped without sacrificing the numerical accuracy. Since the neural network is trained only in the training phase, the computational burden in the inference phase is negligible.

It is noted that skipping two or three Newton-Raphson iterations for the nonlinear Poisson equation does not save the computation time significantly. However, the present approach has great potential for realistic situations. Although we have considered only the equilibrium case in this feasibility study, it can be extended to non-equilibrium cases [6]. By training the deep neural network with the simulated electrostatic potential profiles even at non-equilibrium cases, the neural network can provide an appropriate potential profile for a biased pn diode. With the predicted potential profile under the given bias point, the numerical simulation can be performed directly. On the other hand, in the conventional device simulation, several intermediate bias points should be simulated to cope with the system nonlinearity.

Therefore, the speed-up ratio for the Newton-Raphson method can be expressed as follows:

$$SpeedUp \approx \frac{N_{Bias}^{Conv} \times N_{Newton}^{Conv}}{N_{Newton}^{NN}}, \qquad (2)$$

where N_{Bias}^{Conv} is the number of bias points to be simulated and N_{Newton}^{Conv} is the average number of the Newton-Raphson iterations per bias point. These quantities are for the conventional bias ramping scheme. On the other hand, N_{Newton}^{NN} is the number of the Newton-Raphson iterations when the predicted profile is used as the initial guess. Two numbers, N_{Newton}^{Conv} and N_{Newton}^{NN}, may be comparable. In such a case, the simulation speed-up is mainly achieved by skipping intermediate bias points. Further results will be reported elsewhere [6].

IV. Conclusion

In conclusion, the deep neural network can generate the initial electrostatic potential profile reasonably well. Compared with the local charge balance approximation, the number of the Newton-Raphson iterations required to get the converged solution is considerably reduced. It is expected that this work can be expanded to non-equilibrium cases with additional efforts.

Acknowledgment

This work was supported by GIST Research Institute (GRI) grant funded by the GIST in 2019.

References

[1] F. Benistant, "Future of TCAD: A foundry perspective," International Conference on Simulation of Semiconductor Processes and Devices, 2018.
[2] U. Kwon, T. Okagaki, S. Ahn, J. Shin, A.-y. Kim, Y.-s. Song, S. Kim, J. Kim, D. S. Kim, W. Qi, Y. Lu, H.-H. Park, W. Choi, "Intelligent DTCO (iDTCO) for next generation logic path-finding," International Conference on Simulation of Semiconductor Processes and Devices, pp. 45–48, 2018.
[3] G. Hegde and R. C. Bowen, "Machine-learned approximations to density-function theory Hamiltonians," Scientific Reports, vol. 7, p. 42669, 2017.
[4] S. J. Russell and P. Norvig, Artificial Intelligence: A Modern Approach, Third Edition, Prentice Hall, 2010.
[5] Keras library. [Online]. Available: https://keras.io
[6] S.-C. Han and S.-M. Hong, unpublished.

Multiscale Modeling of Charge Trapping in Molecule Based Flash Memories

Oves Badami
School of Engineering
University of Glasgow
Glasgow, United Kingdom
oves.badami@glasgow.ac.uk

Toufik Sadi
School of Science
Aalto University
Aalto, Finland
toufik.sadi@aalto.fi

Vihar Georgiev
School of Engineering
University of Glasgow
Glasgow, United Kingdom
vihar.georgiev@glasgow.ac.uk

Fikru Adamu-Lema
School of Engineering
University of Glasgow
Glasgow, United Kingdom
fikru.adamu-lema@glasgow.ac.uk

Vasanthan Thirunavukkarasu
School of Engineering
University of Glasgow
Glasgow, United Kingdom
vasanthan.thirunavukkarasu@glasgow.ac.uk

Jie Ding
College of Electrical and Power Engineering
Taiyuan University of Technology,
030024 China
dingjie2015@foxmail.com

Asen Asenov
School of Engineering
University of Glasgow
Glasgow, United Kingdom
asen.asenov@glasgow.ac.uk

Abstract—To keep up with the increase in demand for storing data, flash memories have been scaled down dramatically and stacked by the semiconductor industry. Furthermore, processing large data has highlighted the limitations of the von Neumann architecture. To overcome this, different types of memory devices like Resistive Random-Access Memories (RRAMs) have also gained a lot of importance. Hence, carrier dynamics in oxides has gained significant traction in recent years. In this work, we discuss the kinetic Monte Carlo methodology as implemented in our integrated simulation environment NESS (Nano-Electronic Simulation Software) that allows us to study carrier transport in the oxide using accurate physics based models. As an example, we study the retention characteristics in a molecule based flash memory.

Index Terms—Kinetic Monte Carlo, POM molecule, Trap Assisted Tunneling

I. INTRODUCTION

In the last two decades, the SiO_2 based gate oxide in metal-oxide-semiconductor field-effect transistors (MOSFETs) has been replaced by high-κ dielectrics in order to maintain the down-scaling trend of these devices. Nonvolatile memory devices likewise have also undergone remarkable changes to increase their packing density, to keep up with the increase in the demand to store data. Hence, charge trap and nanocrystal based flash memories have attracted substantial attention because of their ability to overcome the intercell capacitance problem of the floating gate based flash memories [1]. Furthermore, oxide-based Resistive Random Access Memory (RRAM) devices have gained recently a huge interest, as they are extremely well-suited for in-memory computing [2], [3]. All these advancements in memory technology necessitate the development of reliable tools for the modeling of carrier dynamics in oxides.

The research is funded by the EPSRC (UK) grants no. EP/S000224/1 and EP/S001131/1. This work is partly supported by NSFC project (Project No. 61604105)

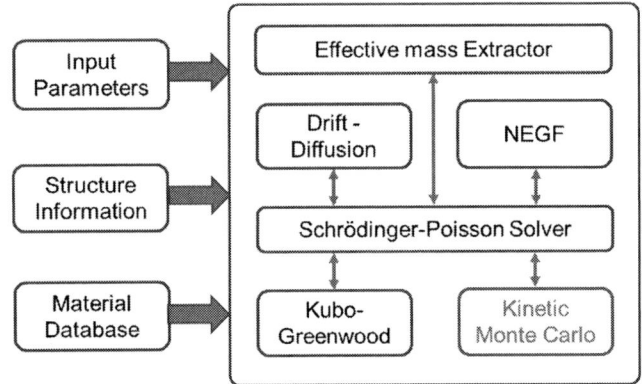

Fig. 1. Capabilities of the Nano-Electronic Simulation Software (NESS) framework.

Fig. 1 illustrates the different modules available in our in-house comprehensive 3D Nano-Electronic Simulation Software (NESS) [4], which can be employed to study different aspects of the carrier transport in modern day transistors. In this work, we report on the extended the capabilities of NESS to simulate charge transport in oxide layers using the kinetic Monte Carlo (kMC) simulation methodology [5]. Here, we illustrate the applicability of the new kMC module by performing a simulation study of the retention time in molecule based flash memory.

II. METHODOLOGY

Fig. 2 shows the flowchart of the kMC module implemented in NESS. After reading and initializing the simulation details, such as the trap type (neutral or positive), trap status (filled or empty) at time (0 sec), Poisson's equation is solved self-consistently in the entire 3D domain along with the equilibrium carrier statistics. This is followed by randomly generating the initial electron state (kinetic energy of the electron)

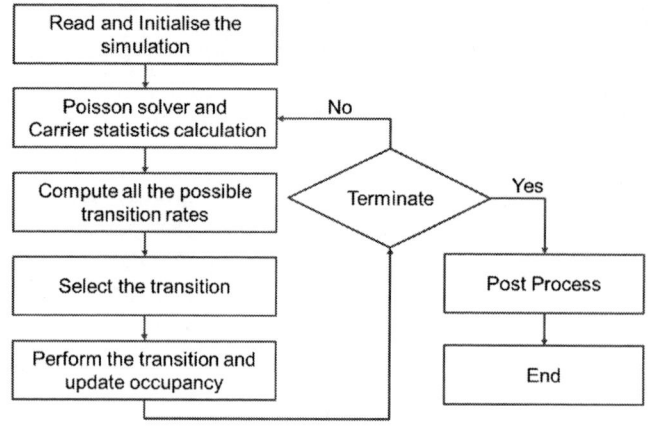

Fig. 2. A flowchart of the implementation of the kinetic Monte Carlo module in NESS.

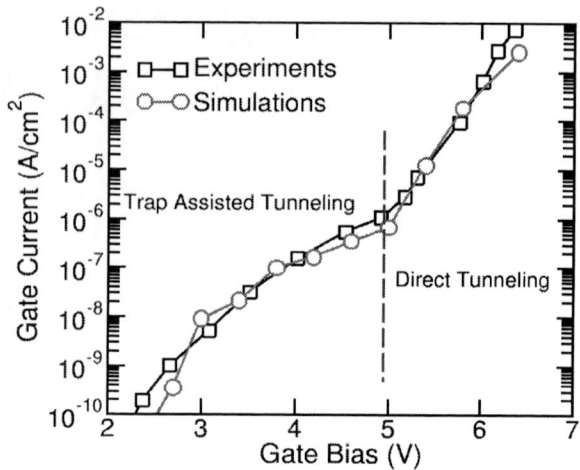

Fig. 3. Comparison of the gate current calculated in this work and experiment. The semiconductor-oxide barrier and electron mass in the oxide were tuned to 3 eV and 0.42 m_0, respectively, to calibrate to the experimental data which agree with those reported in the literature [9].

at the silicon-oxide and metal-oxide interfaces according to the Fermi-Dirac distribution. Following this, all the possible transition rates for the elastic and inelastic trap assisted tunneling, Poole-Frenkel emission, defect-to-defect tunneling mechanisms are computed [5]–[7]. In order to calculate the tunneling rates, the transmission coefficients were evaluated using the Wentzel-Kramers-Brillouin (WKB) approximation assuming a parabolic effective mass. This is followed by randomly selecting a transition, μ, from the cumulative ladder constructed from all the possible tunneling rates according to [5]:

$$\frac{\Sigma_{\mu'=1}^{\mu-1} R_{\mu'}}{R_{TOT}} < r \leq \frac{\Sigma_{\mu'=1}^{\mu} R_{\mu'}}{R_{TOT}}, \quad (1)$$

where r is a uniformly distributed random number between 0 and 1, R_{TOT} is the total transition rate and $R_{\mu'}$ is the transition rate due to the process indexed by μ'. The trap is either populated or depopulated by an electron depending on whether the transition is of an electrode-to-defect nature or vice-versa. The time step, τ, can then be calculated as [5] :

$$\tau = -\frac{ln(r')}{R_{TOT}} \quad (2)$$

where r' is a uniformly distributed random number between 0 and 1 and R_{TOT} is the total transition rate for different tunneling processes. After checking the termination condition, Poisson's equation is solved again to update the electrostatic potential, to account for the rearrangement of the charge in the simulation domain.

The implemented kMC module is flexible allowing different looping conditions, such as terminating when all the traps are filled or empty in addition to the conventional check on the total simulation time. This is extremely useful in analyzing retention, programming and erase times for the memory devices.

III. RESULTS

In order to validate the implementation, we compared the gate current calculated using the methodology discussed in

Section II along with the experimental data published in the literature [8]. The simulated structure corresponded to a MOS capacitor having an oxide thickness of 5 nm, having a trap density of about 3×10^{19} cm^{-3} which agrees with those reported in the literature [9]. In this work, the traps were assumed to be neutral when empty and uniformly distributed in the oxide. The average energy of trap was assumed to be 2.7 eV which also agrees well with those reported in literature [9]. The substrate doping was taken to be 10^{18} cm^{-3}. The relaxation energy, $E_{REL} = S\hbar\omega_0$ was calibrated to 0.6 eV which agrees well with the value for SiO$_2$ reported in the literature [10]. The direct tunneling current component was calculated using the Tsu-Esaki formulation [5].

Fig. 4(a) shows a schematic of the charge trap flash memory cell with a W$_{18}$O$_{54}$(SO$_3$)$_2$ molecule (polyoxymetalate, POM) layer. Fig. 4(b) shows the kMC simulation domain adopted here. The Lowest Unoccupied Molecular Orbital (LUMO) energy level was set to 3.67 eV (with respect to the conduction band edge of the SiO$_2$) as calculated from the first-principle simulations [11]. The POM molecules were approximated by point defects with a cross-section area of $10^{-14}cm^2$ [11]. The p-type doping concentration in the substrate is 10^{18} cm^{-3}. We assume that POM molecules are arranged in a 3×3 configuration embedded in the SiO$_2$ layer. The total oxide thickness ($T_{TUN}+T_{CON}$) was set to 14.5 nm. Fig. 5 shows the threshold voltage (V_T) as a function of the number of electrons trapped in the POM molecules. The threshold voltage is evaluated at the center of the device. V_T was defined as the gate bias at which the electron density at the oxide-semiconductor interface becomes 10% of the p-type doping. As expected the presence of the electrons in the POM molecules increases the threshold voltage.

Fig. 6 shows the conduction band edge, E_C, in the plane of the POM molecules at different time instants as the electrons

978-1-7281-0941-1/19 $31.00 © 2019 IEEE

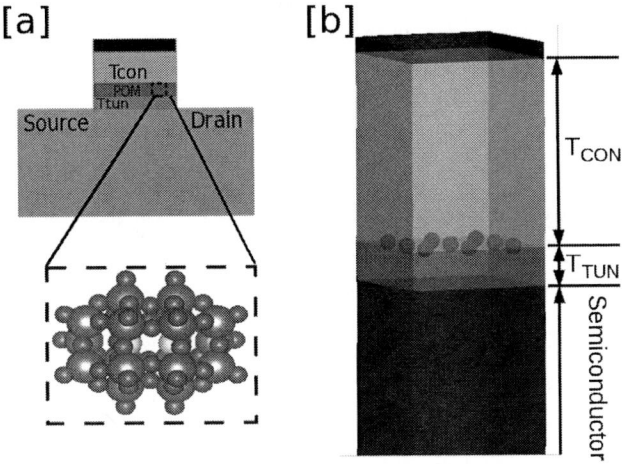

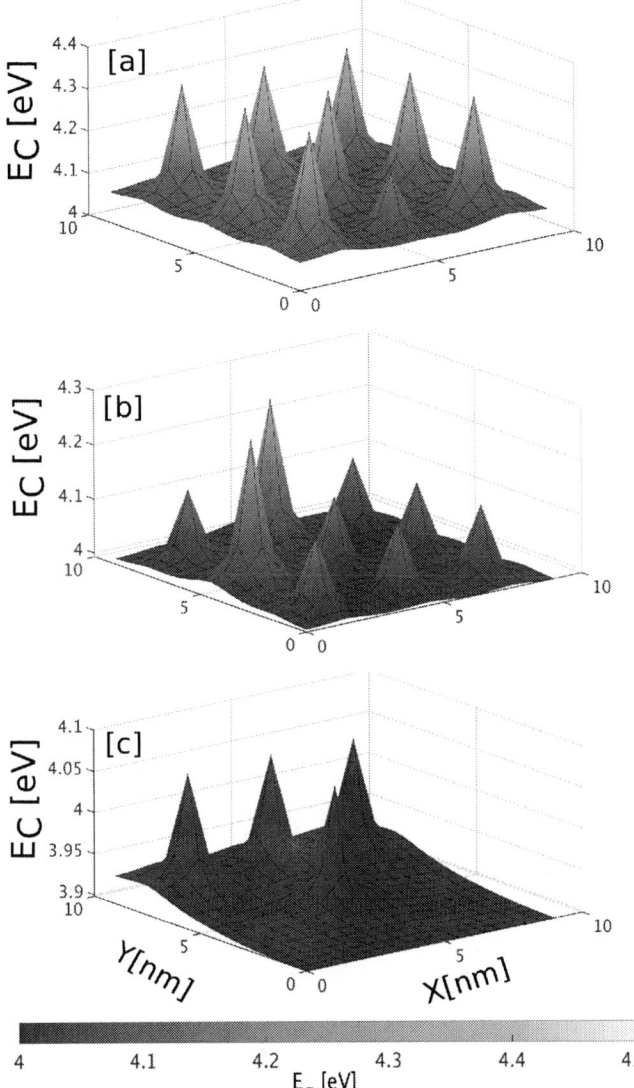

Fig. 4. (a) Illustration of a charge trap flash memory cell with a POM layer. A POM molecule (without counter-cations) is also illustrated in the inset. (b) Schematic of the kMC simulation domain. The 3×3 configuration of the POM molecules is also illustrated (red spheres).

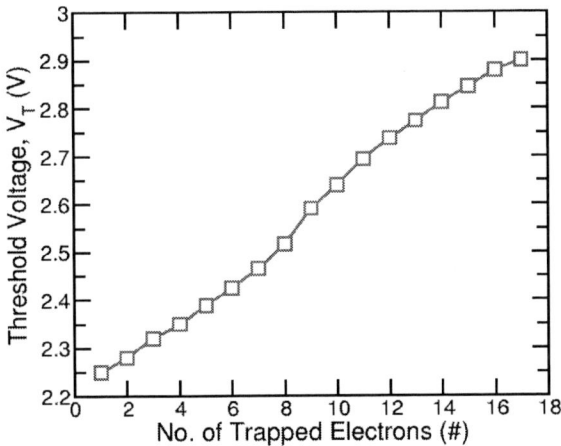

Fig. 5. Calculated threshold voltage (V_T) as a function of the number of electrons trapped in the oxide.

Fig. 6. The evolution of the conduction band edge in the plane of the POM molecules as the electrons tunnel out of the POM molecules with (a) 17 (b) 11 and (c) 4 electrons trapped in the oxide.

from the POM molecules tunnel out through the tunnel oxide. As it is evident, the landscape of the E_C changes dramatically at these time instants because of the different number of the electrons present in the oxide. The presence of the electrons in the POM molecule affects the local potential and hence the electric field. This has a significant impact on the electron tunneling rate out of the doubly charged POM molecules.

Fig. 7 shows the variation of V_T as a function of time where the tunneling rates are calculated self-consistently with the Poisson's equation solution and in the case where the self-consistency is not considered. For the purpose of simulations, $E_{REL} = 0.36eV$ was used [10]. In the absence of the self-consistency, the electrostatic potential in the oxide is not updated and hence all the tunneling rate calculations do not

account for the correct potential. When self-consistency is applied, then during the time period $t < t_1$, all the POM molecules are significantly more likely to undergo transformation from doubly charged (2× reduced) to singly charged (1× reduced) state. This is followed by a significant reduction in the tunneling rates as the local electrostatic potential is modified (as shown in Fig. 6) out of the POM molecules which gives the distinct plateau in Fig. 7. Then, subsequently, POM molecules undergo a transition from 1× reduced state to their parent (neutral) state. Fig. 8 compares the impact of the E_{REL} on the retention behaviour of the flash memory cell shown in Fig. 4. It can seen that as the E_{REL} is increased the retention time increases because of the reduction in the multiphonon

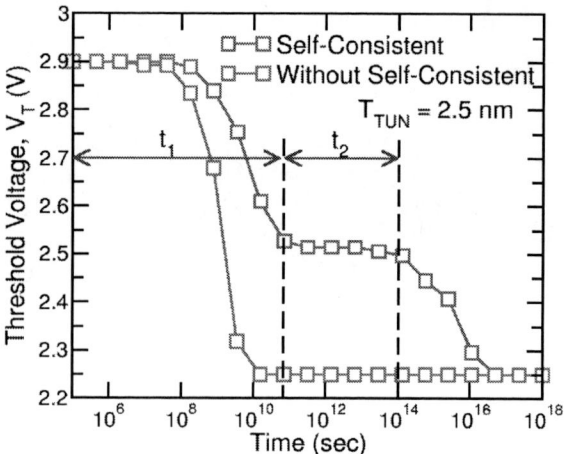

Fig. 7. The impact of the self-consistency between the tunneling rates and Poisson equation on the threshold voltage evolution. For the purpose of simulations, it was assumed that initially all the POM molecules contain two electrons.

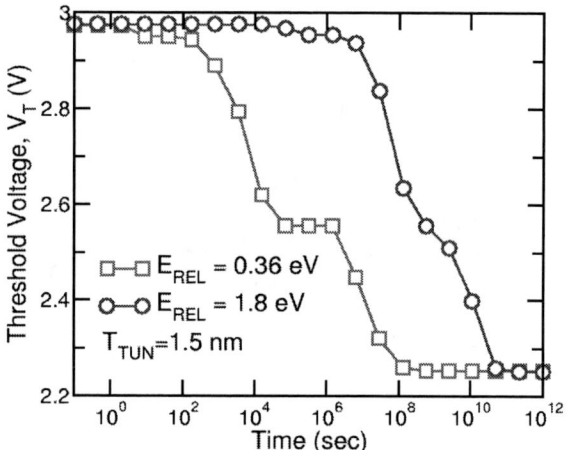

Fig. 8. Impact of the $E_{REL} = S\hbar\omega$ on the retention behaviour. For the purpose of simulation, the phonon energy was assumed to be 0.06 eV while the Huang-Rhys factor (S) was varied.

transition probability. In addition to this, with the increase in the E_{REL} the distinct plateau like feature is also softened. Comparison of Fig.7 and Fig.8 highlights an important fact that with reduction in the T_{TUN} the retention time decreases significantly.

IV. CONCLUSION

We have developed a kMC module within the in-house simulation framework NESS. This module enables us to perform a complete 3D analysis of carrier transport in oxide layers using a comprehensive set of physics-based models. We have bench-marked the module by comparing it with the experimental results for gate tunneling currents. The module allows the user to specify the individual trap location in the

3D space, its type (positive or neutral), the energy depth and cross-section area. Furthermore, by employing a very simple user interface, we can switch on or off different mechanisms, such as electrode-to-defect, defect-to-electrode and defect-to-defect tunneling, enabling the study of their individual impact.

REFERENCES

[1] S. Tiwari *et al.*, "Volatile and non-volatile memories in silicon with nano-crystal storage," Proceedings of International Electron Devices Meeting, Washington, DC, USA, 1995, pp. 521-524.

[2] L. Larcher *et al.*, "Multiscale modeling of neuromorphic computing: From materials to device operations," 2017 IEEE International Electron Devices Meeting (IEDM), pp. 11.7.1-11.7.4, 2017.

[3] M. A. Zidan *et al.*, "A General Memristor-Based Partial Differential Equation Solver," Nature Electronics, 1 (7), 411420, 2018.

[4] S. Berrada *et al.*, "NESS: new flexible Nano-Electronic Simulation Software" SISPAD 2018.

[5] G. C. Gunther, "Modeling of leakage currents in high-κ dielectrics," PhD Thesis, Technical University Munich, Germany 2015.

[6] T. Sadi *et al.*, "Investigation of resistance switching in SiOx RRAM cells using a 3D multiscale kinetic Monte Carlo simulator," Jour. Phys.: Condens. Matter, 30, pp. 084005, 2018.

[7] M. Herrmann *et al.*, "Field and high-temperature dependence of the long term charge loss in erasable programmable read only memories: Measurements and modeling," Jour. of Appl. Phys. vol. 77 no. 9, 1995.

[8] B. Ricco *et al.*, " Modeling and simulation of stress-induced leakage current in ultrathin SiO2 films," Trans. on Electron Devices, vol. 47, no. 7, 1998.

[9] F. Jimenez-Molinos *et al.*, "Physical model for trap-assisted inelastic tunneling in metal-oxide-semiconductor structures," Jour. of Appl. Phys., vol. 90 no. 7, 2001.

[10] L. Vandelli *et al.*, " Physical Model of the Temperature Dependence of the Current Through SiO2/HfO2 Stacks," Trans. on Electron Devices, vol. 58, no. 9, 2011.

[11] V. P. Georgiev *et al.*, "Comparison Between Bulk and FDSOI POM Flash Cell: A Multiscale Simulation Study," Trans. on Electron Devices, vol. 62, no. 2, Feb. 2015.

TCAD Model for Ag-GeSe₃-Ni CBRAM Devices

Kiraneswar Muthuseenu
School of Electrical,
Computer and Energy
Engineering
Arizona State University
Tempe, AZ, USA
kmuthuse@asu.edu

E. Carl Hylin
Silvaco Inc.,
North Chelmsford, MA,
USA
carl@hylin.net

Hugh J. Barnaby
School of Electrical,
Computer and Energy
Engineering
Arizona State University
Tempe, AZ, USA
hbarnaby@asu.edu

Priyanka Apsangi
School of Electrical,
Computer and Energy
Engineering
Arizona State University
Tempe, AZ, USA
papsangi@asu.edu

Michael N. Kozicki
School of Electrical,
Computer and Energy
Engineering
Arizona State University
Tempe, AZ, USA
michael.kozicki@asu.edu

Garrett Schlenvogt
Silvaco Inc.,
North Chelmsford, MA,
USA
garrett.schlenvogt@silvaco
.com

Mark Townsend
Silvaco Inc.,
North Chelmsford, MA,
USA
mark.townsend@silvaco.c
om

Abstract— A model for Ag-GeSe₃-Ni Conductive Bridge Random Access Memory (CBRAM) device is developed using Technology Computer-Aided Design (TCAD) simulations. A new field-dependent ion mobility saturation model that combines Mott-Gurney ionic transport and a high-field saturation ionic drift velocity model is implemented. Also, an electron mobility model for charge transport through the conductive filament is presented. The model simulates forming and dissolving of the filament at different bias conditions. The simulation results of CBRAM I-V hysteresis curves match well to the experimental data.

Keywords— *CBRAM, TCAD, Mott-Gurney model*

I. INTRODUCTION

CBRAM is an ultra-low power non-volatile memory technology that can be integrated into conventional back-end-of-line (BEOL) CMOS processes [1], [2]. CBRAM have been the subject of extensive research since they provide an ultra-low power alternative to conventional non-volatile memory technologies. When functioning as digital memory, they represent stored data as either a high or low resistance across anode and cathode terminals. CBRAM can also be programed in multi-level or even analog states which makes it a potential technology for artificial synapses in neuromorphic applications [3]. The CBRAM device is a two-terminal metal-electrolyte-metal stack. For the material system considered in this work, the active metal (top) anode is silver (Ag) and the inert metal (bottom) cathode is nickel (Ni). The electrolyte layer is a silver doped chalcogenide glass (Ag-Ge₃₀Se₇₀), with a thickness of 70 nm, length of 1μm and width of 10 μm. A thin layer of Ag is dissolved into chalcogenide glass (ChG) using a photo-diffusion process to facilitate ion migration through the glass [4]. The doping profile of Ag in the ChG follows a Gaussian distribution with maximum concentration at the anode/ChG interface. An experimental device with the features above was fabricated at the Arizona State University.

The TCAD device simulator that implements our new CBRAM model is Silvaco's Victory Device Simulator with the Electrochemistry Module. A two-dimensional TCAD structure representative of the fabricated device is shown in Fig. 1. A 2 nm width seed element is added to the top of the Ni cathode in TCAD simulation structure. This enables the formation of the filament from the seed. Without the addition of this seed, a conductive metal sheet will be formed instead of the filament during transient simulation.

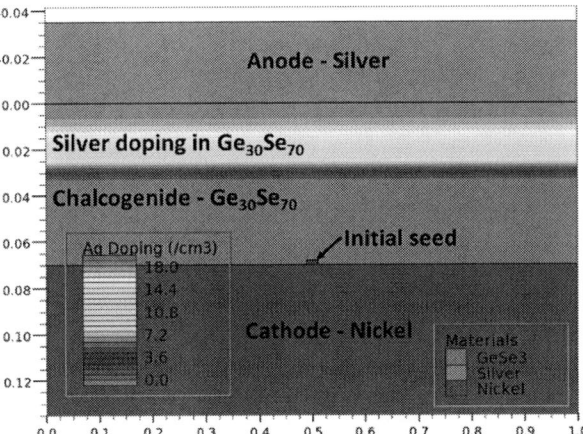

Fig. 1. CBRAM device structure used in this work.

Fig. 2. shows the current-voltage (I-V) hysteresis characteristics of the Ag-GeSe₃-Ni CBRAM device fabricated at Arizona State University. The black square curve shows the response of the device for positive anode bias (relative to the cathode potential) and the blue circle curve is the response for negative anode bias.

CBRAM is a type of electrochemical memory that operates via oxidation-reduction reactions. The data stored in the device is set by the resistance between the anode and cathode contacts. In binary operation, the resistance can be switched between a High Resistance State (HRS) and a Low Resistance State (LRS) [1], [2]. Assuming an initial HRS condition, there exists no filament or low resistance pathway between the terminal. When a positive bias is applied to the anode relative to the cathode, the following sequential processes occur: 1) Ag cations (Ag^+) are created through oxidation at the anode/ChG interface (Eq. 1), 2) the cations drift through the electrolyte in the direction of the applied electric field, 3) Ag cations are reduced by gaining an electron at the cathode (Eq. 2). This corresponds to SET process shown in Fig. 2 (inset A).

$$Ag \rightarrow Ag^+ + e^- \qquad (1)$$

$$Ag^+ + e^- \rightarrow Ag \qquad (2)$$

On continued application of positive bias, a conductive Ag filament grows from the cathode back to the anode. After the application of positive voltage over an appropriate time, the Ag filament will create a conducting bridge between the

978-1-7281-0941-1/19 $31.00 © 2019 IEEE

opposite electrodes through the electrolyte layer. The voltage at which this occurs is called the SET voltage. At this stage, the device switches into LRS (ON state) as shown in Fig. 2 (inset B). The maximum current reached in LRS is set by an externally programmed compliance current that prevents damage to the cell. The device retains the ON state until a sufficient negative bias is applied to the anode. When the polarity of the bias is reversed, Ag ions will drift back to the anode and dissolve the filament. This is called the RESET process (Fig. 2 inset C). Finally, when the filament is fully dissolved the device switches back to HRS (OFF state) as shown in Fig. 2 inset D.

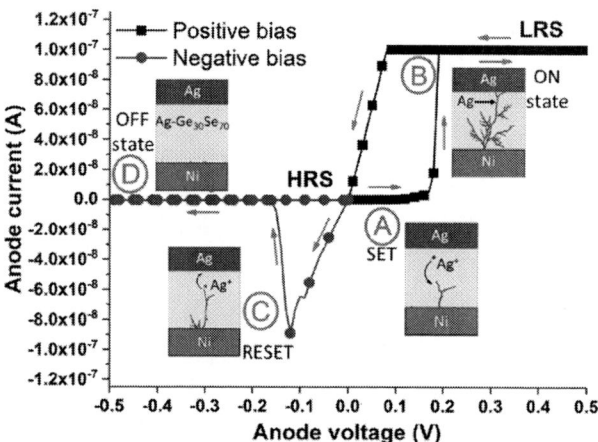

Fig. 2. I-V characteristics of the CBRAM device fabricated at Arizona State University. The insets show different stages of the switching process during the I-V sweep. The compliance current is set at 100 nA.

II. PHYSICAL MODEL

A. Field Dependent Ion Mobility Saturation Model

The field dependent ion mobility of a chemical species is given by the Mott-Gurney model [5]. According to Mott-Gurney, the transport of chemical species in the solid electrolyte (ChG) occurs through ion hopping. The diffusivity, D_S of the chemical species can be expressed as

$$D_S = d_{hop}^2 * v_a * exp\left(\frac{-E_A}{kT_L}\right) * \left(\frac{sinh(\eta_S)}{\eta_S}\right), \quad (3)$$

where d_{hop} and v_a is the hopping distance and the hopping frequency of the species in the ChG layer, respectively [1], [6]. The exponential term corresponds to the probability that the attempted hopping of the species succeeds in overcoming the energy barrier represented by the activation energy E_A. In Eq. 3, k is the Boltzmann constant and T_L is the lattice temperature of the ChG layer. The term η_s can be expressed as [1], [6]

$$\eta_S = \left(\frac{d_{hop}}{2}\right) * z_S * \frac{q}{kT_L} * |E|, \quad (4)$$

where, $|E|$ is the magnitude of the electric field and z_S corresponds to the magnitude and sign of the electric charge on the chemical species S. At large values of η_s, the probability of ion hopping, P_{hop}, in the Mott-Gurney model can be expressed as [6],

$$P_{hop} = exp\left(\eta_S - \frac{E_A}{kT_L}\right). \quad (5)$$

For the CBRAM device discussed here, two chemical species are present: Ag with neutral charge and Ag^+ with a single positive charge. The diffusivity of Ag in ChG layer is negligible compared to Ag^+. The diffusivity parameters of Ag^+ in ChG layer are obtained from [1].

It should be noted that at very high fields, the Mott-Gurney model becomes invalid because the P_{hop} cannot exceed unity. Therefore, at high electric fields the ionic drift velocity saturates at a fraction f_{vsat}, of the thermal velocity. The field dependent ion mobility saturation model implemented in this paper combines the Mott-Gurney model with the high field saturation of the ionic drift velocity. Thus, the diffusivity of the chemical species is revised as [7]

$$D_S = d_{hop}^2 * v_a * exp\left(\frac{-E_A}{kT_L}\right) * \{[\frac{Sinh(\eta_S)}{\eta_S}]^{-2} +$$
$$[f_{vsat} * \frac{sinh\left(\frac{E_A}{kT_L}\right)}{|\eta_S|}]^{-2}\}^{-1/2} . \quad (6)$$

The mobility of an ionic species, μ_S, can be related to diffusivity of the species using Einstein relation [6], and is given as

$$\mu_S = z_S * \frac{q}{kT_L} * D_S. \quad (7)$$

Fig. 3. shows the mobility of Ag^+ ions in the ChG layer as a function of electric field. It compares the Ag^+ ion mobility calculated using the Mott-Gurney model and the field dependent ion mobility saturation model. The black square and blue circle symbols correspond to Mott-Gurney model and field dependent ion mobility saturation model, respectively. From the figure it can be observed that the mobility of Ag^+ ions in the ChG layer calculated using Mott-Gurney model will increase rapidly when the magnitude of electric field is approximately 1.1×10^6 V/cm. The hopping probability exceeds unity at this electric field and therefore Mott-Gurney model is no longer valid above this field. Thus, for high fields the mobility of Ag^+ ion is calculated based on the new field dependent ion mobility saturation model. Using this new model, the mobility of Ag^+ ions peaks at 3×10^{-4} cm²/V.s.

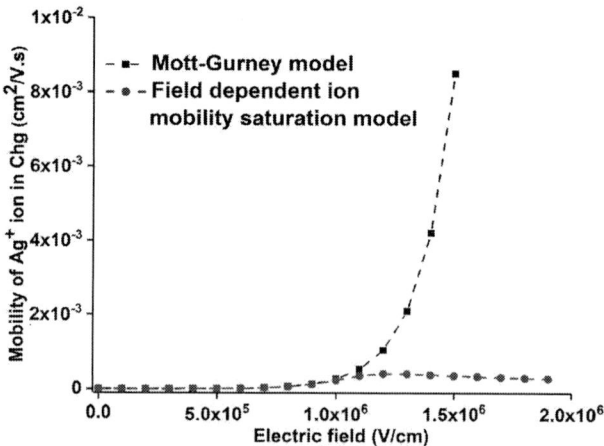

Fig. 3. Plot of mobility Vs electric field for Ag^+ ion in ChG, calculated using Mott-Gurney model and field dependent ion mobility saturation model.

B. Electron Mobility Model for Filament

The primary electrochemical reactions occurring in the CBRAM device are given by Eqs. 1 and 2. When the filament starts to form between the cathode and anode, the conductivity of the device increases. As the concentration of the Ag in the filament increases, the electron mobility in the filament increases. Therefore, the total effective electron mobility (μ_{eff}) depends on the electron mobility in ChG layer, electron mobility in the filament and the concentration of the silver in the filament. This electron mobility model is given as [6]

$$\mu_{eff} = \mu_s + w * (\mu_c - \mu_s), \qquad (8)$$

where μ_s is the electron mobility in the host layer (ChG), μ_c is the electron mobility in the conductive material (Ag filament) and w is a weight factor that depends on the concentration of the Ag in the conductive filament that can vary between 0 to 1. The weight factor formula is

$$w = \varepsilon + \frac{\left\{ (f - f_0) + \delta * ln \left\{ cosh \left[\frac{(f - f_0)}{\delta} \right] \right\} \right\}}{2 * (1 - f_0)}, \qquad (9)$$

where f is the ratio of local concentration of the conductive bridging species relative to its maximum concentration in the ChG, f_0 is the threshold of bridging species fraction at which the electron mobility starts to increase, δ is the half width of the transition region around f_0, ε is a constant required to make $w = 0$ when $f = 0$.

Fig. 4 compares the electron mobility in the filament as a function of bridging species concentration for different values of f_0 with all the other parameters kept constant. The filled black square, blue circle and red triangle symbols correspond to f_0=0.1, 0.3 and 0.5 respectively. It can be observed that the value of f_0 directly correlates to the local concentration fraction and the conductivity through the filament increases rapidly after this fraction is exceeded. The theory behind this model is that for a local concentration value less than f_0, silver in ChG layer exists in disconnected islands and there is no long-range conduction path between the anode and cathode.

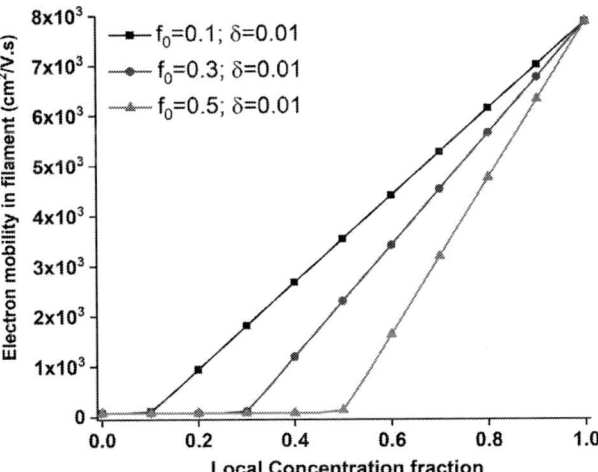

Fig. 4. Mobility of electron in the filament as a function of concentration for different values of f_0 when δ=0.01.

Fig. 5 compares the electron mobility in the filament for different values of δ and f_0=0.3. The unfilled black square, blue circle and red triangle symbols correspond to δ=0.01, 0.05 and 0.1, respectively. As can be observed from Fig. 5, the electron mobility in the filament increases more gradually as the δ value is increased.

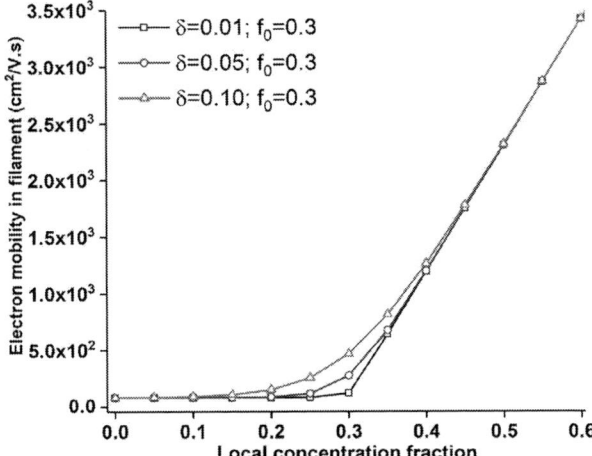

Fig. 5. Mobility of electron in the filament as a function of concentration for different values of δ=0.01 when f_0=0.3.

III. SIMULATION RESULTS

Fig. 6 shows TCAD simulations of the forming and dissolving of the Ag filament at different biases during the I-V hysteresis sweep. It can be observed that the concentration of Ag is being modulated in the horizontal center of the device where the seed element was added. In the initial condition with 0V, no filament is present in the device and the device is in HRS as shown in Fig. 6 (inset A). When a 0.5V is applied to the anode, a full Ag filament is formed between the cathode and anode as can be seen in Fig. 6 (inset B). The concentration of Ag throughout the filament is on the order of 10^{20}-10^{21}cm^{-3}. When the voltage applied is ramped down to 0V (inset C), the device retains the filament in the ChG layer, though the concentration of Ag near the anode drops to 10^{19}cm^{-3}. When -0.5V is applied to the anode (inset D), the filament is fully dissolved, and the device goes to HRS.

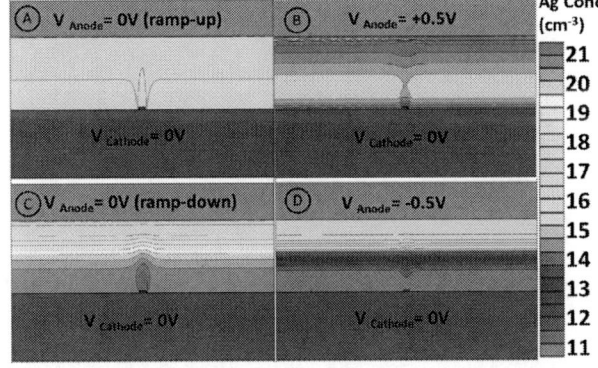

Fig. 6. Concentration of silver at different biases during the I-V hysteresis sweep.

Fig. 7 compares experimental data and simulation results of the I-V hysteresis sweep performed on a Ag-Ge$_{30}$Se$_{70}$ CBRAM device. Unlike Fig. 2, the curves in Fig. 7 plot log current magnitude vs. linear voltage. The filled black squares and unfilled blue circle symbols correspond to the experimental data and TCAD simulation results, respectively.

In Fig. 7 the region of the sweep labeled (1) corresponds to ramping up of the anode voltage from 0V to 0.5. Sweep region (2) corresponds to ramping down the anode voltage from 0.5V to 0V. Region (3) corresponds to the ramping down of voltage from 0V to -0.5. Region (4) corresponds to ramping up of voltage from -0.5V to 0V. The silver filament is formed in region (1), i.e., the device is switched from HRS to LRS. The filament is dissolved in region (3), i.e., the device is switched from LRS to HRS. A reasonably close fit between the experimental data and TCAD simulation can be observed.

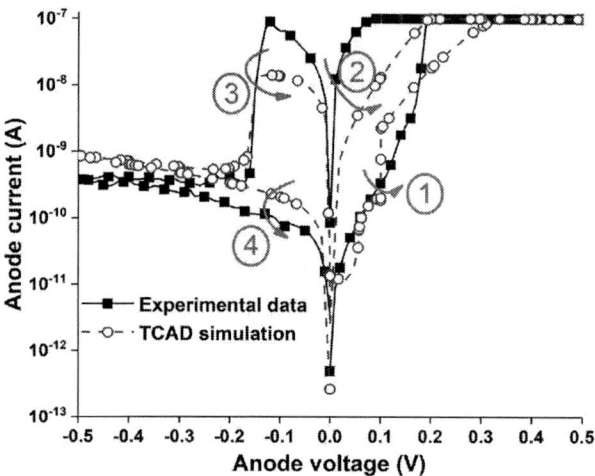

Fig. 7. Comparison of experimental data and simulation results of an I-V hysteresis sweep. The compliance current is set at 100 nA.

Fig. 8. plots the SET voltage for different ramp-rates, simulated using the TCAD model. For our purposes, we define the SET voltage as the voltage at which the current reaches 100 nA during I-V forward sweep. This relationship between ramp-rate and SET voltage was reported experimentally in [8], providing further validation of the model.

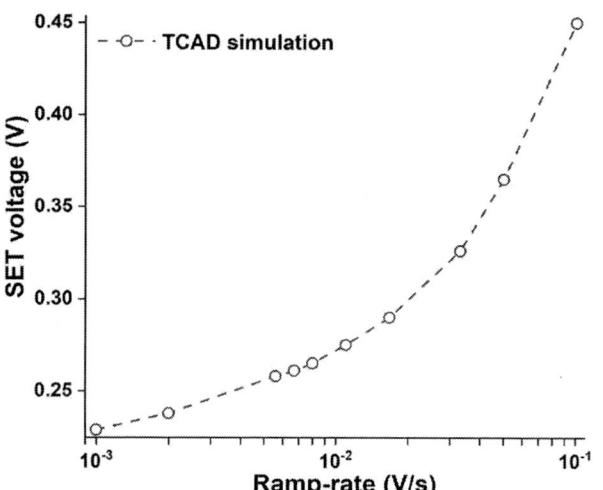

Fig. 8. Simulation results of SET voltage vs. ramp-rate.

IV. CONCLUSION

This paper presents modeling results which reproduce the formation and dissolution of filaments in CBRAM devices. This is the first time the CBRAM has been modelled in a commercial TCAD device simulator tool. The simulator implements new model for the transport of charged species in the ChG layer. The field dependent ion mobility saturation model presented in this paper combines the Mott-Gurney model and high field saturation of the ionic drift velocity model. At high fields, when the hopping probability in Mott-Gurney model exceeds unity, the diffusivity of the chemical species saturates at a fraction of thermal velocity. The second model described in this paper calculates the mobility of the electrons in the filament as a function of bridging species concentration. This model assumes that below a certain threshold of bridging species concentration, the species are present in the ChG layer as disconnected islands and no long-range conduction path exists between anode and cathode. After the species concentration exceeds the threshold, a linear increase in the conductivity through the filament is implemented by the model. The above two models have been implemented using the electrochemistry module in Silvaco's Victory device simulator. Comparison of I-V hysteresis characteristics between the simulations and the experimental data shows good agreement.

ACKNOWLEDGMENT

This work was funded in part by DTRA under grant no. HDTRA1-17-1-0038 and Arizona State University Nanofab which is supported by NSF award number NNCI – 1542160.

REFERENCES

[1] M. N. Kozicki and H. J. Barnaby, "Conductive bridging random access memory - Materials, devices and applications," *Semicond. Sci. Technol.*, vol. 31, no. 11, 2016.

[2] M. N. Kozicki, M. Park, and M. Mitkova, "Nanoscale memory elements based on solid-state electrolytes," in *IEEE Transactions on Nanotechnology*, 2005.

[3] G. Indiveri, B. Linares-Barranco, R. Legenstein, G. Deligeorgis, and T. Prodromakis, "Integration of nanoscale memristor synapses in neuromorphic computing architectures," *Nanotechnology*, 2013.

[4] M. N. Kozicki and M. Mitkova, "Mass transport in chalcogenide electrolyte films – materials and applications," *J. Non. Cryst. Solids*, vol. 352, no. 6–7, pp. 567–577, May 2006.

[5] N. F. Mott and R. W. Gurney, *Electronic processes in ionic crystals; 2nd ed.* Oxford University Press, 1948.

[6] *Victory Device User's Manual.* Silvaco Inc. USA.

[7] J. Habasaki, C. Leon, and K. L. Ngai, *Dynamics of Glassy, Crystalline and Liquid Ionic Conductors*. 2017.

[8] C. Schindler, G. Staikov, and R. Waser, "Electrode kinetics of Cu–SiO2-based resistive switching cells: Overcoming the voltage-time dilemma of electrochemical metallization memories," *Appl. Phys. Lett*, vol. 94, p. 72109, 2009.

Comprehensive Comparison of Switching Models for Perpendicular Spin-Transfer Torque MRAM Cells

Simone Fiorentini
Christian Doppler Laboratory for Magnetoresistive Nonvolatile Memory and Logic
Institute for Microelectronics, TU Wien
Vienna, Austria
fiorentini@iue.tuwien.ac.at

Roberto Orio
Christian Doppler Laboratory for Magnetoresistive Nonvolatile Memory and Logic
Institute for Microelectronics, TU Wien
Vienna, Austria
orio@iue.tuwien.ac.at

Wolfgang Goes
Silvaco Europe Ltd
Cambridge, United Kingdom
wolfgang.goes@silvaco.com

Johannes Ender
Christian Doppler Laboratory for Magnetoresistive Nonvolatile Memory and Logic
Institute for Microelectronics, TU Wien
Vienna, Austria
ender@iue.tuwien.ac.at

Viktor Sverdlov
Christian Doppler Laboratory for Magnetoresistive Nonvolatile Memory and Logic
Institute for Microelectronics, TU Wien
Vienna, Austria
sverdlov@iue.tuwien.ac.at

Abstract—Simulations of free-layer switching in spin - transfer torque MRAM are usually performed with the torque computed approximately by assuming a position-independent electric current density through the structure. For high values of the tunneling magnetoresistance, this description is not accurate anymore, and one needs to solve the spin and charge drift-diffusion equations in the whole structure self-consistently. We compute the switching time distribution obtained by the self-consistent model and compare it to the switching times from the fixed current density approach. We show that, provided the current is appropriately adjusted, the simplified model can mimic the correct switching time distribution even in the case of high TMR.

Keywords—Spin-transfer torque, MRAM, perpendicular magnetization, tunneling magnetoresistance

I. INTRODUCTION

In recent years, the development of computer memory has been focused on miniaturization and scaling of semiconductor devices. This, however, has increased the stand-by power and leakages in modern integrated circuits, due to the volatile nature of SRAM and DRAM. An attractive path to dramatically reduce the power consumption is to introduce non-volatility.

One possible path to achieve this goal is to consider the spin degree of freedom of electric charges. For this, non-volatile spin-transfer torque (STT) magnetoresistive random access memory (MRAM) is a viable candidate. STT-MRAM combines higher speed, superior endurance and lower costs as compared to flash memories. In addition, STT-MRAM is compatible with CMOS technology and can be straightforwardly embedded in circuits. The potential market for STT-MRAM is ranging from IoT and automotive applications to embedded DRAM and L3 caches [1][2].

In this work we present a way of computing the magnetization dynamics in a STT-MRAM by solving the spin drift-diffusion equations with non-uniform current density, and compare it to the standard approach which prescribes the current density to be fixed and position-independent.

II. STT-MRAM

The key element of modern MRAM cells consists of a magnetic tunnel junction (MTJ), formed by two ferromagnetic layers separated by a thin oxide layer, where the latter provides the tunnel barrier (see Fig. 1 for reference). The magnetization of the layers has two possible configurations: parallel (P) and anti-parallel (AP). The magnetization in one of the layers is free to switch (free layer), while the magnetization in the second is fixed (reference layer). This can be achieved by tuning the geometry of the layers or by antiferromagnetically coupling it to a pinned layer [3].

Due to the tunneling magnetoresistance effect, tunneling electrons polarized by the reference layer are easily accommodated by the free layer when the magnetization vectors of the layers are parallel, so the resistance R_P is lower than the resistance R_{AP}. This resistance difference is usually characterized by the tunneling magnetoresistance ratio (TMR), defined as

$$TMR = \frac{R_{AP} - R_P}{R_P}. \qquad (1)$$

Achieving a high TMR ratio is important in order to reliably discern between the P and AP configurations. The use of a CoFeB/MgO/CoFeB MTJ can grant a TMR ratio of up to 600% [4].

The commercial application of MRAM in embedded circuits requires a reliable switching process between the two magnetization configurations of the MTJ. Such a switch can be efficiently realized by letting an electric current flow through the structure: the spins of the electrons get polarized by the reference layer, and the exchange coupling with the magnetization provides the torque (STT) necessary to rotate

978-1-7281-0941-1/19 $31.00 © 2019 IEEE

the magnetization in the free layer and achieve switching [5][6].

In order to design more efficient memories, it is necessary to introduce methods of simulating the switching process in realistic structures.

III. MODELS

The equation that describes magnetization dynamics is the Landau-Lifshitz-Gilbert (LLG) equation. When introducing a term describing STT ($\mathbf{T_S}$), the final magnetization equation is [7]

$$\frac{\partial \mathbf{m}}{\partial t} = -\gamma \mathbf{m} \times \mathbf{H}_{\text{eff}} + \alpha \mathbf{m} \times \frac{\partial \mathbf{m}}{\partial t} + \frac{1}{M_S} \mathbf{T_S}$$

$$\mathbf{T_S} = \gamma \frac{\hbar}{2e} \frac{0.5\, Jc\, P}{d\left(1 + P^2 \cos\theta\right)} \mathbf{m} \times (\mathbf{m} \times \mathbf{x}),$$

(2)

where $\mathbf{m} = \mathbf{M}/M_S$ is the position-dependent normalized magnetization in the free layer, M_S is the saturation magnetization, α is the Gilbert damping constant, γ is the gyromagnetic ratio, $\hbar$ is the reduced Plank constant, e is the electron charge, Jc is the current density, P is the spin current polarizing factor [8] assumed to be equal in both ferromagnetic layers, d is the thickness of the free ferromagnetic layer, θ is the angle between local magnetization vectors in the free and fixed layer, and $\mathbf{x}$ is the unit vector along the fixed layer magnetization (Fig. 1). The effective magnetic field $\mathbf{H}_{\text{eff}}$ includes the external field, the magnetic anisotropy field, the demagnetizing field and the stray field from the reference layer/magnetic stack. It also includes the thermal field, a randomly fluctuating magnetic field used to simulate the effects of finite temperature on the magnetization dynamics.

The usual approach in micromagnetic simulations of STT switching is to assume that the current density J_C is position-independent [9]. For low TMR and in-plane MTJs, where the resistance difference between the low and high resistance configuration is small, this assumption can be justified [10]. However, modern MTJs are perpendicularly magnetized (p-MTJs) and possess a large TMR above 200%. In this case the simplified description offered by (2) may not be accurate anymore. Indeed, if the current is running, the local magnetization vectors at every point are not collinear. This results in position-dependent current density, which in turn results in position-dependent spin currents and spin torques at every time step. The validity of (2) must be justified by a complete computation of spin accumulation and spin torques in p-MTJs due to the position dependent current density coupled to the magnetization dynamics.

In order to achieve this, we need to take into consideration the spin and charge drift-diffusion equations [11][12]:

$$\mathbf{J_S} = \frac{\mu_B}{e} \beta_\sigma \sigma \left(\mathbf{J_C} + \beta_D D_e \frac{e}{\mu_B} [(\nabla \mathbf{S})\mathbf{m}] \right) \otimes \mathbf{m} - D_e \nabla \mathbf{S} \quad (3)$$

$$\frac{\partial \mathbf{S}}{\partial t} = -\nabla \mathbf{J_S} - D_e \left(\frac{\mathbf{S}}{\lambda_{sf}^2} + \frac{\mathbf{S} \times \mathbf{m}}{\lambda_J^2} + \frac{\mathbf{m} \times (\mathbf{S} \times \mathbf{m})}{\lambda_\varphi^2} \right), \quad (4)$$

where $\mathbf{J_S}$ is the spin current density, μ_B is the Bohr magneton, β_σ and β_D are polarization parameters, σ is the electron conductivity, D_e is the electron diffusivity constant, $\lambda_{sf}, \lambda_J, \lambda_\varphi$ are scattering lengths, and $\otimes$ stands for the tensor product.

We compute the electric current density $\mathbf{J_C}$ by taking into consideration the dependence of the tunnel barrier resistance on the relative angle of magnetization vectors in the reference and free layer. In order to do this, we solve the Laplace equation $-\nabla^2 V = 0$, where V is the electric potential, in the ferromagnetic contacts. The non-uniform resistance of the barrier, described as [10]

$$R(\theta) = R_P \left(1 + \left(\frac{\text{TMR}}{2}\right) \cdot (1 - \cos(\theta)) \right), \quad (5)$$

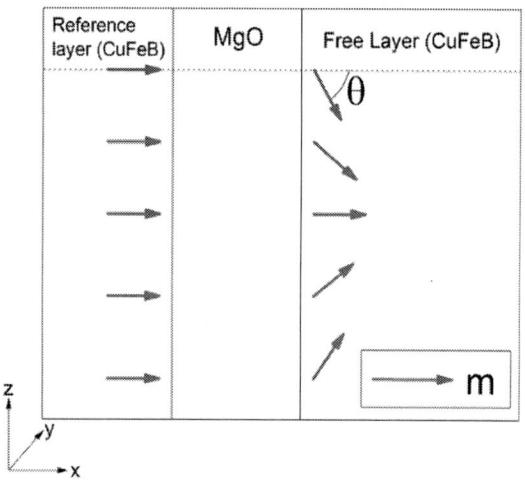

Fig. 1. Schema of the magnetization configuration in a mid-switch scenario.

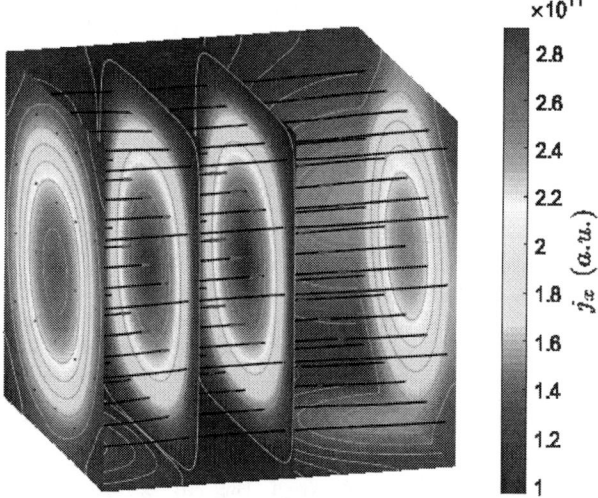

Fig. 2. Current density distribution through a square MTJ in a mid-switch scenario. The current flow is higher where the magnetization is aligned, due to lower resistance.

is imposed by Neumann boundary conditions at the interface between the ferromagnetic and the oxide layers. Dirichlet boundary conditions are imposed on the electrodes, thus fixing the potential. The current density $\mathbf{J_C}$ across the structure is then given by

$$\mathbf{J_C} = -\sigma\nabla V. \qquad (6)$$

In Fig. 2 we report the computed current density for the magnetization configuration schematized in Fig. 1. The current density is highly nonhomogeneous and redistributed in order to accommodate the varying resistance across the barrier.

With $\mathbf{J_C}$ known, (3) and (4) can be solved to compute the spin accumulation $\mathbf{S}$. The torque term to be inserted in the LLG equation is thus:

$$\mathbf{T_S} = -\frac{D_e}{\lambda_J^2}\mathbf{m}\times\mathbf{S} - \frac{D_e}{\lambda_\varphi^2}\mathbf{m}\times(\mathbf{m}\times\mathbf{S}). \qquad (7)$$

IV. RESULTS

We compare the model described by (3-7) (Model 1) with a *fixed voltage* across the MTJ, to the model with the *constant current* density described by (2) (Model 2) and to a reference model [10] generalized to p-MTJs, in which the total *current* is fixed but redistributed at every time step according to the local resistance value (Model 3). In order to account for the differences between the models, we set the total currents for Model 2 and Model 3 to the value of the current in Model 1 at the beginning of the switching process. The simulations are performed for a p-MTJ. The system's main parameters, which were set to typical experimental values [13], are summarized in Table 1. The stack is of a circular pillar shape of 40 nm in diameter. The thicknesses of the free and the reference CoFeB layers are 1.7 nm and 1 nm, respectively. The thickness of the MgO layer is 1 nm.

With this choice of parameters, the TMR of the structure is 200%, and the thermal stability factor, given by

$$\Delta = \frac{M_S H_K V}{2k_B T}, \qquad (8)$$

is equal to 68 at room temperature, in agreement with the minimum factor of 60 required for standalone memories. In (8), H_K is the anisotropy field, V is the volume of the free layer, k_B is the Boltzmann constant and T is the temperature.

Examples of switching realizations in the structure described above are shown in Fig. 3, for both AP→P and P→AP processes. The fluctuating thermal field guarantees a unique switching path every time the simulation is run. Thus, the switching time can vary slightly between different switching realizations. We note that the switching time required to switch from P to AP is higher than from AP to P. This is due to the uncompensated stray field of the reference layer, which tends to keep the magnetization of the two ferromagnets aligned, helping the switching to the parallel configuration and opposing the switching to the antiparallel one. By using a pinned layer antiferromagnetically coupled to the reference layer, we can compensate the total stray field acting on the free layer. This allows to have more symmetric switching times from P to AP and from AP to P, correspondingly.

Fig. 4 reports the switching times (ST) as a function of uncompensated stray field, where the stray field is modeled by the total saturation magnetization of the antiferromagnetically coupled layers. The results are averaged over 30 realizations, in order to take into account the effects of the thermal field. As expected, compensating the stray field results in a higher ST for the AP→P configuration and a lower ST for P→AP. However, results for STs are ~15% higher for AP→P and ~10% lower for P→AP for Model 2 and 3 as compared to Model 1. This

TABLE I.   SYSTEM PARAMETERS

Parameter	Value
Gilbert damping, α	0.02
Gyromagnetic ratio, γ	$2.3\cdot10^5$ m/(A·s)
Saturation magnetization, Ms	$1.2\cdot10^6$ A/m
Free layer thickness, d	1.7 nm
Perpendicular anisotropy energy, K	$9.0\cdot10^5$ A/m
Resistance for P state, R_P	14 kΩ
Resistance for AP state, R_{AP}	42 kΩ
Voltage, V	±2 Volts
Switching current for AP→P, $J_{AP->P}$	$3.8\cdot10^{10}$ A/m^2
Switching current for P→AP, $J_{P->AP}$	$-1.1\cdot10^{11}$ A/m^2

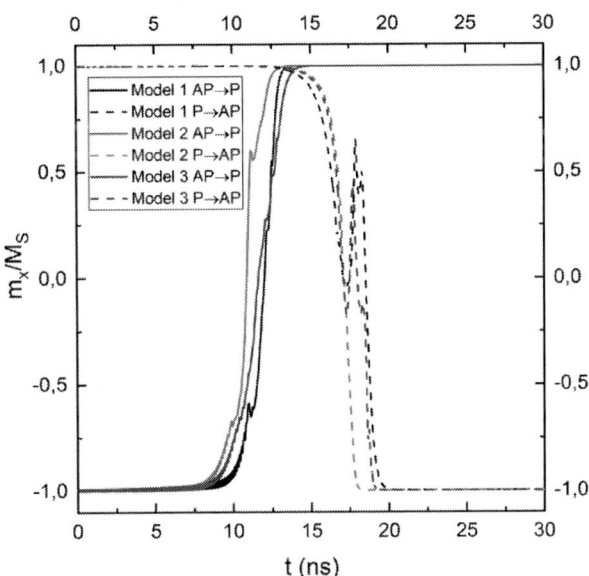

Fig. 3. AP→P and P→AP switching for the 3 different models. In the P→AP switching attempts, the stray field from the fixed layer is opposing the switch, creating the observable oscillations.

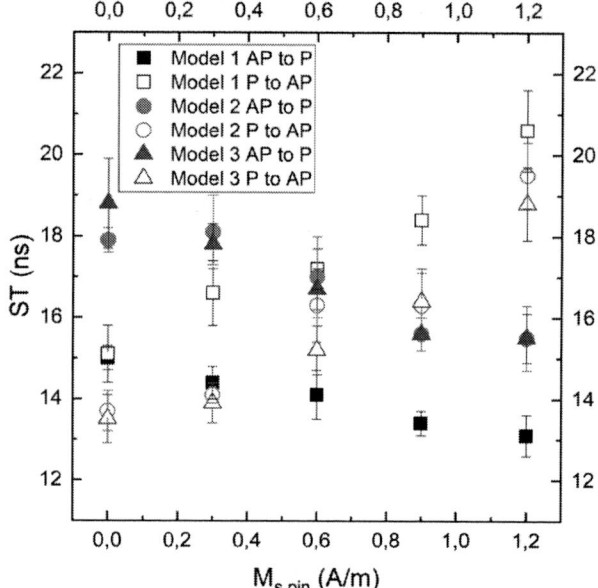

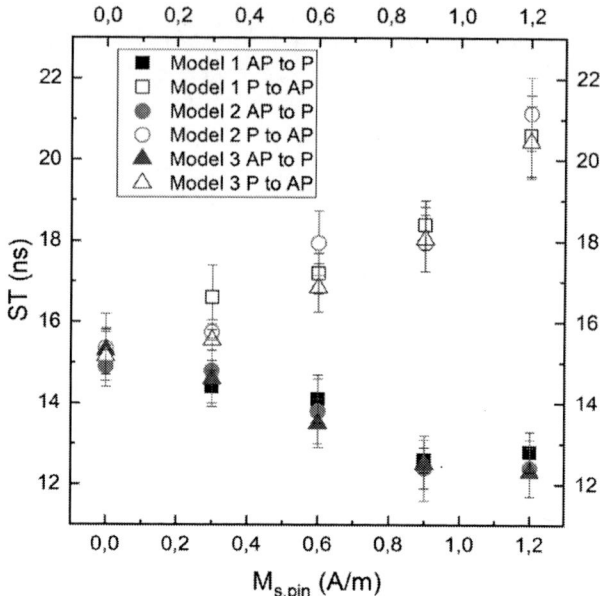

Fig. 4. Comparison between AP→P and P→AP switching for various levels of the uncompensated stray field. Filled symbols represent P→AP switching, empty ones AP→P. The bars show ST variations due to thermal fluctuations for 30 realizations.

Fig. 5. Comparison of switching times for the tuned values of input currents for models 2 and 3. The switching times of all 3 models are compatible within the thermal spread.

discrepancy is attributed to the fact that in Model 1 the *voltage* is fixed, and the total current is allowed to vary according to the MTJ resistance, while Model 2 and 3 are at a fixed total *current* during the whole process.

In order to compensate the effect of the varying resistance on the switching time, we adjust the currents in Model 2 and 3 to make them ~10% higher for AP→P and ~5% lower for P→AP switching. The resulting STs are shown in Fig. 5: with this tuned choice of currents, all the models produce compatible results within the thermal spread. The proposed current adjustment, albeit providing a feasible way of reproducing Model 1 STs distribution using the simplified description offered by (2), depends on the system parameters, especially on the TMR ratio, the dimensions of the stack, and the voltage. Further analysis and simulations are required in order to quantitatively determine the current correction needed to employ the simplified and therefore more computationally efficient model (2) with constant current density.

V. CONCLUSION

We presented a numerical implementation of the coupled spin and charge drift-diffusion equations for the simulation of magnetization dynamics in a p-MTJ based, high TMR STT-MRAM structure. We compared the results for switching time distribution obtained within the application relevant condition of fixed voltage across the structure to two simplified models of switching under fixed current and fixed current density constraints. The switching times obtained by the three models were compared in the presence of different values of the uncompensated stray magnetic field acting on the free layer. We showed that a TMR and voltage-dependent adjustment of the fixed current values is required for the approximated models to correctly reproduce the switching time distribution in the case of high TMR.

ACKNOWLEDGMENT

This work was supported by the Austrian Federal Ministry for Digital and Economic Affairs and the National Foundation for Research, Technology and Development.

REFERENCES

[1] D. Apalkov, B. Dieny, J. M. Slaughter, "Magnetoresistive Random Access Memory", Proc. IEEE 104, 1796 (2016).

[2] O. Golonzka, J. G. Alzate, U. Arslan, M. Bohr, P. Ba *et al.*, "MRAM as Embedded Non-Volatile Memory Solution for 22FFL FinFET Technology", Proc. IEDM 2018, 18.1.1 (2018).

[3] S. Bhatti, R. Sbiaa, A. Hirohata, H. Ohno, S. Fukami *et al.*, "Spintronics Based Random Access Memory: a Review", Mat. Today 20, 530 (2017).

[4] S. Ikeda, J. Hayakawa, Y. Ashizawa, Y. M. Lee, K. Miura *et al.*, "Tunnel Magnetoresistance of 604% at 300 K by Suppression of Ta Diffusion in CoFeB/MgO/CoFeB Pseudo-Spin-Valves Annealed at High Temperature", Appl. Phys. Lett. 93, 082508 (2008).

[5] J.C. Slonczewski, "Current-Driven Excitation of Magnetic Multilayers", J. Magn. Magn. Mat. 159, L1 (1996).

[6] L. Berger, "Emission of Spin Waves by a Magnetic Multilayer Traversed by a Current", Phys. Rev. B 54, 9353 (1996).

[7] A. Makarov, "Modeling of Emerging Resistive Switching Based Memory Cells", Ph.D. thesis, Institute for Microelectronics, TU Wien, Vienna (2014).

[8] J. Slonczewski, "Currents, Torques, and Polarization Factors in Magnetic Tunnel Junctions", Phys. Rev. B 71, 024411 (2005).

[9] A. Makarov, T. Windbacher, V. Sverdlov, S. Selberherr, "CMOS-Compatible Spintronic Devices: a Review", Semic. Sci. Tec. 31, 113006 (2016).

[10] D. Aurelio, L. Torres, G. Finocchio, "Magnetization Switching Driven by Spin-Transfer-Torque in High-TMR Magnetic Tunnel Junctions", J. Magn. Magn. Mat. 321, 3913 (2009).

[11] S. Lepadatu, "Unified Treatment of Spin Torques using a Coupled Magnetisation Dynamics and Three-Dimensional Spin Current Solver", Sci. Rep. 7, 12937 (2017).

[12] C. Abert, M. Ruggeri, F. Bruckner, C. Vogler, G. Hrkac *et al.*, "A Three-Dimensional Spin-Diffusion Model for Micromagnetics", Sci. Rep. 5, 14855 (2015).

[13] S. Ikeda, K. Miura, H. Yamamoto, K. Mizunuma, H. D. Gan *et al.*, "A Perpendicular-Anisotropy CoFeB-MgO Magnetic Tunnel Junction", Nat. Mater. 9, 721 (2010).

3D TCAD Model for Poly-Si Channel Current and Variability in Vertical NAND Flash Memory

D. Verreck*, A. Arreghini*, F. Schanovsky[†], Z. Stanojević[†], K. Steiner[†], F. Mitterbauer[†], M. Karner[†],
G. Van den bosch*, A. Furnémont*,
*imec, 3001 Leuven, Belgium, email: devin.verreck@imec.be
[†] Global TCAD Solutions GmbH., 1010 Vienna, Austria

Abstract—The polycrystalline nature of state-of-the-art 3D NAND flash channels complicates on-current and variability modeling. We have therefore developed a 3D TCAD model that captures percolating current behavior and the resulting variability, and implemented it into the Global TCAD Solutions software package. In our simulation flow, we model the channel transport through the randomly generated grain structure with thermionic emission modulated by discrete traps at the grain boundaries, combined with a crystal orientation dependent mobility model inside the grains. We show that this approach can reproduce experimentally observed on-current temperature dependence and variability and use it to investigate the influence of defect density levels and average grain size.

I. INTRODUCTION

Flash memory scaling has turned vertical, with devices stacked onto each other in strings with a cylindrical channel geometry to enhance bit density and improve memory operation [1]. In such a structure, the channel is created through crystallization of as-deposited amorphous silicon. Numerous grains with different crystal orientations are thereby formed, as the crystallization initiates at various random nucleation sites throughout the structure. The current conduction through such a polycrystalline channel leads to particular experimental observations that require a different modeling approach from the monocrystalline case to explain. First, the cell on-current (I_{ON}) is strongly reduced, as it is hampered by the increased scattering in the grain boundary regions that results from the change in crystal orientation [2]. Second, unwanted threshold voltage fluctuations are observed due to the trapping of charge carriers in the highly defective regions at the grain boundaries and at the oxide interfaces [3]. Third, inter-device variability in I_{ON} is significantly increased because of the random nature of the grain structure and the location of the defects. The degree of variability is strongly determined by the percolating nature of the current flow, which results from an interplay between the grain and defect configurations [4]. And finally, a positive temperature dependence of the current is typically observed, which is a sign of conduction by thermionic emission [5], [6].

To gain understanding of these intricate experimental observations, we present a 3D TCAD model implemented in the Global TCAD Solutions (GTS) package. We first detail the model and then show its use in reproducing I_{ON} temperature (T) dependence and variability of experimental macaroni channel NAND devices.

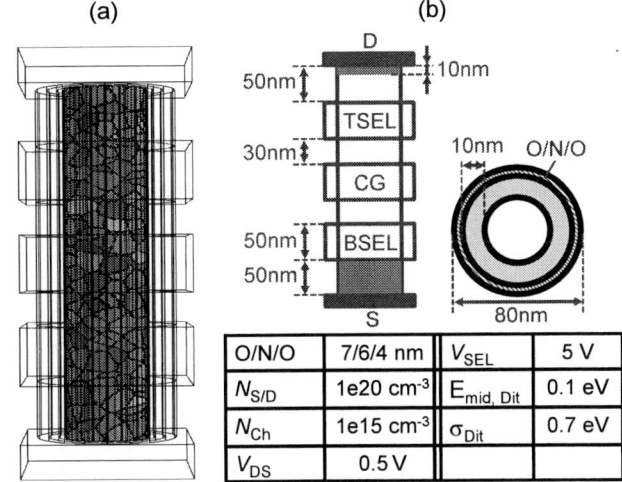

Fig. 1: Investigated macaroni channel 3D NAND device structure. (a) Grain structure in the channel for average L_{grain} of 12 nm. The gates and other layers have been rendered transparent. (b) Configuration details and parameters. The red areas indicate the doping extensions at source and drain. V_{SEL} is applied at both BSEL and TSEL and is chosen large enough to ensure the SEL transistors are in the linear regime.

O/N/O	7/6/4 nm	V_{SEL}	5 V
$N_{S/D}$	1e20 cm^{-3}	$E_{mid, Dit}$	0.1 eV
N_{Ch}	1e15 cm^{-3}	σ_{Dit}	0.7 eV
V_{DS}	0.5 V		

II. MODELS AND SIMULATION FLOW

The first step in the simulation flow is the definition of the device structure, followed by the Voronoi-based growth of grains in the channel region from randomly placed nucleation sites (see Fig. 1) [7]. The number of sites is determined by the specified average grain size (L_{grain}). Besides a random location and shape, each grain is also assigned a random crystal orientation. Where the grains meet, grain boundaries are defined as 2D interfaces to prevent a mesh dependence that occurs for boundaries with a finite volume.

At the grain boundaries, we model two main physical effects, based on previous ab-initio research [2]: discrete trapping and carrier velocity reduction. For the former, acceptor-type discrete traps are placed randomly at the boundary mesh points. The traps interact with the charge carriers through a Shockley-Read-Hall (SRH) process, which governs their occu-

978-1-7281-0941-1/19 $31.00 © 2019 IEEE

pancy. Once charged, they form local potential barriers that are key to reproducing the experimental temperature behavior. The random nature of these local trap potentials also renders the channel current non-uniform, with the charge carriers flowing in percolation paths. The amount of traps that are placed is determined by the specified trap density ($N_{t,GB}$), which serves as a calibration parameter. The traps are also placed at the channel-oxide interfaces ($N_{t,ox}$). The second effect that is modeled at the grain boundaries, carrier velocity reduction, results from increased carrier scattering due to the breaking of crystal periodicity in the grain boundary and is captured with a thermionic emission model. The current through a boundary between two grains (labeled with subscripts 1 and 2) is expressed as [8]:

$$J_{n1} = J_{n2} = q\left(v_b n_2 - v_b n_1 \exp\left(-\frac{\Delta E_b}{k_B T}\right)\right) \quad (1)$$

with q the elementary charge, v_b the barrier velocity, n the electron density, ΔE_b the uniform energy barrier height, k_B the Boltzmann constant and T the temperature. v_b acts as a collection velocity of carriers from one grain to the other [9]:

$$v_b = \alpha v_{th} = \alpha\sqrt{\frac{2k_B T}{\pi m^*}} \quad (2)$$

with v_{th} the thermal velocity, m^* the electron effective mass and α the scattering factor. Through α, v_b can be reduced to represent increased scattering.

Inside the grains, carrier transport is governed by standard drift-diffusion continuity equations, but with a crystal orientation dependent mobility. The random orientation means the oxide interfaces and the local electric field break the band structure symmetry differently in each grain [10]. We therefore extract the mobility at each mesh point from a table, based on the local crystal orientation relative to the nearest oxide interface (see Fig. 2). This table is calculated separately on a 2D slice of the device. On the slice, an effective mass Poisson-Schrödinger system is solved self-consistently, after which the obtained carrier subband wave functions and energy levels are fed into a Kubo-Greenwood based formula [11]. Combined with relaxation times from various scattering mechanisms (listed in Fig. 2), this module calculates a mobility value. A table of such mobilities is constructed by repeating this procedure for a representative sample of crystal orientations and for each V_{CG}. Fig. 3 shows the full simulation flow.

III. SIMULATION RESULTS AND CALIBRATION

We first confirm in Fig. 4 that the implemented model qualitatively reproduces the expected T dependence of I_{ON} for a single device (see details in Fig. 1(b)) with varying L_{grain} and $N_{t,GB}$. The spread with T in I_{ON} indeed increases and becomes more positive for smaller grains and larger $N_{t,GB}$, while I_{ON} itself decreases. This is a result of the increased presence of trap-induced energy barriers in the current paths, which hinder carrier transport and require thermal energy to overcome. The uniform barrier height at the grain boundaries

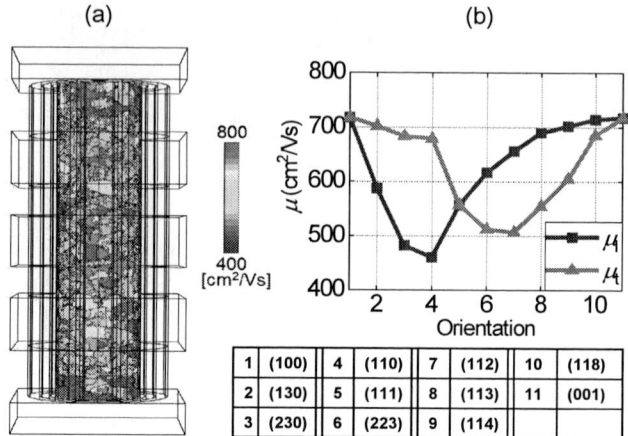

Fig. 2: (a) Calculated intrinsic non-uniform mobility in the macaroni channel and (b) longitudinal (μ_l) and transversal (μ_t) mobility variation with the surface orientations listed in the table. Considered scattering mechanisms are optical and acoustic phonon scattering, inter-valley and surface roughness scattering [11]. V_{CG} is 1 V. Other configuration details are listed in Fig. 1.

1	(100)	4	(110)	7	(112)	10	(118)
2	(130)	5	(111)	8	(113)	11	(001)
3	(230)	6	(223)	9	(114)		

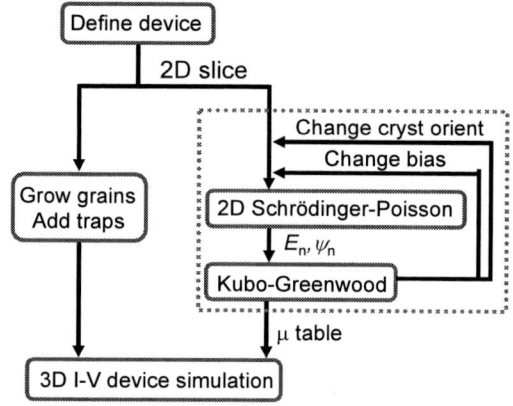

Fig. 3: Flowchart of the simulation procedure implemented in the GTS software package. E_n and ψ_n are energy eigenvalues and state wave functions respectively, obtained from the converged self-consistent Schrödinger-Poisson loop.

ΔE_b is set to zero, which means the presence of traps alone suffices to explain the positive T dependence.

We see the same trends in the I_{ON} distributions in Fig. 5. Here, $N_{t,ox}$ and $N_{t,GB}$ are varied independently for 10 random seeds of the trap and grain configurations. The average L_{grain} is assumed to be close to the channel thickness and is therefore fixed at 12 nm [4]. Both increasing oxide and grain boundary trap density are shown to invert the T dependence from negative to increasingly positive, while I_{ON} is decreased. At the same time, the distributions grow wider, pointing to stronger inter-device variability due to enhanced current percolation. Increasing $N_{t,GB}$ has a stronger effect than $N_{t,ox}$: as the grain

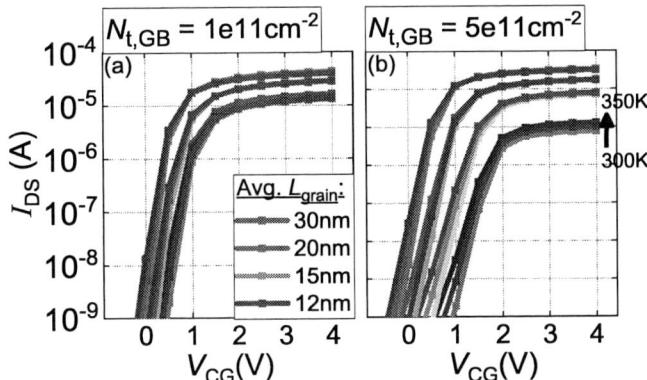

Fig. 4: Simulated transfer characteristics for a single macaroni device for varying grain size and temperature (300K, 325K, 350K) and two values for $N_{t,GB}$. $N_{t,ox}$ is fixed at 1e12 cm^{-2}. Other configuration details are listed in Fig. 1(b).

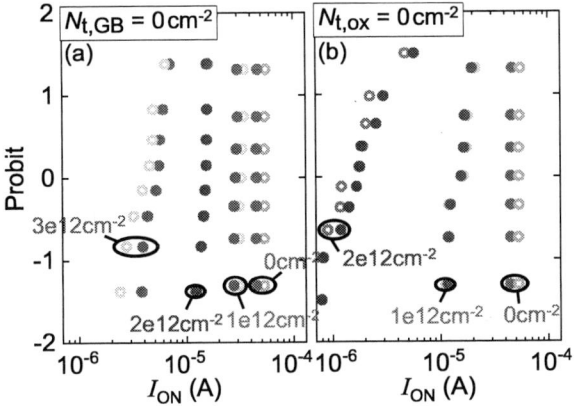

Fig. 5: Simulated I_{ON} distributions for random grain and trap configurations, extracted at $V_{CG}=V_T + 2$ V with V_T at I_{OFF}=1e-9 A/μm, for varying (a) $N_{t,ox}$ and (b) $N_{t,GB}$. Average L_{grain} is 12 nm. Open symbols correspond to a temperature of 300 K, closed to 350 K. Other configuration details are listed in Fig. 1(b).

boundaries run through the full thickness of the channel, traps at those boundaries are located more directly in the current path of the charge carriers.

Finally, we calibrate the model to experimental I_{ON} distributions in Fig. 6. The measured devices are three-gate test vehicles fabricated on a 300 mm platform with a process flow inspired by Bit-Cost Scalable technology [12], [13]. The target dimensions and doping levels are the same as those listed for the simulated configurations in Fig. 1(b). After processing, the experimental devices have been treated with a high pressure hydrogen anneal, which is known to improve performance in terms of I_{ON} and STS by curing defects [14]. Again, we assume that the grain size (L_{grain}) is of the same order as the channel thickness (12 nm). Fig. 6 shows a good agreement between experimental and simulated

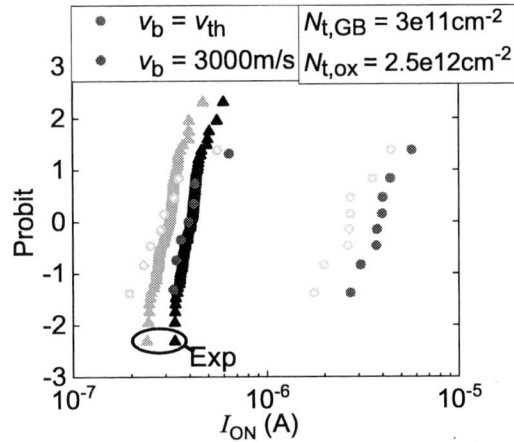

Fig. 6: Calibration of simulated I_{ON} distributions and temperature dependence to experimental results. Average L_{grain} is 12 nm. Open symbols correspond to a temperature of 300 K, closed to 350 K. Other configuration details are listed in Fig. 1(b).

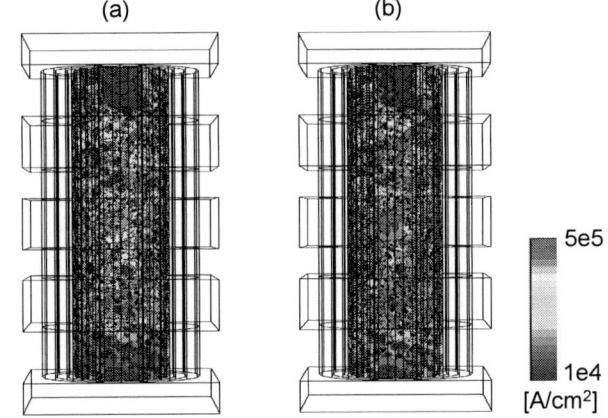

Fig. 7: Electron current density in the channel for a T of (a) 300 K and (b) 350 K. V_{CG} is 4 V. Average L_{grain} is 12 nm. $N_{t,ox}$ and $N_{t,GB}$ have the calibrated values of 2.5e12 cm^{-2} and 3e11 cm^{-2}, respectively. Other configuration details are listed in Fig. 1.

distributions. The temperature dependence and distribution spread is matched by adjusting $N_{t,ox}$ and $N_{t,GB}$, while the I_{ON} magnitude is captured by lowering the scattering factor α in Eq.(1). Fig. 7 illustrates the current density for one of these simulated configurations at the measurement temperatures: the current percolation caused by the grain boundaries and traps is clearly visible. As the temperature is increased, the percolation paths grow wider and the flow of current increases as a result of enhanced thermionic emission.

IV. Conclusion

We reported on a polycrystalline channel current model and simulation flow for 3D NAND flash strings that has been implemented in the GTS TCAD package. This model captures the particular conduction in a random grain structure with discrete traps and a reduced carrier velocity at grain boundaries, combined with an orientation dependent intra-grain mobility. With this approach, we showed that an increased grain boundary trap density and a decreased grain size both not only deteriorate I_{ON}, but also render the I_{ON} temperature dependence more positive. Looking at statistical distributions, we showed a larger effect of grain boundary traps on the distribution spread than oxide interface traps. Finally, we reproduced experimental temperature behavior and statistical distribution of I_{ON} through a calibration of trap densities and carrier velocity at the grain boundaries. For future research, the presented simulation flow can be a valuable tool in the investigation of current conduction mechanisms and variability in scaled 3D NAND devices.

Acknowledgments

This work was supported by imec's Industrial Affiliation Program for advanced flash memory. The authors acknowledge R.Degraeve and B.Lee for useful discussions.

References

[1] S.-H. Lee, "Technology scaling challenges and opportunities of memory devices," in *2016 IEEE International Electron Devices Meeting (IEDM)*, Dec 2016, pp. 1.1.1–1.1.8.

[2] R. Degraeve, S. Clima, V. Putcha, B. Kaczer, P. Roussel, D. Linten, G. Groeseneken, A. Arreghini, M. Karner, C. Kernstock, Z. Stanojevic, G. Van den bosch, J. Van Houdt, A. Furnemont, and A. Thean, "Statistical poly-Si grain boundary model with discrete charging defects and its 2D and 3D implementation for vertical 3D NAND channels," in *2015 IEEE International Electron Devices Meeting (IEDM)*, Dec 2015, pp. 5.6.1–5.6.4.

[3] G. Nicosia, A. Mannara, D. Resnati, G. M. Paolucci, P. Tessariol, A. S. Spinelli, A. L. Lacaita, A. Goda, and C. Monzio Compagnoni, "Characterization and Modeling of Temperature Effects in 3-D NAND Flash Arrays—Part II: Random Telegraph Noise," *IEEE Transactions on Electron Devices*, vol. 65, no. 8, pp. 3207–3213, Aug 2018.

[4] R. Degraeve, M. Toledano-Luque, A. Arreghini, B. Tang, E. Capogreco, J. Lisoni, P. Roussel, B. Kaczer, G. Van Den Bosch, G. Groeseneken, and J. Van Houdt, "Characterizing grain size and defect energy distribution in vertical SONOS poly-Si channels by means of a resistive network model," in *Technical Digest - International Electron Devices Meeting, IEDM*, 2013.

[5] D. Resnati, A. Mannara, G. Nicosia, G. M. Paolucci, P. Tessariol, A. S. Spinelli, A. L. Lacaita, and C. Monzio Compagnoni, "Characterization and Modeling of Temperature Effects in 3-D NAND Flash Arrays Part I: Polysilicon-Induced Variability," *IEEE Transactions on Electron Devices*, vol. 65, no. 8, pp. 3199–3206, Aug 2018.

[6] A. Subirats, A. Arreghini, E. Capogreco, R. Delhougne, C. . Tan, A. Hikavyy, L. Breuil, R. Degraeve, V. Putcha, G. Van den bosch, D. Linten, and A. Furnémont, "Experimental and theoretical verification of channel conductivity degradation due to grain boundaries and defects in 3D NAND," in *2017 IEEE International Electron Devices Meeting (IEDM)*, Dec 2017, pp. 21.2.1–21.2.4.

[7] C. Yang and P. Su, "Simulation and Investigation of Random Grain-Boundary-Induced Variabilities for Stackable NAND Flash Using 3-D Voronoi Grain Patterns," *IEEE Transactions on Electron Devices*, vol. 61, no. 4, pp. 1211–1214, April 2014.

[8] T. Simlinger, *Simulation von Heterostruktur-Feldeffekttransistoren.* PhD Thesis, TU Wien, 1996.

[9] C. Crowell and M. Beguwala, "Recombination velocity effects on current diffusion and imref in schottky barriers," *Solid-state electronics*, vol. 14, no. 11, pp. 1149–1157, 1971.

[10] T. Satô, Y. Takeishi, H. Hara, and Y. Okamoto, "Mobility anisotropy of electrons in inversion layers on oxidized silicon surfaces," *Phys. Rev. B*, vol. 4, pp. 1950–1960, Sep 1971.

[11] Z. Stanojević, O. Baumgartner, L. Filipović, H. Kosina, M. Karner, C. Kernstock, and P. Prause, "Consistent low-field mobility modeling for advanced MOS devices," *Solid-State Electronics*, vol. 112, pp. 37 – 45, 2015.

[12] A. Arreghini, K. Banerjee, D. Verreck, S. V. Palayam, E. Rosseel, L. Nyns, G. Van den bosch, and A. Furnémont, "Improvement of conduction in 3-D NAND memory devices by channel and junction optimization," in *2019 IEEE 11th International Memory Workshop (IMW)*, May 2019, pp. 1–4.

[13] A. Subirats, A. Arreghini, L. Breuil, R. Degraeve, G. Van den bosch, D. Linten, and A. Furnemont, "Impact of discrete trapping in high pressure deuterium annealed and doped poly-Si channel 3D NAND macaroni," in *2017 IEEE International Reliability Physics Symposium (IRPS)*, April 2017, pp. 5A–2.1–5A–2.6.

[14] L. Breuil, J. G. Lisoni, R. Delhougne, C. L. Tan, J. Van Houdt, G. Van den bosch, and A. Furnemont, "Improvement of Poly-Si Channel Vertical Charge Trapping NAND Devices Characteristics by High Pressure D2/H2 Annealing." in *2016 IEEE 8th International Memory Workshop (IMW)*, May 2016, pp. 1–4.

Optimization of select gate transistor in advanced 3D NAND memory cell

Jin Cho, Derek Kimpton, Eric Guichard
Silvaco, Inc, Santa Clara, CA, USA
jin.cho@silvaco.com

Abstract—**There are several device challenges unique to the select gate transistor in 3D NAND memory cell. It requires low leakage current to prevent read and program disturb problem and it needs to provide enough current during read and erase operation. In this paper, we examined the design optimization of select gate transistor with respect to various device elements including work-function, S/D overlap, and trap density. Finally, we reviewed the path to reduce the channel length of the select gate transistor in conjunction with the role of dummy cells.**

Keywords—*3D NAND memory, program/erase operation, program disturb, self-boosting effect*

I. INTRODUCTION

Demand for aggressive bit density scaling of 3D NAND memory device is driving more cells per string as well as more string per block. These multi layers of materials such as oxide and polysilicon introduce manufacturing complexity in various NAND process steps including memory hole, stair step, and slit etch process. It is desirable, therefore, to reduce the layer thickness while increasing the number of memory cells per string. In addition, the number of dummy cells and gate length of select gate device at the end of memory string play a big role on overall stack thickness. In this paper we

studied the role of select gate and dummy cells on memory operation including program disturb and cell operation speed.

II. PROCESS SIMULATION

In this paper, BICS type 3D NAND memory device [1] was examined. The process flow of memory array was simulated as shown in Fig. 1. In this task, special 3D process simulation was used in order to accommodate challenges unique to 3D NAND process flow simulation; process simulation must encompass memory array, interface region and peripheral circuits. As the number of stacks is increased, cell array to circuit interface structure become larger and more complex. From the 3D simulator perspective of view this is a daunting task: simulator need to use fine grid space for memory array simulation e.g. nm resolution, yet it needs to cover 10x10um area. Usually, simulating such a large area with fine resolution hamper the overall performance. In order to overcome this problem, we used Cell-mode 3D simulation. It is a geometry tetrahedral based mesh structure with polygonal algorithm for etch and deposition process. This mode automatically adjust the grid space without presetting of the resolution. Fig. 2 shows the close-up image of memory cell constructed by conventional and Cell-mode. It is apparent that the core cell structure simulated by Cell-mode show superior quality. It is noted that the Cell-mode is capable of performing conventional 3-D process simulations such as implantation and diffusion.

Note that the NAND cell is comprised of polysilicon core with P+ doped poly-Si gate material. The bottom of the memory cell is connected to NWell through Bottom Select Gate (BSG) and top of the memory cell is connected to Bit-line through Top Select Gate (TSG) device.

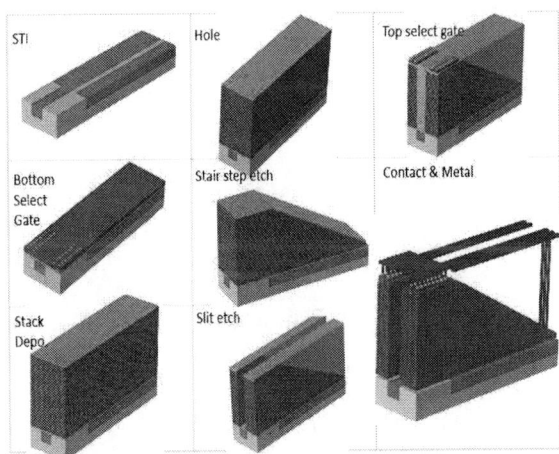

Fig. 1. Process flow of BICS type 3D NAND memory.

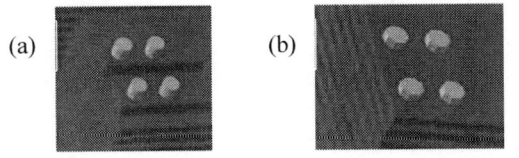

Fig. 2. Close-up image of memory cell region simulated by two different mode in 3D process simulation. (a) Cell-Mode, (b) Process-Mode with 5nm resolution.

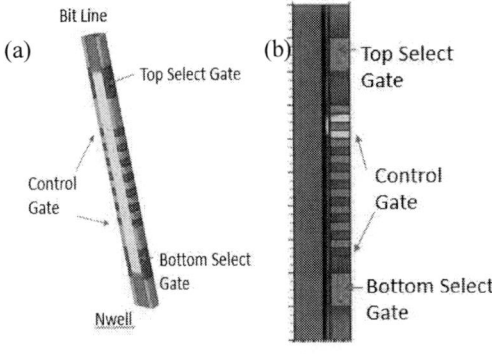

Fig. 3. Structure used in device simulation (a) 3D image (b) 2D cut image of potential after programing of one cell.

978-1-7281-0941-1/19 $31.00 © 2019 IEEE

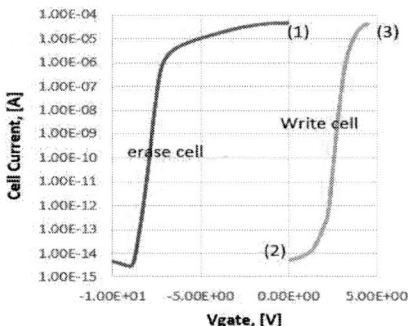

Fig. 4. Cell current with control gate bias for (a) erase cell and (b) programed cell

III. DEVICE SIMULATION

A. Erase/Program/Read simulation

Basic cell operation, program/erase/read, simulation is conducted. The bias condition of cell operation used in this simulation is described in Table 1.

TABLE I.   CELL OPERATION CONDITION

	Read	Program	Erase
Word Line (select)	0	10	0
Word Line (unselect)	4.5	3	0
TSG (select)	4.5	3	Floating
TSG (unselect)	4.5	0	Floating
BSG transistor	4.5	0	Floating
Bit Line (select)	1	0	Floating
Bit Line (unselect)	0	3	Floating
Nwell	0	0	20

For the cell operation simulation, we examined one string of memory structure which is consisted of 7 storage cells and Bottom and Top select transistor as shown in Fig. 3(a). Fig. 4 shows the Bit Line current with respect to the gate voltage of selected cell after (a) erase and (b) write operation, which correspond to Fig. 3(b). For read operation, all unselected cells are biased to 4.5V and TSG/BSG is turned ON. Good erase operation is demonstrated and achieved Bit Line current level of 50uA at Vgate=0V (1). For the simulation of write operation, we first erase entire cells and immediately write one cell. Good blocking state is obtained as the Bit line current of string is less than 1e-12A at Vgate=0V (2). It is also noted that we achieved good pass current level of 40uA at gate bias of 4.5V (3), high enough to establish basic NAND operation.

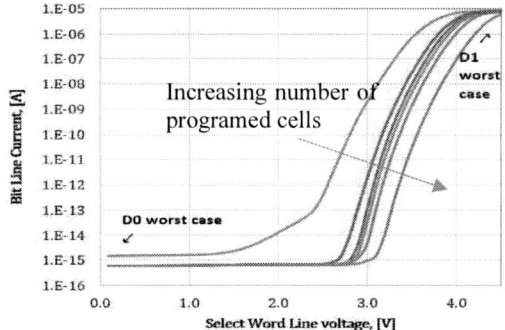

Fig. 5. Bit line current with respect to control gate voltage of single cell, with different number of programmed cells.

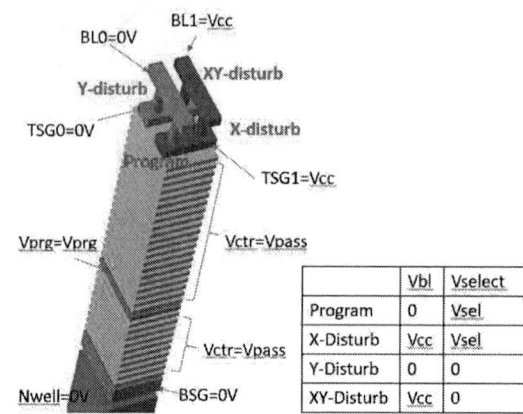

	Vbl	Vselect
Program	0	Vsel
X-Disturb	Vcc	Vsel
Y-Disturb	0	0
XY-Disturb	Vcc	0

Fig. 6. Bias condition for programming operation used in 3D device simulation for 32-layer NAND device.

So far we demonstrated one pattern of erase/program operation consisting of block erase and single cell program and read. It is important to test the worst case or pathological case. Data1 worst case is to read the erased cell when all other cells in the string were programed. Data0 worst case is to access the programed cell when all other cells are in the erase state. Fig. 5 is the read current for both data1 and data0 in various data patterns. For D1/D0 read, the current level is decreased/increased as the number of program cell is increased. Good ON and OFF ratio is achieved considering all data patterns. The worst case of D1/D0 simulation is used in

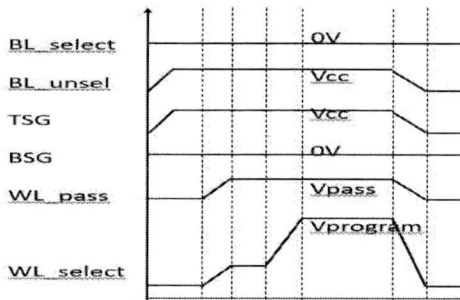

Fig. 7. Timing diagram of program operation

all subsequent simulations shown this paper.

B. Program disturb simulation

The basic cell configuration of 3D NAND memory is that Bit Line is connected to each string and Word Line of each cell is connected at a common point in the block. This shared word line topology creates efficient block erase scheme but generate program disturb problem. Fig. 6 illustrate how programing of one selected cell produces various bias conditions on unselected cell. In order to prevent unwanted programing on unselected cell, all bias conditions applied to unselected cell must fit the condition of minimum electric field between body and control gate. TCAD device simulation of program operation was investigated capturing all four cells.

978-1-7281-0941-1/19 $31.00 © 2019 IEEE   66

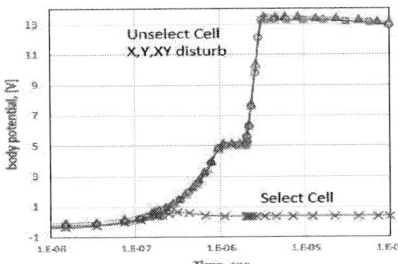

Fig. 8. Time dependent body potential from four cells defined in Fig. 6 with Lg of select gate transistor of 150nm.

This method enables to monitor program cell and all three disturb cells (X, Y, XY) at the same time. The timing diagram used on the program operation is shown in Fig. 7 as in [2].

We monitored the potential at the center of the cell during the program operation, Fig. 8. The body potential of select cell maintains ground and this ground potential is supplied by Bit Line through TSG device, which is in ON state. In the case of unselected cell, there is no path to make cell body ground since either both side of select transistor is OFF (Y, XY-disturb) or Bit Line is not grounded (X-disturb). Therefore the body potential level shoots up at the rising edge of the gate bias: self-boosting effect. The boosted potential will gradually drop and eventually reset to ground by leakage current of two select transistors. In order to avoid program disturb, this boosted body bias need to stay in high level during the program operation. One of the example of demonstrating poor self-boosting effect is when the TSG channel length is small (Fig. 9), which degrade the leakage control. In this case, boosted potential at the body drop fast creating higher e-field between body to control gate causing unwanted programing for the unselected cell. Note that the bias condition for Y-disturb is the worst among three disturb modes: the Vgs(gate to source voltage difference) level is zero, which is less favorable compare to other disturb modes, in which the Vgs level become negative [3,4].

Memory cell disturb behavior was examined in respond to various device parameters of select gate transistor (Fig. 10). In this study, we introduced the "critical potential", body potential at fixed time (10us), as a measure for disturb behavior. Effects of various transistor parameters on disturb margin are investigated, which includes (a) Work Function, (b) Source/Drain overlap, (c) interface charge density, and (d) bulk trap density. It is apparent that the ability to control

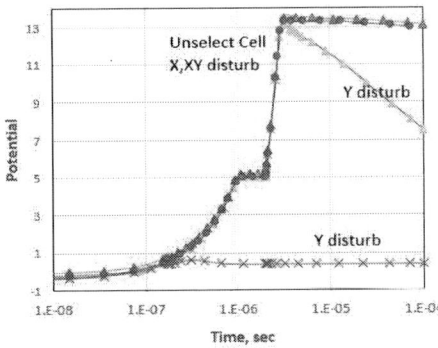

Fig. 9. Time dependent body potential from four cells with Lg of select gate transistor of 100nm.

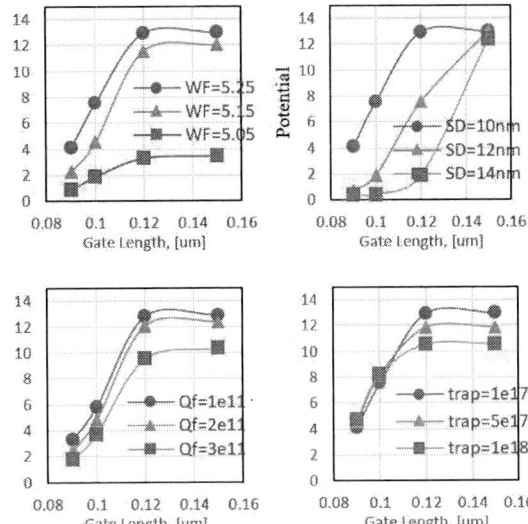

Fig. 10. Critical potential in various device parameters. (a) Workfunction (b) S/D overlap (c) Interface charge (d) Trap density

leakage current, either by increasing Vth (WF, Qf) or improving short channel control (Lg, S/D overlap) is the primary way to prevent the program disturb.

C. Effect of traps

Effect of bulk traps on disturb behavior is investigated Fig. 10(d). The long channel device containing high defect density exhibit poorer disturb margin. Interestingly this trend reversed for the short channel device: higher traps device produces better disturb margin. To understand this behavior, we investigated select gate transistor focusing on trap effects. Fig. 11 depicts the Id_Vg curve of select transistor with various trap densities showing traps pushing Vth higher. We believe this to be due to the extra charges coming from traps and contribute to the charge balance. The Poisson's equation in cylindrical coordinates become [5],

$$\frac{1}{r}\frac{\partial}{\partial r}\left(r\frac{\partial}{\partial r}\Psi(r)\right) = \frac{q}{\varepsilon_{si}}(N_a + N_T\ \Psi(r))$$

In our simulation traps are assumed to be of acceptor type, i.e., positively charged when occupied and neutral when not occupied. This equation implies that traps (N_T) are contributing to effective charge like dopant and subsequently

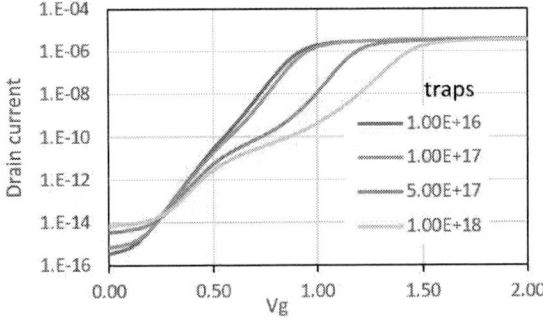

Fig. 11. Id_Vg curve of select gate transistor with various trap concentrations.

978-1-7281-0941-1/19 $31.00 © 2019 IEEE   67

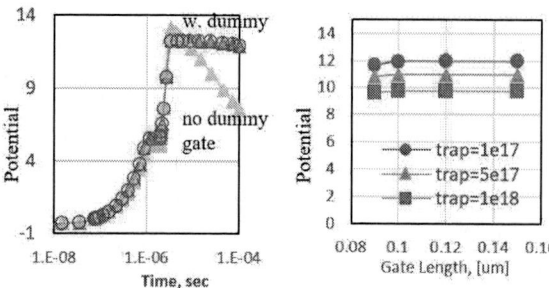

Fig. 12. Time dependent body potential with (a) with/w.o. dummy (b) critical potential in various Lg and traps with dummy cell.

modulating Vth level. It is also observed that the off-state leakage current is increased with traps. Unlike workfunction effect where higher Vth provides lower leakage floor. Effect of traps is different; it increases the threshold voltage and leakage current at the same time. When the channel length is long, the leakage current is the main contributor of the disturb problem whereas for the short channel device, higher Vth attributed by high trap density plays a bigger role.

D. Impact of dummy cell

It is common practice to place dummy cells right next to both select gate transistors to alleviate any edge cell adverse effect. We investigated the role of dummy cell on disturb behavior. Fig. 12(a) shows the Y-disturb potential plot with/without dummy devices. The presence of dummy cell improves self-boosting effect significantly. Fig. 12(b) is the same plot of Fig. 10(d) except that it contains dummy cell. The disturb margin is improved with dummy cell across all gate lengths.

Fig. 13(a) shows the conduction band energy diagram at the center of the cell. When there is no dummy, the energy barrier was lowered at the select gate device causing in the injection of electrons into the cell body, which reduces the boosted potential. Fig. 13(b) illustrate the 3D image of potential at programed cell showing reduction of potential at the center of the cell, where the influence of control gate is minimal.

E. Impact of GIDL on erase operation

For erase operation, it requires high positive bias at the core of the cell, which produces high electric field between

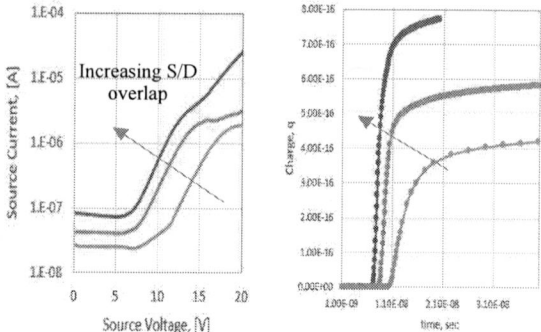

Fig. 14. (a) Source voltage-current curve at various s/d overlap. (b) Charge storage level during erase operation, with various s/d overlap. (S/D overlap: 60,80,100nm)

cell body to control gate enabling charge storage. This body bias is supplied from Nwell in the form of block erase scheme. Since the junction is revers biased, it takes time to raise the body potential. The conventional reverse junction leakage current is too small to raise the potential in time. Therefore, it requires GIDL current [6].

The GIDL current effect on erase operation is investigated. Fig. 14(a) shows the simulation result of source current for the device with various S/D junction overlap. The results show that current level is increased with larger junction overlap due to higher GIDL current. The rate of charge storage in the cell, which tracks the speed of the erase operation, is improved with increasing S/D overlap, Fig. 14(b). It is noted that larger junction overlap is adverse effect on disturb problem as observed previously in Fig.10 (b). Adding that the dopant diffusion in poly-Si material is highly variable, it is proper to have the channel length of the BST longer.

IV. CONCLUSION

We investigated the device characteristics of select gate transistor on 3D NAND structure. We found that there is a room to shrink the channel length of the Top select gate device providing that there is a dummy cell. We project that Bottom Select Gate device has a limit to shrink the channel length considering program disturb and erase speed.

REFERENCES

[1] H. Tanaka, et al., "Bit Cost Scalable Technology with Punch and Plug Process for Ultra High Density Flash Memory" pp. 14-15, VLSI 2017.

[2] Myounggon Kang and Yoon Kim, "Natural Local Self-Boosting Effect in 3D NAND Flash Memory" IEEE Electron Device Letters, Vol. 38, No. 9, pp. 1236-1239, Sep. 2017.

[3] Eun-Seok Choi and Sung-Kye Park, "Device Considerations for High Density and Highly Reliable 3D NAND Flash Cell in Near Future" pp. 211-214, IEDM 2012.

[4] Keon-Soo Shim, et al., "Inherent Issues and Challenges of Program Disturbance of 3D NAND Flash Cell" IMW 2012.

[5] Hyun-Jin Cho, James Plummer, "Modeling of Surrounding Gate MOSFETs with Bulk Trap States", IEEE Tansactions on Electron Devices, Vol. 54 No. 1, pp 166-169, Jan. 2007.

[6] R. Yosuke Komori, et al., "Disturbless Flash Memory due to High Boost Efficiency on BICS Structure and Optimal Memory Film Stack for Ultra High Density Storage Device" pp. 851-854, IEDM 2008.

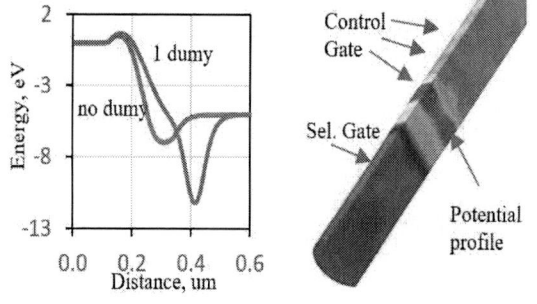

Fig. 13. (a) Conduction band diagram of the device with and w.o. dummy cell. (b) Potential distribution profile at programing operation.

A Hybrid Mode-Space/Real-Space Scheme for DFT+NEGF Device Simulations

F. Ducry[*], M. H. Bani-Hashemian, and M. Luisier

Integrated Systems Laboratory (ETH Zurich), Email: fabian.ducry@iis.ee.ethz.ch

Abstract—Density functional theory based simulation techniques enable thorough investigation of the operational characteristics of nanoscale devices regardless of their configurational complexity. However, this flexibility comes with considerable computational cost. In this work, we present a hybrid mode-space/real-space scheme that utilizes a mode-space basis to represent periodic contacts while maintaining the real-space representation of the central device region. Reducing the size of the contact blocks via mode-space approximation speeds up the calculation of the open boundary conditions and reduces the overall size of the Hamiltonian and overlap matrices, which leads to significant improvements in the computational efficiency of simulations. Keeping the real-space representation of the device blocks preserves the versatility and accuracy of the *ab-initio* approach. The merits of the proposed method are demonstrated with the simulation of an amorphous device with metallic contacts.

I. INTRODUCTION

Following the continuous decrease of the device dimensions, atomistic quantum mechanical simulation approaches have become essential to accurately predict the performance of nanoscale components. Complex material stacks, metallic contacts, and amorphous layers further call for an *ab-initio* treatment of their electronic properties. Coupling density functional theory (DFT) with the Non-equilibrium Green's Function (NEGF) formalism [1,2] meets this requirement. Consequently, DFT+NEGF has established itself as a reference to shed light on the behavior of nano-devices. However, it is typically limited to small atomic systems because of its heavy computational burden. This issue can be addressed by reducing the dimensionality of the real-space (RS) Hamiltonian matrix. Among the techniques proposed to do so, the so-called mode-space (MS) method [3,4] stands out. The MS transformation uses a reduced basis set to accurately reproduce the band structure of a periodic cell within a restricted energy interval only. As such, it does not lend itself to structures made of non-repeatable unit cells, which is true for most realistic components featuring interfaces and amorphous phases. This is the case, for example, of conductive bridging random access memories (CBRAMs), where an amorphous oxide is surrounded by two electrodes, as illustrated in Fig. 1(a) [5].

Here, we propose a hybrid MS/RS scheme that decouples the contact extensions, composed of periodic cells, from the central oxide region with random atomic placement. While the former parts are converted to MS, the latter remains expressed in RS. This approach offers two key advantages over pure RS simulations. By reducing the size of the contact blocks within the Hamiltonian matrix H, we significantly decrease the time

to calculate the open boundary conditions (OBCs) and to solve the NEGF equations. At the same time, the memory required to store H drastically goes down.

Beside this innovation, we also successfully apply an MS transformation to metallic electrodes, fully capturing their band structure and accurately calculating the ballistic current flowing through the resulting device.

II. APPROACH

The MS transformation is used to precisely reproduce the RS band structure within the energy window of interest. This transformation usually introduces unphysical energy states that need to be removed. The contact RS-to-MS transformation matrix U can be obtained and refined using the method of Refs. [3,4] with a few modifications to process large metallic blocks. First, an initial guess for U is created based on the eigenvectors of the contact unit cell. Previous works disregarded interactions beyond nearest neighbors [4], which is acceptable for semiconductors. Metals however, usually feature delocalized electrons resulting in long range connections that cannot be ignored without loss of accuracy. To account for these, the following adaptations were developed. In a system periodic along the x-axis, the E-k relation is given by

$$H_{k_x}\Psi_{k_x} = E(k_x)S_{k_x}\Psi_{k_x}, \tag{1}$$

where H_{k_x} and S_{k_x} are the k_x-dependent Hamiltonian and overlap matrices, respectively, $E(k_x)$ the energy at k_x, and Ψ_{k_x} the corresponding wave function. For unit cells interacting with N_N neighbors along the $\pm x$-direction H_{k_x} is obtained according to

$$H_{k_x} = H_0 + \sum_{n=1}^{N_N} \left(H_n e^{ink_x} + H_n^\dagger e^{-ink_x} \right), \tag{2}$$

where H_n connects one periodic unit cell of width Δ_x at x_i to another one at $x_i + n\Delta_x$, as illustrated in Fig. 1(b). Note that S_{k_x} has the same structure as H_{k_x}. Based on H_{k_x} and S_{k_x} the initial guess is calculated according to [4].

Once U is obtained, it needs to be iteratively refined to eliminate unphysical energy states. The elimination of these energies is achieved by iteratively adding additional normalized basis vectors in the form of $\Xi \cdot C$ to U. Ξ is a trial basis and C the vector that minimizes the following expression:

$$\mathcal{F}(C) = \frac{1}{n_z} \sum_q \sum_z \left[\frac{C^T A(q,z)C}{C^T B(q,z)C}(z-\epsilon_c) \right] + (C^T C - 1)^2, \tag{3}$$

978-1-7281-0941-1/19 $31.00 © 2019 IEEE

where q and z are a set of trial k_x and energies, respectively, A and B are matrices defined as in [4] and depending on U, H_q, S_q, and Ξ. The ability of C to effectively remove unphysical bands strongly depends on the choice of Ξ. Previous works used the commutator $[S_{k_{x1}}^{-1} H_{k_{x1}}, S_{k_{x2}}^{-1} H_{k_{x2}}]$, with $k_{x1} = 0$ and $k_{x2} = \pi$. We found a better performance of the removal procedure using the k_x with the largest number of unphysical energies, in our case $k_{x1,x2} = \pm\pi/5$ such that

$$\Xi = (1 - UU^\dagger)[S_{k=\pi/5}^{-1} H_{k=\pi/5}, S_{k=-\pi/5}^{-1} H_{k=-\pi/5}]U. \quad (4)$$

Finding the global minimum of Eq. (3) is a challenging problem as the function has many local minima. Depending on the initial guess, an optimizer not always finds a vector C that removes an unphysical band from the MS band structure. To remedy this shortcoming, we applied multiple initial guesses and kept the result with the lowest value of $\mathcal{F}(C)$. The rows of $\Re\{U\}$ proved to be the best initial guesses for C.

Convergence of the optimization process was improved using the analytical first derivative:

$$\frac{\partial \mathcal{F}(C)}{\partial C} = \sum_q \sum_z \left[\frac{C^T A(q,z)}{C^T B(q,z)C} - \frac{C^T A(q,z)C}{(C^T B(q,z)C)^2} C^T B(q,z) \right] + 4(C^T C - 1)C, \quad (5)$$

which is a vector of the same length as C. Note that A and B are assumed to be symmetric without loss of generality and the scalar factors of Eq. (3) are absorbed into A.

The transformation of H_n (S_n) from RS to MS requires to project each contact block onto the created MS basis. The RS→MS coupling is achieved by partial, one-sided projection of the RS block onto MS. The MS blocks $\widetilde{H}_n$ and hybrid blocks $\mathcal{H}_n$ are then defined as

$$\widetilde{H}_n = U^\dagger H_n U, \ \mathcal{H}_{n,RM} = U^\dagger H_n \text{ or } \mathcal{H}_{n,MR} = H_n U. \quad (6)$$

The $\mathcal{H}_{n,RM/MR}$ matrices are the hybrid blocks at the RS-MS interface. Eq. (6) is only applied to the contacts, not to the central oxide and the metallic interface layers attached to it.

III. RESULTS

As benchmark example, a CBRAM cell made of 3870 atoms, with two Cu contacts separated by amorphous silicon dioxide (a-SiO$_2$) is considered [6]. The H and S matrices were computed from DFT with CP2K [7] using double-zeta valence (DZVP) basis sets, and their contact extensions converted into MS. In RS each Cu unit cell is made of 240 atoms with 25 orbitals per atom summing up to 6000 orbitals per block. The RS→MS transformation reduced the original block size to 692 (11.5% of the original value), as listed in Fig. 2(a). The success of the transformation is illustrated with the RS and MS contact band structures in Fig. 2(b). The MS band structure is shown both before and after the refinement procedure, in the left and right half respectively. After refinement, the RS and MS results almost perfectly agree with a maximum error of 0.047 meV. Moreover, MS does not exhibit spurious states anymore within the energy interval of interest.

The sparsity pattern of the device Hamiltonian matrices before and after Eq. (6) are plotted in Fig. 3(a-b) for a subset of the CBRAM in Fig. 1(a) (left contact and Cu/a-SiO$_2$ interface). Five Cu contact blocks are converted to MS, while one Cu block at the interface and the a-SiO$_2$ remain in RS. It can be observed that the off-diagonal blocks coupling the MS to RS domains change shape and become thin rectangles.

To validate our approach, H and S were passed to our in-house quantum transport solver [8]. The energy-resolved transmission and low field IV-curve through the investigated cell, calculated with the RS and MS/RS matrices, are reported in Figs. 3(c) and 4(a): the MS/RS transmission overlaps almost perfectly with the RS one, and the IVs agree very well, confirming the strength and validity of the hybrid scheme.

For improved simulation speed, the metal blocks at the interface can also be converted into MS at the cost of accuracy. The relative error of the IVs calculated from hybrid matrices with zero and one RS metal blocks is shown in Fig. 4(b). With a single Cu block in RS representation the current is within 1% of the original value. Transforming all electrode blocks to MS, however, causes an underestimation of the current by up to 20%. It is apparent that one RS block at the interface is required to capture all necessary interactions.

The computational efficiency of the proposed method was also tested on the Piz Daint supercomputer at CSCS by measuring the time to calculate the OBCs and NEGF equations and the total time needed per energy point. Results are presented in Fig. 5. The evaluation of the OBCs on CPUs could be accelerated by a factor of 55, irrespective of the number of RS blocks at the interface. Solving the NEGF system on GPUs is accelerated by a factor 40 (65) for one (zero) RS block, for an overall speed up of 20 (40) per energy point. The reduced memory consumption helped decrease the required number of nodes per energy point from 20 with only RS blocks to 4 with the hybrid approach. Thus, the total cost is lowered by a factor of at least 20 (speed up) x 5 (node reduction) = 100.

As a result of these improvements, at a fixed computational cost, the change in the resistance state of a CBRAM cell can be studied at more steps during the formation of filaments (see Fig. 6), as same MS transformation can be applied to the contact region of each case.

IV. CONCLUSION

We have developed a method that combines real-space and mode-space representations of non-homogeneous device structures, demonstrating both excellent physical accuracy and enhanced computational efficiency. By creating one transformation matrix U, we will be able to more rapidly simulate electron transport through dynamically evolving structures like CBRAM cells with different amorphous oxide layers and nano-filament configurations, but (almost) identical contacts.

ACKNOWLEDGMENT

This work was supported by the Werner Siemens Stiftung, by SNF under Grant No. PP00P2 159314, by ETH Research Grant ETH-35 15-2, and by a grant from the Swiss National Supercomputing Centre (CSCS) under Project s714.

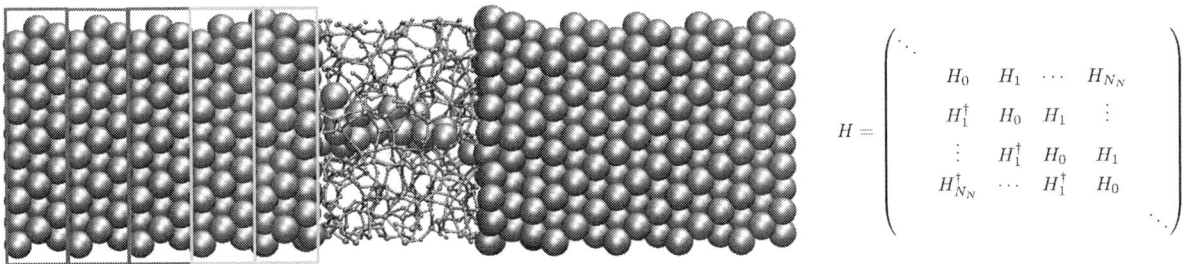

(a) 3D CBRAM Structure

(b) Block N_B-diagonal Pattern of H

Fig. 1. (a) Atomistic Cu/a-SiO$_2$/Cu CBRAM cell structure used in this work to benchmark the proposed hybrid mode-space/real-space approach. It is composed of 3870 atoms. Grey spheres represent Cu atoms (25 orbitals per atom), the orange ones Si and O (both 13 orbitals per atom). The Cu contacts are in fact longer than shown here. The red rectangles mark the contact blocks, while the green ones refer to the interface blocks. (b) Typical block N_B-diagonal pattern of the real-space Hamiltonian matrix H corresponding to the structure in (a). Each block represents a unit cell. Due to long-range interactions, N_B=5 for N_N=2. The mode-space Hamiltonian has the same pattern with much smaller blocks.

RS block size:	6000
Initial guess size:	538
Refinement iterations:	154
MS block size:	692
Size reduction:	88.5%
Energy window:	2 eV
Number of k-points:	128
Max. error band structure:	0.047 meV

(a) Mode-Space Transformation

(b) Cu Contact Band Structure

Fig. 2. (a) Table summarizing the real-space to mode-space transformation parameters. For an energy window of 2 eV around the Fermi level, 128 k-points are necessary to sample the Cu contact band structure and obtain accurate results, leading to a reduction of the block size by 88.5%. The largest error in the MS band structure is well below 10^{-4} eV. (b) Band structure of the Cu contact. Large red dots refer to the real-space. On the left branch the black crosses correspond to the MS band structure before refinement. The right-hand-side shows with small black dots the final mode-space results. The unphysical bands have been removed and all spurious states have been successfully eliminated.

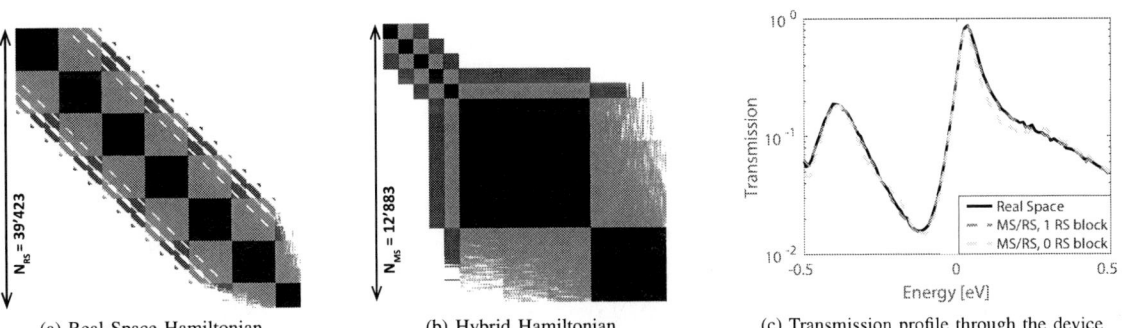

(a) Real-Space Hamiltonian

(b) Hybrid Hamiltonian

(c) Transmission profile through the device

Fig. 3. (a) Sparsity pattern of the real-space, penta-diagonal Hamiltonian matrix (left contact and Cu/a-SiO$_2$ interface). Black blocks correspond to the different Cu contact and a-SiO$_2$ (last block, bottom-right) unit cells, red (blue) blocks to their (second) nearest-neighbor connections. (b) Same as (a) after transformation into mode-space. Note the rectangular shape of the RS→MS coupling blocks and the significant size reduction from $N_{RS} = 39'423$ to $N_{MS} = 12'883$. (c) Transmission function through the structure in Fig. 1(a) as a function of energy around the Fermi energy (0 eV). Real-space (solid black line) and mode-space (dashed lines) are shown. The red (blue) dashed line corresponds to hybrid Hamiltonian with one (zero) RS metal block. While the red and black curve perfectly agree, the blue one is slightly off.

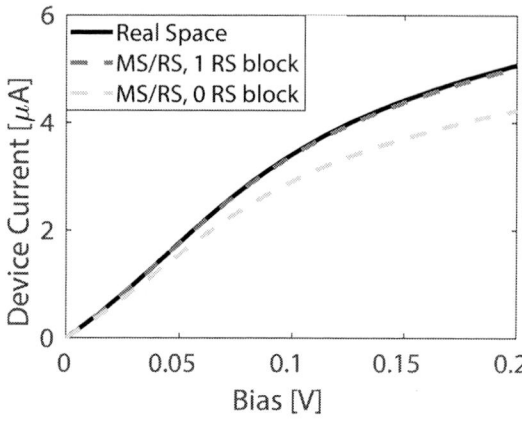
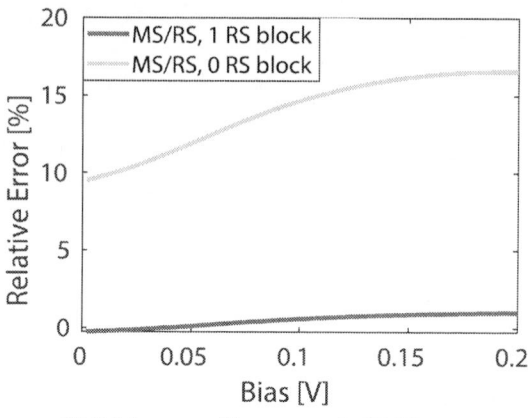

(a) Low-field IV characteristics of the device

(b) Relative error with respect to the RS IV curve

Fig. 4. (a) Low-field IV curves for the same simulations as in Fig. 3(c). The RS and MS curves agree given that at least one interface block remains in RS. (b) Relative error of the MS IVs shown in (a) with respect to RS. With one RS metal block the relative error of the current remains below 1%. With all metal blocks transformed to MS, however, the current is underestimated by almost 20% at a bias of 0.2 V.

	RS	MS/RS Hybrid 1 RS block	Gain vs RS	MS/RS Hybrid 0 RS blocks	Gain vs RS
Matrix size:	78'846	25'766	3.1	15'150	5.2
Non-zero elements:	673.9e6	132.0e6	5.1	47.5e6	14.2
Time OBC [s]:	308.5	5.3	57.8	5.8	53.2
Time linear system [s]:	273.9	6.4	42.5	4.1	66.8
Total time [s]:	336.0	16.5	20.4	8.2	41.0
Nodes:	20	4	5	2	10
Cost [time*nodes]:	6720	66	101.8	16.2	414.8

Fig. 5. Computational benchmark: Table summarizing the numerical problem size and computing times obtained for both the real-space and mode-space Hamiltonians on the Piz Daint supercomputer at CSCS (http://www.cscs.ch). Each node of this machine features 12 Intel Haswell cores (64 GB RAM) and 1 Pascal 100 GPU (16 GB). All benchmarks were run on the same number of cores (20) and GPUs (4), but different number of nodes, 20 for RS and 4 (2) for MS/RS with one (zero) RS blocks. Significant speed up factors of 20 (40) and reductions of memory footprint (using 4 (2) instead of 20 nodes) were achieved by applying the proposed hybrid MS/RS scheme.

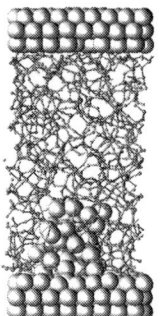
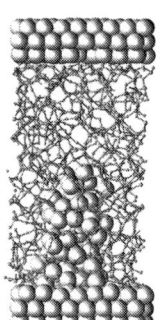
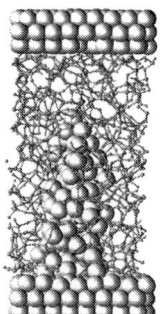
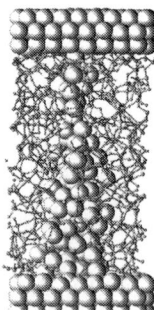

Fig. 6. Different CBRAM cells with the same cross section dimensions, but different nano-filament states. In the left-most structure, the central oxide is almost free of Cu atoms, while the top and bottom contacts are short-circuited by a filament in the right-most one. As the electrodes remain identical for all configurations, the same mode-space transformation can be applied to all of them to reduce their block size.

References: [1] M. Brandbyge et al., *Phys. Rev. B* 65, 165401 (2002). [2] M. V. Fernández-Serra et al., *Nano Lett.* 6, 2674 (2006). [3] G. Mil'nikov et al., *Phys. Rev. B* 85, 035317 (2012). [4] M. Shin et al., *J. Appl. Phys* 119, 154505 (2016). [5] I. Valov et al., *Nanotechnology* 22, 254003 (2011). [6] F. Ducry et al., in *Proceedings of the 2017 IEEE IEDM*, 4.2.1 (2017). [7] J. Hutter et al., *Comp. Molec. Sci.* 4, 15 (2014). [8] M. Luisier et al., *Phys. Rev. B* 74, 205323 (2006).

Full band quantum transport modelling with EP and NEGF methods: application to nanowire transistors

M. Pala
C2N, Université Paris-Saclay, Palaiseau, France
Email: marco.pala@c2n.upsaclay.fr

D. Esseni
DPIA, University of Udine, Udine, Italy
Email: david.esseni@uniud.it

I. INTRODUCTION

The active region of many modern electron devices consists of semiconductors structured at truly nanometric dimensions, either as ultra-thin-body FETs (UTB-FETs), or as 3D architectures such as Fin-FETs, multi-gate FETs (MuGFETs), and nanowire (NW) FETs [1]. Quantum mechanical effects have thus become prominent not only in terms of subband splitting [2], but also in terms of source-drain tunnelling in CMOS FETs [3], [4], [5], and band-to-band-tunnelling (BTBT) in Tunnel FETs (TFETs) [6], [7]. The relevance of quantum effects in nanoscale FETs is also witnessed by the fact that CMOS based quantum dots have been proposed as a platform for quantum computing [8].

The empirical tight-binding (TB) method is the most mature method for full-band quantum transport simulations based on the non-equilibrium Green's function (NEGF) formalism [9], [10], however an approach based on an Empirical Pseudopotentials (EP) Hamiltonian and a plane-waves basis has recently raised substantial interest, with contributions reported for carbon nanotubes [11], ultra-thin-body FETs [12], [13], [14], [15], and graphene nanoribbon transistors [16], [17].

We have recently reported improved methods for full band, EP based NEGF simulations of UTB-FETs [18]. In this paper we first extend the methodology to nanowires and then present complete, self-consistent simulations for nanowire Tunnel FETs with a few nanometer diameter.

II. QUANTUM CONFINEMENT AND TRANSPORT MODELLING

The formulation of the EP method for a bulk semiconductor has been discussed by many authors [19]. We here focus on the device structure shown in Fig. 1, namely a gate-all-around (GAA) InAs NW Tunnel-FET, and our goal is to express quantum confinement as a local operator in real space, because our previous non local formulation set a lower limit to the size of the blocks of the Hamiltonian matrix [15]. To this purpose we introduce a pseudopotential model Hamiltonian for a pseudo-oxide region, whose only purpose is to set a conduction and valence band discontinuity with respect to the semiconductor that effectively confines electrons in the semiconductor region.

To this purpose we let $V_{sc}(\mathbf{r})$ and $V_{ox}(\mathbf{r})$ denote the pseudopotentials describing respectively the actual semiconductor

and the pseudo-oxide, and then define the overall pseudopotential $V_{1D}(\mathbf{r})$ for the 1D eletron gas in the NW as

$$V_{1D}(\mathbf{r}) = V_{sc}(\mathbf{r}) + V_{cnf}(\mathbf{r})\,\theta_{1D}(y,z)\,, \qquad (1)$$

where $V_{cnf}(\mathbf{r})$ is defined as $V_{cnf}(\mathbf{r}) = [V_{ox}(\mathbf{r}) - V_{sc}(\mathbf{r})]$, and $\theta_{1D}(y,z)$ is a box function that is 1 in the oxide region and 0 in the semiconductor.

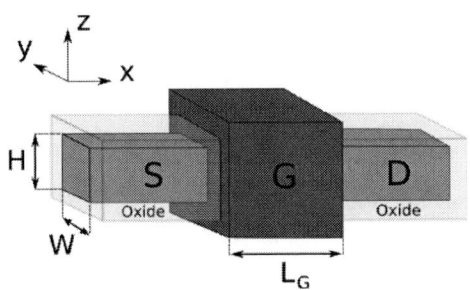

Fig. 1: Sketch of the simulated gate-all-around nanowire FET, where x is the transport direction and (y,z) the plane of quantum confinement.

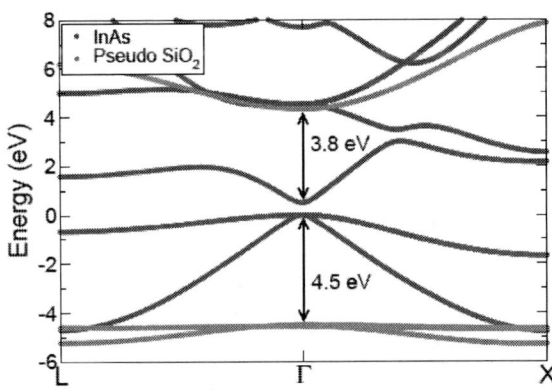

Fig. 2: Bandstructure of the pseudo-SiO$_2$ (red) compared to that of bulk InAs (blue). The corresponding material parameters for the EP model are reported in Tab. I.

We here assumed that the pseudo-oxide has the same lattice constant a_0 as the semiconductor (hence also same reciprocal lattice), and adjusted its EP parameters in order to

978-1-7281-0941-1/19 $31.00 © 2019 IEEE

	U_{S3}	U_{S8}	U_{S11}	U_{A3}	U_{A4}	U_{A11}
ψ-SiO$_2$	-0.64	0.0	0.14	0.225	0.14	0.08
InAs	-0.220	0.0	0.050	0.080	0.050	0.030

TABLE I: EP parameters (Ry) for InAs and for the pseudo-SiO$_2$.

obtain the desired values for the conduction and valence band discontinuity with respect to the semiconductor. Fig. 2 reports the energy dispersion of InAs (in blue) and of a pseudo-SiO$_2$ material (in red), showing that the pseudo-SiO$_2$ has a direct bandgap of about 9 eV, and that the conduction and valence band discontinuity with InAs have the desired values of about 3.8 eV and 4.5 eV [2].

An important feature of the $V_{1D}(\mathbf{r})$ defined in Eq. (1) is that it is by definition local in real space, and its $\mathbf{K}$ space representation can be obtained recalling that the real space product $V_{cnf}(\mathbf{r})\theta_{1D}(y,z)$ in Eq. (1) transforms into a convolution in reciprocal space. We here employ standard notation for wave-vectors $\mathbf{K}=[(k_x, k_{yz})+\mathbf{G}]$, $\mathbf{K}'=[(k_x, k'_{yz})+\mathbf{G}']$, where $\mathbf{k}=(k_x, k_{yz})$ belongs to the reduced zone (same for semiconductor and pseudo-oxide), and $\mathbf{G}=(G_x, G_{yz})$ is a reciprocal lattice vector [19]. In such notation $V_{1D}(\mathbf{K}-\mathbf{K}')$ is given by

$$V_{sc}(\mathbf{G}-\mathbf{G}')\delta_{\mathbf{k},\mathbf{k}'} + \sum_{\mathbf{G}''_{yz}} V_{cnf}(G_x - G'_x, G_{yz} - G'_{yz} - G''_{yz})$$
$$\times \theta_{1D}(\mathbf{K}_{yz} - \mathbf{K}'_{yz} + \mathbf{G}''_{yz})\delta_{k_x, k'_x}. \quad (2)$$

The energy dispersion of the 1D electron gas is obtained by the eigenvalues of the Hamiltonian matrix

$$\mathbf{H}_{k_x}(\mathbf{K}, \mathbf{K}') = T(\mathbf{k}+\mathbf{G})\delta_{\mathbf{G},\mathbf{G}'}\delta_{\mathbf{k}_{yz},\mathbf{k}'_{yz}} + V_{1D}(\mathbf{K}-\mathbf{K}') \quad (3)$$

where k_x varies in the 1D reduced zone and $T(\mathbf{k}+\mathbf{G})$ is the well known kinetic energy term.

A real space discretization is indispensable for transport modelling with the NEGF approach, and in this work we use a simple second order, centered difference discretization of the kinetic energy operator given by [15]

$$T(\mathbf{k}+\mathbf{G}) = 2t_0 \sum_{s=x,y,z} \{1 - \cos[(k_s + G_s)d]\} \quad (4)$$

where $t_0 = \hbar^2/2m_0 d^2$. In all spatial directions $s=\{x,y,z\}$ we employ the same discretization step $d = a_0/N_d$. As shown in Fig. 3, in order to attain an accetable agreement of the bandstructures obtained with a continuous and with a discretized kinetic operator, we used a large $N_d = 30$, which also implies a larger number of Hamiltonian blocks. However, this drawback is compensated by the reduction of the size of the single block, which is the most relevant scaling parameter describing the computational burden.

The use of a second order discretization and of the local formulation of quantum confinement are the two key points that allowed us to reduce the size of the blocks of the block tridiagonal Hamiltonian matrix (compared to the formulation

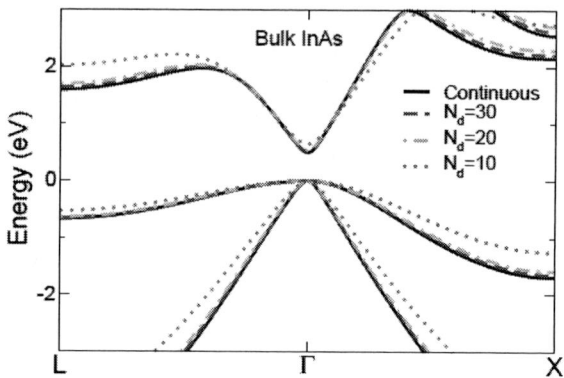

Fig. 3: Bandstructure of bulk InAs obtained with a continuous and a discretized kinetic operator with different values of N_d.

in [15]), which is a crucial parameter for NEGF based simulations. More specifically, the rank of the Hamiltonian blocks for the nanowire is $M_{1D} = (2N_G/N_d)N_{k_z}N_{k_y}$, where N_{k_y} and N_{k_z} are the number of the wavevectors in the reduced zone and are equal to N_{cy} and $2N_{cz}$, respectively, with N_{cy} (N_{cz}) being the number of unit cells along y (z) in the simulation domain. Hence, M_{1D} increases proportionally to the number of unit cells in the confinement directions.

A further reduction of the computational complexity can be achieved by employing a mode-space transformation [20], [21], and then by keeping only the lowest energy transverse modes, which are the most relevant for bandstructure and transport calculations in semiconductors. The mode-space Hamiltonian is obtained by means of a unitary transformation for each section of the system along x, namely for a single discretization point for the methodology of this work. At any discretization point x_l, the transformation matrix is equal to $\mathbf{U}^{(l)} = \left[\xi_1^{(l)} \cdots \xi_{N_{mod}}^{(l)}\right]$, where $\xi_n^{(l)}$ is the eigenvector of the eigenvalue problem

$$\left[\mathbf{H}_{l,l} + \mathbf{H}_{l,l+1} + \mathbf{H}_{l,l+1}^\dagger\right]\xi_n^{(l)} = E_n^{(l)}\xi_n^{(l)} \quad (5)$$

where $\mathbf{H}_{l,l'}$ are the block matrices of the Hamiltonian in the hybrid basis consisting of real space in the transport direction x and plane waves in the (y,z) directions [15].

We found that the mode space approximation works well and it allows one to reduce significantly the size of Hamiltonian blocks for the 1D electron gas. This is not surprising because the off-diagonal blocks, $\mathbf{H}_{l,l+1}$, of the Hamiltonian matrix are diagonal matrices in the hybrid basis with a constant term $-t_0$ on the diagonal. In particular, the $\mathbf{H}_{l,l+1}$ blocks are independent of the transverse Bloch wave-vector k_z (or (k_y, k_z)), so that the transverse modes $\left[\xi_n^{(l)}\right]$ obtained by Eq. (5) are also eigenfunctions of the diagonal blocks $\mathbf{H}_{l,l}$.

We verified that, thanks to the mode space approach, the size of the Hamiltonian blocks can be reduced to $M_{1D} = N_{mod}N_{k_z}N_{k_y}$ for a 1D gas, where, for the materials and devices analyzed in this paper, an N_{mod} of about 12 is sufficient to have an agreement within a few percent between the mode space results and the results obtained without introducing the mode-space approximation.

In our model the charge and current density are computed in terms of the retarded, $[\mathbf{G}_{x\xi}]$, and lesser-than, $[\mathbf{G}_{x\xi}^{<}]$, Green's functions in the hybrid basis consisting of real space in the transport direction x and transverse modes in the lateral directions. At a given energy E, Green's functions are defined as

$$[\mathbf{G}_{x\xi}(E)] = [E\mathbf{I} - [\mathbf{H}_{x\xi}] - [\boldsymbol{\Sigma}(E)]]^{-1} \quad (6)$$

and

$$[\mathbf{G}_{x\xi}^{<}(E)] = [\mathbf{G}_{x\xi}(E)][\boldsymbol{\Sigma}^{<}(E)][\mathbf{G}_{x\xi}(E)]^{\dagger} \quad (7)$$

where the retarded, $[\boldsymbol{\Sigma}]=[\boldsymbol{\Sigma}_L]+[\boldsymbol{\Sigma}_R]+[\boldsymbol{\Sigma}_{ph}]$, and lesser-than self-energy $[\boldsymbol{\Sigma}^{<}]=[\boldsymbol{\Sigma}_L^{<}]+[\boldsymbol{\Sigma}_R^{<}]+[\boldsymbol{\Sigma}_{ph}^{<}]$ describe the connection to contacts (i.e. left, L, and right lead, R), or a possible interaction with phonons [22]. The self-energies for inelastic electron-phonon in mode-space were discussed in [23], but in this paper we do not address incoherent transport.

Moreover, because the periodicity of the unit cell is a_0, but two adjacent unit cells are connected only by the kinetic operator through the first and last discretization point, we developed a new approach to compute the contact self-energies that, for the case at study, is more effective than the standard Sancho-Rubio algorithm [24]. More precisely, instead of computing the surface Green's function corresponding to a the entire unit cell of length a_0, as prescribed by the Sancho-Rubio algorithm, we focus on the surface Green's function corresponding to a single discretization point along x. Again, this allows us to deal with Hamiltonian blocks of reduced size and to significantly improve the computational efficiency. Details on this iterative procedure are given in Ref. [25].

The correct electrostatics of the InAs NW Tunnel-FET was simulated by self-consistently coupling the solutions of NEGF equations (6-7) with that of the 3D Poisson equation

$$\nabla \cdot [\epsilon(\mathbf{r})\nabla\phi(\mathbf{r})] = -e\left[p(\mathbf{r}) - n(\mathbf{r}) + N_D(\mathbf{r}) - N_A(\mathbf{r})\right] \quad (8)$$

where $\phi(\mathbf{r})$ is the electrostatic potential, $\epsilon(\mathbf{r})$ is the material-dependent permittivity, and $N_A(\mathbf{r})$, $N_D(\mathbf{r})$ are the acceptor and donor concentration, respectively. Before entering in the r.h.s. of Eq. (8), electron and hole concentrations were first computed on a fine discretization grid with step $d = a_0/N_d$, then, thanks to the fact that the electrostatic potential has fairly slow spatial variations on the scale of the lattice constant, they were interpolated on a coarser mesh with a discretization step $d_c = a_0/2$.

III. NUMERICAL RESULTS AND DISCUSSION

The treatment of strain and arbitrary crystal orientations in our EP based model has been discussed in [18]. Fig. 4 illustrates the bandstructure for an InAs nanowire either relaxed or subject to a tensile biaxial stress and having a square cross-section with a 3.04 nm side. As expected the biaxial strain reduces significantly the energy gap [26], which is approximately 0.95 eV in the unstrained system.

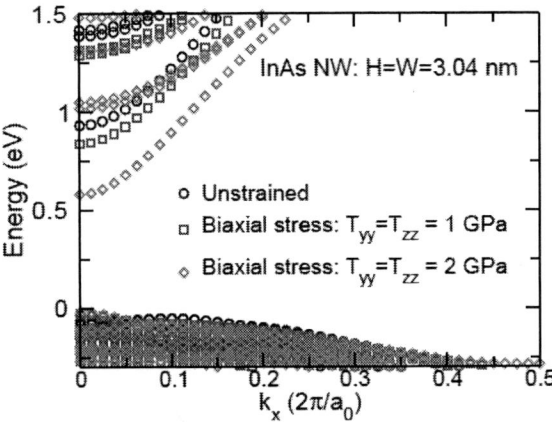

Fig. 4: Bandstructure in the 1D Brillouin zone of an InAs nanowire with different tensile biaxial stress values in the (y,z) plane normal to transport dire

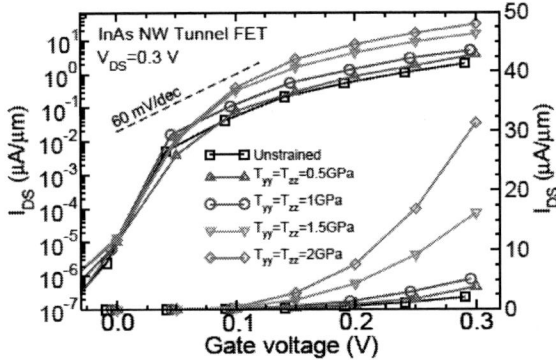

Fig. 5: Transfer characteristics of an InAs nanowire Tunnel FET either unstrained or for a tensile biaxial stress in the (y,z) plane normal to the transport direction. The gate length is L_G=15.2 nm and the cross section of the semiconductor region is a square with a 3.04 nm side.

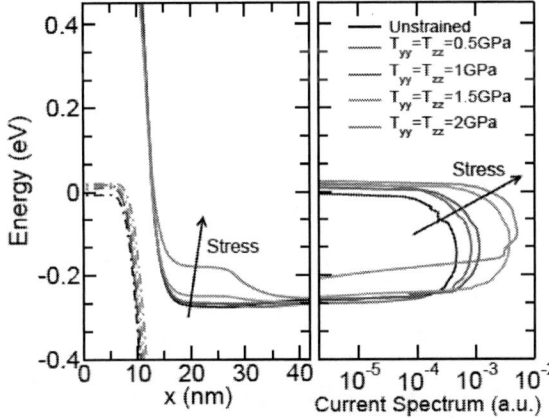

Fig. 6: Spatial profiles of the lowest conducation and highest valence subbands along the source to drain direction and current spectra for the InAs nanowire in Fig. 5 and for different stress levels.

Fig. 5 reports the corresponding I_{DS} versus V_{GS} characteristics at V_{DS}=0.3 V and obtained by means of self-consistent NEGF simulations. The metal gate workfunction was adjusted so as to have approximately the same off current I_{OFF}=$I_{DS}[V_{GS}$=0]=10 pA/μm in all cases. As it can be seen the biaxial tensile stress improves the on state I_{DS} at fixed I_{OFF}, without any sizeable change of the sub-threshold swing. These features are consistent with previous results obtained with a **k·p** Hamiltonian [26], which reinforces the confidence in the results of this paper that, for the first time, were obtained with a full band EP Hamiltonian.

Fig. 6 illustrates the subband profiles along the device channel and the current spectra for the InAs nanowire in Fig. 5 at V_{GS}≈V_{DS}=0.3 V. As it can be seen the biaxial stress greatly increases the transmission across the channel region and consequently the on current of the Tunnel FET.

IV. CONCLUSION

This paper has shown that the EP-NEGF methodology is a viable approach to simulate narrow NW FETs, and may thus deliver a good balance between physical accuracy and numerical burden for electron devices analysis. We argue that our transport formalism can be directly applied to plane-waves ab-initio Hamiltonians, hence it may be an alternative approach to the methods based on maximally localized Wannier functions.

REFERENCES

[1] (2015) The International Technology Roadmap for Semiconductors (ITRS).

[2] D. Esseni, P. Palestri, and L. Selmi, *"Nanoscale MOS Transistors - Semi-Classical Transport and Applications"*, 1st ed. Cambridge University Press., 2011.

[3] R. Kim, U. E. Avci, and I. A. Young, "Comprehensive Performance Benchmarking of III-V and Si nMOSFETs (Gate Length = 13 nm) Considering Supply Voltage and OFF-Current," *IEEE Trans. on Electron Devices*, vol. 62, no. 3, pp. 713–721, 2015.

[4] D. Esseni, M. Pala, and T. Rollo, "Essential Physics of the OFF-State Current in Nanoscale MOSFETs and Tunnel FETs," *IEEE Trans. on Electron Devices*, vol. 62, no. 9, p. 30843091, 2015.

[5] C. Grillet, D. Logoteta, A. Cresti, and M. G. Pala, "Assessment of the Electrical Performance of Short Channel InAs and Strained Si Nanowire FETs," *IEEE Trans. on Electron Devices*, vol. 64, no. 5, pp. 2425–2431, 2017.

[6] A. Seabaugh and Q. Zhang, "Low-Voltage Tunnel Transistors for Beyond CMOS Logic," *Proceedings of the IEEE*, vol. 98, no. 12, pp. 2095–2110, dec. 2010.

[7] D. Esseni, M. Pala, P. Palestri, C. Alper, and T. Rollo, "A review of selected topics in physics based modeling for tunnel field-effect transistors," *Semiconductor Science Technology*, vol. 32, p. 083005, 2017.

[8] S. D. Franceschi, L. Hutin, R. Maurand, L. Bourdet, H. Bohuslavskyi, A. Corna, D. Kotekar-Patil, S. Barraud, X. Jehl, M. S. Y.-M. Niquet, and M. Vinet, "SOI technology for quantum information processing," *in IEEE International Electron Devices Meeting*, pp. 339–342, 2016.

[9] M. Luisier, A. Schenk, and W. Fichtner, "Atomistic simulation of nanowires in the sp3d5s* tight-binding formalism: From boundary conditions to strain calculations," *Phys. Rev. B*, vol. 74, p. 205323, 2006.

[10] G. Klimeck, S.S. Ahmed, H. Bae, N. Kharche, S. Clark, B. Haley, S. Lee, M. Naumov, H. Ryu, F. Saied, M. Prada, M. Korkusinski, and T.B. Boykin, "Atomistic Simulation of Realistically Sized Nanodevices Using NEMO 3D Part I: Models and Benchmarks," *IEEE Trans. on Electron Devices*, vol. 54, no. 9, pp. 2079–2089, 2007.

[11] H. J. Choi and J. Ihm, "Ab initio pseudopotential method for the calculation of conductance in quantum wires," *Phys. Rev. B*, vol. 59, no. 3, pp. 2267–2275, 1999.

[12] Xiang-Wei Jiang, Shu-Shen Li, Jian-Bai Xia, and Lin-Wang Wang, "Quantum mechanical simulation of electronic transport in nanostructured devices by efficient self-consistent pseudopotential calculation," *Journal of Applied Physics*, vol. 109, p. 054503, 2011. [Online]. Available: http://dx.doi.org/doi/10.1063/1.3208067

[13] A. Garcia-Lekue, M. Vergniory, X. Jiang, and L. Wang, "Ab initio quantum transport calculations using plane waves," *Progress in Surface Science*, vol. 90, no. 3, pp. 292 – 318, 2015. [Online]. Available: http://www.sciencedirect.com/science/article/pii/S0079681615000209

[14] M. Pala, O. Badami, and D. Esseni, "NEGF based transport modeling with a full-band, pseudopotential Hamiltonian: Theory, Implementation and Full Device Simulations," *in IEEE International Electron Devices Meeting*, pp. 35.1.1–35.1.4, 2017.

[15] M. G. Pala and D. Esseni, "Full-band quantum simulation of electron devices with the pseudopotential method: Theory, implementation, and applications," *Phys. Rev. B*, vol. 97, p. 125310, Mar 2018. [Online]. Available: https://link.aps.org/doi/10.1103/PhysRevB.97.125310

[16] J. Fang, W. G. Vandenberghe, B. Fu, and M. V. Fischetti, "Pseudopotential-based electron quantum transport: Theoretical formulation and application to nanometer-scale silicon nanowire transistors," *Journal of Applied Physics*, vol. 119, p. 035701, 2016. [Online]. Available: http://dx.doi.org/doi/10.1063/1.3208067

[17] J. Fang, S. Chen, W. G. Vandenberghe, B. Fu, and M. V. Fischetti, "Theoretical Study of Ballistic Transport in Silicon Nanowire and Graphene Nanoribbon Field-Effect Transistors Using Empirical Pseudopotentials," *Electron Devices, IEEE Transactions on*, vol. 64, no. 6, pp. 2758 – 2764, 2017.

[18] M. Pala, O. Badami, and D. Esseni, "Transport models based on NEGF and empirical pseudopotentials: a computationally viable method for self-consistent simulation of nanoscale devices," *in IEEE International Electron Devices Meeting*, pp. 33.1.1–33.1.4, 2018.

[19] M.L. Cohen and J.R. Chelikowsky, *Electron Structure and Optical Properties of Semiconductors*. Springer Series in Solid-State Sciences. Springer-Verlag Berlin Heidelberg New York London Tokyo, 1988.

[20] R. Venugopal, Z. Ren, S. Datta, and M. Lundstrom, "Simulating quantum transport in nanoscale transistors: Real versus mode-space approaches," *J. Appl. Phys.*, vol. 92, no. 7, pp. 3730–3739, 2002.

[21] S. Poli, M. G. Pala, T. Poiroux, S. Deleonibus, and G. Baccarani, "Size Dependence of Surface-Roughness-Limited Mobility in Silicon-Nanowire FETs," *IEEE Transactions on Electron Devices*, vol. 55, no. 11, pp. 2968–2976, Nov 2008.

[22] G.D. Mahan, *Many-Particle Physics*. New York: Plenum Press, 1988.

[23] K. Rogdakis, S. , E. Bano, K. Zekentes, and M. Pala, "Phonon-and surface-roughness-limited mobility of gate-all-around 3C-SiC and Si nanowire FETs," *Nanotechnology*, vol. 20, no. 29, p. 295202, 2009.

[24] M. P. L. Sancho, J. M. L. Sancho, and J. Rubio, "Quick iterative scheme for the calculation of transfer matrices: application to Mo (100)," *Journal of Physics F: Metal Physics*, vol. 14, no. 5, p. 1205, 1984. [Online]. Available: http://stacks.iop.org/0305-4608/14/i=5/a=016

[25] M. G. Pala and D. Esseni, "Quantum transport models based on NEGF and empirical pseudopotentials for accurate modelling of nanoscale electron devices," *to be published*, 2019.

[26] F. Conzatti, M. Pala, D. Esseni, E. Bano, and L. Selmi, "Strain-Induced Performance Improvements in InAs Nanowire Tunnel FETs," *Electron Devices, IEEE Transactions on*, vol. 59, no. 8, pp. 2085–2092, 2012.

978-1-7281-0941-1/19 $31.00 © 2019 IEEE

A Quantum Element Reduced Order Model

Ming-C. Cheng , Dept. of ECE, Clarkson University, Potsdam, NY 13699-5720, USA, mcheng@clarkson.edu

Abstract—A reduced-order model for quantum eigenvalue problems developed previously is revised and combined with the domain decomposition method to construct the quantum *element* method (QEM). The basic idea of the QEM is to partition a quantum domain structure into several subdomains or *elements*. Each element is projected onto a functional space using the proper orthogonal decommission. These elements are then combined together to construct the whole domain structure. The proposed QEM has been demonstrated in 2 quantum well structures constructed with several elements. The study illustrates that the QEM is capable of offering accurate prediction of wave functions and quantum eigenenergies with a substantial reduction in the numerical degrees of freedom compared to direct numerical simulation of the Schrödinger equation.

Keywords—Proper orthogonal decomposition; Schrödinger equation; domain decomposition; discontinuous Galerkin; quantum wells; nanostructures

I. INTRODUCTION

This study presents the quantum element method (QEM) that implements the quantum reduced-order model [1] based on the proper orthogonal decomposition (POD) [2]-[4] in the domain decomposition. The basic idea of the QEM is to partition a domain of a quantum eigenvalue problem into smaller *subdomains* (or elements). Each element is projected onto a functional space represented by a small set of basis functions (or modes) extracted from the POD. To construct a POD model for a large domain, the projected elements are glued together, and the interior penalty discontinuous Galerkin method [5],[6] is applied to stabilize the numerical solution and to achieve the continuity across the element interfaces. The POD is able to optimize the basis functions (or POD modes) specifically tailored to the geometry and parametric variations of the problem and can therefore substantially reduce the degrees of freedom (DoF) needed to solve the Schrödinger equation with a high accuracy.

A multi-element POD approach offers many advantages. First, to generate POD modes for a quantum eigenvalue problem, collection of wave function (WF) data in the simulation domain subjected to enough parametric variations is needed to "train" the modes [1]. In a large simulation domain especially with fine resolution, generation of POD modes of smaller subdomains is certainly more efficient. Second, domain decomposition possesses a nature advantage for parallel and/or distributed computing [7]-[10]. Finally, many quantum structures contain repeating substructures. If the selected elements appear frequently in a group of quantum structures, the quantum POD modes of these generic elements can be generated and stored in a technology library that can then be used for cost-effective simulation and design of large quantum structures. The POD approach has been successfully applied to thermal simulations of devices and integrated circuits with a high resolution [11]-[13]. It has been shown that the approach offers a reduction in numerical degrees of freedom (DoF) by 3-4 orders and 5-6 orders of magnitude in 2D and 3D domains, respectively, compared to direct numerical simulation if solution with a high resolution is needed.

II. QUANTUM ELEMENT METHOD

A. Formulation of the Single-Element POD Model

To generate an optimal set of modes, POD on WF data of electrons/holes for the quantum domain structure is applied based on the Fredholm equation of the second kind [2],[3],

$$\int_{\Omega'} \left\langle \vec{\psi}(\vec{r}) \otimes \vec{\psi}^T(\vec{r}') \right\rangle \vec{\eta}(\vec{r}') d\vec{r}' = \lambda \vec{\eta}(\vec{r}), \tag{1}$$

where $\otimes$ is a tensor product and λ is the POD eigenvalue of the data representing the mean squared WFs captured by the corresponding POD mode η. Once the POD modes are determined, the WF $\psi(\vec{r})$ in the domain is given by a linear combination of these POD modes,

$$\psi(\vec{r}) = \sum_{j=1}^{M} a_j \eta_j(\vec{r}). \tag{2}$$

where M is the selected number of modes or the DoF for representing the WF and a_j are weighting coefficients. To derive an equation for $\vec{a}$, the Galerkin projection is applied to the Schrödinger equation along η_i in the its POD space,

$$\int_{\Omega} \nabla \eta_i(\vec{r}) \cdot \frac{\hbar^2}{2m^*} \nabla \psi(\vec{r}) d\Omega - \int_s \eta_i(\vec{r}) \frac{\hbar^2}{2m^*} \nabla \psi(\vec{r}) \cdot \hat{n} dS \\ + \int_{\Omega} \eta_i(\vec{r}) U(\vec{r}) \psi(\vec{r}) d\Omega = E \int_{\Omega} \eta_i(\vec{r}) \psi(\vec{r}) d\Omega \tag{3}$$

where m^* is the electron/hole effective mass, $\hbar$ the reduced Planck's constant, U the potential energy, E the quantum-state (QS) energy, and $\hat{n}$ the outward normal vector of the boundary surface of the domain Ω. The parametric variations are accounted for via U, which may be induced by external electric fields and/or the charge distributions in the structure.

Using (2) in (3), an equation for $\vec{a}$ in terms of the $M \times M$ Hamiltonian matrix $\mathbf{H}_\eta$ in POD space can be derived,

$$\mathbf{H}_\eta \vec{a} = E_\eta \vec{a}, \tag{4}$$

where $\vec{a}$ is the eigenstate vector of $\mathbf{H}_\eta$ that is expressed as

$$\mathbf{H}_\eta = \mathbf{T}_\eta + \mathbf{U}_\eta \tag{5}$$

with the kinetic energy matrix given as

$$T_{\eta i,j} = \int_{\Omega} \nabla \eta_i(\vec{r}) \cdot \frac{\hbar^2}{2m^*} \nabla \eta_j(\vec{r}) d\Omega \tag{6}$$

and the potential energy matrix given as

$$U_{\eta i,j} = \int_{\Omega} \eta_i(\vec{r}) U(\vec{r}) \eta_j(\vec{r}) d\Omega. \tag{7}$$

WFs near the boundary in the single-element domain is assumed small and the surface integral vanishes. In multi-element cases, the surface integral in each element is coupled with the surrounding elements, which is presented below.

In the previous quantum POD model [1], decomposition was performed on WF data in each QS. In order to extend the QSs across multi-elements in the QEM, a global approach for all QSs is proposed. More specifically, WFs in all the selected QSs subjected to N_s different applied electric fields are collected numerically from the Schrödinger equation. Using these N_s sets of WF data in (1), N_s sets of POD eigenvalues and eigenmodes are generated. The coefficients in (6) and (7) are then evaluated and used to solve $\vec{a}$ in (4).

B. Quantum Element Model

For a large structure that are partitioned into multiple elements, each element is projected to POD space using (1) represented by its POD modes. The QEM constructs the large structure using these elements and then projects the quantum eigenvalue problem of the large multi-element structure onto a POD space described by multiple sets of POD modes. To arrive at an equation for the eigenvector $\vec{a}$ of the multi-element structure, (3) is modified to project the Schrödinger equation along the ith mode of the pth element as

$$\int_{\Omega_p} \nabla \eta_{p,i} \cdot \frac{\hbar^2}{2m_p^*} \nabla \psi_p d\Omega + \int_{\Omega_p} \eta_{p,i} U \psi_p d\Omega - \sum_{q=1,q\neq p}^{N_{el}} \int_{S_{pq}} \left[\Delta \left(\frac{\hbar^2}{2m^*} \right)_{pq} \langle \nabla \eta_i \rangle_{pq} + \left\langle \frac{\hbar^2}{2m^*} \nabla \psi \right\rangle_{pq} \Delta(\eta_i)_{pq} \right] \cdot d\vec{S} - \mu \sum_{q=1,q\neq p}^{N_{el}} \int_{S_{pq}} \Delta \left(\frac{\hbar^2}{2m^*} \psi \right)_{pq} \Delta(\eta_i)_{pq} dS = E \int_{\Omega_p} \eta_{p,i} \psi_p d\Omega,$$

$$(8)$$

where N_{el} is the total number of elements in the domain, μ is the penalty constant defined as N_μ/dr with dr as the local numerical mesh size, and $\Delta(*)_{p,q}$ and $\langle * \rangle_{p,q}$ are the difference and average in the surface integral, respectively, across the interface between the pth and qth elements. In (8), the interior penalty discontinuous Galerkin method [5],[6] is applied at the surfaces to enforce the interface continuity. For the pth element projected along its ith mode, (8) can be rewritten as

$$\sum_{j=1}^{M_p} \left(T_{\eta_p,ij} + U_{\eta_p,ij} \right) a_{p,j} + \sum_{q=1,q\neq p}^{N_{el}} \sum_{j=1}^{M_p} B_{p,pq,ij} a_{p,j} + \sum_{q=1,q\neq p}^{N_{el}} \sum_{j=1}^{M_q} B_{pq,ij} a_{q,j} = E a_{p,i},$$

$$(9)$$

where M_p and M_q are the selected numbers of modes in the pth and qth elements, the interior kinetic and potential energy matrices are given in (6) and (7), the diagonal boundary kinetic energy matrix is given as

$$B_{p,pq,ij} = -\frac{1}{2} \int_{S_{pq}} \frac{\hbar^2}{2m_p^*} \left[\left(\nabla \eta_{p,i} \right) \eta_{p,j} + \eta_{p,i} \left(\nabla \eta_{p,j} \right) \right] \cdot d\vec{S} + \mu \int_{S_{pq}} \frac{\hbar^2}{2m_p^*} \eta_{p,i} \eta_{p,j} dS,$$

$$(10)$$

and the off-diagonal boundary kinetic energy matrix is

$$B_{pq,ij} = \frac{1}{2} \int_{S_{pq}} \frac{\hbar^2}{2m_q^*} \left[\left(\nabla \eta_{p,i} \right) \eta_{q,j} - \eta_{p,i} \left(\nabla \eta_{q,j} \right) \right] \cdot d\vec{S} - \mu \int_{S_{pq}} \frac{\hbar^2}{2m_q^*} \eta_{p,i} \eta_{q,j} dS.$$

$$(11)$$

A multi-element POD Hamiltonian matrix equation for the N_{el}-element domain can be derived from (9),

$$\begin{bmatrix} \mathbf{H}_1 & \mathbf{H}_{1,2} & \cdots & \mathbf{H}_{1,N_{el}-1} & \mathbf{H}_{1,N_{el}} \\ \mathbf{H}_{2,1} & \mathbf{H}_2 & \cdots & \mathbf{H}_{2,N_{el}-1} & \mathbf{H}_{2,N_{el}} \\ \vdots & \vdots & \ddots & \vdots & \vdots \\ \mathbf{H}_{N_{el}-1,1} & \mathbf{H}_{N_{el}-1,2} & \cdots & \mathbf{H}_{N_{el}-1} & \mathbf{H}_{N_{el}-1,N_{el}} \\ \mathbf{H}_{N_{el},1} & \mathbf{H}_{N_{el},2} & \cdots & \mathbf{H}_{N_{el},N_{el}-1} & \mathbf{H}_{N_{el}} \end{bmatrix} \cdot \begin{bmatrix} \vec{a}_1 \\ \vec{a}_2 \\ \vdots \\ \vec{a}_{N_{el}-1} \\ \vec{a}_{N_{el}} \end{bmatrix} = E \begin{bmatrix} \vec{a}_1 \\ \vec{a}_2 \\ \vdots \\ \vec{a}_{N_{el}-1} \\ \vec{a}_{N_{el}} \end{bmatrix}, (12)$$

where the diagonal block matrix $\mathbf{H}_p$ for the pth element is

$$H_{ij} = \sum_{j=1}^{M_p} \left(T_{\eta_p,ij} + U_{\eta_p,ij} \right) + \sum_{q=1,q\neq p}^{N_{el}} \sum_{j=1}^{M_p} B_{p,pq,ij} \quad (13)$$

and the off-diagonal block matrix $\mathbf{H}_{pq}$ is

$$H_{pq,ij} = \sum_{j=1}^{M_q} B_{pq,ij} \quad \text{with } p \neq q. \quad (14)$$

With $\vec{a}_p$ solved from (12) for $p = 1$ to N_{el}, the WF in each state for the whole structure can then be constructed by combining WFs in space based on (2) over N_{el} elements. If the pth and qth elements do not neighbor each other, surface integrals in (8) vanish; $B_{p,pq,ij} = B_{pq,ij} = 0$ and $H_{pq,ij} = 0$. With a large number of elements, most of matrix entries in (12) are zeros. The block matrices in (12) can be pre-evaluated from the integrals of $\eta_{p,i}$ and $\nabla \eta_{p,i}$. The matrices and modes are then stored in a library for simulation/design of large quantum structures.

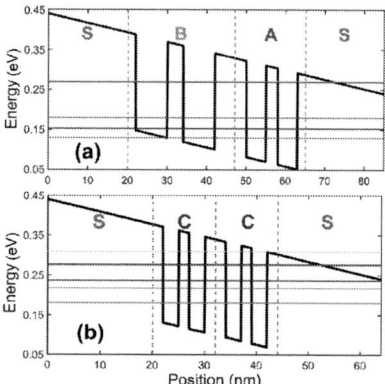

Fig. 1. Structures for generation of POD modes for each of Elements A, B, C, SC and CS. The slope of each band indicate the maximum electric field in the simultions for the data collection.

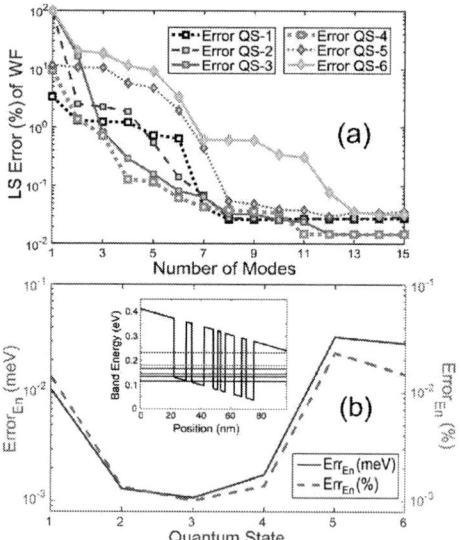

Fig. 2. (a) LS errors of WFs in QSs 1-6 derived from the 3-element POD simulations of the SB-C-AS structure at -18kV/cm. (b) The error of predicted eigenenergy with an inset showing the first 6 QS energies in the energy band at -18 kV/cm.

III. DEMONSTRATION OF THE QEM

Two QW structures given in Fig. 1 are used to collect WF data over 6 QSs in order to generate POD modes for Elements A, B, C, SC and CS, as labeled in the band diagrams. 15 electric fields from -24 to 24 kV/cm are applied to the structure in Fig. 1(a) and from -30 to 30 kV/cm to the one in Fig. 1(b). Simulations with such setups allow each projected element to experience different boundary conditions (BCs)

978-1-7281-0941-1/19 $31.00 © 2019 IEEE

induced by the adjacent elements due to electric field variations. These projected POD elements are applied to construct 2 multi-element structures to validate the QEM against detailed numerical solutions of the Schrödinger equation. The test QW structures include a 3-element structure with Elements SB, C and AS and a 7-element structure with Elements SC, B, A, B, C, A and CS.

Theoretically, the training described above allows the generated POD modes of each element to predict WFs with a high accuracy only if the element is neighbored by those encountered during the training. However, most of the elements in the test QW structures are neighbored by those not encountered in the training. As will be seen below, the generated POD modes appear to be more robust than expected. In Fig. 1(b), 2 possible sets of POD modes can be generated for Element C. The POD modes for Element C on the left are used in this study.

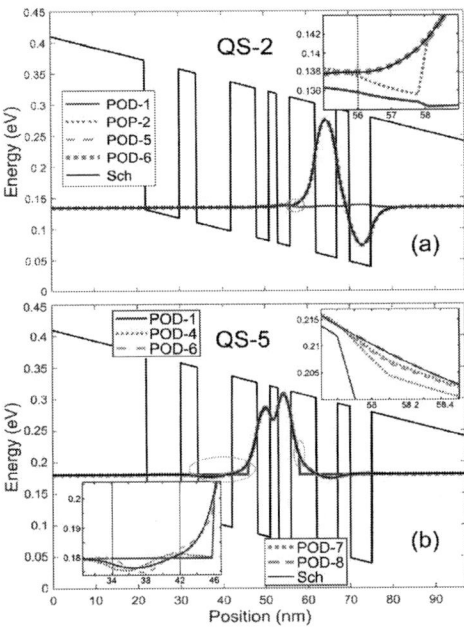

Fig. 3 Energy band diagram and WFs in QSs 2 and 5 of the 3-element structure at -18 kV/cm with a different numbers of modes. The insets show the interface discontinuities influenced by the number of modes.

Using the projected POD elements, 3-element POD simulation of the SB-C-AS structure with an equal number of modes in each element and $N_\mu = 20$ is performed. Compared to numerical simulation of the Schrödinger equation, errors of WFs and POD eigenergies are illustrated in Fig. 2, and its WFs in QSs 2 and 5 are included in Fig. 3. In general, a small error of WF (near or below 2%) in each state can be reach with 2 or 3 modes for each element except for QSs 5 and 6 that require more modes to minimize the large discontinuities at interfaces between elements. With a small number of modes the discontinuity is considerably smaller in QS-2 than in QS-5. The least square (LS) error in QS-2 is thus much smaller and its error drops from 2% to 0.52% when the discontinuity near 58nm shown in the inset of Fig.3 (a) is successfully minimized with 5 modes. On the other hand, the large discontinuities in QS-5 are gradually smoothed while the number of modes increase from 4 to 6, as seen in the insets of Fig. 3(b). The LS error of the QS-5 WF in Fig. 2(a) thus gradually reduces with 4 to 6 modes, and suddenly drops from 2% to an error less than 0.1% with 7th and 8th modes included.

For the 7-element structure, simulation at an applied voltage of 0.25V across the 166nm-long QW structure (electric field ≈ 15kV/cm) is performed to demonstrate effectiveness of the QEM. LS errors of the WFs in the first 14 QSs are displayed in Figs. 4(a) and 4(b). The band diagram, the first 14 eigenenergies and the errors of these eigenenergies are displayed in the insets. Some predicted WFs are illustrated in Figs. 3(a)-3(d), compared to numerical solution from the Schrödinger equation.

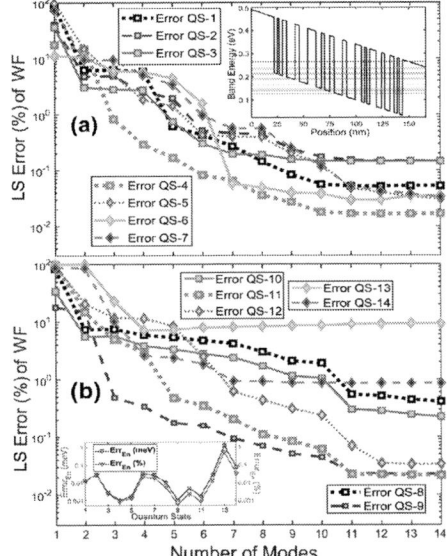

Fig. 4. LS errors of WFs in (a) QSs 1-7 and (b) QSs 8-14 derived from the 7-element POD simulation of the SC-B-A-B-C-A-CS structure at -15 kV/cm. The insets display the predicted 14 QS energies in the energy band diagram in (a) and the error of the predicted eigenenergy in each QS in (b).

POD prediction of the QS energy is very accurate, as seen in the inset of Fig. 4(b), with an error less than 0.05% (< 0.1meV) for all QSs except for QS 13 whose error is 0.6% (< 1meV). Results reveal that the predicted WFs in most of states are able to reach a good accuracy with just 3 to 6 modes and the errors mostly are induced by the interface discontinuity. Once the discontinuity is suppressed using more modes, high accuracy (with an error far below 1%) can be achieved. For example, when the discontinuity in QS 1 is minimized with 5 modes (see Fig. 5(a)), a sudden drop of its error to 0.6% is found in Fig. 4(a). Except for QS 13, this is also observed in other states; some need 3 to 6 modes to minimize the discontinuity (QSs 1-7, 9 and 11) but some need 10 or 11 modes (QSs 8 and 10). Except for QSs 13 and 14, the error far below 1% can be achieved if enough modes are included. The error of the QS-14 WF stays near 1% with 7 or more modes. Unlike the other states, QS 13 appears to be an unbound state, which may not be accounted for thoroughly in the generated POD modes. As seen in Fig. 5(d), the predicted QS-13 WF is not able to reach a high accuracy and a minimum error of 7.2% is achieved with 4 modes. Fig. 4(b) shows that its error actually increases slowly as more modes are added.

The demonstrations have shown that the QEM offers a very accurate approach for the well-bounded WFs in the 2 test structures. The POD modes appear to be more *intelligent* than expected in the multi-element simulation. The POD modes are able to offer very accurate prediction even when they experience elements that were not included in the training process. However, larger errors are observed for the unbounded WFs perhaps due to incomplete data of the unbounded WFs provided in the training process.

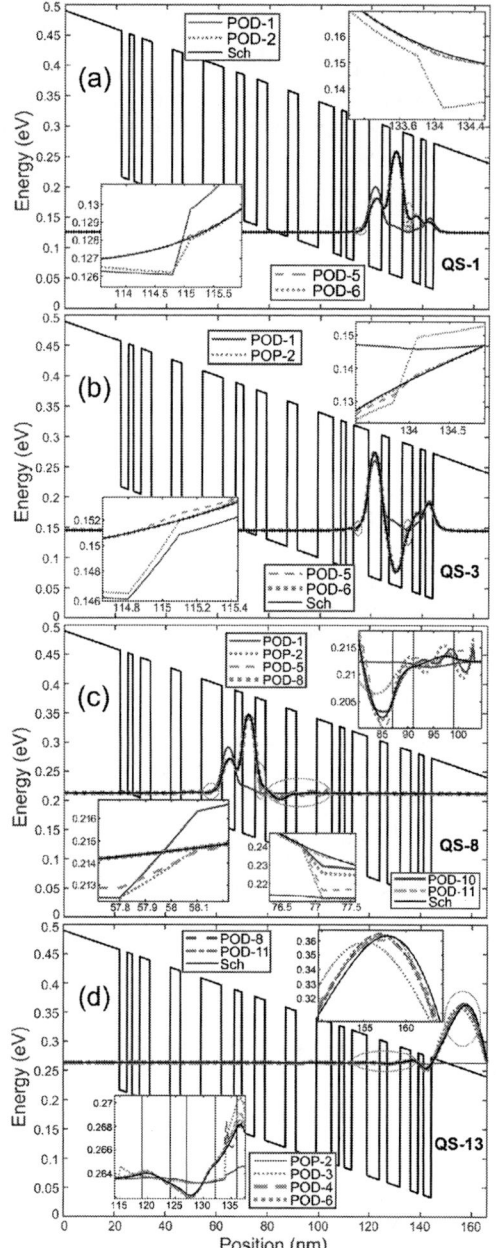

Fig. 5. Energy band diagram and WFs in QSs 1, 3, 8 and 13 of the 7-element structure at -15kV/cm derived from the QEM with a different numbers of modes. The insets show the interface discontinuities influenced by the number of modes.

IV. CONCLUSIONS

The QEM proposed in this work combines the POD quantum model [1] with the domain decomposition method to offer an efficient approach for simulation of large domain structures for quantum eigenvalue problems. The approach partitions a large domain into smaller elements, each of which is projected to a POD functional space represented by a set of POD modes. The large quantum structure is then constructed using the projected POD elements. The developed QEM has been demonstrated in 2 QW structures, including a 3-element

structure with 6 QWs and a 7-element structure with 14 QWs. With the projected multi-element QW structure onto a POD space represented by the several sets of POD modes, the QEM is able to predict well-bounded WFs and QS energies with high accuracy with a small number of DoF.

This study presents the first application of the QEM. The developed approach will be useful for simulations/design of nanostructures or materials that require solution of the Schrödinger equation or Schrödinger-like equation. It is particularly useful for quantum structures with a high degree of geometrical repetition, including periodic lattice structures in materials. For simulation of a periodic structure, the general practice is to simulate a small basic element with periodic BCs, which offers a feasible computational time. In reality, there are always desired or undesired localized imperfections and/or non-uniformity in crystals or nanostructure in which periodic BCs cannot be used. To understand these types of structures, computationally intensive simulation of a large domain with a large number of basic elements, together with imperfections and/or nonuniformity, is needed. With the novel concept of the QEM, POD modes for a collection of elements, including imperfection and/or nonuniformity, can be generated first. The QEM would be able to offer accurate simulation of a large domain structure, constructed using these POD elements, to account for realistic BCs and nonuniformity at a reasonable computational time.

REFERENCES

[1] M. C. Cheng, "A Reduced-Order Representation of the Schrödinger Equation," AIP Advances, 6, 095121, 2016.

[2] J. L. Lumley, *Atmospheric Turbulence and Wave Propagation*, Moscow, Russia, Nauka publisher, 166 (1967).

[3] J. L. Lumley, *Stochastic Tools in Turbulence*, Academic, New York, 1970; reprint, Dover publisher 2007.

[4] Y. Maday and E. M. Rønquist, "A reduced-basis element method," *Comptes Rendus Math.*, 335, pp. 195–200, 2002.

[5] D.N. Arnold, F. Brezzi, B. Cockburn, and D. Marini. "Discontinuous Galerkin methods for elliptic problems," Lecture Notes in Comp. Sci. Eng., 11, pp. 89-101, 2000.

[6] D. N. Arnold, F. Brezzi, B. Cockburn, and L. Donatella Marini. "Unified analysis of discontinuous galerkin methods for elliptic problems," SIAM, 39, pp. 1749-1779, 2002.

[7] J. Mandel, "Balancing domain decomposition," Commun. Numer. Methods Engrg. 9 (1993) 233–241.

[8] V. Vondrák, T. Kozubek, A. Markopoulos, Z. Dostál, "Parallel solution of contact shape optimization problems based on Total FETI domain decomposition method," Struct. Multidiscip. Optim. 42, 955–964, 2010.

[9] A. Mobasher Amini, D. Dureisseix, P. Cartraud, N. Buannic, "A domain decomposition method for problems with structural heterogeneities on the interface: Application to a passenger ship," Comput. Methods Appl. Mech. Engrg. 198, 3452–3463, 2009.

[10] M. Papadrakakis, G. Stavroulakis, A. Karatarakis,, "A new era in scientific computing: Domain decomposition methods in hybrid CPU–GPU architectures," Computer Methods in Applied Mechanics and Engineering, Vol. 200, Issues 13–16, pp. 1490-1508, 2011.

[11] W. Jia, B. Helenbrook, M. C. Cheng, "Fast Thermal Simulation of FinFET Circuits Based on a Multi-Block Reduced-Order Model", IEEE Trans. ICs & Systems, vol. 35, no. 7, pp. 1114-1124, 2016.

[12] W. Jia, B. Helenbrook, M. C. Cheng, "Thermal Modeling of Multi-Fin Field Effect Transistor Structure Using Proper Orthogonal Decomposition", IEEE Trans. Electron Devices, Vol. 61, No. 8, pp. 2752-2759, 2014.

[13] R. Venters, B. Helenbrook, K. Zhang, M. C. Cheng, "Proper Orthogonal Decomposition Based Thermal Modeling of Semiconductor Structures," IEEE Trans. Electron Devices, Vol.59, No. 11, pp. 2924-2931, 2012.

Quantum Mechanical Simulations of the Impact of Surface Roughness on Nanowire TFET performance

Yunhe Guan
School of Microelectronics
Xi'an Jiaotong University
Xi'an, China
guanyunhe@stu.xjtu.edu.cn

ZunChao Li
School of Microelectronics
Xi'an Jiaotong University
Xi'an, China
zcli@xjtu.edu.cn

Hamilton Carrillo-Nuñez
School of Engineering
University of Glasgow
Glasgow, United Kingdom
hacarrillo@gmail.com

Vihar P. Georgiev
School of Engineering
University of Glasgow
Glasgow, United Kingdom
vihar.georgiev@glasgow.ac.uk

Asen Asenov
School of Engineering
University of Glasgow
Glasgow, United Kingdom
asen.asenov@glasgow.ac.uk

Abstract—In this work, the impact of the surface roughness (SR) on the variability in p-type InAs nanowire Tunnel FET (TFET) has been investigated. Using the Non-Equilibrium Green's Function (NEGF) module implemented in the University of Glasgow quantum transport simulation tool, called NESS, we have simulated a statistical ensemble of 200 TFETs with unique SR profiles. The SR in each device is defined by the characteristic values of the SR root mean square amplitude (RMS) and correlation length. Our results show that the larger the RMS, the stronger the variability. We find that the SR-induced variability is reduced in InAs-Si heterostructure TFETs when comparing with their homogenous InAs counterpart. The impacts of both metal grain granularity and random discrete dopants on InAs TFETs are also studied. Our finding suggests that SR is the weakest source of statistical variability.

Keywords—surface roughness (SR), variability, tunnel field-effect transistor, quantum simulation.

I. INTRODUCTION

With the continuous scaling of the MOSFETs, the power consumption has become a critical concern since the subthreshold swing (SS) in MOSFETs is limited to 60mV/decade at room temperature. Based on band-to-band tunneling (BTBT) the tunneling FET (TFET) is a promising candidate allowing to achieve SS<60mV/decade and thus offering reduction of leakage or supply voltage at constant drive current [1].

Although TFETs have been widely studied experimentally and through simulation in order to optimize performance [2]-[4], the variability induced by the surface roughness (SR) is still not very well investigated [5], [6]. Considering that the SR has a strong impact on MOSFETs performance [7], it is crucial to investigate its impact on TFETs. F. Conzatti *et al.* [5] reported simulation of statistical ensemble with 50 SR realizations in n-type TFET in comparison with MOSFETs. H. Carrillo-Nuñez *et al.* [6] studied the impact of SR in p-type TFET by carrying out atomistic simulations of only few devices. Therefore, there is a need of thorough investigation of the SR in TFETs on a proper statistical scale.

In this work, the variability induced by SR is investigated in ensembles 200 p-type InAs nanowire TFETs by using the quantum transport solver module in NESS [8]. The dependence of the variations on the values of the root mean square (RMS) amplitude has been studied. We also report that

This work was supported by NSFC (Project No. 61176038).

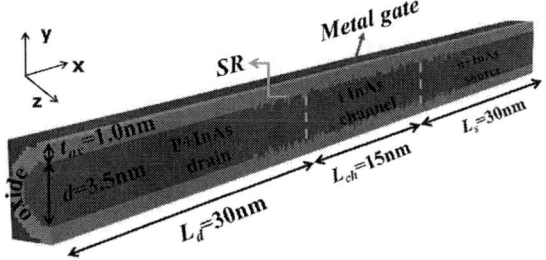

Fig. 1. 3D view of the p-type InAs nanowire TFET with SR. For all devices, the diameter of nanowire is $d = 3.5$nm, and the channel length is $L_{ch} = 15$nm. The SR is between the oxide and InAs semiconductor, covering the whole channel and 10nm source and drain regions. The left 20nm source/drain without SR is for better convergence. The source-drain voltage is fixed at $V_{ds} = -0.5$V.

InAs-Si heterojunction not only can improve the current but also can alleviate the variability caused by SR. Physical insight into the influence of SR is also provided. Besides, a comparison between the variability induced by SR, random discrete dopants (RDD), and metal grain granularity (MGG) is given.

II. SIMULATION METHODOLOGY

Fig. 1 shows the sketch of a p-type InAs nanowire TFET with SR. The nanowire diameter is $d = 3.5$nm, the gate length is $L_{ch} = 15$nm, and the oxide thickness is $t_{ox} = 1$nm with dielectric constant $\varepsilon_{ox} = 9.0$. The doping level in the n+-source and p+-drain is $N_S = N_D = 5 \times 10^{19}$cm^{-3}, and the channel is left intrinsic. The SR at the interface between InAs and gate oxide is introduced by means of an autocorrelation function $C(x) = \Delta_m^2 \exp(-\sqrt{2}x/L_m)$, which is characterized by the RMS Δ_m and correlation length (CL) L_m parameters [9]. The SR region covers the whole channel and 10nm source and 10nm drain regions. The smooth part in source/drain region is required for numerical stability. The source-to-drain voltage is fixed at $V_{ds} = -0.5$V, and all simulations are performed at room temperature (T = 300K).

The Flietner model [10], [11], combined with the non-equilibrium Green's function (NEGF) within an effective mass approximation, is used to calculate the BTBT current. The accuracy of this approach, implemented in NESS, has been validated by comparison with atomistic simulation results showing a very good agreement [11], [12]. The BTBT

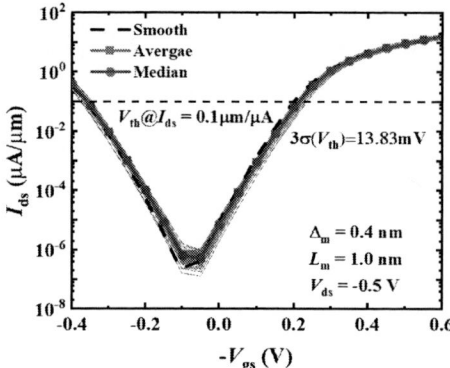

Fig. 2. $I_{ds} - V_{gs}$ characteristics of an ensemble of 200 p-type InAs nanowire TFETs with SR (gray curves). $\Delta_m = 0.40$nm, and $L_m = 1$nm. The threshold voltage is defined as the V_{gs} yielding $I_{ds} = 0.1\mu A/\mu m$, and the off-current is $I_{off} = 10$pA/μm at $V_{gs} = 0$V. The current is normalized by πd. The ambipolar current is also plotted.

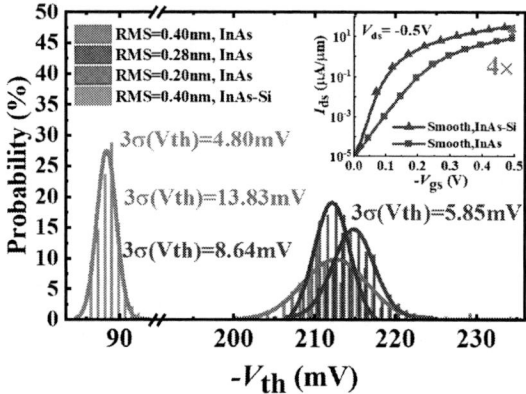

Fig. 3. The distribution of V_{th} with 0.20, 0.28 and 0.40nm RMS amplitude of SR variation of an ensemble of 200 InAs and InAs-Si TFETs. The inset shows the $I_{ds} - V_{gs}$ characteristic of the smooth InAs and InAs-Si TFETs. The off-current at $V_{gs} = 0$V is defined as 10pA/um. In InAs-Si TFET, Si (InAs) is used as the channel/drain (source) material. With the increase of RMS, the variability induced by SR increases. The InAs-Si heterostructure not only can improve the current but also can alleviate the variation caused by SR. $L_m = 1$nm in all devices.

model is detailly explained in [11]. The inclusion of the electron-phonon coupling would improve the accuracy of our results. However, it has been reported phonon scattering to only have a small influence on the OFF-state in TFETs [6]. The inclusion of phonon scattering is also very computational intense, particularly, when considering a large number of samples. The latter would not significantly change the conclusions of this study and thus phonon scattering is neglected in this work.

III. RESULTS AND DISCUSSIONS

Fig. 2 shows the $I_{ds} - V_{gs}$ characteristics of an ensemble of 200 nanowire TFETs with random SR configurations determined by $\Delta_m = 0.40$nm and $L_m = 1$nm. The smooth, average, and median curves are also presented. The off-current of smooth case is $I_{off} = 10$pA/μm at $V_{gs} = 0$V, and the threshold voltage (V_{th}) is defined as the V_{gs} yielding $I_{ds} = 0.1\mu A/\mu m$. It can be seen that the ambipolar current can also be predicted by NESS. In this case, the standard deviation of threshold voltage is $3\sigma(V_{th}) = 13.83$mV. I_{off} of the

ensemble ranges from $1.92 \times 10^{-6}\mu A/\mu m$ to $1.25 \times 10^{-5}\mu A/\mu m$, and the on-current ($I_{on}$) defined at $V_{gs} = 0.5$V varies from $7.56\mu A/\mu m$ to $11.34\mu A/\mu m$, indicating a smaller variability than I_{off}. The smaller variability in I_{on} than I_{off}, also found in MOSFETs [9], is attributed to the smaller SS in on-state.

In Fig. 3, the threshold voltage probability distribution (PD) is plotted for different values of RMS, i.e. 0.20, 0.28, and 0.40nm. It can be found that the variability increases with the increase of RMS, which is in agreement with [13]. The PD of the V_{th} in case of the InAs(source)-Si(channel) heterojunction nanowire TFET is also shown in Fig. 3. In the latter, the doping levels in the source and drain are 5×10^{19}cm^{-3} and 2×10^{20}cm^{-3}, respectively. The $3\sigma(V_{th})$ with $\Delta_m = 0.40$nm in the InAs-Si TFET is reduced by 17.95%, 44.44%, and 65.29% when comparing with that in InAs TFETs with $\Delta_m = 0.20$nm, 0.28nm, and 0.40nm. The inset in Fig. 3 compares the $I_{ds} - V_{gs}$ characteristics of the smooth InAs and InAs-Si TFETs under the same $I_{off} = 10$pA/μm. The InAs-Si TFET provides an on-current four times higher than the InAs TFET. Therefore, the heterojunction not only can improve the current but also can reduce the SR-induced variability. In order to gain a physical insight into our findings, the current spectra of both TFETs with the SR are shown in Fig. 4. The regions from left to right are drain, channel, and source regions, respectively. The red lines denote the highest valence and the lowest conduction subbands. Observe the fluctuations in the subbands due to the SR influence on the confinement. Fluctuations in the valence subband in the InAs TFET are more pronounced in comparison to InAs-Si

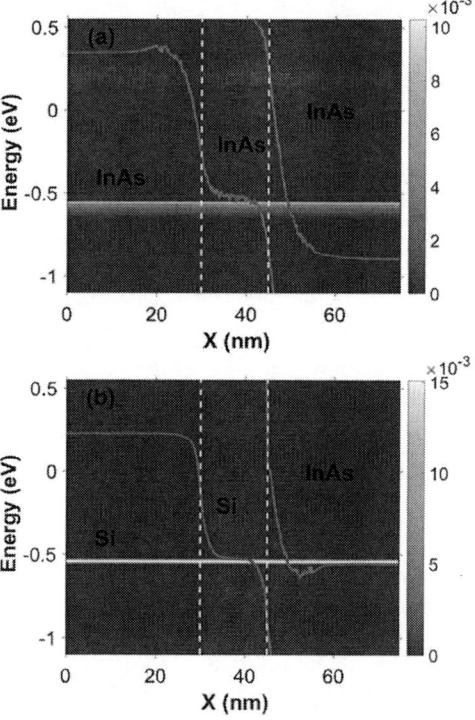

Fig. 4. The current spectra of the (a) InAs TFET and (b) InAs-Si TFET at the maximum V_{th} with 0.40nm RMS amplitude of SR variation. $L_m = 1$nm. The unit is μA/eV. The red lines denote the highest valence and the lowest conduction subbands. Due to the fluctuation in thickness from SR, the subbands are not smooth anymore which results to the variation of performance. Compared with the case in InAs-Si TFET, the fluctuation of the valence subband in InAs TFET is much stronger.

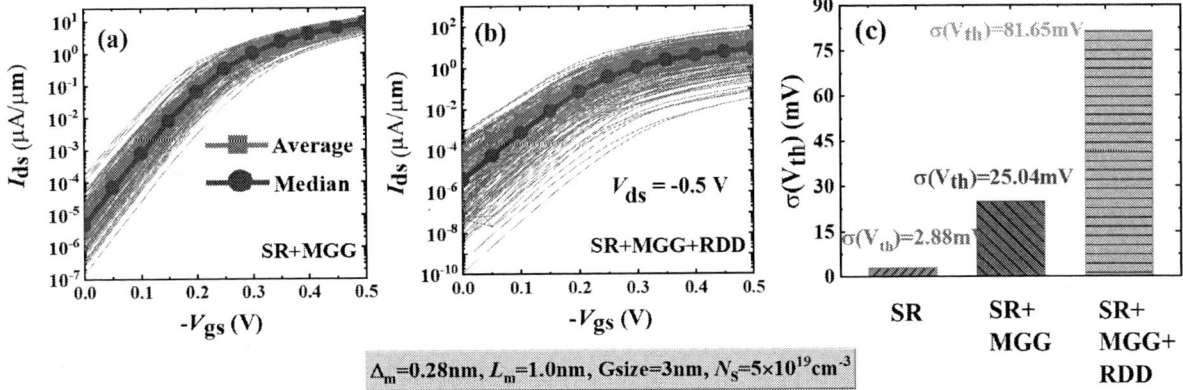

Fig. 5. The variability of V_{th} induced by SR compared to other major sources of variability (MGG and RDD) at $V_{ds} = -0.5$V of an ensemble of 200 InAs TFETs. MGG is generated from the realistic Voronoi pattern with average grain size (Gsize) of 3.0nm. RDD is considered in the source and drain regions, and covers 10nm length in each region as SR does. The number of dopants in RDD working regions obeys the Poisson distribution. For SR, $\Delta_m = 0.28$nm and $L_m = 1$nm. Compared with the influence from MGG and RDD, SR shows the least impact on the variability of the p-type InAs TFET.

transistors, resulting from the stronger quantum confinement effect in InAs channel due to its lighter hole effective mass than Si material. As a result, the SR-induced variability is more pronounced in InAs TFETs.

Except SR, there are still other source of variabilities, such as MGG, RDD, and trap-assisted tunneling (TAT). The latter strongly affects the subthreshold region by increasing the off-state BTBT current. TAT is not considered in this work since its impact is already well known [14]-[16]. We rather focus on the MGG- and RDD-induced variability. The MGG is generated by using the realistic Voronoi pattern [9], [17]. Here, we have considered a TiN metal gate which has two possible grain orientations (<200> and <111>) with a work function difference of 0.2 eV [18]. The occurrence probabilities of each orientation are 60% and 40%, respectively. The default average grain size of the metal gate is Gsize = 3nm. In case of RDD, the number of dopants in each of the TFETs is randomly chosen from a Poisson distribution, placing them by means of a probability rejection technique. The mean is determined by the doping concentration multiplied by the volume of the RDD region [9]. In InAs TFETs, the RDD regions cover 10 nm length in the source and drain, i.e. the same SR regions. For numerical purposes, uniform doping is adopted at the endings of source and drain regions to guarantee the numerical convergence.

Fig. 5 compares the impact of the different aforementioned variability sources, including SR, on the characteristics of TFETs. Compared with SR (Fig. 2), MGG has stronger influence on the $I_{ds} - V_{gs}$ characteristics, as observed in Fig. 5(a). When considering MGG, I_{off} ranges from $1.35 \times 10^{-7}\mu$A/μm to $1.64 \times 10^{-4}\mu$A/μm, larger than 3 orders of magnitude. I_{on} ranges from 3.84μA/μm to 13.54μA/μm, and its variation is still much weaker than I_{off} case. However, we find that RDD-induced variability is the dominant source of variability. This is shown in Fig. 5(b) where one can observe that the on-current now spreads over 4 orders of magnitude. The same effect is found for the off-current which spreads over 7 orders of magnitude. This strong impact is attributed to the position and number of the dopants that directly affect the electrical field across the tunnel junction [12], [19]. The impact on the V_{th} variation is summarized in Fig. 5(c). It is observed that the $\sigma(V_{th})$ in the overall combined case is nearly 30× larger than that in SR case.

IV. CONCLUSION

The impact of the SR on p-type InAs nanowire TFET is analyzed through quantum transport simulations of ensembles of 200 samples. The statistical analysis shows that the increase of RMS enhances the SR induced variability. Heterojunction TFETs, such as InAs-Si TFETs, not only can improve the performance of TFET but also could effectively reduce the influence of SR, due to the heavier hole effective mass in the Si channel and thus the weaker quantum confinement effect. It was also found that SR is the weakest source of variability in TFETs when compared to other variability sources such as MGG and RDD. Whereas, the RDD-induced variability is found to be the strongest.

ACKNOWLEDGMENT

Yunhe Guan is indebted to the China Scholarship Council (CSC) for its financial support as a visiting student at the University of Glasgow, U.K.

REFERENCES

[1] A. M. Ionescu and H. Riel, "Tunnel field-effect transistors as energy-efficient electronic switches," *Nature*, vol. 479, no. 7373, pp. 329-337, Nov. 2011, doi: 10.1038/nature10679.

[2] E. Memisevic, M. Hellenbrand, E. Lind, A. R. Persson, S. Sant, A. Schenk, J. Svensson, R. Wallenberg, and L. E. Wernersson, "Individual Defects in InAs/InGaAsSb/GaSb Nanowire Tunnel Field-Effect Transistors Operating below 60 mV/decade," *Nano lett.*, vol. 17, no. 7, pp. 4373–4380, Jul. 2017, doi: 10.1021/acs.nanolett.7b01455.

[3] H. Riel, K. E. Moselund, C. Bessire, M. T. Björk, A. Schenk, H. Ghoneim, and H. Schmid, "InAs-Si heterojunction nanowire tunnel diodes and tunnel FETs," in IEDM Tech. Dig., Dec. 2012, pp. 16.6.1–16.6.4, doi: 10.1109/IEDM.2012.6479056.

[4] W. Li and J. C. S. Woo, "Optimization and Scaling of Ge-Pocket TFET," IEEE Trans. Electron Devices, vol. 65, no. 12, pp. 5289-5294, Dec. 2018, doi: 10.1109/TED.2018.2874047.

[5] F. Conzatti, M. G. Pala and D. Esseni, "Surface-Roughness-Induced Variability in Nanowire InAs Tunnel FETs," IEEE Electron Device Lett., vol. 33, no. 6, pp. 806-808, June 2012, doi: 10.1109/LED.2012.2192091.

[6] H. Carrillo-Nuñez, R. Rhyner, M. Luisier and A. Schenk, "Effect of surface roughness and phonon scattering on extremely narrow InAs-Si Nanowire TFETs," 2016 46th European Solid-State Device Research Conference (ESSDERC), Lausanne, 2016, pp. 188-191, doi: 10.1109/ESSDERC.2016.7599618.

[7] A. Cresti, M. Pala, S. Poli, M. Mouis, and G. Ghibaudo, "A comparative study of surface-roughness-induced variability in silicon nanowire and double-gate FETs," IEEE Trans. Electron Devices, vol.

58, no. 8, pp. 2274–2281, Aug. 2011, doi: 10.1109/TED.2011.2147318.

[8] S. Berrada, H. Carrillo-Nuñez, J. Lee, C. Medina-Bailon, T. Dutta, M. Duan, F. Adamu-Lema, V. Georgiev, and A. Asenov, "NESS: New Flexible Nano-Transistor Simulation Environment,". International Conference on Simulation of Semiconductor Processes and Devices (SISPAD), Sept. 2018, pp. 24–26, doi: 10.1109/SISPAD.2018.8551701.

[9] J. Lee, O. Badami, H. Carrillo-Nuñez, S. Berrada, C. Medina-Bailon, T. Dutta, F. Adamu-Lema, V. P. Georgiev, and A. Asenov, "Variability Predictions for the Next Technology enerations of n-type SixGe1-x Nanowire MOSFET", Micromachines, vol. 9, no. 12, pp. 643-1-11, Dec. 2018, doi: 10.3390/mi9120643.

[10] H. Flietner, "The E(k) relation for a 2-band scheme of semiconductors and the application to metal-semiconductor contact," Phys. Stat. Sol. (b), vol. 54, pp. 201–208, 1972.

[11] H. Carrillo-Nuñez, J. Lee, S. Berrada, C. Medina-Bailon, M. Luisier, A. Asenov and V. P. Georgiev, "Efficient Two-Band based Non-Equilibrium Green's Function Scheme for Modeling Tunneling Nano-Devices," 2018 International Conference on Simulation of Semiconductor Processes and Devices (SISPAD), Austin, TX, 2018, pp. 141-144. doi: 10.1109/SISPAD.2018.8551629.

[12] H. Carrillo-Nuñez, J. Lee, S. Berrada, C. Medina-Bailon, F. Adamu-Lema M. Luisier, A. Asenov and V. P. Georgiev ., "Random Dopant-Induced Variability in Si-InAs Nanowire Tunnel FETs: A Quantum Transport Simulation Study," IEEE Electron Device Lett., vol. 39, no. 9, pp. 1473-1476, Sept. 2018, doi: 10.1109/LED.2018.2859586

[13] N. Agrawal, H. Liu, R. Arghavani, V. Narayanan and S. Datta, "Impact of Variation in Nanoscale Silicon and Non-Silicon FinFETs and Tunnel FETs on Device and SRAM Performance," IEEE Trans. Electron Devices, vol. 62, no. 6, pp. 1691-1697, June 2015, doi: 10.1109/TED.2015.2406333.

[14] R. N. Sajjad, W. Chern, J. L. Hoyt, and D. A. Antoniadis, "Trap assisted tunneling and its effect on subthreshold swing of tunnel FETs," IEEE Trans. Electron Devices, vol. 63, no. 11, pp. 4380–4387, Nov. 2016. doi: 10.1109/TED.2016.2603468.

[15] S. Sant and A. Schenk, "Trap-Tolerant Device Geometry for InAs/Si pTFETs," IEEE Electron Device Lett., vol. 38, no. 10, pp. 1363-1366, Oct. 2017, doi: 10.1109/LED.2017.2740262

[16] S. Sant and A. Schenk, "The effect of density-of-state tails on band-to-band tunneling: Theory and application to tunnel field effect transistors", Journal of Applied Physics, vol. 122, no. 13, pp. 135702, 2017, doi: 10.1063/1.4994112

[17] C. Hsu, M. Fan, V. P. Hu and P. Su, "Investigation and Simulation of Work-Function Variation for III–V Broken-Gap Heterojunction Tunnel FET," IEEE J. Electron Devices Soc., vol. 3, no. 3, pp. 194-199, May 2015, doi: 10.1109/JEDS.2015.2408356.

[18] K. M. Choi and W. Y. Choi, "Work-Function Variation Effects of Tunneling Field-Effect Transistors (TFETs)," IEEE Electron Device Lett., vol. 34, no. 8, pp. 942-944, Aug. 2013, doi: 10.1109/LED.2013.2264824.

[19] S. S. Sylvia, K. M. M. Habib, M. A. Khayer, K. Alam, M. Neupane, and R. K. Lake, "Effect of random, discrete source dopant distributions on nanowire tunnel FETs," IEEE Trans. Electron Devices, vol. 61, no. 6, pp. 2208–2214, Jun. 2014, doi: 10.1109/TED.2014.2318521.

978-1-7281-0941-1/19 $31.00 © 2019 IEEE

Efficient Coupled-mode space based Non-Equilibrium Green's Function Approach for Modeling Quantum Transport and Variability in Vertically Stacked SiNW FETs

V. Thirunavukkarasu[*1], H. Carrillo-Nunez[1], F. D. Alema[1], S. Berrada[1], O. Badami[1], C. Medina-Bailón[1], T. Datta[1], J. Lee[1], Y. Guen[2], V. Georgiev[1] and A. Asenov[1]

[1]School of Engineering, University of Glasgow, Glasgow G12 8LT, Scotland, UK.
[2]Department of Electrical Engineering, Xian Chiao Tung University, People Republic of China.
e-mail* nanodhasan@gmail.com

Abstract— In this paper we present state of the art coupled-mode space based Non-Equilibrium Green Function approach for modeling quantum transport accurately in the vertically stacked Silicon nanowire (SiNW) FETs. Random discrete dopants (RDD) and metal grain granularity (MGG) induced variability in stacked SiNW FETs are also investigated. Furthermore, charge spectrum, current spectrum w.r.t. sub bands and the space-resolved Local Density of States (LDOS) corresponding to the location of band edge are analyzed in detail. The newly developed flexible and computationally efficient models implemented in quantum transport simulation tool NESS provides valuable insights on the effect of RDD and MGG variability on Sub-Threshold Swing (SS), Threshold Voltage (V_{TH}) shift, On/Off Current (I_{ON}/I_{OFF}) ratio and quantum confined charge transport mechanism.

I. INTRODUCTION

As CMOS technology advances, logic devices are aggressively scaled [1]. It is very crucial to develop accurate models to investigate the ultra-scaled nano devices as the quantum charge transport is highly influenced by the strong confinement effects in quantum regime [2-4]. Stacked Gate All Around (GAA) transistors promises better gate electrostatics and high performance in sub 7nm advanced technology nodes [5-7]. To investigate short channel effects, quantum charge confinement, ballistic and diffusive transport, tunneling and variability induced effects accurately advanced tools with quantum NEGF approach is a must. In this work, we demonstrated quantum simulation tool NESS *(i)* capability to handle complex device structures and sophisticated geometries; *(ii)* extract and investigate the quantum transport with electron-phonon interactions and *(iii)* use the fully 3D self-consistent couple-mode space NEGF formalism to model the impact of the process induced variation such as MGG or RDD using statistical ensembles of 100 devices. The impact of phonon scattering on the performance of staked-nanowire FETs is also investigated. Details on the simulation procedure is given below.

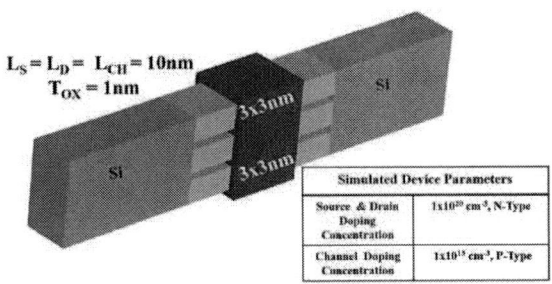

Fig. 1 Device structure of simulated Vertically Stacked Gate-All-Around Transistor with three 3 x 3nm stacked Channels. Inset shows the doping concentration used in the simulation.

II. SIMULATION METHODOLOGY

Assuming steady-state conditions, quantum transport is described by means of the Non-Equilibrium Green's Function (NEGF) formalism within the coupled mode-space representation [8][9], as implemented in the quantum transport simulator NESS [10] from the University of Glasgow. The NEGF solver is based on the effective-mass approximation and is self-consistently coupled with Poisson's equation. Both acoustic and optical phonon scattering are tackled within the self-consistent Born approximation (SCBA) [11]. Using the latter, the electron-phonon interactions are incorporated in the NEGF formalism through self-energies in terms of the retarded/lesser/greater-than Green's function $\Sigma^{R/</>}=M_{ep} G^{R/</>}$, where M_{ep} is the electron-phono coupling strength. Exact expression of the electron-phonon self-energies can be found in Refs. [12] [13]. In practice, both NEGF approach and SCBA are solved self-consistently using a recursive algorithm [11] until the criteria of convergence for both electron and current densities are reached.

We examined the quantum charge transport properties in stacked GAA transistors with three stacked GAA 3x3 channels as shown in Fig. 1 and compared the transfer characteristics of statistical simulations for 100 samples with *(i)* only RDD, *(ii)* only MGG and *(iii)* both RDD and MGG variability sources. Dopants in the source and drain regions are randomly chosen from a Poisson distribution, placing them by means of a probability rejection technique. The mean is determined by the

978-1-7281-0941-1/19 $31.00 © 2019 IEEE

doping concentration multiplied by the volume of the RDD region. Regarding MGG, the grains in the metal gate region are generated by using the Voronoi algorithm [14]. The work-function for each grain can be either 4.4 or 4.6 eV with the probability of 40% or 60% based on previous experimental results [15]. It was reported that, as the grain size increases, the more significant variability is observed, meaning that the small average grain size causes less variability [14]. Therefore, the average grain size of 3.0 nm used in this paper is small enough to expect a relatively less MGG-induced variability. The aforementioned models and an efficient NEGF recursive algorithms are implemented in NESS to the compute physical quantities like carrier density and current. Our main findings are summarized as follows.

III. RESULTS AND DISCUSSIONS

Fig. 2 compares the ballistic and scattering current at low bias V_{DS}= 0.05V with and without RDD+MGG for a 3x3 nanowire GAA device. The impact of scattering with RDD+MGG is not very significant. Hence to reduce computational cost, for 3x3 stacked nanowire statistical simulation of 100 devices ballistic simulations are considered. Fig. 3 shows the transfer characteristics (I_d-V_g) of the 100 three-channels stacked 3x3nm nanowire devices. I_d obtained is normalized by the effective width of the three stacked 3x3nm nanowires.

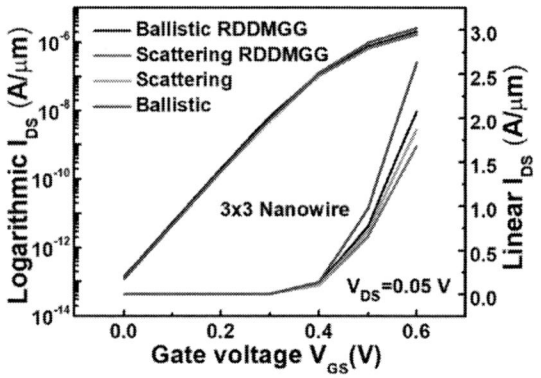

Fig. 2 Id-Vg Transfer Characteristics of 3x3 square nanowire devices in ballistic and considering phonon scattering with and without RDD+MGG. Impact of scattering is less severe in extremely-scaled NW GAA devices. Vds of 0.05V is applied.

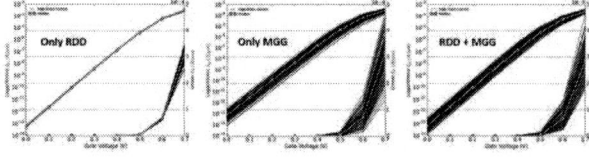

Fig. 3 Id-Vg Transfer Characteristics of statistical ensemble of 100 devices with only RDD, only MGG and both RDD+MGG. Median is highlighted as a line with red circles. Vds of 0.6V is applied.

The devices are simulated in three different setups. Devices with only RDD, only MGG and both RDD+MGG are considered. RDD has less impact on I_{on} current. Clearly the RDD is less significant compared to huge variability due to MGG which is evident from the transfer characteristics at V_{ds}=0.6V. The mean values of the drain current obtained is indicated in red.

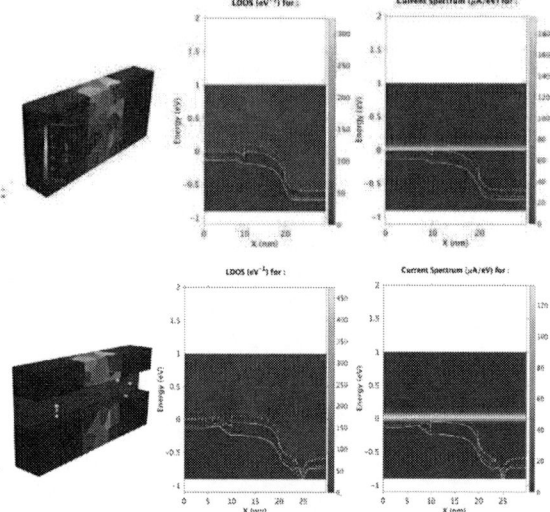

Fig. 4 Illustration of simulated Stacked Gate-All-Around Transistor with MGG and RDD+MGG along three 3 x 3nm stacked Channels. Space-resolved LDOS and Current Spectrum shows the impact of RDD and MGG on quantum-confined charge transport in sub bands.

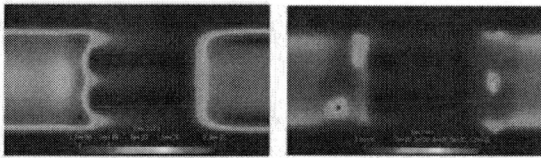

Fig. 5 Electron Density distribution in Stacked SiNW GAA without and with RDD. Random Dopants impacts the charge distribution.

Fig. 4 illustrates the MGG and RDD+MGG devices simulated. The current spectrum clearly indicates the impact of RDD in S/D region and MGG generated in the device along the transport direction. LDOS corresponding to the local band edge shows effect of potential barrier w.r.t. applied gate voltage. The space-resolved local density of state (LDOS) exhibits strong correlation between different electron densities which is shown in Fig. 5 for a device without RDD and with RDD. The impact of random dopants on the band edges can be seen. The Vth, Ion, Ioff probability distributions for the MGG, RDD and both RDD+MGG devices are shown in Fig. 6. The best-fit Gaussian function curve is indicated in the probability distribution plot. For devices with only RDD a standard deviation of 0.00136V,

978-1-7281-0941-1/19 $31.00 © 2019 IEEE

Vth and SS determines the performance of the stacked nanowire. Hence it is important to understand the impact metal grain granularity in stacked nanowire transistors to circumvents the performance degradation due to MGG.

IV. CONCLUSION

We examined the quantum charge transport in stacked GAA transistors and compared the transfer characteristics from RDD and MGG induced statistical variability in Stacked SiNW GAAFETs. The insights from the current spectrum and LDOS results sheds light on electron density and quantum transport which will help us to design ultra-scaled stacked GAA CMOS FETs for high performance logic devices in advanced nodes.

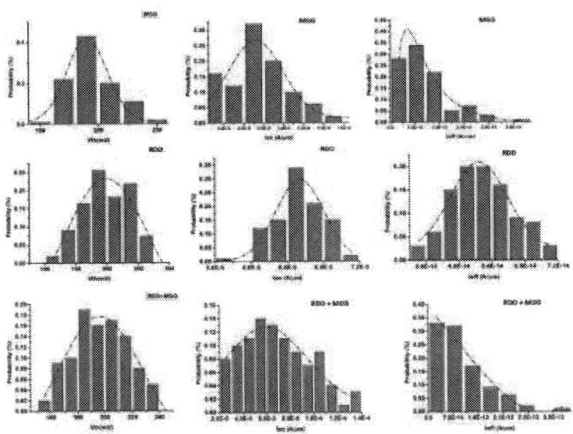

Fig. 6 Threshold Voltage (Vth), On current (Ion), Off current (Ioff) probability density distribution for three different variability samples along with best-fit Gaussian curve in black dotted-line.

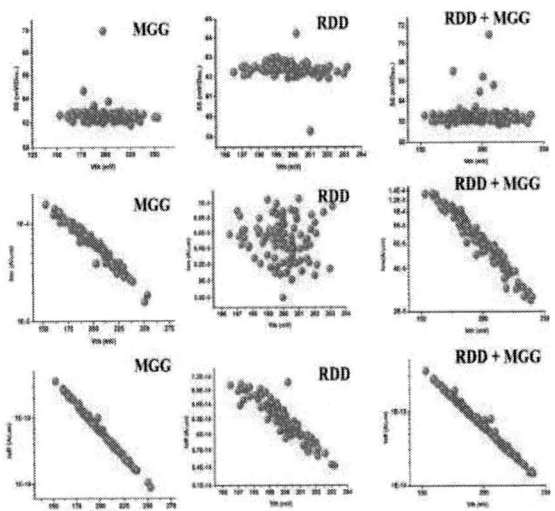

Fig. 7 Threshold Voltage (Vth) versus Subthreshold swing (SS), On current (Ion) and Off current (Ioff). RDD has significant impact on Ion. Devices with both RDD+MGG exhibit large variation in Ioff and SS.

$3.77E^{-14}$ A/μm, $2.60E^{-5}$ A/μm and 0.42mV/dec. are observed in Vth, Ioff, Ion and SS respectively. In RDD+MGG devices a standard deviation of 0.02015V, $6.78E^{-14}$ A/μm, $2.79E^{-5}$ A/μm is observed in threshold voltage (Vth), Off current (Ioff), on current (Ion) and subthreshold swing (SS) respectively. Fig. 7 shows the scatter plot of SS, Ion and Ioff variation w.r.t. threshold voltage Vth. SS can be fine-tuned by modulating the Vth further more. For RDD only devices SS_{max} of 64.5 mV/dec. and SS_{mean} of 62.36mV/dec. and for RDD+MGG devices SS_{max} of 70.94mV/dec. and SS_{mean} of 63.60mV/dec. is observed. The correlation between

References

[1] C. Hu, "Device Challenges and Opportunities," in VLSI Symp. Tech. Dig., pp. 4-5, Jun. 2004. doi: 10.1109/VLSIT.2004.1345359

[2] S. Migita, Y. Morita, M. Masahara, H. Ota, "Fabrication and Demonstration of 3-nm-Channel-Length Junctionless Field-Effect Transistors on Silicon-on-Insulator Substrates Using Anisotropic Wet Etching and Lateral Diffusion of Dopants "in Jpn. J. Appl. Phys., 52 pp.04CA01-01 - 04CA01-5, Feb. 2013. doi: 10.7567/JJAP.52.04CA01

[3] V. Thirunavukkarasu, Y. R. Jhan, Y. B. Liu, E. D. Kurniawan, Y. R. Lin, S. Y. Yang, C. H. Cheng, Y. C. Wu, "Gate-all-around junctionless silicon transistors with atomically thin nanosheet channel (0.65 nm) and record sub-threshold slope (43 mV/dec)" in Applied Physics Letters, vol. 110, no.3, pp. 032101-1 - 032101-5, Jan. 2017. doi: 10.1063/1.4974255

[4] M. Brandbyge, J. L. Mozos, P. Ordejón, J. Taylor, and K. Stokbro, "Density-functional method for nonequilibrium electron transport" *Phys. Rev. B* 65, 165401, March 2002. doi: 10.1103/PhysRevB.65.16540.

[5] C. H. Yu, Y. S. Wu, V. P. H. Hu and P. Su, "Impact of Quantum Confinement on Backgate-Bias Modulated Threshold-Voltage and Subthreshold Characteristics for Ultra-Thin-Body GeOI MOSFETs," in IEEE Transactions on Electron Devices, vol. 59, no. 7, pp. 1851-1855, July 2012. doi: 10.1109/TED.2012.2194499

[6] X. Jiang, S. Guo, R. Wang, X. Wang, B. Cheng, A. Asenov, and R. Huang, "A device-level characterization approach to quantify the impacts of different random variation sources in FinFET technology," IEEE Electron Device Lett., vol. 37, no. 8, pp. 962–965, Aug. 2016, doi: 10.1109/LED.2016.2581878

[7] M. Rau, E. Caruso, D. Lizzit, P. Palestri, D. Esseni, A. Schenk, L. Selmi, and M. Luisier, "Performance projection of III-V ultra-thin-body, FinFET and nanowire MOSFETs for two next-generation technology nodes," in IEDM Tech. Dig., Dec. 2016, pp. 30.6.1–30.6.4, doi: 10.1109/IEDM. 2016.7838515.

[8] D. Ferry and C. Jacobini, "Quantum transport in semiconductors", Springer Science, 1992.

[9] L.S.D. Esseni and P. Palestri, "Nanoscale MOS transistors: Semi-classical transport and applications", Cambridge University Press, 2011.

[10] S. Berrada, H. Carrillo-Nunez, Tapas Dutta, Meng Duan, Fikru Adamu-Lema, Jehyun Lee, Vihar Georgiev, Cristina Medina-Bailon, Asen Asenov, "NESS: new flexible Nano-Electronic Simulation Software," *2018 International Conference on Simulation of Semiconductor Processes and Devices (SISPAD)*, Austin, TX, 2018, pp. 22-25. doi: 10.1109/SISPAD.2018.8551701

[11] A. Svizhenko and M. P. Anantram, IEEE Trans. Electron Devices 50, 1459 (2003).

[12] H. Carrillo-Nuñez, M. Bescond, N. Cavassilas, E. Dib, and M. Lannoo, J. Appl. Phys. **116**, 164505 (2014) ; https://doi.org/10.1063/1.4898863

[13] C. Jacoboni and L. Reggiani, Rev. Mod. Phys. 55, 645 (1983).

[14] X. Wang, A. R. Brown, N. Idris, S. Markov, G. Roy, ans A. Asenov, Statistical Threshold-Voltage Variability in Scaled Decananometer Bulk HKMG MOSFETs: A Full-Scale 3-D Simulation Scaling Study. *IEEE Trans. Electron Devices* **2011**, *58*, 2293–2301

[15] H. Dadgour, K. Endo, V. De, ans K. Banerjee, Modeling and analysis of grain-orientation effects in emerging metal-gate devices and implications for SRAM reliability. In Proceedings of the 2008 IEEE International Electron Devices Meeting, San Francisco, CA, USA, 15–17 December 2008; pp. 1–4.

Surface Roughness Scattering in NEGF using self-energy formulation

Oves Badami
School of Engineering
University of Glasgow
Glasgow, United Kingdom
oves.badami@glasgow.ac.uk

Salim Berrada
School of Engineering
University of Glasgow
Glasgow, United Kingdom
salim.berrada@glasgow.ac.uk

Hamilton Carrillo-Nunez
School of Engineering
University of Glasgow
Glasgow, United Kingdom
hacarrillo@gmail.com

Cristina Medina-Bailon
School of Engineering
University of Glasgow
Glasgow, United Kingdom
cristina.medinabailon@glasgow.ac.uk

Vihar Georgiev
School of Engineering
University of Glasgow
Glasgow, United Kingdom
vihar.georgiev@glasgow.ac.uk

Asen Asenov
School of Engineering
University of Glasgow
Glasgow, United Kingdom
asen.asenov@glasgow.ac.uk

Abstract—**The microelectronic industry has moved from matured bulk planar transistor to three-dimensional (3D) architectures with small non-trivial cross-sections and short channel lengths requiring quantum simulation techniques. In addition, novel materials, which enhance the transistor performance, are considered as silicon channel replacement. This necessitates the efficient inclusion of surface roughness scattering in quantum transport simulations. In this work, we report an approximate methodology to include surface roughness scattering in 3D Non-Equilibrium Green's Function (NEGF) simulations using self-energy formulation within the self-consistent Born approximation (SCBA). The method is validated with the well established methodology of treating surface roughness as a variability source. We also extract the mobility from our simulations and then compare with to those reported in the literature.**

Index Terms—**surface roughness, NEGF, mobility, scattering**

I. INTRODUCTION

The progress of the semiconductor industry in the last 50 years has been governed by scaling the physical dimensions of the metal-oxide-semiconductor field effect transistor (MOSFET). In order to continue this scaling trend, the industry has adopted 3D multi-gate architecture [1]. This led to a better control of short channel effects that limit the scaling of the conventional (bulk) MOSFETs. The attention of both industry and academia is also focused on the novel materials to boost the performance of the transistor without scaling of the MOSFETs [2]. These factors make the interface between the oxide and semiconductor even more critical in the working of the MOSFET. Because of these reasons, modeling of surface roughness scattering is important for predicting the performance of the contemporary and future transistors.

The non-equilibrium Green's Function (NEGF) formalism has become the method of choice to model ultra-scaled transistors. NEGF can accurately capture the confinement

Part of this research was funded by the EPSRC (UK) grant no. EP/S001131/1

in the direction normal to the electron transport as well as quantum transport effects, such as source to drain tunneling [3]. However, the efficient modelling of the surface roughness (SR) scattering in NEGF for realistic current simulations has not been addressed yet in the literature [4]. Therefore, in this paper, we report a new approximate methodology to include SR scattering in the 3D NEGF simulation through the inclusion of a self-energy expression for the SCBA. This is implemented in our comprehensive in-house simulation framework, NESS – Nano-Electronic Simulation Software [5].

II. METHODOLOGY

A. Surface Roughness Scattering

Traditionally, SR scattering in the NEGF formalism has been modeled by generating an ensemble of devices with statistically different surface roughness configurations and by performing transport simulation for each one of them [6]–[8]. Then the statistical average of the current is computed for the ensemble in order to estimat the effect of surface roughness scattering. Due to the need of large simulation ensemble to reduce the statistical noise, this method is computationally expensive. In this work, the interface between the semiconductor and oxide is characterized by an auto-correlation function

$$C(r) = \Delta_{rms}^2 e^{-\sqrt{2}r/L_C} \tag{1}$$

where Δ_{rms} is the root mean square of the surface roughness, L_C is the correlation length, and r is the distance between two mesh points.

The electron dynamics in the coupled mode space NEGF implementation is modeled by [9]:

$$\left[E - H_{MS} - \Sigma_{MS}^R \right] G_{MS}^R = I \tag{2}$$

$$G_{MS}^< = G_{MS}^R \Sigma_{MS}^< \left(G_{MS}^R \right)^\dagger \tag{3}$$

where H_{MS} is the Hamiltonian, Σ_{MS}^R and $\Sigma_{MS}^<$ are retarded and lesser self energies (comprising of both the contacts and interactions), respectively, and I is the identity matrix. In this work, we have employed parabolic effective mass approximation, which has been shown to give reasonably accurate results when compared to the full-band approach [10]. For elastic interactions, the self-energies [11] can be written as:

$$\Sigma_{MS}^{<} = M^2 G_{MS}^{<} \tag{4}$$

$$\Sigma_{MS}^{R} = \frac{M^2}{2}\left(\Sigma_{MS}^{>} - \Sigma_{MS}^{<}\right) \tag{5}$$

where M is the matrix element. For SR scattering, the ensemble average squared matrix element (using exponential power spectrum of the surface roughness) can be written as [12]:

$$
\begin{aligned}
\langle|M_{m,n}(k,k')|^2\rangle &= |\langle m,k'|H_{pert}|n,k\rangle|^2 \\
&= \frac{e^2 N_E^2(m,n)\Delta_{rms}^2\sqrt{\pi}L_C}{L_x}e^{-\frac{q^2 L_C^2}{4}}
\end{aligned} \tag{6}
$$

where H_{pert} is the perturbed Hamiltonian (SR), L_x is the normalization length, q is the difference of magnitudes of the wavevectors before and after scattering ($|k_x| - |k_x'|$), m and n are the subband indices. The effective electric field, $N_E(m,n)$, can be calculated as:

$$N_E(m,n) = \int d\mathbf{A}\zeta_n(\mathbf{r})\mathbf{E}_{y,z}(x)\zeta_m(\mathbf{r}) \tag{7}$$

where $\zeta_m(\mathbf{r})$ is the wavefunction corresponding to the mth subband in the cross-section, and $\mathbf{E}_{y,z}$ is the electric field along the confinement directions, which correspond to 'y' and 'z' directions in this work. In the limit $q \to 0$ (and making the self-energies local in real space) makes the matrix element independent of the wave-vector and thus allows us to express the matrix elements in coupled mode space as:

$$
\begin{aligned}
\langle|M_{m,n}(x,x')|^2\rangle &= |\langle m,x'|H_{pert}|n,x\rangle|^2 \\
&= e^2 N_E^2(m,n)\Delta_{rms}^2\sqrt{\pi}L_C\delta(x-x')
\end{aligned} \tag{8}
$$

B. Mobility Calculation

We simulate three nanowires transistors with gate lengths, L_{G1}, L_{G2} and L_{G3}, and with a very low drain bias ($V_{DS} = 1$ mV) [13]. Then, using their drain current I_{D1}, I_{D2} and I_{D3}, the mobility (μ) and the necessary condition for the diffusive transport (C_D) can be calculated. The mobility is given by

$$\mu = \frac{L_{G2} - L_{G1}}{R_2 - R_1}\frac{1}{e\rho_{1D}} \tag{9}$$

where $R_1 = V_{DS}/I_{D1}$ and $R_2 = V_{DS}/I_{D2}$ are the resistances of the nanowires, e is the unit electron charge and ρ_{1D} is the 1D charge density evaluated at the center of the device. The diffusive transport condition (variation in the resistance is linearly proportional to the change in the gate length) can be evaluated using

$$C_D = \frac{R_3 - R_2}{L_{G3} - L_{G2}}\frac{L_{G2} - L_{G1}}{R_2 - R_1} \approx 1 \tag{10}$$

where $R_3 = V_{DS}/I_{D3}$.

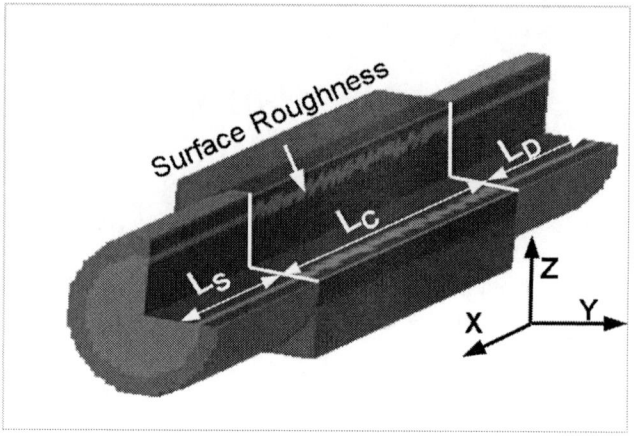

Fig. 1. Schematic of the circular nanowire and illustrating the surface roughness in gated region that is used in the statistical simulations. For the simulation gate length, L_G, was set to 20 nm, and source (L_S) and drain (L_D) length are set to 10 nm. In this study, the diameter, D, of the device was varied.

III. RESULTS

Fig. 1 shows the schematic of the silicon nanowire transistor (NWT) used in the simulations. The transport orientation of the nanowire was [100]. The source and drain doping is 10^{20} cm^{-3} and the channel doping is 10^{15} cm^{-3}. The effective oxide thickness is 0.8 nm. In this work, surface roughness was considered only in the gated region. This is because the electric field normal to the transport direction is highest under the gate and thereby it is expected to dominate the transport. The oxide-semiconductor interface in the source and drain region is assumed to be smooth.

Fig. 2 shows the comparison of the band structure between the tight-binding method (calculated using QuantumATK [14] and parameters published in [15]) and the parabolic effective mass method. The spin-orbit coupling was ignored in the tight-binding band structure calculations as it is expected to have negligible impact on the conduction band. The comparison highlights an excellent agreement between the tight-binding and effective mass (with calibrated effective masses) approximation in the energy range relevant for the MOSFET in terms of the subband minimas and band curvature around the minimas. In this paper, the effective masses for different diameters were calibrated to reproduce the band-structure calculated with the tight-binding method. The effective masses for different cross-sections are reported in [16].

Fig. 3 compares the transfer characteristics for the circular nanowire with diameter of 4 nm. The result shows a good agreement between the transfer characteristic for a nanowire MOSFET obtained with mean of the statistical simulations and self-energy simulations. For statistical simulations, an ensemble of 140 instances were used to obtain a correct statistical average. The root mean square of the surface roughness, Δ_{rms}, for statistical simulation was taken to be 0.21 nm which is consistent with the experimental data for bulk Si

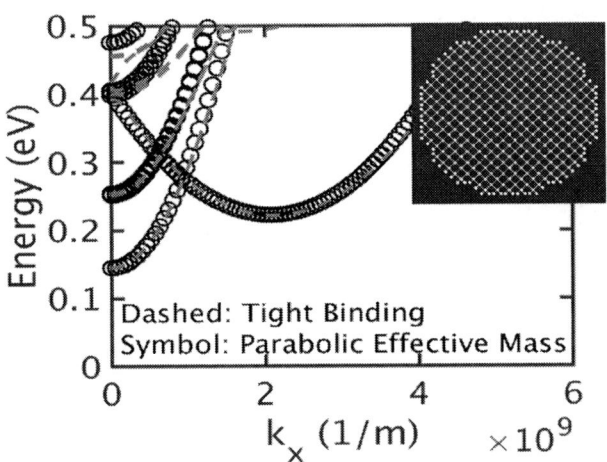

Fig. 2. Comparison between the tight-binding and parabolic effective mass approximation for a circular nanowire with the diameter, D = 4 nm. The inset shows the arrangement of atoms in the cross-section of the nanowire. The bottom of the bulk conduction band edge was taken to be the reference (0 eV).

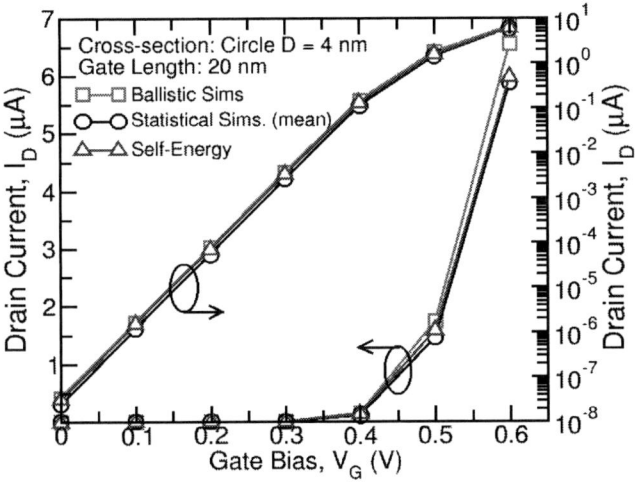

Fig. 3. Comparison of the transfer characteristics calculated with the self-energy formulation and mean of the statistical simulations is reported. The diameter of the device was set to 4 nm. A good comparison between the two methods is obtained.

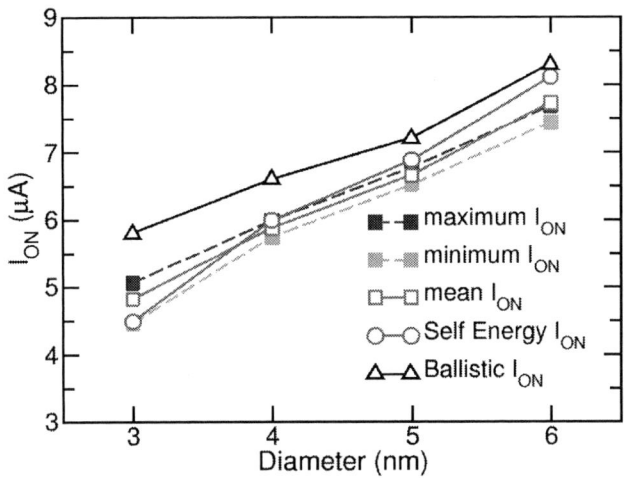

Fig. 4. Comparison of the mean on-current calculated with the statistical simulations and the self-energy method discussed. For comparison, we have also plotted the maximum and minimum on-currents calculated with the statistical simulations. The Δ_{rms} used for the statistical simulation was 0.21 nm while for the self-energy it was taken to be 0.18 nm. The correlation length, L_c, was set to 1.4 nm for all the simulations.

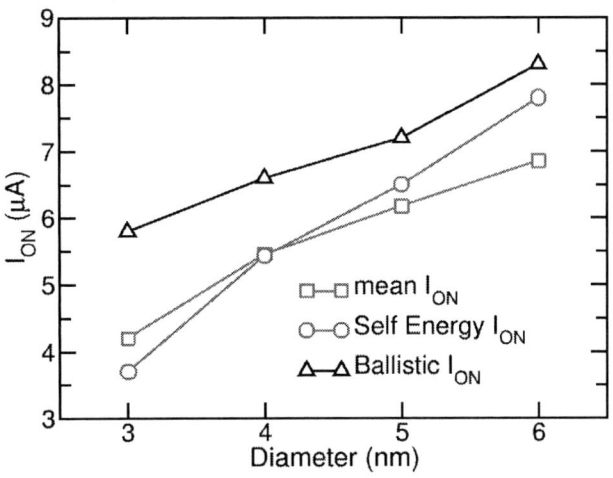

Fig. 5. Comparison of the mean on-current calculated with the statistical simulations and the self-energy method discussed. The Δ_{rms} used for the statistical simulation was 0.31 nm while for the self-energy it was taken to be 0.3 nm. The correlation length, L_c, was set to 1.4 nm.

based transistors [17]. Δ_{rms} for self-energy formulation was calibrated to 0.18 nm to match the statistical simulation. Fig. 4 and Fig. 5 shows the comparison between the mean on-current (I_{on}) calculated with the statistical simulations and with the self-energy for diameters ranging from 3 nm to 6 nm and for different surface roughness values. The I_{on} was defined as the drain current at $V_{DS} = V_{GS} = 0.6$ V. The results highlight a fair comparison of the I_{on} between the statistical and self-energy based methodologies with some discrepancy at larger diameter (D = 6 nm).

Fig. 6 compares the mobility calculated with NEGF formulation and those reported in the literature [18]. The mobility in this work was calculated using the method discussed Section II-B. For the purpose of the mobility calculations the gate length of 20, 30 and 40 nm were used. The effective electric field (E_{eff}) was calculated at the center of the device and it is given by:

$$E_{eff} = \frac{1}{n_s} \int \int dy \; dz \; n(\mathbf{r}) \left[E_y^2 + E_z^2 \right]^{0.5} \quad (11)$$

where n_s is the total electron density in the cross-section, E_y and E_z are the electric field along the y and z directions and $n(\mathbf{r})$ is the electron density in the cross-section. In order

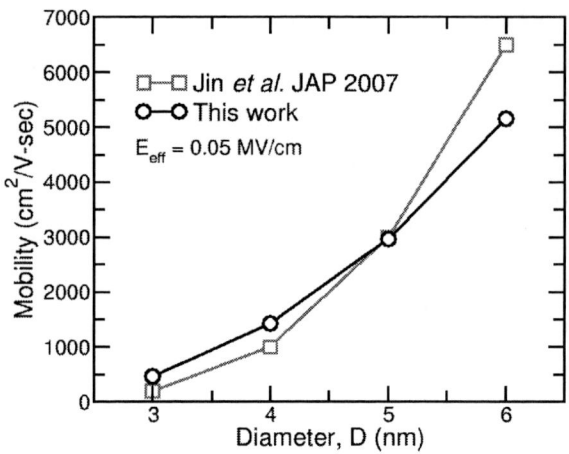

Fig. 6. Comparison between the mobilities calculated from NEGF and ones reported in [18] for circular nanowires with different diameters and for an effective field, $E_{eff} = 0.05$ MV/cm. The effective mass, oxide-semiconductor barrier are consistent with those used in the Ref. [18].

to match the mobility, $\Delta_{rms} = 0.21$ nm and $L_c = 1.4$ nm were used. A fair agreement is observed between the mobility calculated with NESS-NEGF simulations and the results reported in the literature obtained using the Kubo-Greenwood formulation.

IV. CONCLUSION

In this paper, we propose a self-energy correction, Σ_{SR}, that allows a perturbative treatment of the surface roughness scattering in 3D NEGF simulations, which greatly reduces the computational burden compared to statistical simulations. The locality of Σ_{SR} in real space makes it easy to implement. The ON-current obtained by the perturbative treatment showed a fair agreement with statistical simulations for different cross-section areas for circular nanowire transistors. The mobility values that we extracted using self-energy formulation in NESS-NEGF follow the same trend as the ones available in the literature.

REFERENCES

[1] C.-H. Jan *et al.*, "A 22nm SoC Platform Technology Featuring 3-D Tri-Gate and High-k/Metal Gate, Optimized for Ultra Low Power, High Performance and High Density SoC Applications," 2012 International Electron Devices Meeting, San Francisco, CA, 2012, pp. 3.1.1-3.1.4.

[2] G. Doornbos *et al.*, "Benchmarking of III-V n-MOSFET Maturity and Feasibility for Future CMOS," IEEE Electron Device Lett., vol. 31, no. 10, pp. 1110 1112, Oct. 2010.

[3] C. Grillet *et al.*, "Assessment of the Electrical Performance of Short Channel InAs and Strained Si Nanowire FETs," Trans. on Elec. Devices, vol. 64, no. 5, June 2017.

[4] G. Kannan *et al.*, "The impact of surface-roughness scattering on the low-field electron mobility in nano-scale Si MOSFETs," Jour. of Appl. Phys., vol. 122, 2017.

[5] S. Berrada *et al.*, "NESS: new flexible Nano-Electronic Simulation Software," SISPAD 2018.

[6] A. Martinez *et al.*, "Variability in Si nanowire MOSFETs due to the combined effect of interface rogughness and random dopants: A fully three-dimensional NEGF simulation study," Trans. on Electron Devices vol. 57, no. 7, 2010.

[7] F. Conzatti *et al.*, "Surface-Roughness-Induced Variability in Nanowire InAs Tunnel FETs," Trans. on Elec. Devices, vol. 3, no. 6, June 2012.

[8] J. Wang *et al.*, "Theoretical investigation of surface roughness scattering in silicon nanowire transistors, " Applied Physics Letters, vol. 87, no. 4, 2005.

[9] M. Lusier *et al.*, "Quantum transport in two- and three-dimensional nanoscale transistors: Coupled mode effects in the non-equilibrium Greens function formalism," Jour. of Appl. Phys, vol. 100, 2006.

[10] Jan-Laurens P. J. van der Steen *et al.*, "Validity of the Parabolic Effective Mass Approximation in Silicon and Germanium n-MOSFETs With Different Crystal Orientations," Trans. on Elec. Devices, vol. 54, no. 8, August 2007.

[11] Roger Lake EE 212 Notes: Quantum Devices Modeling with Non Equilibrium Green Function.

[12] T. Sadi *et al.*, "Simulation of the Impact of Ionized Impurity Scattering on the Total Mobility in Si Nanowire Transistors," Materials , 12(1), 2019.

[13] Reto Rhyner, "Quantum Transport Beyond the Ballistic Limit," PhD Thesis, ETH 2015.

[14] QuantumATK version O-2018.06. 2018. [Synopsys, inc., 2018].

[15] T.B. Boykin *et al.*, "Valence band effective-mass expressions in the $sp^3 d^5 s^*$ empirical tight-binding model applied to a Si and Ge parameterization," Phys. Rev. B 69, 115201, 2004.

[16] O. Badami *et al.*, "Comprehensive study of cross-section dependent effective masses for silicon based gate-all-around transistors," Applied Sciences, vol. 9, no. 9, April 2019.

[17] O. Badami *et al.*, "Improved surface roughness modeling and mobility projections in thin film MOSFETs," in Solid State Device Research Conference (ESSDERC), 2015 45th European, 2015, pp. 306 309

[18] S. Jin *et al.*, "Modeling of electron mobility in gated silicon nanowires at room temperature: Surface roughness scattering, dielectric screening, and band nonparabolicity," Jour. of Appl. Phys., vol. 102, 2007.

Impact of BEOL Design on Self-heating and Reliability in Highly-scaled FinFETs

Jaehee Choi, Udit Monga, Yonghee Park, Hyewon Shim*, Uihui Kwon, Sangwoo Pae*, Dae Sin Kim
Semiconductor R&D Center, *Foundry Business,
Samsung Electronics Co., Ltd.,
Hwasung-si, Gyeonggi-do, Korea.
email:joyful.choi@samsung.com, udit.monga@samsung.com

Abstract— This paper investigates the impact of BEOL design on device and backend reliability – HCI, BTI, EM – due to dependence of self-heating on BEOL in highly-scaled FinFETs. Our analysis indicates that due to poor thermal coupling to substrate – in the thin fin body devices – a large part of heat flows out of BEOL. This makes self-heating, and thus device (FEOL) temperature, very sensitive to BEOL design. The heat flow through BEOL also significantly increases the metal and via temperatures. The increased temperature negatively affects the overall reliability, and one of the ways to mitigate device degradation is optimization of BEOL design.

Keywords— *Impact of BEOL design, Self-heating effect, Aging, HCI, BTI, EM, Reliability, FinFET*

I. INTRODUCTION

Bulk FinFETs started replacing planar CMOS at 22nm because of its superior electrical characteristics [1]. Self-heating is one of the major reliability concerns in FinFETs [2], and continued scaling is only going to aggravate this issue. The large self-heating in FinFETs compared to planar devices is attributed to poor thermal coupling and reduced thermal conductivity in highly-scaled FinFETs – arising from the effects of size and surface roughness [3]. Lower thermal conductivity of thin silicon due to size effects significantly inhibits the thermal conduction to substrate, and back-end-of-line (BEOL) offers relatively lower resistive path. This means the dominant heat dissipation path is through the BEOL from the device to ambient. Because BEOL serves as a thermal path, the reliability mechanism of the device is affected by the BEOL configuration, even if it is thought to be independent. In this paper, we analyze the impact of BEOL design variation on self-heating in sub-10nm FinFETs. Our analysis indicates that BEOL design has strong impact on both device and backend temperature. And this also affects the device aging due to BTI (Bias Temperature Instability), HCI (Hot Carrier Injection) and EM (Electro-Migration).

II. METHODOLOGY

A. Device-Level Simulation

Well-calibrated TCAD and 3D FEM solver (Sentaurus Interconnect) were used to simulate SH in FEOL and BEOL. In nano-scale devices, most of the heat is generated deep inside the drain junction [4], and location of maximum heat generation is quite important to calculate accurate temperature rise. To accurately estimate the location of maximum heat generation, spherical harmonics expansions (SHE) solver of Boltzmann transport equation (BTE) was used in the device simulation. [5]

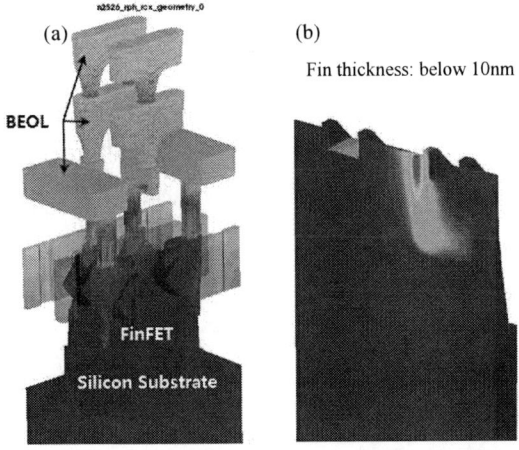

Fig. 1. (a) SI structure with BEOL (b) TCAD device structure

Moreover, in highly-scaled FinFETs, the thermal conductivity of Fin is significantly lower than its bulk counterpart due to phonon confinement [3], and thus a thickness-dependent thermal conductivity model was also employed. Fig. 1 illustrates one of the generated TCAD structures. The analysis present in this paper is for a large inverter cell – multi-fin and multi-finger inverter – at high bias values. The structure was simulated with three different top metal levels – M1, M2, and M3. Dirichlet boundary condition of T=300K was assumed at top-metal and substrate.

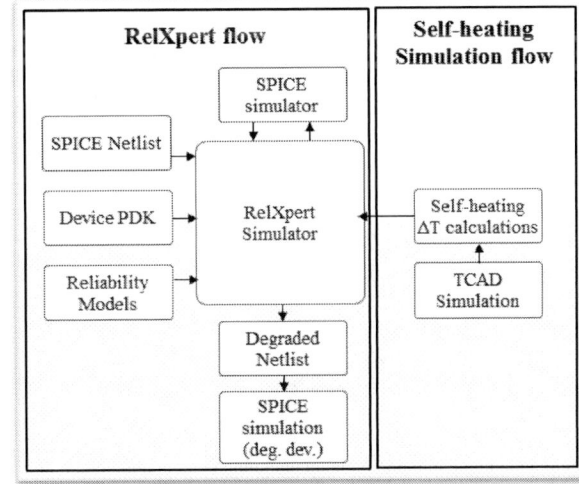

Fig. 2. Cadence-RelXpert simulation flow with self-heating.

B. Circuit-Level Simulation

As the circuit ages, the performance degrades at some point until unacceptable. Therefore, it is important to evaluate the end-of-life (EOL) of the circuit during the initial circuit design to enable design optimization. To analyze the impact of self-heating on device aging (BTI and HCI), Cadence-RelXpert was used with modified simulation flow. Fig. 2 shows the RelXpert's simulation flow [6] augmented with self-heating simulations.

III. BEOL IMPACT AND ITS IMPLICATIONS

A. BEOL Impact : self-heating effect in device temperature

This study focuses on highly-scaled FinFETs as shown in Fig. 1 where the heat flows over backend. Heat dissipation happens mainly by conduction of metal interconnect. A significant point is that heat transfer efficiency from the device to the top metal level is dominated by on the differences in the number of vias. Fig. 3 illustrates the 2D device temperature profile in metal-via plane. Fig. 4 shows the variation of average device temperature with number of vias (normalized), for three different top metal levels. The result shows that the device temperature exponentially increases with decreasing the number of vias. As an example, we observed a 33% reduction of device temperature by increasing the number of vias seven times. It also observed a 20% reduction while changing top metal level.

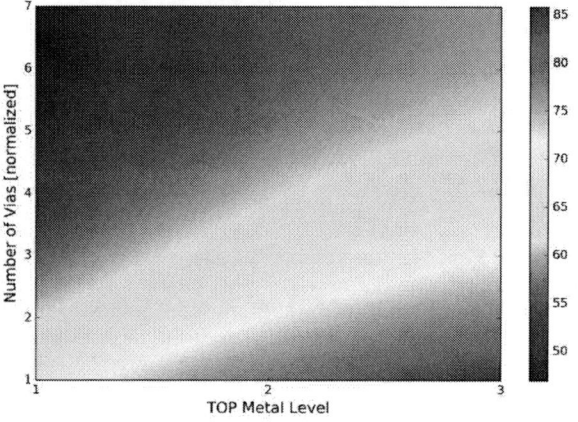

Fig. 3. Temperature profile. Surface plot of 2D thermal gradient.

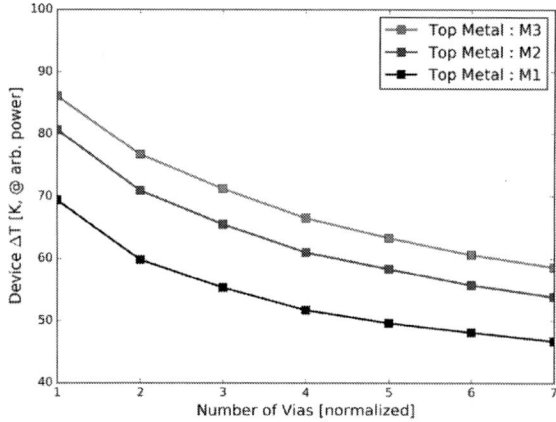

Fig. 4. Temperature with number of vias for three top metal levels.

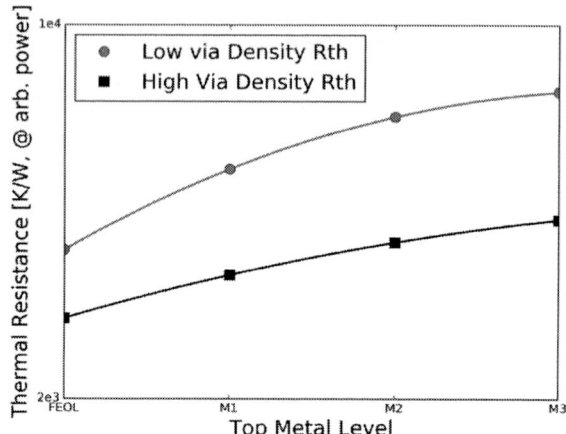

Fig. 5. Thermal resistance for three top metal levels

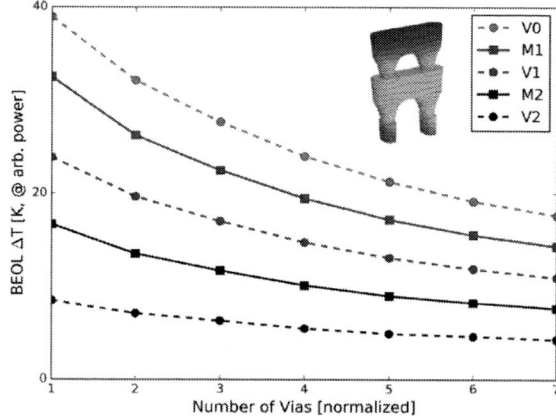

Fig. 6. BEOL Temp. of Mx and Vx layers with number of vias.

Fig. 5 shows the extracted thermal resistance value according to the stacked BEOL layer and via density, which will be used in circuit simulation. As evident from the figure, self-heating significantly reduces with number of vias and increases with top metal level. As aforementioned (in introduction section), device scaling significantly reduces the thermal conductivity of Silicon and thus heat flows mainly through BEOL since it offers lower thermal resistance. Adding additional metal layer increase the thermal resistance while increasing the via number results in the its reduction. Our analysis indicates that vias have very significant impact on self-heating, and a substantial reduction in device temperature is possible by increasing via density in the design. Since inter-layer-dielectric (ILD) has very low thermal conductivity, in highly-scaled FinFETs, vias will become "thermal-bottleneck".

B. BEOL Impact : self-heating effect in backend temperature

In addition, to studying the vertical variations in temperature at a certain inverter in the 3D structure, the temperature gradients were analyzed at different vias, as shown in Fig. 6. This figure shows the variation of average temperature of metal and via layer with number of vias. Due to heat flow through backend, metal and via temperatures also increase with increase in self-heating.

C. BEOL Impact : reliability

The most widely used method for the degradation of metal wire is Mean-Time-To-Failure (MTTF), which in case of electro-migration is modeled by Black's law:

$$MTTF = \frac{A}{J^n} \exp\left(\frac{E_a}{kT}\right) \qquad (1)$$

Therefore, increase in backend temperature can lead to serious issues, since MTTF due to electro-migration exponentially decreases with increase in BEOL temperature [7]. According to (1), Fig. 7 shows that MTTF increases with increase in via density. It is thus evident that the reduction in BEOL thermal resistance is imperative for the mitigation of electro-migration.

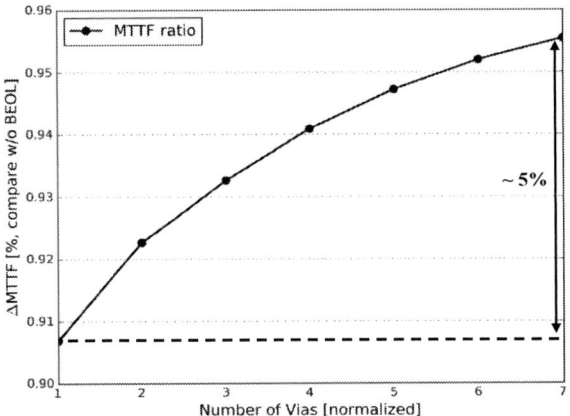

Fig. 7. MTTF degradation with number of vias for M1.

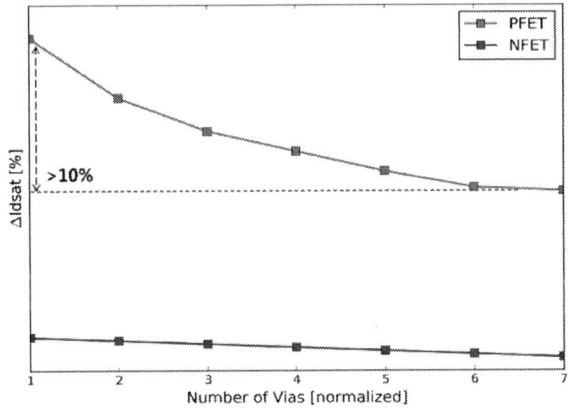

Fig. 8. Idsat degradation variation due to BEOL. NFET shows negligible variation.

Increase in temperature due to self-heating is also detrimental to device aging due to BTI and HCI. Since the device temperature is dependent upon BEOL, we studied the impact of BEOL on device aging Cadence-RelXpert [6]. Fig. 2 shows the RelXpert flow used for the simulation, the self-heating simulation results for different BEOL configurations were fed to RelXpert. The aging simulation was done for both BTI and HCI and the simulation was done for 1 year at room temperature. The temperature rise was only due to self-heating. Ideally, BEOL variations wouldn't impact the device-level reliability. But since the device temperature varies strongly with backend variation, so does the device

degradation. As evident from Fig.8, more than 10% reduction in PFET drain-current degradation is possible by increasing vias in the design. For NFET, the variation is insignificant due to much lower degradation as compared to PFET.

D. BEOL Impact : self-heating effect in circuit temperature

From a thermal point of view, the main difference between the D.C. and A.C. operations is the power consumption of the internal device of the circuit, which allows the different temperature ranges with respect to the frequency. Fig. 9 and Fig. 10 show the SH simulation results of an inverter chain. The solid lines represent the device temperature with via density and the dotted lines represent the M1 temperature with via density. So, by increasing the via density and kept the top metal to M3 as shown in Fig. 10, we observed that the device temperature and M1 temperature decreased. The self-heating during A.C. circuit operation is much lower than SH during D.C. operation due to finite thermal capacity. Notice that D.C. means that current continues to flow on the device. However, heat accumulation still raises the device and backend temperature. From the figure, it's evident that the device temperature and metal layer temperature during the AC operation depends upon BEOL design.

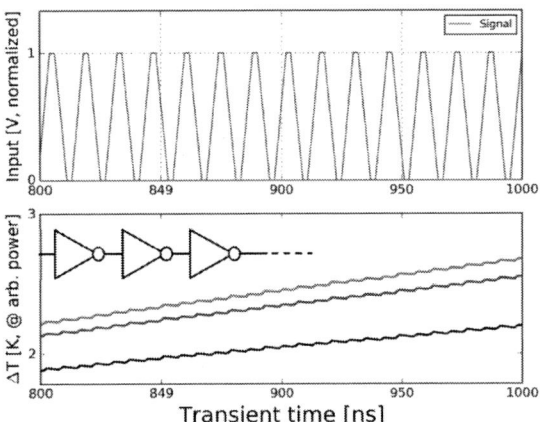

Fig. 9. Variation of the AC and self-heating over time at the inverter.

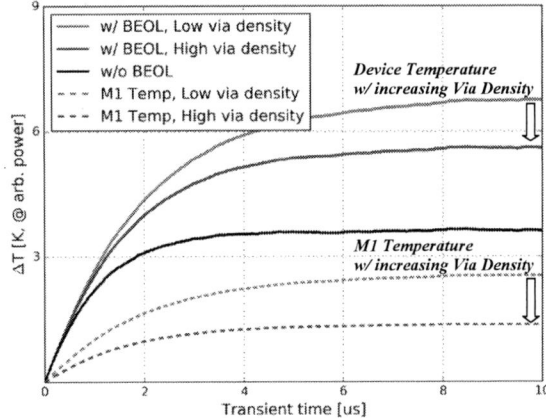

Fig. 10. Self-heating of the inverter. NFET.

IV. CONCLUSIONS

We have shown the importance of BEOL optimization for self-heating reduction in highly-scaled technologies. Our analysis indicates that significant portion of heat generated due to self-heating flows through vias and large via density is important for reduction of device temperature. We also presented that the reduction of backend thermal resistance is important for mitigation of electro-migration. Finally, we have analyzed the impact of BEOL dependent self-heating on device aging due to BTI and HCI. Thus, one way to effectively suppress increasing temperatures to improve reliability is to increase via density. As discussed earlier, the optimization of BEOL designs increases the reliability margin.

ACKNOWLEDGMENT

The authors would like to acknowledge Dr. Jeon (Asst. Prof., Kon-kuk University) and Dr. Harsono Simka (Samsung Advanced Logic Lab) for helpful suggestions, and Samsung San Jose R&D team for TCAD solution development.

REFERENCES

[1] C. Auth, et al, "A 22nm high performance and low-power CMOS technology featuring fully-depleted tri-gate transistors, self-aligned contacts and high density MIM Capacitors," Proc. VLSI. Tech., pp. 131-132, 2012

[2] C. Prasad, et al, "Self-heat reliability considerations on Intel's 22nm tri-gate technology, " Proc. IRPS 2013, pp. 5.D.1.1-5.D.1.5

[3] M.Kazan, et al, "Thermal conductivity of silicon bulk and nanowires: Effects of isotopic composition, phonon confinement, and surface roughness," J. Appl. Phys (108), 2010

[4] Eric pop, "Self-heating and scaling of thin body transistors", PhD thesis, Stanford University, 2004.

[5] S. Jin, A. Wettstein, W. Choi, F. M. Bufler and E. Lyumkis, "Gate current calculations using spherical harmonic expansion of Boltzmann equation," 2009 International Conference on Simulation of Semiconductor Processes and Devices, San Diego, CA, 2009, pp. 1-4. doi: 10.1109/ SISPAD.2009.5290216

[6] Virtuoso RelXpert reliability simulator user guide, ver. 12.1.1, 2013

[7] J. R. Lloyd, C. E. Murray, T. M. Shaw, M. W. Lane, X.-H. Liu, and E. G. Liniger, "Theory for electromigration failure in Cu conductors," in Proc. AIP Conf., 2006, vol. 817, pp. 23–33.

Modeling 1/f and Lorenzian noise in III-V MOSFETs

E. Caruso*, F. Bettetti, L. Del Linz, D. Pin, M. Segatto, P. Palestri

Tyndall National Institute, University College Cork, Cork, Ireland
DPIA, University of Udine, Via delle Scienze 206, 33100, Udine, Italy

Abstract— **We present an approach to model 1/f and random telegraph noise in TCAD combining the models for non-local tunneling to traps and generation/recombination noise. The TCAD results are compared with simple numerical expression to understand the influence of the device and trap parameters on the noise spectrum. The simulation deck is then used to compute the low-frequency noise spectrum in III-V MOSFETs using traps distributions extracted from multi-frequency C-V measurements.**

Keywords—Lorenzian, 1/f, border traps, modelling, TCAD, noise

I. INTRODUCTION AND MODEL DESCRIPTION

1/f noise in MOSFETs has a significant impact on the performance of RF circuits such as oscillators and mixers due to up-conversion [1]. The power spectral density goes as the reciprocal of the area and so becomes critical with device scaling. Furthermore, for short channel devices single traps increase the variability of the noise spectrum and result in random-telegraph noise (RTN) [2][3].
Models for 1/f noise are usually based on the assumption of uniform trap distribution over space and energy in the gate oxide [4][5] while multi-frequency C-V and G-V experiments [6] have shown complex border traps distributions in high-*k* dielectrics.

We use TCAD simulations [7] implementing the model for generation/recombination noise of [8]. Coupled with the non-local model tunneling to and from traps, our approach describes the carrier number fluctuations due to trapping/de-trapping in the gate insulator with arbitrary energy and space distributions of donors and acceptors. Mobility fluctuations [5] are not accounted for by the present implementation.

II. RESULTS FOR UNIFORM TRAP DISTRIBUTION

Before considering complex trap distributions, we validate our approach by comparing the simulation results with the prediction of the well known formula for drain current noise due to carrier number fluctuation [4]:

$$S_{Id}(f) = \frac{q^2 KT g_m^2 N_{BT}}{\alpha_t WLC_{OX}^2 f} \quad (1)$$

N_{BT} is the concentration of traps per unit volume and unit energy. The transconductance g_m is extracted from the I-V curves. The factor $\alpha_t = (2/\hbar)\sqrt{2m_{eff}\phi_B}$ is computed using the TCAD tunneling mass (m_{eff}) and barrier (ϕ_B) between semiconductor and insulator.
Fig. 1 compares Eq. 1 and TCAD results for Si/SiO$_2$ and In$_{0.53}$Ga$_{0.47}$As/Al$_2$O$_3$ large area MOSFETs at low V$_{DS}$. The noise is larger in the In$_{0.53}$Ga$_{0.47}$As device due to the larger mobility (we used bulk mobility just to simplify the comparison with the analytical model).
The agreement between Eq. 1 and the TCAD results is very good and somewhat surprising: Eq. 1 is obtained

assuming that the tunneling rate decreases exponentially when penetrating into the oxide and that the decay rate is the same for all energies, i.e. $P_{tun}=exp(-\alpha_t x)$. On the other hand, in TCAD the devices show a triangular tunneling barrier and, furthermore, the trapping rate depends on energy [7]. This point will be discussed in the next section.
Fig. 2 shows the effect of the gate bias on the spectrum, considering elastic and inelastic tunneling. Above threshold (not shown) all curves stay very close (same g_m), while below threshold we see a decrease of the current spectrum. This decrease is even larger when only elastic tunneling is used: in such case, there are very few electrons at the interface able to tunnel into the traps. We thus propose a correcting factor to Eq. 1 in order to reproduce the TCAD results:

$$S_{Id}(f) = \frac{q^2 KT g_m^2 N_{BT}}{\alpha_t WLC_{OX}^2 f} \frac{1}{1+exp(-\frac{qE_F}{KT})} \quad (2)$$

where E_F is the distance between the Fermi level and the conduction band minimum at the interface. Eq. 2 captures the decrease of S_{Id} at low V$_{GS}$ for elastic traps.

III. INTERPRETATION WITH NUMERICAL MODELS

To understand why Eq. 1 hold also with triangular tunneling barrier, we have worked out the different dependences related to the tunneling probability and trap capture/emission rates and see that they tend to compensate each other. First of all, we model the fluctuation of the trap occupation due to elastic tunneling with a Lorentzian function with inverse time constant:

$$\frac{1}{\tau} = \frac{\sqrt{8m_{eff}}m_0 g_c v_T}{\hbar^4 \pi} E_T^2 \sqrt{\phi_B - E_T} P_{tun} \quad (3)$$

where the meaning of the terms and the energy references can be seen in Fig. 3 and g_C=1. The expression is the same used in [7] for elastic tunneling into traps. The tunneling probability P_{tun} across the trapezoidal barrier is computed with WKB approximation from x=0 to x_T. Ramo's theorem is used to compute the current fluctuation induce by the trap as qv/L. At low V$_{DS}$ the velocity is the mobility μ multiplied by V_{DS}/L. We thus get:

$$S_{Id}(f) = \left(\frac{q\mu V_{DS}}{L^2}\right)^2 4f_T(1-f_T)\frac{\tau}{1+(2\pi f\tau)^2} \quad (4)$$

where f_T is the trap distribution (Fermi-Dirac statistic). Fig. 4 compares Eq. 4 with TCAD results: the agreement is quite good and tends to be almost perfect when using T_{OX}>30nm (not shown) where the assumption of the Ramo weighting field equal to $1/L$ is more appropriated. The noise is larger for V$_{GS}$ values such that the trap energy gets aligned with the Fermi level.
When considering multiple traps, Eq. 4 should be integrated on the trap distribution:

978-1-7281-0941-1/19 $31.00 © 2019 IEEE

$$S_{Id}(f) = \left(\frac{q\mu V_{DS}}{L^2}\right)^2 WL \iint N_{BT} \frac{4f_T(1-f_T)\tau}{1+(2\pi f\tau)^2} dE_T dx_T \quad (5)$$

where f_T, τ and N_{BT} depend on the trap position and energy. If N_{BT} is constant over energy and space, and the electric field in the oxide is almost null, we get $\tau = \tau_0\, exp(\alpha_t x_T)$ and Eq. 5 leads to Eq. 2.

Fig. 5 compares Eq. 5 with TCAD results for a case with F_{OX}=1.44 MV/cm. We see that even if the oxide field is not null, the agreement between Eq. 5 and Eq. 2 is very good. We have verified that this holds even for higher F_{OX}. Differences between Eq. 5 and Eq. 2 appear for example at 7 MV/cm as in Fig. 6. The influence of F_{OX} is thus visible only at low frequency and makes S_{Id} deviate from the 1/f behavior.

In addition, from Figs. 5 and 6 we see that the agreement between Eq. 5 and the TCAD results is good, in line with Figs. 1 and 2. It improves when using T_{OX}>30 nm (not shown) for the same reasons discussed for Fig. 4.

To motivate the weak influence of F_{OX} on the results, we first simplify the problem by noticing that $f_T(1-f_T)$ resembles a Dirac-delta at energy E_F. We thus can eliminate the integral over energy in Eq. 5 and get:

$$S_{Id}(f) = \left(\frac{q\mu V_{DS}}{L^2}\right)^2 WL4KT \int N_{BT} \frac{\tau}{1+(2\pi f\tau)^2} dx_T \quad (6)$$

Where τ and N_{BT} are compute for the E_T level aligned with E_F. Fig. 5 shows that the Eq. 6 is very close to Eq. 5 over the whole frequency range. Differences between Eq. 5 and Eq. 6 appear at high F_{OX} as can be seen in Fig. 6, but in any case Eq. 6 is a good starting point for a simplified analysis.

Fig. 7 plots the term $\tau/[1+(2\pi f\tau)^2]$ inside the integral of Eq. 6 for 3 different F_{OX} values: the shape of the curves and their integral is the same for 1MV/cm and 7 MV/cm. The function indeed peaks at $\tau = 1/(2\pi f)$ with a peak value of $1/(4\pi f)$; for large F_{OX} the tunneling probability (Fig. 8) increases with respect to the simple exponential term related to rectangular barrier, reducing τ and moving the peak of $\tau/[1+(2\pi f\tau)^2]$ further inside the oxide. When F_{OX} is too large, the function $\tau/[1+(2\pi f\tau)^2]$ gets cut off because there is no position where $\tau = 1/(2\pi f)$, as can be seen in Fig. 7 for the case with F_{OX}=9 MV/cm and the integral is lower than at low F_{OX}. Note that in Fig. 7 we see a weaker effect of F_{OX} compare to Fig. 6. This is due to the fact that in Fig. 7 we set a fixed E_F=0.1 eV while in Fig. 6 the high F_{OX} is associated to a large E_F due to the low DoS of the channel material. The energy of the tunneling path is thus higher in Fig. 6 and the classical turning point is moved to the left compared to. Fig. 7.

IV. REALISTIC TRAP DISTRIBUTION

We now consider a realistic trap distribution obtained by fitting multi-frequency C-V (Fig. 9a) and G-V (not shown) curves for InGaAs/Al$_2$O$_3$ capacitors [6]. The acceptor and donor profiles (plots b and c) are not uniform in energy and are positioned close to the interface. The resulting noise spectra at different biases are reported in Fig. 10. In strong inversion, tunneling is more localized and takes place with traps closer to the interface. Since 1/f noise is the combination

RTN over the different tunneling rates, this results in a flat spectra up to a knee frequency related to the slowest tunneling rate. Above that, we have 1/f dependence. In accumulation, instead, the slowest tunneling rate is much smaller and we see 1/f noise even at low frequency, tending to 1/f^2 when approaching the largest tunneling rate.

We now consider the influence of an additional uniform trap distribution inside the oxide. This additional charge does not affect the C-V curves simulated in Fig. 9 up to N_{BT} ~10^{19} cm^{-3}eV^{-1}, and thus is out of the detection limit of the multi-frequency C-V experiment (from 1 kHz up to 1 MHz). Fig. 11 shows that this additional contribution results in 1/f noise at low frequency in inversion, demonstrating that low-frequency noise measurements can probe traps deep in the oxide that are not probed by C-V at typical frequency range of measurement (1 kHz up to 1 MHz). Fig. 12 shows that the impact of bulk traps is much weaker in depletion/accumulation, where the D_{BT} is higher and goes deeper into the oxide than in inversion (Fig. 9 b-c), so that the additional uniform N_{BT} does not play a significant role.

V. CONCLUSIONS

We have shown that TCAD simulations can be used to analyze low frequency noise in MOSFETs considering complex trap distributions. The simple formula for carrier number fluctuation holds also when the tunneling barrier is not rectangular and the trapping rate changes with energy. On the other hand, a correction to that formula has been proposed to describe the lower of the noise due to elastic tunnel in the subthreshold regime.

ACKNOWLEDGMENT

The authors would like to thank Paolo Scarbolo, Luca Selmi and Paul Hurley for many helpful discussions and encouragement.

REFERENCES

[1] T.Lee, "The Design of CMOS Radio-Frequency Integrated Circuits", Cambridge university press, 2003

[2] H. Tuinhout and A. Z. Duijnhoven, "Evaluation of 1/f noise variability in the subthreshold region of MOSFETs", Conf. Proc. ICMTS, pp. 87-92, 2013

[3] M. Si, N. J. Conrad, S. Shin, J. Gu, J. Zhang, M. A. Alam and P. D. Ye, "Low-Frequency Noise and Random Telegraph Noise on Near-Ballistic III-V MOSFETs", IEEE Trans. Electron Devices, vol. 62, no. 11, pp. 3508-3515. 2015

[4] S.ChristenssonI, Lundström and C.Svensson, "Low frequency noise in MOS transistors—I theory", Solid-State Electronics, vol. 11, no. 9, pp. 797-812. 1968

[5] G. Ghibaudo, O. Roux, Ch. Nguyen - Duc, F. Balestra and J. Brini, "Improved analysis of low frequency noise in field - effect MOS transistors", Phys. Stat. Sol. (a), vol. 124, no 2, pp. 571-581, 1991

[6] E. Caruso, J. Lin, K. F. Burke, K. Cherkaoui, D. Esseni, F. Gity, S. Monaghan, P. Palestri, P. Hurley and L. Selmi, "Profiling border-traps by TCAD analysis of multifrequency CV-curves in Al$_2$O$_3$/InGaAs stacks," Conf. Proc. EUROSOI-ULIS, pp. 1-4, 2018

[7] Synposys Inc., Sentaurus DeviceTM, v. L-2016.03-SP2, 2016

[8] F. Bonani and G. Ghione. "Generation–recombination noise modelling in semiconductor devices through population or approximate equivalent current density fluctuations", Solid-State Electronics, vol. 43, no. 2, 1999

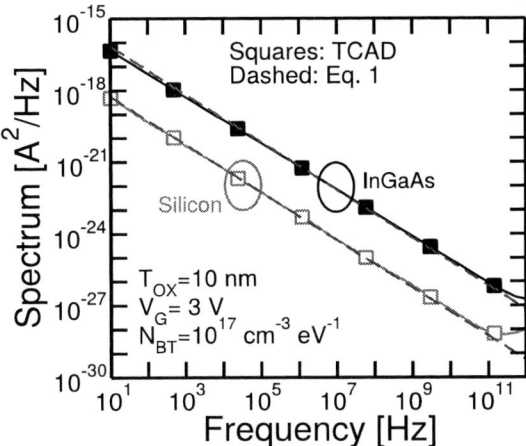

Figure 1 Comparison between the TCAD simulations employ an inelastic tunneling model for traps, using a trap volume $V_T=10^{-11}$ μm^3 and a phonon energy $E_{ph}=48$ meV. W= 1 μm, L=0.2 μm, $V_{DS}=25$ mV.

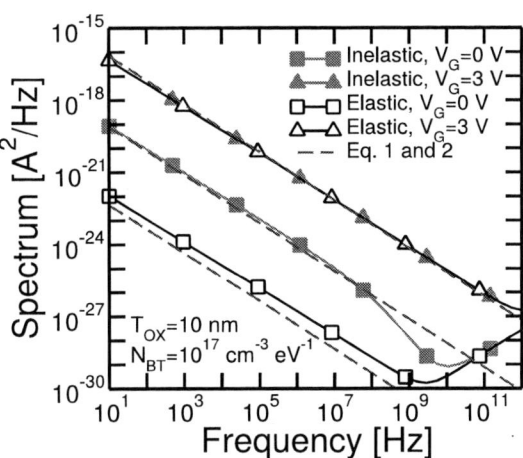

Figure 2 Comparison between TCAD simulations and the analytic model of Eq. 1-2 (dashed line) for an InGaAs/Al$_2$O$_3$ MOSFET simulated at 0 V (squares) and 3 V (triangles). TCAD simulations use inelastic (closed symbols) and elastic (open symbols) tunneling models for traps.

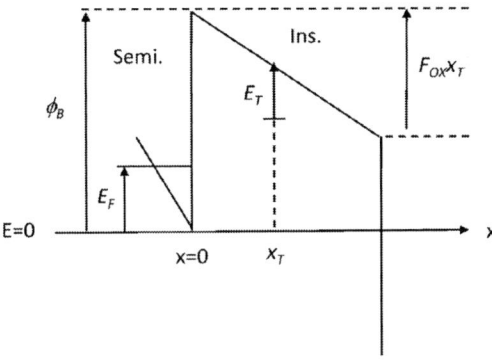

Figure 3 Band-diagram of the semiconductor/insulator system in the case of a single trap. The energy reference is the bottom of the conduction band at the interface.

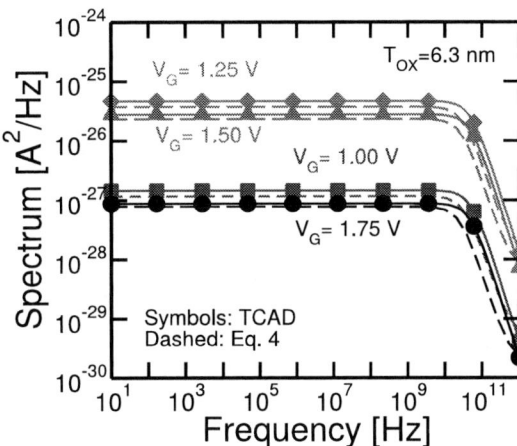

Figure 4 Comparison between the analytical model based on Eqs.3,4 and TCAD simulations for a single trap at $x_T=1$nm from the interface with an energy $E_T=3.05$eV. Device parameters are the same as in Fig.1 (InGaAs device) excpet for $T_{OX}=6.3$nm.

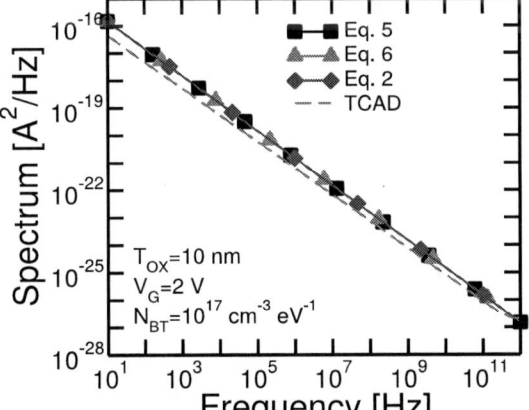

Figure 5 Comparison between the various numerical models and TCAD simulations for distributed traps and $F_{OX}=1.44$ MV/cm. The device is the same as in Fig.1 (InGaAs device) but an uniform trap distribution with $N_{BT}=10^{17}$ cm^{-3}eV^{-1} is used.

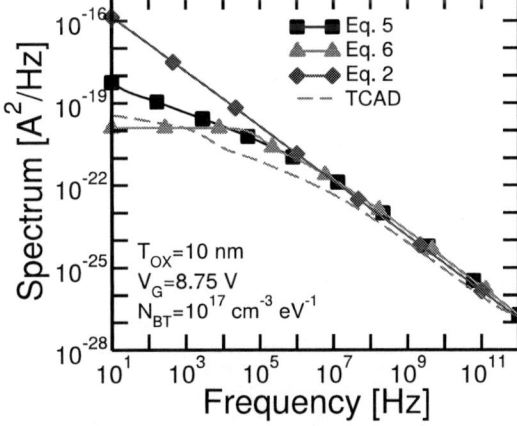

Figure 6 Comparison between the various numerical models and TCAD simulations for distributed traps and $F_{OX}=7$ MV/cm. The device is the same as in Fig.1 (InGaAs device) but an uniform trap distribution with $N_{BT}=10^{17}$ cm^{-3}eV^{-1} is used.

978-1-7281-0941-1/19 $31.00 © 2019 IEEE

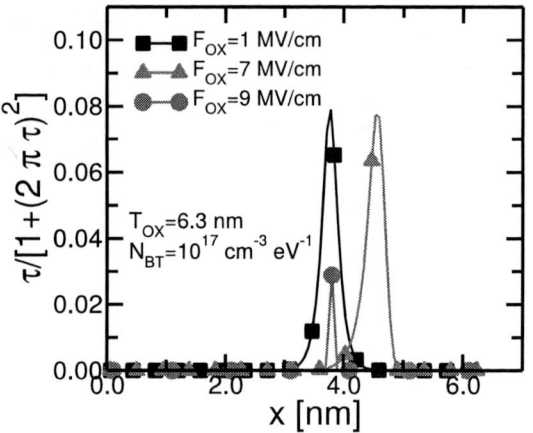

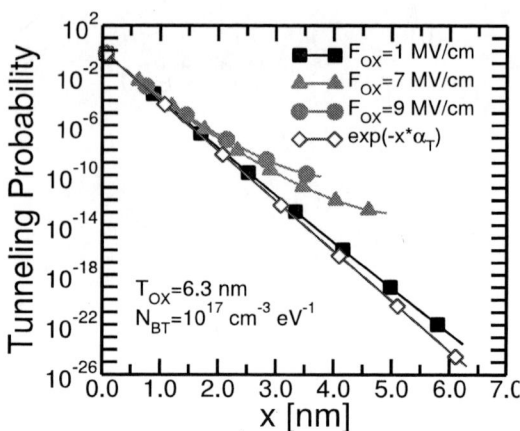

Figure 7 Term $\tau/[1+(2\pi f\tau)^2]$ evaluated inside the insulator for an energy equal to the Fermi level in the substrate (set to 0.1eV). The frequency is set to f=1Hz.

Figure 8 Tunneling probability along the path of Figure 7 for differernt values of the oxide field F_{OX}.

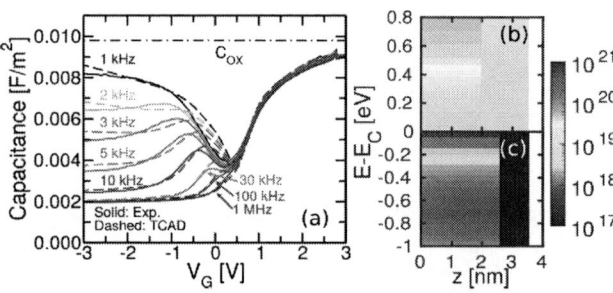

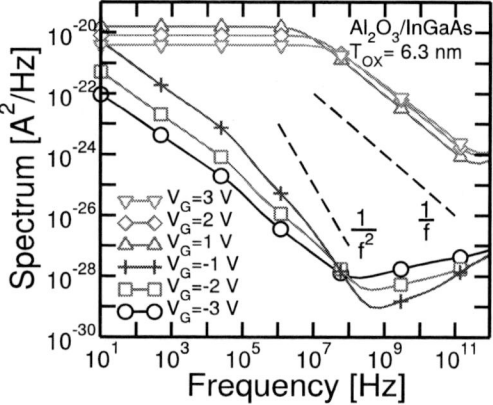

Figure 9 Experimental (solid lines) multi-frequency C-V (a) compared with simulated data (dashed lines). Simulations use (b) acceptor and (c) donor $D_{BT}(z,E)$ inside the 6.3 nm Al_2O_3. Doping is $N_D=3.0\cdot10^{17}$ cm^{-3} and SRH time is $\tau_g=80$ ps. The energy distributions are referred to the InGaAs conduction band (E_C).

Figure 10 Noise spectrum at different V_G using the $D_{BT}(z,E)$ of Fig. 9. W= 1 μm, L=0.2 μm, V_{DS}=25 mV.

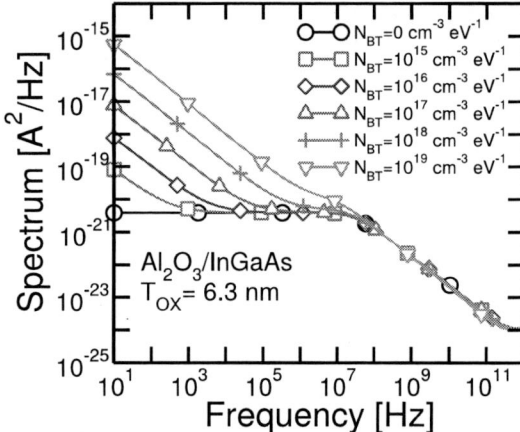

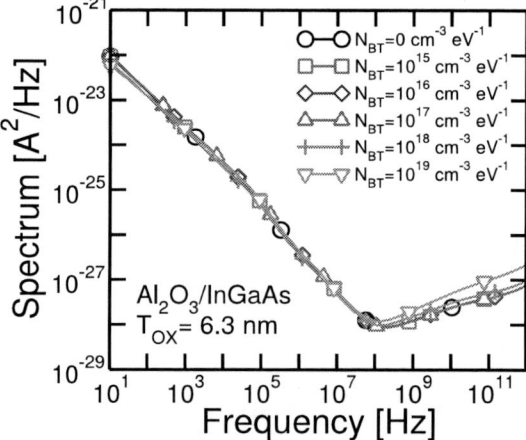

Figure 11 Noise spectrum using a border traps distribution obtained from the sum of the $D_{BT}(z,E)$ in Fig. 9 and an additional uniform acceptor distribution with density N_{BT}. In all simulations V_{GS} is 3 V, except for the one with $N_{BT}=10^{19}$ cm^{-3} eV^{-1}, where V_{GS} is 5 V to compensate for the threshold shift due to the traps.

Figure 12 Noise spectrum using a border traps distribution obtained from the sum of the $D_{BT}(z,E)$ in Fig. 9 and an additional uniform donor distribution with density N_{BT}. The spectrum is extracted at the same E_C-E_F=-0.5 eV (corresponding to negative V_{GS}).

978-1-7281-0941-1/19 $31.00 © 2019 IEEE

Investigation of TCAD Calibration for Saturation and Tail Current of 6.5kV IGBTs

Takeshi Suwa and Shigeaki Hayase
Toshiba Electronic Devices & Storage Corporation,
Kawasaki, Japan
Email:takeshi.suwa@glb.toshiba.co.jp

Abstract—In this work we focus on two calibration methods to clarify a key point of TCAD calibration for turn-off waveforms and IV characteristics including the saturation currents of IGBTs at the same time. Simulated results with the method based on adjustments of the surface N+ and P+ depth ratio reproduce measured results of all calibration targets reasonably in terms of time and accuracy. On the other hand, simulated results by mainly calibrating parameters of the velocity saturation model for saturation currents hardly reproduce all calibration targets simultaneously. We explain the reason using the roughly approximated criteria of the dynamic punch-through oscillation and dynamic avalanche. We also analyze the temperature dependence of the tail current briefly which is one of the important design items of IGBTs.

Keywords—IGBT, tail current, saturation current, calibration

I. INTRODUCTION

When optimizing and designing IGBTs for new products, it is very useful to have well calibrated TCAD tools for that generation that cover the range needed for the design [1], [2]. Here, well calibrated tools means that the tools has good predictability and convergence and it takes only a reasonable calculation time. Good predictability is achieved by selecting suitable calibration targets for the design and calibrating the tools to reproduce the target's measurements well. It is also important that calibration can be performed in a short time and easily.

The key point of the IGBT calibration is that the electrical characteristics crucially depend on the horizontal channel layout on the silicon surface side [3]. Fig. 1 shows schematic IGBT and diode structures for device simulations. The IGBT channel length in the pitch direction of Fig. 1 is shortened to reduce the saturation current so that the IGBT does not break at the time of short circuit. Further, in order to lower the on-resistance by raising the carrier density under the trenches in conduction state, the number of channels in the lateral direction is also reduced. For these reasons, the IGBT is basically a three-dimensional (3D) device, and the number of cells in one unit structure is large unfortunately.

When simulating IGBTs using Sentaurus TCAD tools, we often use following three basic ways:(1) two-dimensional (2D) device simulations using 2D structures made from 2D process simulations, (2) 3D device simulations using 3D structures made from 2D process simulations, and (3) 3D device simulations using 3D structures made from 3D process simulations. Although the method using 2D structures has the shortest calculation time, it is necessary to reproduce the effect of the 3D channel current flow of the IGBT by contriving process and device simulations. For that purpose, a method is generally used that reproduces both the measured on-state voltage and the saturation current by adjusting the

model parameters of channel mobility and velocity saturation in the drift region. Making a structure using 3D process simulations is the most correct and interesting way, however for IGBTs the computation time is relatively long and calibration is a little more difficult than the other ways. As high concentration boron and phosphorus diffuse and mix up in narrow silicon regions sandwiched by gate trenches, Secondary Ion Mass Spectrometry (SIMS) data are not effective for calibrations. Further we have to adjust the profile finely, because electrical characteristics are very sensitive to the depth ratio of the each profile. The low temperature process also makes calibration a bit more difficult. This method is often used when we do not need to create many types of structures or when we have a lot of time for calibrations. On the other hand, 3D device simulations have become relatively easy due to improvements in the software, and it is most efficient especially for large structures such as IGBTs to carry out 3D device simulations with 3D structures made by combining 2D process simulation results. When we simulate the reliability of IGBTs by increasing the number of cells, it is preferable to be able to reduce the number of meshes as much as possible [4]. In this work diodes (Free-Wheeling Diodes) are 2D structures, so device simulation carried out using the results of two-dimensional process simulation for diodes.

II. TARGETS AND SIMULATION APPROCH

Calibration targets are the IGBT turn-off waveforms and collector currents Ic as functions of collector-emitter voltage Vce and gate-emitter voltage Vge. To reproduce turn-on and short-circuit waveforms in the future, carrier lifetime dependence of IV-characteristics of diodes and saturation currents of IGBTs are also included in the calibration targets. The calibration ranges are as follows. The ambient temperature ranges from 298K to 398K, and there are three conditions for p base concentration and two N+ width ratios for IGBTs and four carrier life time conditions for diodes. In this work we adopt Shockley–Read–Hall (SRH) model as a model of carrier recombination through deep defect levels for device simulation, and carrier lifetimes of IGBTs in the drift region are obtained by extrapolating from calibrated lifetimes of diodes. We use Spreading Resistance (SR) data to make p collector and n buffer profiles on the back side, and the thickness of the trench oxide film is adjusted with reference to the cross-sectional Scanning Electron Microscope (SEM) image. The threshold voltage characteristics may be calibrated by adjusting the parameters of the segregation model or the work function difference.

In terms of cost and ease, we often adopt a method of mainly adjusting parameters of the velocity saturation model

978-1-7281-0941-1/19 $31.00 © 2019 IEEE

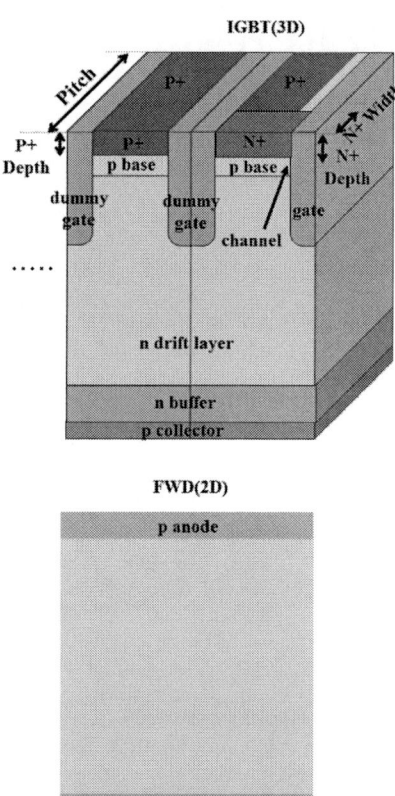

Fig. 1: Schematic view of the simulated structures (IGBT and diode).

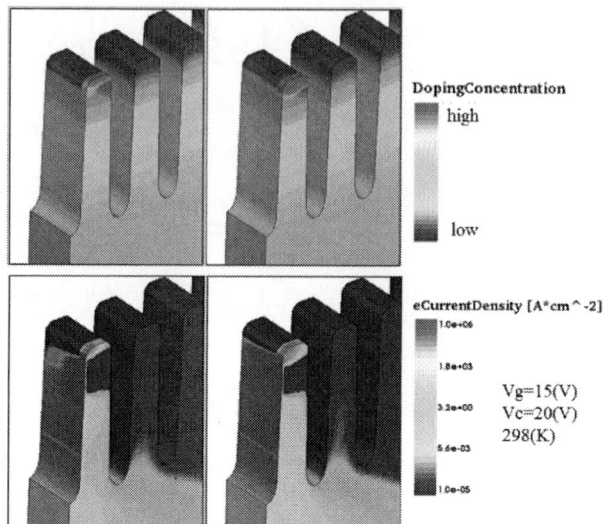

Fig. 2: Examples of the 3D current paths depending on the N+ and P+ depth ratio. The left figure is a current path of a structure having a shallow N + profile compared to P +, and the right figure is a current path of a structure having a deep N + profile.

without correcting the surface profiles to reproduce the saturation current, even if the process or physical basis are scarce (hereinafter abbreviated as Method-1). Although it is preferable to firstly calibrate the surface profile, it takes time to calibrate many type of structures in the 3D process simulation. We propose another method that we mainly adjust the surface N+ and P+ profiles to reproduce the saturation current. When we create a 3D structure by extending and bonding two 2D impurity profiles in the N+ and P+ regions, we calibrate the acceleration voltage of each surface N+ and P+ implantation of 2D structures and adjust the depth ratio of a 3D structure to reproduce the saturation currents (hereinafter abbreviated as Method-2, newly proposed method). Examples in which each current path spreading is different depending on the depth ratio are shown in Fig. 2. At this time, the on-voltages differ by about 10%, and the saturation currents differ by about three times. As is well known, in on-state calculations using the channel mobility degradation model, it is necessary to pay attention to the mesh of channel interfaces because the effect of the discretization error of the first layer mesh at the interface is large [5].

III. RESULTS AND DISCUSSION

Fig. 3 shows some examples of calibrated IV results with Method-2. These are in good agreement with the measurement results. This is because IV characteristics of IGBTs strongly depend on the 3D current path of the channel region, which is mainly determined by depth and width ratios of surface N+ and P+ profiles. If we sufficiently lower the

parameters of carrier velocity saturation model in the drift region and mobility degradation model at channel interfaces without adjustments of N+ and P+ profiles (Method-1), simulated results can also reproduce the measured results of the IV-characteristics to the same extent.

Fig. 4(a) shows comparison between simulated results using Method-1 (IV-characteristics have been well calibrated) and measured results of IGBT turn-off switching waveforms with an inductive load for an example condition. It can be seen that the simulated result of the slope of Vce and the time when tail current suddenly drops are not well reproduced. The timing at which the dynamic avalanche occurs can be calibrated by adjusting the parameters of avalanche model, however then the simulation results of breakdown voltage become unreasonable obviously. The relationship between tail currents and saturation currents is also the same. It is difficult to calibrate all targets at the same time by only adjusting these parameters. As shown in Fig. 4(b) and (c), the tail current suddenly drops when the depletion layer hits the n buffer layer. Although it is difficult to see from the figure, Vce jumps up a little at that time. The main reason why no vibration is seen in the simulation results unlike the measurement results is that the first-order approximation is used in the time discretization and strong numerical damping occurs. The relationship between these phenomena and the physical model parameters becomes clear from following two simple analytical expressions obtained in [6], [7].

$$Ic = qV_S A\left(\frac{2\varepsilon Vce}{qW_B^2} - N_B\right) \qquad (1)$$

where q, V_S, A, e, W_B and N_B are electronic charge, saturation velocity of hole, device active area, silicon dielectric constant, n drift thickness and n drift donor concentration, respectively. This equation is derived from a very simple approximation (Poisson equation considering the charge with the hole transportation in the depletion region) and gives a rough indication of the relationship between Ic and Vce when

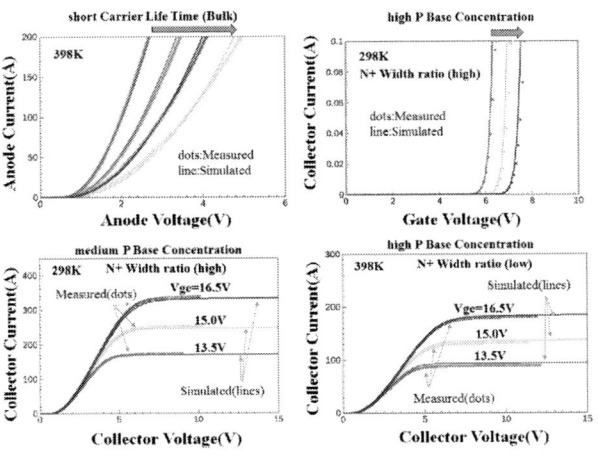

Fig. 3: Some examples of the calibrated IV-characteristics with Method-2.

the depletion layer hits the buffer (dynamic punch-through condition).

$$Ic = qV_S A\left(\frac{\varepsilon E_{crit}^2}{2qVce} - N_B\right) \qquad (2)$$

where Ecrit is the critical electric field. This equation represents the relationship roughly between Ic and Vce when dynamic avalanche occurs with almost no channel current flowing. The relationships between Ic and Vce represented by these expressions are shown by dashed lines in the Fig. 5(a). Curves in this figure are locus curves created from results of turn-off waveform in Fig. 5(b). The symbols (A) to (C) in the figure correspond to the points on Fig. 5(a). Since carrier saturation velocity is lowered in Method-1, two criteria in Fig. 5(a) indicate that dynamic avalanche occurs from a lower Vce, and the depletion layer reaches n buffer layer later. In other words, the smaller the carrier saturation velocity, the slower the discharge of the carrier. This points to that the calibration method of reproducing the saturation current by mainly adjusting velocity saturation model parameters is not so good for the purpose of this calibration. Fig. 6 shows the results of the turn-off switching waveforms of the IGBT for three example conditions with Method-2. The simulated results reproduce the measured results better than Method-1 with respect to the tail current length and the timing of the dynamic avalanche occurring. Using Method-2, we can reproduce saturation and tail currents of IGBTs simultaneously and easily. Therefore, Method-2 is appropriate for the purpose of this calibration, and we reaffirmed that it is important that calibrations are based on physical considerations as much as possible.

Analyzing the simulated results from the viewpoint of TCAD models, it was found that the temperature dependence of the mobility and band gap models dominate the difference in length between the tail currents of each temperature (approximately the turn-off time difference). Specifically, in this device and condition, the lattice temperature dependence of phonon scattering in the n drift region contributes about 70%, the lattice temperature dependence of hole velocity saturation contributes about 13%, and the lattice temperature dependence of the band gap contributes about 13%. On the other hand the temperature dependence of SRH recombination model (carrier lifetimes) is not a dominant

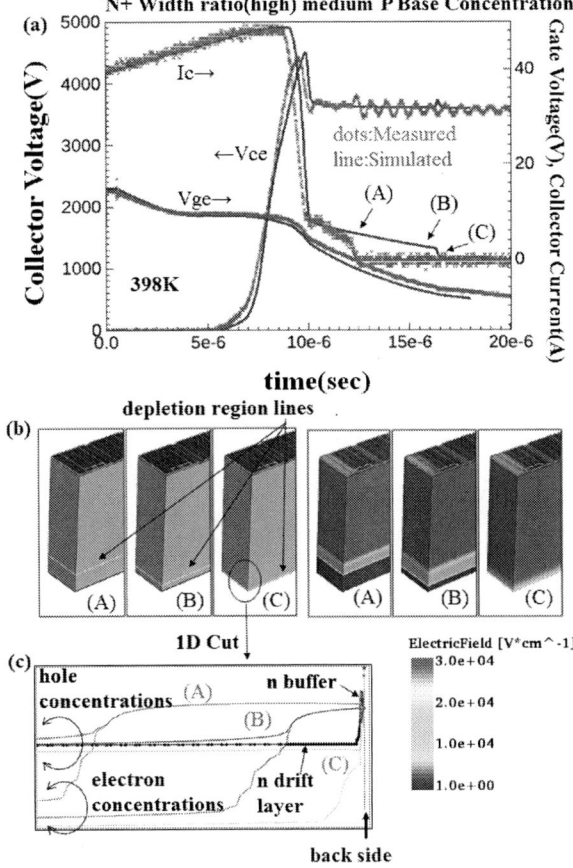

Fig. 4: Results of turn-off simulation with Method-1. (a) Comparison of simulated and measured waveforms. (b) Calculated movement of the depletion region line and distribution of electric field when tail current is flowing. Here, (A)-(C) correspond to the points on Figure (a), respectively. These figures are enlarged in the planar direction. (c) Distributions of carrier concentrations and impurities cut in the vertical direction of the circled area.

effect on the difference because of the low impurity concentrations and very long carrier lifetimes in the n drift layer of 6.5kV IGBTs.

IV. CONCLUSION

To clarify the key points of calibration for saturation currents and turn-off wave form of IGBTs, we calibrated with two methods and analyzed the results. In particular, using simple analytical equations, we explained that it is difficult to reproduce the electrical characteristics of the target simultaneously by the traditional calibration method of mainly adjusting the parameters velocity saturation model. In IGBT calibrations, the N + and P + profiles on the device surface side are particularly important, and we need to properly determine these profiles in 3D structures for simulations. To simplify calibration and reduce the number of meshes, we recommend a method to create 3D structures by adjusting the each depth of the N + and P + profiles in 2D process simulations so that the saturation currents reproduce the measured values. Of course, when there are obvious process or physical causes, it is also necessary to properly adjust the model parameters of channel mobility and velocity saturation.

(a)

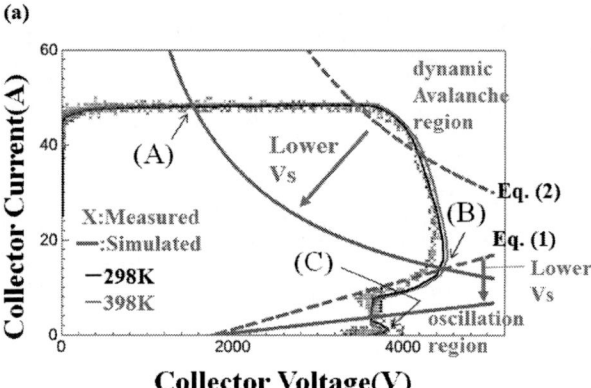

(b)   N+ Width ratio(high) medium P Base Concentration

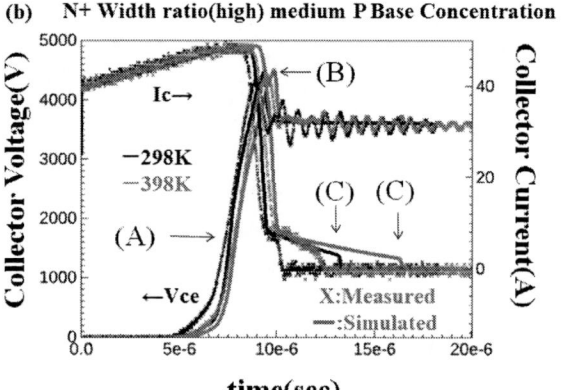

Fig. 5: (a) Locus curves of the collector current vs. the collector voltage are plotted to analyze the simulated turn-off results with Method-1 by the criteria of Ref. 1. Here, Vs represents saturation velocity of holes. (b) Turn-off waveforms that are the source of the locus curve.

REFERENCES

[1] A. Philippou, M. Bina and F.-J. Niedernostheide, "Automated Vertical Design Optimization of a 1200V IGBT," Proc. SISPAD, 2015, pp. 72-75.

[2] M. Bina, A. Philippou, M. Hauf, Ch. Sandow and F.-J. Niedernostheide, "Automated Vertical Design Co-Optimization of a 1200V IGBT and Diode," Proc. SISPAD, 2016, pp. 185-188.

[3] M. Watanabe et al., "Impact of three-dimensional current flow on accurate TCAD simulation for trench-gate IGBTs," Proc. ISPSD 2019, pp. 311-314.

[4] M. Tanaka and A. Nakagawa, "Growth of short-circuit current filament in MOSFET-Mode IGBTs," Proc. ISPSD 2016, pp. 319-322.

[5] T. Enda and N. Shigyo, "Grid size independent model of inversion layer carrier mobility," Proc. SISPAD, 1997, pp. 319-321.

[6] T. Ogura, H. Ninomiya, K. Sugiyama and T. Inoue, "Turn-Off Switching Analysis Considering Dynamic Avalanche Effect for Low Turn-Off Loss High-Voltage IGBTs," IEEE Transactions on Electron Devices, Vol. 51, No.4, April 2004, pp. 629-635.

[7] M. Tsukuda, I. Omura, Y. Sakiyama, M. Yamaguchi, K. Matsushita and T. Ogura, "Critical IGBT Design Regarding EMI and Switching Losses," Proc. ISPSD 2008, pp. 185-188.

(a)   N+ Width ratio(high), medium P Base Concentration

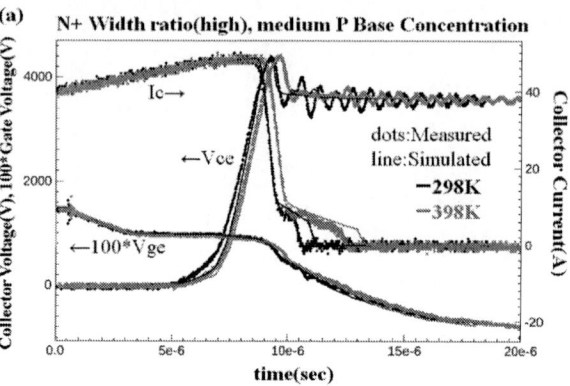

(b)   N+ Width ratio(low), medium P Base Concentration

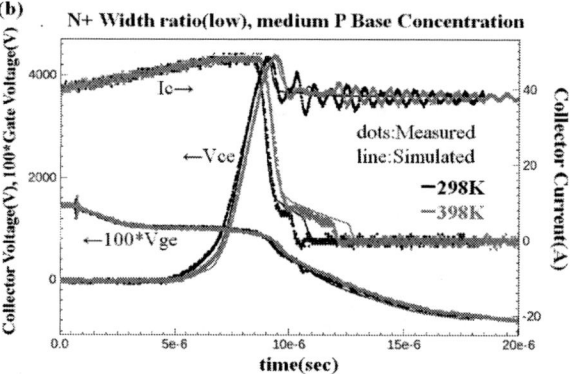

(c)   N+ Width ratio(low), high P Base Concentration

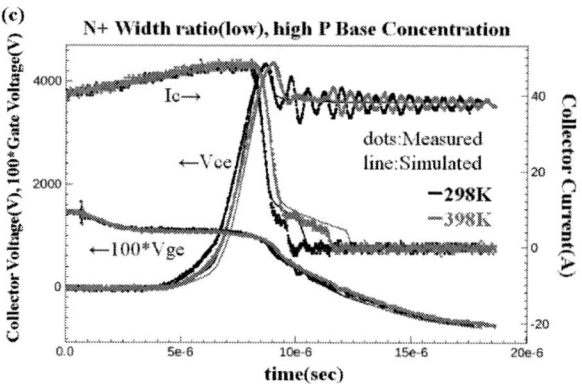

Fig. 6: Examples of turn-off simulation results with Method-2 in three conditions. The simulated results well reproduce the measured results with respect to the lengths of tail currents and the timing of dynamic avalanche occurring at each temperature compared to Method-1.

Simulation of Statistical NBTI Degradation in 10nm Doped Channel pFinFETs

F, Adamu-Lema, V. Georgiev, *Member, IEEE*, and A. Asenov, *Fellow, IEEE*

Device Modeling Group, University of Glasgow, Glasgow, G12 8LT, UK: Fikru.adamu-lema@glasgow.ac.uk

Abstract— In this paper, by means of simulations, we have studied the impact of Negative Bias Temperature Instability (NBTI) in bulk silicon FinFETs suitable to the 10nm CMOS technology generation. Different levels of channel doping are considered in controlling the threshold voltage and the leakage of the FinFETs for SoC applications. The interplay between the initial statistical variability introduced by random discrete dopants, line edge roughness and metal gate granularity and the statistical variability introduced by different level of trapped charges resulting from NBTI degradation is studied in details. Results related to the time dependent variability and the correlation of key transistor figures of merit are also presented.

Keywords—Atomistic doping; NBTI; FinFETs; statistical simulations; statistical variability; MOSFET FOMs correlations

I. INTRODUCTION

Performance limitations and increasing statistical variability [1][2] have prompted the end of bulk planar MOSFET scaling at 28/20nm CMOS technology generation [3]-[5]. FinFETs, with improved electrostatic integrity, that tolerate low channel doping, have been introduced by Intel at the 22nm CMOS technology generation [6] to sustain the benefits from increased power/performance with the continuation of CMOS scaling and to reduce the statistical variability. The rest of the industry follow suit at 14/16nm CMOS and now 10nm and 7nm FinFET CMOS are in production.

The low channel doping that FinFETs tolerate improves the performance due to reduced impurity scattering in the channel and steeper subthreshold slope (SS) [7][8] and could almost completely eliminate the random discrete dopant (RDD) induced variability in the threshold voltage (V_T) [9]. However, for System on Chip (SoC) design FinFETs with different V_T will be needed [10]. Gate work-function engineering via gate implantation or different metal spices can be used to deliver different V_T's at low channel doping, but at 14/16nm and in some 10 nm CMOS implementation doping is used for tuning V_T [11].

Simultaneously Negative Bias Temperature (NBTI) related time dependent statistical variability has been highlighted as a critical issue at 45/40nm CMOS [12]. However, with the further scaling of the bulk CMOS technology the relative importance of the NBTI induced statistical variability has been reduced due to the increasingly dominate role of RDD induced statistical variability associated with the necessary increase in channel doping with scaling. However, the possibility to reduce the channel doping and the related variability in FinFETs brings back the concerns about the NBTI induced statistical variability [9].

In this paper we study and compare statistical NBTI effects associated with the trapping of individual discrete charges in bulk FinFETs designed to meet the requirements of the 10nm CMOS technology generation. Transistors with various level of channel doping, needed to control the V_T and leakage for different aspects of SoC applications, are investigated.

II. DEVICE DESIGN AND SIMULATION METHODOLOGY

The 'template' p-channel FinFET adopted in this study is representative for 10nm technology generation bulk FinFETs introduced by major foundries in 2017 and are close to the FinFETs introduced by Intel in their 14nm CMOS offering [13]. The gate pitch is 64nm, the fin pitch is 40nm, the channel length is 28nm, the spacer is 8nm thick, fin height is 44nm and fin width is 8nm. The device design was initially targeted for high performance applications with leakage current (I_{OFF}) of 100nA/μm using low channel doping of 10^{17}cm^{-3} and work function (WF) engineering. Then the channel doping of 2×10^{18}cm^{-3} and 4.5×10^{18}cm^{-3} has been introduced to shift V_T and to reduce I_{OFF} to 10nA/μm and 1nA/μm respectively. Detailed description of the FinFET template can be found in [14]

To predict accurately the transistor performance, full band (FB) Ensemble Monte Carlo (EMC) simulations are carried out using the EMC module of GARAND [15]. In order to study the statistical variability and reliability effects, drift diffusion (DD) simulations using the DD module of GARAND were calibrated to the results of the EMC simulations.

The resolution of the individual discrete dopants in the random discrete dopants (RDD) simulations employs fine meshing in conjunction with density gradient quantum corrections. This prevents artificial charge trapping in the sharply resolved Coulomb wells of the ionized dopants and avoids acute mesh-spacing sensitivity [16]. Line edge roughness (LER) is modeled based on the assumption that it follows a Gaussian autocorrelation function [17] with three times root-mean-square (Δ) deviation of the gate edge position of LER=3Δ=2 nm and a correlation length Λ=30 nm. Identical LER parameters are used to model both gate edge roughness (GER) and fin edge roughness (FER). The modeling of metal

---

This work was supported in part by the European Commission through the FP7 grant agreement 261868 MORDRED, and FP7 grant SUP
ERTHEME (grant no.318458)

978-1-7281-0941-1/19 $31.00 © 2019 IEEE

gate granularity (MGG) assumes a TiN metal gate with two major grain orientations leading to a WF difference of 0.2 V, with a probability of 0.4/0.6 for the lower/higher WF respectively, and an average grain diameter of 5 nm [18].

Five levels of NBTI degradation corresponding to trapped charge densities ranging from $N_T = 1.0\times10^{11}$ cm^{-2} to $N_T=2.0\times10^{12}$ cm^{-2} are considered. The random trapped charges are introduced at the channel/gate dielectric interface using the methodology described in [19] resulting in random numbers and positions of traps in each individual transistor. The low drain bias statistical current voltage characteristics the highly doped channel FinFETs with 4.5×10^{18} cm^{-3} are illustrated in Fig. 1 as fresh devices (a) and after NBTI degradation (b) resulting in average trapped charge density of 2×10^{12} cm^{-2}. We would like to make some observations based on Fig.1 before proceeding with the detailed analysis of the simulation results at various degradation conditions in the next section. It is clear that the virgin low channel doping FinFET has lower statistical variability compered to virgin high channel doping FinFET. Simultaneously the same level of NBTI degradation resulted in significant increase in the variability in the low channel doping FinFET and much smaller increase of the variability in the high channel doping FinFET.

It has been shown previously that the continuous doping TCAD simulations differ from the average of the atomistic simulations yielding progressively erroneous results in the subthreshold with the reduction of transistor dimensions [20]. The problems in the case of FinFET simulations increase with the increase of the channel doping. Here we investigate to what extent the discrepancy between the continuous doping and the average 'atomistic' simulation is exacerbated with the increase in the average trapped charge density. First we investigate the V_T lowering properties of these devices.

Fig. 2 illustrates the continuously doped device, average and median V_T dependences on the average trapped charge at low channel doping of 10^{17}cm^{-3} and at the highest 4.5×10^{18}cm^{-3}. Cleary the increase of the channel doping results in higher threshold voltage lowering particularly at high drain bias. Simultaneously the increase of the average trapped charge density contributes to a relatively small increase in threshold voltage lowering at both channel-doping concentrations. To study further this effect, we present in Fig. 3 uniform, average and median threshold voltage shift dependence on the average trapped charge at low channel doping of 10^{17} cm^{-2}. There is a clear lowering in the threshold voltage shift (ΔV_T) which is

more pronounced at low drain and can reach more than 5 mV at $N_T = 2\times10^{12}$ cm^{-2}. This newly reported ΔV_T-lowering phenomena reduces the average threshold voltage shift associated with charge trapping compared to the results from uniform simulation.

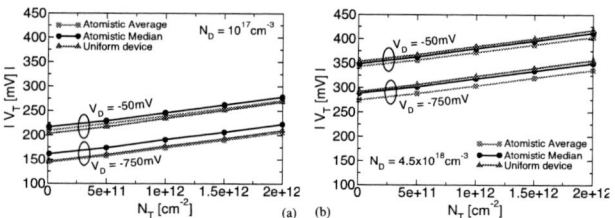

Fig. 2: Uniform, average and median V_T dependence on N_T $N_D=1\times10^{17}$ cm^{-3} and $N_D=4.5\times10^{18}$ cm^{-3}.

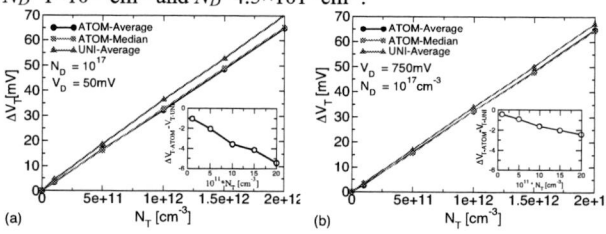

Fig. 3: Uniform, average and median ΔV_Td dependence on N_T at $N_D=1\times10^{17}$ cm^{-3} (a) $N_D=4.5\times10^{18}$ cm^{-3} (b)

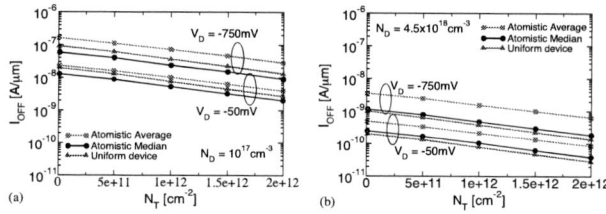

Fig. 4: Uniform, average and median gate leakage dependence on N_T $N_D=1\times10^{17}$ cm^{-3} and $N_D=4.5\times10^{18}$ cm^{-3}.

Figure 4 illustrates the dependence of the uniform, average and median leakage current on the average trapped charge density at low (10^{17} cm^{-3}) and high (4.5×10^{18} cm^{-3}) channel doping. We notice that the percentage discrepancy between the uniform and the average leakage current increase with the increase trapped charge. The effect is stronger at high drain bias and at channel doping of 10^{17} cm^{-3}. The discrepancy increases from 100% to 130% when the trapped charge increases from zero to 2×10^{12} cm^{-2}. At channel doping of 4.5×10^{18}cm^{-3} the discrepancy increases from 300% to 380% when the trapped charge increases from zero to 2×10^{12} cm^{-2}.

It is well understood that the statistical DD simulation accurately captures the statistical variability in the subthreshold region but can underestimate the on state variability, particularly associate with transport variation due to scattering with random discrete dopants and trapped charges [21]. Simultaneously the average on current is lowered due to certain amount of charge trapping in the Coulomb well of individual dopants, even after the introduction of density gradient quantum corrections [16]. Therefore, in the next section we analyze in great details the statistical variability in the subthreshold region.

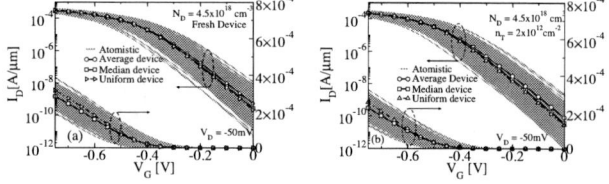

Fig. 1: Statistical Id-Vg of the highly doped FinFET without trap (a), and when interface traps of $N_T = 2 \times 10^{12} cm^2$ are considered in the simulation (b).

978-1-7281-0941-1/19 $31.00 © 2019 IEEE

III. RESULTS AND DISCUSSION

A. Threshold Voltage Variability

In the simulations we have considered two scenarios for the initial statistical variability of the FinFETs. In the first scenario, combined sources of variability, RDD, GER, FER and MGG are considered. In the second scenario we assume that sidewall deposition definition is used for the fin, which results in correlated FER with little impact on statistical variability, and therefore only RDD, GER and MGG are taken into account. The resulting threshold voltage distributions at low drain bias and at different levels of doping and different degrees of degradation are presented in Fig. 5. The corresponding values of the σV_T are plotted as a function of the channel doping concentration in Fig. 6.

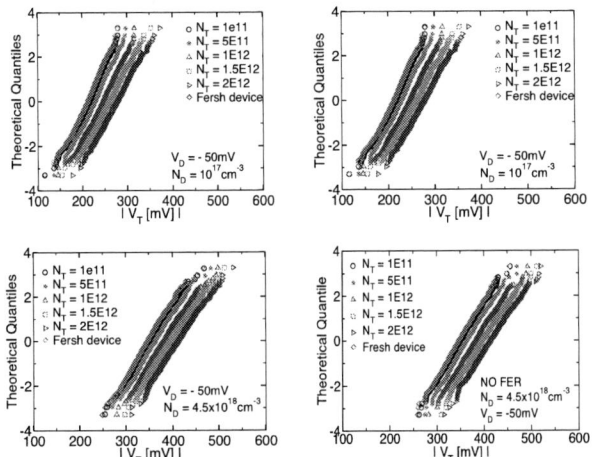

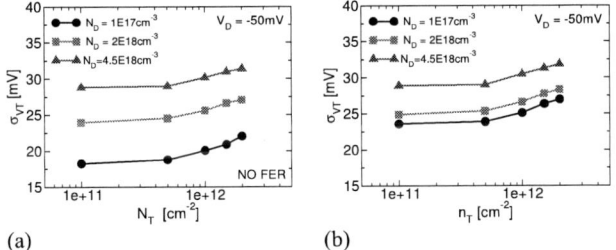

(a)   (b)

Fig. 6 Dependence of the standard deviation of the threshold voltage on the areal density of the trapped charge showing all sources of SV without FER (a) and with (b)

From the visual inspection of Fig. 5 and Fig. 6 it is clear that the trapping of discrete charges in the progressive NBTI degradation result not only in average threshold voltage shift, but also in increase of the statistical variability captured by σV_T. The NBTI induced increase of σV_T at a particular stage of degradation, determined by the areal density of the trapped charge depends on the level of variability in the fresh FinFETs. The impact of the degradation is stronger in the FinFETs with low channel doping and more pronounced in the case without FER at low drain bias conditions. In this case trapped charge of 2.0×10^{12} cm^{-2} increases σV_T from 17.8 mV in the fresh

transistors to 22.2 mV in the degraded one, which is approximately 25% increase.

At high channel doping of 4.5×10^{18} cm^{-3} and all sources of variability present the increase of σV_T is from 28.6 mV to 31.5 mV, which is approximately 10% increase. Also at high drain bias the initial variability is higher due to the LER induced effects and the relative impact of the trapped charge is smaller in all cases. It is clear that the virgin low-doped FinFETs, particularly in the absence of FER variability offer almost 60% improvement in σV_T compared to the highly doped FinFETs. However, this improvement is reduced to 40% after heavy degradation.

B. Subthreshold Figures of Merit Variability

It is also interesting to investigate the impact of the degradation on other important transistor figures of merit like the drain induced barrier lowering (DIBL), SS and I_{OFF}. Fig. 7 illustrated the trapped charge density dependence of the DIBL (a) and SS (b) of the lightly doped (10^{17} cm^{-3}) and the heavily doped (4.5×10^{18} cm^{-3}) FinFETs in the presence of RDD, FER, GER and MGG. The DIBL distribution strongly departs from a normal distribution, which should be represented by a straight line in Fig. 7 (a). It is clear that the charge trapping has a very little impact on the DIBL distribution. However, it should be noted that both in the low add high doped FinFETs the DIBL distribution is very broad, ranging from 30 to 180 mV/V and that the addition of channel doping does not improve DIBL. The SS distribution is also strongly non-Gaussian and has significant upper tail in the case of the heavily doped FinFETs, which are also strongly affected by the charge trapping. The high drain bias trapped charge density dependence of the I_{OFF} is illustrated in Fig. 8 for the lightly doped and the heavily doped FinFETs in the presence of RDD, FER, GER and MGG. Please note that the simulations do not include band-to-band tunneling.

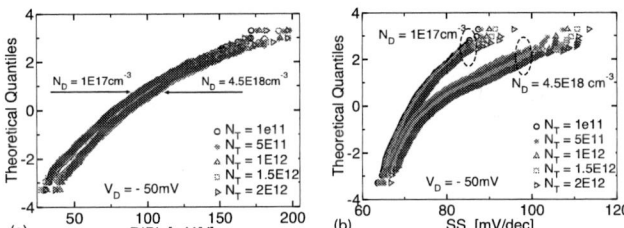

Fig. 7: N_T dependance of the DIBL (a) and SS (b) with $N_D=1\times10^{17}$ cm^{-3} and $N_D=4.5\times10^{18}$ cm^{-3}

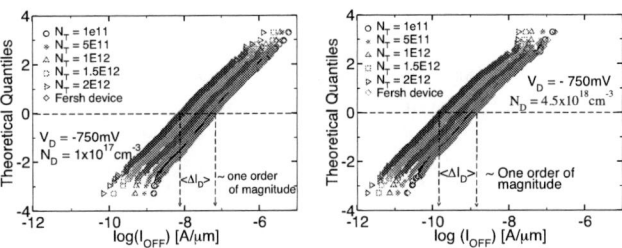

Fig. 8: N_T dependance dependence of the I_{OFF} for $N_D=1\times10^{17}$ cm^{-3} (a) and $N_D=4.5\times10^{18}$ cm^{-3} FinFETs.

Fig. 5: V_T distribution for two cases of SV at V_D=-50 mV. Figures in the left column are for all variability sources while in the right column FER is excluded assuming sidewall Fin definition.

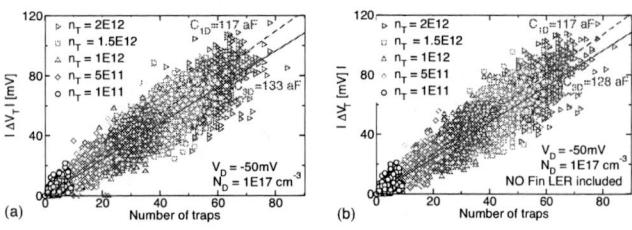

Fig. 9: Fractional ΔV_T as a function of the number of trapped charges at different levels of degradation. The left column is with all relevant variability sources including RDD, GER, FER and MGG (a) and where FER has been excluded by assuming sidewall Fin definition (b). The dashed line represents an ideal capacitance (C_{1D}) from a planar capacitance and the 'effective' capacitance (C_{3D}) derived from the slope of the average curve.

As expected, the progressive charge trapping reduces the average leakage due to the corresponding average increase of the threshold voltage but increases the leakage voltage spread. Analogous to the spread in the V_T the spread in the leakage current increases faster with trapping in the lightly doped channel FinFET.

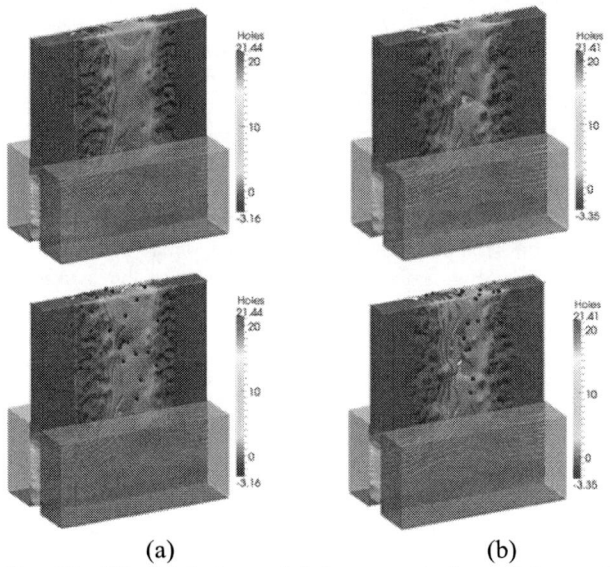

(a)   (b)

Fig. 10: 3D distribution of hole concentration of two extreme transistors in the tail of the distribution. The left images are for $N_D = 1.0 \times 10^{17}$ cm^{-3} with $\Delta V_T = 71.9$ mV. The right images are for $N_D = 8.5 \times 10^{18}$ cm^{-3} with $\Delta V_T = 69.5$ mV.

IV. CONCLUSIONS

The trapping of discrete charges as a result of a progressive NBTI degradation in the investigated 10 nm CMOS technology FinFETs results not only in average ΔV_T, but also in increase of

the statistical variability captured by σV_T and by other important transistor figures of merit. The NBTI induced increase of σV_T at a particular stage of degradation is dominated by the areal density of the trapped charge and depends also on the level variability in the fresh FinFETs. It is higher in the FinFETs with low channel doping and more pronounced in the case without FER at low drain bias conditions. The progressive charge trapping also results in a progressive increase in the discrepancy between the continuous doping and the average results of the 'atomistic' simulations. The charge trapping also results in a de-correlation between the key transistor FOM, particularly in the case of low virgin variability associated at low channel doping and in the absence of FER.

REFERENCES

[1] Dimitri A. Antoniadis and Ali Khakifirooz, in *IEDM*, 2008, pp.1-4.
[2] X. Wang, A.R. Brown, N. Idris, S. Markov, G. Roy and A. Asenov, *IEEE Transactions on Electron Devices*, Vol. 58, No. 8, pp. 2293–2301, Aug. 2011.
[3] H. Fukutome, et al., in *IEDM Tech. Dig.*, 2012. pp.3.5.1-3.5.4.
[4] X. Wang, G. Roy, O. Saxod, A. Bajolet, A. Juge and A. Asenov, *IEEE Electron Device Letters*, Vol.33 No.5, pp.643-645, May 2012.
[5] X. Wang, F. Adamu-Lema, B. Cheng and A. Asenov, *IEEE Trans. Electron Devices*, Vol. 60, No. 5, pp. 1547–1554, May 2013.
[6] C. Auth, et al., in *VLSI Tech. Sym.*, 2012, pp.131-132C.
[7] L. Alexander, G. Roy and A. Asenov. *IEEE Trans. Electron Devices*, Vol. 55, No. 11, pp. 3251–3258, Nov. 2008.
[8] M. Bohr, in in *Proc. IEDM*, 2011, pp.1.1.1-1.1.4.
[9] X. Wang, A.R. Brown, B. Cheng, and A. Asenov, in *Proc. IEDM*, 2011, pp.5.4.1-5.4.4.
[10] C.-J. Han, et al., in Proc. *IEDM*, 2012, pp.3.1.1-3.1.4.
[11] A. Veloso, *et al.*, *Japanese Journal of Applied Physics*, vol. 52, p. 04CA02-1, 2013.
[12] A. R. Brown, V. Huard and A. Asenov, *IEEE Transactions on Electron Devices*, Vol. 57, No. 9, pp. 2320–2323, 2010.
[13] S. Natarajan, et al., in *IEDM* 15-17 Dec. 2014
[14] L. Shifren, R. Aitken, et al., *IEEE Transactions on Electron Devices* Volume: 61, July 2014
[15] Garand, https://www.synopsys.com
[16] G. Roy, A. R. Brown, F. A-Lema, S. Roy, and A. Asenov, *IEEE Trans. Elec. Devices.*, Vol.53, no.12, 2006, pp.3063-3070.
[17] A. Asenov, S. Kaya, and A. R. Brown, *IEEE Trans. Elec. Dev.*, vol.50 no.5, pp.1254-1260, 2003.
[18] A. R. Brown, N. M. Idris, J. R. Watling and A. Asenov, *IEEE Electron Device Letters*, Vol. 31, No. 11, pp. 1199–1201, Nov. 2010.
[19] A. R. Brown, V. Huard and A. Asenov, *IEEE Transactions on Electron Devices*, Vol. 57, No. 9, pp. 2320–2323, 2010.
[20] A. Asenov, F. Adamu-Lema, X. Wang and S. M. Amoroso, *IEEE Transactions on Electron Devices*, Vol. 61, No. 8, pp. 2745–2751, 2014.
[21] C. L. Alexander, G. Roy and A. Asenov, *IEEE Trans. Electron Devices*, Vol. 55, No. 11, pp. 3251–3258, Nov. 2008.
[22] X. Wang, *et al.*, in Proc. *IEEE Silicon Nanoelectronics Workshop*, 2012, pp. 77–78.
[23] X. Wang, *et al.*, in Proc. *SISPAD*, 2012, pp. 296–299.
[24] S. M. Amoroso, *et al.*, *Electron Device Letters*, Vol. PP, No. 99, pp. 1–3, Apr. 2013.
[25] F. Adamu-Lema, *et al.*, *IEEE Trans. Electron Devices*, Vol. 60, No. 2, pp. 833–839, Feb. 201

A SPICE Compatible Compact Model for Process and Bias Dependence of HCD in HKMG FDSOI MOSFETs

Uma Sharma and Souvik Mahapatra*

Department of Electrical Engineering,
Indian Institute of Technology Bombay,
Mumbai 400076, India
*Phone: +91-222-572-0408, Email: souvik@ee.iitb.ac.in

Abstract— **A SPICE compatible model is developed for the time kinetics of linear drain current drift (ΔI_{DLIN}) under Hot Carrier Degradation (HCD) stress in 28 nm Fully Depleted Silicon On Insulator (FDSOI) n-channel FETs having High-K Metal Gate (HKMG) gate stack. The impact of varying the drain (V_D), gate (V_G) and body (V_{BB}) biases is modeled. The framework is also capable of modeling the channel length (L_{CH}) and gate-oxide thickness (T_{OX}) variations. Impact of Self-Heating Effect (SHE) has also been taken into consideration during ΔI_{DLIN} modeling.**

Keywords—HCD, FDSOI, HKMG, SHE, gate-oxide thickness, body bias

I. Introduction

FDSOI MOSFETs are gaining interest for use in IoT and 5G applications. The adaptive body biasing in FDSOI devices can be utilized for dynamic power-performance control [1]. After a brief hiatus, HCD has once again emerged as one of the important and crucial aspects of MOSFET reliability and it impacts both FinFET and FDSOI devices [2]-[9]. HCD causes local damage towards the drain side of the transistor leading to device parametric degradation. Accurate modeling of HCD time kinetics is necessary to extrapolate short time accelerated stress data to determine the HCD magnitude at End-OF Life (EOL) under operating condition. Unlike typical stress condition where V_G and V_D are held fixed, in actual digital circuit V_G and V_D are varied from 0 to V_{DD}. Thus, it is necessary to model the HCD time kinetics under full V_G/V_D space [10]. Unlike FinFET, the FDSOI devices use adaptive V_{BB} to optimize performance and power consumption [1], which also impacts HCD [9] and needs to be modeled. Finally, SHE needs to be carefully modeled in more electrostatically confined architectures (like FinFET and FDSOI) where thermal resistance is much higher than the bulk counterparts [11]. Note that unlike classical HCD, degradation in modern devices show positive temperature (T) activation (increase at larger T) [3]-[5]. The compact model should be SPICE compatible, where the transistor degradation should only depend on V_G, V_D, V_{BB}, T and time, and not on any device parameter (like I_D or saturation drain bias (V_{DSAT}) etc.). Our previous work has shown, using Eq.(1), the model struggles to predict the measured time kinetics across wide V_D and time ranges. Fig.2 (a) plots the prediction of V_D (using Eq.(1)) dependent I_{DLIN} degradation in time from [13]. Therefore, to improve accuracy, the basic Eq.(1) is modified using a time varying *m* Eq.(2). Fig 1(b) and Fig. 2 shows the modeled time kinetics using the modified equation for planar [10] and Fig. 3 for FinFET [12] devices. The framework is robust to model different probes as V_T, I_{DLIN}, I_{DSAT} and CP. Fig. 2 plots the model prediction [10] of time kinetics probed by using (a, b) V_T, (c, d) I_{DSAT} and (e, f) I_{DLIN} methods in shorter L_{CH} planar devices [14]. The V_D dependence at various V_G/V_D combinations is studied in different L_{CH} at room and higher T. The P is listed, which is device and probe specific. Fig. 3 shows the time kinetics of ΔV_T during HCD stress at different V_G and V_D conditions in SiGe channel p-FinFETs measured by an ultra-fast method and explained using the SPICE compatible compact model.

II. Scope of Work

The developed compact model for planar and FinFET devices [10, 12] is suitably modified for FDSOI MOSFET. It can model the ΔI_{DLIN} time kinetics under full V_G / V_D / V_{BB} space and also model the L_{CH} and T_{OX} dependence. In particular, the impact of L_{CH} and T_{OX} on the V_G dependence of HCD can be modeled. The model is validated against measured data from published reports on 28 nm FDSOI technology [6]-[8] and the parameters are listed.

III. Framework

The schematic of a FDSOI MOSFET is shown in Fig. 4 along with localized defects induced by HCD stress in the channel and near drain junction, gate-drain overlap and spacer regions. Threshold voltage and mobility are affected due to the defects in channel, and defects in the overlap and spacer regions impact series resistance. Therefore, defects in all the regions impact drain current. The compact model equations are listed in Table-I. Eq.(1) shows the self-saturating empirical equation to model the HCD time kinetics with a time-varying slope, Eq.(2). Eqs. (3)-(9) govern the modeling of the time-constant (τ) that takes into account all the process parameters (L_{CH} and T_{OX}) and the stress conditions (V_G, V_D, V_{BB}, and T). Eq.(10) is used for SHE. Fig. 5 shows the measured and modeled time kinetics of ΔI_{DLIN} for various V_G/V_D conditions for thin gate oxide at room temperature (25°C).

IV. BODY BIAS DEPENDENCE

In FDSOI structure, forward body biasing can be done to switch the transistor faster. Fig. 6 (a) and (b) show the measured and modeled forward body bias effect on HCD time kinetics in a device having L_{CH} of 0.30 µm and thin gate oxide. Fig. 6 (a) shows the time kinetics with varying V_{BB} and keeping V_G and V_D constant. It is observed that degradation increases with increase in V_{BB}. Fig. 6 (b) shows the fixed time degradation with varying V_G. It can be modeled using a consistent set of parameters, which are also used in Fig. 7 (a).

V. CHANNEL LENGTH DEPENDENCE

Fig. 7 models the impact of L_{CH} variation on the V_G dependence of measured ΔI_{DLIN} (V_D is held fixed) at a fixed stress time for thin and thick oxide devices. Note that the V_G dependence is different for thick and thin oxide devices. Fig. 7 (a) shows the V_G dependence changes from bell shape to monotonic with decreasing L_{CH} for thin gate oxide. However, in Fig. 7 (b) the effect is reversed with L_{CH} for thick gate oxide. It can be seen that the model is capable of explaining both trends with consistent model parameters, which are listed in Table-II. Note that the difference in V_G dependence of ΔI_{DLIN} for thin and thick oxide devices is due to difference in V_G dependence of the sub-components of time constant (τ) as shown in Fig.8 (a) and (b): τ_A monotonically reduces while τ_B increases first and later reduces as V_G is increased and they control the overall τ and HCD lifetime. For thick gate oxide device, τ_A is dominant till $V_G=V_D/2$ while for thin gate oxide it is significant till $V_G= V_D$, which are the well-known worse case conditions for each cases [10]. Also, the model helps to explain the L_{CH} scaling keeping same gate process. Model parameters are listed in Table-II.

VI. OXIDE THICKNESS

Measured HCD data from different T_{OX} devices are modeled using a consistent set of parameters at a fixed L_{CH}. Fig. 9 shows the measured and modeled data for gate oxide thickness variation at 0.15 µm channel length. Thus, the model can help in providing the complete modeling and thus helps in the study of gate-oxide scaling as well.

VII. CONCLUSION

A compact HCD time kinetics model is proposed and validated using measured data from 28 nm FDSOI devices. The model is fully SPICE compatible, as it only uses terminal voltages and T, and does not depend on any internal transistor parameter. It can model data over the full V_G/V_D spectrum (for $V_G <$, = and $> V_D$); different T (including SHE as applicable), V_{BB}, and can also handle technology parameter (L_{CH} and T_{OX}) changes. In particular, it is observed that L_{CH} scaling results in different V_G dependence of measured ΔI_{DLIN} at fixed time for thick and thin T_{OX} devices, which is explained. The model can be incorporated in a SPICE simulator to model HCD in actual circuits under different mission profiles.

VIII. REFERENCE

[1] "FDSOI Technology and Dynamic Body Bias Compensation to Enable Next Generation AI/IoT and Automotive Products – An industrial perspective" V. Huard, tutorial, *2019 IEEE International Reliability Physics Symposium (IRPS)*, Anaheim, CA, 2019.

[2] S. Ramey *et al.*, "Intrinsic transistor reliability improvements from 22nm tri-gate technology," *2013 IEEE International Reliability Physics Symposium (IRPS)*, Anaheim, CA, 2013, pp. 4C.5.1-4C.5.5. doi: 10.1109/IRPS.2013.6532017

[3] H. Jiang, S. Shin, X. Liu, X. Zhang and M. A. Alam, "Characterization of self-heating leads to universal scaling of HCI degradation of multi-fin SOI FinFETs," *2016 IEEE International Reliability Physics Symposium (IRPS)*, Pasadena, CA, 2016, pp. 2A-3-1-2A-3-7. doi: 10.1109/IRPS.2016.7574506

[4] A. Kerber, S. Cimino, F. Guarin and T. Nigam, "Assessing device reliability margin in scaled CMOS technologies using ring oscillator circuits," *2017 IEEE Electron Devices Technology and Manufacturing Conference (EDTM)*, Toyama, 2017, pp. 28-30. doi: 10.1109/EDTM.2017.7947495

[5] Y. Qu et al., "Ultra fast (<1 ns) electrical characterization of self-heating effect and its impact on hot carrier injection in 14nm FinFETs," in Proc. IEEE Int. Electron Devices Meeting (IEDM), San Francisco, CA, USA, Dec. 2017, pp. 39.2.1-39.2.4. doi: 10.1109/IEDM.2017.8268520.

[6] W. Arfaoui, X. Federspiel, P. Mora, M. Rafik, D. Roy and A. Bravaix, "Experimental analysis of defect nature and localization under hot-carrier and bias temperature damage in advanced CMOS nodes," *2013 IEEE International Integrated Reliability Workshop Final Report*, South Lake Tahoe, CA, 2013, pp. 78-83. doi: 10.1109/IIRW.2013.6804163

[7] W. Arfaoui, X. Federspiel, A. Bravaix, P. Mora, A. Cros and D. Roy, "Application of compact HCI model to prediction of process effect in 28FDSOI technology," *2014 IEEE International Integrated Reliability Workshop Final Report (IIRW)*, South Lake Tahoe, CA, 2014, pp. 69-72. doi: 10.1109/IIRW.2014.7049513

[8] W. Arfaoui *et al.*, "Energy-driven Hot-Carrier model in advanced nodes," *2014 IEEE International Reliability Physics Symposium*, Waikoloa, HI, 2014, pp. XT.12.1-XT.12.5. doi: 10.1109/IRPS.2014.6861189

[9] Federspiel, X., Angot, D., Rafik, M., et al. (2012) 28 nm Node Bulk vs FDSOI Reliability Comparison. 2012 IEEE International Reliability Physics Symposium (IRPS), Anaheim, CA, 15-19 April 2012, 3B.1.1-3B.1.4.

[10] U. Sharma and S. Mahapatra, "A SPICE Compatible Compact Model for Hot-Carrier Degradation in MOSFETs Under Different Experimental Conditions," in IEEE Transactions on Electron Devices, 2019. doi: vol. 66, no. 2, pp. 839-846, Feb. 2019.

[11] C. Prasad *et al.*, "Self-heat reliability considerations on Intel's 22nm Tri-Gate technology," *2013 IEEE International Reliability Physics Symposium (IRPS)*, Anaheim, CA, 2013, pp. 5D.1.1-5D.1.5. doi: 10.1109/IRPS.2013.6532036

[12] U. Sharma, N. Parihar and S. Mahapatra, "Modeling of HCD Kinetics for Full V_G/V_D Span in the Presence of NBTI, Electron Trapping, and Self Heating in RMG SiGe p-FinFETs," in *IEEE Transactions on Electron Devices*, vol. 66, no. 6, pp. 2502-2508, June 2019. doi: 10.1109/TED.2019.2911335

[13] J. Goo, Y. Kim, H. L'Yee, H. Kwon and H. Shin, "An analytical model for hot-carrier-induced degradation of deep-submicron n-channel LDD MOSFETs," in *1995 Solid-State Electronics*, 38(6), pp.1191-1196. DOI: 10.1016/0038-1101(94)00221-Z

[14] D. Varghese, M. A. Alam and B. Weir, "A generalized, IB-independent, physical HCI lifetime projection methodology based on universality of hot-carrier degradation," in *Proc. Int. Rel. Phys. Symp* 2010, pp. 1091-1094. **DOI:** 10.1109/IRPS.2010.5488666

978-1-7281-0941-1/19 $31.00 © 2019 IEEE

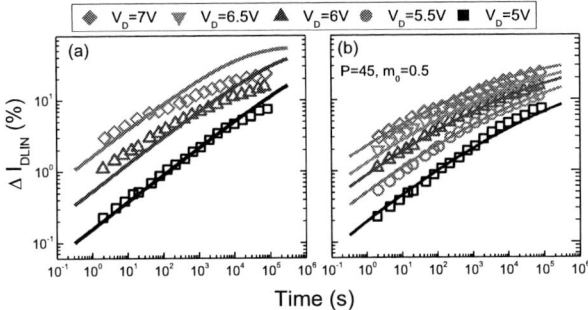

Fig.1. Modeling of measured ΔI_{DLIN} time kinetics at different V_D for on-state HCD stress (data from [13]) using standard and modified compact models. Further details [10].

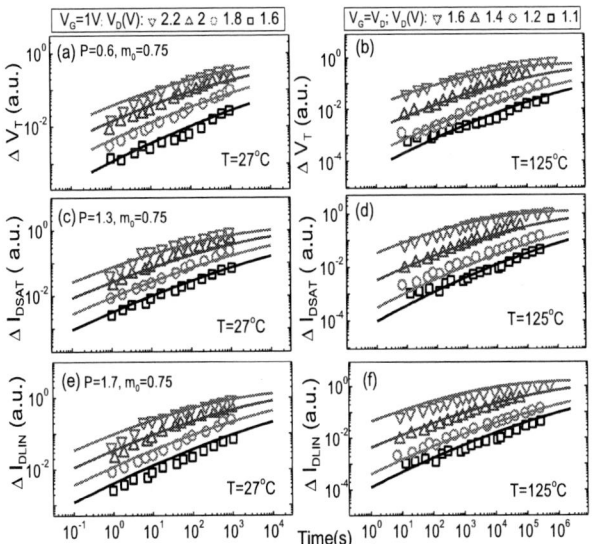

Fig.2. Modeling of V_D dependence of measured (a, b) ΔV_T, (c, d) ΔI_{DSAT} and (e, f) ΔI_{DLIN} time kinetics at (a, c, e) room and (b, d, f) high T from planar devices at low V_D stress [14]. Further details [10].

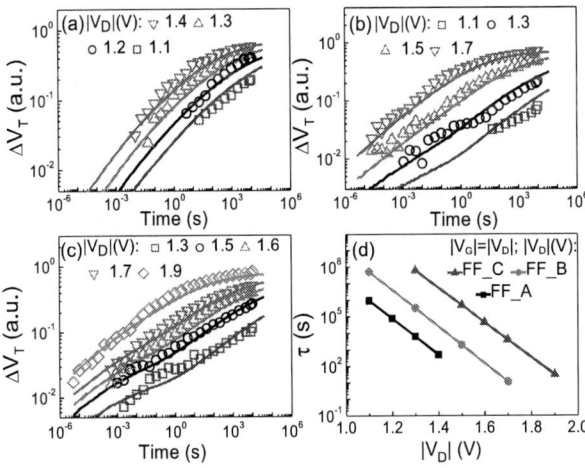

Fig.3. Time kinetics of measured and modeled ΔV_T at different stress V_D but for $|V_G| = |V_D|$ condition in SiGe devices (a) FF_A (Low N%, L_{CH}= 20 nm), (b) FF_B (High N%, L_{CH}= 20 nm) and (c) FF_C (High N%, L_{CH}= 60 nm) devices. Symbols: data, lines: model. In (d), the V_D dependence of the parameter τ (see Eq. (1) and Eq. (2), Table-I) is identical for all three devices. Further details [12].

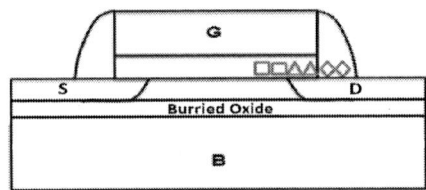

Fig 4. Schematic of a FDSOI MOSFET showing regions along the channel that get degraded during HCD: square (channel), triangle (gate/drain overlap) and diamond (spacer).

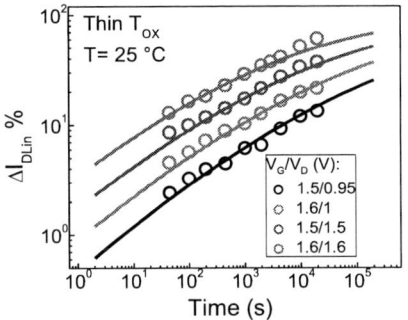

Fig. 5. Time evolution of experimental ΔI_{DLIN} (in %) for DC stress, with model prediction at different V_G and V_D

TABLE I. COMPACT MODEL EQUATIONS

$$\Delta P = P[1 - e^{\left\{-\left(\frac{t}{\tau}\right)^m\right\}}] \tag{1}$$

$$m = m_0 e^{[-(\frac{t}{\tau_m})^k]} \tag{2}$$

$$\tau = (\tau_A + \tau_B) \tag{3}$$

$$\tau_A = \frac{A}{e^{\Gamma_{A2}(V'_D)}} e^{\Gamma_{A1}(V'_D - \alpha V_G)} \tag{4a}$$

$$\tau_B = \frac{B}{C} e^{\Gamma_{B1}(V_G - \beta V'_D)} \tag{4b}$$

$$C = [1 + C_1 e^{\Gamma_{C1}(V_G - \sigma V'_D)}] e^{(\Gamma_{C2} V'_D)} \tag{5}$$

$$V'_D = V_D + (\chi * V_{BB}); \tag{6}$$

$$\alpha = \alpha_0 * T_{ox}; \tag{7a}$$

$$\Gamma_{B1} = \Gamma_{B10} * e^{-(Tox*s1)}; \tag{7b}$$

$$\Gamma_{C1} = \Gamma_{C10} * T_{ox}^{(s2)}; \tag{7c}$$

$$A = A_1 e^{\left(\frac{-Ea_A}{kT}\right)}, B = B_1 e^{\left(\frac{-Ea_B}{kT}\right)} \tag{8}$$

$$A_1 = A_2 L_{CH}^{(\lambda_A)}, B_1 = B_2 L_{CH}^{(\lambda_B)}, C_1 = C_2 L_{CH}^{(\lambda_c)} \tag{9}$$

$$T = T_{SHE} + T_{CHUCK}, T_{SHE} = \theta V_G V_D \tag{10}$$

ΔP: device parameter shift,
P: Model parameter related to maximum degradation,
t: stress time, τ : related to bond dissociation rate,
m: governs the kinetics at early time before the onset of saturation

978-1-7281-0941-1/19 $31.00 © 2019 IEEE

TABLE II. MODEL PARAMETERS

Fixed Parameters:
$m_0 = 0.5$; τ_m (s) = 1e4; k = 0.036; χ = 0.08;
s1 = 0.1; s2 = 0.5; Γ_{A1} (V^{-1}) = 4.5; Γ_{A2} (V^{-1}) = 17.6;
E_{A_HC} (eV) = 0.4; β = 0.5; E_{B_HC} (eV) = 0.4; σ = 1;
Γ_{C2} (V^{-1}) = 6.96;

Adjustable Parameters:

	Fig 4a	Fig 4b	Fig 5
A2*	1	0.4	1e-5
α/α_0	2	2.5	0.027
B2*	1	1e3	1e-1
Γ_B / Γ_{B0} (V^{-1})	2	1.7	60
C2*	1	10	10
Γ_{C1}/Γ_{C10} (V^{-1})	6	5.8	35

*values are normalized to thin gate oxide parameters.
Data for thick and thin oxide are normalized to
different values in referenced paper.

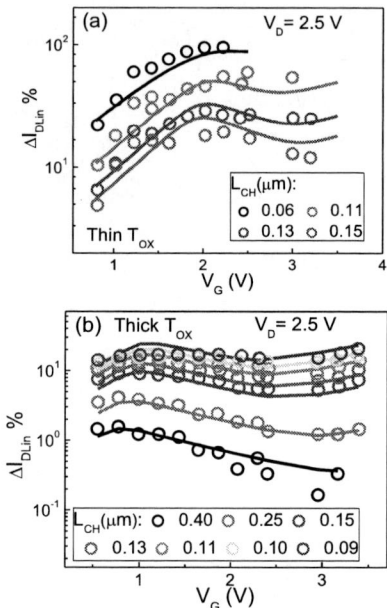

Fig. 7. Modeling of channel length dependence. (a) thin gate-oxide
and (b) thick gate oxide. (a, b) shows varying channel length across V_G.
Parameters are shown in Table II.

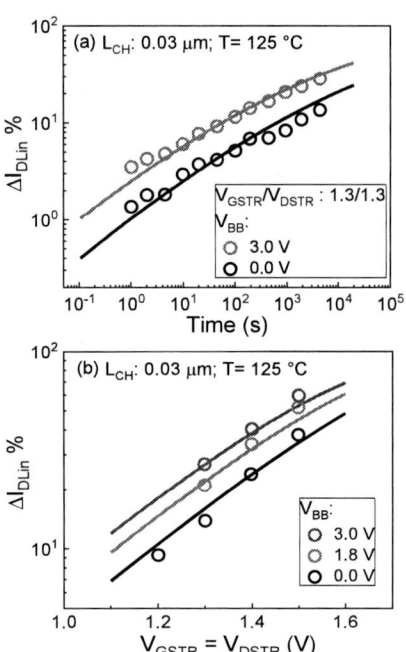

Fig. 6. Modeling of V_{BB} stress impact over I_{DLIN} drift in thin gate-oxide
(a) time kinetics over classical HCI stress with FBB stress for (b) over
voltage stress where $V_G = V_D$ with and without FBB at fixed time.

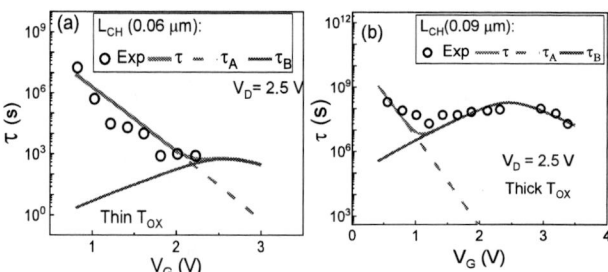

Fig. 8. Contribution from different τ subcomponents. (a) thin gate-
oxide and (b) thick gate oxide. (b) shows varying channel length across
VG. Parameters are shown in Table II.

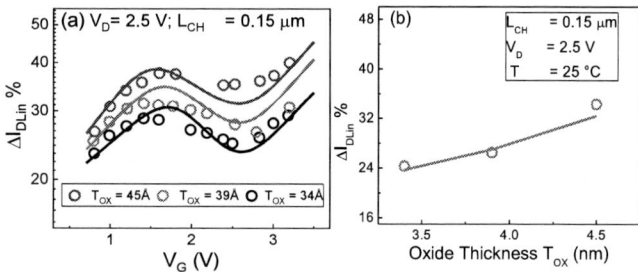

Fig. 9. I_{DLIN} drift modeling over gate voltage stress for (a) several
gate-oxide thickness at fixed channel length (b) several gate-oxide
thickness at fixed condition $V_G = V_D$.

978-1-7281-0941-1/19 $31.00 © 2019 IEEE

Quantum Transport Simulations of the Zero Temperature Coefficient in Gate-all-around Nanowire pFETs

Hyeongu Lee
School of Electrical Engineering
Korea Advanced Institute of Science and Technology
Daejeon, Republic of Korea
hyeongulee@kaist.ac.kr

Junbeom Seo
School of Electrical Engineering
Korea Advanced Institute of Science and Technology
Daejeon, Republic of Korea
jbseo@kaist.ac.kr

Mincheol Shin*
School of Electrical Engineering
Korea Advanced Institute of Science and Technology
Daejeon, Republic of Korea
mshin@kaist.ac.kr

Abstract— **We present a full quantum transport study of the zero-temperature coefficient (ZTC) point for sub-10 nm gate-all-around nanowire p-type field effect transistors (GAA NW pFETs). The phonon scattering effects are included through the self-consistent Born approximation in the non-equilibrium Green's function framework. The main findings are that the ZTC point can be present in GAA NW pFETs in sub-10 nm regime and the gate voltage at the ZTC point shows an opposite trend and has an upper limit at a certain gate length. This is due to the interplay between the ballisticity ratio and the ballistic current ratio, which can be explained only by the quantum transport simulations.**

Keywords—zero-temperature coefficient (ZTC), quantum transport, non-equilibrium Green's function (NEGF), k·p method, gate-all-around (GAA) nanowire field effect transistor

I. INTRODUCTION

Zero-temperature coefficient (ZTC) point has attracted attention to design devices immune to process, voltage, and temperature variation [1]. The ZTC point, corresponding to the bias voltage where current-voltage (I-V) characteristics show little variation with temperature, can play an important role in designing thermally stable CMOS integrated circuits [2]. For example, the current reference circuit exploiting bias voltages near the ZTC point has an advantage with low temperature coefficient [3]. For memory application, 1T-DRAM biased in the ZTC point shows a stable memory window over a wide range of temperature [4].

There are numerous works on the ZTC point for multi-gate structures such as double-gate ultra-thin-body field effect transistors (FETs) [5] and triple-gate FinFETs [6]. These studies have demonstrated the existence of the ZTC point for the FETs with several tens of nanometer. However, few works have been done on the ZTC point for gate-all-around nanowire (GAA NW) FETs with sub-10 nm feature size. In order to utilize the property of the ZTC point, it is necessary to evaluate the behavior of the ZTC point in sub-10 nm scale GAA NW FETs where the thermal reliability is a critical issue in particular. At such small dimensions, quantum effects such as tunneling and confinement strongly affect the device performance. Since it is insufficient to analyze the ZTC point in sub-10 nm scale through classical drift-diffusion transport model, quantum transport simulation should be considered to

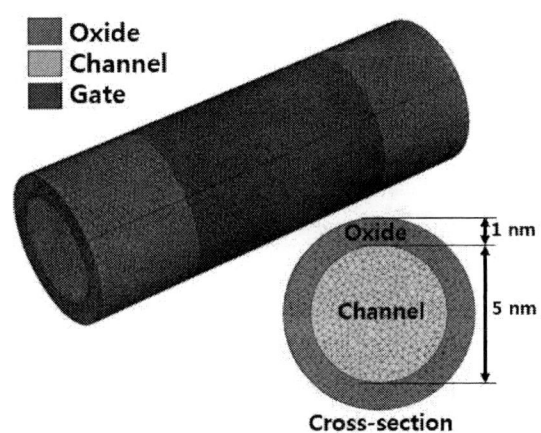

Fig. 1: Schematic structure of Si GAA NW and the circular cross-section with the FEM mesh configuration.

correctly investigate the behavior of the ZTC point for GAA NW FETs in sub-10 nm size.

In this paper, we present a theoretical study of the ZTC point for Si GAA NW p-type FETs (pFETs) using a full three-dimensional (3-D) quantum transport simulation. To the best of our knowledge, it is the first time to study the ZTC point through the quantum transport simulations. We investigate the behavior of the ZTC point for a small size GAA NW pFETs and the scaling effects of L_g on the drain current (I_d) and the gate voltage (V_g) at the ZTC point.

II. SIMULATION METHOD

We have simulated NWs with a circular cross-section as shown in Fig. 1. The diameter of the NWs is 5 nm, which are surrounded by an oxide layer with an equivalent oxide thickness of 0.5 nm. L_g is scaled from 32 nm to 3 nm and the current flows along [100] direction. The channel is undoped and the source and drain are p-type doped with a concentration of 1×10^{20} cm^{-3}. All the simulations are conducted in the temperature range of 200-400 K with a supply voltage of 0.4 V.

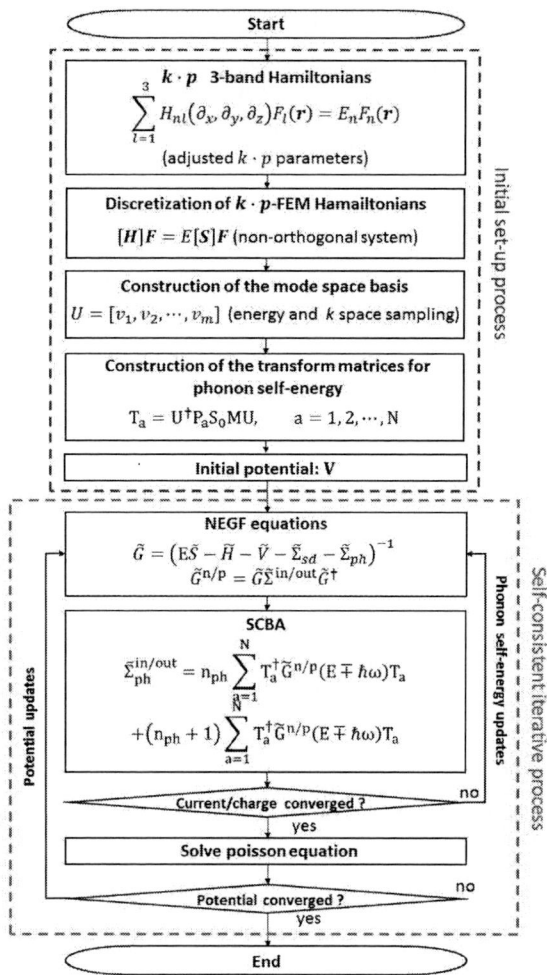

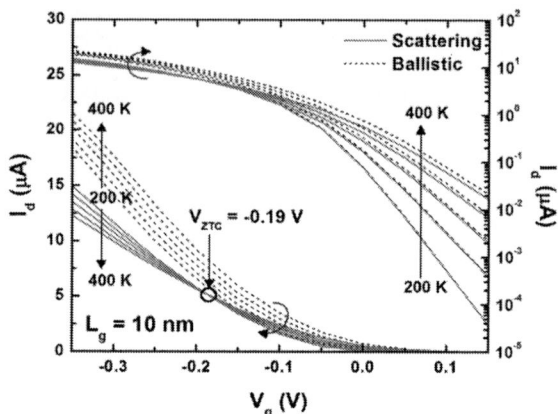

Fig. 3: I-V characteristics of the ballistic and the phonon scattering transport for L_g of 10 nm. V_{ZTC} is V_g at the ZTC point. The ballistic current was calculated using the electrostatic potential converged in the phonon scattering transport simulation.

above 77 K [9], we have only considered phonon scattering in this work. The hole-phonon interaction has been rigorously treated through the self-consistent Born approximation (SCBA). We have assumed bulk phonon scattering as the mobility variation due to phonon confinement is weak for NWs with diameters larger than 5 nm [10]. The optical and acoustic deformation potential parameters are taken from [11]. For efficient NEGF simulations, we have reduced the size of the $k \cdot p$ Hamiltonians using the mode space method [12] where the Bloch modes are sampled in energy and k space. We have developed an efficient scheme to calculate the hole-phonon scattering self-energy in the mode space, which greatly reduces the computational cost of SCBA.

III. RESULTS AND DISCUSSION

The GAA NW pFET with L_g of 10 nm was simulated to investigate the ZTC point for a small size FET. Fig. 3 shows the I-V curves for the ballistic and scattering transport. It is noted that I_d is insensitive to temperature variation at $V_g = -0.19$ V, indicating the ZTC point even in a small size FET. In order to clearly understand the ZTC point, we have compared the ballistic transport with the phonon scattering transport. We calculated the ballistic transport using the electrostatic potential of the scattering transport to exclude an artificial band edge shift [13].

Fig. 3 shows that in the subthreshold region, a similar trend is observed for ballistic and dissipative cases, where I_d is increased with the temperature and consequently V_{th} is decreased. This is due to the injection of carriers at high energy states, excited by the broadening effect of the Fermi-Dirac distribution. However, in the high V_g region, I_d of the scattering transport is decreased as the temperature increases, while I_d of the ballistic transport follows the same increasing trend as in the subthreshold region. The reason for the former is that the strength of the phonon scattering is enhanced at the elevated temperature. We therefore note that the ZTC point occurs even in sub-10 nm scale GAA NW FETs by the phonon scattering effects and the broadening effect of the Fermi-Dirac distribution.

In order to investigate how the scaling of L_g affects the ZTC point, we calculated the V_g (V_{ZTC}) and I_d (I_{ZTC}) at the

Fig. 2: Flowchart of the overall simulation procedure. m and N are the number of modes and FEM nodes, respectively. P_a is the orthogonal projection onto the subspace spanned by the $k \cdot p$ basis vectors of the ath node. S_0 is the overlap matrix and $n_{ph}(\omega)$ is the Bose-Einstein distribution of the phonon with frequency ω. The tilde represent the mode space quantities. In this work, 39 modes are used.

Fig. 2 shows the overall simulation flow chart. To faithfully represent a circular cross-section of the NWs, we have employed the finite element method (FEM) for 3-D discretization of the $k \cdot p$ Hamiltonians and the Poisson equation (see Fig. 1). The 3-band $k \cdot p$ model has been used to describe the valence bands. Since the bulk $k \cdot p$ parameters are inadequate to describe the band structure of the NW with a diameter of 5 nm, we used the adjusted $k \cdot p$ parameters in [7] to calculate the $k \cdot p$ Hamiltonians, whose the band structure matches well with tight-binding ones.

We have solved 3-D Poisson equation self-consistently with the non-equilibrium Green's function (NEGF) equations. As the ZTC point results from mutual compensation of mobility and threshold voltage (V_{th}) temperature effects [8], scattering mechanisms such as phonon scattering, surface roughness scattering, and Coulomb scattering should be included in the NEGF framework. Since, however, the most dominant scattering mechanism is the phonon scattering

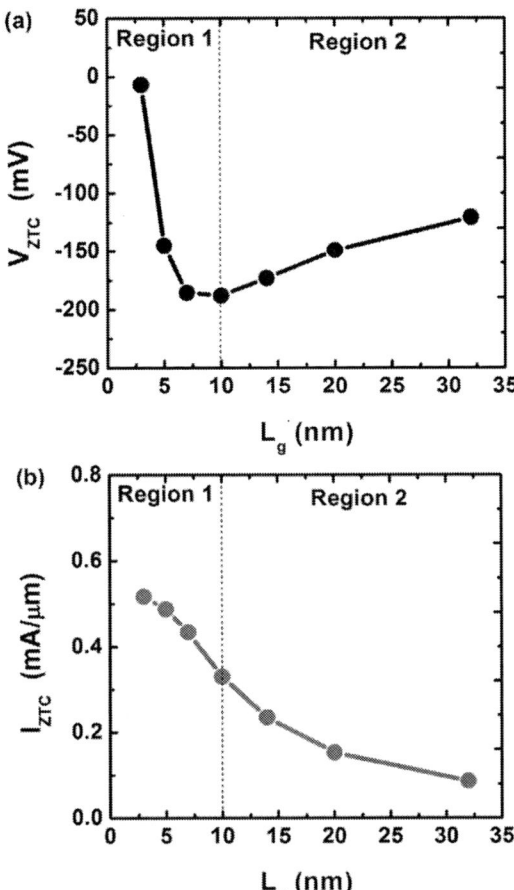

Fig. 4: V_{ZTC} (a) and I_{ZTC} (b) as a function of L_g, where V_{ZTC} and I_{ZTC} are V_g and I_d at the ZTC point, respectively. L_g of GAA NW FETs are 32, 20, 14, 10, 7, 5, and 3 nm.

ZTC point for various L_g from 32 to 3 nm as shown in Fig. 4. It is interesting to note that the $|V_{ZTC}|$ shows an opposite trend about a specific L_g and has a maximum value as illustrated in Fig. 4(a). As L_g scales down from 32 to 10 nm (see region 2 in Fig. 4(a)), the $|V_{ZTC}|$ is moderately increased by about 67 mV. On the other hand, when L_g shrinks down from 10 to 3 nm in region 1, the $|V_{ZTC}|$ is rapidly decreased by approximately 181 mV. The trend of the $|V_{ZTC}|$ in the region 2 is compatible with that for n-type FDSOI devices [14] where the $|V_{ZTC}|$ monotonically increases as gate length decreases from 69 to 29 nm. What is notable is the inverted trend of the $|V_{ZTC}|$ below L_g of 10 nm. For the case of I_{ZTC}, it has a different trend with the $|V_{ZTC}|$ as shown in Fig. 4(b). As L_g decreases, I_{ZTC} monotonically increases.

The contrary trends of the $|V_{ZTC}|$ can be explained by ballisticity and ballistic current, which are useful for describing the behavior of the $|V_{ZTC}|$ as it is determined by the interplay between those two values. The ballisticity is defined as the ratio of the scattering current to the ballistic current, expressing the strength of backscattering. Since the ZTC point occurs when I_d of 400 K ($I_{400 K}$) is the same as I_d of 200 K ($I_{200 K}$), the following equation (1) is satisfied at the ZTC point,

$$\frac{I_{400 K}}{I_{200 K}} = \frac{B_{400 K}}{B_{200 K}} \times \frac{I_{400 K}^{Ballistic}}{I_{200 K}^{Ballistic}} = 1, \quad (1)$$

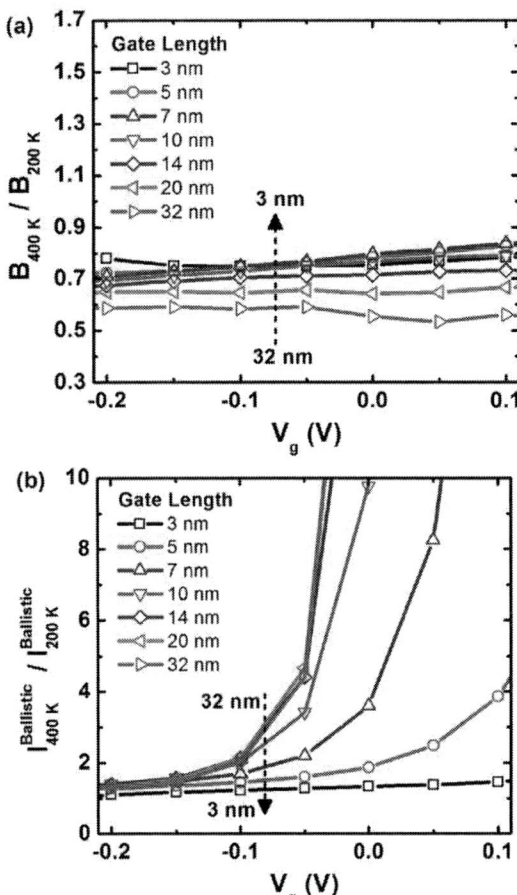

Fig. 5: Ballisticity ratio ($B_{400 K}$ / $B_{200 K}$) (a) and ballistic current ratio ($I_{400 K}^{Ballistic}/I_{200 K}^{Ballistic}$) (b) as a function of V_g. The ballisticity ratio represents the dependence of scattering strength on the variation of temperature. The ballistic current ratio reflects the intensity of source-to-drain direct tunneling (SDDT) current and the gate controllability.

where B and $I^{Ballistic}$ are the ballisticity and the ballistic current, respectively.

Fig. 5 shows the ballisticity ratio ($B_{400 K}$ / $B_{200 K}$) and the ballistic current ratio ($I_{400 K}^{Ballistic}/I_{200 K}^{Ballistic}$) for GAA NW FETs with varying L_g in a range of 3 to 32 nm. It is helpful to describe the scaling dependences of these two ratios, before moving on to the analysis of the V_{ZTC} and I_{ZTC} variation. In Fig. 5(a), the ballisticity ratio is smaller than one for all devices simulated, meaning that the phonon scattering is enhanced at high temperature. It is noted that while the ballisticity ratio is moderately increased as L_g decreases from 32 to 10 nm, a further decrease of L_g does not significantly change the ratio. It is because the L_g is too short for the backscattering to take effect except the source/drain extension region.

Unlike the ballisticity ratio, the ballistic current ratio is larger than one as illustrated in Fig. 5(b). It is because of the broadening effect of the Fermi-Dirac distribution. We note that the ballistic current ratios of devices with L_g from 32 to 14 nm show almost similar behavior over V_g. The reason is that the thermionic current is dominant over source-to-drain

978-1-7281-0941-1/19 $31.00 © 2019 IEEE

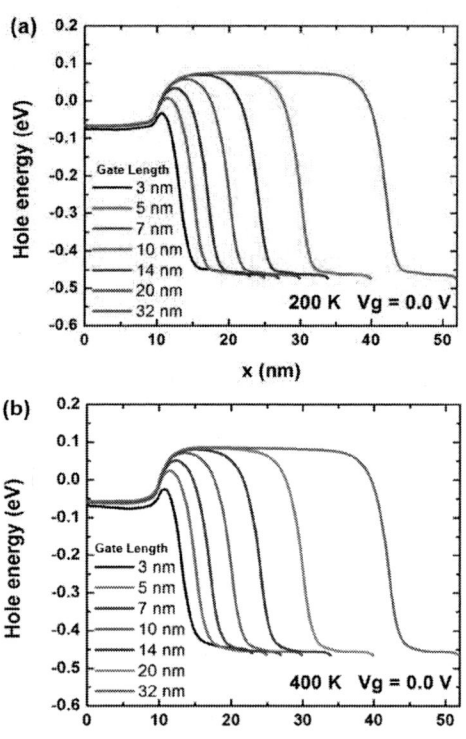

Fig. 6: Valence band edge profiles for 200 K (a) and 400 K (b) at Vg = 0.0 V.

direct tunneling (SDDT) current. When L_g is large enough for the SDDT current to be negligible, the ballistic current ratio is mainly determined by the difference of the amount of injected carriers over the band edge. Fig. 6 shows the valence band edge profiles at Vg = 0.0 V for the temperatures of 200 K and 400 K. In Fig. 6, for L_g from 32 to 14 nm, the valence band barrier heights are nearly same for all temperatures. Thus, the Fermi-Dirac broadening factor solely determines the ballistic current ratio. However, as the SDDT becomes dominant and the gate controllability degrades with reduction of L_g, the ballistic current ratio is rapidly decreased (see Fig. 5(b)).

As mentioned above, the trend of the $|V_{ZTC}|$ with L_g is easily described by examining the tendencies for the ballisticity ratio and the ballistic current ratio. For the devices with $L_g < 10$ nm, as L_g decreases, the ballisticity ratio is not significantly changed in Fig. 5(a). By contrast, the ballistic current ratio rapidly decreases with L_g in Fig. 5(b). Thus, to satisfy (1), the $|V_{ZTC}|$ should sharply decrease with L_g. In case of $L_g > 10$ nm, the trend of the $|V_{ZTC}|$ can be explained in the same manner but with the roles interchanged. As L_g decreases, the ballistic current ratio slightly changes as shown in Fig. 5(b). However, the ballisticity ratio increases so that the $|V_{ZTC}|$ is moderately increased.

When it comes to I_{ZTC}, it is attributed to the increase of the SDDT and the degradation of the gate controllability. Fig. 6 confirms this behavior. As L_g decreases, the tunneling length is decreased and the valence band barrier height is reduced. This enhances the SDDT and thermionic current, leading to the larger I_{ZTC}.

IV. CONCLUSION

In this paper, we have investigated the ZTC point of GAA NW pFETs by using the quantum transport simulations. We theoretically demonstrated that the ZTC point can appear in sub-10 nm scale. We discussed the L_g scaling effects on the ZTC point. It is found that as L_g decreases, $|V_{ZTC}|$ has an inverted trend at a specific L_g, where $|V_{ZTC}|$ moderately increases for $L_g > 10$ nm and rapidly decreases for $L_g < 10$ nm. Unlike the trend of $|V_{ZTC}|$, I_{ZTC} is monotonically increased with L_g. The results will help design the thermally stable CMOS circuit using GAA FETs with future technology nodes.

ACKNOWLEDGMENT

This research was supported by Basic Science Research Program through the National Research Foundation of Korea (NRF) funded by the Ministry of Science and ICT (2017R1A2B2005679).

REFERENCES

[1] A. R. Mohamed, M. Chen, and G. Wang, "Untrimmed CMOS Nano-Ampere Current Reference with Curvature-Compensation Scheme," 2019 IEEE International Symposium on Circuits and Systems, pp. 1-4, May. 2019.

[2] M. A. Alim and A. A. Rezazadeh, "Device behaviour and zero temperature coefficients analysis for microwave GaAs HEMT," Solid State Electron., vol. 147, pp. 13-18, Sept. 2018.

[3] P. Toledo et al., "MOSFET ZTC condition analysis for a Self-biased current reference design," Journal of integrated Circuits and Systems, vol. 10, no. 2, pp. 103-112, Dec. 2015.

[4] K. R. A. Sasaki, L. M. Almeida, J. A. Martino, M. Aoulaiche, E. Simoen, and C. Claeys, "Temperature influence on UTBOX 1T-DRAM using GIDL for writing operation," 2012 International Caribbean Conference on Devices, Circuits and Systems, pp. 1-4, Mar. 2012.

[5] S. K. Mohapatra, K. P. Pradhan, and P. K. Sahu, "Temperature dependence inflection point in Ultra-Thin Si directly on Insulator (SDOI) MOSFETs: An influence to key performance metrics," Superlattices Microstrut., vol. 78, pp. 134-143, Feb. 2015.

[6] C. W. Lee, et al., "High-Temperature Performance of Silicon Junctionless MOSFETs," IEEE Trans. Electron Devices, vol. 57, no. 3, pp. 620-625, Mar. 2010.

[7] M. Shin, S. Lee, and G. Klimeck, "Computational Study on the Performance of Si Nanowire pMOSFETs Based on the k · p Method," IEEE Trans. Electron Devices, vol. 57, no. 9, pp. 2274-2283, Sept. 2010.

[8] I. M. Filanovsky and A. Ahmed, "Mutual compensation of mobility and threshold voltage temperature effects with applications in CMOS circuits," IEEE Trans. Circuits Syst. I. Fundam. Theory Appl., vol. 48, no. 7, pp. 876-884, Jul. 2001.

[9] S. I. Takagi et al., "On the universality of inversion layer mobility in Si MOSFET's: Part I-effects of substrate impurity concentration," IEEE Trans. Electron Devices, vol. 41, no. 12, pp. 2357-2362, Dec. 1994.

[10] E. B. Ramayya et al., "Electron transport in silicon nanowires: The role of acoustic phonon confinement and surface roughness scattering," J. Appl. Phys., vol. 104, no. 6, p. 063711, Sept. 2008.

[11] N. Neophytou and K. Kosina, "Large enhancement in hole velocity and mobility in p-type [110] and [111] silicon nanowires by cross section scaling: an atomistic analysis," Nano Lett, vol. 10, no. 12, pp. 4913-4919, Nov. 2010.

[12] M. Shin, W. J. Jeong, and J. Lee, "Density functional theory based simulations of silicon nanowire field effect transistors," J. Appl. Phys., vol. 119, no. 15, p. 154505, Apr. 2016.

[13] M. Luisier and G. Klimeck, "Atomistic full-band simulations of silicon nanowire transistors: Effects of electron-phonon scattering," Phys. Rev. B, vol. 80, no. 15, p. 155430, Oct. 2019.

[14] T. Nicoletti, et al., "The impact of gate length scaling on UTBOX FDSOI devices: The digital/analog performance of extension-less structures," 2012 13th International Conference on Ultimate Integration on Silicon, pp. 121-124, Mar. 2012.

978-1-7281-0941-1/19 $31.00 © 2019 IEEE

Electro-Thermal Analysis and Edge Termination Techniques of High Current β-Ga₂O₃ Schottky Rectifiers

Ribhu Sharma
Material Science and Engineering
University of Florida
Gainesville, Fl, USA
https://orcid.org/0000-0001-5754-7873

Erin Patrick
Electrical and Computer Engineering
University of Florida
Gainesville, Fl, USA
erin.patrick@ece.ufl.edu

Jiancheng Yang
Chemical Engineering) University of Florida
Gainesville, Fl, USA
yjcallen@ufl.edu

Fan Ren
Chemical Engineering
University of Florida
Gainesville, FL, USA
https://orcid.org/0000-0001-9234-019X

Mark Law
Electrical and Computer Engineering
University of Florida
Gainesville, Fl, USA
law@ece.ufl.edu

Stephen Pearton
Material Science and Engineering
University of Florida
Gainesville, Fl, USA
https://orcid.org/0000-0001-6498-1256

Abstract—The performance and limitations of β-Ga₂O₃ Schottky rectifiers is studied via simulation using the Florida Object Oriented Device and Process (FLOODS) TCAD simulator. The effect of forward bias and power is examined for various bulk and epitaxial layer thicknesses as well as for heat sink geometries. Thicker bulk/substrate results in higher maximum temperature values whereas a thinner epitaxial-layer results in higher forward currents and hence a higher maximum temperature values via Joule heating. A Cu finned heat sink geometry results in a 26.76% reduction in the maximum temperature. Edge termination techniques are examined for β-Ga₂O₃ Schottky rectifiers in order to maximize the breakdown voltage, identify the location of breakdown and mitigate the maximum electric field. Best results have been observed for Al₂O₃ as the dielectric material in a field-plate structure while the effect of field-plate dimensions is also studied.

Keywords—Gallium oxide, semiconductors – II-VI, Rectifiers, electro-thermal, edge-termination

I. INTRODUCTION

Gallium oxide has recently emerged as the prime candidate for applications in power electronics due to its large bandgap, high electric breakdown strength, low on-resistance and low charge storage times [1][2][3]. The suitability of semiconductors as electronic power switches is evaluated by calculating various figure-of-merits (FOM). The large bandgap (~4.85 eV) of Ga₂O₃ translates into a high breakdown field (E_{br}) creflected by a high Baliga's figure-of-merit which is about 4-7 times higher than that for SiC, leading to lower conductance losses at a low manufacturing cost in comparison to 4H-SiC diodes [4]. Notably, the electric field breakdown strength (E_{br}) of Ga₂O₃ is more than double the theoretic limits of SiC and GaN.

Schottky rectifiers have been adopted for power electronics applications due to its fast switching speed, which is required to improve the efficiency of inductive motor controllers and power supplies [5][6] as well as low forward voltage drop and high-temperature operability. Along with the high E_{br}, Ga₂O₃ also possesses an on-state resistance value of over 10 times larger than conventional Si rectifiers.

An ongoing issue with Ga₂O₃ is its thermal conductivity (21 W/m.K) [7][8], which is an order of magnitude lower than its commercial counterparts, i.e. SiC and GaN. Recent studies [9][10][11] performed on vertical Schottky rectifiers have achieved high forward-currents and high breakdown voltages,

due to the rapid progress in the epitaxial growth methods. These studies focus on the realization of large dimension devices; hence it is important to understand the rise and inefficient dissipation of heat from the channel resulting in a degradation of electron transport properties. The thermal properties of Gallium oxide devices have been studied recently with a focus on thermoreflectance-based thermography and imaging of Ga₂O₃ Schottky diodes [4][12], Ga₂O₃ Field effect transistors [13] and Ga₂O₃ MOSFETs [14]. Edge termination techniques for GaN and SiC devices to maximize breakdown voltage (V_{br}) have been well established [15][16][17][18][19]; however, such methods have not yet been fully developed for Ga₂O₃. One of the biggest challenges for Ga₂O₃ technology is the relative absence of p-type epitaxy or p-type doping which rules out the 'guard ring structure' method of edge termination [20][21]. This leaves methods like field plate structures [21][22] and high resistivity layers created by ion implantation [23] for edge termination.

In this paper, Ga₂O₃ Schottky diodes are examined to assess the limitations stated above and to improve the design considerations in terms of dimensions, structure, and the materials used. We report on the effect of dimensionality on the forward current and heat generation by modeling the current density and heat flow in the device. Edge termination techniques are also evaluated in a separate set of simulations.

II. SIMULATION METHODS

The partial and total differential equations governing the physics of the electrical and thermal domains are solved self-consistently with the FLOODS TCAD simulator. FLOODS/FLOOPS is a partial differential equation (PDE) solver, written as an extension to the Tcl language for easy specification of PDEs and boundary conditions [24].

A. Device structure

The device structure as given in Fig. 1a is used as the reference for the study as a two-dimensional model and designed to achieve both high forward currents and good reverse breakdown characteristics [9]. The device consists of a lightly doped epitaxial layer grown on highly doped bulk β-Ga₂O₃ with [001] surface orientation and E-beam evaporated Ni/Au is used to form the Schottky (top) contact, while Ti/Au is used to form the Ohmic contact. In order to accurately model the electro-thermal profile the experimental IV curves are fitted via modeling [29]. The doping concentrations for the electro-thermal simulations are 1.3×10^{16} and 3.6×10^{18}

U. S. Government work not protected by U. S. copyright.

978-1-7281-0941-1/19 $31.00 © 2019 IEEE

cm^{-3} for the epi and bulk layer respectively, and for the edge termination simulations are 2.8×10^{16} and 4.8×10^{18} for the epi and bulk layer respectively.

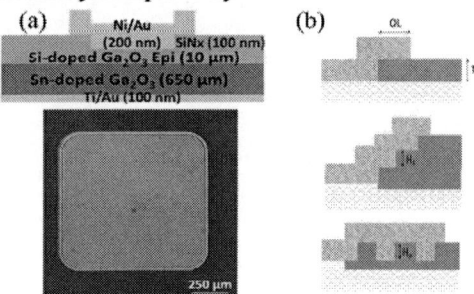

Figure 1. Schematic of a) vertical geometry Ni/Au Schottky structure (top) and top view with a contact area of 0.01 cm^{-2}, and b) field plate structures showing field plate overlap (OL), dielectric thickness (t), dielectric step height (Hs), and pillar height (Hp).

B. Electrical domain

Equations used the model include the common device equations [25] including the Poisson's, continuity and current density equations. By using the Boltzmann relation, the quasi-Fermi level from the current density equation can be related to the electrostatic potential. For the electric field analysis, the gradient of the electrostatic potential is then used to obtain the electric field, $|\vec{E}|$. Equation 1 describes the formulation of ionized donor trap density,

$$\frac{N_D^+}{N_{tot}} = \int \left(\frac{1}{1+2e^{\frac{E_F-E}{kT}}} \right) \left(\frac{1}{\nabla E\sqrt{2\pi}} e^{\frac{(E-E_T)^2}{2\nabla E^2}} \right) dE, \quad (1)$$

where N_{tot} is the total donor trap concentration, N_D^+ is the ionized donor density, E_F and E_T are the electron quasi-Fermi levels and trap levels, respectively, and ∇E is the energy spread of the traps. The total trap densities and energy levels of the traps associated with β-Ga$_2$O$_3$ [26] have been added into the above model. The mobility model used has been incorporated using the equations derived by Ma et al. [27].

Table 1. Material (electronic and thermal) properties of Ga$_2$O$_3$ vs other semiconductors.

	Si	4H-SiC	GaN (Wurzite)	Ga$_2$O$_3$
E$_g$ (eV)	1.1	3.3	3.4	4.6-4.9
μ$_n$ (cm^2/V-s)	1400	1000	1200	100-200
E$_{cr}$ (MV/cm)	0.3	2.5	3.3	8
ε	11.8	9.7	9.0	10.0
Normalized BFOM	1	340	870	1100-2250
κ at 300K (W/m-K)	150	270	210	27 [010] 11 [100]

C. Thermal Domain

Using the electron and hole current densities, heat generation (Q) is incorporated into the model via Joule heating [28][29]. Heat transport is modeled as a function of heat generated and temperature change with respect to time and space via the following equation:

$$C\frac{\partial T}{\partial t} - \nabla . K\nabla T = Q \quad (2)$$

where C is the specific heat capacity, T is temperature and K is the thermal conductivity. The thermal conductivity for Ga$_2$O$_3$ has been studied in detail, while Table I also shows the anisotropic nature of the thermal conductivity. Due to the surface orientation along the [001] direction and heat moving along the [001] direction, we assume an average value of 21 ± 2 W/mK. In coherence with previous thermal studies [31] an isotropic value for the thermal conductivity has been considered. However, future work with anisotropic thermal conductivity and the temperature dependence on the conductivity is ongoing. We use a convective heat transfer [32] equation as the boundary condition for heat transfer, given by Newton's law of cooling:

$$q'=hA(T_s-T_\infty) \quad (3)$$

where q' is the heat flux, h is the heat transfer coefficient, A is the area of the surface, T_s and T_∞ (300K) are the surface and ambient temperatures respectively. The heat transfer coefficient is expressed as a function of three dimensionless numbers, the Nusselt (Nu), Grashof (Gr) and the Prandtl (Pr) numbers.

$$Nu(Gr, Pr) = hL/k \quad (4)$$

$$Gr = \frac{g\beta (T_s-T_\infty)L^3}{v^2} \quad (5)$$

$$Pr = \frac{\mu C}{k} \quad (6)$$

In Eqs 4-6, L is the characteristic length, k is the thermal conductivity, g is the gravitational force, β is the volume coefficient of expansion (β = 1/T; for an ideal gas), μ and υ are the dynamic and kinematic viscosities. In order to get an expression for the heat transfer coefficient for different surfaces a simpler expression gives a better understanding:

$$\overline{Nu} = \frac{\bar{h}k}{L} = c(Gr\ Pr)^m \quad (9)$$

where c and m are constants, which depend on the type of surface in question (i.e., a vertical surface or a horizontal surface). A Dirichlet boundary condition is applied for the temperature at the bottom side of the wafer of T=300K, representing a perfect heat sink. While modeling the convective heat transfer from the top and sides of the device, Neumann boundary conditions have been set for the temperature.

III. RESULTS AND DISCUSSIONS

A. Electro-Thermal management

The effect of forward (0-2.5 V) and power generation (0-5.5 W) is examined for the device structure. The effect of the voltage bias is reflected in heat generation in the drift region near the epi-metal interface and results in highest temperature values near this interface. A temperature profile is simulated in the device cross-section as seen in Fig 2a, and the effect of dimensionality is seen via this temperature profile and peak temperature (T_{max}) values. The bulk thickness is varied from 100 μm to 1000 μm while the epi-layer thickness and contact area were kept constant. This resulted in T_{max} values from 313 K for the 100 μm bulk to 417 K for 1000 μm bulk thickness. The rise in temperature also causes a drop in the electron mobility, while a thinner bulk also causes a higher effective heat dissipation via the bottom contact. Thinning the substrate thickness shows a significant reduction in heating in the device as seen in Fig 3.

Similarly, the epitaxial layer thickness was varied from 3 μm to 20 μm while the bulk-layer thickness and contact area

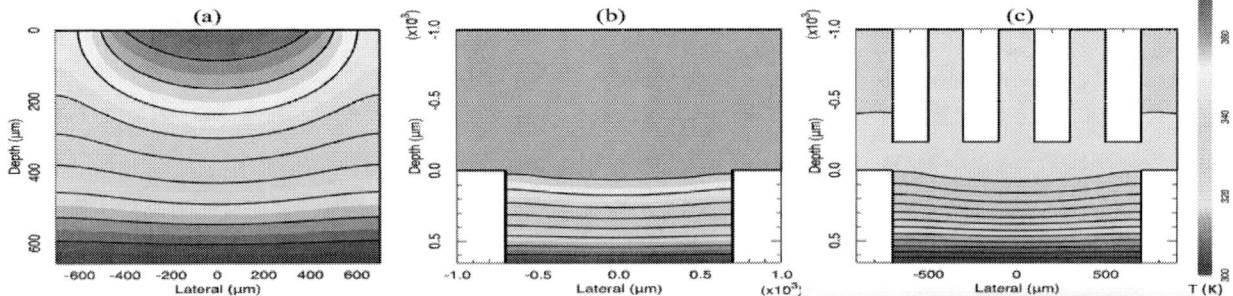

Figure 2. Cross-sectional temperature profile for a diode structure with free/natural convection from (a) device top and side surfaces peak temperature 369 K), (b) a copper top-side heat sink block (peak temperature 359 K), and (c) a copper top-side finned heat sink (peak temperature 344 K

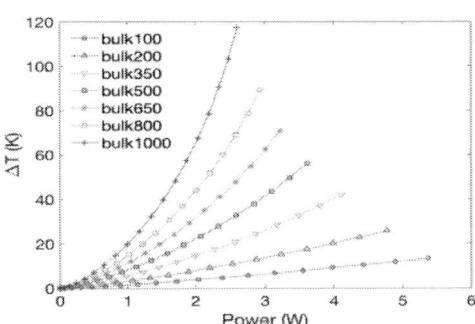

Figure 3. The temperature rises as a function of Power ($P=V^2/R$) while varying the bulk layer thickness from 100 μm to 1000 μm with epi-layer thickness constant at 7 μm.

were kept constant. As the epi-layer is thinned, the resistance decreases which results in higher currents in the device, (Fig 4); however, this results in a trade-off for peak temperature as the T_{max} ranges from 464 K for the 3 μm epi layer to 324 K for the 20 μm epi-layer thickness diode. The self-heating effects are evident, and the importance of efficient heat dissipation is realized.

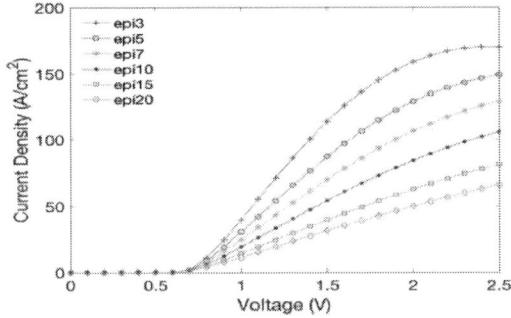

Figure 4. The dependence of the forward J-V characteristics on the epi-layer thickness from 3 μm to 20 μm with bulk thickness constant at 650 μm.

Initial studies show the importance of dimensions [4] and the simulations show considerable amounts of heat being dissipated from the Schottky (top) contact, a requirement for top-side passive cooling methods like nanocrystalline diamond capping layers and flip-chip hetero-integration. In electronics, copper is one of the most widely used material as a heat sink and different structures have been investigated thoroughly. Two top-side cooling cap heat-sink designs have been considered: a solid copper block and a finned copper structure. Heat loss from the heat sinks to the atmosphere is modeled as free convection via a steady flux. Fig 2 shows the temperature profile for three test simulations: a) natural

convection with bulk thickness of 650 μm and epi thickness of 7 μm, b) the same device structure with a 1 mm thick solid Cu heat sink block with a width of 2 mm on top, c) the same device structure with a 1 mm thick finned copper heat sink and 200 μm fin thickness. As seen in the figure, the solid heat sink reduces the T_{max} value by ~10 K, while the finned heat sink reduces T_{max} by ~25 K compared to the device with no heat sink. The reduction is device temperature causes the mobility to increase which results in higher forward currents.

B. Edge termination techniques

The field plate (FP) structural variance is shown in Fig 1b. The electric field distribution is simulated and breakdown locations are observed which helps in concluding where the device is breaking down. Traditionally device breakdown happens near the contact edges due to field crowding near the edge, and edge termination is important to mitigate these effects. The electric field is mapped out as a function of reverse voltage and the breakdown voltage is noted as the critical electric field of the material is reached. The simulations are performed for each structure using the same mesh and the same technological parameters in order to maintain homogeneity in the results. Four dielectric materials have been studied as the FP dielectric; SiN_x (ε=7.0, E_{br}=6.7 MV/cm), SiO_2 (ε=3.9, E_{br}=10 MV/cm), Al_2O_3 (ε=8.0, E_{br}=8.68 MV/cm), and HfO_2 (ε=15.5, E_{br}=5.3 MV/cm). Figure 1.(b) shows the different FP structures used in the

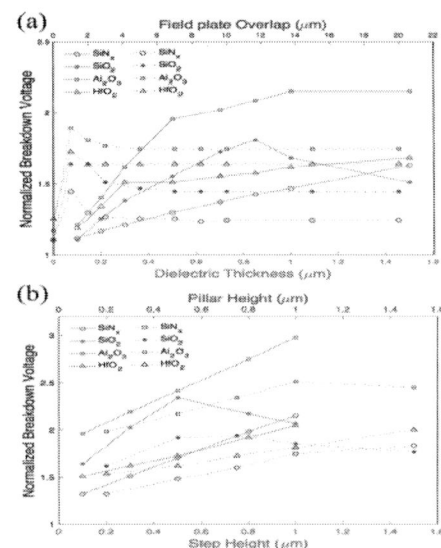

Figure 5. Plots showing the normalized breakdown voltage as a function of (a) field plate overlap (blue) and dielectric thickness (red), and (b) dielectric pillar height (blue) and dielectric step height (red).

978-1-7281-0941-1/19 $31.00 © 2019 IEEE

simulations and denotes the variables in our simulations, i.e. dielectric layer thickness, filed plate overlap, step height, and pillar height. The results have been represented in Fig 5, the best results have been achieved for Al_2O_3 due to its high critical field strength and high relative permittivity, with a normalized breakdown field (V_{Nbr}) of more than 2 for the dielectric thickness equal to or greater than 0.5 μm. Furthermore, using Al_2O_3 dielectric step FP structure, a V_{Nbr} or more than 3 is achieved, while an Al_2O_3 pillar height of 1 μm achieves the highest V_{Nbr}. Future work on ion implantation to create highly resistive area near the contact is currently being developed.

IV. CONCLUSIONS

This paper conducted a thorough investigation via simulations into the effect of device design on β-Ga_2O_3 Schottky rectifiers to aid device optimization. Two pressing concerns of Ga_2O_3 as a power-electronics materials are addressed and the results predict efficient design considerations. Industry desires devices that generate high power and low self-induced heating. The results predict that a thin epitaxial layer and a thin substrate would achieve the desired result. The simulations predict a significant amount of heat being dissipated neat the Schottky contact and warrant the requirement of top-side passive cooling. In order to maximize the breakdown voltage, edge termination techniques were investigated and field plate designs and dielectric materials were studied.

ACKNOWLEDGMENT

The work at UF is partially supported by HDTRA1-17-1-0011 (Jacob Calkins, monitor). The project or effort depicted is sponsored by the Department of the Defense, Defense Threat Reduction Agency. The content of the information does not necessarily reflect the position or the policy of the federal government, and no official endorsement should be inferred.

REFERENCES

[1] G. Jessen, K. Chabak, A. Green, J. McCandless, S. Tetlak, K. Leedy, R. Fitch, S. Mou, E. Heller, S. Badescu, A. Crespo, and N. Moser. *75th IEEE Device Research Conference (DRC)* (2017).

[2] K. Momma and F. Izumi, *J. of Appl. Crys.*, **41**, 653–658 (2008).

[3] H. Wang, thesis, (2007) http://hdl.handle.net/10919/27517.

[4] B. Chatterjee, A. Jayawardena, E. Heller, D. W. Snyder, S. Dhar, and S. Choi, *Review of Scientific Instruments*, **89**, 114903 (2018).

[5] S. J. Pearton, J. Yang, P. H. Cary, F. Ren, J. Kim, M. J. Tadjer, and M. A. Mastro. *Appl. Phy. Rev.*, **5**, 11301 (2018).

[6] M. A. Mastro, A. Kuramata, J. Calkins, J. Kim, F. Ren, and S. J. Pearton, ECS J. of Sol. State Sci. and Tech. **6**,356 (2017).

[7] Z. Guo *et al.*, "Anisotropic thermal conductivity in single crystal β-gallium oxide," *Appl. Phys. Lett.*, vol. 106, no. 11, p. 111909, Mar. 2015.

[8] M Handwerg and R Mitdank and Z Galazka and S F Fischer, "Temperature-dependent thermal conductivity in Mg-doped and undoped β -Ga 2 O 3 bulk-crystals," *Semi. Sci. and Tech.*, vol. 30, no. 2, p. 24006, 2015.

[9] J. Yang, F. Ren, S. J. Pearton, and A. Kuramata. *AIP Advances*, **8**, 55026 (2018).

[10] J. Yang, S. Ahn, F. Ren, S. J. Pearton, S. Jang, and A. Kuramata, *IEEE Elec. Dev. Lett.*, **38**, 906–909 (2017).

[11] J. Yang, F. Ren, M. Tadjer, S. J. Pearton, and A. Kuramata, *ECS J. of Sol. State Sci. and Tech.*, **7**, Q92–Q96 (2018).

[12] P. E. Raad *et al.*, "Thermoreflectance Temperature Mapping of Ga2O3 Schottky Barrier Diodes," in *H01 - Wide Bandgap Semiconductor Materials and Devices 20*, Dallas, TX, 2019.

[13] J. Chen, Z. Xia, S. Rajan, and S. Kumar, "Analysis of Thermal Characteristics of Gallium Oxide Field-Effect-Transistors," presented at the 2018 17th IEEE Intersociety Conference on Thermal and Thermomechanical Phenomena in Electronic Systems (ITherm), 2018.

[14] B. Chatterjee, K. Zeng, C. D. Nordquist, U. Singisetti, and S. Choi, "Device-Level Thermal Management of Gallium Oxide Field-Effect Transistors," *Trans. on Comp., Pack. and Manu. Tech.*, June, 2019.

[15] V. Soler *et al.*, "Planar edge terminations for high voltage 4H-SiC power MOSFETs," *Semi. Sci. and Tech.*, vol. 32, no. 3, p. 35007, Jan. 2017.

[16] M. C. Tarplee, V. P. Madangarli, Quinchun Zhang, and T. S. Sudarshan, "Design rules for field plate edge termination in SiC Schottky diodes," *IEEE Tran. on Elec. Dev.*, vol. 48, no. 12, pp. 2659–2664, Dec. 2001.

[17] J. R. Laroche, F. Ren, K. W. Baik, S. J. Pearton, B. S. Shelton, and B. Peres, "Design of edge termination for GaN power Schottky diodes," *J. of Electronic Materials*, vol. 34, no. 4, pp. 370–374, Apr. 2005.

[18] A. M. Ozbek and B. J. Baliga, "Planar Nearly Ideal Edge-Termination Technique for GaN Devices," *IEEE Electron Dev. Lett.*, vol. 32, no. 3, pp. 300–302, Mar. 2011.

[19] A. M. Ozbek and B. J. Baliga, "Finite-Zone Argon Implant Edge Termination for High-Voltage GaN Schottky Rectifiers," *IEEE Elec. Dev. Lett.*, vol. 32, no. 10, pp. 1361–1363, Oct. 2011.

[20] J. B. Varley, A. Janotti, C. Franchini, and C. G. Van de Walle, "Role of self-trapping in luminescence and p-type conductivity of wide-band-gap oxides," *Phys. Rev. B*, vol. 85, no. 8, p. 81109, Feb. 2012.

[21] J.-H. Choi, C.-H. Cho, and H.-Y. Cha, "Design consideration of high voltage Ga2O3 vertical Schottky barrier diode with field plate," *Results in Physics*, vol. 9, pp. 1170–1171, Jun. 2018.

[22] K. Konishi *et al.*, "1-kV vertical Ga2O3 field-plated Schottky barrier diodes," *Appl. Phys. Lett.*, vol. 110, no. 10, p. 103506, Mar. 2017.

[23] Y. Gao *et al.*, "High-Voltage β-Ga2O3 Schottky Diode with Argon-Implanted Edge Termination," *Nano. Res. Lett.*, vol. 14, no. 1, p. 8, Jan. 2019.

[24] M. E. Law and S. M. Cea, "Continuum based modeling of silicon integrated circuit processing: An object-oriented approach," *Computational Materials Science*, vol. 12, no. 4, pp. 289–308, Nov. 1998.

[25] R. Sharma, E. Patrick, M. E. Law, J. Yang, F. Ren, and S. J. Pearton, "Thermal Simulations of High Current β-Ga2O3 Schottky Rectifiers," *ECS J. of Sol. State Sci. and Tech.*, vol. 8, no. 7, pp. Q3195–Q3201, Jan. 2019.

[26] Z. Zhang, E. Farzana, A. R. Arehart, and S. A. Ringel, "Deep level defects throughout the bandgap of (010) β-Ga2O3 detected by optically and thermally stimulated defect spectroscopy," *Appl. Phys. Lett.*, vol. 108, no. 5, p. 52105, Feb. 2016.

[27] N. Ma *et al.*, "Intrinsic electron mobility limits in β-Ga2O3," *Appl. Phys. Lett.*, vol. 109, no. 21, p. 212101, Nov. 2016.

[28] G. K. Wachutka, "Rigorous thermodynamic treatment of heat generation and conduction in semiconductor device modeling," *IEEE Trans. on Computer-Aided Design of Int. Circ. and Sys.*, vol. 9, no. 11, pp. 1141–1149, Nov. 1990.

[29] E. Patrick, D. Horton, M. Griglione, and M. E. Law, "A Self-Consistent Electro-Thermo-Mechanical Device Simulator based on the Finite-Element Method," *SISPAD 2012*, Sep. 2012.

[30] T.-S. Kang *et al.*, "Thermal Simulation of 193 nm UV-Laser Lift-Off AlGaN/GaN High Electron Mobility Transistors Mounted on AlN Substrates," *ECS Transactions*, vol. 41, no. 6, pp. 129–136, Oct. 2011.

[31] E. A. Douglas, F. Ren, and S. J. Pearton, "Finite-element simulations of the effect of device design on channel temperature for AlGaN/GaN high electron mobility transistors," *J. of Vac. Sci. & Tech. B*, vol. 29, no. 2, p. 20603, Mar. 2011.

[32] F. P. Incropera, D. P. Dewitt, T. L. Bergman, and A. S. Lavine, *Fundamentals of heat and mass transfer*, 6th ed. John Wiley and son.

Thermal Conductivity of Silicon Nanowire Using Landauer Approach for Thermoelectric Applications

Ming-Yi Lee*, Min-Hui Chuang*, Yiming Li*‡, Seiji Samukawa†‡

*Institute of Communications Engineering, National Chiao Tung University,
1001 University Road, Hsinchu 300, Taiwan
e-mail: ymli@faculty.nctu.edu.tw
†Institute of Fluid Science and WPI-AIMR, Tohoku University,
2-1-1 Katahira, Aoba-ku, Sendai, 980-8577, Japan
‡Center for mmWave Smart RadarSystems and Technologies, National Chiao Tung University,
1001 University Road, Hsinchu 300, Taiwan

Abstract—The electronic and phononic band structure of silicon nanowires embedded in SiGe$_{0.3}$ is calculated and used to investigate its effect on the thermoelectric properties by Landauer approach. The contribution from elec-tron/hole on power factor and electronic thermal con-ductance is less than that from phonons on lattice ther-mal conductance.

Index Terms—Landauer Approach, Silicon Nanowire, Thermoelectric

I. Introduction

Thermoelectric (TE) energy conversion materials have been attracting attention for use in solid-state power generation devices. The dimensionless figure of merit (ZT) is the parameter used to indicate the performance of TE energy conversion materials. ZT is given by the equation:

$$ZT = \frac{S^2 \sigma T}{\kappa_{ph} + \kappa_{el}},\qquad(1)$$

where S is the Seebeck coefficient, σ the electrical conduc-tivity, κ_{ph} the lattice thermal conductivity from phonon and κ_{el} the electronic thermal conductivity from electron. To achieve $ZT > 1$, the first is to reduce the lattice or electronic thermal conductivity in the denominator of Eq. (1) and the other is to enhance the power factor, $S^2\sigma$, in the numerator. These properties are determined by the details of the electronic and phonon structure with the scattering of charge carriers so that they are not inde-pendently controlled. Silicon (Si)-based nanostructure is one of attractive materials to realize low-cost TE devices. For example, Si nanowires (SiNWs) with a diameter of 50 nm have a much lower thermal conductivity of 1.6 W/mK than bulk Si (around 150 W/mK) [1].

In this paper, a high density array of silicon nanowires (SiNWs) with a 10-nm diameter embedded in matrix of SiGe$_{0.3}$ as shown in Fig. 1 is considered. We first calculate the electron band structure by solving the Schrödinger equation with Bloch theorem [2] and phonon energy dispersion by solving the Elastodynamic equation [3]. Then, the Landauer approach [4], which works in ballistic limit as well as quasi-ballistics and diffusive regimes, is used to investigate the quantum effect of nanostructure

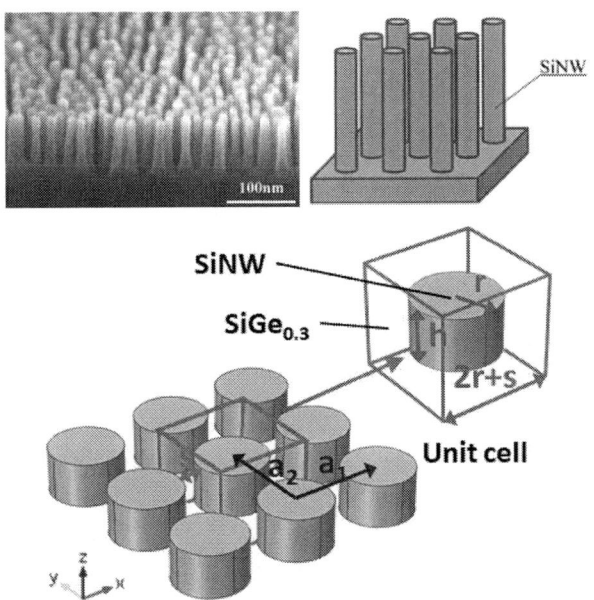

Fig. 1. A high-density array of SiNWs with a 10-nm diameter in SEM image is formed after neural beam etching using polyethylene glycol-modified ferritin as a mask, which is schematically illustrated as a square superlattice of SiNWs. The SiNWs is then embedded in SiGe$_{0.3}$ by thermal CVD techniques [5]. The SiNWs-SiGe$_{0.3}$ composite is simulated by a square superlattice with cuboid unit cell of radius r, height h and varied space s from 2 to 15 nm.

on the thermoelectric performance with different density of SiNWs by tuning space between SiNWs.

II. Modeling and Simulation Methodology

For the periodic SiNWs as shown in Fig. 1, the phonon energy dispersion is numerically solved by the elastody-namic wave equation as [6]

$$\nabla \cdot [C \nabla u(\boldsymbol{r})] = -\rho \omega^2 u(\boldsymbol{r}),\qquad(2)$$

where u is the displacement vector, ρ is the mass density, ω is the eigenfrequency and C is the elastic constant

978-1-7281-0941-1/19 $31.00 © 2019 IEEE

TABLE I
List of parameters used in the simulation of electronic band
structure and phononic dispersion [8].

Materials	Electron mass (m_e)		Hole mass (m_e)		Bandgap
	m_l^*	m_t^*	m_{hh}^*	m_{lh}^*	eV
Si	0.98	0.19	0.49	0.16	1.12
SiGe$_{0.3}$	1.14	0.12	0.41	0.10	1.00

Materials	Elastic constants (GPa)		
	C_{11}	C_{12}	C_{44}
Si	165.8	63.9	79.6
SiGe$_{0.3}$	154.6	59.2	75.8

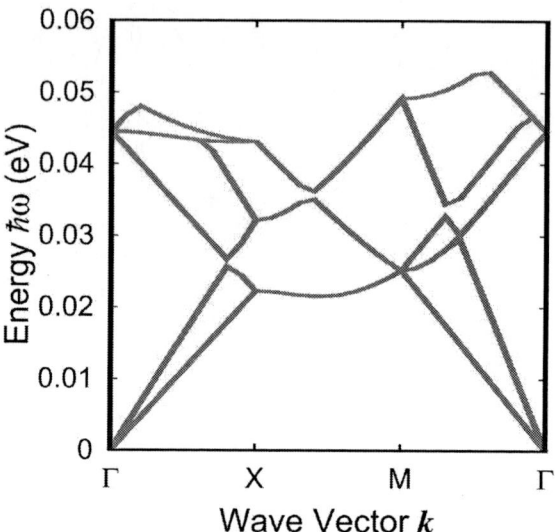

Fig. 2. Phonon Energy dispersion for bulk silicon along with the specific symmetric k-points.

matrix which describes second-order strain energy density [3]. Here C is a 6×6 symmetric matrix that has 21 independent elements. Since the silicon have a cubic symmetry, the number of independent elastic constants reduced to three, C_{11}, C_{12}, and C_{44}.

On the other hand, the electron band structure is numerically solved by the Schrödinger equation with the effective mass approximation under the Bloch theorem [2] as

$$\nabla \left[-\frac{\hbar}{2m^*} \nabla u_{\boldsymbol{k}}(\boldsymbol{r}) \right] - \frac{i\hbar}{m^*} \boldsymbol{k} \cdot \nabla u_{\boldsymbol{k}}(\boldsymbol{r})$$
$$+ \left[V(\boldsymbol{r}) + \frac{\hbar^2 k^2}{2m^*} \right] u_{\boldsymbol{k}}(\boldsymbol{r}) = E_{n,k} u_{\boldsymbol{k}}(\boldsymbol{r}), \quad (3)$$

where $\hbar$, m^*, $V(\boldsymbol{r})$, $E_{n,k}$, $u_{\boldsymbol{k}}(\boldsymbol{r})$ are the reduced Plank's constant, the effective mass, the position-dependent potential energy, quantum energy levels, and the corresponding wave function respectively.

Since SiNW-SiGe$_{0.3}$ composite is modeled by a square superlattice with cuboid unit cell as Fig. 1, both Eq. (2) and Eq. (3) follow specific boundary conditions based on the periodicity. The displacement vector u in Eq. (2) satisfies the Floquet periodic condition, $u(\boldsymbol{r}) = u_{\boldsymbol{q}} \exp(\boldsymbol{q} \cdot \boldsymbol{r})$, at boundary of a unit cell [7]. Here $\boldsymbol{q}$ is a wave vector. Meanwhile, the wave function $u_{\boldsymbol{k}}$ in Eq. (3) satisfies the periodic condition, $u_{\boldsymbol{k}}(\boldsymbol{r} + \boldsymbol{R}) = u_{\boldsymbol{k}}(\boldsymbol{r})$, where $\boldsymbol{R} = n_1 \boldsymbol{a}_1 + n_2 \boldsymbol{a}_2$ is the lattice vector with integer n_1 and n_2. Thus, Both Eq. (2) and Eq. (3) are discretized within a unit cell formed by the primitive vectors ($\boldsymbol{a}_1$ and $\boldsymbol{a}_2$) and then solved by a finite element method (FEM) solver for each sampling q-point or k-point in the irreducible Brillouin zone. The parameters used to calculate the electron band structure and phonon energy dispersion are listed in Table I.

With the calculated electron band structure $E(\boldsymbol{k})$ from Eq. (3) and phonon energy dispersive relation $\omega(\boldsymbol{q})$ from Eq. (2), the Landauer approach [9], [10] is adopted to describe the electron and phonon transport in nanostructures and investigate the quantum effect of nanostructure on the thermoelectric performance [4] because of its physical insight in ballistic limit as well as quasi-

ballistic and diffusive regimes. The transport coefficients for thermoelectricity in Eq. (1) are expressed as

$$\sigma = \int_{-\infty}^{\infty} \sigma'(E) dE$$
$$S = \frac{k}{q\sigma} \int_{-\infty}^{\infty} (E - E_F) \sigma'(E) dE$$
$$\kappa_0 = T \left(\frac{k}{q} \right)^2 \int_{-\infty}^{\infty} (E - E_F)^2 \sigma'(E) dE' \quad (4)$$
$$\kappa_{el} = \kappa_0 - S^2 \sigma T$$
$$\kappa_{ph} = \int_{-\infty}^{\infty} \kappa_{ph}'(E) dE$$

where E_F is the Fermi level, σ' and κ_{ph}' are differential conductivity for electron and phonon, respectively. For bulk materials [11], [12]

$$\sigma'(E) = q^2 \Sigma(E) \left(-\frac{\partial f_0}{\partial E} \right)$$
$$\kappa_{ph}'(E) = \frac{E^2}{T} \Sigma_{ph}(E) \left(-\frac{\partial n_0}{\partial E} \right)' \quad (5)$$

where $\Sigma(E)$ and $\Sigma_{ph}(E)$ are are transport distribution functions for electron and pho-non respectively that depends on both band structure or phonon dispersion (the number of modes) and scattering (mean free path).

III. Results and Discussion

Figure 2 shows the calculated phonon energy dispersion for bulk silicon using Eq. (2) and parameters in Table I. The result approximates the experimental data [3] well under the same order with only three independent elastic constants in Table I.

Figure 3 shows the calculated conductance as function of the Fermi level using Eq. (4). P-type SiNW has larger

978-1-7281-0941-1/19 $31.00 © 2019 IEEE

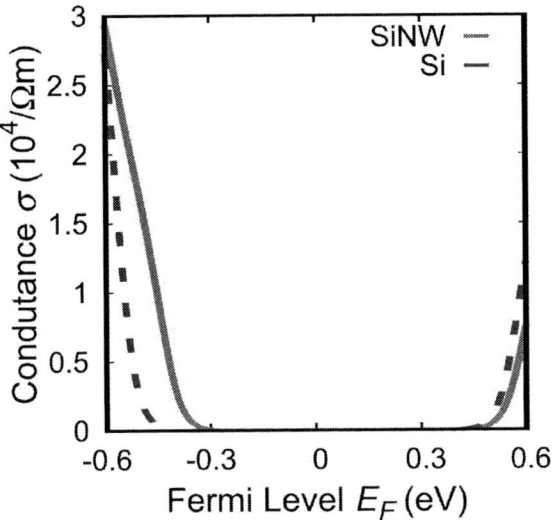

Fig. 3. Calculated conductance as function of the Fermi level for SiNW (solid red line) and bulk silicon (dash blue line).

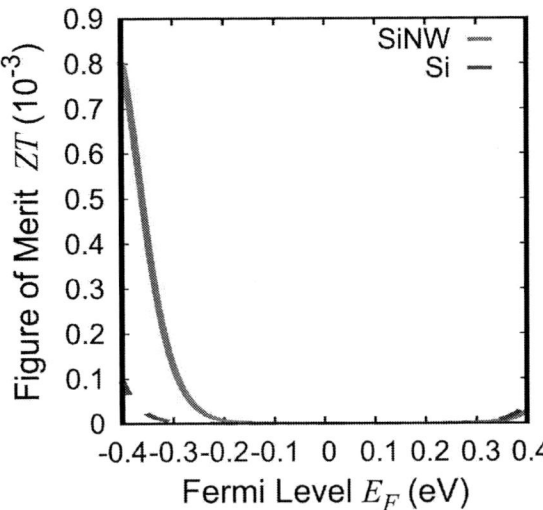

Fig. 5. ZT as function of the Fermi level for SiNW-SiGe$_{0.3}$ superlattice (solid red line) and bulk silicon (dash blue line).

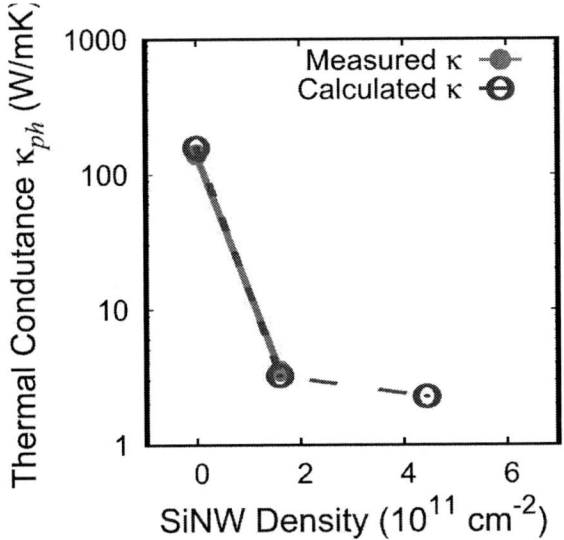

Fig. 4. Thermal conductance contributed from lattice dynamic for bulk silicon (zero SiNW density at x-axis) and SiNW-SiGe$_{0.3}$ superlattice from experiment (solid symols) [5] and simulation (open symols).

mal conductivity by Eq. (4) and Eq. (5). Fig. 4 shows the lattice thermal conductivity with a comparison between bulk silicon (zero SiNW density at x-axis) and SiNW-SiGe$_{0.3}$ superlattice. The simulation shows reduction of thermal conductivity around almost two orders for SiNW-SiGe$_{0.3}$ superlattice as the experiment results [5].

Figure 5 shows the calculated ZT as function of the Fermi level with contribution of electronic and lattice thermal conductivity. The main difference comes from the lattice thermal conductivity so that the ZT can reach around 0.01 for SiNW-SiGe$_{0.3}$ superlattice (at the Fermi level around -0.4 eV) and ZT is in value of less than 10^{-4} for bulk silicon, which is also two order smaller than that SiNW-SiGe$_{0.3}$ superlattice as Fig. 4.

IV. Conclusion

In this paper, we take use of the Landauer approach to investigate the quantum effect on thermoelectric properties for SiNWs embedded in matrix of SiGe$_{0.3}$. The impact of SiNWs on thermoelectric from electron, such as power factor and electronic thermal conductance, is less than that that from phonon, which is consistent with the observation in the experiment.

Acknowledgment

This work was supported in part by the Ministry of Science and Technology (MOST), Taiwan, under grants MOST 106-2221-E-009-149, 106-2622-8-009-013-TM, 107-3017-F-009-001, and 107-2221-E-009-094, and the "Center for mmWave Smart Radar Systems and Technologies" under the Featured Areas Research Center Program within the framework of the Higher Education Sprout Project by the Ministry of Education (MOE) in Taiwan.

conductance than bulk p-type silicon since the SiGe$_{0.3}$ has lower bandgap energy than Si as list in Table I and more energy band offset between SiNW and SiGe$_{0.3}$ for hole that induces quantum confinement and higher tunneling probability. However, the energy band offset between SiNW and SiGe$_{0.3}$ for electron is small so that the difference to bulk silicon for the conductance is little.

Based on the Elastodynamic wave equation Eq. (2), the phonon energy dispersion for SiNW-SiGe$_{0.3}$ superlattice is also calculated and then used to simulate the lattice ther-

978-1-7281-0941-1/19 $31.00 © 2019 IEEE

References

[1] A. I. Hochbaum, R. Chen, P. D. Delgado, W. Liang, E. C. Garnett, M. Najarian, A. Majumdar, and P. Yang, "Enhanced thermoelectric performance of rough silicon nanowires," Nature, vol. 451, pp. 163–167, 2008.

[2] M.-Y. Lee, Y. Li, and S. Samukawa, "Miniband calculation of three-dimensional nanostructure array for solar cell applications," IEEE Trans. Elec. Devices, vol. 62, pp. 3709–3714, 2015.

[3] W.-W. Zhang, H. Yu, S.-Y. Lei, and Q.-A. Huang, "Modelling of the elastic properties of crystalline silicon using lattice dynamics," J. Phys. D, vol. 44, p. 335401, 2011.

[4] J. Maassen and M. Lundstrom, "The landauer approach to electron and phonon transport," ECS Trans., vol. 69, pp. 23–36, 2015.

[5] A. Kikuchi, A. Yao, I. Mori, T. Ono, and S. Samukawa, "Composite films of higly ordered Si nanowires embedded in $SiGe_{0.3}$ for thermoelectric applications," J. Appl. Phys., vol. 122, p. 165302, 2017.

[6] R. Anufriev and M. Nomura, "Thermal conductance boost in phononic crystal nanostructures," Phys. Rev. B, vol. 91, p. 245417, 2015.

[7] M. Collet, M. Ouisse, M. Ruzzene, and M. N. Ichchou, "Floquet–bloch decomposition for the computation of dispersion of two-dimensional periodic, damped mechanical systems," Int. J. Solids and Structures, vol. 48, pp. 2834–2848, 2011.

[8] F. Schaffler, "Silicon-Germanium," in Properties of Advanced Semiconductor Materials: GaN, AIN, InN, BN, SiC, SiGe, M. E. Levinshtein, S. L. Rumyantsev, and M. S. Shur, Eds. New York: Wiley, Inc., 2001, pp. 149–188.

[9] R. Landauer, "Spatial variation of currents and fields due to localized scatterers in metallic conduction," IBM J.Res. Dev., vol. 1, no. 3, pp. 223–231, 1957.

[10] S. Datta, Electronic Transport in Mesoscopic Systems. New York: Cambridge University Press, 1997.

[11] C. Jeong, S. Datta, and M. Lundstrom, "Full dispersion versus Debye model evaluation of lattice thermal conductivity with a Landauer approach," J. Appl. Phys., vol. 109, p. 073718, 2011.

[12] ——, "Thermal conductivity of bulk and thin-film silicon: A Landauer approach," J. Appl. Phys., vol. 111, p. 093708, 2012.

Simulation of Chemically Reacting Flow in Plasma Native Oxide Cleaning Process

Seung-Min Ryu
Dept. of Aerospace Engineering and
Engineering Mechanics
The University of Texas at Austin
Austin, U.S.
smryu@utexas.edu

Yunho Kim
Dept. of Aerospace Engineering and
Engineering Mechanics
The University of Texas at Austin
Austin, U.S.
ykim96@utexas.edu

Dylan Pederson
Dept. of Aerospace Engineering and
Engineering Mechanics
The University of Texas at Austin
Austin, U.S.
dpederson@utexas.edu

Jonghyun Lee
Memory Technology Innovation Team
Samsung Electronics
Hwaseong-si, South Korea

Youngkwon Kim
Memory Technology Innovation Team
Samsung Electronics
Hwaseong-si, South Korea

Laxminarayan L. Raja
Dept. of Aerospace Engineering and
Engineering Mechanics
The University of Texas at Austin
Austin, U.S.
lraja@mail.utexas.edu

Jiho Uh
Memory Technology Innovation Team
Samsung Electronics
Hwaseong-si, South Korea

Sang-Jin Choi
Memory Technology Innovation Team
Samsung Electronics
Hwaseong-si, South Korea

Abstract—**A plasma native oxide cleaning process is widely used on the semiconductor production line to remove oxide impurities on silicon surfaces of an wafer. In this study, a flow simulation with microwave plasma species has been conducted to analyze the flow characteristics in a showerhead that affect the batch uniformity in the process. In particular, the distributions of temperature and mass flow rate of the gas as well as the number density of hydrogen radicals at the showerhead hole outlets were compared for different showerhead designs by using computational fluid dynamics. The distribution of gas temperature at the hole outlets was found to be inversely proportional to one of gas mass flow rate by the simulation results. However, mass flow rate distribution for the total gas shows a different trend from one of hydrogen radicals in the showerhead hole outlets. The showerhead design with low temperature gradient also showed a more uniform mass flow rate profile at the hole outlets, which was validated by the simulation results.**

Keywords—*Multi-physics, Simulation, Plasma, Species, Flow*

I. INTRODUCTION

A key issue for semiconductor manufacturing plants is to increase the productivity and yield of high quality chips simultaneously [1]. Semiconductor companies have tried to obtain better uniformity of the pattern features as well as the electrical characteristics of the wafer. Specifically, a Si thermal deposition process is one of the most challenging areas in the semiconductor fabrication because it is difficult to achieve pattern uniformity due to a narrow range of acceptable process parameters and sensitivity to pre-cleaned surface conditions. For instance, the Si layer is deposited uniformly on the super-clean Si surface of the wafer, especially in the Si epitaxial growth process, but the oxygen atoms on the Si surface interfere with the growth of the Si layer, which results in nonuniform pattern height. To resolve this issue, it has become critical to effectively remove the oxide layer from the Si wafer for the advanced devices, which has motivated the semiconductor industry to optimize various native oxide removal methods to obtain high performance of the devices. As a chemical dry cleaning process, batch type plasma native oxide cleaning (PNC) equipment has been widely used to treat

and remove a native oxide layer on the Si surface due to its high throughput and superior cleaning efficiency. However, it has been difficult to obtain even distribution of gas temperature as well as species concentration in the batch chamber. The key requirement of the process is to make the silicon surface ultra clean uniformly across the wafers in the batch chamber [2].

In this paper, firstly we study the flow mechanism by computational fluid dynamic (CFD) simulation to elucidate potential causes related to the batch uniformity. Then we analyze characteristics of gas flow in a different showerhead design affecting uniform cleaning quality in the PNC equipment.

II. BACKGROUND AND NUMERICAL PROCEDURE

A. PNC process

The schematic of a batch-type PNC equipment is as shown in Fig. 1. There are two types of gas inlet ports in the chamber. A mixture of N_2 and NH_3 gas enters the two applicators and is dissociated by a continuous wave source driven at 2.45 GHz. Hydrogen radicals are generated in the applicator and supplied to the process chamber through a showerhead inlet. An NF_3 gas directly enters the chamber through straight nozzles. Production and dummy wafers are loaded together on a wafer-boat in the vertical type chamber at one time. The exhaust port is located in the center of the chamber opposite to showerhead and the outlet pressure of the chamber is controlled by automatic pressure valve. Halogen lamps are installed to heat the wafers to vaporize and remove the by-product on the wafers after the etch step.

B. Simulation procedure

The simulation consists of three different parts. First, the plasma discharge is simulated to obtain the mole fractions of plasma generated species inside a microwave waveguide applicator. Second, a steady-state fluid flow simulation is performed for each test case to obtain the distribution of pressure, temperature and mass flow rates at the hole outlets of the showerhead. Finally, reacting flow simulations are conducted with the inclusion of the neutral plasma generated species. The main goal of these simulations is to obtain the transport characteristics of the plasma generated radical species to the showerhead hole outlets of the gas. For this study, reactive hydrogen radicals are of main interest as they become an important source for the formation of etchants ($NH_x F_y$) [3-6].

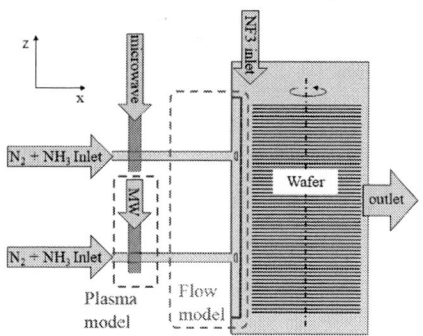

Fig. 1. Chamber structure for PNC batch process.

C. Plasma and reacting flow chemistry

For the simulation of plasma discharge, a kinetic mechanism that includes 23 Species and 152 reactions is used. The database is provided by VizGlow [7]. The species include: e, NH_3, NH_3^+, NH_4^+, NH_2, NH_2^+, NH_2^-, NH, NH^+, N, NH^+, N, N^+, N_2, N_2^+, H, H^+, H^-, H_2, H_2^+, H_3^+, N_2H_2, N_2H_3, N_2H_4, $NH_3(v)$. For the reacting flow simulation, all charged species and reactions that involve them are ignored. This assumption is reasonable since the recombination of charged species occurs in μs timescale or less, which is much smaller than the flow timescale (ms). To model the loss of atomic hydrogen during its transport to the outlet holes, the surface recombination coefficient is set as 0.01 [8].

D. Microwave plasma simulation

a) Microwave plasma discharge: Plasma discharge simulation is performed to obtain the mole fraction of each species generated by the plasmas. The mole fractions are used as the inlet boundary condition of the reacting flow simulation in the next section. Based on the pressure and temperature at the inlet of the showerhead obtained from the fluid flow simulation, a plasma discharge simulation is conducted to find out the species number densities at quasi-steady state. At a fixed mass flow rate of $2.0 \times 10^{-4} \ kg/s$ and the temperature of $700K$, the inlet pressures are $300Pa$ and $600Pa$ for the Case *I* and Case *II*, respectively. These temperature and pressures are chosen as the operating conditions of an NH_3/N_2 mixture. Numerical models developed for high frequency plasma discharges are used for this section [9-10].

b) Computational domain: To model the microwave discharge, a downstream plasma source built by MKS is considered [11]. It employs the TE_{10} mode of a rectangular waveguide whose side view of the schematic is shown in Fig. 2. In the figure, D (25mm) is the diameter of the quartz tube, L (30.6mm) is the distance between the tube and the short plunger at the end of the waveguide, and H (33mm) is the height of the waveguide. L is the quarter wavelength of the operating frequency (2.45GHz) which indicates the standing wave formation inside. The maximum wave amplitude occurs

near the surface of the dielectric shown as the thin grey area in Fig. 2. The dielectric quartz thickness is 2mm and its relative dielectric permittivity is 3.8. We model the electromagnetic wave propagation as a simple 1D waveguide as shown in the Fig. 2 by impinging the plane wave from the left. The Perfectly Matched Layer (PML) is implemented on the left to effectively absorb the reflected waves from the short plunger. While the Maxwell's equations are solved in the entire computational domain, the self-consistent plasma governing equations are solved only inside the plasma subdomain.

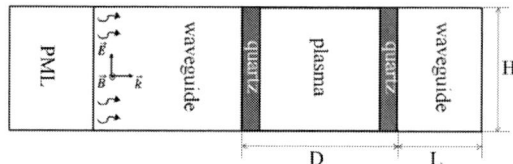

Fig. 2. Schematic of the 1D waveguide.

E. Reacting flow simulation

Two showerhead designs were modeled to check the distribution of mass flow rate including radical species and gas temperature in the hole outlets of the showerhead. The simulation consists of a two-step process. First, steady state 2D analysis of the gas flow inside the showerhead was conducted using only the fluid flow solver in VizGlow. The models were meshed with dominant quadrilateral and triangular elements. The total mass flow rate of the process gases was used as an inlet condition of the model. The gases were assumed to be ideal gas. Major boundary conditions for the models are shown in Table 1. The temperature at the gas inlet was set as 700K. The static pressure 200 Pa was set as outlet boundary condition of the chamber. The constant temperature on the backside surface of the showerhead is applied to the wall condition because the chamber wall in contact with the showerhead is cooled and kept at 300K during the process. The other surfaces of the showerhead were assumed to have thermally insulated walls. Then, a transient simulation of the gas flow including radical species generated by microwave plasma in the applicator was conducted using a species density solver in VizGlow. In particular, the species density solver includes bulk chemical reactions as well as surface reactions on the wall. The time step was set to 1.0×10^{-6}s to resolve the chemical reaction timescale. Two types of showerhead design with different gas flow paths are shown in Fig. 3. Especially, mass flow rate, temperature and the number density of the hydrogen radical species were monitored at the hole outlets of the showerhead.

TABLE I.   BOUNDARY CONDITIONS FOR GAS FLOW SIMULATION MODEL

Process parameter	Mass flow rate at inlet(kg/sec)		Pressure at the outlet(Pa)	Wall conditions	
	N_2	NH_3		*Showerhead backside surface*	*Other surfaces*
value	1.63×10^{-4}	3.29×10^{-5}	200	300K Constant Temp.	Thermally Insulated

978-1-7281-0941-1/19 $31.00 © 2019 IEEE   126

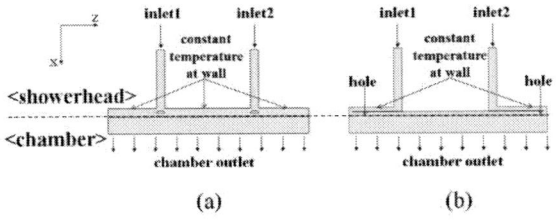

Fig. 3. Showerhead designs of (a) Case I and (b) Case II.

III. RESULT AND DISCUSSION

A. Microwave plasma formation

At each operating pressure, the ratio of the seed number densities of NH_3 and N_2 is given as $1:3$ according to the inlet flow condition, respectively. The order of magnitude of the number densities are $10^{22} m^{-3}$. For all other species, the seed number density is set as $10^{12} m^{-3}$. The incident wave amplitude is set as $50 kV/m$. The transient plasma formation is shown in Fig. 4 where the tracepoint is $2.5 mm$ away from the dielectric surface on the right. Here, the generation of radical species H, NH, and NH_2 is important as they mainly participate in the formation of etchants ($NH_x F_y$) [3]. The order of magnitude of these species densities is comparable to or larger than the electron number density ($10^{18} m^{-3}$). The mole fraction of each species is obtained from these results. For the operating pressure of $300 Pa$, the mole fractions of H, NH, and NH_2 are found to be 1.98×10^{-4} 4.07×10^{-5} and 1.30×10^{-4}, respectively. For $600 Pa$, they are 1.93×10^{-4}, 3.38×10^{-5}, and 1.39×10^{-4}, respectively. The mole fractions of the other neutral species are also obtained and used as the inlet boundary condition for the reacting flow simulation in the next section.

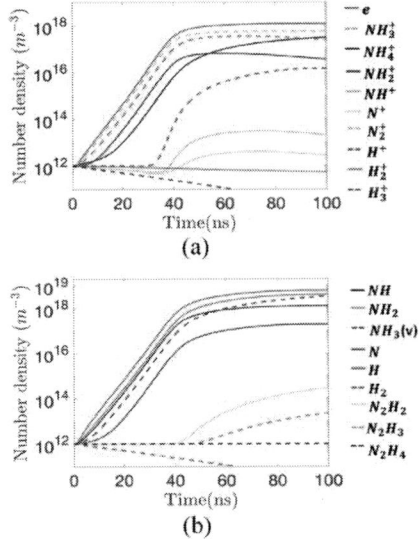

Fig. 4. Transient evolution of species number density. (a) Charged species and (b) neutral species.

B. Fluid characteristics in the showerhead

Case I shown in Fig. 5(a) has diffusers to disperse the gas throughout the showerhead and obtain flow uniformity in the showerhead. However, Case II without any diffuser in Fig.5(b) has returning paths of the gas at both sides to maintain continuous flow in the showerhead. Inlet pressure values of Case I and Case II model used in the microwave plasma simulation were calculated by the CFD models to be 300 Pa and 600 Pa, respectively, as shown in Fig. 5. Return paths of gas flow in the Case II made pressure drop larger in the showerhead, whose inlet pressure is about twice larger than Case I. The pressure inside both chambers were nearly 200Pa which is similar to the outlet boundary condition.

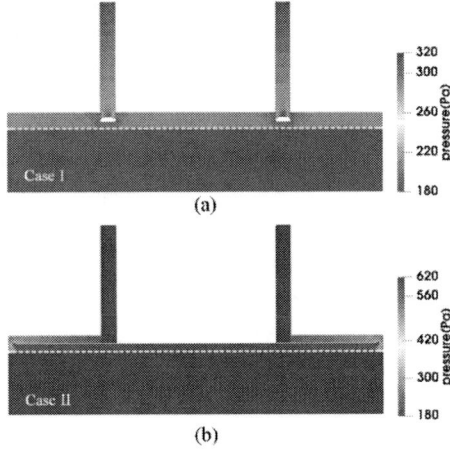

Fig. 5. Pressure distribution of showerhead models; (a) Case I and (b) Case II.

The distribution of temperature normalized by inlet gas temperature 700K at each hole of the showerhead is plotted in Fig. 6. The gas temperature in Case I gradually decreases with distance from the applicators because of conductive heat loss from the showerhead to the chamber. Therefore, the temperature distribution of the gas in the showerhead shows an 'M'-shaped profile as shown in Fig. 6. Case II exhibits relatively flat shape in the temperature profile compared to Case I. Case II shown in the Fig. 6, however, has no diffusers but instead there is a thermally isolated vacuum barrier to change the gas flow path and prevent heat flux from the showerhead to process chamber. Here it is important to reduce the heat flux from the hot gas to the chamber through the showerhead. The temperature distribution of the gas at the hole outlets of the showerhead was affected by the showerhead backside area in contact with the chamber.

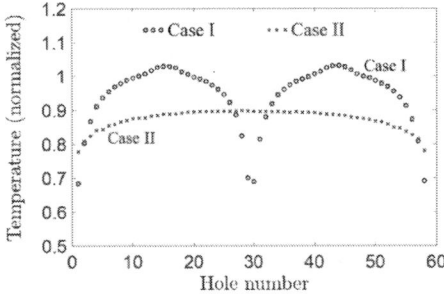

Fig. 6. Normalized temperature distribution at the hole outlets of the showerheads.

978-1-7281-0941-1/19 $31.00 © 2019 IEEE

The distributions of total gas mass flow rate and number density for hydrogen radicals normalized by inlet flow condition for each showerhead design are shown in Fig. 7 and Fig. 8, respectively. It is found on the distribution that the mass flow rate is inversely dependent on the gas temperature. For both models, the distribution of hydrogen radicals has a different trend compared to the mass flow rate. It is noticeable that it depends on the flow path. Case *II* shows 16% and 23% better uniformity in the process zone than Case *I* in terms of the number density of hydrogen radical species and temperature, respectively. However, the radical species loss in Case *I* is smaller than Case *II* due to its shorter flow path which yields less amount of surface recombination. The simulation results imply that Case *II* can show the better the etch rate uniformity but lower silicon oxide etch amount on the wafer.

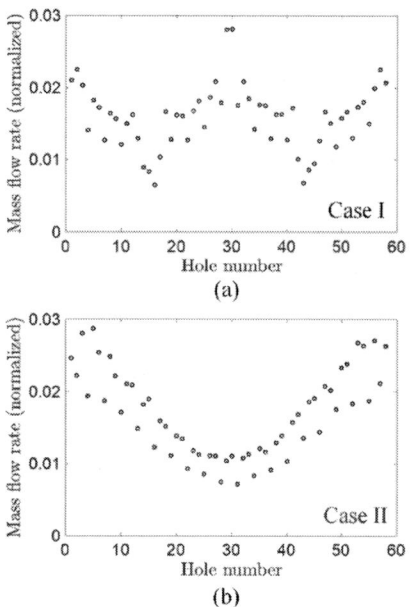

Fig. 7. Distribution of normalized mass flow rates at the hole outlets of the showerheads: (a) Case I and (b) Case II.

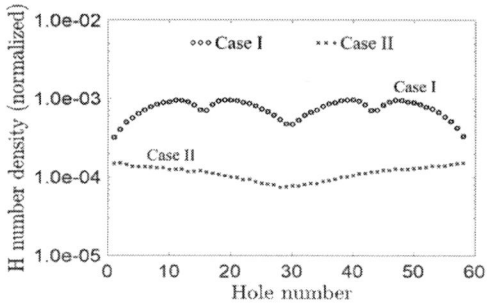

Fig. 8. Distribution of normalized number densities for hydrogen radical species at the hole outlets of the showerheads.

IV. CONCLUSIONS

The effects of showerhead design on flow characteristics are studied in this work to estimate their correlation to the process uniformity. High frequency plasma discharge model and chemically reacting flow model are employed to find out the distribution of temperature and plasma generated radical species. Mass flow rate at the hole outlets of the showerheads was inversely dependent on the gas temperature, which is affected by the heat flux to the showerhead backside surface. However, the distribution of number density for hydrogen radicals was affected by the flow path of the gas in the showerhead. By altering the flow path, the gas temperature gradient at the hole outlets was changed to yield better uniformity. CFD simulation results show the showerhead design with a vacuum barrier has 23 % lower temperature gradient in the process zone than without the thermal barrier. It is also shown that the showerhead with the thermally isolated barrier had improved 16% in hydrogen radical flow uniformity. However, the radical species loss was larger for the case with a longer flow path.

ACKNOWLEDGMENT

This work was supported by Samsung Electronics Co., Ltd.

REFERENCES

[1] K. Kim, "Silicon technologies and solutions for the data-driven world", IEEE International Solid-State Circuits Conference, 2015, pp. 8-14.

[2] W. S. Kim, W. G. Hwang, I. Kim, K. Yun, K. M. Lee and S. K. Chae, "Development of New Batch-Type Plasma Assisted NOR (Native-Oxide-Removal) Dry Cleaning Equipment", Solid State Phenomena, 2005, pp. 63-66, vol. 103-104.

[3] T. Hayashi, Recent Development of Si Chemical Dry Etching Technologies", J. of Nanomedicine & Nanotechnology, 2012, s15:001.

[4] N. Posseme, V. Ah-Leung, O. Pollet, C. Arvet and M. Garcia-Barros, "Thin layer etching of silicon nitride: A comprehensive study of selective removal using NH3/NF3 remote plasma", J. Vac. Sci. Technol. A, 2016, vol. 34, no. 6, pp. 061301.

[5] H. Nishino, N. Hayasaka, and H. Okano, "Damage‐free selective etching of Si native oxides using NH3/NF3and SF6/H2O down‐flow etching", 1993, J. Appl. Phys., vol. 74, no. 2, pp.1345–1348.

[6] H. J. Oh, J. H. Lee, M. S. Lee, W. G. Shin, S. Y. Kang, G. D. Kim and D. H. Ko, "NF3/NH3 Dry Cleaning Mechanism Inspired by Chemical and Physical Surface Modification of Si, SiO2, and Si3N4", ECS Transactions, 2014, vol. 61, pp. 1–8.

[7] https://esgeetech.com/

[8] B. J. Wood and H. Wise, " the Kinetics of Hydrogen Atom Recombination on Pyrex Glass and Fused Quartz 1 ," J. Phys. Chem., 2007, vol. 66, no. 6, pp. 1049–1053.

[9] P. P. Chelvam and L. L. Raja, "Modeling of gas breakdown and early transients of plasma evolution in cylindrical all-dielectric resonators," J. Phys. D. Appl. Phys., 2017, vol. 50, no. 47.

[10] Y. Kim and L. L. Raja, "Modeling of microwave surface plasmas on the meta-surface at atmospheric pressure," in AIAA Scitech 2019 Forum, 2019, no. January, pp. 1–13.

[11] M. Mehdizadeh. "Microwave/RF Applicators and Probes : for Material Heating, Sensing, and Plasma Generation", 2nd ed., Elesevier, 2015, pp. 348-351.

978-1-7281-0941-1/19 $31.00 © 2019 IEEE

Simulation of deep level transient spectroscopy using circuit simulator with deep level trap model implemented by Verilog-A language

Koichi Fukuda
National Institute Of Adivanced
Industrial Science and Technology
(AIST)
Tsukuba, Japan

Junichi Hattori
National Institute Of Adivanced
Industrial Science and Technology
(AIST)
Tsukuba, Japan

Hidehiro Asai
National Institute Of Adivanced
Industrial Science and Technology
(AIST)
Tsukuba, Japan

Mitsuaki Shimizu
National Institute Of Adivanced
Industrial Science and Technology
(AIST)
Tsukuba, Japan

Tamotsu Hashizume
Research Center for Integrated
Quantum Electronics (RCIQE)
Hokkaido University
Hokkaido, Japan

Abstract— A modeling method of deep level transient spectroscopy (DLTS) using circuit simulation with a MOS capacitor compact model which takes into account influences of deep level traps is proposed. In the proposed method, DLTS measurement procedures are described by transient analysis of circuit simulation. Stable numerical convergence is obtained even for the case in which carrier traps with wide range of time scales are included. Through case studies, it is proved that this method is a robust and versatile theoretical tool to predict DLTS signals, which helps to understand DLTS results and to optimize DLTS measurement conditions. Furthermore, the method is applied to several capacitance measurement methods discussed in literatures concerning GaN MIS capacitors, which ensures the practical ability of the proposed simulation approach.

Keywords—deep level transient spectroscopy, circuit simulation, compact model)

I. INTRODUCTION

Deep trap levels are still a major issue for power devices [1] and also a major technical barrier for two-dimensional materials [2]. Identification of trap species is important in order to overcome these deep trap level problems. Among various trap measurement methods, deep level transient spectroscopy (DLTS) is one of the most detailed physical measurement methods for traps [3]. Using DLTS, it is possible to separate out the information of traps with energy distributions and different time constants. On the other hand, quantitative modeling of the DLTS measurement is a problem in which various traps with a wide range of time constants are obstacles and therefore it is difficult to make them converged even using numerical simulation, and a versatile theoretical tool is strongly required. Recently, the authors used circuit simulation as an analysis tool of MOS capacitors containing various traps by implementing the physical behavior of traps in the Verilog-A language [4]. In this paper, a simple DLTS modeling approach using the circuit simulation with the MOS capacitor model is proposed, and its usefulness is shown through simulation case studies.

II. SIMULATION METHOD

The circuit simulation used for DLTS modeling utilizes the compact model of a MOS capacitor in which the influence of trap levels is taken into account [4]. In-side the compact model implemented using Verilog-A language [5], Poisson equation, current continuity equations, and one rate equation for each trap species are solved self-consistently by an iterative method for a time step given by the circuit simulator. The rate equation for each trap is as in (1),

$$\partial N_{TA} / \partial t = E_n N_{TA}^- - C_n n (N_{TA} - N_{TA}^-) \qquad (1)$$

where N_{TA} and N_{TA}^- are total and negatively charged acceptor-like traps, C_n is the electron capture rate, n is the electron concentration, and E_n is the electron emission rate as in (2),

$$E_n = v_{th} \sigma_n N_C \exp(-\Delta E / k_B T) \qquad (2)$$

where v_{th} is the thermal velocity σ_n is the capture cross section, N_C is the effective density of states of the conduction band, and ΔE is the trap energy depth from the conduction band.

The schematic explanation of the trap dynamics is shown in Fig. 1 in which an example of acceptor type trap is explained. The acceptor trap fastly captures an electron and is negatively charged, but the captured electron is slowly emitted by obtaining the thermal energy corresponding to the energy depth from the conduction band. The negatively charged acceptor trap fastly captures a hole and becomes neutral charge condition. The equation (2) shows that the time scale of the trap behavior strongly depends on the trap energy depth. These set of equations are created for each trap, and solved self-consistently all in one device instance of the circuit.

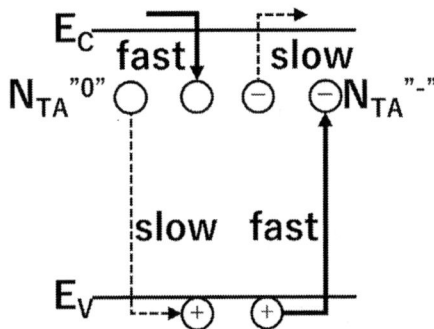

Fig. 1.   A schematic explanation of the dynamics of the acceptor-like traps.

The circuit simulator controls the time steps and manages the Newton convergence including the consistency with the other components of the circuit. This nested algorithm is shown as a flowchart-like viewgraph in Fig. 2. Poisson equation, carrier continuity equation and rate equations for each trap species are solved by decouple iterations in order to obtain the self-consistent solutions of all equations. The time step discretization is treated by the implicit method. The self-consistency with all other circuit components is automatically ensured by the convergence loop of SPICE. The Newton Jacobian matrices are automatically created by the Verilog-A language system.

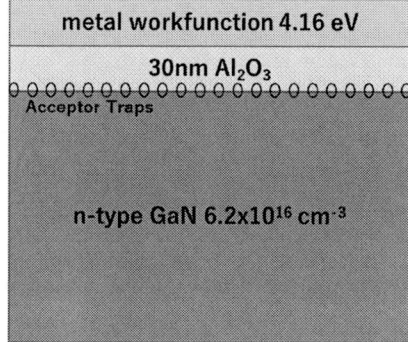

Fig. 2. A flow-chart like viewgraph of the proposed and implemented method. Poisson equation, carrier continuity equations, and rate equations of various traps are self-consistently solved in the internal decouple iterations written by Verilog-A language. The time steps are controlled by SPICE convergence loop governed by Newton-Raphson method.

All circuit simulations for C-V measurements and DLTS measurements are performed as transient mode analyses as shown in the schematic viewgraph Fig. 3. The control of time-dependent voltage is described in the input deck of the circuit simulator, and in particular, each capacitance measurement is performed by giving sine-curve signals with the given amplitude and frequency. The signal part of the simulation input deck can easily be obtained by using a shell script or a small program. The capacitance values are obtained from the time dependent gate currents which include time dependence of carrier concentrations in the semiconductor, trap charges at the insulator semiconductor interface, and also the displacement currents arising from the sinusoidal input voltages. Additionally, when the conversion is obtained for each time step, electron and trap charges are output to the standard output of the program, controlled by Verilog-A systems. This helps to differentiate the displacement current of the gate bias and charge current in the semiconductor. It should be mentioned that time steps differ by orders of magnitude depending on the input deck and conversion status, which is well managed by SPICE time step control. Compared with device simulation approach as in [6], input deck for circuit simulation is much simpler, because the purpose of the simulation is well focused.

III. RESULTS AND DISCUSSIONS

A. A case study.

Fig. 4 shows the assumed GaN MOS capacitor with n-type substrate dopant density of 6.2×10^{16} /cm^3 and with acceptor type traps at the semiconductor-insulator interface. Two cases of assumed energy distribution of trap densities are shown in Fig. 5. Both cases consist of three density peaks at the energy depth from the conduction band of 0.2, 0.4, and 0.6 eV. For

each peak, the accumulated trap density is 2×10^{12} /cm^2, which results in 6×10^{12} /cm^2 for each case. The difference of these two cases is the standard deviation of the trap energy, 10 and 40 meV for each peak. The circuit simulation is performed with Silvaco SmartSpice [7].

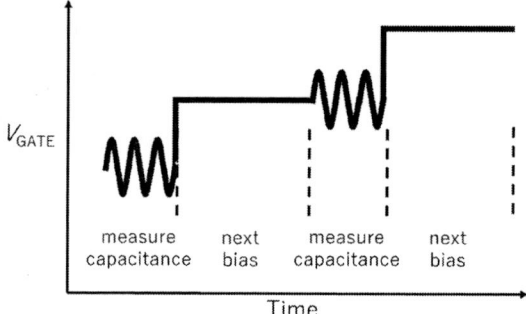

Fig. 3. Schematical description of time dependent bias control for both C-V and DLTS measuremets. Both are simulated by full transient analysis mode of the circuit simulator. Gate voltage is controlled through SPICE input decks.

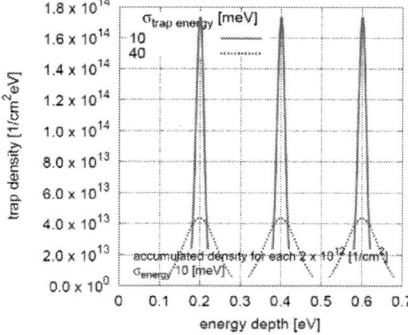

Fig. 4. Assumed device structure of GaN MOS capacitor with deep level traps at the semiconductor insulator interface.

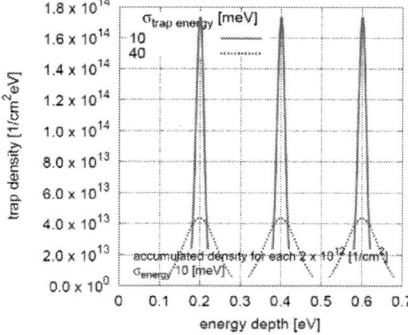

Fig. 5. Two trial cases of trap density energy distribution for demonstration. Three peaks at the energy depth of 0.2, 0.4, and 0.6 eV with accumulated trap density of 2×1012 /cm2 for each peak. Two cases are different in standard deviation of the trap energy, 10 and 40 meV.

Fig. 6 shows the C_G-V_G characteristics of assumed two cases for frequency of 1 MHz. Although the difference in energy distribution in these cases affects the C-V characteristics, the difference is minor and it might be difficult to determine the difference in the actual measurements. Fig. 6 shows the DLTS signal for these two cases. The DLTS signal is the difference of 1 MHz capacitance at 1 msec. and 10 msec. after the gate bias is changed from the initial gate bias of 3 V to the target gate bias of -1.5 V. In the case of energy standard

978-1-7281-0941-1/19 $31.00 © 2019 IEEE   130

deviation of 10 meV, three peaks corresponding to 0.2, 0.4, and 0.6 eV are clearly observed, but for the case of 40 meV, the peaks for 0.4 and 0.6 eV are ambiguous. As easily imagined from the examples, the proposed method is an optimal tool to optimize the detailed measurement conditions.

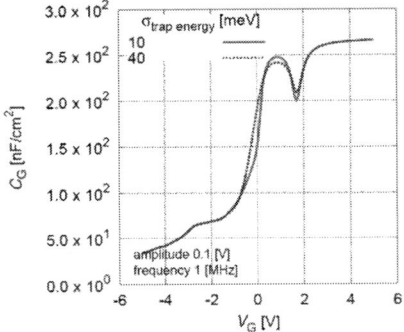

Fig. 6. *C-V* curves obtained for the two cases in Fig. 5. The frequency is 1 MHz. The circuit simulations are well converged even for trap energy distribution with wide range time scale.

B. Comparison with the literature

Comparison with experimental values in the literature is important to verify the usefulness of this method. We will focus on the article on GaN in particular as a semiconductor with various applications in the near future and suffering from trap behavior. In particular, Nakano and Jimbo have shown detailed and respectable measurement results regarding the GaN MIS structure [8], and we confirmed the usefulness of this method by simulating their conditions. The metal is Al, the insulator is sputtered 100 nm SiO_2, and the GaN is n-type with 6.8×10^{17} /cm^3 confirmed by secondary ion mass spectrometry.

They combined several measurement methods to characterize the MIS interface traps, the conventional *C-V*, the pulsed *C-V*, transient capacitance *C-t*, and deep level transient spectroscopy DLTS. Fig. 7 is the pulsed *C-V* curves for several delay time td conditions. The pulsed *C-V* is used to evaluate time dependence of charging and discharging at the SiO_2 / GaN interface.

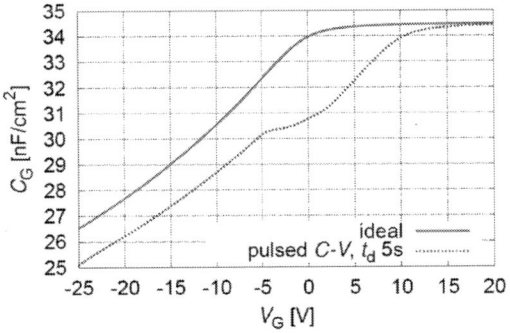

Fig. 7. Pulsed *C-V* characteristics for delay time of 1s, 5s discussed in [8]. The ideal *C-V* curve is also shown for comparison. This measurement is used to evaluate time dependence of charging and discharging at the SiO_2 / GaN interface.

Fig. 8 shows temperature dependence of measured *C-t* curves using transient capacitance measurement method as described in [8]. The method is used in order to examine the capacitance transient observed in deep depletion. The relaxation time for both capacitance transients becomes shorter in accordance with the rising of the temperature, which is in reasonable agreement with the theory of thermal carrier emission based on the Shockley–Read–Hall statistics, as discussed in [8]. Three types of traps are identified in [8], however detailed energy peak information is not available. In the simulations, three energy peaks of 0.55, 0.77 and 0.9 eV are assumed for trial. By comparing simulation results and measured results, it is possible to discuss more detailed information of the energy distributions of the interfacial traps. In addition, a capacitance transient with a fast relaxation time of 10^{-5} s is observed in the *C-t* curves at temperature higher than 330 K.

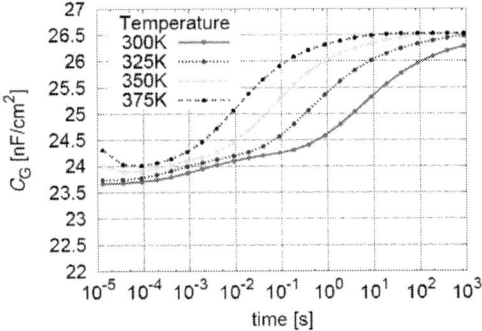

Fig. 8. Simulation results of temperature dependence of *C-t* curves for target V_G of -25 V obtained by the capacitance transient measurement method configured as in [8].

Fig. 9 shows the simulated DLTS curves for several t_1 / t_2 configurations of measurements in [8]. The detailed discussions of these dependences are realized by the present theoretical method ensured by sufficient accuracies of the simulations. The rate windows t1 (ms) / t2 (ms) = 4 / 8, 5 / 10, 8 / 16, and 10 / 20 correspond to the emission rates of 173, 139, 87, and 69 s^{-1}. The peak shift depending on the rate window enables the Arrhenius plot and the activation energy is extracted as 0.77 eV. In [8], other two peaks are also discussed but no detailed information is described. Therefore, only the main trap energy peak of 0.77 eV is assumed with accumulated trap area density of 2.2×10^{12} cm^{-2}.

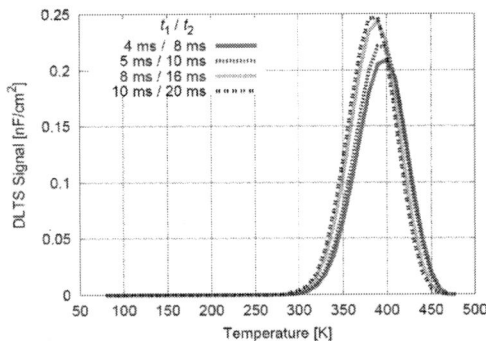

Fig. 9. Simulated DLTS curves for several rate window t_1 / t_2 configurations of measurements in [8]. The peak shift depending on the rate window include the information of the activation energy of the interface trap.

For the comparison to published measurement results, the proposed method covers not only the conventional C-V measurement, but also pulsed C-V measurements, transient C-t measurements, and DLTS measurements for MIS capacitors in order to characterize interface traps. These simulation conditions can be managed changing only the transient input voltage sequences in the input deck of the circuit simulator SPICE. Through the simulation according to the configurations in [8], practical ability of the present method is sufficiently confirmed, and therefore it is expected that the method helps to overcome the trap issues in coming generation of novel semiconductor materials.

IV. CONCLUSIONS

The modeling method of deep level transient spectroscopy by circuit simulation is demonstrated which uses the compact model of the MOS capacitor considering the influence of the deep level traps. In the method, Poisson, carrier continuity equations and rate equations for all traps are self-consistently solved by the decouple method for one time-step of circuit simulation, which is implemented using Verilog-A language. All the measurement procedures are described by the transient input deck of circuit simulations. Through the case study, it has been verified that the C-V measurements, not only the conventional method but also the several types of methods, and the temperature dependence of the DLTS signals can be stably obtained. Influence of subtle differences in the trap level density distribution is quantitatively predicted. It is proved that this method is a strong theoretical tool of DLTS signals to optimize and to understand the measurements, and therefore it helps to overcome the trap issues which block to realize applications of novel semiconductor materials.

ACKNOWLEDGMENT

This article is partially based on the results that were obtained in a project commissioned by the New Energy and Industrial Technology Development Organization (NEDO).

REFERENCES

[1] K. Matocha, T. P. Chow, and R. J. Gutmann, "Positive flatband voltage shift in MOS capacitors on n-type GaN," IEEE Electron Device Letters, vol. 23, pp. 79–81, 2002.

[2] F. Nan, K. Nagashio, and A. Toriumi, "Subthreshold transport in mono and multilayered MOS2 FETs," Applied Physics Express, vol. 8, p 065203, 2015.

[3] D. V. Lang, "Deep level transient spectroscopy: A new method to characterize traps in semiconductors," Journal of Applied Physics, vol. 45, pp. 3023-3032, 1974.

[4] K. Fukuda, H. Asai, J. Hattori, M. Shimizu, and T. Hasihzume, "A time-dependent Verilog-A compact model for MOS capcitors with interface traps," Jpn. J. Appl. Phys., vol. 58, p. SBBD06, 2019.

[5] Verilog-AMS Language Reference Manual, Accellela, 2008.

[6] K. Fukuda, H. Asai, J. Hattori, M. Shimizu, and T. Hashizume, "A transient simulation approach to obtaining capacitance–voltage characteristics of GaN MOS capacitors with deep-level traps," Jpn. J. Appl. Phys., vol. 57, p. 04FG04, 2018.

[7] SmartSpice User's Manual, SILVACO Inc., 2017.

[8] Y. Nakano and T. Jimbo, "Electrical characterization of SiO2/n-GaN metal-insulator-semiconductor diodes," J. Vac. Sci. Technol. B, vol. 21, pp. 1364-1368, 2003.

A New Computer-Aided Calibration Technique of Physics Based IGBT & Power-Diode Compact Models with Verilog-A Implementation

Arnab Biswas, Daniel Ludwig, Maria Cotorogea

Infineon Technologies AG
Am Campeon 1-15, D-85579 Neubiberg, Germany
email: arnab.biswas@infineon.com

Abstract—In this work, we present a new calibration technique of an IGBT and power diode compact model using a commercially available tool optiSLang™ [1]. We show that with such a computer-aided technique, we can get a accurate match in switching transients just by calibrating the static (transfer and output characteristics) and the gate charge curves. Furthermore, we present a Verilog-A implementation of a physics based IGBT and power diode compact model [2,3]. We demonstrate the benefits of a Verilog-A model by comparing the run time and convergence performance with a standard SPICE implementation.

Index Terms—Verilog-A, compact-model, IGBT, power-diode, computer-aided calibration

I. INTRODUCTION

It is often said that a model is only as good as its calibration. Hence, it is equally important to have a good calibrated model as the quality of the model itself. Compact model calibration can be a time consuming process. Manual calibration can be prone to error and also time consuming, often requiring multiple iterations. Computer aided techniques are therefore preferred to reduce manual effort and to ensure quick turnaround time. There are several approaches to this problem using Python, Matlab or other programming languages with varying levels of ease of use and final fit quality [4,5,6].
In this work, a new calibration technique is presented using a commercially available optimizer called optiSLang™ [1]. It supports the following optimization/evaluation routines:

1) Sensitivity analysis
2) Multi objective optimization
3) Robustness evaluation
4) Reliability analysis
5) Robust design optimization

In this work, we only use the first two routines. This tool, once set up as a template, is very easy to use. This can be useful when there are a lot of similar calibration tasks that need to be performed. optiSLang™ can be set up to launch a full set of simulations on circuit simulators like SIMetrix or PSpice depending on the objective criterion and the parameter input range. It can then iterate to a set of parameters which gives the best fit to the target curves.

II. VERILOG-A MODEL IMPLEMENTATION

The Verilog-A compact model developed for this work is based on an existing sub-circuit SPICE model [1,2]. The Verilog-A model uses the same model equations for physical descriptions and parameters are identical or equivalently expressed as in the sub-circuit model. Consequently, this implementation produces the identical results as the sub-circuit SPICE model. This is illustrated in the next sub-section.

The physics-based IGBT and power diode models are based on the analytical solution of the one dimensional drift-diffusion equation. In this way, the model achieves acceptable run time in circuit simulations and at the same time ensures a short model development cycle. Since diodes are highly symmetrical in layout and vertical design, a single 1D layout is sufficient. For IGBTs, the description of the MOS capacitances accounts for 2D effects as well. Gauss's law is used to describe capacitive currents and charges including the p-n junctions. Once the electrostatics are defined, the drift-diffusion equation is solved to calculate the current densities.

In the sub-circuit model, auxiliary circuits are used to solve for the dynamic base charge, base resistance, space charge width etc. This results in a complex circuit description of the model equations. The Verilog-A implementation does not use auxiliary circuits. This results in a significant reduction of internal nodes.

A. Advantages of Verilog-A over SPICE

Verilog-A has been the modeling language of choice for compact model developers in the recent past. This is primarily because of the fact that it has many significant advantages over sub-circuit SPICE model implementation. Verilog-A provides advanced features like looping, conditional statements, arrays and much more. We have exploited this advantages of Verilog-A to have a faster running and better converging model.

Example of junction-width calulation

To illustrate the benefit of Verilog-A in solving implicit equations, here we show the example of the calculation of

978-1-7281-0941-1/19 $31.00 © 2019 IEEE

the p-n junction width (x_j) as it is applied to the diode and IGBT models.

$$x_j(V_j, x_j) = \frac{\Big((V_j - f(x_j)\Big) + x_j f'(x_j)}{f'(x_j)} \quad (1)$$

where $f(x_j)$ is given by equation (2) below and $f'(x_j)$ is the symbolic differentiation of $f(x_j)$.

$$f(x_j) = \int_0^{x_j} \Big(E(x_j) - E(z)\Big) dz \quad (2)$$

where $E(z)$ is the electric field. Equation (1) results from the approximate Taylor's series expansion (cut off after first derivative) of equation (2). It can be seen that the junction width x_j is an implicit equation depending on the voltage V_j across the space charge region and itself. In our SPICE model, the junction width needs to be evaluated using a self-iterating auxiliary circuit as shown in Fig. 1. This may lead to typical convergence problems in transient simulations. In Verilog-A, we can solve such a problem with a simple Newton-Raphson's algorithm using a do-while loop or as an implicit contribution statement as shown in equation (3).

$$V(x_j) \; : \; V(x_j) \; == \; f(x_j) - V_j \quad (3)$$

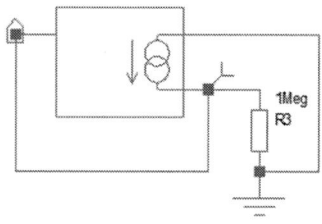

Fig. 1. Self-iterating circuit used for junction width calculation.

This results in less number of circuit equations and hence lower number of iterations are needed to reach convergence in case of Verilog-A as shown in Table I. Due to the lower number of circuit equations less transient/total iterations are needed for the full simulation. This results in a much faster run time with Verilog-A as shown in the following section.

TABLE I
NUMERICAL ITERATIONS AND CIRCUIT EQUATIONS.

	Sub-circuit	*Verilog-A*
Total iterations	4224	2009
Transient iterations	2331	1989
Circuit equations	155	30

B. Comparison with SPICE sub-circuit model

A simple chopper circuit shown in Fig. 2 was simulated to capture the collector current and collector-emitter voltage with respect to time during switching events. As shown in Fig.3, the results from Verilog-A (dashed line) and sub-circuit model (solid line) simulations are almost equivalent.

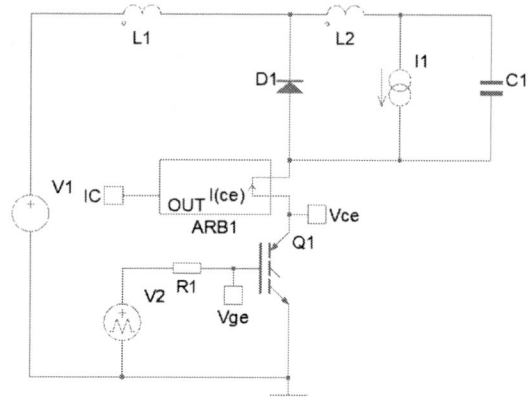

Fig. 2. Schematic of a chopper circuit used as an example.

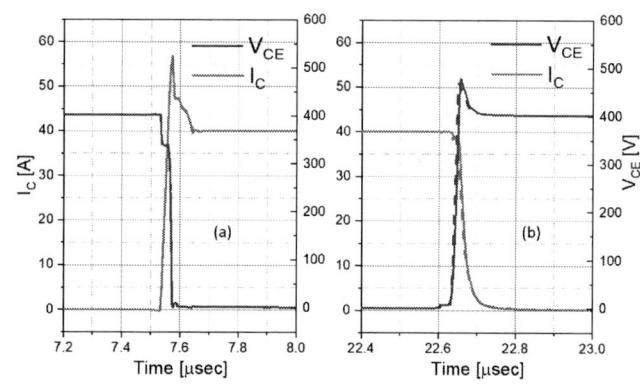

Fig. 3. Comparison of SPICE and Verilog-A simulation output in a chopper circuit (a) turn-on (b) turn-off

Minor differences come from the fact that Verilog-A uses a true time differential operator (ddt), whereas in the sub-circuit model, a resistor-capacitor network is used as an equivalent differentiator circuit. We now compare the run time of sub-circuit versus Verilog-A in Table II for different time steps in the same chopper circuit. As explained in the previous section, the reduced number of circuit equations results in much faster run time in Verilog-A. The speed differences are more noticeable for smaller time steps ($<= 10ns$).

TABLE II
COMPARISON OF RUN-TIME BETWEEN SUB-CIRCUIT AND VERILOG-A

	Sub-circuit	*Verilog-A*
300 ns time step	3.62 sec	0.78 sec
20 ns time step	4.48 sec	1.98 sec
10 ns time step	33.2 sec	5.05 sec

III. COMPUTER-AIDED CALIBRATION

A. Methodology

In a first step, we fit the static transfer and output characteristics for the IGBT and the forward characteristic for the diode at two different temperatures. Further, we use the best-fit IGBT model from this stage to calibrate the gate-charge

978-1-7281-0941-1/19 $31.00 © 2019 IEEE

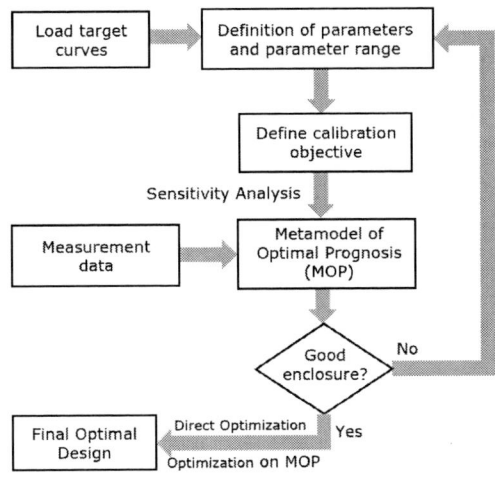

Fig. 4. Algorithm for the calibration methodology.

In the next stage, the criterion for the optimization is defined. We use a quadratic difference method to compare the reference and simulation results. This method is basically the summation of the squared differences between each segment of the reference and simulation curves. The optimizer is then tasked to minimize this quadratic difference. A zero difference would mean that the target and simulations are overlapping.

B. Sensitivity analysis and meta-model generation

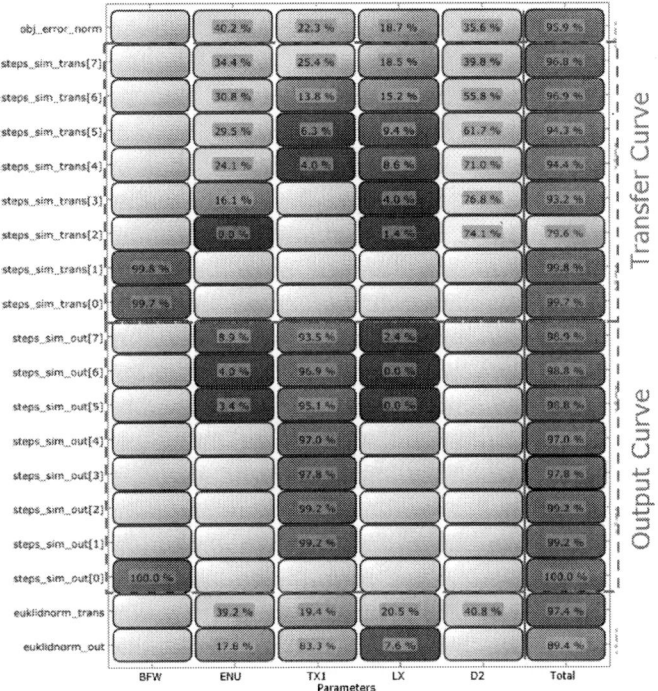

Fig. 6. Meta-model generated after the sensitivity analysis.

curve from transient simulations. The simulations are done in a commercially available circuit simulator supporting Verilog-A. Such a simulator can be easily coupled to optiSLang™ using windows command line interface. optiSLang™ designs the simulation runs, calls the simulator for the simulations, then reads back the results and thereafter tweaks the parameters and launches further set of simulations.

The overall process is summarized in a flow-chart in Fig. 4 and will be explained further on for the IGBT model. In the first stage, the compact model parameters to be optimized are identified in the model and their initial range of variation are determined. The target curves are divided into multiple segments (10-12 for our example) of equal length. The optimizer fits each of these discrete segment together in parallel. This approach has two advantages. Firstly, the segments of the target curves (measurements) where a good fit is important (for example threshold region in transfer characteristic) are identified and can be weighted higher if a higher accuracy is needed in that region. Secondly, as we will see later, this results in a meta-model of the system where each segments dependence on each parameter can be visualized in a convenient way.

At this stage, the tool has all the required information about the reference curves and the compact model parameters to be optimized. The optimizer now does a sensitivity analysis to evaluate the dependence of each parameter on each section of the target curve. The number of total simulation runs are decided on the number of parameters varied and their

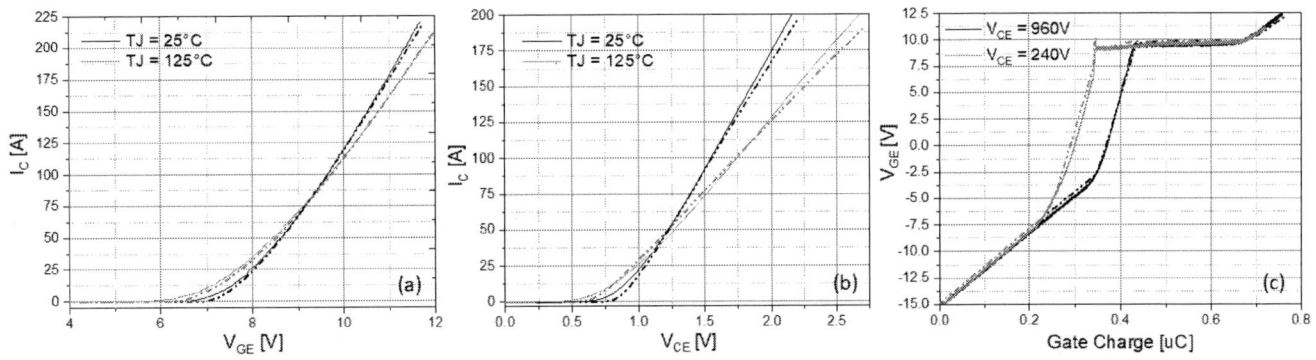

Fig. 5. Comparison of simulated and measured curves after final fit (a) transfer (b) output and (c) gate-charge characteristics.

978-1-7281-0941-1/19 $31.00 © 2019 IEEE

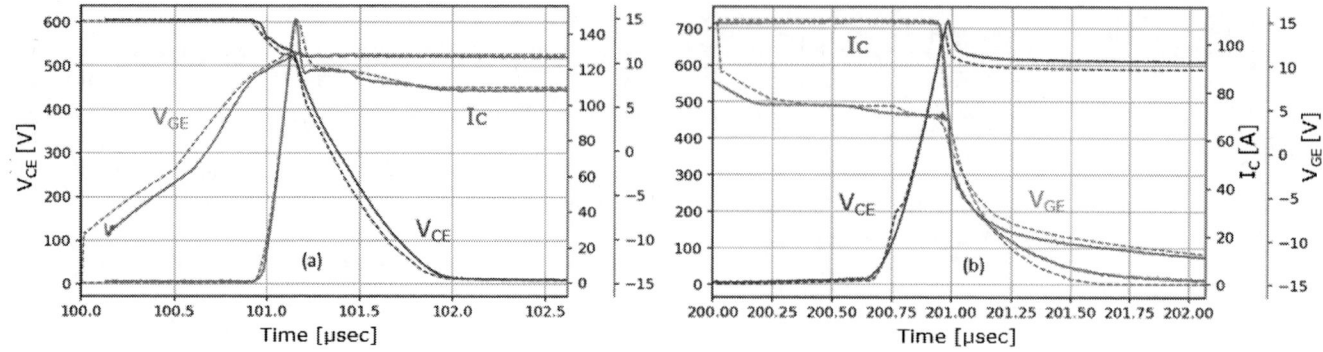

Fig. 7. Comparison of simulated and measured switching transient curves after final fit (a) turn-on (b) turn-off.

variation range. The optimizer can run a set of simulations, analyze the difference between the curves and then launch further simulations in the range where it estimates the objective criterion will be even lower.

As a result of the sensitivity analysis, a meta-model of prognosis (MoP) is created. This meta-model can be visualized as shown in Fig. 6. The five parameters being varied are on the x-axis and the discrete segments (0 being the first segment of the reference curve and 7 being the last) of the transfer and output curves are on the y-axis. The impact of each parameter on each segment of the target curves is shown with a color plot with red being the highest impact and blue the smallest. For example, parameter D2 has a strong impact in segments 2 to 4, which is the threshold region of the transfer curve. Greyed boxes have zero impact. The last two rows summarize this contributions into an overall score for the transfer and output characteristics. A higher score implies a higher confidence of the optimizer in representing the simulated curve.

Various steps can be taken to improve the overall score before going to the final optimization stage. This steps include, removing simulations from the design which have large deviation from the reference, redefining the parameter ranges or removing some parameter dependence from the system. The optimizer may also automatically neglect a certain parameter if it has negligible impact on the target curves. The tool checks if the family of performed simulations has a full enclosure of the target curve. If not, this would mean the parameter ranges were not correct to begin with. In this case, the sensitivity has to be repeated with corrected ranges until full enclosure is achieved.

C. Final optimization and best fit

In the final stage, a further optimization step is done. This can be achieved in the following two ways.

Direct optimization: Direct optimization method would run further simulations in a smaller or same parameter range using the best result of the sensitivity stage as a starting point.

Optimization on MoP: MoP optimization on the other hand, does not need to run new simulations. Instead, it uses the meta-model created from the sensitivity stage to evaluate a best combination of parameters, which results in the best fit

of the simulation data to the measurements.

The final stage then gives the best-calibrated model to the reference curves. The results of the optimization process can be seen in Fig. 5 for (a) transfer (b) output and (c) gate-charge characteristic simulations (dashed line) compared to measurements (solid lines). With this best-fit model, we now check the transient switching behavior of the IGBT with a free wheeling diode. Fig. 7 shows that after such an optimization process, even switching transient simulations (dashed line) fit very well to measurements (solid line). This proves the quality of the physics based model approach.

IV. CONCLUSION

We have demonstrated a new computer-aided calibration technique applied to physics based compact models for IGBTs and power diodes. Just a good calibration of the static curves and the gate-charge curves leads to excellent match with measurements in switching transients as well. In addition, we have shown how Verilog-A compact models can significantly improve the run time and convergence over sub-circuit models.

ACKNOWLEDGMENT

The authors would like to thank Dr. P. Tuerkes for support during the Verilog-A model development and Mr. R. Kallmeyer from Dynardo in setting up the calibration flow.

REFERENCES

[1] optiSLang user manual: www.dynardo.de/software/optislang.html.
[2] R. Kraus, P.Türkes, J.Sigg, "Physics-Based Models of Power Semiconductor Devices for the Circuit Simulator SPICE," PESC 98 Record. 29th Annual IEEE Power Electronics Specialists Conference, May 1998.
[3] R. Kraus, K. Hoffmann and H. J. Mattausch, "A precise model for the transient characteristics of power diodes," PESC '92 Record. 23rd Annual IEEE Power Electronics Specialists Conference, Toledo, Spain, 1992, pp. 863-869 vol.2.
[4] D. Cavaiuolo et al., "An effective parameters calibration technique for PSpice IGBT models application," 2014 International Symposium on Power Electronics, Electrical Drives, Automation and Motion, Ischia, 2014, pp. 133-138.
[5] A. C. Olivieri, Hai-Long Wu, Ru-Qin Yu, "Chemometrics and Intelligent Laboratory Systems," ISSN: 0169-7439, Vol: 96, Issue: 2.
[6] F. Pianosi, F. Sarrazin, T. Wagener, "A Matlab toolbox for Global Sensitivity Analysis," Environmental Modelling & Software, Volume 70, August 2015, Pages 80-85.

978-1-7281-0941-1/19 $31.00 © 2019 IEEE

First-principles investigation of paramagnetic centers in P_2O_5 based glasses

1st Luigi Giacomazzi
Materials Research Laboratory
(University of Nova Gorica)
Nova Gorica, Slovenia
luigi.giacomazzi@ung.si

4th Matjaz Valant
Materials Research Laboratory
(University of Nova Gorica)
Nova Gorica, Slovenia
mvalant@ung.si

2nd Layla Martin-Samos
CNR-IOM Democritos
Trieste, Italy
msamos@sissa.it

5th Nadege Ollier
LSI-IRAMIS, CEA,
Palaiseau, France
nadege.ollier@polytechnique.edu

3rd Nicolas Richard
DAM-DIF CEA
Arpajon, France
nicolas.richard@cea.fr

Abstract—We present a first-principles investigation of paramagnetic centers in P_2O_5 based on the calculation of electron paramagnetic resonance (EPR) parameters (*g*-tensor and Fermi contacts). Calculations of the EPR parameters for the P_1 configuration in crystalline *o'*(P_2O_5) support the previous attribution of the P_1 to a P-defect structurally analogous to the Si-*E'* center. As far as concerns the P_1 center in glassy P_2O_5, the present work suggests the possible occurrence of another configuration besides the analogue of the Si-*E'* center. Such an alternative P_1 center configuration is likely to be relevant for those P_2O_5 based glasses featuring a considerable fraction of Q_1 and Q_2 tetrahedral units besides the Q_3 unit which dominates the structure of pure P_2O_5.

Keywords—P_2O_5, P_1 center, Si-E' center , first-principles

I. Introduction

Phosphate glasses, due to their low Tg (~300 to 500 °C) and melting temperature (~800 to 1300 °C) facilitating the final formation process, are used in various industrial applications. In nuclear waste storage, in particular by means of vitrification processes, iron-phosphate glasses are regarded as a stable storage medium for high-level nuclear waste. In microelectronics, phosphosilicate glass (PSG) is employed to stabilize the field effect transistor (FET) device, since the PSG traps impurity ions, namely Na^+, which would be detrimental for the correct functioning of the device [1]. Furthermore the PSG is used for photovoltaic applications as a dopant source in the fabrication of crystalline silicon solar cells [2]. For optical applications, phosphate glasses are widely used, in particular, doped with rare-earth elements. For instance, phosphate glasses doped with Yb give highly efficient emission at ~1000 nm. As the dissolution of rare-earth ions is much more easier in a phosphate matrix than in silicate glasses, phosphate glasses can be used as laser glass for the production of high-power solid-state lasers systems [3]. Similarly, in silica glass, it has been shown that P-codoping is more efficient to dissolve a rare-earth element such as Yb, than Al-codoping. In fact the P-codoping allows to avoid the Yb cluster formation, which limits the efficiency of Yb^{3+} emission [4].

Ionizing irradiation induces many point defects in phosphate and phospho-silicate glasses [5,6,7]. Several paramagnetic point defects have been detected: the so-called P_1, P_2, P_3, P_4 and r-POHC, and l-POHC centers. The radiation induced generation of the latter phosphorus oxygen hole centers (POHC) has an almost linear dependence on the dose, so that phosphate glasses are nowadays considered also for dosimetry applications [7]. The P_1 center gives rise to an absorption band peaking at about 0.79 eV, which is detrimental for transmission in the infrared domain [6].

Although the P_1 center is commonly accepted to be analogous to the Si-*E'* center [6,8,9], its experimental characterization in phosphate glasses, and in particular in pure P_2O_5 glass, is rather poor (as compared to Si-*E'*) and mainly based on a 30 years old data analysis [5,6]. The present work aims, mainly by means of theoretical first-principles techniques for the calculation of electron paramagnetic resonance (EPR) parameters, to improve our understanding and modelling of point defects, namely of the P_1 center, in P_2O_5 containing glasses.

II. Methods & Models

The calculations presented in this work are based on density functional theory (DFT). The codes we used are freely available with the Quantum-Espresso (QE) package [10]. We used the QE-GIPAW code that exploits the gauge including projector augmented wave (GIPAW) method for the calculation of the EPR parameters [10,11] The Perdew-Burke-Ernzerhof exchange correlation functional (PBE) has been adopted for the present calculations [12]. Norm-conserving Trouiller-Martins gipaw pseudopotentials are used and Kohn-Sham wavefunctions are expanded in a basis of plane waves up to a kinetic cutoff of 70 Ry. The defect configurations here analyzed have been obtained by using a o'(P_2O_5) crystal model and a recently generated vitreous P_2O_5 model both consisting of a 112 atoms supercell with 32 regular corner-sharing PO_4 tetrahedral units. Regular PO_4 units exhibit three normal P-O bonds and one P=O double bond, and are sometimes labelled as Q_3 units [13]. Configurations of paramagnetic centers (POHC, P_1 P_2) are obtained by removing a terminal oxygen from one PO_4 tetrahedron and then by performing a first-principles relaxation of the atomic structure which is put in a

978-1-7281-0941-1/19 $31.00 © 2019 IEEE

positive charge state. Next EPR parameters (g-tensors and Fermi-contacts) are calculated by using the QE-GIPAW code. The calculations of the computationally expensive g-tensor have been carried out only for a selected number of configurations.

III. RESULTS

A. g-tensor and Fermi-contact analysis of selected paramagnetic configurations in P_2O_5

Despite we are not aware of any EPR measurements on irradiated crystalline o'(P_2O_5) we still begin our investigation by considering oxygen vacancies, at terminal oxygen atoms, in o'(P_2O_5) where there is no issue concerning the structural arrangement of atoms. The spin density of our P_1 defect configuration in o'(P_2O_5) is shown in Fig. 1: the spin-density mainly localizes on a sp^3 dangling bond at the three-fold P atom, similarly to a Si-E' center. However, terminal oxygen atoms of nearby [(O-)$_3$P=O] tetrahedra also show some spin-density localization. For the three-fold P atom shown in Fig. 1, the average O-P-O bond angle is 106.3° and P-O bond length is 1.57 Å. The calculated g-principal values and Fermi contacts are g_1 =2.0023, g_2 =2.0038, g_3 =2.0063, and A$_{iso}$(^{31}P) = 105.8 mT. Such values are in a reasonable agreement with the available experimental data [5]: $<g>$=2.005, A$_{iso}$(^{31}P) =95 mT, and provide a solid ground for further investigations in P_2O_5 based glasses. Note that although only the average value $<g>$ was given in [5] a certain anisotropy could be expected on the basis of the experimental EPR parameters of the P_1 center in P-doped silica: $g_{//}$=2.002, $g_\perp$=1.999, and A$_{iso}$ (^{31}P)=91 mT [6]. As previously shown by Abarenkov *et al* [14], the top of the valence band of P_2O_5 mainly comprises of O $2p$ states which, given the large number of terminal oxygen atoms, implies an ease of formation of phosphorus-oxygen hole centers (POHC). In fact, by applying the same procedure used to generate a P_1 center in the o'(P_2O_5) crystal (i.e. removing one terminal oxygen atom and one electron and subsequently carrying out an *ab-initio* relaxation of the atomic structure, see also the following discussion on the Fermi contacts distribution) we have generated P_1, P_2 and also POHC-like center (Fig. 2) in our glass model of P_2O_5. The g principal values calculated for the configuration shown in Fig. 2 are g_1 =2.013, g_2 =2.018, g_3 =2.019, which indeed speak for a center resembling a POHC [6]. Note that this configuration features a (neutral) three coordinated P atom with average P-O bond length of 1.67 Å and O-P-O angle of 96.7°. Only a few terminal oxygen sites in the glass allowed for the generation of a P_1 center configuration similar to the one discussed here above in the crystalline o'(P_2O_5) model structure. Moreover, from a direct inspection of the spin-density, only one configuration shows a spin-density very close to the ideal sp^3 typical of E' centers, whereas the others all shows spin-density localization on a fourth nearby oxygen atom, thus reminiscing of the spin-density of a P_2 center. The calculated g-principal values and Fermi contacts of our "ideal" P_1 configuration are g_1 =2.0018, g_2 =2.0026, g_3 =2.0036, and Aiso(P) = 98.5 mT. The formation energy [9] calculated for the neutral oxygen vacancy at this site is about 2.7 eV (for a few other oxygen vacancy sites we found formation energies to be in the range 2.3 to 2.8 eV), about 2 eV less than reported for neutral oxygen vacancies in silica [9]. In the neutral charge state, the average P-O bond length is 1.67 Å, and O-P-O angle is 97.7°, while once it is positively charged the P-O bond length becomes 1.57 Å and O-P-O 108.8°.

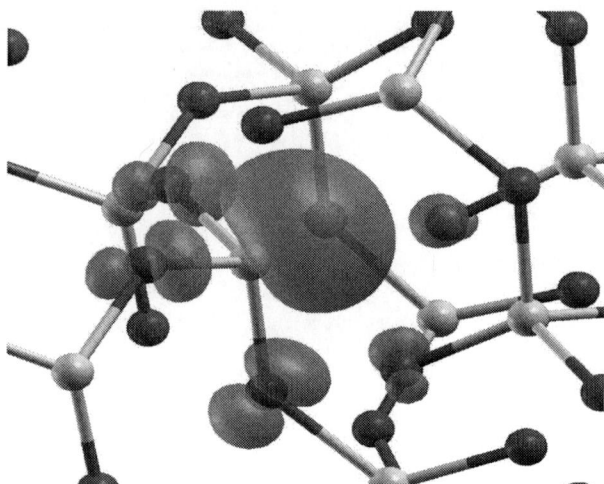

Fig. 1: Spin-density (shaded) of a P_1 configuration (positively charged oxygen vacancy obtained from removal of a terminal oxygen) in o'(P_2O_5).

In the glass, we also generated a considerable number of P_2 center configurations. The calculated g-principal values and Fermi contacts of a typical P_2 configuration are g_1 =2.0002, g_2 =2.0027, g_3 =2.0038, and A$_{iso}$(^{31}P) = 142.5 mT which is in reasonable agreement with [5]. The PO$_4$ tetrahedron of this P_2 configuration is remarkably distorted, as previously found for P_2 centers in P-doped silica [15], and shows a wide O-P-O angle of 163.6° between two long P-O bonds (1.69 and 1.95 Å), while the other two P-O bonds are shorter (1.61 Å) and form a O-P-O angle of 101.1°.

An alternative model for the P_1 center, not structurally analogous to the Si-E' center, was obtained by adding a PO$_2$ unit nearby a terminal oxygen atom and then by carrying out a first principles relaxation of the structure. The spin-density of the final configuration is shown in Fig. 3.

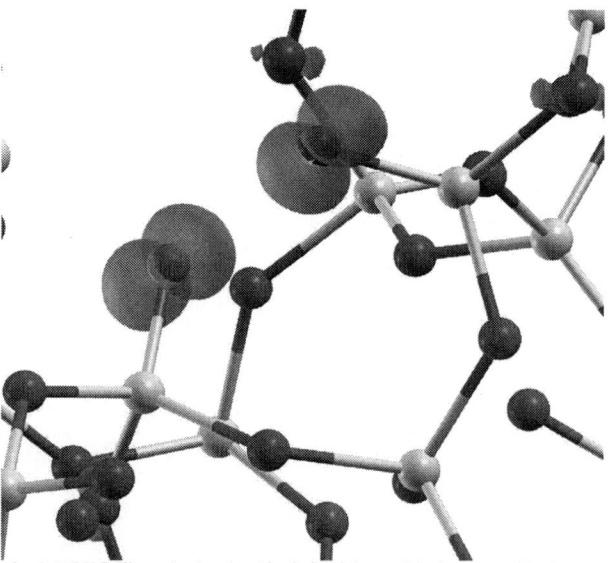

Fig. 2: POHC-like spin-density (shaded) of the positively charged P_2O_5 glass model containing an oxygen vacancy obtained from removal of a terminal oxygen atom.

Fig. 3: Spin-density (shaded) of the P_1-like configuration obtained after placing an extra PO_2 unit nearby a terminal oxygen in a P_2O_5 glass model.

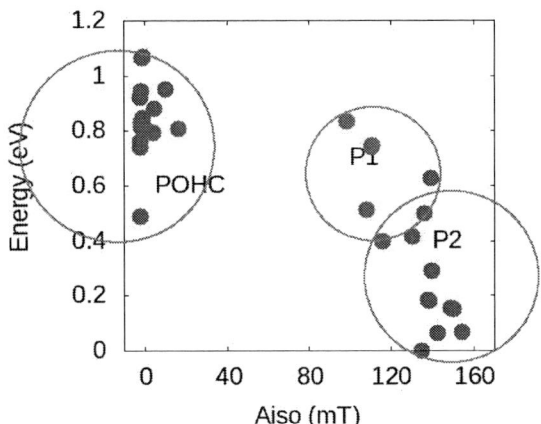

Fig. 5: Fermi-contacts $A_{iso}(^{31}P)$ distribution of oxygen vacancies (at a non-bridging oxygen atom) in a P_2O_5 glass model.

The P atom bearing the unpaired spin in Fig. 3 forms one double bond with a terminal oxygen (1.49 Å) and two bonds with bridging oxygen atoms (1.87 Å and 1.55 Å). The spin-density shown in Fig. 3 is very similar to the one shown in Fig. 1, thus suggesting its classification as a P_1 center configuration. Moreover, the Fermi contact A_{iso} (^{31}P)=101.3 mT and the g principal values g_1 =1.9969, g_2 =2.0023, g_3 =2.0049 calculated for the configuration shown in Fig. 3 further support its attribution to a paramagnetic defect of the P_1 center type [6]. In the table Tab. 1, we summarize the results of our EPR calculations for the three P_1-like configurations generated in the present work, according to the number of non-bridging oxygen (NBO) atoms.

B. Fermi contacts $A_{iso}(^{31}P)$ distribution of oxygen vacancies in P_2O_5 glass

The Weeks & Bray paper [5] is to our knowledge, the only paper which addresses the irradiation induced generation of paramagnetic centers in pure P_2O_5. However, also because the information on g-tensors given in [5] is rather scarce, we have dedicated more attention to the hyperfine splittings and in particular we have obtained the distribution of Fermi contacts $A_{iso}(^{31}P)$ calculated for all the paramagnetic configurations generated at a terminal oxygen site in our model of vitreous P_2O_5. In agreement with Weeks & Bray [5], the present investigation shows (Fig. 4) that in pure P_2O_5 the presence of oxygen vacancies (at the terminal oxygen) leads to the three kinds of EPR centers mentioned here above (P_1, P_2, POHC), so that the occasional presence of centers other than the three under discussion should be ascribed to the occurence of impurities in P_2O_5 [5]. Among the three kinds of centers which are emphasized in Fig. 4 with circles, the lowest

(relative) energy one corresponds to P_2 centers which exhibit Fermi contacts in the range ~130-150 mT. P_1 centers and POHC centers have a larger relative energy 0.6-1.0 eV. However these differences, when considering a non-equilibrium condition (irradiation) are rather small and do not imply necessarily a preference for the generation of P_2 centers. Moreover, we here only generated a few P_1-like configurations so that, with a more statistically significant set of configurations, their average relative energy difference with respect to P_2 could become smaller.

IV. CONCLUSIONS

We performed first-principles calculations of the structure and of the EPR parameters for the P_1, P_2 and POHC paramagnetic centers in o'(P_2O_5) and P_2O_5 glass. The results are satisfactory and confirm the validity of the adopted theoretical approach to model paramagnetic defects in phosphate glasses and support for the P_1 a structural analogy to the Si-E' center. It should be noted however that only a minor fraction (~15%) of the oxygen vacancy sites in the glass here considered allow for the generation of a P_1-like configuration. In fact, the vast majority of the terminal oxygen vacancy sites allows for the generation of POHC and P_2 centers in a similar amount. As far as concerns the P_1 center in glassy P_2O_5, the present work suggests the possible occurrence of another configuration besides the previous one, i.e. a three-fold P atom featuring a P=O double bond and two normal P-O bonds. Such an alternative P_1 center configuration is likely to be relevant for those glasses featuring a considerable fraction of Q_1, Q_2 tetrahedral units besides the Q_3 unit which dominates the structure of pure P_2O_5 glass [13].

ACKNOWLEDGMENT

We acknowledge support through the ARRS-CEA Bilateral Project "REPHLES" NC-0008.

REFERENCES

[1] P. Balk and J. M. Eldridge, "Phosphosilicate glass stabilization of FET devices," Proceed. IEEE, vol. 57, no. 9, pp. 1558-1563, September 1969.
[2] H. Wagner et al, "Optimizing phosphorus diffusion for photovoltaic applications: peak doping, inactive phosphorus, gettering, and contact formation," J. Appl. Phys, vol. 119, no. 18, pp. 185704, May 2016.

TABLE I.   G-TENSORS AND FERMI CONTACTS (mT) OF P_1 CONFIGURATIONS IN P_2O_5 AS CALCULATED BY FIRST-PRINCIPLES

Number of NBO	structure	g_1	g_2	g_3	$A_{iso}(P)$
0	o'(P_2O_5)	2.0023	2.0038	2.0063	105.8
0	glass	2.0018	2.0026	2.0036	98.5
1	glass	1.9969	2.0023	2.0049	101.3

[3] J. H. Campbell, J. S. Hayden, and A. Marker, "High-power solid-state lasers: a laser glass perspective,"J. Appl. Glass Sci. vol. 2, no. 1, pp. 3-29, February 2011.

[4] B. Schaudel, P. Goldner, M. Prassas, and F. Auzel, "Cooperative luminescence as a probe of clustering in Yb^{3+} doped glasses," J. Alloys Compd, vol. 300-301, pp. 443-449, April 2000.

[5] R. A. Weeks and P. J. Bray, "Electron spin resonance spectra of gamma-ray-irradiated phosphate glasses and compounds: oxygen vacancies," J. Chem. Phys, vol. 48, no. 1, pp. 5-13, January 1968.

[6] D.L. Griscom, E. J. Friebele, K. J. Long, and J. W. Fleming, "Fundamental defect centers in glass: Electron spin resonance and optical absorption studies of irradiated phosphorus doped silica glass and optical fibers,"J. Appl. Phys, vol. 54, no. 7, pp. 3743, March 1983.

[7] Z. M. Da Costa, W. M. Pontuschka, J. M. Giehl, and C. R. Da Costa, "ESR dosimeter based on P_2O_5-CaO-Na_2O glass system," J. Non-Cryst. Solids, vol. 352, no. 32-35, pp. 3663-3667, September 2006.

[8] L. Giacomazzi et al., "EPR parameters of E' centers in v-SiO_2 from first-principles calculations," Phys. Rev. B, vol. 90, no. 1, pp. 014108, July 2014.

[9] L. Giacomazzi et al., "Photoactivated processes in optical fibers: generation and conversion mechanisms of twofold coordinated Si and Ge atoms," Nanotechnology, vol. 28, no. 19, pp. 195202, April 2017.

[10] P. Giannozzi et al., "QUANTUM ESPRESSO: a modular and open-source software project for quantum simulations of materials," J. Phys. Condens. Matter, vol. 21, no. 39 pp. 395502, September 2009.

[11] C. J. Pickard and F. Mauri, "First-principles theory of the EPR g-tensor in solids: defects in quartz," Phys. Rev. Lett, vol. 88, no. 8-25, pp. 086403, February 2002.

[12] J. P. Perdew, K. Burke, and M. Ernzerhof, "Generalized gradient approximation made simple," Phys. Rev. Lett., vol. 77, no. 18, pp. 3865-3868, October 1996.

[13] N. Shcheblanov et al., "Vibrational and structural properties of P_2O_5 glass: advances from a combined modeling approach," unpublished.

[14] I. V. Abarenkov, I. I. Tupitsyn, V. G. Kuznetsov, and M. C. Payne, "Electronic structure of crystalline phosphorus pentoxide and the effect of an Ag impurity," Phys. Rev. B, vol. 60, no. 11, pp. 7881-7885, September 1999.

[15] G. Pacchioni, D. Erbetta, D. Ricci, and M. Fanciulli, "Electronic structure of defect centers P_1,P_2, and P_4 in P-doped SiO_2", J. Phys. Chem. B, vol. 105, no. 26, pp. 6097-6102, March 2001.

Theoretical Study of the Edge Effect of Dumbbell-shape Graphene Nanoribbon with a Dual Electronic Properties by First-principle Calculations

1st Qinqiang Zhang
Dept. of Finemechanics
Tohoku University
Sendai, Japan
zhang.qinqiang@rift.mech.tohoku.ac.jp

2nd Takuya Kudo
Dept. of Finemechanics
Graduate School of Engineering
Sendai, Japan
takuya.kudo@@rift.mech.tohoku.ac.jp

3rd Jowesh Gounder
Dept. of Finemechanics
Graduate School of Engineering
Sendai, Japan
jowesh@rift.mech.tohoku.ac.jp

4th Ying Chen
Fracture and Reliability Research
Institute
Tohoku University
Sendai, Japan
ying@rift.mech.tohoku.ac.jp

5th Ken Suzuki
Fracture and Reliability Research
Institute
Tohoku University
Sendai, Japan
kn@rift.mech.tohoku.ac.jp

6th Hideo Miura
Fracture and Reliability Research
Institute
Tohoku University
Sendai, Japan
hmiura@rift.mech.tohoku.ac.jp

Abstract—The electronic band structure (band gap) and electronic transmission properties of dumbbell-shape graphene nanoribbons (DS-GNRs), which consists of a thinner semiconductive GNR and two wider metallic GNRs at its both ends, was theoretically investigated using first-principles calculation to clarify the dominant controlling factors of their electronic performance for their applications to various smart sensors. The electronic properties of the DS-GNR was found to vary drastically depending on the combination of the total number of carbon atoms along the width direction of each portion, the length of the semiconductive portion, the width of the metallic portion, and so on.

Keywords—*graphene nanoribbon, dumbbell-shape structure, band gap, electronic transmission property*

I. INTRODUCTION

After the discovery of graphene in 2004, many researchers have been dedicated to explicate the basic characteristics of this outstanding material which has a great potential to substitute the conventional semiconductor material such as silicon. Previous studies have shown that a single layer of graphene sheet has been proved to show super metallic properties [1]. However, when a graphene sheet is cut into nanoscale stripe with a width thinner than 70 nm, so-called graphene nanoribbon (GNR), it starts to exhibit semiconductive properties [2], and the semiconductive properties vary a lot depending on its size. Even various graphene-based electronic applications have been proposed, however, the high performance electronic devices based on GNR have not been successfully fabricated in mass production.

It was experimentally validated that the sub-50-nm GNR showed clear semiconductive properties [3]. In addition, the semiconductive GNRs showed wide variation of the strain sensitivity of their electrical resistance under the application of uniaxial strain [4]. However, their electronic properties varied widely from metallic-like ones to semiconductive ones. There are two essential problems which caused the variation of the electrical properties. One is the periodic change in the band gap of GNRs smaller than 70 nm which appear to have a strong function of the total number of carbon atoms along the width direction of GNRs. The other is electronic junction between the semiconductive GNR and metallic GNRs. Therefore, it is very important to control the effective band gap of the DS-GNR structure and a stable electronic junction between the semiconductive GNR and metallic GNRs.

In this study, the electronic band structure and electronic transmission properties of a novel proposed dumbbell-shape structure of GNRs were investigated by using first-principle calculations. The dumbbell-shape structure consists of different sizes of single GNRs with different edge sizes at both end portion of single GNRs. The simulation results were used for providing theoretical data and guides to fabricate highly reliable, stable and sensitive GNR-based electronic devices.

II. ANALYTICAL MODEL OF DUMBBELL-SHAPE STRUCTURE

The synthesized 40-nm wide monolayer graphene nanoribbons periodically patterned on a Si/SiO_2 substrate has been fabricated successfully as shown in Fig. 1 [3]. To indicate a more intuitional illustration on the simulation model of novel proposed dumbbell-shape graphene nanoribbon (DS-GNR), the structure is highlighted by blue and red color as shown in Fig. 1. The blue portions consist of the large size monolayer graphene with metallic properties, and the red ones are composed by 40-nm wide GNR stripes with semiconductive properties. The light grey areas in this figure are the deposited electrodes on the top of the DS-GNRs and the black area represents the etched area where no graphene exists on Si/SiO_2 substrate. It was confirmed that the DS-GNR showed semiconductive photosensitivity and the sensitivity was more than ten times higher than that of conventional silicon-base devices. However, the sensitivity of each GNR showed wide variation. Therefore, in order to fabricate highly reliable, stable and sensitive GNR-based electronic devices, a

Fig. 1. GNR-based electronic device with multi-channel dumbbell-shape structure highlighted by blue and red bars. The scale is 5 micro meter.

978-1-7281-0941-1/19 $31.00 © 2019 IEEE

further theoretical study is indispensable for finding out the dominant controlling factors of their electronic performance.

A simple schematic image of the developed dumbbell structure of GNR is shown in Fig. 2(a) and the simulated dumbbell-shape GNR (DS-GNR) model is also shown in Fig. 2(b). The large graphene sheet area of DS-GNR, where the graphene sheet shows metallic properties, was covered by metal electrodes. In this simulation, only the wide GNRs at both ends were considered to be electrodes. This wide GNRs with metallic properties were assumed that it had an ohmic contact with any common deposited metal electrodes, and the energy barrier between the metallic electrodes and the wide GNR was negligible. The electronic properties of a single semiconductive GNR between the two GNR electrodes are categorized into three-families: $3p$, $3p+1$, $3p+2$ (p is an integer) by a total number of carbon atoms along width direction. The groups of $3p$ and $3p+1$ show semiconductive properties, and the group of $3p+2$ shows semi-metallic properties as the same explanation with previous studies [5]. In this study, the $3p+2$ type of GNR was selected as an example for representing the large area graphene/GNR which shows semi-metallic properties. The $3p/3p+1$ types of GNR were represented for semiconductive GNRs whose width was less than or much less than 70 nm in practical application. Identically, it was regarded as narrow portion GNRs in the simulation. The model structure was assumed to be periodic along the vertical direction and has Armchair GNR (AGNR) with hydrogen termination. Hence, the pure DS-GNR has Zigzag GNR (ZGNR) with hydrogen termination perpendicular to the periodic direction. The electronic properties of pristine ZGNR and AGNR have already been studied [5]. Therefore, in this study, the electronic properties of pristine ZGNR and AGNR were recalculated as a reference to that of the novel proposed DS-GNR. However, previous studies have not shown the combined structure which has interesting electronic properties.

III. ANAYSIS OF ELECTRONIC BAND STURCTURE OF DS-GNR

The SIESTA package which was developed based on the density functional theory (DFT) was used for the calculations of electronic band structure in this study. The post-processing calculation was performed by using the TranSIESTA utility which was using nonequilibrium Green's function (NEGF) method based on DFT for calculating the current-voltage (I-V) characteristics of DS-GNR [6]. The generalized gradient approximation (GGA) in the Perdew-Burke-Ernzerhof (PBE) form was used to describe the exchange and correlation energy of electron-electron interactions. The conjugate gradient (CG) method was used to fully optimize all the atomic positions without any geometric constraints until the total maximum force became less than 0.02 eV Å^{-1}. An energy mesh cutoff was set at 500 Ry. The tolerance energy of convergence was set to 0.001 eV. The k-point sampling for unit cells was constructed with a $1 \times 20 \times 20$ Monkhorst-Pack grid. The electronic temperature was fixed at 450 K to increase the convergence speed. The vacuum area was 10 Å on the left and right side along horizontal direction and 7.5 Å on the top and bottom of the DS-GNR. Also, based on the frontier molecular orbital theory, the top of the valence band was also referred to as the highest occupied molecular orbitals (HOMO), and the bottom of the conduction band was referred to as the lowest unoccupied molecular orbitals (LUMO). All the orbital distribution figures shown in this study is at HOMO energy

level since that of figures at LUMO energy level is also symmetric but shows opposite phase color.

Since it was hard to analyze the electronic properties of the actual DS-GNR structure shown in Fig.1 due to the limitation of the memory capacity of used supercomputer, a small model structure was analyzed to explicate the dominant structural factors of the dumbbell-shape structure for its electronic properties. At first, the effect of the length of the semiconductive narrow GNR on the electronic properties of the DS-GNR structure was analyzed. Because the quantum tunneling effect should appear and the electrons can jump between two wide metallic GNRs without extra energy field, when the length of narrow portion is too short.

The structures with the same width of narrow portion of DS-GNR ($N_N = 7$, semiconductive type) and the same width of wide portion ($N_W = 17$, semi-metallic type) were calculated with different length portion from $N_L = 1$ to $N_L = 20$. Two examples of dumbbell-shape structures with the length portion $N_L = 3$ and $N_L = 7$ are shown in Fig. 3(a) on the left and right hand side, respectively. The orbital distribution of DS-GNR with $N_L = 3$ at HOMO illustrated the quantum tunneling effect clearly. The colored area appeared throughout the whole structure. When the length of narrow portion was increased from 3 to 7, however, the localized orbital distribution started

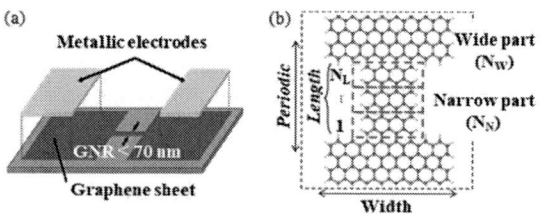

Fig. 2. (a) Simple schematic image of dumbbell-shape GNR base electronic device. (b) Simulated DS-GNR model. It consists of wide portions at both ends of narrow portion. N_L and N_W is the number of total continuous carbon atoms along the width direction. N_L is the six member ring of carbon atoms along width direction as one group which is represented by red dash line box.

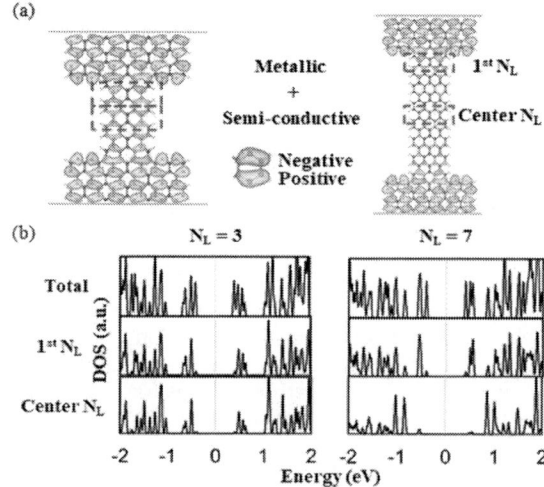

Fig. 3. (a) Simulation model with 17 at wide portion, 7 at narrow portion. Left side has $N_L = 3$ and right side has $N_L = 7$. (b) Density of states of DS-GNR at different areas which are corresponding to each upper structure, respectively. Carbon atoms included in red dash line box are counted for DOS analysis.

to appear around the interface between wide portion and narrow portion. The different distributions are highlighted by red dash lines. As shown in the right hand side of Fig. 3(a), there was no orbital distribution at the center of the narrow portion. It shows the hint that when the 0 eV or higher bias voltage applied to this structure, it may show semiconductive properties which will be explained later.

The local density of states (LDOS) with different length of narrow portion of DS-GNR are shown in Fig. 3(b). The first row panel is the total density of states (DOS) of DS-GNR, and the second row panel indicates the summation of LDOS of each carbon atom included in the red-dashed box in Fig. 3(a) near the junction between wide portion and narrow portion (1st N_L). Similarly, the third row panel shows the summation of LDOS of each carbon atom exists in the red-dashed box at the center of narrow portion (center N_L). The vertical axis of the figure is the arbitrary unit of DOS and horizontal axis is the energy. The arbitrary unit was rescaled to compare the peaks near the Fermi level. On the left side of Fig. 3(b), the LDOS doesn't show much difference among the total, first N_L, and center N_L. But the most inner peak of LDOS at center N_L with the length of narrow portion equaling 3 ($N_L = 3$) was disappeared. However, on the right side of Fig. 3(b), the LDOS shows a large difference among the total, first N_L and center N_L. There were no high inner peaks around the 0.5 eV to the Fermi level which was different from the results obtained from the shorter DS-GNR. This result clearly indicates that the localized energy barriers exist in the DS-GNR.

In order to simulate the practical condition and dismiss the effect of two wide portions as described above, the length increased up to $N_L = 20$ to find the length dependence of the narrow portion of DS-GNR. It was confirmed that the band gap increased monotonically with the length of the narrow portion as shown in Fig. 4. The band gap of DS-GNR was converged to one specific value when the length of narrow portion reached 20. Thus, the performance of the proposed DS-GNR structure is close to that of single pristine GNR when the length is long enough. This result provides the reasonable guide for experiments. As shown in Fig. 4, the curve becomes saturated when N_L is larger than 10. Since the length of carbon to carbon bonding is still around 1.42 Å after fully optimized, the width of narrow portion was about 7.2 Å and the length of narrow portion was around 42.6 Å when $N_L = 10$. Therefore, when the length is at least 6 times larger than the width, the performance of DS-GNR is close to that of pristine GNR at the narrow portion. This result provides the instruction to fabricate the DS-GNR for experiment that the length to width ration should be larger than 6.

Furthermore, the width dependence of DS-GNR at wide portion was also considered as edge effect. Since the wide portion should be a large graphene sheet area or GNR, the width of the wide portion was increased from 17 to 23, 29, and 41 to reach the specific value that total DOS started to show no band gap around the Fermi level. The simulated models are shown in Fig. 5(a). Two structures have the same length and width of the narrow portion which $N_L = 7$ and $N_N = 7$ but with different width of the wide portion which $N_W = 17$ (left side) and $N_W = 29$ (right side). Even when the width of the wide portion was increased to 29, the orbital distribution still showed the localized pattern. It implies the same hint of electronic transmission properties which described above. The figures of LDOS of DS-GNR are shown in Fig. 5(b). The left

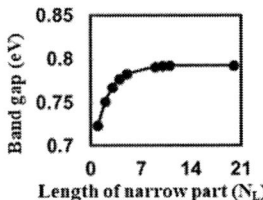

Fig. 4. Tendency of the band gap of DS-GNR with increasing length of N_L

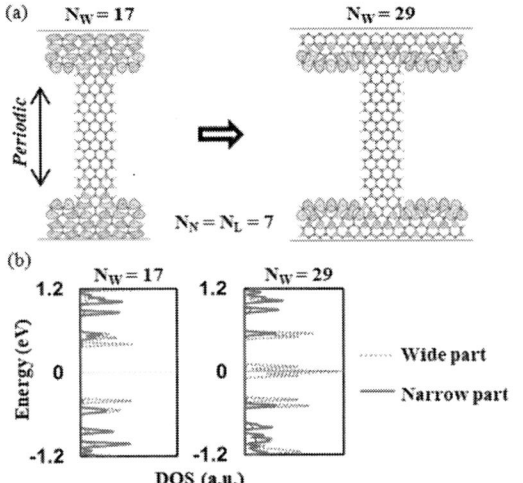

Fig. 5. (a) Simulation model with different width of wide portion. The orbital distribution of two DS-GNRs shows localized distribution. (b) Local density of states of each upper DS-GNR, respectively. Blue dashed line is the LDOS of wide portion and red line is that of narrow portion.

side is the LDOS of DS-GNR with $N_W = 17$ and the right side is that of with $N_W = 29$, respectively. The vertical axis is the energy and the horizontal axis is the intensity of DOS. The blue dashed lines are the LDOS of carbon atoms in the wide portions and the red lines indicate the LDOS of carbon atoms in the narrow portion. It is clear that the band gap becomes 0 eV in the wide portion around Fermi level but the band gap does not change too much in the narrow portion. The total DOS was the summation of blue dashed and red lines, hence, the performance of the total DS-GNR structure shows metallic properties when the width of the wide portion increased to specific value, though the LDOS in the wide portion is different from that in the narrow portion. This result indicates that there exists an energy barrier around the junction area between the wide and narrow portions, even the component of this structure consists of only carbon atoms. Usually, this property appears in the structure at the boundary with different component or element. In order to further investigate the electronic properties, the post-processing was also implemented to obtain the electronic transmission properties of DS-GNR. The result is also known as the current-voltage (I-V) characteristics.

IV. ANALYSIS OF ELECTRNOIC TRANSIMISSION PROPERTIES OF DS-GNR

The schematic image of DS-GNR implemented by TranSIESTA is shown in Fig. 6(a). The scattering portion of DS-GNR was not calculated periodically in TranSIESTA, and the electrode portions were added at the both ends with periodical open boundaries. The current-voltage (I-V) curve of two structures are shown in Fig. 6(b) and Fig. 6(c), respectively. The y-axis is the current through the DS-GNR

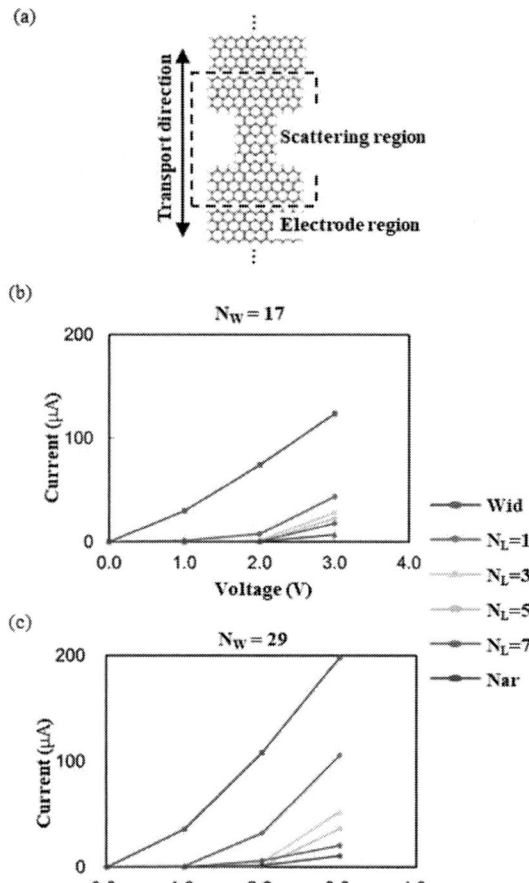

(a)

(b)

(c)

Fig. 6. (a) Schematic image of DS-GNR for electronic transmission calculation. Bias voltage applied on electrode regions and scattering region is enclosed by dash black line. (b), (c) are the current-voltage curve of DS-GNR with different length of narrow portion and that of single pristine GNR at wide and narrow portion, respectively.

along the periodical direction and the x-axis is the bias voltage applied on the structure along the periodical direction. The unit of current is micro ampere and that of voltage is volt. The line of label Wid indicates the I-V curve of pristine single GNR in the wide portion. Similarly, the line of label Nar implies the I-V curve of pristine single GNR at narrow portion. $N_L = 1$ to 7 are the I-V curves of DS-GNR with different length of the narrow portion, respectively. Fig. 6(b) shows the tendency of I-V curve of DS-GNR with the width of wide portion 17 ($N_W = 17$) and also Fig. 6(c) represents that of DS-GNR with the width of wide portion 29 ($N_W = 29$). When the length of the narrow portion increased, the I-V curve became more flat due to the increase of band gap of DS-GNR. And when the length of narrow portion was long enough, the I-V curve became close to that of pristine single GNR with the same width of DS-GNR in the narrow portion.

When the width of the wide portion increased from $N_W = 17$ to $N_W = 29$, however, the tendency of the length dependence didn't change too much as shown in Fig. 6(c), even when the band gap of DS-GNR with $N_W = 29$ was zero.

Thus, the electronic transmission properties of DS-GNR was semiconductive. It also means that the localized energy barrier exists in the DS-GNR which consists of only carbon atoms. It can be explained by the LDOS of DS-GNR in the wide and narrow portions. However, the electronic transmission properties through the DS-GNR along the periodical direction have no significant relationship with the width of DS-GNR in the wide portion. The width of the wide portion of DS-GNR, therefore, doesn't affect the electronic transmission properties of DS-GNR significantly. This is because the zigzag edge effect in the wide portion of DS-GNR mainly dominates the band gap of the total system, but no significant effect on the electronic transmission properties perpendicular to the zigzag edge direction.

V. CONCLUSIONS

In this study, the effect of the structure of DS-GNR such as the length of the semiconductive portion and the width of the wide metallic portion on its electron transmission properties wsa analyzed by using first principle calculation. It was found that the electronic band structure was converged to a specific value when the length of the narrow portion was at least 6 times larger than its width. Thus, the performance of the proposed dumbbell-shape structure is close to that of pristine single GNR when the length of the narrow portion is long enough. Furthermore, when the width of the wide portion is wide enough, for instance, the whole structure starts to show a localized energy barrier around the junction between the wide and narrow portions. Therefore, the electronic properties of the wide and narrow portions of DS-GNR keep their intrinsic properties even after they are jointed together. The energy barrier between them still remains even the whole structure consists of only carbon atoms. These results give basic idea of the structural design of GNR-based two dimensional electronic devices.

ACKNOWLEDGMENT

This work was supported by JSPS KAKENHI Grant Number JP16H06357. The authors would like to express their sincere thanks to the crew of Center for Computational Materials Science of the Institute for Materials Research, Tohoku University for their continuous support of the supercomputing facilities.

REFERENCES

[1] Novoselov, Kostya S., et al. "Electric field effect in atomically thin carbon films." Science 306.5696 (2004): 666-669.

[2] Yang, Meng, et al. " Electronic properties and strain sensitivity of CVD-grown graphene with acetylene " Japanese Journal of Applied Physics 55, 04EP05(2016), pp.04EP05-1~04EP05-8.

[3] JA Goundar, et al. " Photosensitivity of Monolayer Graphene-Base Field Effect Transistor " Proc. of ASME IMECE2019, No. 87245, (2018), pp. 1-6.

[4] Ryohei Nakagawa, et al. "Area-arrayed graphene nano-ribbon-base strain sensor." Proc. of ASME IMECE2019, No. 87277, (2018), pp. 1-6.

[5] Son, Young-Woo, Marvin L. Cohen, and Steven G. Louie. "Energy gaps in graphene nanoribbons." *Physical review letters* 97.21 (2006): 216803.

[6] Soler, José M., et al. "The SIESTA method for ab initio order-N materials simulation." *Journal of Physics: Condensed Matter* 14.11 (2002): 2745.

978-1-7281-0941-1/19 $31.00 © 2019 IEEE

A new wet etching method for black phosphorus layer number engineering: experiment, modeling and DFT simulations

Teren Liu
Department of Materials Engineering
University of British Columbia
Vancouver, Canada
ORCID: 0000-0003-1060-3350

Tao Fang
[1]Department of Materials Engineering,
[2]Department of Physics and Astronomy
[3]Stewart Blusson Quantum Matter
Institute
University of British Columbia
Vancouver, Canada

Karen Kavanagh
Department of Physics
Simon Fraser University
Vancouver, Canada
ORCID: 0000-0002-3059-7528

Hongyu Yu
School of Microelectronics
line 3: *Southern University of Science*
of Technology
Shenzhen, China

Guangrui (Maggie) Xia [*]
Department of Materials Engineering
University of British Columbia
Vancouver, Canada
guangrui.xia@ubc.ca

Abstract—This paper reports the successful atomic layer patterning of 2-dimensional Black Phosphorus (BP) and the simulation of the etching process by Density Functional Theory (DFT) method. The wet etching process can etch selected regions of few-layer black phosphorous with an atomic layer accuracy, which provides a feasible patterning approach for large-scale manufacturing of few-layer BP materials and devices. Absorption energies of iodine atoms/molecules at different location of BP layer edge were also calculated by DFT method, shown a vertical etching direction preference which was important for achieving high quality patterns.

Keywords—Black Phosphorus, atomic layer etching, Density Functional Theory, etching process simulation, absorption energy

I. INTRODUCTION

Two-dimensional (2D) materials have drawn high interests for applications in electronic and photonic devices. They have unique properties such as high carrier mobilities, high on-off ratios, anisotropy, lack of surface dangling bonds, low defect concentrations, high optical absorption, and tunable bandgaps. [1-5] For most 2D materials, the atom layer number greatly affects their energy band structures and electronic transport properties. This made many band-gap-engineering by layer numbers possible. Devices such as Tunneling Field Effect Transistor (TFET) [2] or wavelength-tunable photodetectors/emitters can be relativity easy to fabricate [3-4]. However, on the other side, layer number variation in device fabrication can cause undesired device performance variations. Thus, atom layer number controlling in 2D material devices is a critical issue. Currently, experimental or theoretical reports on atom layer etching processes for the 2D materials are still very rare. [3]

2D black phosphorus (BP) can be viewed as a stack of phosphorus atomic planes with interlayer Van der Waals force, whose direction is along [010] direction. BP has a direct bandgap, which is tunable by its atomic layer number. Here,

we present a wet etching method for BP, which could achieve atom layer accuracy and controllability along [010] direction.

II. ETCH EXPERIMENTS AND RESULTS

Iodine was chosen as the etchant in the experiment due to its compatibility with mainstream photoresist and the suitable etch rate range. Since oxygen and water may introduce oxygen defects in BP, we used Isopropyl Alcohol (IPA) and methanol mixture as solvents [5]. Etching depth was measured by an Atomic Force Microscope (AFM). Etching depths with different times and/or concentrations are shown in Figure 1. The relations between the etch depths and etch time/etchant concentrations are linear. The linear fitting lines were shown in Figure 1. From the fitting results, the etching rate along [010] direction was around 0.081-0.101 nm/(min*g/L). Considering that one BP layer is 1.0473 nm thick [6], this etch rate is slow enough to have an atomic layer accuracy. The etch rate can be then modelled as:

$$R = 0.091 * \frac{C_{I2}}{\frac{g}{L}} \frac{nm}{min}.$$

In the above equation, R is the etch rate of BP, and C_{I2} is the concentration of iodine.

It should be noticed that the etch rate was thickness-independent. For example, before etching, the BP sample shown in Figure 2 (a) was not uniform. After etching, the thickness was reduced by 11 nm uniformly across the 20 microns keeping the same profile shape, as measured by AFM (Figure 2 (b)). Due to the very large area to measure and the AFM tip size (around 10 nm), the edge steepness is very hard to be measured by AFM.

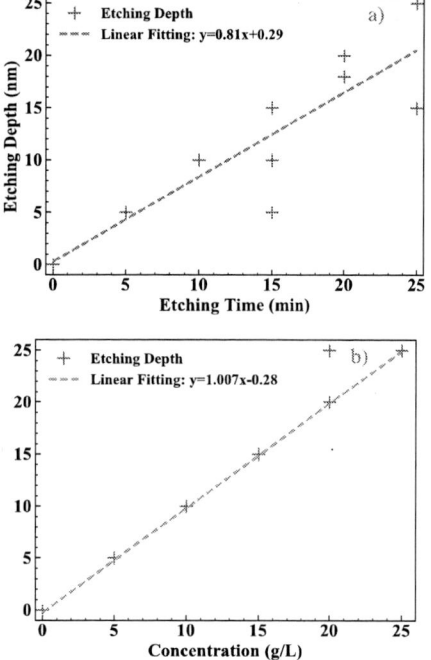

Figure 1. (a) Etching depths vs. different etch times using 10g/L iodine/IPA-methanol solutions. (b) Etching depths after 10 minutes etching vs. iodine concentrations of iodine/IPA-methanol solutions.

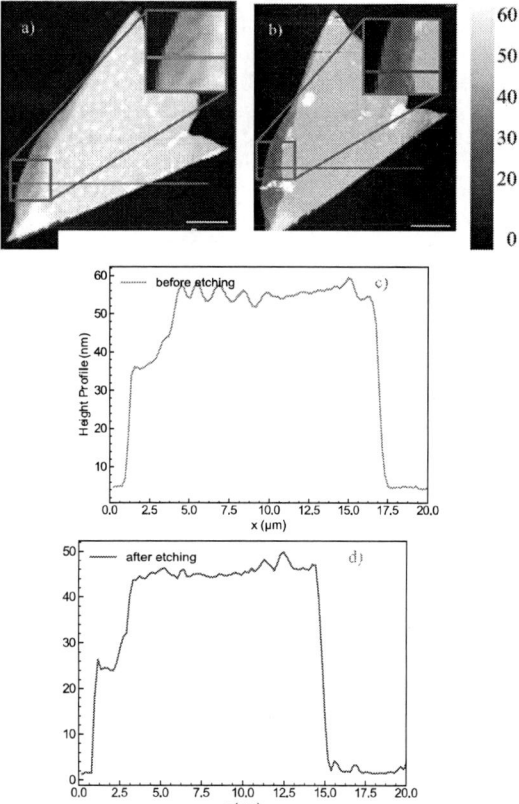

Figure 2. AFM thickness characterization for a BP sample. (a) and (c): 2D and 1D thickness profiles before etching. (b) and (d): 2D

and 1D thickness profiles after etching. The etch solution was 5g/L iodine/(IPA-methanol) solution, and the etch time was 20 min.

Samples etched by 10g/L iodine solutions for 10 mins were characterized by Scanning Transmission Electron Microscope (STEM), Raman microscopy, energy-dispersive X-ray spectroscopy (EDS) and electron energy loss spectroscopy (EELS) spectra (Figure 3), which showed that there were no iodine remains after the etching process, and the crystal structure were not changed by the etching (Figure 3). Raman spectrum showed the good crystallinity (figure not shown here).

After we developed an etch recipe to control the atom layer numbers, BP patterning with deep ultra-violate lithography and wet etching was investigated. Polymethyl methacrylate (PMMA) was used as the photoresist. A TEM copper grid of 6 μm bar width was used as the photomask. After 12 min etching, we achieved 15 nm thickness difference between the exposed and unexposed region as shown in Figure 4.

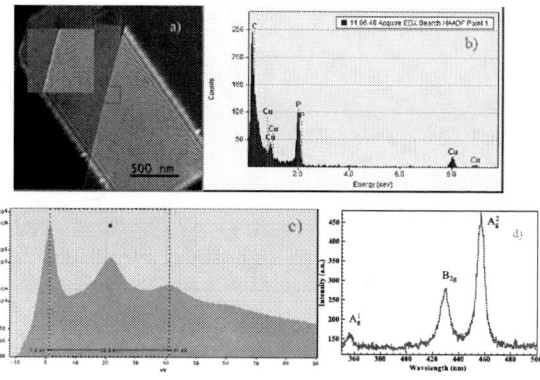

Figure 3. Material characterizations of a sample after etching by a 10 g/L iodine/IPA-methanol solution for 10 mins. a) STEM image and d) Raman spectrum   showing a good BP crystallinity after etching. b) EDS data, c) EELS spectrum: showing that no iodine residues left. The carbon peak is from the polycarbonate (PC) film remain on the copper TEM grid.

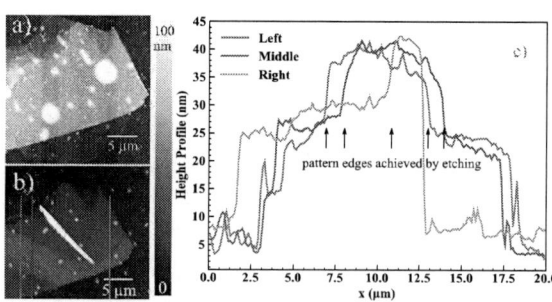

Figure 4. (a), (b) AFM thickness data of the sample before and after patterning. (c) 1D thickness profiles after etching at three locations indicated by the purple, blue and green cutlines shown in (b).

978-1-7281-0941-1/19 $31.00 © 2019 IEEE

III. DFT SIMULATION FOR THE ETCHING PROCESS

Both exfoliation methods and etching can introduce thickness differences thus plane edges. Different from bulk materials which can have defected surfaces, 2D materials normally have defects or dangling bonds on the edges of the layers due to the Van der Waals force structure. Therefore, etching behavior might be different at edge regions compared to other regions. To investigate this problem, Density Functional Theory (DFT) method was used to simulate the etching behavior around plane edges. Specifically, absorption energies for iodine atoms and molecules were calculated as both may exist during the etching processes. It was also of our interest to calculate the absorption energies of iodine atoms/molecules with P atoms laterally and those in the layer below to see which direction is preferred in the reactions.

We used a 4×4×3 BP supercell with half of the first layer removed to emulate a surface edge. Quantum-Espresso [7] software package was used for the calculations.

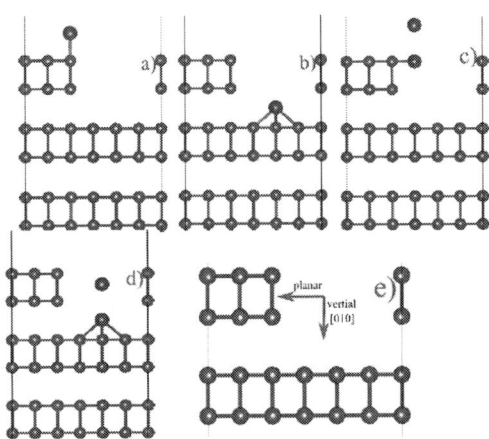

Figure 5. Schematic view from a [001] direction of an ideal BP lattice and the iodine atom/molecule before geometry relaxation. (a): iodine atom on top layer edge. (b): iodine atom on second layer. (c): I_2 molecule on top layer edge. (d): I_2 molecule on second layer edge. (e) Schematic view of the two etching directions.

After a full relaxation, we obtained the lowest total energy for each case. For top layer edge, the absorption energy for iodine atom or molecules were $E_a = 1.06eV$, $E_c = 0.96eV$. While the second layer were $E_b = 1.69eV$, $E_d = 1.52$ eV.

The absorption energies for iodine atom or molecule on the top layer edges are much lower than those on the second layer. This means that iodine atom or molecule tends to attach on the exposed second layer region instead of the exposed edges, which means that the etching direction is [010] direction, the surface normal direction.

The DFT calculations support a vertical etch direction instead of the possibility of an area shrinkage from the edges. This is desired for the use in lithography and patterning of BP, and is consistent with the patterning results shown in Figure 4 (b).

IV. CONCLUSIONS

We report a new iodine-based wet etching method for black phosphorus, which could achieve an atom layer accuracy along [010] direction. The etch rate is thickness-independent, and has a linear relationship with the etchant concentration. With this method, a BP pattern with a 15 nm etch depth was successfully fabricated. DFT calculations were used to study the etch behavior around edges. Absorption energy calculations show that iodine atoms/molecules prefer to be absorbed by the layer below the edges, predicting an [010] etch direction.

ACKNOWLEDGMENT

The DFT calculations in this research were enabled by the High Performance Computing center at Southern University of Science and Technology (SUSTech), China, and Westgrid, Compute Canada.

REFERENCES

[1] Tran V., Soklaski R., Liang Y. and Yang L. "Layer-controlled band gap and anisotropic excitons in few-layer black phosphorus". Physical Review B. 2014, vol 89, pp. 235319.

[2] Chen, F., Ilatikhameneh, H., Ameen, T. A., Klimeck, G. Rahman, R. "Thickness Engineered Tunnel Field-Effect Transis-tors Based on Phosphorene", IEEE Electron Device Letters. 2017, 38, pp. 130-133

[3] Stanford M. G., Rack P. D. and Deep Jariwala. "Emerging nanofabrication and quantum confinement techniques for 2D materials beyond graphene", npj 2D Materials and Applications. 2018, vol 2:20.

[4] Tan, W., Wang, L., Feng, X., Chen, L., Huang, L., Huang, X., and Ang, K. "Recent Advances in Black Phosphorus-Based Electronic Devices", Advanced Electronic Materials. 2019, vol 5(2), 1800666.

[5] Yang, S. Zhang, K., Ricciardulli, A. G., Zhang, P. Liao, Z., Lohe, M., Zschech, E., Blom, P., Pisula, W. Mullen, K., Feng, X. "A Delam ination Strategy for Thinly Layered Defect-Free High-Mobility Black Phosphorus Flakes", Angewandte Chemie-International Edition. 2018, vol 57, 4677.

[6] Cartz, L.; Srinivasa, S. R.; Riedner, R. J.; Jorgensen, J. D.; Worlton, T. G. "Effect of pressure on bonding in black phospho-rus Locality: synthetic Note: pressures calculated from the meas-ured unit cell volume". Journal of Chemical Physics, 1979, vol 71, pp. 1718-1721

Giannozzi P., Andreussi O., Brumme T., Bunau O., Buongiorno N. M., Calandra M., Car R., Cavazzoni C., Ceresoli D., Cococcioni M., Colonna N., Carnimeo I., Dal Corso A., de Gironcoli A., Delugas P., J. DiStasio R., Ferretti A., Floris A., Fratesi G., Fugallo G., Gebauer R., Gerstmann U., Giustino F., Gorni T., Jia J., Kawamura M., Ko H., Kokalj A., Küçükbenli E., Lazzeri M., Marsili M., Marzari N., Mauri F., Nguyen L. N., Nguyen H. V., Otero-de-la-Roza A., Paulatto L., Poncé S., Rocca D., Sabatini R., Santra B., Schlipf M., Seitsonen A.P., Smogunov A., Timrov I., Thonhauser T., Umari. P, Vast N., Wu X. and S Baroni, "Advanced capabilities for materials model-ling with Quantum ESPRESSO", Journal of Physics : Con-densend Matter. 2017. 29, 465901.

978-1-7281-0941-1/19 $31.00 © 2019 IEEE

978-1-7281-0941-1/19 $31.00 © 2019 IEEE   148

OFF Current Suppression by Gate-gontrolled Strain in The N-type GaAs Piezoelectric FinFETs

Yuxiong Long[1], Jun Z. Huang[2,*], Zhongming Wei[1], Jun-Wei Luo[1], Xiangwei Jiang[1,†]

[1]Institute of Semiconductors, Chinese Academy of Sciences, Beijing 100083, China, †email: xwjiang@semi.ac.cn
[2]MaxLinear Inc., Carlsbad, CA 92008, USA, *email: junhuang1021@gmail.com

Abstract—The gate-controlled compressive strain induced by piezoelectric layers (piezo-layers) is used to suppress the OFF current of n-type GaAs piezoelectric FinFETs (Piezo-FinFETs). Quantum ballistic transport of n-type GaAs Piezo-FinFETs is modeled by the self-consistent Schrödinger–Poisson system. Our results suggest that n-type GaAs Piezo-FinFETs reduce OFF current by an order of magnitude for both high performance and low power applications compared with their counterparts without piezo-layers. The influences of device orientations on device performance is also investigated. The optimal device orientation of n-type GaAs Piezo-FinFETs is on the crystal surface (111).

Keywords—FinFET, Piezoelectric, Steep slope, strain modulation

I. INTRODUCTION

Silicon complementary metal–oxide–semiconductor (CMOS) scaling is approaching its limits, and the reduction of operating voltage (V_{DD}) become very difficult due to the fundamental thermal limit, for which the scaling of subthreshold swing (SS) is limited to 60 mV/decade [1], [2].

Therefore, as a new channel material, the III–V group compound semiconductors is introduced in which the velocity of charge carriers is much higher than in silicon. This would cause a reduction in operating voltage without a loss of performance [3].

However, n-type III-V transistors suffer from a large OFF current (I_{OFF}) leakage as a result of low effective mass. As an intriguing device concept, the Piezo-FinFETs with piezoelectric layers in gate stacks has been proposed to enhance ON current and meanwhile achieve a steeper SS [4], [5]. Furthermore, Piezo-FinFETs can be redesigned by simply changing the piezo-gate voltage to V_{DD} to suppress OFF current and also achieve a steeper SS.

In this work, n-type GaAs Piezo-FinFETs has been studied. The device transport characteristics is obtained by self-consistently solving the open-boundary Schrödinger equation and Poisson equation. An asymptotic waveform evaluation (AWE) technique combined with complex frequency hopping (CFH) is used to speed up the numerical calculation [6]. The influence of device orientations on device performance is also systematically discussed.

II. DEVICE CONCEPT AND SIMULATION APPROACH

A. Device structure

The device structure of n-type GaAs Piezo-FinFETs is shown in Fig. 1. The Fin width is 5nm and channel length is 14 nm. The PZT-5H is chosen as a piezoelectric material in this work, and the piezo-layers have a thickness of 5 nm which can sustain a 0.5 V supply voltage [7]. The piezoelectric layer is placed between the control-gate and

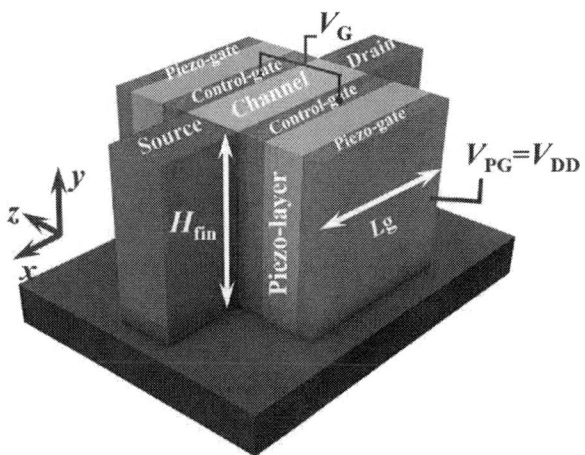

Fig. 1: Device structure of n-type GaAs Piezo-FinFETs. x is the transport direction, y is device height direction, and z is the confined and also periodic direction. The piezo-gate is set to supply voltage V_{DD}.

piezo-gate. In order to suppress OFF current, the voltage of control-gate equals gate voltage, while the voltage of piezo-gate is set to supply voltage (V_{DD}), so that there is an electric filed applied to the piezo-layers in the OFF-state, and meanwhile the piezo-layers are expanded which causes a large compressive strain to the channel in the OFF-state. Furthermore, this dynamic strain reaches its maximum in the OFF-state. However, there is no strain induced by piezo-layers in the ON-state due to the equal voltage of control-gate and piezo-gate. Therefore, I_{OFF} is suppressed that will be explained next, while I_{ON} remains the same in Piezo-FinFETs

TABLE I. MATERIAL PARAMETER

Material	Thickness	Elastic compliance constant $\left[\frac{10^{-3}}{GPa}\right]$					Piezoelectric strain constants $[10^{-12}C/N]$
		S_{11}	S_{12}	S_{13}	S_{33}	S_{44}	d_{33}
GaAs	5 nm	11.6	-3.68	Negligible	Negligible	1.67	Negligible
GaSb	5 nm	15.8	-4.94	Negligible	Negligible	23.1	Negligible
PZT-5H	5 nm	16.5	-4.78	-84.5	20.7	4.35	593

compared with their counterparts without piezo-layers.

B. Simulation approach

Some simplifying assumptions have been embraced in this work [7]: (i) T_{xx}, T_{yy} components have been assumed to be negligible compared to T_{zz} that results in a 1D stress model, because there is no directly applied stress in the x and y directions as shown in Fig. 1; (ii) the device structure repeats

978-1-7281-0941-1/19 $31.00 © 2019 IEEE

periodically in the z-axis, so that the outer PE-gates can be assumed to be mechanically firm; (iii) Stress induced by layers is uniformly distributed in the fin region. The 1D stress model is here used for simplicity, and we have previously shown that its results compare fairly well with 3D numerical simulations [7].

The gate tunable stress in the z-axis, T_{zz}, is analytically derived from the principle of piezoelectric effect as shown in Eq. (1) [7], where $D \rightarrow C$ is the rotation matrix from the Device Coordinate System (DCS) to the Crystal Coordinate System (CCS), and W and S is the abbreviation of width and elastic compliance constant respectively.

The 3D FinFETs structure is simplified to a quasi-2D one to save computation time. To further accelerate the simulation, asymptotic waveform evaluation (AWE) technique combined with complex frequency hopping (CFH) is used in this work. The accuracy and efficiency of this method have been demonstrated by simulation of a triple gate MOSFET in the presence of surface roughness [6]. Moreover, material parameters used in this work are listed in Table I [5], [8].

$$T_{zz} = -d_{33}V_{gs}\left[\frac{1}{0.5W_{fin}R_{D\rightarrow C}^{-1}S_{fin}R_{D\rightarrow C}+W_{ox}S_{ox,11}+W_G S_{G,11}+W_{pie}S_{pie,33}}\right]$$

(Eq.1).

III. RESULT AND DISCUSSION

A. OFF current suppression

The polarization direction of piezo-layers is chosen to introduce a large compressive strain along the z-axis to the channel in the OFF-state. The variations of gate-controlled stress (T_{zz}) and channel conduction band minimum (CBM) along with control-gate voltage are shown in Fig. 2. It is found that a compressive strain applied to the channel of n-type GaAs Piezo-FinFETs lifts its channel CBM and the variations of both stress and channel CBM reach their maximum in the OFF-sate, namely $V_{GS} = 0$ V. An upward shift of the channel CBM results in a raise of the top of the barrier for electron injection in the OFF-state, which causes an OFF current suppression.

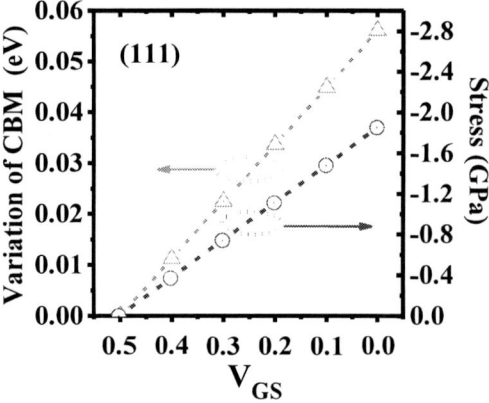

Fig. 2: the variations of stress introduced by piezo-layers and conduction band minimum (CBM) along with control-gate voltages (V_{GS}) for the optimal device orientation on the crystal surface (111).s

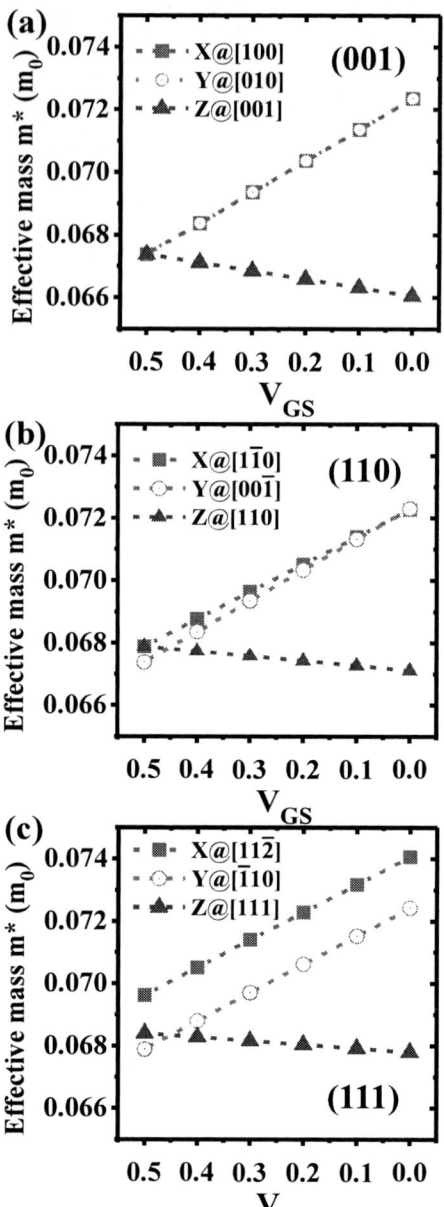

Fig. 3: The dependences on control-gate voltages (VGS) of Effective masses m* for different device orientations. (a): (001); (b): (110); (c): (111); Note that (001) means that the channel surface perpendicular to z-axis is the crystal surface (001) and the transport direction is along the crystal orientation [100].

The variations of effective mass (m*) along with control-gate voltages for different surface orientations are shown in Fig.3 (a), (b), and (c), where (001) indicates that the channel surface of n-type GaAs Piezo-FinFETs perpendicular to z-axis is the crystal surface (001), and the transport direction is along the crystal orientation [100]. What's more, the transport effective mass along the x-axis marked by solid green square increases for all three different surface orientations in the OFF-state, which also helps to reduce the OFF-state current.

978-1-7281-0941-1/19 $31.00 © 2019 IEEE   150

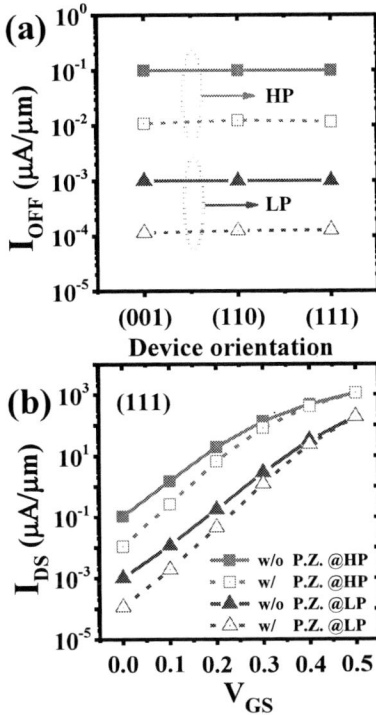

larger degradation of SS than for HP applications, and the SS improvement of n-type GaAs Piezo-FinFETs for HP applications is larger than for LP applications. Our simulations suggest that n-type GaAs Piezo-FinFETs achieve about 14 *mv/decade* SS improvement for HP applications and 10 *mv/decade* SS improvement for LP applications in the optimal device orientation on surface (111). Moreover, the influence of device orientations on OFF current is not so obvious.

IV. CONCLUSION

Piezo-FinFETs has been redesigned by simply changing the piezo-gate voltage to V_{DD} to introduce a large compressive into channel in the OFF-state. This gate-controlled compressive strain induced by expanded piezoelectric layers lifts channel conduction band edge and meanwhile increases the transport effective mass in the OFF-state, which suppresses the OFF current and meanwhile improves the SS of n-type GaAs Piezo-FinFETs. Our results show that n-type GaAs Piezo-FinFETs achieve an OFF current reduction by an order of magnitude and a SS improvement by 14 *mv/decade*. The device orientation on surface (111) has the smallest SS, While the influence of device orientations on I_{OFF} is very small.

ACKNOWLEDGMENT

This work was supported by China Key Research Development Program (2018YFA0306101), NSFC (Nos. 11774338, 11574304), CAS-PKU Pioneer Cooperation team, and the Youth Innovation Promotion Association CAS.

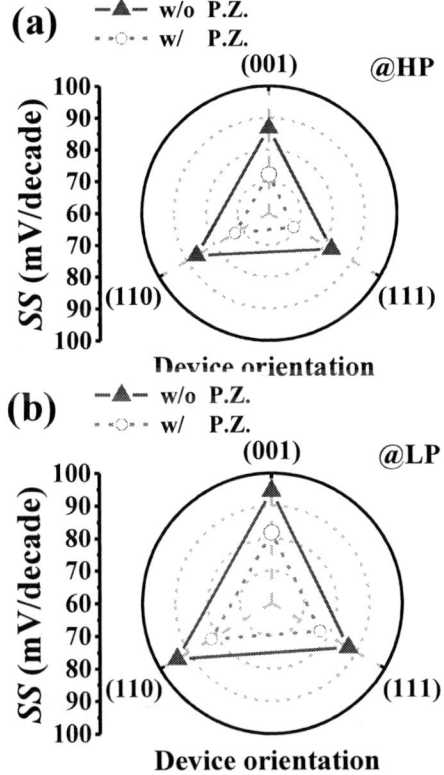

Fig. 5: The subthreshold swing (SS) of n-type GaAs Piezo-FinFETs (red empty circle) and their counterparts without piezo-layers (blue solid triangle). (a): for high performance applications; (b): for low power applications.

Fig. 4: (a) shows the OFF current of n-type GaAs Piezo-FETs (dash lines) and their counterparts without piezo-layers are compared (solid line). The OFF current of their counterparts without piezo-layers is fixed to 10^{-1} μ A/μm and 10^{-3} μ A/μm respectively for both high performance (red solid lines) and low power (blue solid lines) applications. $I_{DS} - V_{GS}$ curves of n-type GaAs Piezo-FinFETs (dash lines) for both high performance and low power applications are shown in (b), and their counterparts (solid lines) without piezo-layers are also compared. Note that the optimal device orientation on surface (111) is selected.

The OFF current suppression of n-type GaAs Piezo-FinFETs is shown in Fig. 4 (a) marked by dash lines, where the OFF current of their counterparts without piezo-layers is marked by solid lines and targeted at 10^{-1} μA/μm and 10^{-3} μA/μm respectively for high performance (HP) and low power (LP) applications.

Our results suggest that the n-type GaAs Piezo-FinFETs reduce the OFF current by an order of magnitude for both high performance and low power applications. The corresponding I_{DS}-V_{GS} curves are drawn in Fig. 4 (b) for the optimal device orientation on surface (111). It is obviously observed that n-type GaAs Piezo-FinFETs results in a large reduction of I_{OFF} and meanwhile a steeper SS.

B. Device orientation dependence

The subthreshold swing of n-type GaAs Piezo-FinFETs for different device orientations is shown in Fig. 5, where (a) is for high performance applications and (b) is for low power applications, and their counterparts without piezo-layers are also compared. The device orientation on surface (111) has the smallest SS for both HP and LP applications due to its largest transport effective mass shown in Fig. 3. However, n-type GaAs Piezo-FinFETs for LP applications suffer from a

REFERENCES

[1] J. Meindl, Q. Chen, and J. Davis, "Limits on silicon nanoelectronics for terascale integration," Science, vol. 293, no. 5537, pp. 2044–2049, Sept. 2001.

[2] A. M. Ionsecu and H. Riel, "Tunnel field-effect transistors as energy efficient electronic switches," Nature, vol. 479, pp. 329–327, Nov. 2011.

[3] J. A. Alamo, "Nanometre-scale electronics with $III - V$ compound semiconductors" Nature, vol. 479, pp. 317-323, 2011.

[4] T. V. Hemert and R. J. E. Hueting, "Active strain modulation in field effect devices," in Proc. 42nd Eur. Solid-State Device Res. Conf., Sep. 2012, pp. 125–128.

[5] T. V. Hemert and R. J. E. Hueting, "Piezoelectric strain modulation in FETs," IEEE Trans. Electron Devices, vol. 60, no. 10, pp. 3265–3270, Oct. 2013.

[6] J. Z. Huang, W. C. Chew, M. Tang, and L. Jiang, "Efficient simulation and analysis of quantum ballistic transport in nanodevices with AWE," IEEE Trans. Electron Devices, vol. 59, no. 2, pp. 468–476, Feb. 2012.

[7] H. Wang, X. Jiang, N. Xu, G. Han, Y. Hao, S. Li, and D. Esseni, "Revised Analysis of Design Options and Minimum Subthreshold Swing in Piezoelectric FinFETs," IEEE Electron Device Lett., vol. 39, no. 3, pp. 444–447, Mar. 2018.

[8] I. Vurgaftman, J. R. Meyer, and L. R. Ram-Mohan, "Band parameters for III-V compound semiconductors and their alloys," J. Appl. Phys., vol. 89, p. 5815, 2001.

Numerical Investigation of the Photogating Effect in MoTe₂ Photodetectors

J.M. Gonzalez-Medina[†‡], E.G. Marin[†*], A. Toral-Lopez[†‡], F. G. Ruiz[†‡], A. Godoy[†‡]

[†]Dpto. Electrónica, Universidad de Granada. Av. Fuentenueva S/N, 18071, Granada, Spain
[‡]Pervasive Electronics Advanced Research Laboratory, CITIC, Universidad de Granada, 18071, Granada, Spain
[*]Dipartamento di Ingegneria dell'Informazione, Università di Pisa, 56122 Pisa, Italy.
Email: jmgonzalme@ugr.es

Introduction

The necessity of overcoming the limitations (e.g. weight, cost and brittleness) of traditional bulk semiconductors employed to build conventional photodetectors, has fueled the interest of the scientific community towards two-dimensional crystals. Its most representative member, graphene [1], with outstanding electrical and mechanical properties, has however a severely limited photoresponsivity due to 1) the lack of bandgap and 2) a reduced carrier lifetime that hardly reaches a few picoseconds [2]. Greater expectations lay on Transition Metal Dichalcogenides (TMDs) [3], the bandgap of which is sensitive to the number of layers. Moreover, TMDs can be stacked forming vertical or lateral heterojunctions [4] giving rise to structures similar to field-effect transistors (FETs) that behave as photodetectors. In this work we theoretically study the optoelectronic properties of a back-gated phototransistor, with its channel formed by few-layer MoTe₂, and we focus on the role played by the charges trapped at the channel-insulator interface through the photogating effect [5,6].

Device description and simulation

The device considered here is inspired by the experimental realization described in [7] and is schematically depicted in Fig. 1. The p-type, 8.4 nm-thick MoTe₂ lays on top of a bulk SiO₂ substrate (280 nm thick), back-gated by a p-type doped Si. During the fabrication process, impurities, defects and imperfections can be located at the interface between the SiO₂ substrate and the MoTe₂, resulting in the presence of a noticeable density of interface traps that are also included in the numerical model. The few-layer MoTe₂ flake is ohmically contacted by two metal electrodes acting as source and drain, respectively.

To analyze this structure we have used the SAMANTA code suite [8], which solves self-consistently the 2D Poisson and Drift-Diffusion equations including the effect of light-induced generation, Shockley-Read-Hall recombination and interface and/or bulk traps. The bandgap of MoTe₂ has been set to 0.8 eV, corresponding to its bulk form, with an acceptor doping density of $N_A = 1.5 \times 10^{18}$ cm⁻³. Electron and hole mobilities have been fixed to 0.3 cm²/Vs and 5.9 cm²/Vs, and effective masses to $0.5m_0$ and $0.6m_0$, respectively [7]. The dielectric constant of MoTe₂ is set to $10.4\varepsilon_0$. In order to analyze the impact of the photogating effect in 2D-based photodetectors we have considered the presence of hole deep traps following a Gaussian energetic distribution centered at mid-gap and with standard deviation $\sigma = 0.05$ eV. The value of the maximum trap density N_{max} is varied in order to evaluate its effect. The traps are spatially located at the interface between the insulator substrate and the MoTe₂ layer (see Fig. 1).

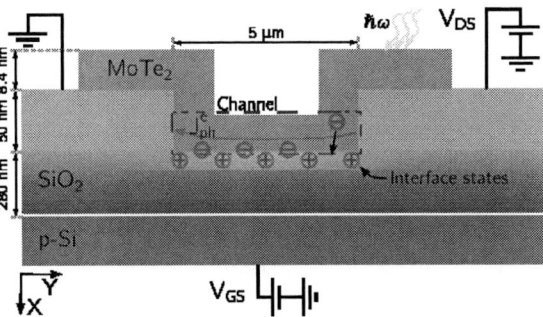

Figure 1 Schematic of the MoTe₂ phototransistor with a layer of interface states located between the MoTe₂ and the SiO₂.

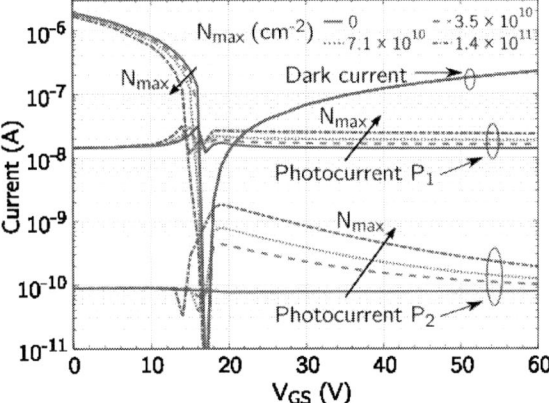

Figure 2 Dark current and photocurrent *vs.* gate voltage for different values of N_{max}, at $V_{DS} = 1$ V, for light power densities: $P_1 = 1$ W/cm² and $P_2 = 6$ mW/cm².

We have considered two light power densities, $P_1 = 1$ W/cm² and $P_2 = 6$ mW/cm², so to analyze the traps role and the photogating effect in the 2D-based photodetector under different illumination conditions. The resulting transfer characteristics of the photodetector can be seen in Fig. 2, where the dark current is plotted together with the photocurrents achieved for P_1 and P_2 light power densities, and for different N_{max} values. The influence of the traps is negligible in dark conditions. However, when the device is illuminated, they increase significantly the photocurrent due to the so-called photogating effect [9,10]; in particular, for the lower value of the light power density. In this phenomenon, the trapped holes act as a local gate that augments the channel conductance, resulting into a photoconductive gain. When a high light power density (P_1) is considered, the photocurrent is mainly due to the photoconductive effect, being only slightly enhanced due to the photogating effect. In both high and low light power density cases, the photocurrent remains independent of N_{max} for $V_{GS} < 10$ V, as the hole traps only favor the electron conductivity.

978-1-7281-0941-1/19 $31.00 © 2019 IEEE

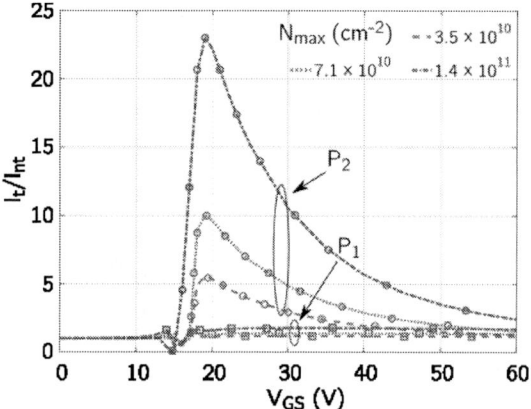

Figure 3 Ratio of the photocurrent calculated for different N_{max} values (I_t) with respect to the case without traps (I_{nt}) for P_1 and P_2 vs. V_{GS}.

In the scenario of lower light power density, the contribution of the photoconductive current is smaller due to the reduced amount of photogenerated carriers, and the photogating effect becomes dominant. This contribution is especially remarkable for gate voltages higher than 10 V, where the electron conductivity increases and holes are not strongly repelled from the lower part of the channel by the gate voltage, increasing the probability to be trapped. The relative role of the photogating on the total photocurrent is better appreciated from Fig. 3, that depicts the ratio between the photocurrent when interface traps are considered I_t and when they are not I_{nt}. For the P_2 case, the photocurrent can increase more than one order of magnitude due to the effect of the hole traps. For P_1, on the contrary, the photoconductive contribution is dominant and the effect of the traps is much more smaller. The effect of the charged traps can be better understood in Fig. 4, where the electron and hole photocurrent distributions in the channel are plotted (see inset) for two illumination power densities. The hole current density flows in a similar fashion regardless the light power density. However, the electron photocurrent distribution is different for P_1 and P_2. For P_2 it mainly flows in a sharp region close to the charged interface, due to the photogating effect. On the other hand, for the higher light power density, P_1, the photoconductive effect is dominant due to the large density of photogenerated carriers and the current flow is more homogeneously distributed.

Conclusions

The optoelectronic properties of a MoTe$_2$ phototransistor have been analyzed making use of detailed numerical simulations. The different contributions to the photocurrent has been assessed as a function of the light power density and the hole trap density. In particular, for high light power densities the photoconductive effect dominates the photocurrent, whereas for low power densities the photogating effect has demonstrated its importance, enhancing the overall photocurrent more than one order of magnitude compared with the situation where a perfect interface is assumed.

Acknowledgment

This work has been partially supported by the project TEC2017-89955-P (MINECO/AEI/FEDER, EU) and the grants FPU014/02579 and FPU016/04043. E. G. Marín acknowledges Juan de la Cierva Incorporación IJCI-2017-32297 (MINECO/AEI).

References

[1] A. K. Geim et al. Nature Materials, 6(3):183-191, 2007.
[2] M. M. Furchi et al. Nano Letters, 14, 6165−6170, 2014.
[3] Yi Ding et al. Physica B, 406:2254-2260, 2011.
[4] A. K. Geim et al. Nature, 499:419-425, 2013.
[5] M. Buscema. Chem. Society Rev., 44(11):3691, 2015.
[6] F. H. L. Koppens et al. Nat. Nano., 9(10):780-793, 2014.
[7] H. Huang et al. Nanotechnology, 27, 445201, 2016.
[8] S. Riazimehr et al. ACS Photonics, 6(1):107-115, 2019.
[9] H. Fang et al. Advanced Science, Wiley, 4, 1700323, 2017.
[10] B. Miller et al. Applied Physics Letters, 106, 122103, 2015.

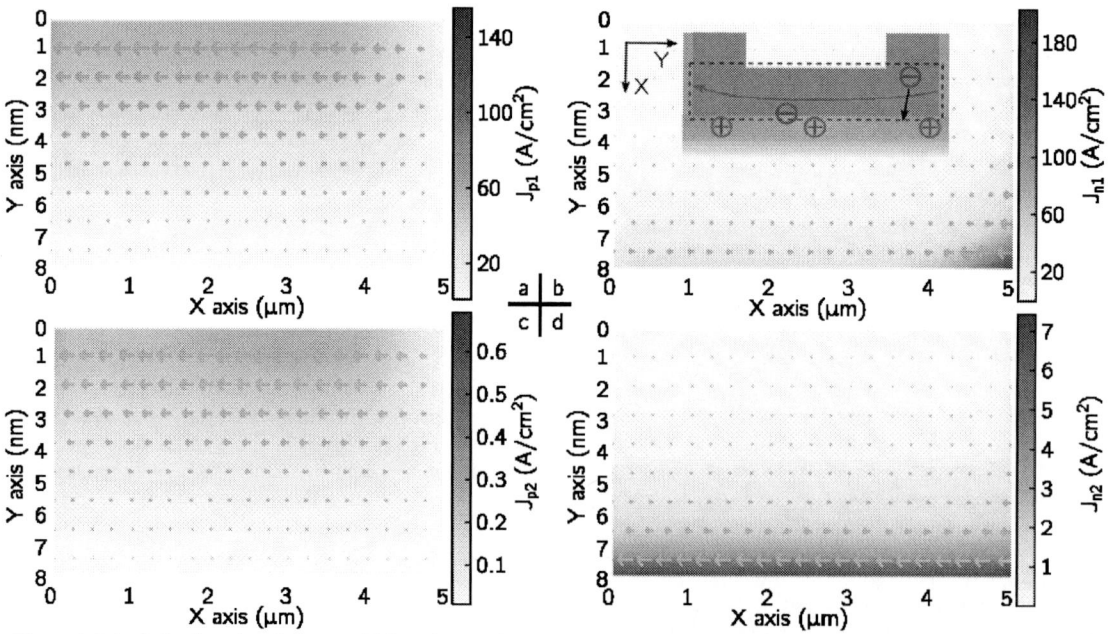

Figure 4 Hole (left column) and electron (right column) photocurrent density distribution in the channel region for $P_1 = 1\,W/cm^2$ (top) and $P_2 = 6\,mW/cm^2$ (bottom), $V_{DS} = 1\,V$ and $V_{GS} = 19V$. Inset: schematic of the device illustrating the zoomed region and the photogating effect .

Physical Insights into the Transport Properties of RRAMs Based on Transition Metal Oxides

Toufik Sadi[1], Oves Badami[2], Vihar Georgiev[2], Jie Ding[3] and Asen Asenov[2]

[1]*Engineered Nanosystems Group, School of Science, Aalto University, PO Box 12200, 00076 AALTO, Finland*
[2]*School of Engineering, Electronic and Nanoscale Engineering, University of Glasgow, Glasgow G12 8LT, Scotland, UK*
[3] *College of Electrical and Power Engineering, Taiyuan University of Technology, 030024 China.*
toufik.sadi@aalto.fi

Abstract—Nowadays, resistive random-access memories (RRAMs) are widely considered as the next generation of non-volatile memory devices. Here, we employ a physics-based multi-scale kinetic Monte Carlo simulator to study the microscopic transport properties and characteristics of promising RRAM devices based on transition metal oxides, specifically hafnium oxide (HfO_x) based structures. The simulator handles self-consistently electronic charge and thermal transport in the three-dimensional (3D) space, allowing the realistic study of the dynamics of conductive filaments responsible for switching. By presenting insightful results, we argue that using a simulator of a 3D nature, accounting for self-consistent fields and self-heating, is necessary for understanding switching in RRAMs. As an example, we look into the unipolar operation mode, by showing how only the correct inclusion of self-heating allows the proper reconstruction of the switching behaviour. The simulation framework is well-suited for exploring the operation and reliability of RRAMs, providing a reliable computational tool for the optimization of existing device technologies and the path finding and development of new RRAM options.

Index Terms—Kinetic Monte Carlo (KMC), resistive random-access memories (RRAMs), multi-scale models, transport phenomena.

I. Introduction

For several decades, the semiconductor industry experienced a strong growth, thanks to device downscaling, leading to increased functionality and performance. However, as this miniaturization trend is maintained and Moore's law is approaching its limits, undesirable effects, such as excessive power dissipation and self-heating, hinder the performance of microchips. This has forced the industry to re-evaluate the von-Neumann architecture by moving towards in-memory computing. In this paradigm shift, devices based on resistive random access memories (RRAMs) are expected to play an important role, which necessitates the development of advanced physics-based simulators to understand better RRAM operation and provide optimal device designs.

The idea of memristor devices, such as RRAMs, was put forward theoretically almost 50 years ago [1]. Since their experimental demonstration 11 years ago [2], the interest in RRAMs has been increasing exponentially [3]–[5], being considered as the next generation of non-volatile memories.

The research is funded by the EPSRC (UK), under grants no. EP/S000224/1 and no. EP/S001131/1.

Indeed, the 'International Technology Roadmap for Semiconductors' (ITRS) cites a multitude of incentives for developing RRAMs, such as low cost and power dissipation, high endurance and three-dimensional (3D) crossbars integration [6]. The applications of RRAMs are also innumerable, ranging from high-density memories and novel processor architectures to neuromorphic computing and artificial intelligence [4].

In this work, we analyze the switching behaviour and certain interesting features of RRAM structures based on hafnium oxide (HfO_x), using a kinetic Monte Carlo (KMC) simulation framework. In Sec. II, we discuss the main attributes of the simulator and describe its original aspects. In Sec. III, we discuss the basic switching behaviour of the simulated devices, and highlight the importance of including coupled electro-thermal transport to capture correctly switching.

II. Simulation Methodology

Most previous work on the simulation of RRAMs relied mostly on phenomenological models, such as the resistor breaker network [5], [7], which do not account accurately for self-heating and self-consistent fields. In addition, most existing models use two-dimensional (2D) approximations [9], [10] which may produce less reliable and insightful results [11]. The 3D KMC simulator used in this work is capable of providing a complete picture of particle dynamics in oxide based RRAMs. It incorporates several features that distinguish it from established phenomenological models [5], [9], [10], as discussed in Ref. [3].

We employ an in-house 3D device simulator, which has been previously used for gaining insight into the operation of SiO_x structures [3], [8], to study HfO_x-based RRAMs, a widely used transition metal oxide (TMO) in memristor technology. Hafnia is highly suitable for high-density CMOS integration due to their high dielectric constants. Figure 1(a) illustrates the simulation framework. Unlike previously used 2D and phenomenological models [5], [9], [10], our simulator uses a powerful combination of tools, describing accurately electron-ion interactions and reconstructing realistically the electroforming and rupture of conductive filaments in the 3D real space. It couples, in a self-consistent manner, electron and oxygen ion KMC trajectory simulations to the electric field and temperature distributions determined from the solution of Poisson's and the time-dependent heat diffusion equations.

978-1-7281-0941-1/19 $31.00 © 2019 IEEE

(a)

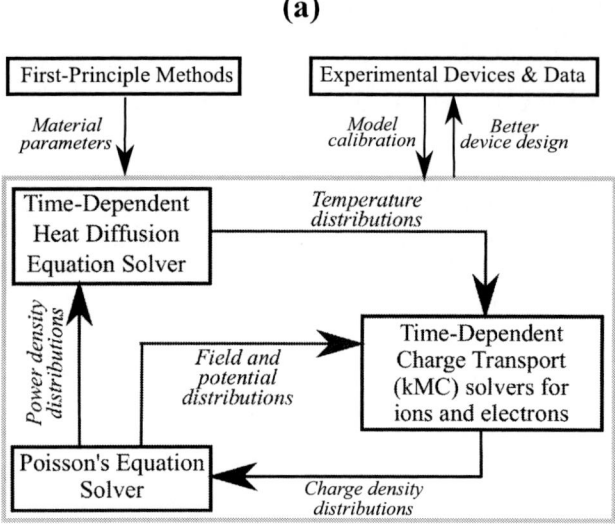

(b)

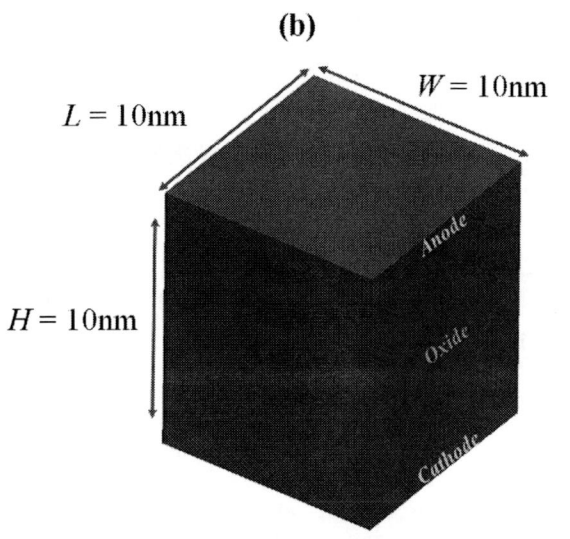

Fig. 1. (a) The simulation framework, coupling the KMC description of charge transport to the local temperature and electric field distributions in the oxide. Relevant material parameters, e.g. the activation energy, are obtained using first-principle methods. In general, the simulator is calibrated with experiments for enhanced predictivity power. (b) The simulated two-terminal RRAM structure, consisting of an oxide (HfO$_x$ in this case) volume (thickness $T = 10$nm) sandwiched between the cathode and the anode. Realistic experimental structures may have electrode areas as large as 100μm$\times 100\mu$m [14], but it is sufficient to limit our study to a small contact area ($L \times W$=10nm$\times$10nm here), to minimize computational cost.

The dynamic nature of the vacancy formation and annihilation, and electron trapping is considered accurately, as discussed rigorously in Refs. [3], [8], [12]. The ion and vacancy time-dependent dynamics (drift, diffusion, generation and recombination) are also modeled carefully, as discussed in Refs. [3], [8]. The effect of all the dominant electron transport mechanisms are carefully considered, including trap-assisted tunneling, trap-to-trap tunneling, Fowler-Nordheim tunneling,

Poole-Frenkel emission, and direct tunneling mechanisms [3]. Electron and oxygen ion movements as well as ion-vacancy generation and recombination events are tracked down in time via the stochastic KMC algorithm, providing a realistic picture of the interplay between electrons, ions and vacancies as influenced by the evolving local electrostatic and temperature effects. More details about the simulation methodology and the included physical processes are given in Ref. [3].

III. RESULTS AND DISCUSSION

A. Simulated Structure and Practical Considerations

Here, we illustrate how 3D electrothermal modelling, accounting for self-heating effects and self-consistent fields, as neglected in other KMC simulation models (see for example Ref. [9]), provides a deep physical insight into RRAM switching. The studied devices simply consist of the oxide (HfO$_x$ in this case) layer (thickness$\sim$ 10nm) located between two electrodes (the cathode and the anode), as illustrated in Fig. 1(b). As discussed extensively in literature, the memristive behaviour of oxide-based RRAMs is a direct consequence of the forming and destruction of conductive filaments, which are formed by direct electrical conductive paths between the cathode and the anode [3], [8], [13]. These filaments are created by the generation of oxygen ion-vacancy pairs, whose rates and transport are in general governed by the electric field (potential) and temperature distribution within the oxide. While the experimental structures used to validate the simulator may have electrode areas as large as 100μm$\times 100\mu$m [14], the simulations can be limited to a small contact area ($L \times W$=10nm$\times$10nm here), which can represent a region incorporating e.g. a grain boundary. This is common practice in Monte Carlo modeling methods, aiming to reduce the computational cost while allowing reasonable numerical simulation accuracy [3].

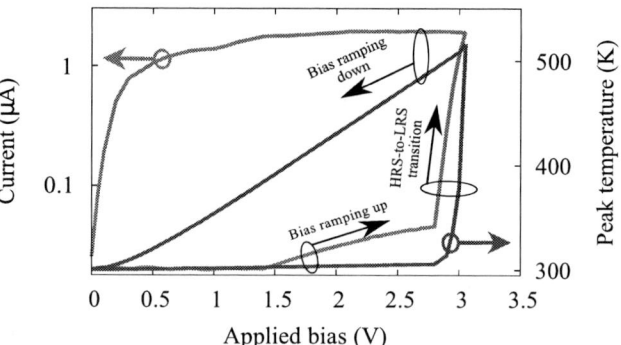

Fig. 2. The $I-V$ and peak temperature curves, as bias is ramped up towards the CF forming and then lowered down to 0V.

B. Basic Characteristics

Figure 2 shows the $I - V$ and peak temperature curves obtained during the electroforming of the conductive filament, using an electric current compliance limit of 2μA. Figure 3 shows the distribution of the oxygen vacancies created, and

978-1-7281-0941-1/19 $31.00 © 2019 IEEE   156

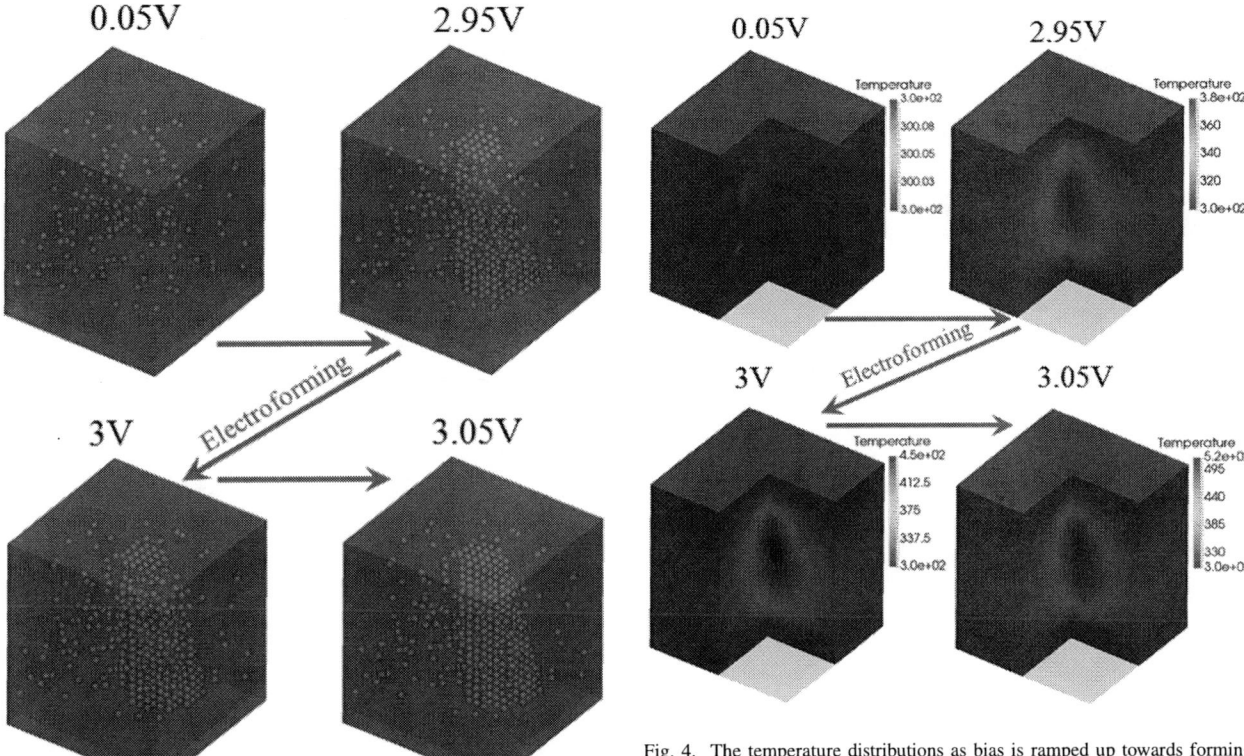

Fig. 3. The generated vacancy distributions as bias is ramped up towards forming.

Fig. 4. The temperature distributions as bias is ramped up towards forming.

Fig. 4 shows the corresponding local temperature distributions, as bias is increased and the CF is gradually created. Figure 2 illustrates the expected memristive characteristics of the RRAM device. Figure 3 highlights the three-dimensional nature of conductive filaments. At biases below 2.5V, very few vacancies are generated. As bias is increased, more vacancies are generated and filament seeds start to appear (e.g. at 2.9V). Such seeds start to grow as bias is further increased. At around 3V, an accelerated generation of vacancies occurs, leading to the creation of a full conductive filament, linking both electrodes; at this condition, percolation paths are created, as an abrupt jump in the device current is observed.

Figure 4 shows how the oxide temperature reaches values beyond 500K during the electroforming process. The peak temperature tends to occur within the filament volume, where the combination of the elevated current densities and the low oxide thermal conductivity can lead to such very high values. In general, the elevated local temperatures, resulting mainly from Joule heating, affect significantly the device behavior, as they can boost the probability of vacancy generation and ion hopping, but also electron transport via trap-assisted tunneling and other relevant mechanisms [3], [8], [12], [15].

C. Switching and Self-Heating

The critical role of device self-heating can be illustrated by looking into the reset process considering the unipolar RRAM switching mode. Unlike the bipolar switching mode, where the bias is further reduced, from 0V to negative values, to realize the reset process (after the CF is formed) [3], this regime is achieved in the unipolar mode by increasing the bias from 0V to positive values. Figure 5 shows the oxygen vacancy and temperature distributions just before and after the filament is ruptured, at a bias voltage of around 2.4V, during the reset process for a unipolar mode. It can be seen that the fully-conductive filament is broken near the top surface (near the anode), resulting in the device switching from a high-current (ON) low-resistance state (LRS) to a low current (OFF) high-resistance state (HRS). As expected, the ON-to-OFF transition also results in the peak temperature dropping considerably, as current densities are reduced. The RRAM device experiences such transition because the oxygen ions in the anode contact move back to the oxide volume, recombining with the nearby vacancies and breaking the CF. This phenomenon occurs thanks to ion diffusion, which is facilitated by the elevated temperatures in the oxide just before the transition. It has been verified that such transition cannot be easily reconstructed, using simulations, in the unipolar RRAM operation mode without the inclusion of thermal effects. Self-heating in the bipolar switching mode is not of critical importance, as ions easily drift back to the oxide as a certain reset bias (field) is

978-1-7281-0941-1/19 $31.00 © 2019 IEEE

reached.

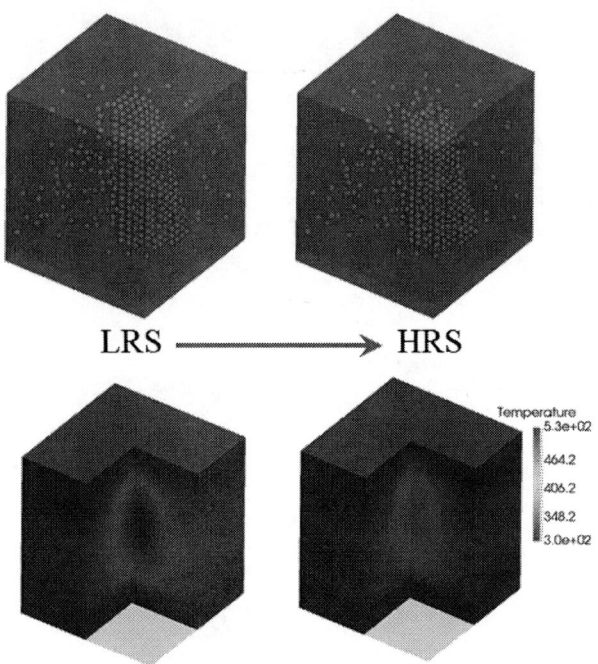

Fig. 5. The vacancy and temperature distributions, in the reset process for a unipolar mode, just before and after the CF is ruptured (bias ~ 2.4V).

IV. OUTLOOK

We applied a kinetic Monte Carlo simulation framework to study the operation and switching physics of hafnia-based RRAM devices. We discussed the need for using 3D models accounting for self-heating and self-consistent fields to capture carefully the expected switching behaviour. In addition to exploring RRAM physics and operation, the model can be applied to investigate reliability issues and bottlenecks of RRAM technology development, such as the design of reliable oxygen storage or supply systems to increase the RRAM reliability and endurance. Physics-based modelling is only the initial step in designing high-performance RRAMs. Data from the physical simulation of experimental devices will contribute to the development of analytical (compact) models, which will be integrated into circuit simulators to design cross-bar circuits for interesting applications.

REFERENCES

[1] L. Chua, "Memristor-the missing circuit element," IEEE Trans. Circuit Theory, vol. 18, pp. 507–519, 1971.
[2] D. B. Strukov, G. S. Snider, D. R. Stewart, and R. S. Williams, "The missing memristor found," Nature, 453, pp. 80–83, 2008.
[3] T. Sadi et al., "Investigation of resistance switching in SiOx RRAM cells using a 3D multiscale kinetic Monte Carlo simulator," Jour. Phys.: Condens. Matter, 30, pp. 084005, 2018.
[4] A. Mehonic and A. J. Kenyon, "Emulating the electrical activity of the neuron using a silicon oxide RRAM cell," Front. Neurosci., 10, pp. 1–10, 2016.
[5] S. C. Chae et al., "Random circuit breaker network model for unipolar resistance switching," Adv. Mater., 20, pp. 1154–1159, 2008.

[6] The ITRS Report 2013, http://www.itrs2.net/2013-itrs.html. Accessed 5 Jun 2018.
[7] S. Brivio and S. Spiga, "Stochastic circuit breaker network model for bipolar resistance switching memories," J. Comput. Electron., vol. 16, pp. 1154–1166, 2017.
[8] T. Sadi et al., "Advanced Physical Modeling of SiO$_x$ Resistive Random Access Memories," In Proc. SISPAD, pp. 149–152, 2016.
[9] S. Yu, X. Guan, and H.-S. P. Wong, "On the stochastic nature of resistive switching in metal oxide RRAM: physical modeling, Monte Carlo simulation, and experimental characterization," In: 2011 IEEE Int. Electron Devices Meeting (IEDM), p. 17.3.1., 2011.
[10] S. Kim et al., "Physical electro-thermal model of resistive switching in bi-layered resistance-change memory," Sci. Rep., 3, p. 1680, 2013.
[11] T. Sadi and A. Asenov, "Microscopic KMC Modeling of Oxide RRAMs," In: Nikolov G., Kolkovska N., Georgiev K. (eds) Numerical Methods and Applications. NMA 2018. Lecture Notes in Computer Science, vol. 11189. Springer, Cham, pp. 290–297, 2019.
[12] L. Vandelli et al., "A physical model of the temperature dependence of the current through SiO$_2$/HfO$_2$ stacks," IEEE Trans. Electron Devices, 58, pp. 2878–2887, 2011.
[13] M. Buckwell, L. Montesi, S. Hudziak, A. Mehonic, A. J. Kenyon, "Conductance tomography of conductive filaments in intrinsic silicon-rich silica RRAM," Nanoscale, vol. 7, pp. 18030–18035, 2015.
[14] A. Mehonic et al., "Resistive switching in silicon sub-oxide films," J. Appl. Phys., vol. 111, p. 074507, 2012.
[15] G. C. Jegert, "Modeling of leakage currents in high-k dielectrics," PhD Dissertation, Tech. Univ. Munich, Germany, 2011.

Molecular Dyanamics Simulation of Thermal Chemical Vapor Deposition for Hydrogenated Amorphous Silicon on Si (100) Substrate by Reactive Force-Field

1st Naoya Uene
Graduate school of engineering
Tohoku university
Sendai, Japan
uene@nanoint.ifs.tohoku.ac.jp

2nd Takuya Mabuchi
Frontier research institute for
interdisciplinary sciences
Tohoku university
Sendai, Japan
mabuchi@tohoku.ac.jp

3rd Masaru Zaitsu
Rsearch & development
Japan advanced chemicals ltd.
Sagamihara, Japan
masaru.zaitsu@japanadvancedchemical
s.com

4th Shigeo Yasuhara
Rsearch & development
Japan advanced chemicals ltd.
Sagamihara, Japan
shigeo.yasuhara@japanadvancedchemi
cals.com

5th Takashi Tokumasu
Institute of fluid science
Tohoku university
Sendai, Japan
tokumasu@ifs.tohoku.ac.jp

Abstract—We calculate a deposition process of hydrogenated amorphous silicon (a-Si:H) films on a silicon (100) substrate by reactive force-field molecular dynamics simulations. The influences of (a) substrate temperatures and (b) coverage of hydrogen atoms on the substrate on the adsorption probability are investigated, and it is found out that (a) the adsorption probability is almost constant for SiH_2 and SiH_3, but decrease with increase in the substrate temperature for SiH_4, (b) it decreases with the increase in hydrogen coverage.

Keywords—Chemical Vapor Deposition, Reactive Force-Field Molecular Dynamics Simulation, Reactive Sticking Coefficient, Surface-Covering Bonded Hydrogen, Surface Reaction

I. INTRODUCTION

Chemical vapor deposition (CVD) is one of the common and powerful methods to form a high-quality thin film for advanced semiconductor devices. The film properties strongly depend on the process parameters such as substrate temperature, gas composition, gas pressure, and deposition time etc., therefore, understanding these influences is essential to obtain the desired physical properties. However, deposition phenomena in CVD process are complicated and difficult to understand, and trial and error by experiments are repeated. Predicting the microstructure and properties of thin films using a simulation is one of the most challenging project in material science [1][2]. The CVD process simulation should include all physical phenomena (mass transport) and chemical reactions (gas phase reaction and surface reaction) to analyze the phenomena over wide length and time scale accurately. Fast elementary processes (e.g. adsorption and diffusion of atoms on the surface) occur $10^{-12} - 10^{-15}$ s and involve displacements of approximately 10^{-10} m. The time for the deposition of approximately 10^{-6} m thick in reaction chamber of approximately 10^{-1} m in diameter requires usually $10^2 - 10^4$ s [1]. It is not practical to simulate the development of film formation with capturing microscopic scale behavior because there are large gap between them even regarding the ever-increasing power of computer.

Multiscale modeling strategies are now becoming available to simulate entire CVD processes. Multiscale modeling strategies are realized by coupling together individual models at different length scales, i.e. the reactor scale model (macroscale) and the feature scale model (microscale).

1) Reactor scale model: The reactor scale model solves the governing equation of mass, momentum, heat, species transport to compute the flow, temperature, and species concentrations as functions of positions and time. Recent advances of computer performance have made it possible to solve it using a computational fluid dynamics (CFD) code.

2) Feature scale model: Various models have been proposed as a feature scale model [3][4][5]. Two approaches have been adapted (i.e. numerical and stochastic approach).

a) numerical approach: The numerical approach consists of ballistic model, a surface model, and a profile evolution algorithm. The transport of species has been dealt with by numerically solving continuum ballistic models when $Kn > 1$, continuum diffusion models when $Kn \ll 1$, and Boltzmann equations when $Kn \sim 1$, depending on the Knudsen number (Kn). For the surface model, empirical expressions [6] for the surface reaction rate have been exploited. This approcch is not universal because it is necessary to change the equation empirically depending on the composition of the gas.

b) Stochastic approach: Monte Carlo (MC) based models have been used for the transport of species and their interaction with the surface when $Kn > 1$. The concentrations of the various species, then the corresponding impingement rates at the surface, are estimated from a reactor scale model. The impingement rates are used as a input in feature scale model, which tracks surface events at the atomic scale (e.g. adsorption, desorption, reflection, and relaxation of surface atoms). As a result of MC simulation, the reactive sticking coefficient (RCS) can be obtained from statistical averages

978-1-7281-0941-1/19 $31.00 © 2019 IEEE

for the reactivity of the various species. The RSCs are fed back into the reactor scale model and the process is repeated until convergence is obtained. In general, this approach has a larger computational cost than the numerical one because of time consuming MC simulations. It may become a more promising approach by improving the computer performance in the future because it can takes into account molecular knowledge on the surface. However, surface reconstructions and multiple bonds between atoms were ignored because of the hypothesis [7].

We consider a deposition process of hydrogenated amorphous silicon (a-Si:H) films on a silicon (100) substrate by the thermal CVD using SiH_4 as a precursor. a-Si:H are highly expected as potential materials applicable to electronic or optoelectronic thin-film devices such as thin-film transistors (TFT), photoreceptors, color sensors, and solar cells. The surface is covered with many hydrogen atoms before and during deposition processes. Understanding relationships between the structure, hydrogen coverage, morphology, and reactivity is very important as mentioned in many papers [8][9][10]. Here, we investigated influence of hydrogen atoms covering the substrate as a hydrogen coverage θ and the substrate temperature T_S on the adsorption behaviors for each chemical species. SiH_4, SiH_3, and SiH_2 are selected as a chemical species because they have a possibility to form during the deposition processes using SiH_4. Many studies have reported that SiH_3 is the dominant chemical species in the a-Si:H deposition process [8][11], but other chemical species cannot be ignored under high substrate temperatures [9].

Our study focuses on investigations of accurate molecular knowledge on the surface, and provides the data necessary for the construction of the feature scale model. For that purpose, we perform a reactive force-field molecular dynamics (ReaxFF MD) simulation that incorporates chemical reactions and molecular transport. Since the ReaxFF can reproduce the reaction path by the relation between bond order and bond distance more accurately, the CVD process can be handled with higher accuracy than MC and classical MD simulations. In addition, it is possible to perform calculations considering dynamics due to the smaller calculation cost compared with the quantum chemical simulations [12][13]. It is important for CVD processes to include dynamics knowledge such as molecular transport form.

II. METHODS

A. Reactive Force-Field (ReaxFF)

Molecular dynamics simulations were performed with the large-scale atomic/molecular massively parallel simulator (LAMMPS) MD packages using reactive force-field (ReaxFF) [14]. ReaxFF can describe bond-breaking and bond-formation events unlike general force-fields. In the ReaxFF formalism, each element is described by only one single atom type even within different chemical environment. In addition, the information of reactive sites or the bond connectivity is not defined before simulations. Instead, this information is derived from bond orders (BOs) that are calculated from interatomic distances. The BOs are repeatedly updated at every MD step. Our MD simulations were based on the ReaxFF potential which parameters are taken from the Si/Ge/H system [15]. We confirmed that inter- and intra-molecular potential agree well with results from the density functional theory (DFT) calculation. DFT calculations are

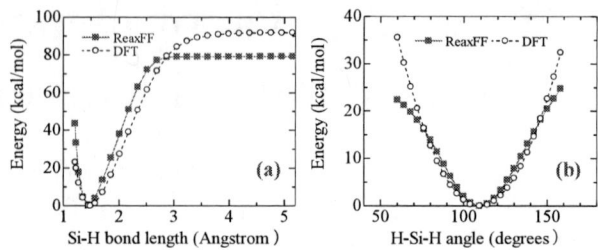

Fig. 1. Comparison between DFT (black open circles) and ReaxFF (blue solid squares) results for: (a) Si-H bond dissociation in SiH_4 and (b) H-Si-H angle bending in SiH_4.

performed by using commercial software package BIOIVA Materials Studio. Results of inter-molecular potential in SiH_4 obtained from DFT and ReaxFF are shown in Fig.1.

B. Calculation System

We report results of simulations that each chemical species impact on the hydrogenated Si (100) substrates. The substrate was composed of (2×2×1) Si lattice and has about

15 Å vacuum region above the substrate. Beyond the distance gas molecules are not affected by the surface. The periodic boundary conditions are applied for the x, y directions, and the free boundary condition are applied for the z direction. The velocity Verlet algorithm with a step of 0.25 fs that can describe the fast-chemical reaction was used. The number of surface-covering bonded hydrogen atoms on the Si substrate were chosen to 0, 9, 18, 27, and 36, which corresponds to the hydrogen coverage θ of 0.00, 0.25, 0.50, 0.75, and 1.00, respectively. To mimic the bulk behavior of Si substrate, the one bottom layer was fixed and the other layers were maintained in a constant temperature of 500-1000 K in 100 K intervals using the Nose-Hoover thermostat. The initial position of each chemical species is set at the height of 10 Å above hydrogen atoms on the substrate (x and y coordinates are random). The incident velocity of each chemical species is set at 0.00394094, 0.004004278, and 0.004070771 Å/fs for SiH_4, SiH_3, and SiH_2 respectively (corresponding to the most probable speed of 300 K for each chemical species). Each chemical species has only an incident velocity normal to the surface. Our calculation system with chemical species and surface conditions of each hydrogen coverage are shown in Fig. 2 and 3. After the 15 ps NVT simulation from the start of incidence, we judged the adsorption according to the bonding state. The case where Si atom constituting a SiH_x forms bonds on the substrate (i.e. Si-Si bond) until the end of simulations is defined as the adsorption. 2000 times simulations are performed for each condition and then, an adsorption probability is calculated. In general, the more we calculate, the

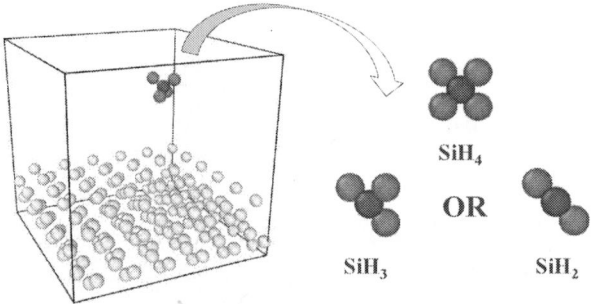

Fig. 2. Calculation system with chemical species.

978-1-7281-0941-1/19 $31.00 © 2019 IEEE

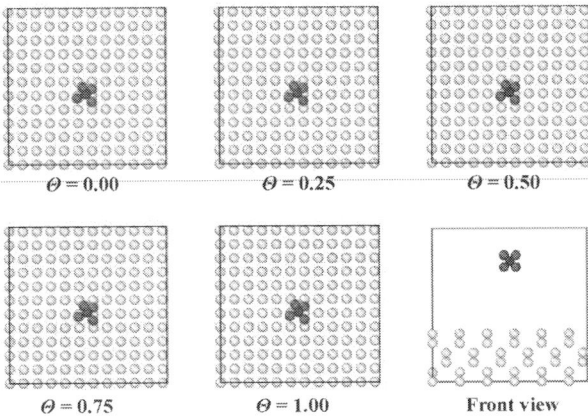

Fig. 3. Surface conditions of each hydrogen coverage ($Si_{substrate}$: yellow, $H_{adsorption}$: white, $Si_{precursor}$: blue, and $H_{precursor}$: red).

more accurate results will be obtained. However, long time and many times simulation lead to high costs. We increased the number of calculations by 1 ps and determined the calculation conditions so that the change in adsorption probability converges within ±5.00 %.

III. RESULTS AND DISCUSSION

A. Substrate Temperature

Fig. 4 (a), (b) shows the result of the substrate temperature dependence on the adsorption probability when the hydrogen coverage = 0.00 and 0.25, respectively. The adsorption probabilities of SiH_2 and SiH_3 are almost constant regardless of the substrate temperature. This trend is consistent with the experimental values of the surface reaction probability β obtained from Time-resolved cavity ringdown spectroscopy (τ-CRDS) [16]. In addition, the values at the hydrogen coverage of 1.0 agree well with the previously reported experimental values (0.3 for SiH_3 and 0.6 for SiH_2) [16]. For SiH_4, it tends to decrease as the substrate temperature increases. These results suggest that an increase in substrate temperature has the effect of preventing adaptation to the incident precursor. It is related to the fact that the thermal accommodation coefficient in kinetics theory of gases decreases with the increase in wall temperature [17]. However, the degree of influence differs depending on the chemical species. SiH_2 and SiH_3 which have relatively strong reactivity are less affected by the decrease in adsorption probability as the substrate temperature increases. SiH_4 is strongly affected because it is a relatively stable chemical species compared with SiH_2 and SiH_3.

B. Hydrogen Coverage

In Fig. 5 (a), (b) show the result of the hydrogen coverage dependence on the adsorption probability when the substrate temperature = 500 K and 1000 K. The adsorption probabilities decreased with increasing the hydrogen coverage. These results indicate that the hydrogen coverage inhibits the reactivity. It is considered that the adsorption probability decreases with the increase of the hydrogen coverage because the dangling bond is occupied by the surface-covering bonded hydrogen atom.

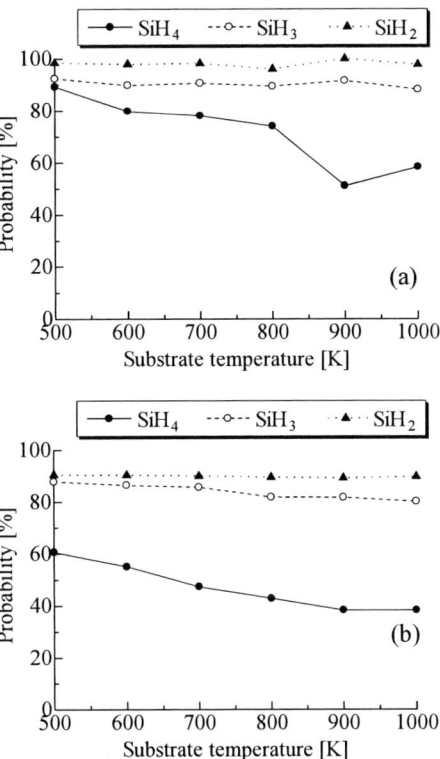

Fig. 4. Substrate temperature dependence of the adsorption probability.
(a) Hydrogen coverage = 0.00 and (b) Hydrogen coverage = 0.25

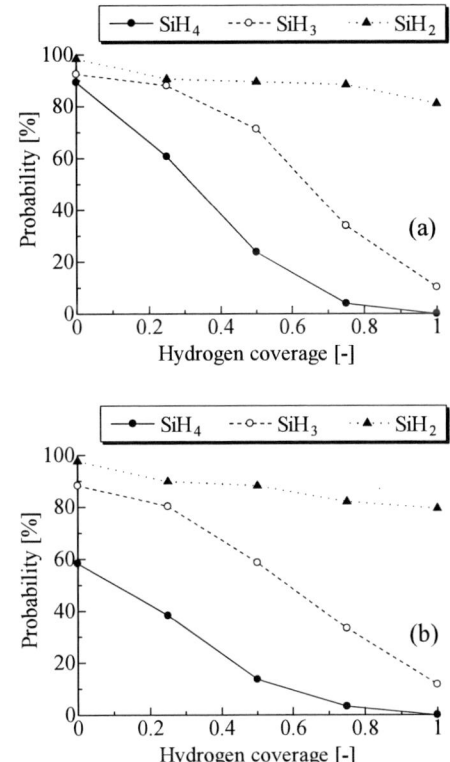

Fig. 5. Hydrogen coverage dependence of the adsorption probability.
(a) Substrate temperature = 500K and (b) Substrate temperature = 1000K

IV. CONCLUSIONS

An influence of (a) substrate temperatures, (b) hydrogen coverage on the adsorption probability are investigated by using ReaxFF MD simulations.

We found that (a) the adsorption probability is almost constant for SiH_2 and SiH_3, but decrease with increase in the substrate temperature for SiH_4, (b) it decreases with the increase in hydrogen coverage. It is well known that the deposition rate generally increases as the substrate temperature increases. However, our results show that increasing the substrate temperature leads to a decrease in reactivity depending on the chemical species. We think that it is related to the thermal accommodation coefficient decreases with the increase in wall temperature. From these results, it is suggested the reason why the reactivity increases with increasing the substrate temperature is that the surface-covering bonded hydrogen atoms are eliminated on the substrate.

ACKNOWLEDGMENT

Numerical simulations were performed on the Supercomputer system "AFI-NITY" at the Advanced Fluid Information Research Center, Institute of Fluid Science, Tohoku University.

REFERENCES

[1] A. Dollet, "Multiscale modeling of CVD film growth-a review of recent works," *Surf. Coatings Technol.*, vol. 177–178, pp. 245–251, 2004.

[2] P. Smereka, X. Li, G. Russo, and D. J. Srolovitz, "Simulation of faceted film growth in three dimensions: Microstructure, morphology and texture," *Acta Mater.*, vol. 53, no. 4, pp. 1191–1204, 2005.

[3] A. Dollet, S. de Persis, M. Pons, and M. Matecki, "Simulation of SiC deposition from SiH4/C3 H8/Ar/H2 mixtures in a cold-wall CVD reactor," *Surf. Coatings Technol.*, vol. 177–178, pp. 382–388, 2004.

[4] M. Grujicic and S. G. Lai, "Multi-length scale modeling of CVD of diamond," *J. Mater. Sci.*, vol. 35, no. 21, pp. 5359–5369, 2000.

[5] T. S. Cale, M. O. Bloomfield, D. F. Richards, K. E. Jansen, and M. K. Gobbert, "Integrated multiscale process simulation," *Comput. Mater. Sci.*, vol. 23, no. 1–4, pp. 3–14, 2002.

[6] C. R. Kleijn and C. J. Hoogendoorn, "Transport Phenomena in Tungsten LPCVD in a Single-Wafer Reactor," *J. Electrochem. Soc.*, vol. 138, no. 2, pp. 509–517, 1991.

[7] G. Chaix and A. Dollet, "Three dimensional Monte Carlo simulation of β-SiC deposition from the vapor phase," in *Proceedings of the 199th meeting of the Electrochemical Society, Symposium: Fundamental Gas Phase and Surface Chemistry of Vapor Deposition II*, 2001, pp. 33–40.

[8] A. Matsuda, K. Nomoto, Y. Takeuchi, A. Suzuki, A. Yuuki, and J. Perrin, "Temperature dependence of the sticking and loss probabilities of silyl radicals on hydrogenated amorphous silicon," *Surf. Sci.*, vol. 227, no. 1–2, pp. 50–56, 1990.

[9] S. Ramalingam, S. Sriraman, E. S. Aydil, and D. Maroudas, "Evolution of structure, morphology, and reactivity of hydrogenated amorphous silicon film surfaces grown by molecular-dynamics simulation," *Appl. Phys. Lett.*, vol. 78, no. 18, pp. 2685–2687, 2001.

[10] A. Matsuda, "Microcrystalline silicon. Growth and device application," *J. Non. Cryst. Solids*, vol. 338–340, pp. 1–12, 2004.

[11] W. M. M. Kessels, Y. Barrell, P. Van Den Oever, J. P. M. Hoefnagels, and M. C. M. van de Sanden, "The a-Si: H Growth Mechanism: Temperature Study of the SiH3 Surface Reactivity and the Surface Silicon Hydride Composition During Film Growth," *Symp. A – Amorph. Nanocrystalline Silicon-Based Film. – 2003*, vol. 762, 2003.

[12] I. I. Oleinik, D. G. Pettifor, A. P. Sutton, and J. E. Butler, "Theoretical study of chemical reactions on CVD diamond surfaces," *Diam. Relat. Mater.*, vol. 9, no. 3–6, pp. 241–245, 2000.

[13] S. Canovic *et al.*, "TEM and DFT investigation of CVD TiN/κ-Al2O3 multilayer coatings," *Surf. Coatings Technol.*, vol. 202, no. 3, pp. 522–531, 2007.

[14] A. C. T. Van Duin, S. Dasgupta, F. Lorant, and W. A. Goddard, "ReaxFF: A reactive force field for hydrocarbons," *J. Phys. Chem. A*, vol. 105, no. 41, pp. 9396–9409, 2001.

[15] G. Psofogiannakis and A. C. T. Van Duin, "Development of a ReaxFF reactive force field for Si/Ge/H systems and application to atomic hydrogen bombardment of Si, Ge, and SiGe (100) surfaces," *Surf. Sci.*, vol. 646, pp. 253–260, 2016.

[16] J. P. M. Hoefnagels, Y. Barrell, W. M. M. Kessels, and M. C. M. van de Sanden, "Time-resolved cavity ringdown study of the Si and SiH3 surface reaction probability during plasma deposition of a-Si:H at different substrate temperatures," *J. Appl. Phys.*, vol. 96, no. 8, pp. 4094–4106, 2004.

[17] H. Yamaguchi, K. Kanazawa, Y. Matsuda, T. Niimi, A. Polikarpov, and I. Graur, "Investigation on heat transfer between two coaxial cylinders for measurement of thermal accommodation coefficient," *Phys. Fluids*, vol. 24, no. 6, 2012.

Device-to-circuit modeling approach to Metal – Insulator – 2D material FETs targeting the design of linear RF applications

Alejandro Toral-López
Departamento de Electrónica y Tecnología de Computadores
Universidad de Granada
Granada, Spain
atoral@ugr.es

Francisco Pasadas
Departament d'Enginyeria Electrònica
Universitat Autònoma de Barcelona
Barcelona, Spain
francisco.pasadas@uab.es

Enrique G. Marín
Departamento de Electrónica y Tecnología de Computadores
Universidad de Granada
Granada, Spain
egmarin@ugr.es

Alberto Medina-Rull
Departamento de Electrónica y Tecnología de Computadores
Universidad de Granada
Granada, Spain
amedinarull@correo.ugr.es

Francisco J. G. Ruiz
Departamento de Electrónica y Tecnología de Computadores
Universida de Granada
Granada, Spain
franruiz@ugr.es

David Jiménez
Departament d'Enginyeria Electrònica
Universitat Autònoma de Barcelona
Barcelona, Spain
david.jimenez@uab.es

Andrés Godoy
Departamento de Electrónica y Tecnología de Computadores
Universidad de Granada
Granada, Spain
agodoy@ugr.es

Abstract—We present a physics-based device-to-circuit modeling approach to metal – insulator – 2D material based field-effect transistors (2DFETs). Starting from numerical simulations based on the self-consistent solution of the 2D Poisson and 1D Drift-Diffusion equations, we obtain the electrostatics and current-voltage characteristics of such devices. Then, assuming small-signal operation, a charge-based equivalent circuit is fed with the small-signal parameters computed from the numerical results and then it is implemented in a standard circuit simulator. This framework enables the design and assessment of linear radio-frequency applications based on novel and emergent 2DFETs. The approach has been applied to an experimental MoS₂ transistor by benchmarking the transfer characteristics and then predicting the expected performance of such device as a common-source power amplifier, for instance a power gain of 8.6 dB at 2.45 GHz.

Keywords—2D material, drift-diffusion, field-effect transistor, molybdenum disulfide, power amplifier, radio-frequency, small-signal

I. INTRODUCTION

Since the emergence of graphene, a wide number of two-dimensional (2D) materials have been experimentally demonstrated in an impressive short period of time [1]. Indeed, some of them have already shown potential as channels in field-effect transistors (FETs), being promising candidates to replace and/or augment conventional CMOS technology in a near future. In this context, it becomes essential to develop a methodology able to interpret the electrical measurements of novel 2D material based FETs (2DFETs), to expedite fast prototyping and enabling the design and assessment of linear radio-frequency (RF) circuits based on such devices.

While physics-based numerical simulations provide rigorous information about the electrical operation of a device, their computational burden is quite demanding even for the evaluation of the DC operation of a single transistor. As a consequence, such approach is out of the question for general AC circuit-level simulations. Much more efficient models or methodologies are thus needed for the design and optimization of circuits based on emergent 2DFETs.

In this work, our proposal is to take advantage of the accuracy and potential of numerical simulations by computing the DC operation of a selected device to later build, under a small-signal approximation, a charge-conserving equivalent circuit able to carry out the AC circuit design (implying its analysis and optimization) in commercial circuit simulators. A schematic description of the methodology proposed is depicted in Fig. 1. We have applied this device-to-circuit modeling methodology to an experimental MoS₂-FET, demonstrating the design of a microwave power amplifier based on such a device.

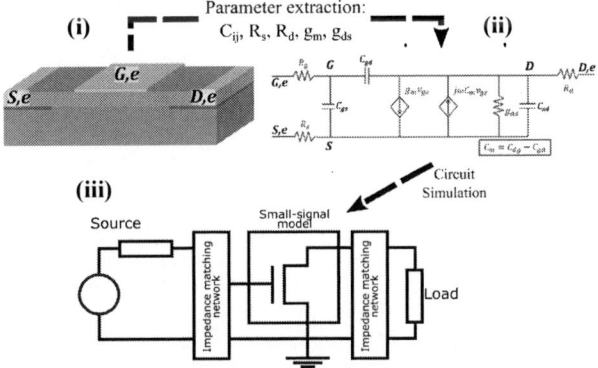

Fig. 1. Schematic description of the approach proposed in this work: from (i) self-consistent simulations of a single device to (ii) the extraction of the small-signal parameters of a charge-based equivalent circuit and (iii) the circuit design, implying its analysis and optimization, in a standard circuit simulator.

The paper is structured as follows. Section II introduces the numerical self-consistent simulator used to compute the electrostatics and the drain-to-source current. Then, Section III presents the small-signal equivalent circuit used in circuit simulators to describe a 2DFET. Section IV encompasses the comparison of the DC electrical simulations of a MoS$_2$-FET with experimental measurements together with the design of a power amplifier based on that device. Finally, the conclusions are drawn in Section V.

II. SELF-CONSISTENT SIMULATOR OF 2DFETs

A. Electrostatics and drain current

The numerical simulation comprises the 2D Poisson equation self-consistently coupled with the 1D Drift-Diffusion (DD) transport equation. Under equilibrium conditions, the electron (hole) density profile, n_L (p_L), is calculated using the 2D density of states. The 2D electron (hole) distribution is obtained combining the longitudinal charge profile with a fixed sinusoidal profile (φ_L) that defines the charge spread along the semiconductor thickness. As emerging 2DFETs suffer from interface traps that degrade the device performance, we have included this effect distributing the trapped charge density in a Gaussian spatial profile along the direction normal to the 2D material – insulator interface. Out of equilibrium, n_L (p_L) is calculated using the 1D DD transport equation which reads as:

$$J_n(x) = qn_L(x)\mu(x)\frac{\partial V(x)}{\partial x} + qD_n\frac{\partial n_L(x)}{\partial x}$$
$$J_p(x) = -qp_L(x)\mu(x)\frac{\partial V(x)}{\partial x} + qD_p\frac{\partial p_L(x)}{\partial x} \quad (1)$$

where J_n (J_p) stands for the 1D electron (hole) current density which is given as a function of position x along the channel (positive x direction is from source, $x = 0$, to drain, $x = L$; where L is the channel length); V is the potential profile along the 2D channel; D_n (D_p) is the electron (hole) diffusion coefficient; q is the electron charge; and μ is the carrier mobility which has been assumed the same for both type of carriers and dependent on the longitudinal electric field as follows [2]:

$$\mu = \frac{\mu_0}{\left[1 + \left(\frac{\mu_0}{v_{sat}}|\vec{E}_x|\right)^\beta\right]^{\gamma_\beta}} \quad (2)$$

where μ_0 is the low-field mobility; v_{sat} is the carrier saturation velocity; β is the saturation coefficient and E_x is the electric field along the longitudinal direction.

B. Charge-based capacitance model

An accurate modeling of the intrinsic capacitances of FETs requires the consideration of any change in the channel charge distribution as a response to the changes of terminal voltages. Thus, once the device electrostatic description is completed, the charge density, $Q_{net} = q(p_L - n_L)$, is integrated along the channel (Q_g) and then split into the drain (Q_d) and source (Q_s) components by making use of the Ward-Dutton's linear charge partition scheme [3], which guarantees charge conservation, as follows:

$$Q_g = -W\int_0^L Q_{net}(x)dx$$
$$Q_d = W\int_0^L \frac{x}{L} Q_{net}(x)dx \quad (3)$$
$$Q_s = -(Q_g + Q_d) = W\int_0^L \left(1 - \frac{x}{L}\right)Q_{net}(x)dx$$

In particular, a three-terminal device can be modelled with only four independent capacitances by assuming that the dynamic description is charge-conservative and reference-independent [4]. Each element C_{ij} describes the dependence of the charge at terminal i with respect to a varying voltage applied to terminal j, assuming that the voltage at any other terminal remains constant as follows:

$$C_{ij} = -\frac{\partial Q_i}{\partial V_j} \quad i \neq j \qquad C_{ij} = \frac{\partial Q_i}{\partial V_j} \quad i = j \quad (4)$$

where i and j stand for g, d, and s.

III. SMALL-SIGNAL DESCRIPTION OF 2DFETs

When electronic devices are employed in analog and RF circuits, their terminals are biased with a DC voltage over which a time-varying signal is superimposed. If the amplitude of that time-varying excitation is small enough, the resulting small current and charge variations can be expressed in terms of it using linear relations. This way a non-linear device can be treated as a linear circuit with conductance, inductance and capacitance elements forming a lumped network [5]. The appropriate small-signal equivalent circuit of a 2DFET is shown in Fig. 2 [4]. It presents two important features: (i) it guarantees charge conservation paying attention to the nonreciprocity of the capacitances and (ii) it includes the metal contact and access resistances, which are of upmost importance when dealing with low dimensional FETs.

Accordingly, our methodology is based on the accurate calculation of the different small-signal elements contained in the equivalent circuit depicted in Fig. 2 with our self-consistent simulator. Specifically, the capacitances are numerically computed by evaluating (3) and (4); the conductances by calculating: $g_m = \partial I_{ds}/\partial V_{gs}$ and $g_{ds} = \partial I_{ds}/\partial V_{ds}$; and the resistances $R_s = R_{s,acc} + R_c$; $R_d = R_{d,acc} + R_c$ are split into the series combination of the access and metal contact resistances, respectively. $R_{s,acc}$ and $R_{d,acc}$ are extracted from the numerical simulations, while R_c and the gate resistance, R_g, are treated as fitting parameters.

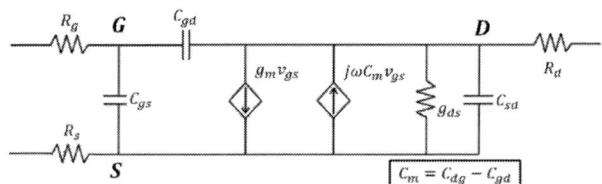

Fig. 2. Charge-based small-signal model suited to 2DFETs. The small-signal elements are: g_m transconductance, g_{ds} output conductance and C_{gs}, C_{gd}, C_{sd} and C_{dg} independent intrinsic capacitances. R_g is the gate resistance and R_d, R_s account for the contact and access resistances of the drain and source, respectively.

IV. RESULTS

A. MoS₂-based FET

In this section, we validate the proposed approach against the measured characteristics of an experimental 2DFET [6]. The device consists of an *n*-type monolayer CVD-grown MoS₂ (length of $L = 450$ nm) deposited on 285-nm thick SiO₂, and covered by a 30-nm thick HfO₂ controlled electrostatically by a metal gate of 250-nm length. We calibrate our simulator by comparing the experimental DC transfer characteristic (symbols) with our simulations (solid line) as shown in Fig. 3. The accurate fit is achieved assuming $R_c = 200\Omega$. Then we extract the intrinsic small-signal parameters, summarized in Table I, for an operation bias point of $V_{GS} = -3.9$V and $V_{DS} = 0.7$V corresponding to the saturation regime.

TABLE I. EXTRACTED SMALL-SIGNAL PARAMETERS OF THE MoS₂-FET [6]

Element	Value	Element	Value
C_{gs}	13.317 fF	g_m	10.8 mS
C_{gd}	1.745 fF	g_{ds}	0.58 mS
C_{dg}	7.039 fF	$R_{d,acc}$	4.37 kΩ
C_{sd}	-0.415 fF	$R_{s,acc}$	1.04 kΩ

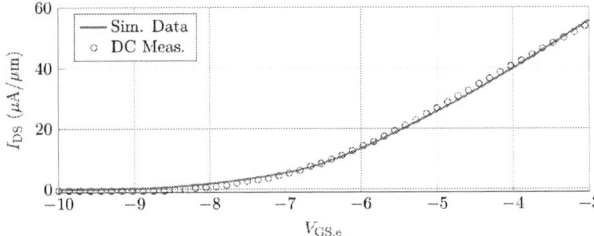

Fig. 3. Fitting of the measured transfer characteristic reported in [6] ($V_{DS,e} = 3.5$V).

B. Design of a power amplifier based on a MoS₂-FET

Once the device has been thoroughly characterized, the performance of a power amplifier operating at the ISM band is assessed by taking advantage of the Keysight© ADS design tools. The design of the RF linear circuit is based on the MoS₂-FET (described in Section IV.A) but considering an optimized device in which the underlapped access regions have been removed ($R_s = R_d = R_c$). Finally, $R_g = 10\Omega$ is assumed.

First, the main RF figures of merit of the device under test are analyzed; which in turn are the cutoff frequency, f_T, and the maximum frequency of oscillation, f_{max}. f_T stands for the frequency at which the current gain (h_{21}) of the transistor drops to unity and the f_{max} is the frequency at which the power gain (U, Mason's invariant [7],[8]) becomes unity. Such gains are evaluated from the short-circuit admittance parameters as follows [4]:

$$h_{21}(\omega) = -\frac{y_{21}}{y_{11}}$$

$$U(\omega) = \frac{|y_{12} - y_{21}|^2}{4\left(\text{Re}[y_{11}]\text{Re}[y_{22}] - \text{Re}[y_{12}]\text{Re}[y_{21}]\right)} \quad (5)$$

Fig. 4 shows the h_{21} and U of the MoS₂-FET as estimated in the circuit simulator by connecting 50Ω-ports to the gate-

source and drain-source terminals of the equivalent circuit depicted in Fig. 2. The maximum stable gain / maximum available gain (MSG/MAG), calculated as in [9], are also shown in Fig. 4. Both U and MSG/MAG becomes unity at the same frequency, $f_{max} = 109$ GHz. In practice, as a rule of thumb, the operating frequency of the power amplifier should be lower than ~20% of the transistor's f_{max} to guarantee sufficient power gain [10]; thus, the selected design frequency is 2.45 GHZ where a MSG of 19.8 dB is available.

Considering the tradeoff between operating far from the unstable region and getting high power gain, the design of an Input Matching Network (IMN) and an Output Matching Network (OMN) has been carried out to properly adapt the input and output impedances to the source (Z_s) and load (Z_l) impedances, respectively, at the frequency of operation. Fig. 5a shows the schematics of the proposed power amplifier where the OMN and IMN have been included as a network of lumped elements. Table II contains the magnitude of the S-parameters after the matching. Fig. 5b shows the input voltage (V_{in}) and the amplified output voltage (V_{out}) when an incident input power (P_{in}) of -20 dBm (small-signal operation) is delivered by the power source. The proposed design results in a power gain of 8.6 dB for a frequency of operation of 2.45 GHz demonstrating its potential application in RF electronics.

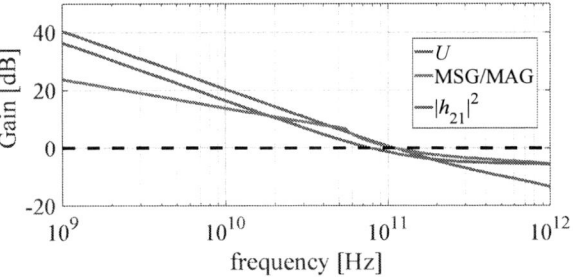

Fig. 4. Radio-frequency performance of the MoS₂-FET under test described by the small-signal parameters provided in Table I. Specifically, the small-signal current gain ($|h_{21}|$), the unilateral power gain (U) and the maximum stable gain / maximum available gain (MSG/MAG) are plotted versus frequency.

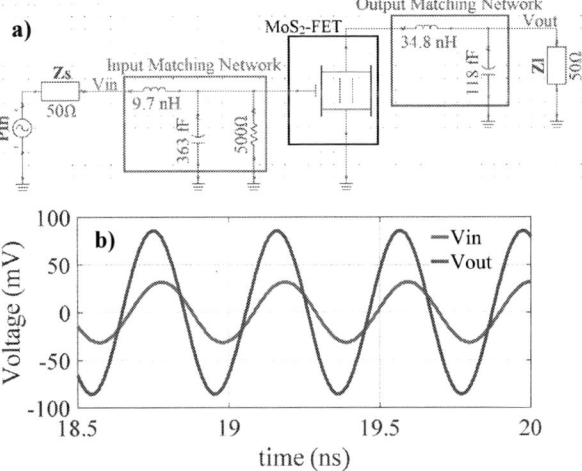

Fig. 5. a) Microwave power amplifier design based on a novel 2D technology. The MoS₂-FET symbol contains the network depicted in Fig. 2 and the small-signal elements described in Table I. b) Input and output voltages showing an amplification of 8.6 dB at 2.45 GHz.

TABLE II. MAGNITUDE (IN dB) OF THE S-PARAMETERS AT 2.45 GHz AFTER INPUT AND OUTPUT MATCHING

S_{11}	S_{12}	S_{21}	S_{22}
-73.45	-31.03	8.65	-24.54

V. CONCLUSIONS

A device-to-circuit modeling approach to 2DFETs is proposed, allowing the design, optimization and analysis of circuits for linear RF applications. In a first step, the electrostatics and DC current-voltage characteristics of such devices are obtained by means of a robust and accurate self-consistent simulator. Then, a small-signal equivalent circuit suited to 2DFETs is used for circuit simulation feeding its elements with the results achieved in the previous step. The design of a matched power amplifier based on an optimized experimental MoS$_2$-FET is predicted to operate at 2.45 GHz showing a power gain of 8.6 dB.

ACKNOWLEDGMENTS

The authors would like to thank the financial support of Spanish Government under projects TEC2017-89955-P (MINECO/AEI/FEDER, EU) and TEC2015-67462-C2-1-R (MINECO). A. Toral-López acknowledges the FPU program (FPU16/04043). E. G. Marín acknowledges Juan de la Cierva Incorporación IJCI-2017-32297 (MINECO/AEI). F. Pasadas and D. Jiménez also acknowledge the support from the European Union's Horizon 2020 Grant Agreement No. 785219 GrapheneCore2.

REFERENCES

[1] F. Schwierz, R. Granzner, and J. Pezoldt, "Two-dimensional materials and their prospects in transistor electronics," *Nanoscale*, pp. 8261–8283, 2015.

[2] P. C. Feijoo, D. Jiménez, and X. Cartoixà, "Short channel effects in graphene-based field effect transistors targeting radio-frequency applications," *2D Mater.*, vol. 3, no. 2, p. 025036, Jun. 2016.

[3] D. E. Ward and R. W. Dutton, "A charge-oriented model for MOS transistor capacitances," *IEEE J. Solid-State Circuits*, vol. 13, no. 5, pp. 703–708, Oct. 1978.

[4] F. Pasadas, W. Wei, E. Pallecchi, H. Happy, and D. Jiménez, "Small-Signal Model for 2D-Material Based FETs Targeting Radio-Frequency Applications: The Importance of Considering Nonreciprocal Capacitances," *IEEE Trans. Electron Devices*, vol. 64, no. 11, pp. 4715–4723, Nov. 2017.

[5] Y. Tsividis, Operation and modeling of the MOS transistor, 2nd ed. New York ; Oxford : Oxford University Press, 1999.

[6] A. Sanne, R. Ghosh, A. Rai, M. N. Yogeesh, S. H. Shin, A. Sharma, K. Jarvis, L. Mathew, R. Rao, D. Akinwande and S. Banerjee, "Radio Frequency Transistors and Circuits Based on CVD MoS2," *Nano Lett.*, vol. 15, no. 8, pp. 5039–5045, Aug. 2015.

[7] S. Mason, "Power Gain in Feedback Amplifier," *Trans. IRE Prof. Gr. Circuit Theory*, vol. CT-1, no. 2, pp. 20–25, Jun. 1954.

[8] M. S. Gupta, "Power Gain in Feedback Amplifiers, a Classic Revisited," *IEEE Trans. Microw. Theory Tech.*, vol. 40, no. 5, pp. 864–879, 1992.

[9] P. C. Feijoo, F. Pasadas, J. M. Iglesias, M. J. Martín, R. Rengel, C. Li, W. Kim, J. Riikonen, H. Lipsanen and D. Jiménez, "Scaling of graphene field-effect transistors supported on hexagonal boron nitride: Radio-frequency stability as a limiting factor," *Nanotechnology*, vol. 28, no. 48, 2017.

[10] F. Schwierz, R. Granzner, and J. Pezoldt, "Two-dimensional materials and their prospects in transistor electronics," *Nanoscale*, pp. 8261–8283, 2015.

Simulation and Investigation of Electrothermal Effects in Heterojunction Bipolar Transistors

Xujiao Gao
electrical models and simulation
sandia national laboratories
Albuquerque, USA
xngao@sandia.gov

Gary Hennigan
electrical models and simulation
sandia national laboratories
Albuquerque, USA
glhenni@sandia.gov

Lawrence Musson
electrical models and simulation
sandia national laboratories
Albuquerque, USA
lcmusso@sandia.gov

Andy Huang
electrical models and simulation
sandia national laboratories
Albuquerque, USA
ahuang@sandia.gov

Mihai Negoita
electrical models and simulation
sandia national laboratories
Albuquerque, USA
mnegoit@sandia.gov

Abstract—We present a comprehensive physics investigation of electrothermal effects in III-V heterojunction bipolar transistors (HBTs) via extensive Technology Computer Aided Design (TCAD) simulation and modeling. We show for the first time that the negative differential resistances of the common-emitter output responses in InGaP/GaAs HBTs are caused not only by the well-known carrier mobility reduction, but more importantly also by the increased base-to-emitter hole back injection, as the device temperature increases from self-heating. Both self-heating and impact ionization can cause fly-backs in the output responses under constant base-emitter voltages. We find that the fly-back behavior is due to competing processes of carrier recombination and self-heating or impact ionization induced carrier generation. These findings will allow us to understand and potentially improve the safe operating areas and circuit compact models of InGaP/GaAs HBTs.

Index Terms—heterojunction bipolar transistor, TCAD, self-heating, impact ionization, mobility reduction, hole back injection

I. Introduction

Heterojunction bipolar transistors (HBTs) based on the InGaP/GaAs material system are widely used in wireless communication systems, due to their higher power, faster switching speed, and higher efficiency, when compared to silicon devices. It is generally recognized [1] [2] that the electrothermal effect (a.k.a. self-heating effect) has a strong influence on device performance of III-V HBTs. The interplay of self-heating with impact ionization (a.k.a. avalanche breakdown) limits their safe operating areas (SOAs) [3] [4]. Several papers [3] [4] [5] have reported the measured SOAs of

This work is funded by the Advanced Scientific Computing (ASC) program at Sandia National Laboratories. Sandia National Laboratories is a multimission laboratory managed and operated by National Technology and Engineering Solutions of Sandia, LLC., a wholly owned subsidiary of Honeywell International, Inc., for the U.S. Department of Energy's National Nuclear Security Administration under contract DE-NA-0003525. The views expressed in the article do not necessarily represent the views of the U.S. Department of Energy or the United States Government.

InGaP/GaAs HBTs in non-radiation environments. Accompanying the experimental work, there exists a number of papers [1] [2] [6] [7] that employed Technology Computer Aided Design (TCAD) device codes to model the self-heating effect in III-V HBTs. Although Rinaldi *et al.* [8] [9] were able to simulate the complex device failure characteristics due to self-heating and impact ionization in silicon bipolar transistors, modeling and understanding of device failure mechanisms in III-V HBTs is quite limited. In fact, the physics of self-heating and its interplay with impact ionization in III-V HBTs is not well understood. This is evidenced by the fact that existing compact HBT models often do not work well [10] when modeling the self-heating effect even in the operating regimes far from device failure.

In this paper, we present physics-based TCAD simulation and modeling of self-heating and impact ionization effects in an InGaP/GaAs HBT. We present a detailed physics understanding of the self-heating effect and its interplay with impact ionization. This understanding would enable us to potentially improve the SOAs and circuit compact models of these devices.

II. Modeling Approach

Simulations were done using Charon [11], a multi-dimensional, Messaging Passing Interface (MPI) based parallel TCAD device code, which we developed at Sandia National Laboratories. Charon supports isothermal drift diffusion (DD) modeling and coupled electrothermal (i.e., DD + lattice heating) simulation. Our device of interest is an emitter-up $In_{0.5}Ga_{0.5}P$/GaAs NP^+N HBT [12]. Figure 1 shows the simulated two-dimensional (2D) structure, which represents a truncated half-finger cross-section of a real device. The coupled model is applied to all the semiconductor regions, while the lattice heat equation is solved in the metal and nitride regions. To properly simulate device self-heating, temperature dependencies were incorporated in all important material mod-

978-1-7281-0941-1/19 $31.00 © 2019 IEEE

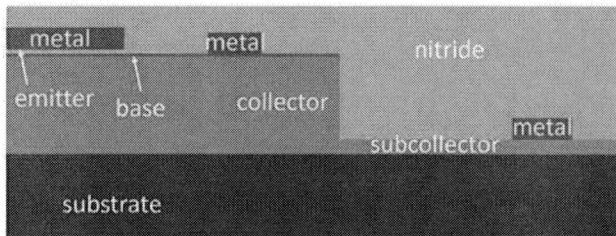

Fig. 1. Simulated 2D structure for a lattice-matched In$_{0.5}$Ga$_{0.5}$P/GaAs HBT. The emitter is n-type In$_{0.5}$Ga$_{0.5}$P, the base is p$^+$-type GaAs, while the collector and subcollector are n-type GaAs. The metal regions provide electrical contacts. To properly simulate the lattice heating, the capping nitride region and a small part of the semi-insulating GaAs substrate are also included.

els, including band gap, carrier mobility, thermal conductivity, and heat capacity [13]. The carrier mobility models are similar to those given in [14]. For impact ionization (II), several GaAs II models [15] [16] [17] were studied and found to produce similar II coefficient vs. field curves. The Plimmer II model [16] was used in this work. Carrier transport across the emitter-base (E-B) heterojunction (HJ) is governed by thermionic emission (TE) and tunneling processes [18]. The net hole TE current density across the HJ is given by

$$J_{TE,p} = A_p^* T^2 \left[\frac{p_E}{N_{V,E}} - \frac{p_B}{N_{V,B}} \exp\left(-\frac{\Delta E_V}{k_B T} \right) \right]. \quad (1)$$

Here A_p^* is the Richardson coefficient and other symbols have their usual meanings. Our simulation results show that the tunneling current across the HJ contributes only about 10% of the total current in the HBT.

The choice of thermal boundary conditions plays a vital role in determining the temperature profile. The bottom of the simulated structure had a thermal conductance of 10 W/(K.cm^2) estimated from the substrate thickness and GaAs thermal conductivity. The top surface had a thermal conductance of 200 W/(K.cm^2) obtained by fitting simulated results to measured output response data at a given base current.

III. RESULTS AND DISCUSSION

We first investigate the self-heating (SH) effect in the InGaP/GaAs HBT without II. Figure 2 shows the simulated common-emitter results under constant base currents (I_B). The black solid curves in the top figure are widely observed in simulation and experimental results [1] [4] for III-V HBTs. The negative slopes in these curves are often attributed to the reduction of carrier mobility with increasing temperature [1]. However, this explanation is only one part of the puzzle, because the simulated collector current (I_C) vs. collector voltage (V_{CE}) curves still show strong negative slopes, even when the temperature dependencies were removed from the mobility models, as shown by the red curves in the top of Fig. 2. We discovered that the other dominant mechanism is the hole back injection from the base to the emitter. This can be seen from the blue curves, which show a much smaller decrease in I_C at high V_{CE}, when the temperature in the exponential term of the

hole TE model (1) was replaced by 300 K. The observation becomes more evident from the bottom plot of Fig. 2. In the low temperature regime, I_C shows a power-law reduction with increasing temperature, indicating the reduced carrier mobility is responsible for the I_C reduction. In the high temperature regime, I_C shows an exponential reduction with increasing temperature, indicating the base-to-emitter hole back injection dominates the decreasing in I_C. For medium temperatures, both mechanisms play an important role in reducing I_C.

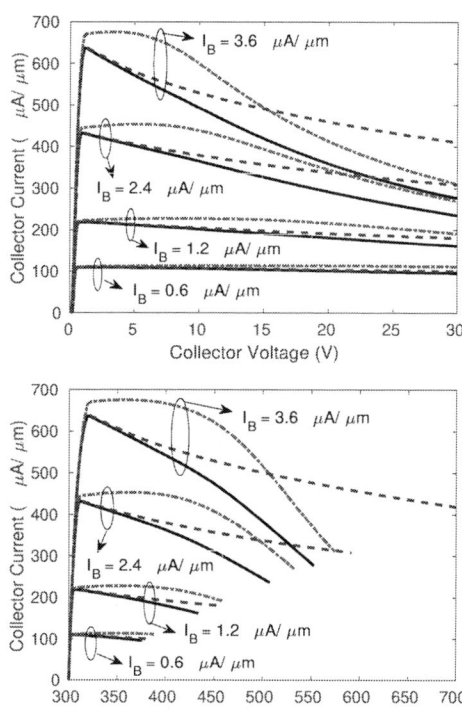

Fig. 2. Simulated I_C vs. V_{CE} (top) and I_C vs. device peak temperature (bottom) curves under constant I_B. Black solid curves were obtained with all temperature dependencies included. Red dot-dashed curves were obtained with the temperature dependencies removed from the mobility models. Blue dashed curves were obtained with the temperature in the exponential term of the hole TE model (1) replaced by 300 K, whereas keeping all other temperature dependencies in the simulation. The device peak temperature was taken to be the maximum temperature in the simulation domain at a given bias condition.

In the case of constant base-emitter voltages (V_{BE}), the I_C vs. V_{CE} curves are very different from those of the constant I_B case. Figure 3 shows the simulated I_C (top) and device peak temperature (bottom) vs. V_{CE} curves using the SH model without II under constant V_{BE}. We observe that the simulated device is significantly heated up, especially in the high V_{BE} and/or high V_{CE} regimes. The simulated temperature profiles corresponding to the turning point and the peak current for the case of $V_{BE} = 1.26$ V are plotted in Fig. 4. We see that the temperature has a strong non-uniform shape and the device is heated up mostly in the active region below the emitter.

The I_C-V_{CE} curves in Fig. 3 clearly show fly-back behavior

978-1-7281-0941-1/19 $31.00 © 2019 IEEE

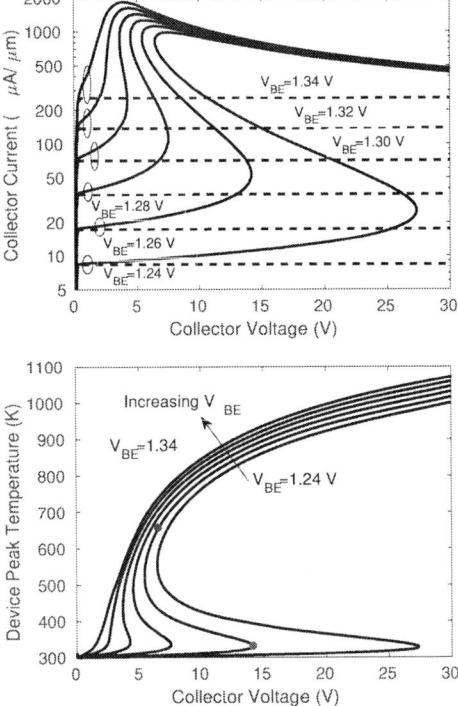

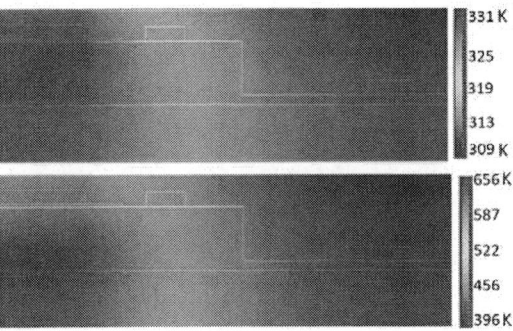

Fig. 4. Simulated device temperature spatial profiles at the turning point (top), indicated by the red dot in the bottom panel of Fig. 3, and at the peak current (bottom), indicated by the blue dot in the bottom panel of Fig. 3, for the case of $V_{BE} = 1.26$ V.

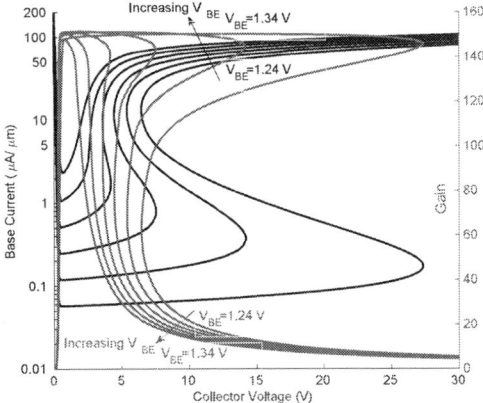

Fig. 3. Simulated I_C (top) and device peak temperature (bottom) vs. V_{CE} curves under constant V_{BE}. Black solid curves were obtained using the SH model without II, whereas black dashed curves were obtained using the isothermal DD model without II.

Fig. 5. Simulated I_B (black) and gain (red) vs. V_{CE} curves that correspond to the black solid curves in Fig. 3.

at low V_{BE} values, whereas this SH induced fly-back does not occur in the constant I_B case (Fig. 2). Though similar phenomena were observed by other authors [8] [9], the underlying physics was still unclear. We found that, during the fly-back process, I_B increases and the transistor gain decreases (Fig. 5), while the device temperature keeps increasing (Fig. 3). This indicates that fly-back occurs due to competing mechanisms of SH induced carrier generation and standard carrier recombination processes. At a given low V_{BE}, as the device heats up with increasing V_{CE}, thermal generation increases carrier densities and hence increases the overall carrier injection from the emitter, leading to larger I_C on the fly-back curve; on the other hand, carrier recombination in the base also increases due to excess carriers, which leads to larger I_B and smaller gain. At high V_{BE}, the fly-back does not occur, because carrier injection from the emitter electrode is so high that it obscures the effect of thermal carrier generation.

Furthermore, we notice that, at high I_C, independent of V_{BE}, all the I_C-V_{CE} curves show negative slopes and converge to more or less the same curve. This is because the base-to-emitter hole back injection dominates at high temperatures and the device loses its transistor action. It is worth pointing out that the simulated I_C-V_{CE} curves without the fly-backs at high V_{BE} are qualitatively very similar to the measured curves by Lee *et al.* [4] for a different InGaP/GaAs HBT design.

To investigate the impact ionization effect in the HBT, we simulated the I_C-V_{CE} curves using the isothermal DD model with the Plimmer II coefficients [16]. As shown in Fig. 6, the I_C-V_{CE} curves also show fly-back behavior using the II model alone. The II induced fly-back is due to competing processes of avalanche carrier generation and standard carrier recombination. However, the fly-back occurs at a much higher voltage for a given V_{BE}, even when the critical fields in the II model were reduced by 25%. In addition, unlike in the SH case where the I_B are always positive (i.e., the electrode provides holes to the device), the I_B values here are negative (i.e., the device supplies holes to the electrode). We note that the II induced fly-back behavior was experimentally measured in InGaP/GaAs HBTs by R. Jin [5].

Using the SH and II models together resulted in the I_C-V_{CE} curves plotted in Fig. 7 for constant V_{BE}. Clearly, the I_C-V_{CE} curves are determined by SH and slightly affected by II when the II critical fields were set to 1.25 times the Plimmer values. As the critical fields were reduced to 75%, the I_C-V_{CE} curves (blue) show more fly-backs and fly-forwards than the SH case.

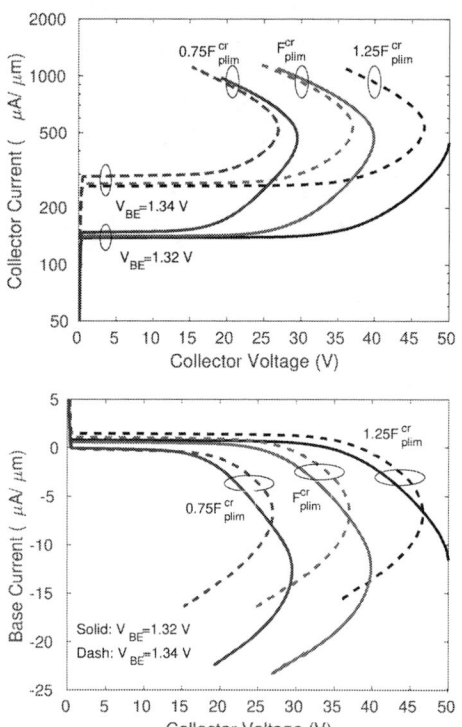

Fig. 6. Simulated I_C (top) and I_B (bottom) vs. V_{CE} under constant V_{BE}, obtained using the isothermal DD model with the Plimmer II parameters. The critical fields in the II model were modified to 1.25 (black) and 0.75 (blue) times the original values (red).

However, when measuring a real HBT, the device would burn out due to SH long before it could reach the second fly-back (II induced). It indicates that the device failure in an InGaP/GaAs HBT under quasi-DC operation is mainly caused by SH and little affected by II.

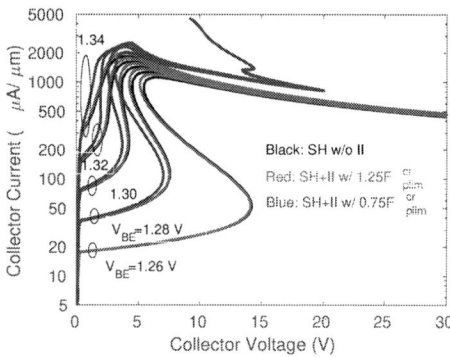

Fig. 7. Simulated I_C vs. V_{CE} curves under constant V_{BE}, obtained using different model configurations. The black curves were obtained using the SH model without II. The red and blue curves were obtained using the SH and II models with the critical fields set to 1.25 (red) and 0.75 (blue) times the original Plimmer values.

IV. CONCLUSION

We have shown the negative slopes of common-emitter output responses in InGaP/GaAs HBTs are not only due to carrier mobility reduction, but more importantly also due to the increased base-to-emitter hole back injection, with increasing temperature. Both SH and II can cause fly-backs in the output responses of HBTs under constant base-emitter voltages. The fly-back behavior are caused by competing processes of carrier recombination and thermal or avalanche induced carrier generation. Device failure under quasi-DC operation is dominated by self-heating because the impact ionization induced fly-back occurs at a much higher voltage.

REFERENCES

[1] L. L. Liou, J. L. Ebel, and C. I. Huang, "Thermal effects on the characteristics of AlGaAs/GaAs heterojunction bipolar transistors using two-dimensional numerical simulation," *IEEE Trans. Electron Devices*, vol. 40, no. 1, pp. 35–43, 1993.

[2] A. Kager and J. J. Liou, "Two-dimensional numerical analysis of AlGaAs/GaAs heterojunction bipolar transistors including the effects of graded layer, setback layer and self-heating," *Solid-State Electronics*, vol. 39, no. 2, pp. 193–199, 1996.

[3] C. P. Lee, F. H. F. Chau, W. Ma, and N. L. Wang, "The safe operating area of GaAs-based heterojunction bipolar transistors," *IEEE Trans. Electron Devices*, vol. 53, no. 11, pp. 2681–2688, 2006.

[4] C. P. Lee, N. G. M. Tao, and B. J. F. Lin, "Studies of safe operating area of InGaP/GaAs heterojunction bipolar transistors," *IEEE Trans. Electron Devices*, vol. 61, no. 4, pp. 943–949, 2014.

[5] R. Jin, "Sub-nanosecond pulse characteristics of InGaP/GaAs HBTs," Lehigh Univ. Lehigh Pres., Tech. Rep., 2011.

[6] B. R. Lin, N. G. M. Tao, C. P. Lee, T. Henderson, and B. J. F. Lin, "2D numerical simulation for InGaP/GaAs HBT safe operating area," in *The 9th European Microwave Integrated Circuits Conference.*, 2014, pp. 1–4.

[7] C. Mukerjee and C. K. Maiti, "Simulation and modeling of self-heating effects in heterojunction bipolar transistors," *J. Basics and Appl. Phys.*, vol. 3, no. 1, pp. 16–25, 2014.

[8] N. Rinaldi and V. d'Alessandro, "Theory of electrothermal behavior of bipolar transistors: Part I - single-finger devices," *IEEE Trans. Electron Devices*, vol. 52, no. 9, pp. 2009–2021, 2005.

[9] ——, "Theory of electrothermal behavior of bipolar transistors: Part III - impact ionization," *IEEE Trans. Electron Devices*, vol. 53, no. 7, pp. 1683–1697, 2006.

[10] P. J. Robertson, "HBT compact thermal model," Sandia National Laboratories, Tech. Rep., 2018.

[11] https://charon.sandia.gov/.

[12] S. Choi and G. M. P. et al., "Thermal design and characterization of heterogeneously integrated InGaP/GaAs HBTs," *IEEE Trans. Compon. Packag. Manuf. Technol.*, vol. 6, no. 5, pp. 740–748, 2016.

[13] X. Gao, G. L. Hennigan, L. Musson, A. Huang, and M. Negoita, "Simulation and investigation of electrothermal and dose rate effects in heterojunction bipolar transistors," *submitted to J. radiation effects research and engineering*, 2019.

[14] W. R. Wampler and S. M. Myers, "Model for transport and reaction of defects and carriers within displacement cascades in gallium arsenide," *J. Appl. Phys.*, vol. 117, no. 045707, 2015.

[15] G. E. Stillman, V. M. Robbins, and K. Hess, "Impact ionization in InP and GaAs," *Physica*, vol. 134B, pp. 241–246, 1985.

[16] S. A. Plimmer, J. P. R. David, G. J. Rees, and P. N. Robson, "Ionization coefficients in $Al_x Ga_{1-x} As$ ($x = 0 - 0.60$)," *Semicond. Sci. Technol.*, vol. 15, pp. 692–699, 2000.

[17] C. Groves, R. Ghin, J. P. R. David, and G. J. Rees, "Temperature dependence of impact ionization in GaAs," *IEEE Trans. Electron. Devices*, vol. 50, no. 10, pp. 2027–2031, 2003.

[18] X. Gao, B. Kerr, and A. Huang, "Analytic band-to-trap tunneling model including band offset for heterojunction devices," *J. Appl. Phys.*, vol. 125, no. 054503, 2019.

978-1-7281-0941-1/19 $31.00 © 2019 IEEE

Variability of Threshold Voltage Induced by Work-Function Fluctuation and Random Dopant Fluctuation on Gate-All-Around Nanowire nMOSFETs

Wen-Li Sung
Department of Electrical and
Computer Engineering,
Center for mmWave Smart Radar
System and Technologies,
National Chiao Tung University,
Hsinchu 30010, Taiwan.

Min-Hui Chuang
Department of Electrical and
Computer Engineering,
Center for mmWave Smart Radar
System and Technologies,
National Chiao Tung University,
Hsinchu 30010, Taiwan.

Yiming Li
Department of Electrical and
Computer Engineering,
Center for mmWave Smart Radar
System and Technologies,
National Chiao Tung University,
Hsinchu 30010, Taiwan.
Email: ymli@faculty.nctu.edu.tw

Abstract—We advance the localized work-function fluctuation (LWKF) method to examine the variability of threshold voltage (V_{th}) induced by titanium nitride (TiN) metal-gate work-function fluctuation (WKF) and combined the WKF with the random dopant fluctuation (RDF) for various grain sizes on Si gate-all-around (GAA) nanowire (NW) MOSFETs. Our results show that the WKF-induced variability of V_{th} will be dominated by bamboo-type TiN grains and its impact is larger than that induced by the RDF with doped channel (RDF (doped)). Additionally, the variability of V_{th} induced by the WKF and the RDF (doped) could be treated as independent fluctuation sources because the channel dopants are away from the metal-gate/high-κ interface. Consequently, statistical models are further proposed for the σV_{th} induced by the WKF and the combined WKF with RDF (doped) by considering position effect of nanosized TiN grains.

Keywords—GAA; Nanowire; TiN; Work function fluctuation; Random dopant fluctuation; Grain size; Statistical model.

I. INTRODUCTION

Nano-sized nanowire (NW) metal-oxide semiconductor field-effect transistors (MOSFETs) with high-κ/metal gate technology become an attractive device to replace the fin field-effect transistors (FinFETs) for sub-5-nm technological nodes due to better electrical characteristics [1]-[3]. However, the crystal orientation of nanosized metal grains is hard to control during the high temperature process in relative smaller gate area [4]-[5]. This will cause the variability of V_{th} owing to different value of work function at the metal-gate/high- interface. It was reported that the WKF-induced V_{th} variability for various metal materials can be modeled below by using the average WKF (AWKF) method [6]:

$$\sigma_{th} = A_{VT}(GS/A)^{0.5}, \qquad (1)$$

where GS is the average grain size and A is the metal gate area, and A_{VT} (mV) is a fitting factor depending on various metals. However, the AWKF method did not consider the random position effect of metal grains and it will underestimate the

impact of WKF-induced variability of V_{th} [7]. Thus, in this study, the LWKF method is further advanced to estimate the variability of V_{th} considering the location effect for exact grain size ratio (GSR = GS/A). Explicit statistical expressions for the σV_{th} are then discussed.

II. STATISTICAL DEVICE SIMULATION METHOD

Figure 1(a) illustrates the schematic of the GAA NW device with various types of RDDs that are statistically generated by the Monte Carlo (MC) method [9]. The simulated structure consists of a metal-gate (TiN)/high-κ (HfO₂) stack with cylindrical Si channel, where the effective work-function (WK) of the TiN metal is set to be 4.552 eV, the gate length is 10 nm, the radius of cylindrical channel is 5 nm, and the effective oxide thickness is 0.6 nm. The various RDDs are generated by the equivalent channel doping concentration with 5×10^{17} cm^{-3}, the equivalent S/D extension doping concentration with 5×10^{18} cm^{-3}, and dopants concentrations with 3.36×10^{17} cm^{-3} resulting from the penetration of the S/D extension, respectively [9]. Fig. 1(b) shows a flow chart of LWKF method. First, metal gate is partitioned by different grain sizes. Then, we set different WK and probability (p) in the MC program for different metal grain orientations: TiN<200> (p = 0.6, WK = 4.63 eV) and TiN<111> (p = 0.4, WK = 4.43 eV), respectively [10]. Finally, we generate 200 fluctuated 3D devices with different grain sizes by the MC program to explore the effect of the WKF and combined the WKF and RDF (doped) on the device variability.

III. RESULTS AND DISCUSSION

Figure 2 presents the cumulative probability of V_{th} induced by the RDF, the WKF, and the combined WKF and RDF with doped channel. It is obvious that the distributions of V_{th} with these simulation conditions were close to Gaussian and continuous distribution. However, it differs from other researcher's results with uniformly square grain pattern method

978-1-7281-0941-1/19 $31.00 © 2019 IEEE

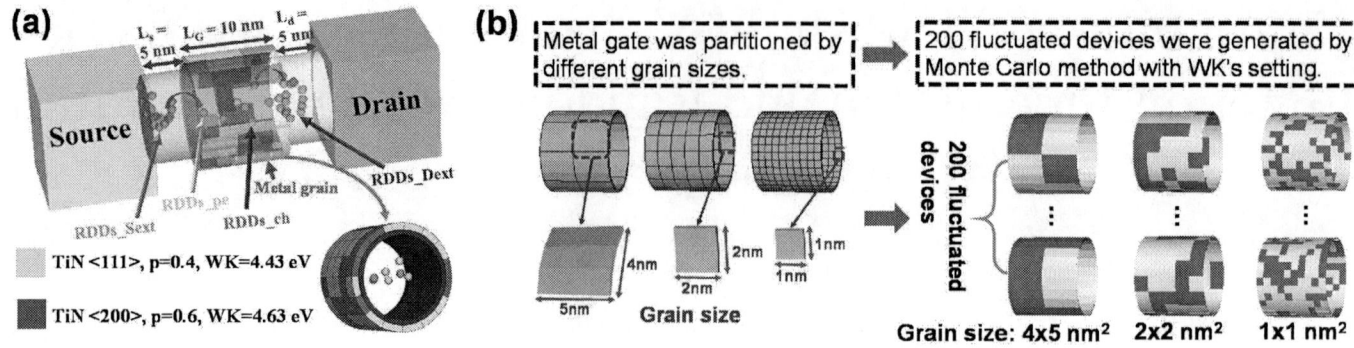

Fig. 1. (a) Schematic of the GAA NW device with various types of RDDs and metal grains, where the gate length is 10 nm, the radius of cylindrical channel is 5 nm, and the effective oxide thickness is 0.6 nm [9]. (b) Flow chart of LWKF method, which partitioned by three grain sizes in the same metal gate area.

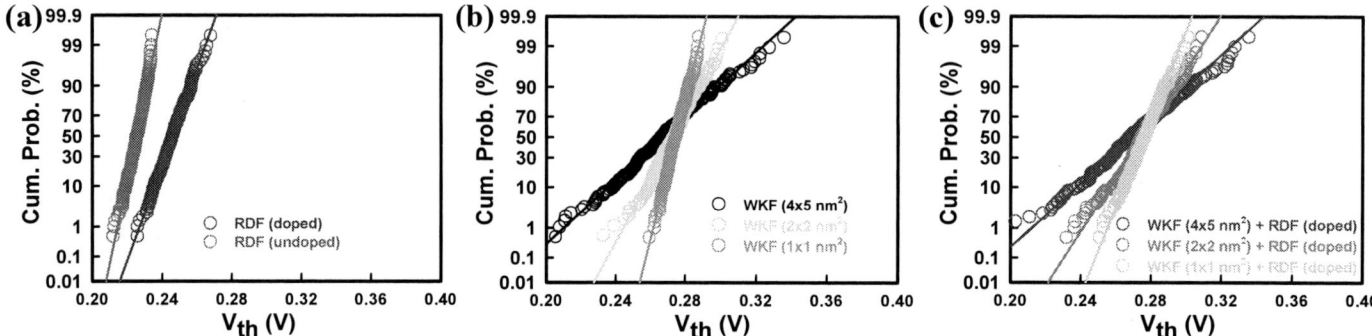

Fig. 2. The cumulative probability of V_{th}. (a) RDF, (b) WKF, and (c) Combined WKF with RDF (doped).

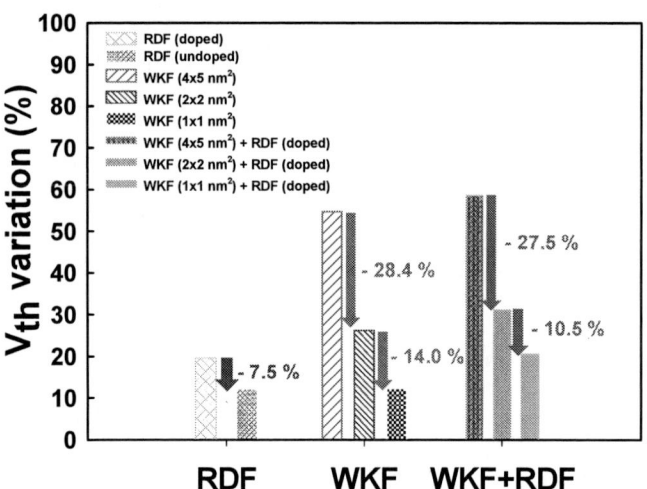

Fig. 3. The V_{th} variation induced by the RDF, the WKF and the combined WKF with RDF (doped), respectively.

[11]. As indicated in Fig. 2(a), the slope of the V_{th} distribution induced by doped channel was smaller than that of the undoped channel condition. The results indicate that the V_{th} variation can be improved without channel doping for the RDF simulation. Fig. 2(b) shows the distribution of V_{th} induced by the WKF with different grain size patterns. It is observed that the grain size increases, the slope of the V_{th} distribution increases in value. It means that the reduction of the V_{th} variation can be controlled by smaller grain size. We will further discuss the grain size

effect by the GSR from statistical perspective later. Similarly, the distribution of V_{th} induced by the combined WKF and RDF with doped channel, as shown in Fig. 2(c), has the same trend as compared to Fig. 2(b).

To estimate and compare the impact of the V_{th} variation induced by the RDF, the WKF and the combined WKF and RDF with doped channel, we normalized the V_{th} variation by $(6\sigma/mean) \times 100\%$, where σ is the standard deviation of the characteristic parameters, and mean is the average of the characteristic parameters. Fig. 3 shows the V_{th} variation induced by the RDF, the WKF and the combined WKF and RDF with doped channel, respectively. The reduction of the V_{th} variation induced by the RDF with undoped channel is 7.5%, as compared to the RDF with doped channel. It is smaller than the cases of the WKF (the total reduction rate is the sum of 28.4% and 14.0%) and the combined WKF and RDF with doped channel (the total reduction rate is the sum of 27.5% and 10.5%). In contrast, the V_{th} variation (54.66%) induced by the WKF with largest grain size (bamboo-type grain structure) in our simulation has larger variation than that induced by the RDF with doped channel (19.63%). While the WKF with grain size is below 4 nm², the V_{th} variation induced by the RDF with doped channel will be comparable to that induced by the WKF (12.24% ~ 26.2%). This implies that the grain size effect induced by the WKF may be larger than that the random-discrete doping effect in the channel.

To analyze the grain size effect and statistically empirical equations, we employ the GSR to an x-axis [12]-[13] for the plot

978-1-7281-0941-1/19 $31.00 © 2019 IEEE   172

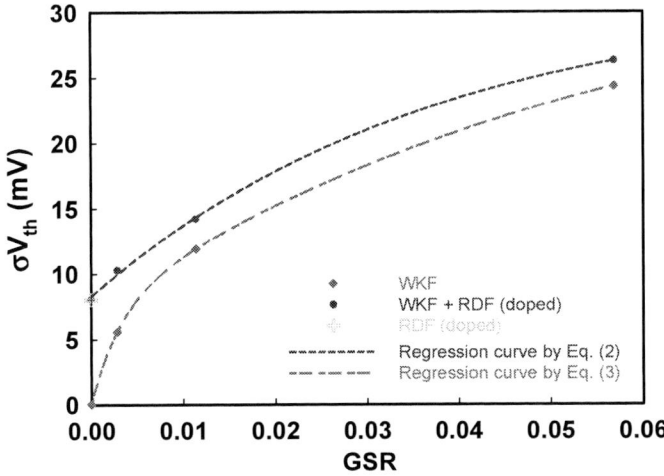

Fig. 4. Plot of the σV_{th} versus GSR, where the blue circle dot is the σV_{th} induced by WKF + RDF (doped channel), the red diamond dot is the σV_{th} induced by WKF, the green cross is the σV_{th} induced by the RDF (doped), the blue short-dash line is the regression curve by Eq. (2) and the red short-long line is the regression curve by Eq. (3).

of the standard deviation of the V_{th} (σV_{th}) versus GSR, as shown in Fig. 4. We consider the range of the GSR in the x-axis from 0 to 0.06 (i.e, from the amorphous-type to the bamboo-type grain structure) in our simulation. First, we apply the non-linear regression curves to fit the simulation results by the optimized method. Because of uniformly square grain size method in this simulation, the value of the GSR is exactly plot in the x-axis. Thus, we only consider the sum of squares of residuals for y-axis (σV_{th}). Moreover, we assume that the data points are independent and statistical noise-free. Then, we try to employ the following empirical Eq. (2) and Eq. (3) to fit the curves of σV_{th} versus GSR induced by the combined WKF and RDF with doped channel, and the WKF, respectively, due to best correlation coefficient (the values of the R^2 are close to 1).

$$\sigma V_{th\_RDF+WKF} = \sigma V_{th\_RDF} + a \times (1 - e^{-b \times GSR}) \quad (2)$$

$$\sigma V_{th\_WKF} = a \times (1 - e^{-b*GSR}) + c \times (1 - e^{-d \times GSR}), \quad (3)$$

where the a,b,c, and d are the extracted parameters.
At the GSR = 0, as shown in Table I, the value of the σV_{th_RDF} in Eq. (2) is close to the value of the σV_{th} (8.05 mV) induced by the RDF with the doped channel (green cross). To analyze the different regions of the GSR for grain size effect, we differentiate the Eq. (2) and Eq. (3) to obtain the slopes of the curves for the whole GSR, respectively, as following equations:

$$\frac{d(\sigma V_{th\_RDF+WKF})}{d(GSR)} = a \times b \times \left(e^{-b \times GSR}\right) \quad (4)$$

$$\frac{d(\sigma V_{th\_WKF})}{d(GSR)} = a \times b \times (e^{-b \times GSR}) + c \times d \times (e^{-d \times GSR}) \quad (5)$$

For smaller grain size (i.e, the GSR is below 0.01), we apply the Taylor's expansion to Eq. (2) and Eq. (3), respectively. We can find the linear relationship between the σV_{th} and the GSR, where the parameters {a×b} for Eq. (2) and {a×b + c×d} for Eq. (3) are the slopes of the linear equations, respectively. Many research groups in the world-wide also found this linear behavior [11]-[13]. Furthermore, the slopes will determine the

Table I. The extracted parameters from Eq. (2) and Eq. (3).

Equation	σV_{th_RDF}	a	b	c	d
Eq. (2)	8.27	23.0	26.8	0	0
Eq. (3)	0	72.6	31.6	27.5	17.1

changing rate of the V_{th} variation according to Eq. (4) and Eq. (5). As the GSR increases, the slopes decrease due to exponential decay factor. In the smaller GSR, the slope of the combined WKF and RDF with doped channel is smaller than that of the WKF. It implies that the impact of the V_{th} variation induced by the RDF with doped channel is comparable to that of the WKF for smaller grain size. The slopes may be dominated by the metal orientations of various materials that depend on the different generated probabilities. Thus, various metal materials or composited metal materials may alter the behavior of the V_{th} variation induced by the WKF. For the larger GSR (> 0.01), the curve of the σV_{th} versus the GSR will become linear-like saturation [11]-[13] and it is controlled by the asymptote parameter {a} or {c}. Although the freedom of data points will influence the extracted parameters, fitting curves and 95% confidence intervals, based on our simulation results and other researchers' results [11]-[13], we believe that the non-linear empirical equations in our study is one of best ways to describe the metal grain size effect in our limited knowledge. Finally, the WKF and the RDF can be viewed as independent fluctuation sources since the statistical sum of the WKF with the RDF (doped) can be approximated by Eq. (6) due to the dopants away from the TiN/HfO$_2$ interface.

$$\sigma V_{th\_RDF+WKF} = \sqrt{\left(\sigma V_{th\_RDF}\right)^2 + \left(\sigma V_{th\_WKF}\right)^2}. \quad (6)$$

IV. CONCLUSIONS

In summary, we have found that the grain size/metal gate area and location of metal grains are critical to suppress the variability of V_{th}. The smaller grain size will have a smaller V_{th} variation induced by the WKF. For comparison of the impact of the WKF and the RDF, the results show that the V_{th} variation (54.66%) induced by the WKF with largest grain size (bamboo-type grain structure) has larger impact than that induced by the RDF with doped channel (19.63%). If the GSR is larger than 0.01, the grain size effect induced by the WKF will be more important than that induced by the random-discrete doping effect in the channel. And, according to statistically non-linear empirical equations in our study, in the smaller GSR region, the curves of the σV_{th} versus the GSR will be linear relationship. The changing rate of the σV_{th} are depending on the slopes of the curves. In larger GSR region, the curves of the σV_{th} versus the GSR become linear-like saturation. Moreover, the slopes of the fitting curves may be influenced by the distribution of metal orientations with different materials. The statistical models of σV_{th} offer a new perspective relating to grain size, metal gate area, and different metal-gate materials for sub-5-nm technological nodes.

ACKNOWLEDGMENT

This work was supported in part by the Ministry of Science and Technology, Taiwan, under Grant MOST 106-2221-E-009-149, Grant MOST 106-2622-8-009-013-TM, Grant MOST 107-2622-8-009-011-TM, Grant MOST 107-2221-E-009-094, and Grant MOST 107-3017-F-009-001, in part by the "Center for mmWave Smart Radar Systems and Technologies" under the Featured Areas Research Center Program within the framework of the Higher Education Sprout Project by the Ministry of Education in Taiwan.

REFERENCES

[1] J. Appenzeller, J. Knoch, M. T. Björk, H. Riel, H. Schmid, and W. Riess, "Toward nanowire electronics," IEEE Trans. Electron Devices, vol. 55, no. 11, pp. 2827–2845, Nov. 2008, doi: 10.1109/TED.2008.2008011.

[2] H. Mertens, R. Ritzenthaler, V. Penal, G. Santoro1, K. Kenis, A. Schulze, E. D. Litta, S. A. Chew, K. Devriendt, T. Chiarella, S. Demuynck, D. Yakimets, D. Jang, A. Spessot, G. Eneman, A. Dangol, P. Lagrain, H. Bender, S. Sun, M. Korolik, D. Kioussis, M. Kim, K-.H. Bu, S. C. Chen, M. Cogorno, J. Devrajan, J. Machillot, N. Yoshida, N. Kim, K. Barla, D. Mocuta, N. Horiguchi, "Vertically stacked gate-all-Around Si nanowire transistors: key process optimizations and ring oscillator demonstration," in IEDM Tech. Dig., pp.828–831, 2017, doi: 10.1109/IEDM.2017.8268511.

[3] H. Mertens, R.S. Barraud, V. Lapras, B. Previtali, M.P. Samson, J. Lacord, S. Martinie, M.-A. Jaud, S. Athanasiou, F. Triozon, O. Rozeau, J.M. Hartmann, C. Vizioz, C. Comboroure, F. Andrieu, J.C. Barbé, M. Vinet, and T. Ernst, "Performance and design considerations for gate-all-around stacked-nanowires FETs," in IEDM Tech. Dig., pp.677–680, 2017, doi: 10.1109/IEDM.2017.8268473.

[4] M. M. Hussain1, M. A. Quevedo-Lopez1, H. N. Alshareef, H. C. Wen, D. Larison, B. Gnade and M. El-Bouanani, "Thermal annealing effects on a representative high-k/metal film stack," S. S. Technol., vol. 21, no. 10, pp. 1437-1440, 2006, https://iopscience.iop.org/article/10.1088/0268-1242/21/10/012.

[5] J.L. He, Y. Setsuhara, I. Shimizu, S. Miyake, "Structure refinement and hardness enhancement of titanium nitride films by addition of copper," Surf. Coat. Technol., vol. 137, pp. 38-42, 2001, https://doi.org/10.1016/S0257-8972(00)01089-6.

[6] C.-H. Yu et al., "Statistical Simulation of Metal-Gate Work-function Fluctuation in High-/Metal-Gate Devices," in Proc. SISPAD, 2010, pp. 153–156, doi: 10.1109/SISPAD.2010.5604544.

[7] H.-W. Cheng and Y. Li, "Random Work Functions Induced DC and Dynamic Characteristic Fluctuations in 16-nm High-κ/Metal Gate CMOS Device and Digital Circuit," in ASQED, pp. 203-206, 2011, doi: 10.1109/ASQED.2011.6111745.

[8] Y. Li, H.-T. Chang, C.-N. Lai, P.-J. Chao, and C.-Y. Chen, "Process variation effect, metal-gate work-function fluctuation and random dopant fluctuation of 10-nm gate-all-around silicon nanowire MOSFET devices," in IEDM Tech. Dig., pp. 887–890, Dec. 2015, doi: 10.1109/IEDM.2015.7409827.

[9] W.-L. Sung and Y. Li, "DC/AC/RF characteristic fluctuations induced by various random discrete dopants of gate-all-around silicon nanowire n-MOSFETs," IEEE Trans. Electron Devices, vol. 65, no. 6, pp. 2638-2646, Jun. 2018, doi: 10.1109/TED.2018.2822484.

[10] A. Yagishita, T. Saito, K. Nakajima, S. Inumiya, K. Matsuo, T. Shibata, Y. Tsunashima, K. Suguro, and T. Arikado, IEEE Trans. Electron Devices, vol. 48, no. 8, pp. 1604–1611, 2001, doi: 10.1109/16.936569.

[11] S.-H. Chou, M.-L. Fan, and P. Su, "Investigation and comparison of work function variation for FinFET and UTB SOI devices using a voronoi approach," IEEE Trans. Electron Devices, vol. 60, no. 4, pp. 1485–1489, Apr. 2013, doi: 10.1109/TED.2013.2248087.

[12] D. Nagy, G. Indalecio, A. J. Garcia-Loureiro, M. A. Elmessary, K. Kalna, and N. Seoane, "Metal grain granularity study on a gate-all-around nanowire FET," IEEE Trans. Electron Devices, vol. 64, no. 12, pp. 5263–5269, 2017, doi: 10.1109/TED.2017.2764544.

[13] H. Nam, Y. Lee, J.-D. Park, and C. Shin, "Study of work-function variation in high-κ/metal-gate gate-all-around nanowire MOSFET," IEEE Trans. Electron Devices, vol. 63, no. 8, pp. 3338–3341, Aug. 2016, doi: 10.1109/TED.2016.2574328.

A First Principle Insight into Defect Assisted Contact Engineering at the Metal-Graphene and Metal-Phosphorene Interfaces

Jeevesh Kumar, Adil Meersha, Ansh and Mayank Shrivastava
Department of Electronic Systems Engineering
Indian Institute of Science
Bengaluru, India -560012
Email: jeevesh@iisc.ac.in

Abstract—In this work we have studied bonding nature of Graphene and Phosphorene with metal (Pd) followed by carrier transport behavior and contact resistance engineering across the metal-Graphene and the metal-Phosphorene interfaces using Density Functional Theory (DFT) and Non Equilibrium Green's Function (NEGF) computational methods. We have studied, how carrier transports at the interfaces is limited by van der Waals (vdW) gap across the interfaces and how the gap can be reduced by creating the Carbon vacancy (defect engineering) at the Graphene-Palladium interface. We have seen that the defect engineering enhances the Carbon-Palladium bond at the interface which reduces the van der Walls (vdW) gap, hence contact resistance due to corresponding reduction in the tunneling barrier width at the interface. We have also studied that the defect engineering (Phosphorous vacancy) at the Phosphorene-Palladium interface is not effective as Graphene-Palladium interface because it has less interfacial (vdW) gap than Graphene-Palladium interface intrinsically.

Keywords—DFT; NEGF; vdW

I. INTRODUCTION

Invention of Graphene [1] along with other 2D materials (eg. TMDs, Phosphorene) [2] has opened new door for the semiconductor device scaling research but very soon it has been realized that the contact resistance at the metal-2D material interface [2], [3] is the main bottleneck for their performance improvements. The contact resistance at the interface is high due to Schottky barrier, tunnel barrier and various kinds of trap states (eg. defect assisted, metal induced) at the interface [3]. Defect engineering (creating vacancy at the metal-2D material interface) is a promising technique for contact resistance improvement in the Graphene based transistors [4], [5], but its physics and chemistry at the interface is needed to be explored more. Defect engineering at other 2D elemental material-metal contacts like Phosphorene-metal is not explored yet so it can also be explored to see the impact of the defect engineering on the carrier transport through its interfaces and also corresponding physics and chemistry at the interface.

In this work, we have explored the interface chemistry and carrier transport physics of Graphene and Phosphorene with metal (Pd) using DFT and NEGF methods supported in Quantum ATK package [6], [7], [8]. We have studied how Graphene-metal and Phosphorene-metal bonding properties at their contact interfaces are changed when Carbon (C) and

Phosphorous (P) vacancies are created (defect engineering) at the intrinsic Graphene-Palladium and Phosphorene-Palladium contacts respectively. We have studied how carrier transmission probability and hence contact resistance at the interfaces changes due to change in interfacial bonding distance by these defect engineering. We have explained the improvement in the contact resistance due to defect engineering at the Graphene-Palladium interface by corresponding improvement in the carrier transmission probability. Finally after transport current analysis at the interfaces, we have concluded that defected engineering is a promising way to improve the contact resistance (~2.3 times) at the Graphene-Palladium interface but it is not promising way for the Phosphorene- Palladium interface.

II. COMPUTATION DETAILS

All the bulk structures are optimized with 0.01 eV/Å force tolerance and 0.001 eV/Å^3 energy tolerance. Palladium (Pd) is cleaved in [111] direction before creating Graphene-Palladium (Gr.-Pd) and Phosphorene-Palladium (Ph.-Pd) interfaces for minimum interfacial strain, which is applied on the metal to minimize the strain effect on the Graphene/Phosphorene. The Interface width and length are 8.382 Å and 12.062 Å for Graphene- Palladium and, 16.49 Å and 9.92 Å for Phosphorene-Palladium interfaces respectively. The density mesh cutoff is 75 Hartree with 10 k-points along the width and 200 k-points along the channel of the devices in the calculation with local density approximation (LDA) exchange correlation functional. A source–drain bias of 250 mV is applied to conduct the carrier transport analysis on 300K electron temperature.

III. RESULTS AND ANALYSIS

When the carrier (electron or hole) transports through the metal-2D material interface, it encounters tunnel barrier and Schottky barrier at the interface. There are following four parameters which determines the carrier transport probability at the interface: (i) Tunnel barrier width; (ii) Tunnel barrier height; (iii) Schottky barrier width and (iv) Schottky barrier height (Figure 1). Tunnel barrier width is the van der Waals (vdW) gap between metal and 2D materials which can be reduced by enhancing metal and 2D materials bonding by creating vacancy at the interfaces.

Authors would like to acknowledge Council for Scientific & Industrial Research (CSIR) and Department of Science and Technology (DST), Government of India for providing financial support..

In the First step, we have computed bond energy minima optimization for the Graphene-Pd and Phosphorene-Pd interfaces using DFT for their intrinsic interface and interfaces with vacancy. It is observed that the Carbon (C) vacancy at the Graphene-Pd interface region improves the 'Pd-C' covalent bond while Phosphorous (P) vacancy does not have a major impact on the bonding at Phosphorene-Pd interface (Figure 2).

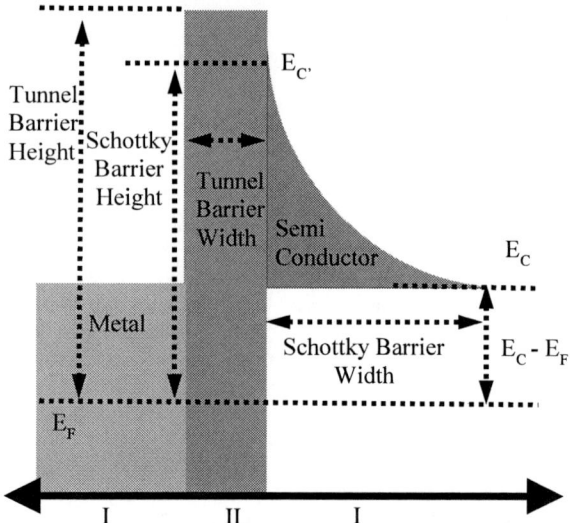

Figure 1: Schematic of different types of barriers in the carrier transport at the metal-2D materials interface.

Graphene has weak vdW interaction with 'Pd' due to unavailability of vacant orbital in the 'C' atom. When a 'C' vacancy is created at the interface, Graphene-Pd bond becomes stronger due to bonding of 'Pd' with vacant orbitals available at the three nearest neighbor 'C' atoms of the vacant site which reduces average bond length from 2.61 Å to 2.50 Å. Phosphorene has relatively stronger bond with 'Pd' intrinsically due to interaction of 'Pd' electron cloud with vacant 3d orbitals available in Phosphorous. However, unlike Graphene-Pd interface, the bond length is not significantly

affected (2.20 Å to 2.18 Å) in the presence of 'P' vacancy at the Phosphorene-Pd interface.

Contact resistance of the metal-semiconductor interface is determined by carrier (electron for this work) transmission probability (CTP) and density of state (DOS) near fermi level of the interface. CTP reflects how easily the electron can move from metal to semiconductor (and vice versa) at the interface. When electron flows from metal to semiconductor across interface, it moves from fermi level state of the metal side to the conduction band minima (CBM) of the semiconductor side. There are mainly two barriers available at the interface to throttle the electron flow, Schottky barrier and tunnel barrier. The tunnel barrier depends on the interfacial distance between metal and semiconductor. Defect engineering on the Graphene-Pd interface reduces the interfacial distance and hence reduces the tunnel barrier (Figure 3). Reduction in the tunneling barrier enhances the electron transmission probability of the electron at the interface (Figure 4).

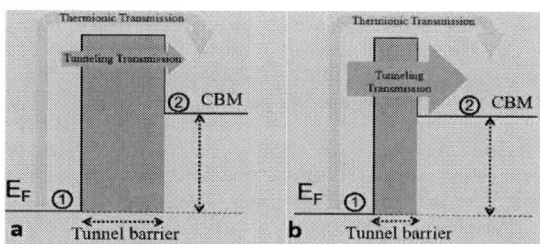

Figure 3: Representation of tunneling distance and hence electron transmission probability in the (a) intrinsic and (b) defect engineered Graphene-Pd interfaces. 'b' has more electron transmission probability than 'a' due to less tunneling distance.

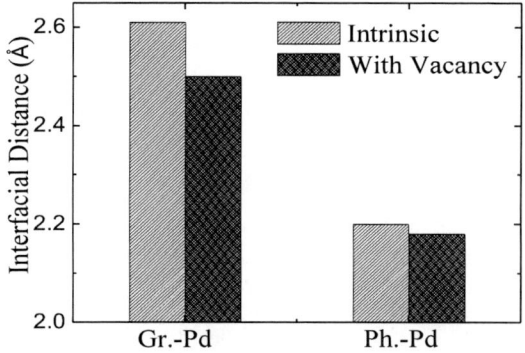

Figure 2: Defect Engineering (Creating Vacancy) reduces interfacial distance significantly (2.61 Å to 2.50 Å) for Graphene-Pd interface but it doesn't have significant impact on the interfacial distance (2.20 Å to 2.18 Å) of the Phosphorene-Pd interface.

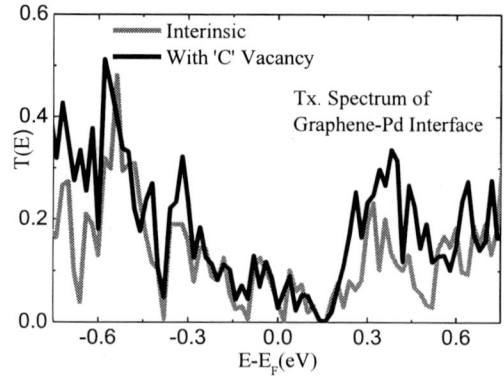

Figure 4: Transmission probability of electron at the intrinsic and defect engineered Graphene-Pd interfaces, computed using DFT and NEGF. Defect engineered interface has better transmission probability than intrinsic interface.

Intrinsic Phosphorene-Pd interface has less interfacial distance and hence tunneling distance than intrinsic Graphene-Pd interface so it has better electron transmission probability than Graphene-Pd interface as shown in Figure 5. Since defect engineering in the Phosphorene-Pd interface hasn't significant impact on the interfacial and hence tunneling distance, so it has similar transmission spectrum as corresponding intrinsic interface (Figure 6)

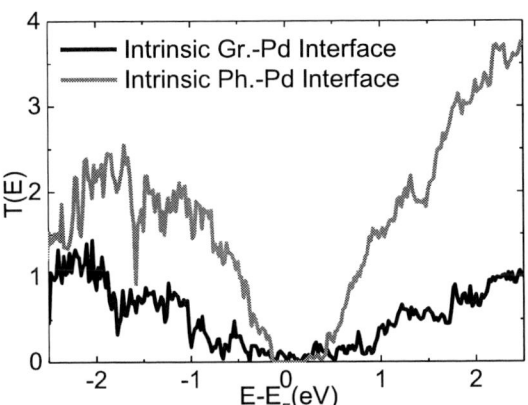

Figure 5: Transmission probability of electron at the intrinsic Graphene-Pd and Intrinsic Phosphorene-Pd interfaces, computed using DFT and NEGF. Intrinsic Graphene-Pd interface has better transmission spectrum than intrinsic Phosphorene-Pd interface.

Enhancement in the electron transmission probability due to defect engineering reduces the contact resistance and hence increases the current flow across the Graphene-Pd interface.

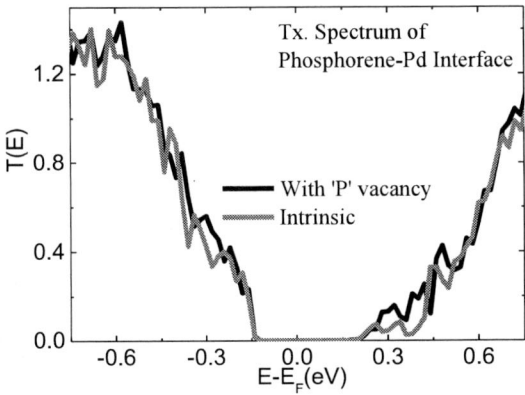

Figure 6: Transmission probability of electron at the intrinsic and defect engineered Pd-Phosphorene interfaces, computed using DFT and NEGF. Both interfaces have similar transmission spectrum

Defect engineering doesn't improve the transmission probability at the Phosphorene-Pd interface hence it doesn't improve the current but decreases due to excess sheet resistance added by Phosphorous vacancy at the interface (Figure 7).

IV. CONCLUSION

We have studied the impact of the defect engineering on the Pd-C and Pd-P bonds at the Graphene-Palladium and Phosphorene-Palladium interfaces respectively. We have seen that defect engineering reduces the average Pd-C bond length but it doesn't change average Pd-P bond length significantly at corresponding interfaces. Reduction in the bond length enhances the carrier transmission probability and hence reduces the contact resistance (~2.3 times) due to reduction in the tunnel barrier at the Graphene-Palladium interface. Reduction in the contact resistance due to defect engineering at the Graphene-Palladium interface is in agreement with the reported experimental work [5]. 'P' vacancy doesn't result in a significant improvement in contact resistance at Phosphorene- Palladium interface due to insignificant changes in the tunnel barrier.

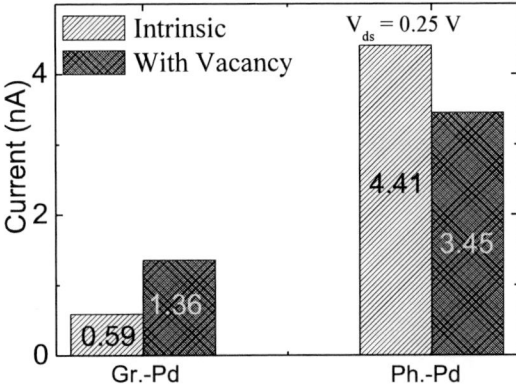

Figure 7: Computed current values across the intrinsic and defect engineers Graphene-Pd and intrinsic and defect engineers Phosphorene-Pd interfaces respectively using DFT and NEGF. Defect engineered Graphene-Pd interface has higher current than corresponding intrinsic interface. Defect engineered Phosphorene-Pd interface has lower current than corresponding intrinsic interface. Phosphorene-Pd interface has higher current than corresponding Graphene-Pd interface.

V. ACKNOLEDGMENT

Student Authors would like to thank the Nano device team of Mayank Shrivastava's Device Physics Lab, DESE, IISc Bangalore, India. For their fruitful discussions.

978-1-7281-0941-1/19 $31.00 © 2019 IEEE

VI. REFERENCES

[1] K. S. Novoselov, A. K. Geim, S. V. Morozov, D. Jiang, Y. Zhang, S. V. Dubonos, I. V. Grigorieva and A. A. Firsov, "Electric Field Effect in Atomically Thin Carbon Films," Science, Vol. 306, No. 5696, 2004, pp. 666-669. doi:10.1126/science.1102896.

[2] F. Schwierz, J. Pezoldt and R. Granzner "Two-dimensional materials and their prospects in transistor electronics", Nanoscale, 2015, 7, 8261. doi: 10.1039/c5nr01052g.

[3] Daniel S. Schulman, a Andrew J. Arnold b and Saptarshi Das, "Contact engineering for 2D materials and devices", rsc.li/chem-soc-rev, 2018, doi: 10.1039/c7cs00828g.

[4] Mayur Ghatge and Mayank Shrivastava, "Physical Insights on the Ambiguous Metal–Graphene Interface and Proposal for Improved Contact Resistance", IEEE Transactions on Electron Devices, 4139 - 4147, Volume: 62 , Issue: 12 , Dec. 2015, doi: 10.1109/TED.2015.2481507.

[5] Adil Meersha, H. B. Variar, K. Bhardwaj, A. Mishra, S. Raghavan, N. Bhat, Mayank Shrivastava, "Record low metal — (CVD) graphene contact resistance using atomic orbital overlap engineering", Electron Devices Meeting (IEDM) 2016 IEEE International, pp. 5.3.1-5.3.4, 2016, doi: 10.1109/IEDM.2016.7838352.

[6] Quantumwise ATK, version 2014.0, Quantumwise, Copenhagen, Denmark, 2014.

[7] Brandbyge M, Mozos J. L, Ordejon P, Taylor J, Stokbro K. "Density-functional method for non-equilibrium electron transport". Physical Review B, 65(16):165401, 2002

[8] Simon M. Sze, Kwok K. Ng. Physics of Semiconductor Devices.

Transient Simulation of Field-Effect Biosensors : How to Avoid Charge Screening Effect

Kyoung Yeon Kim
Department of Electrical and Computer Engineering
Seoul National University
Seoul, South Korea
kim77ky@snu.ac.kr

Byung-Gook Park
Department of Electrical and Computer Engineering
Seoul National University
Seoul, South Korea
bgpark@snu.ac.kr

Abstract—**We developed a numerical simulator to model the operation of a bio-FET in transient state. The simulator takes a realistic device structure as a simulation domain, and it employs the drift-diffusion equation for ion / channel transport, and the Ramo-Shockley theorem for accurate calculation of non-faradaic current. For efficient transient simulation, the implicit time integration scheme is employed where the solution at each time step is obtained from the coupled Newton-Raphson method. Using the simulator, we found that the sensitivity of bio-FET can be improved with transient measurement by redistribution of the mobile ions by an external electric field.**

Keywords—bio-FET, charge screening effect, transient simulation)

I. INTRODUCTION

Field-effect transistor-based biosensors (bio-FET) have been used to detect various biomolecules [1], [2]. In this type of biosensor, the probe molecules are attached to the FET gate, and the probe-target binding event occurs changing conductance of the channel. It is usually measured in the steady state. However, there are research results that higher sensitivity can be obtained by measuring in the transient state after applying the step pulse bias [1], [2]. About this, only one-dimensional conceptual simulation and experimental results were reported; lacking in detailed theoretical study about what happens in the transient state [1]. Therefore, in this study, we have developed a physics-based 3-dimentsional simulator to analyze the terminal current of the realistic device structure including detailed surface physics. Since the buffer solution is used as an electrolyte in many practical experiments, a numerically efficient transient simulation method has been developed to simultaneously include various kinds of ion movements and reactions and couple with the model for the channel transport. Also, we considered the biomolecule layer as an ion-permeable membrane including the partitioning effect from the Born equation. We applied the developed simulator to the actual device structure and showed that the movement of ions in the electrolyte in the transient state affects the channel conductance, which improves the sensitivity.

II. PHYSICS MODELS

The device structure used in the simulation is shown in Fig.1. This structure is based on the fabricated reference-electrode free device structure. The potential of the electrolyte is close to that of the source electrode due to the large capacitive coupling between the source electrode and the electrolyte solution. Therefore, the source electrode acts as a reference electrode and no additional reference electrode is required.

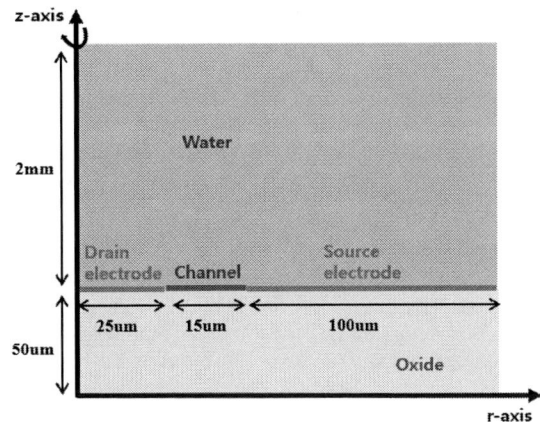

Fig. 1. Schematic of the simulation domain. The cylindrical coordinates are chosen to take advantage of the concentric feature of the electrode

A. The Poisson Equation

The electric potential in the device is governed by the Poisson equation.

$$\nabla^2 V = -\rho/\varepsilon_0 \qquad (1)$$

B. Transport of ions

We considered the phosphate-buffered saline (PBS) buffer solution usually used in experiments [3]. For the charge transport in the solution, we used the drift diffusion equation for each ions ($Na^+, K^+, Cl^-, H^+, OH^-, HPO_4^{2-}, H_2PO_4^-$):

$$J = -zu[n]\nabla V - D\nabla[n] \qquad (2)$$

Where u, D, z and [n] are the mobility, diffusivity, charge state and density of the corresponding ion. We assume a constant mobility and the diffusivity is obtained from the Einstein relation. Since the biomolecules generally react with the

978-1-7281-0941-1/19 $31.00 © 2019 IEEE

hydrogen ion, the pH value should be carefully modeled according to the buffer reactions near the metal electrode and solution interface. It is considered as the net generation and recombination terms of and we express the continuity equation for each ion as follows:

$$d[n]/dt = -\nabla \cdot J + G - R \quad (3)$$

The generation and recombination terms for each ion can be calculated using the detailed balance condition in equilibrium [3].

$$[H^+] : G = a_1 + b_1[H_2PO_4^-],$$

$$R = a_2[H^+][OH^-] + b_2[H^+][HPO_4^{2-}] \quad (4a)$$

$$[OH^-] : G = a_1, R = a_2[H^+][OH^-] \quad (4b)$$

$$[HPO_4^{2-}] : G = b_1[H_2PO_4^-], R = b_2[H^+][HPO_4^{2-}] \quad (4c)$$

$$[H_2PO_4^-] : G = b_2[H^+][HPO_4^{2-}], R = b_1[H_2PO_4^-] \quad (4d)$$

These are expressed in terms of the concentration of several ions. Therefore, if there is buffer reaction (generation and recombination), the continuity equations for each of the seven ions are coupled together and should be considered in the Jacobian matrix.

C. Channel

By default, the drift-diffusion equation and the continuity equation are used for the channel as well as the ions. Also, we assume that the channel is neutral. Ohmic contact was used as boundary condition between metal and semiconductor. To reduce numerical complexity, we assumed that channel is thin film. So thin that the charges in the film are considered as interface (surface) charges and the current in it as surface current.

D. Biomolecule layer

The biomolecule layer is treated as an ion-permeable membrane layer considering the Gibbs free energy barrier using the Born equation [3]. The thickness of this layer is 4nm and the charge is assumed to be uniformly distributed. We assume that the probe molecule is electrically neutral and the concentration of the target molecule after binding event is $10^{14}/cm^2$. When a biomolecule is present in an electrolyte, it causes modulation of permittivity. Therefore, when the mobile ions pass through the membrane layer, they feel a free energy barrier and the drift-diffusion equation in the membrane region is modified as

$$J = -zu[n](\nabla V + \Delta G) - D\nabla[n]. \quad (5)$$

where G is Gibbs free energy difference:

$$\Delta G = G_{membrain} - G_{water} = \frac{N_A z^2 e^2}{8\pi \varepsilon_0 r_0}\left(\frac{1}{\varepsilon_{membrain}} - \frac{1}{\varepsilon_{water}}\right) \quad (6)$$

E. Terminal current

The terminal current consists of direct current flowing through the channel and non-faradaic current due to the movement of the ions in the electrolyte solution. To calculate the non-faradaic current in the numerical simulation, we employed the extended Ramo-Shockley theorem [4]:

$$I_j(t) = -\sum_i^N q_i v_i(t) \cdot \nabla f_j(r_i) \quad (7)$$

where $f_j(r)$ is the electric potential at position r when the j-th electrode is kept at unit potential while all the other electrodes are grounded and all the charges are removed from the simulation domain.

TABLE I.   SIMULATION DEFAULT PARAMETERS

Parameter	Symbol	Default value
Biomolecule layer density	N_p	$10^{14}/cm^2$
Bulk concentration of ions	$[Na^+]$	0.1M
	$[K^+]$	0.022M
	$[Cl^-]$	0.1M
	$[HPO_4^{2-}]$	0.01M
	$[H_2PO_4^-]$	0.22M
Ion mobilites	u_{Na^+}	$5.9 \times 10^{-4} cm^2/V \cdot s$
	u_{K^+}	$8.6 \times 10^{-4} cm^2/V \cdot s$
	u_{Cl^-}	$7.0 \times 10^{-4} cm^2/V \cdot s$
	$u_{HPO_4^{2-}}$	$3.83 \times 10^{-4} cm^2/V \cdot s$
	$u_{H_2PO_4^-}$	$3.03 \times 10^{-4} cm^2/V \cdot s$
	u_{H^+}	$33.3 \times 10^{-4} cm^2/V \cdot s$
	u_{OH^-}	$18.8 \times 10^{-4} cm^2/V \cdot s$
Channel mobility	u_n	$20 cm^2/V \cdot s$
	u_p	$20 cm^2/V \cdot s$
Temperature	T	300K
Stern layer thickness	d_{stern}	5A
Water permittivity	ε_{water}	80
Stern layer permittivity	ε_{stern}	40
Biomolecule layer permittivity	ε_{bio}	3.9
Oxide permittivity	ε_{ox}	3.9

III. NUMERICAL METHOD

Since the size of the region of interest near the electrode is very small, a careful generation of the mesh size is necessary to achieve the numerical efficiency. Since the electrical double layer (EDL) length is usually only a few nanometers and we need to calculate the potential near the channel-electrolyte surface, we used a very small mesh spacing of 2.5A near the surface and a gradually increasing mesh size to bulk. The simulation is based on a cylindrical coordinate and the finite volume (box integration) method was used. We used the Scharfetter-Gummel scheme as the optimal way to discretize the drift-diffusion equation. In transient simulation, the explicit method requires very small time step for numerical stability. Since the required simulation time period is on the order of hundreds of us, the explicit approach would require

978-1-7281-0941-1/19 $31.00 © 2019 IEEE

prohibitively large number of time steps and is therefore not suitable for efficient simulation. Therefore, to freely use the size of the time step, we used the implicit method (Backward Euler method). When using this approach, we also need to minimize the truncation errors due to the large time step. For example, for a step pulse bias, at the moment when the applied voltage is changed, the ions are rapidly redistributed. Thus, the time step of about 10 ns is used at this moment. Then, the time step can be gradually increased as the ion redistribution rate is reduced after EDL charging. At each time step, we need to find the solutions for the ten coupled equations, but the convergence is not good for the Gummel iteration method. Therefore, the fast and robust convergence was obtained through the coupled Newton scheme.

IV. RESULTS

Fig.2 shows the probe-target binding event. When the target material is negatively charged, this negative charge affects the channel conductivity. Therefore, the channel current change and the target material can be measured. However, there is a space charge layer called the electric double layer at the interface between the electrolyte and the solid surface. In typical experimental and simulation environments, this EDL is only a few nanometers thick. Therefore, the charge of a biomolecule is significantly screened by ions and therefore can not achieve high sensitivity. However, as shown in Fig. 3, after a step pulse is applied to the drain electrode, ions are redistributed in the transient state and the EDL is extended. Therefore, the channel can better sense the charge of the target material, and it is confirmed that higher sensitivity was observed at the time of ion redistribution as Fig. 4.

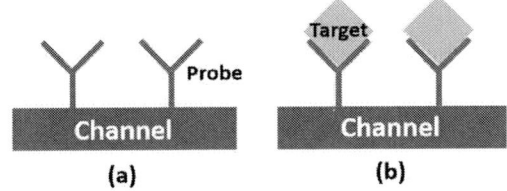

Fig. 2. Schematic of (a) befer and (b) after of the probe-target binding event. .

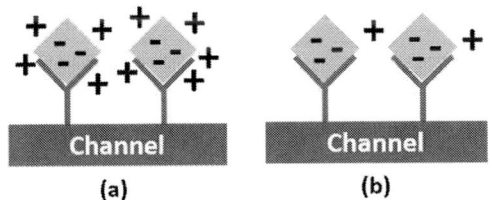

Fig. 3. Distribution of the ions around bio molecule at (a) steady state and (b) transient state.

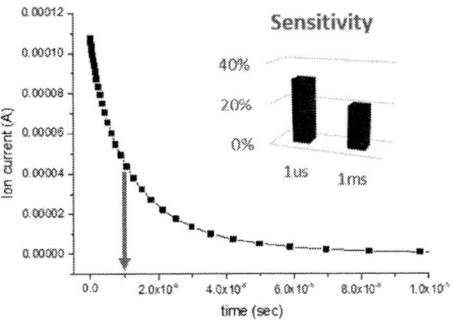

Fig. 4. Non-faradaic current and sensitivity after unit pulse is applied to the drain electrode.

V. CONCLUSION

We developed 3-dimensional physics-based model for simulation for transient simulation of bio-FET. It is confirmed that the sensitivity can be improved by measuring the transient current, which is due to the field effect and redistribution of the mobile ions around the channel.

ACKNOWLEDGMENT

This work was supported by the Brain Korea 21 Plus Project in 2019.

REFERENCES

[1] J. Woo, S. Kim, H. Chun, S. Kim, J. Ahn and Y. Park, "Modulation of molecular hybridization and charge screening in a carbon nanotube network channel using the electrical pulse method", Lab on a Chip, vol. 13, no. 18, p. 3755, 2013.

[2] W. Lee, H. Lee, J. Lim and Y. Park, "An effective electrical sensing scheme using AC electrothermal flow on a biosensor platform based on a carbon nanotube network", Applied Physics Letters, vol. 109, no. 22, p. 223701, 2016.

[3] K. Kim, W. Lee, J. Yun, Y. Lee, S. Choi, S. Jin and Y. Park, "A theoretical study on tunneling based biosensor having a redox-active monolayer using physics based simulation", Journal of Applied Physics, vol. 123, no. 2, p. 024509, 2018.

[4] H. Kim, H. Min, T. Tang and Y. Park, "An extended proof of the Ramo-Shockley theorem", Solid-State Electronics, vol. 34, no. 11, pp. 1251-1253, 1991.

978-1-7281-0941-1/19 $31.00 © 2019 IEEE

Electronic and structural properties of interstitial titanium in crystalline silicon from first-principles simulations

Gabriela Herrero-Saboya
CEA, DAM, DIF
Arpajon, France
LAAS-CNRS
Toulouse, France
gabriela.herrerosaboya@cea.fr

Layla Martin-Samos
*CNR-IOM / Democritos National
Simulation Center, c/o SISSA*
Trieste, Italy

mrsamos@iom.cnr.it

Anne Hemeryck
*LAAS-CNRS,
Université de Toulouse*
Toulouse, France

anne.hemeryck@laas.fr

Denis Rideau
STMicroelectronics
Crolles, France

denis.rideau@st.fr

Nicolas Richard
CEA, DAM, DIF
Arpajon, France

nicolas.richard@cea.fr

Abstract—We demonstrate the presence of small Jahn-Teller distortions for interstitial titanium in silicon at different charge states by performing ground state DFT calculations. We prove the existence of three charged transition levels within the band gap by using a non-empirical parameter-free approach, based on the GW approximation, in agreement with DLTS measurements.

Keywords—titanium, impurity, point defects, silicon, first-principles

I. Introduction

Point-like defects are known to introduce trap levels in the semiconductor band gap allowing carrier recombination and charge capture through non-radiative transitions, modifying/degrading the performance of microelectronic devices. The accurate prediction of point defects is, therefore, a major challenge in microelectronics that has been and is still addressed by both theory and experiments.

Among all defects, Transitions Metals (TMs) impurities are common contaminants in semiconductors. In particular, titanium is believed to cause uncontrolled avalanches in Single-Photon Avalanche Diodes (SPAD). These diodes, by working at the limit of breakdown, are able to detect single photons arriving to the device, triggering an avalanche of carriers. If impurities such as titanium are present, even at low concentrations after the fabrication process, carriers can be generated thermally due to the presence of deep levels and therefore, avalanche processes can be triggered even without external stimulus, i.e. in the dark. We believe that a first approach to the problem is to characterize such traps levels theoretically to correctly identify their experimental signature. Understanding the atomic-scale origin of Ti-related activity is an enabling stage to design mitigation strategies.

TMs impurities in silicon, such as Ti and V, were first characterized by Electronic Paramagnetic Resonance (EPR) [1,2]. The variation of spin between TMs and different charge states was explained through the modification of the electronic structure of 3d metals embedded in silicon, described by the Ludwig-Woodbury model [1,2]. According to such EPR-based model, the initial [Ar] $4s^2 3d^n$ electronic

configuration would be replaced by a $3d^{n+2}$ structure, where the 3d levels are splitted into the so-called $t2$ and e levels with a multiplicity of 6 and 4 respectively (see Fig. 1). The measured spin value was first estimated by filling these electronic levels following Hund's rule. First-principles studies [3] reported, however, deviations from this empirical rule for both Ti and V, after obtaining spin values of 1 and 1/2 for Ti (V) at charge states 0 (-1) and -1 (-2), instead of 2 and 5/2 as predicted by the Ludwig-Woodbury model. They also predicted a 3/2 spin value for Ti at charge +1 and for V at charge +2, in agreement with EPR measurements performed by D. A. Wezep and coworkers [4,5]. Higher charge states have, however, never been observed experimentally. Besides the magnetic properties of TMs in silicon, EPR studies also concluded that both Ti^{1+} and V^{2+} are found in tetrahedral interstitial positions in the lattice with full Td symmetry.

Regarding the electronic properties of TMs, Deep Level Transient Spectroscopy (DLTS) studies [6-11] commonly agree on the presence of three Charge Transition Levels (CTLs) for interstitial Ti within the band gap; an acceptor level at $E_C - 0.09$ eV, a donor level in the range of $E_V + 0.87$-0.92 eV, and a double donor level at $E_V + 0.25$-0.32 eV. References [6,7,9] also reported intermediate band gap levels between $E_V + 0.51$ eV and $E_V + 0.55$ eV. CTLs were also predicted within the framework of Density Functional Theory (DFT) [3,12-14] and the parametrized method DFT+U [15], in fair agreement with most experimental measurements. In the context of standard DFT, however, there is no grounded theoretical foundation for an accurate prediction of CTLs. Indeed, as it is well known, DFT is not Self-Interaction free and it is not piecewise linear against adding/removing an integer charge number.

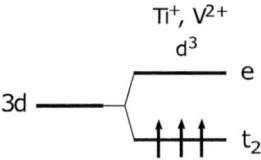

Fig. 1. Electronic configuration of interstitial Ti at charge +1 (Ti^{1+}) and interstitial vanadium at charge +2 (V^{2+}).

In this work, we discuss the properties of interstitial titanium at the atomic scale by performing first-principles simulations within DFT, for the ground state characterization, and Many-Body Perturbation Theory within the GW approximation for CTLs. The followed approach [16] is parameter-free and non-empirical, allowing an unbiased comparison with experimental measurements.

II. METHODOLOGY

Structural and magnetic properties are obtained by means of the Density Functional Theory (DFT) as implemented in the ABINIT code [17]. In order to avoid spurious defect-defect interactions due to the Periodic Boundary Conditions (PBC), a silicon supercell containing 3x3x3 unit cells (216 atoms) and one k-point (gamma) are employed. The calculations are performed using the norm-conserving pseudopotential ONCVPSP [18] and the Perdew-Burke-Ernserhof exchange-correlation functional [19]. A 30 Ha energy cutoff for the plane wave basis set is found to give converged results. The defect geometries at different charge states are obtained by the BFGS algorithm, with a stopping force criteria of 1 meV/Å. Furthermore, we are able to describe the magnetic properties of the system by performing spin-unrestricted calculations.

Many-body corrections are computed on top of the Kohn-Sham energies within the GW method (G_0W_0 as implemented in the ABINIT code [17]) in order to obtain the defect band structure correctly. We employ the Godby-Needs plasmon-pole model and a cutoff energy of 3 Ha to describe the dielectric matrix. In order to assure convergence of the GW exchange-correlation self-energy, we use a very large ratio of 10:1 empty bands versus occupied bands.

III. RESULTS

The capture or the release of a free carrier by a trap induces a structural reorganization of the defect. We therefore start this section by briefly discussing ground states properties of interstitial Ti and V at different charge states. We then present the electronic properties of the defects, focusing on the calculation of thermodynamic transition levels and the comparison with experimental evidence.

A. Structural and magnetic properties

Our ground state calculations also predict spin values of 1/2, 1 and 3/2 for Ti (V) at charge states -1 (0), 0 (+1) and +1 (+2), confirming the break of Hund's rule and the preference for lower spin values. We also obtained the full Td symmetric interstitial configuration, characterized by four equivalent distances between the TM and the host Si atom (TM-Si) of 2.48 Å for Ti^{1+} and 2.46 Å for V^{2+}.

Point-defects in semiconductors often present the so-called Jahn-Teller distortions when varying the trapped charge. The existence of such distortions in the case of interstitial Ti and V was hypothesized previously by [12, 14] but it was never reported as a result. Therefore, in contrast with previous theoretical results, we observed small Jahn-Teller distortions (see Fig. 2), in the order of 0.03 Å. In the case of Ti^0 and V^+, we obtained a flatted tetrahedron, constituted by one short distance TM-Si of 2.45 Å and 2.42 Å respectively, against three longer distances of 2.48 Å and 2.46 Å. Ti^{1-} and V^0, on the contrary, present a slightly elongated configuration, characterized by a long distance of 2.49 Å and 2.47 Å and three shorter distances of 2.45 Å and

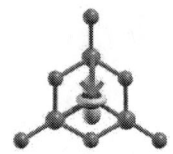

Fig. 2. Jahn-Teller distortions for interstitial Ti in silicon. Red arrows represent the geometry of the elongated (left) and flatted (right) tetrahedron. In yellow, one of the defect *t2* localized electronic states.

2.42 Å respectively. The origin of these slight changes in the defect geometry is the orientation of the *t2* states along the bonding directions as shown in figure 2. The capture of a free carrier starting from a symmetric Td configuration would, therefore, lead to a distorted flatted geometry as the electron is allocated in such *t2* state. Small Jahn-Teller distortions of this order of magnitude should be observed in EPR signals; however, as it was already mentioned above, such signals have not been reported.

B. Electronic properties

Even though DFT correctly describes structural and magnetic properties of defects, the discontinuity of the exchange-correlation functional and the Self-Interaction contribution constitute the origin of the well-known DFT band gap problem, making the description of electronic properties a difficult task. In the present work, for example, we estimate the value of the silicon forbidden gap to be 0.75 eV in DFT, instead of the measured gap equal to 1.2 eV. A better treatment of the electronic correlation is possible by means of Many-Body Perturbation Theory within the GW approximation. The corrected GW energies or quasiparticle energies correspond to ionization potentials, IP, and/or electronic affinities, EA; and therefore, the silicon band gap can be estimated as, $EA_{bulk} - IP_{bulk} = 1.18$ eV. By combining both DFT and GW calculations we are able to compute the quasiparticle band structure of the defect, identifying the trapped states by studying the localization properties of the DFT wavefunctions (see Fig. 3).

In figure 3 we show the quasiparticle band structure for interstitial Ti+. Even though three peaks appear within the silicon forbidden gap, only two of them can be considered as deep defect levels due to the spatial localization of their electronic wavefunctions (states *t2* and *e* in Fig. 3, with a multiplicity of 6 and 4 respectively). The non-labeled state is actually part of the conduction band, since it presents similar delocalization properties as the ones of pure silicon. The small difference in energy is due to the structural distortions around the impurity and it is often referred to as the band gap narrowing effect. From the common structural and magnetic properties presented by interstitial titanium and vanadium, it is not surprising that, for the same number of trapped electrons (e. g. electronic occupation 3, d^3, for Ti^{1+} and V^{2+}), we obtain equivalent quasiparticle band structures (a triply occupied *t2* state). In the case of Ti^{1+}, the occupied *t2* state is located at a distance of $IP_{t2} - IP_{bulk} = 0.41$ eV from the top of the valence band (see Fig. 3). The occupied *t2* state for V^{2+}, on the other hand, is located within the bulk valence band, in consistency with the absence of a triple donor state, (++/+++), in the measured CTLs [20, 21].

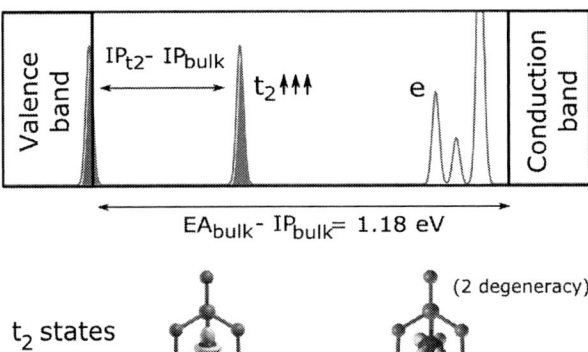

Fig. 3. (Above) Spin-up projection of the quasiparticle density of states for Ti^{1+}. Occupied states are colored and the valence and conduction band are shown for the Si-bulk are shown. An electrostatic correction is applied to the GW energies according to [24]. (Below) Filled electronic defect states, called $t2$ states.

The Charge Transition Levels (CTLs) are defined as the change in total energy between two different charge states at their corresponding relaxed geometries, with respect to the top of the valence band. Such energy difference can easily be computed from DFT ground state calculations; however, the estimated CTLs would suffer from the already mentioned band gap problem. A common strategy to avoid such underestimation [12-15] is the use of an empirical method, known as the Marker Method [22]. The idea of the method is to shift the computed CTLs by the absolute error determined for a "well characterized" defect; the error is defined as the difference between the calculated CTLs and the experimental one. In the present work, CTLs are computed using the non-empirical DFT-GW combine approach [16]. The gain/loose in energy when capturing/releasing an electron is accounted by the GW eigenvalues (first electronic affinity for capture and first ionization potential for releasing). Indeed, by definition, GW eigenvalues are the excitation energies (N+1/N-1) of an interacting N-electron system. The energy exchange during structural reorganization can be safely described by DFT since no change in the particle number occurs in such process. The CTLs are therefore computed as follows,

$$E\ (+/0) = IP^0_{defect} - \Delta E_{relax} - IP_{bulk}, \qquad (1)$$

where E (+/0) represents a single donor level, IP^0_{defect} is the first IP computed from charge zero and ΔE_{relax} the energy difference between the ground state geometries at charges 0 and +1. Furthermore, as it was investigated by [23], the calculation of double acceptor/donor levels is not straight forward because of strong electrostatic interactions between localized charges in neighboring cells (we recall the use of PBC). Even though GW corrections are short ranged, Hartree and ion-electron contributions coming from the DFT eigenvalues are subjected to this spurious electrostatic interaction. In the present work, we use the monopole correction scheme, proposed by G. Makov and M. C. Payne [24], to shift our IP^+/EA^+ in order to calculate the double donor level, (+/++). The obtained CTLs are shown in Table I, as well as previous reported values [13,15] and experimental evidence [6-11].

TABLE I. Charge Transition Levels (CTLs) for interstitial Ti. The acceptor level is computed from the conduction band, $E_C - E(0/-)$, whereas the donor levels are described from the top of the valence band $E_V + E\ (+/0)$. All values are given in eV.

Transition level	DFT@GW	DFT [13]	DFT+U [15]	DLTS [6-11]
E (0/-)	0.19	0.14	0.05	0.09
E (+/0)	0.82	0.51	0.98	0.87-0.92
E (+/++)	0.45	0.10	0.23	0.25-0.32

Previous references employed an empirical method to shift their computed CTLs [13,15], since by using a pure *ab initio* approach based on DFT, the calculated CTLs could not be predicted above the underestimated band gap (as in the case of E (0/-) in references [13,15] and E (+/0) for [15]). As we already showed in Fig. 3, by computing GW corrections on top of DFT, the silicon band gap is in very good agreement with experiments, and therefore, no empirical scheme is needed to shift our values. Furthermore, reference [15] relies not only on an empirical correction, but also on a parametrized first-principles calculation: the DFT+U method. The basic idea behind this method is to describe the strong interaction between localized electrons by an additional Hubbard-like term, characterized by two parameters U and J. LDA+U allows for a recovery of the piecewise linearity for adding/removing electrons from strongly localized frontier orbitals (Top of Valence Band and/or Bottom of Conduction Band). Depending on the CTLs the U/J value might vary significantly (see for instance figure 2 from [15]). Large variations are, however, difficult to meaningful interpret and theoretically ground.

It was established that theoretical CTLs could be assigned to DLTS activation energies [25]. Previous DFT studies already linked the acceptor level to the E40 characteristic peak, the donor level with the E150 signal, and the double donor level with the H180 peak. Due to the proximity of our double donor level to the mid-gap region, we considered the existence of a forth CTL within the values of Ev + 0.51-0.55 eV as suggested by DLTS experiences [6, 7, 9]. In order to clarify this point, CTLs for interstitial vanadium were also computed and compared to DLTS measurements as shown in Fig. 4. As already discussed above, interstitial vanadium presents a similar electronic structure as Ti (splitting and filling of the 3d states). However, due to the difference in atomic number and therefore, electrostatic interaction with the nuclei, levels corresponding to the same change in electronic occupation differ for these two systems. This is the case, for example, of the d^3/d^4 transition, which corresponds to the CTL (+/0) for Ti and (-/0) for V (see Fig. 4). However, by comparing the trend of the theoretical CTLs for both TMs, and specially the double donor levels, it is clear that the E (+/++) level for Ti was correctly assigned to the DLTS peak H180, with activation energies within 0.25-0.32 eV. Since TMs impurities in silicon are known to interact with lighter foreign elements (such as oxygen, hydrogen...), it has been hypothesized in [11] that the origin of the intermediate band gap reported by [6,7,9] is actually a hydrogen-titanium complex. First efforts to support this evidence by the DFT community [12,14] have been made; however, the large number of possible structural configurations for such Ti-H complexes does not allow for a one-to-one assignment.

978-1-7281-0941-1/19 $31.00 © 2019 IEEE

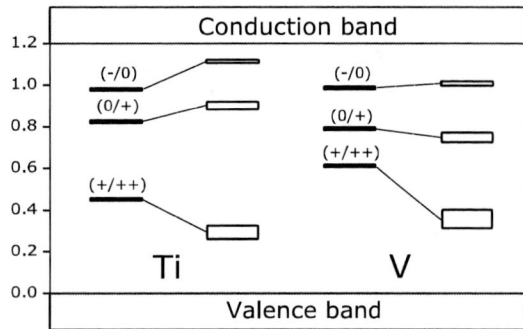

Fig. 4. Charge Transitions Levels (CTLs) for interstitial Ti and V computed with the DFT-GW combine approach (black lines) and measured DLTS activation energies [6-11,20,21] (white rectangles). The width of the white rectangles represents the dispersion of the reported values.

We report a single acceptor, (-/0), a single donor, (0/+), and a double donor, (+/++), levels for both Ti and V, by using state-of-the art first-principles and parameter free approaches, namely DFT+GW. The computed *single* levels are in fair agreement (within 0.1-0.15 eV) with the experimental ones (see Fig. 4). Similar agreements have been reported by [26] for a series of point defects in several types of semiconductors. The reported *double* donor levels, on the other hand, lay a bit further from the experimental references (within 0.15-0.25 eV). The comparison of such absolute error with previous reported double levels in the GW framework is not straight forward since such CTLs usually lay within the bulk valence band for the majority of systems and therefore they are invisible to DLTS measurements.

IV. CONCLUSIONS

Atomic scale calculations accurately predict point defect properties, since we were able to validate our results with experimental evidence (EPR and DLTS studies). The DFT-GW combine approach is the state-of-the art first-principles method to compute CTLs in semiconductors, giving values in fair agreement with the DLTS measurements, without including empirical shifts or fitted parameters. The discussed approach is still limited to small set of systems due to the large number of unoccupied states needed to converge the GW calculation.

The presented work can be considered as a first step into the evaluation of electrical activity of TMs in silicon. Further investigations should include meaningful calculations of the probability of capturing/releasing charges by the already computed CTLs.

REFERENCES

[1] H. H. Woodbury and G. W. Ludwig, "Spin resonance of transition metals in silicon", Phys. Rev., vol. 117, pp. 102, 1960.

[2] G. W. Ludwig and H. H. Woodbury, "Solid States Physics" edited by F. Seitz and D. Turnbull, Vol. 13 (Academic Press, New York, 1962).

[3] F. Beeler, O. K. Anderson, and M. Scheffler, "Theoretical evidences for low-spin ground states of early interstitial and late substitutional 3d transtion-metal ions in silicon", Phys. Rev. Lett., vol. 55, pp. 1498, 1985.

[4] D. A. van Wezep, R. van Kemp, E. G. Sieverts and C. A. J. Ammerlaan, "Electron-nuclear double resonance of titanium in silicon: ^{29}Si ENDOR", Phys. Rev. B, vol 32, pp. 7129, 1985.

[5] D. A. van Wezep and C, A, J, Ammerlaan, "Electron-nuclear double resonance of titanium in silicon: ^{47}Ti and ^{49}Ti ENDOR", Phys. Rev. B, vol. 37 pp. 7268, 1988.

[6] J. W. Chen, A. G. Milnes and A. Rohatgi, "Titanium in silicon as a deep level impurity", Solid State Elect., vol. 22, pp. 801-808, 1979.

[7] J. R. Morante, J. E. Carceller, P. Cartujo and J. Barbolla, "Thermal emision rates and capture cross-section of majority carriers at titanium levels in silicon", Solid State Elect., vol. 26, pp. 1-6, 1983.

[8] A. C. Wang and C. T. Sah, "Complete electrical characterization of recombination propeties of titanium in silicon", J. Appl. Phys., vol. 56, pp. 1021, 1984.

[9] Z. Shurong, L. Yongling and F. Chunyin, "Several physical properties of titanium in silicon", J. of Electronics, vol. 5, pp. 40-46, 1988.

[10] D. Mathiot and S. Hocine, " Titanium-related deep levels in silicon: a reexamination" , J. Appl. Phys., vol. 66, pp. 5862, 1989.

[11] L. Scheffler, VI. Koljovsky and J. Weberm "Identification of titanium-hydrogen complexes with up to four hydrogen atoms in silicon", J. Appl. Phys., vol. 117, pp. 085707, 2015.

[12] D. J. Backlund and S. K. Estreicher, "Structural, electrical, and vibrational properties of Ti-H and Ni-H complexes in Si", Phys. Rev. B vol.81 pp. 155208, 2010.

[13] V. P. Markevich, S. Leonard, A. R. Peaker, B. Halmiton, A. G. Marinopoulos and J. Coutinho, "Titanium in silicon: lattice positions and electronic properties", Appl. Phys. Lett., vol. 104, pp. 152105, 2014.

[14] P. Santos, J. Coutinho, V. J. B. Torres, M. J. Rayson and P. R. Briddon, "Hydrogen passivation of titanium impurities in silicon: effect of doping conditions", Appl. Phys. Lett., vol. 105, pp. 032108, 2014.

[15] A. G. Marianopoulos, P. Santos and J. Coutinho, "DFT+U study of electrical levels and migration barriers of early 3d and 4d transition metals in silicon", Phys. Rev. B, vol. 92, pp. 075124, 2015.

[16] P. Rinke, A. Janotti, M. Scheffler and C. G. Van de Walle, "Defect formation energies without the band-gap problem: combining density-funcation theory and the GW approach for the silicon self-interstitial", Phys. Rev. Lett., vol. 102, pp. 026402, 2009.

[17] X. Gonze et al., "First-principles computation of material properties: the ABINIT software project", Comp. Mat. Sci., vol. 25, pp. 478-492, 2002.

[18] D. R. Hamann, "Optimized norm-conserving Vanderbilt pseudopotentials", Phys. Rev. B, vol. 88, pp. 085117, 2013.

[19] J. P. Perdew, K. Burke and M. Ernzerhof, "Generalized gradiente approximation made simple", Phys. Rev. Lett., vol. 77, pp. 3865, 1996.

[20] K. Graff in "Metals Impurities in silicon-device fabrication", edited by R. Hull, R. N. Osgood, H. Sakaku and R. J. Kriegler and Q. Zunger, Springer, 1981.

[21] D. J. Backlund, T. M. Gibbons, and S. K. Estreicher, "Vanadium interactions in crystalline silicon", Phys. Rev. B, vol. 94, pp.195210, 2016.

[22] J. Coutinho, V. B. Torres, R. Jones and P. R. Briddon, "Electrical activity of chalcogen-hydrogen defects in silicon", Phys. Rev. B, vol. 67, pp. 035205, 2003.

[23] M. Jain, J. R. Chelikowsky and S. G. Louie, "Quasiparticle excitations and charge transition levels of oxygen vacancies in hafnia", Phys. Rev. Lett., vol. 107, pp. 216803, 2011.

[24] G. Makov and M. C. Payne, "Periodic boundary conditions in ab initio calculations", Phys. Rev. B, vol. 51, pp. 4014, 1995.

[25] C. Freysoldt, B. Grabowski, T. Hickel, J. Neugebauer, G. Kresse, A. Janotti and C. G. Van de Walle, "First-principles calculations for point defects in solids", Rev. Mod. Phys., vol. 86, pp. 253, 2014.

[26] W. Chen and A. Pasquarello, "Accuracy of GW for calculating defect energy levels in solids", W. Chen and A. Pasquarello, Phys. Rev. B, vol. 96, pp. 020101, 2017.

Hybrid method for electromagnetic modelling of coherent radiation in semiconductor lasers.

Mateusz Marek Krysicki*, Bartlomiej Salski†, Pawel Kopyt‡

Warsaw University of Technology
Institute of Radioelectronics and Multimedia Technology
Nowowiejska 15/19, 00-665 Warsaw, Poland
*m.krysicki@ire.pw.edu.pl, † b.salski@ire.pw.edu.pl, ‡ p.kopyt@ire.pw.edu.pl

Abstract—**In this paper hybrid method for electromagnetic (EM) modelling of coherent radiation in semiconductor lasers is presented. Described approach consist of drift diffusion (DD) model and electromagnetic simulation. Four-level two-electron atomic system with Pauli Exclusion Principle (PEP) extended by electric pumping ratio has been used as lasing model.**

Index Terms—**simulation, device, FDTD, ADE, PEP, drift diffusion, nonlinear optics**

I. INTRODUCTION

Nowadays laser action can be simulated using various methods, as Finite Difference Time Domain (FDTD) metod with auxiliary differential equation (ADE) which represents classical dispersive Lorentzian gain [1] or with FDTD with rate equations ADE [2]. Another ADE-FDTD technique presented in [3] and extended in [4], uses the density-matrix method for solving Maxwell-Bloch equation.

In this paper ADE-FDTD method for electrically-pumped lasers is reported. Flowchart of algorithm is presented on figure 1. Initial point base on device simulation, computed using drift diffusion method [5] and recalculated in co-processing. Results are computed using electromagnetic simulation.

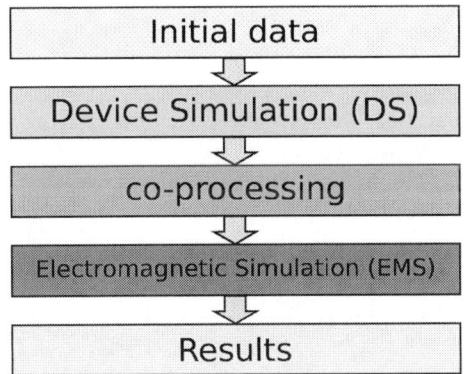

Fig. 1. Flowchart od hybrid method for EM modelling of coherent radiation in semiconductor lasers.

This work was partially supported by the Polish National Science Center within the SONATA project titled "Full-wave electromagnetic modelling of coherent radiation in electrically-pumped metal-clad semiconductor lasers with a folded cavity".

II. DEVICE SIMULATION

First step in the proposed method is the computation of a drift-diffusion model of a device [5], which consists of the Poisson's equation: (3):

$$\nabla \cdot \epsilon \nabla \Psi = -q(p - n + N_D^+ - N_A^-) \tag{1}$$

and continuity equations:

$$\nabla J_n - q\frac{\partial n}{\partial t} = +qR \tag{2}$$

$$\nabla J_p + q\frac{\partial p}{\partial t} = -qR \tag{3}$$

where ϵ is permittivity, Ψ is electrostatic potential, q is elementary charge p, n, N_D^+, N_A^- are densities of electron, hole singly ionized donors and singly ionized acceptors, R is generation/recombination rate, and $J_{n(p)}$ is electron (hole) current density.

As a result of this calculation spatial distribution of electron and hole density (or Fermi-Dirac distributions for electron f_e and holes f_h) is obtained, which will be recalculated to quasi-chemical potentials (4) (5):

$$\mu_e = f_e - E_c \tag{4}$$

$$\mu_h = -f_h + E_v \tag{5}$$

where $E_{c(v)}$ is energy level of conduction (valence) band, and, subsequently, used in co-processing part of algorithm presented in the next Section.

III. LINKING ELECTROMAGNETIC AND DRIFT-DIFFUSION SIMULATIONS (CO-PROCESSING)

Electromagnetic simulation of a semiconductor laser with eq. (12) - (16) requires knowledge on material parameters, which can be determined with the aid of DD. All calculations presented in this section are preformed for one selected voltage. For both type of computing processes the same mesh grid have to be used. In this paper, the study is limited to a 1D structure.

Once DD is computed, gain characteristic is calculated in each cell [6]:

978-1-7281-0941-1/19 $31.00 © 2019 IEEE

$$g(z) = \frac{\nu}{4\pi^2 \hbar \gamma \varepsilon_0 n c} \left(\frac{2m_r}{\hbar^2} \right)^{\frac{3}{2}} \cdots$$

$$\int_0^\infty d\epsilon |\mu_k|^2 \left[f_e(\epsilon) + f_h(\epsilon) - 1 \right] \cdots \qquad (6)$$

$$L(\omega_k - \nu) \left(1 - j \frac{\omega_k - \nu}{\gamma} \right)$$

where ν is angular frequency, $\hbar$ is reduced Planck constant, γ is homogeneous linewidth factor, ε_w is vacuum permittivity, m_r is reduced mass given by:

$$m_r^{-1} = m_e^{-1} + m_h^{-1} \qquad (7)$$

where $m_{e(h)}$ is electron (hole) mass, $d\epsilon$ is part of energy used for integral calculations, $|\mu_k|$ is dipole moment, $L(\omega_k - \nu)$ is Lorentzian lineshape function, ω_k is angular plasma frequency of lorentz model, and f_e, f_h are Fermi-Dirac distributions for electron and holes respectively given by equation:

$$f_\alpha(\varepsilon) = \frac{1}{\exp\left[\beta \left(\varepsilon \frac{m_r}{m_\alpha} - \mu_\alpha \right) \right] + 1} \qquad (8)$$

where α stands as $e(h)$ for electron (hole) distribution, β is coefficient equal to:

$$\beta = \frac{1}{k_B T} \qquad (9)$$

where T is temperature and k_B is Boltzmann coefficient, and μ is quasi-chemical potential.

As an example of those calculation gain curve and imaginary part of electrical susceptibility, which be easy obtained using equation:

$$\chi'' = -2\frac{g}{n K_0} \qquad (10)$$

where K_0 is the wavenumber in vacuum and n is refractive index, is presented on figure 2.

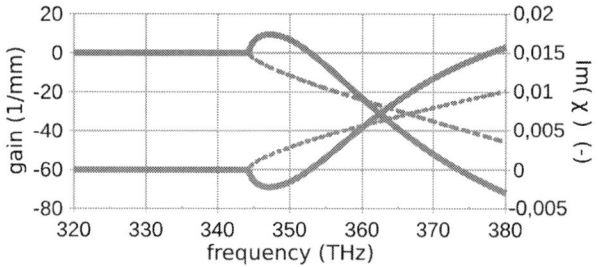

Fig. 2. An example of gain curve (blue lines, left scale) computed using equation (6) and corresponding to it imaginary part of electrical susceptibility (green lines, right scale) computed using equation (10). Solid (dashed) lines are obtained for active (lossy) material

If gain is negative in a given cell, it is assumed to be lossy, which is represented with electric conductivity (11):

$$\sigma = \omega \varepsilon_0 \chi'' \qquad (11)$$

where χ'' is imaginary part of electrical susceptibility.

On the contrary, when gain characteristic is positive at the frequency of interest, pumping rate P_R, is calculated using spatial distribution of current density, by integrating density of states between 0-3 energy levels.

IV. ELECTROMAGNETIC SIMULATION

In the next step, electrodynamic simulation will be undertaken with the aid of a finite-difference time-domain (FDTD) method with auxiliary differential equation (ADE). Rate equations are coupled with Ampère's circuital law (with Maxwell's addition) as density of electric polarization:

$$\nabla \times H = \frac{\partial D}{\partial t} + \frac{\partial P_a(t)}{\partial t} + \frac{\partial P_b(t)}{\partial t} \qquad (12)$$

where i indicates a or b, $P_a(P_b)$ stands for the density of electric polarization between 0-3 (1-2) levels of a four-level atomic model, γ are linewidths, and ω are resonant frequencies, $\Delta N_{a(b)}$ is difference between population densities at 0-3 (1-2 levels) and κ is model parameter.

In addition to a typical form of rate equations [4] [2], electrical pumping rate, P_R, has been added:

$$\frac{\partial N_3}{\partial t} = -\frac{N_3(1 - N_2)}{\tau_{32}} - \frac{N_3(1 - N_0)}{\tau_{30}} + \frac{1}{\hbar \omega_a} E \frac{\partial P_a}{\partial t} + P_R \qquad (13)$$

$$\frac{\partial N_2}{\partial t} = \frac{N_3(1 - N_2)}{\tau_{32}} - \frac{N_2(1 - N_1)}{\tau_{21}} + \frac{1}{\hbar \omega_b} E \frac{\partial P_b}{\partial t} \qquad (14)$$

$$\frac{\partial N_1}{\partial t} = \frac{N_2(1 - N_1)}{\tau_{21}} - \frac{N_1(1 - N_0)}{\tau_{10}} - \frac{1}{\hbar \omega_b} E \frac{\partial P_b}{\partial t} \qquad (15)$$

$$\frac{\partial N_0}{\partial t} = \frac{N_3(1 - N_0)}{\tau_{30}} + \frac{N_1(1 - N_0)}{\tau_{10}} - \frac{1}{\hbar \omega_a} E \frac{\partial P_a}{\partial t} - P_R \qquad (16)$$

where N_i is population density at level $i \in \{0, 1, 2, 3\}$, τ_{ij} is the decay time constant between i and j levels and $\hbar$ is reduced Planck constant

Due to chosen model, wide spectrum of results can be obtain, for example spectral characteristic of generated radiation.

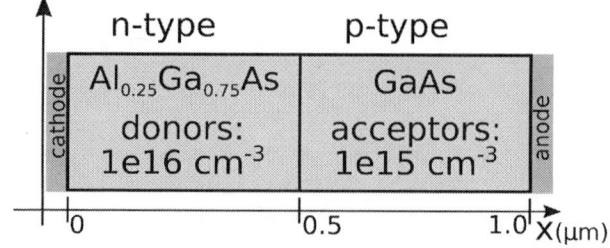

Fig. 3. Geometry of PN junction [7] used for calculations.

V. NUMERICAL RESULTS

As an example PN junction [7] (see fig. 3) has been used. It consist of $0.5\mu m$ n-type $Al_{0.25}Ga_{0.75}As$ and $0.5\mu m$ p-type $GaAs$ blocks with refractive index $n = 3.6$. Firstly, DD simulation is used for computing IV-characteristic (see fig. 4). After that, for chosen voltage (1.4 V) parameters for novel model have been computed using quasi-chemical potentials (see fig. 5).

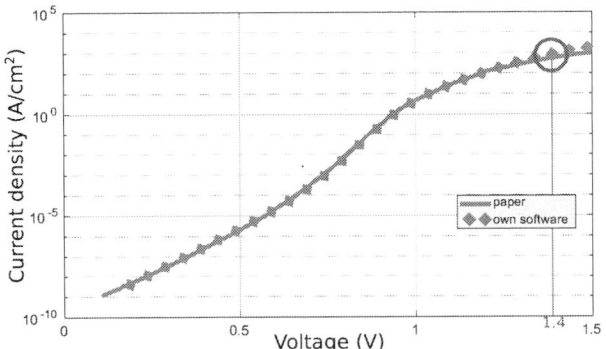

Fig. 4. IV characteristic of PN junction presented on fig. 3).

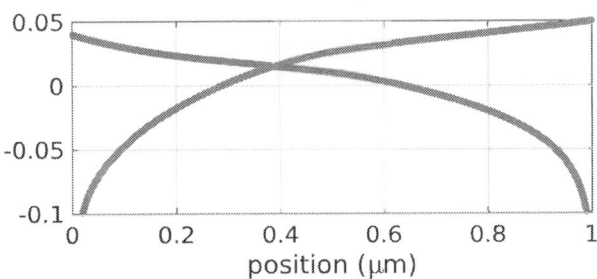

Fig. 5. Spatial distribution of electron (hole) quasi – chemical potential $\mu_n(\mu_p)$ marked by green (blue) line.

As an example of results evolution of population densities between populations densities at levels 2 and 1 (see fig. 6) and value of E-field (see fig. 7) have been presented.

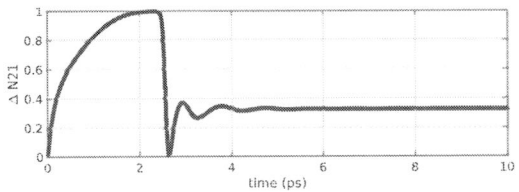

Fig. 6. Time evolution of difference between population densities at levels N_2 and N_1.

For next example of results the simplest optical cavity (Fabry-Perot) of total length $20\mu m$ with similar PN junction as presented on fig. 3 has been used. Two parallel ends are cleaved, and ended with low- and high- reflectivity (LR/HR)

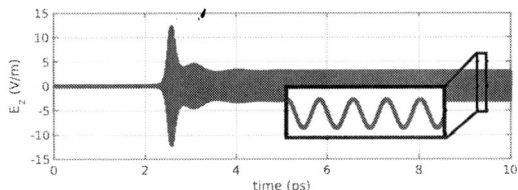

Fig. 7. An example of gain curve computed using equation (6).

facets. Spectral characteristic of radiation from LR facet has been calculated and presented on fig. 8.

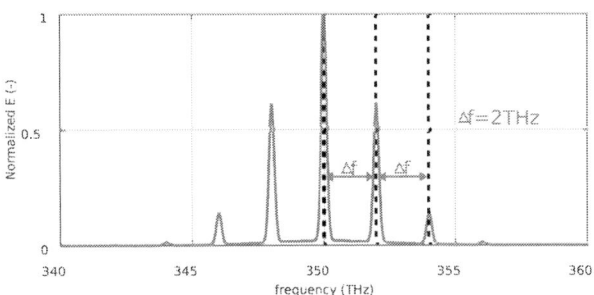

Fig. 8. Normalized E - field spectrum radiated from low-reflectivity facet of FP cavity.

In presented case it is possible to detect many longitudinal modes spaced by $\Delta f = 2THz$ which clearly corresponds to cavity length.

VI. CONCLUSIONS

Using novel method presented in this paper it is possible to calculate insensitivity of generated electromagnetic field and observe dynamics of difference between population densities. Presented methodology will be extended for two dimensional simulations.

REFERENCES

[1] A. Taflove, S. C. Hagness,, *Computational electrodynamics: the finite-difference time-domain method.*, Artech house, 2005.

[2] S. H. Chang, A. Taflove, "'Finite-difference time-domain model of lasing action in a four-level two-electron atomic system'", *Optical and Quantum Electronics*,Optics express, vol. 12, no. 16, pp. 3827–3833, 2004.

[3] A. S. Nagra and R. A. York, "'FDTD analysis of wave propagation in nonlinear absorbing and gain media'", *IEEE Trans. Antennas Propgat*, 46, 334, (1998).

[4] B. Salski, "'Hybrid FDTD Analysis of Two- and Four-Level Atomic Systems'", *Optical and Quantum Electronics*, vol. 47, pp. 1703–1712, 2014.

[5] D. L. Scharfetter, H. K. Gummel, "'Large-signal analysis of a silicon read diode oscillator'", *IEEE Transactions on electron devices*, vol. 16, no. 1, pp. 64–77, 1969.

[6] W. W. Chow, S. W. Koch, *Semiconductor-laser fundamentals: physics of the gain materials.*, Springer Science & Business Media, 2013.

[7] K. Horio, H. Yanai, "'Numerical modeling of heterojunctions including the thermionic emission mechanism at the heterojunction interface'", *IEEE Ttransactions on electron devices*, vol. 37, no. 4, pp. 1093–1098, 1990.

978-1-7281-0941-1/19 $31.00 © 2019 IEEE   190

NEGF simulations of stacked silicon nanosheet FETs for performance optimization

Hong-Hyun Park
Device Laboratory
Samsung Semiconductor Inc.
San Jose, California, USA
honghyun.p@samsung.com

Woosung Choi
Device Laboratory
Samsung Semiconductor Inc.
San Jose, California, USA

Mohammad Ali Pourghaderi
Semiconductor R&D Center
Samsung Electronics
Hwasung-si, Gyeonggi-do, Korea

Jongchol Kim
Semiconductor R&D Center
Samsung Electronics
Hwasung-si, Gyeonggi-do, Korea

Uihui Kwon
Semiconductor R&D Center
Samsung Electronics
Hwasung-si, Gyeonggi-do, Korea

Dae Sin Kim
Semiconductor R&D Center
Samsung Electronics
Hwasung-si, Gyeonggi-do, Korea

Abstract— We present quantum transport simulation results of stacked silicon nanosheet (SiNS) nFETs. Our simulations are based on the non-equilibrium Green's function (NEGF) method which is capable of dealing with all major physical effects necessary for steady-state electron transport in the complex-shaped devices. In order to help find optimal device design many split simulations for various geometry and process conditions were performed as a demonstration.

Keywords—quantum transport, non-equilibrium Green's function (NEGF), stacked nanosheet FET

I. INTRODUCTION

The FinFET technology is the most advanced node for commercial logic device fabrications nowadays. Its success was enabled mainly by the larger effective channel width and stronger gate controllability than previously used planar FETs. As long as we stick to the concept of field-effect transistor, a key technology requirement to boost the device performances will be to achieve better gate controllability. Hence, gate-all-around FETs are being studied as an ultimate structure for future logic devices [1]. Stacked silicon nanosheet FETs are considered as one of the strong candidates which have potential to replace FinFETs [2]. Although the fabrication processes of SiNS FETs are more complex than those of FinFETs, the better gate controllability is more favorable to shorter channel length and better performance. An example of dual SiNS FET is shown in Fig. 1. The number of channel stacks can be more than two in reality.

Considering the size and shape, the stacked SiNS FET is expected to have significant quantum and atomistic effects with its performance. Also, there are many design variables related to geometry dimensions or process conditions, which might not be optimized one by one because of their possible intercorrelations. In this case a tool which can accurately estimate the device performance to the changes of the design variables is required. We employ the self-consistent NEGF-Poisson approach in this work because of its rigorous and general theoretical basis which does not depend on the design variables.

In this paper we explain our simulation models and demonstrate how the NEGF simulation can be used for the optimization of the design of dual SiNS nFETs.

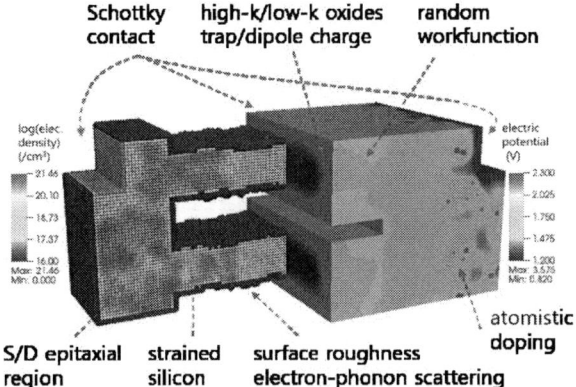

Fig. 1. Structure of simulated dual SiNS nFET and physical models avalable in our in-house NEGF solver. Electron density and electric potential profiles at $V_G = 1$ V and $V_D = 0.01$ V are visualized from atomistic and continuum simulation domains, respectively.

II. SIMULATION METHOD

We use our in-house atomistic and quantum transport simulation tool. For device simulations the NEGF and the Poisson equations are solved self-consistently. The NEGF solver covers the silicon region for carrier transport while the Poisson solver covers the whole region for electrostatic potential. The NEGF solver is equipped with various models to take into account all the major physical phenomena as shown in Fig. 1. The following are explanations of the physical models:

1) The bandstructure of the conduction band of silicon is modelled by the effective-mass approximation (EMA) [3, 4]. Each ellipsoidal valley is characterized by a mass tensor with nonparabolicity correction. The model parameters for each valley are calibrated against tight-binding simulation results to take into account the effects of quantum confinement and strain/stress more accurately [5, 6]. For the sake of numerical efficiency only three valleys out of six are considered without loss of accuracy.

2) Electron-phonon scattering is modelled based on the deformation potential theory which can capture intra- and inter-valley scattering processes [7]. The basic parameters were calibrated to fit bulk mobility of silicon [8, 9]. To fit

978-1-7281-0941-1/19 $31.00 © 2019 IEEE

experimentally observed mobility decrease in the strong inversion regime which could not be captured by the original scattering model, the deformation potential values are adjusted to be enhanced near the surface of the channel [10].

3) Atomistic doping model is used to consider the effects of dopant variation and impurity scattering explicitly in the NEGF simulations [11, 12]. Random and discrete point charges are generated from a given continuous doping profile according to Poisson statistics.

4) Remote Coulomb scattering due to high-k dielectric layer can be modelled by putting static monopole or dipole charges into the source terms of the Poisson solver.

5) Atomistic traps can be generated based on [13]. This effective trap model works well as long as the self-consistent Born approximation (SCBA) does not fail to converge due to any resonance states at the traps.

6) Surface roughness on the channels is generated according to given sets of amplitude and correlation length [14]. Different roughness and random seeds can be applied to different channels. Although the quality of generated surface depends on the lattice constant of the simulation structure, the simulation results were not sensitive to the atomic-scale resolution.

7) To consider the variability effect due to the gate metal phenomenologically, it is assumed that the gate metal consists of grains with random sizes and workfunction values [15]. Some parameters such as the distribution of the grain size and the distribution of workfunction values are necessary.

8) In quantum transport simulations the gate contact is usually modelled as a Dirichlet boundary condition in the Poisson solver with the assumption that there is no current through the contact. To be more realistic, we can treat the contact as an open-boundary lead like the source/drain contacts and calculate the gate leakage current. In order to improve the numerical efficiency, we have improved the conventional recursive Green's function (RGF) technique [16] to be able to simulate arbitrary-shaped devices [17].

9) Schottky contacts are attached on the source/drain epitaxial regions as shown in Fig. 1. In quantum transport simulations, it is usually assumed that a contact is a semi-infinite extension of its connected part of the simulation domain [18, 19], which is not appropriate for realistic metal-semiconductor interfaces. So, in this work the contact model was improved more realistically by two modifications. Firstly, virtual metals are introduced for contacts. Each type of metal requires three parameters under EMA model, i.e. effective mass, workfunction, and band edge. The workfunction parameter is used to give desired Schottky barrier height when a Dirichlet boundary condition is applied to the contact and the other parameters are calibrated to give desired I-V characteristic or Schottky resistance. Secondly, the new contact model assumes that each of the contact atoms is an independent reservoir of carriers and does not interact with the other contact atoms. By this assumption different incident directions can be set for different contact atoms so the shape of a contact can be an arbitrary curved surface. Another advantage is that the calculation of the contact self-energy function is much faster than that of conventional contact models because the function can be evaluated independently for each atom.

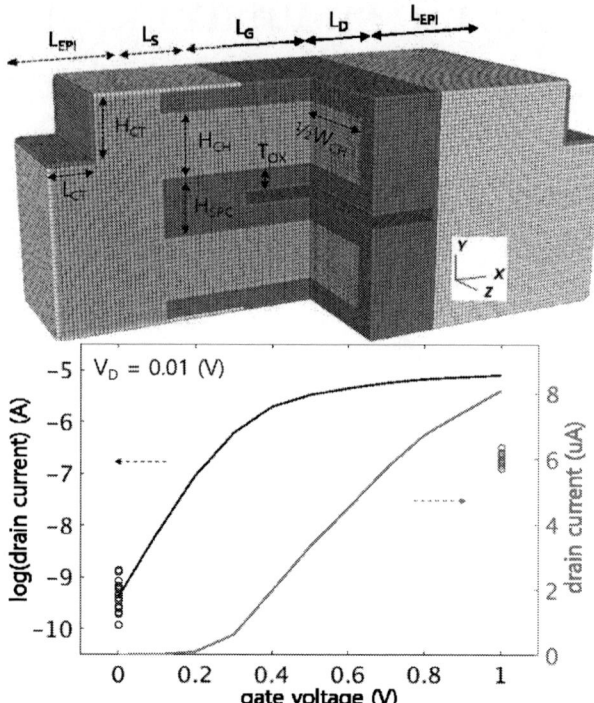

Fig. 2. (TOP) Base simulation structure for the sensitivity analysis of dual SiNS nFETs. Silicon, oxide, and contact metal regions are represented in ivory, red, and orange colors, respectively. The space between the channels is not empty in real devices but filled with gate metal. This particular structure has two SiNSs as its channels and bulky epitaxial regions for source and drain. The dimensions of this base structure are 50 nm, 22 nm, and 24 nm along x, y, and z directions, respectively. The channels are lightly doped with p-type and the source/drain regions are heavily doped with n-type. (BOTTOM) I_D-V_G characteristic of the base simulation structure. The lines are the case of continuum doping model without surface roughness in the channels or random workfunction of metal grains. For the simulations with atomistic dopant, surface roughness, and random workfunction models, only $V_G = 0$ V and $V_G = 1$ V were performed, which are are marked with dots according to different random configurations. The variation in I_{OFF} and the degradation of I_{ON} are confirmed.

III. SIMULATION RESULTS

To show a demonstration how the design variables of stacked SiNS nFETs affect the device performance, we performed many split simulations by changing each design variable one by one from the base device shown in Fig. 2. Please note that the simulated device structures and conditions are not based on real devices but set for the demonstration of this simulation study. All the simulation results are obtained from self-consistent NEGF simulations with the models explained in the previous section. Here, some models like atomistic traps, dipole changes due to remote Coulomb scattering, and gate leakage were not activated because these require statistical analysis based on experimental data or have minor effects on the device performance. All the simulation parameters were calibrated to reproduce basic electrical properties of materials and the same set of parameters were used for all the simulations. The simulation is highly parallelized by MPI, OpenMP, and GPU. A typical simulation time for each bias condition is about 4~5 hours using 80 cores of Intel® Xeon® CPU E5-2699 v3 @ 2.30GHz.

978-1-7281-0941-1/19 $31.00 © 2019 IEEE

TABLE I.   SENSIVIITY OF VTH, SS, AND ION TO THE CHANGES OF DESIGN VARIABLES ($V_{DS} = 0.01$ V)

variable change	Δ Vth (mV)	Δ SS / SS	Δ Ion / Ion
L_G -14 %	-54	15.9 %	4.8 %
EOT 20 %	-17	4.6 %	-2.8 %
W_{CH} -50 %	61.2	-9.8 %	-42.1 %
H_{CH} 33 %	-71	18.5 %	19.1 %
S_{SPC} 33 %	1.1	1.8 %	0.4 %
L_{CT} 100 %	-0.1	0.2 %	9.9 %
H_{CT} 100 %	0.4	0.4 %	14.1 %
L_{OVL} 0 → 0.4 nm	-17	4.1 %	4.2 %
L_{UDL} 0 → 0.4 nm	15.1	-4.7 %	1.1 %
N_{CH} 1E17 → 0 /cm³	-0.8	0.1 %	2.1 %
$N_{S/D}$ 3E20 → 2E20 /cm³	9	-0.8 %	-21.6 %

Table I shows how much threshold voltage (V_{TH}), subthreshold slope (SS), and on-current level (I_{ON}) are affected by the change of design variables. Although just 11 design variables and 3 performance criteria were considered in this paper to find the steepest gradient for better device performance, in principle we can play with any geometry and simulation conditions. Since the model parameters of NEGF solver are only related to the properties of materials, we have high confidence of the accuracy of the simulation results within the limitation of the NEGF formalism itself. In reality, some practical considerations such as fabrication difficulty, cost, yield should be taken into account in the optimization process.

IV. CONCLUSION

In this paper we have shown a demonstration how the self-consistent NEGF-Poisson method can be used to help optimize the design of stacked SiNS nFETs. We could evaluate the sensitivity of device performance to the changes of design variables such as geometry and process conditions.

REFERENCES

[1] B.-H. Lee, M.-H. Kang, D.-C. Ahn, J.-Y. Park, T. Bang, S.-B. Jeon, J. Hur, D. Lee, and Y.-K. Choi, "Vertically Integrated Multiple Nanowire Field Effect Transistor," Nano Lett. 15 (12), pp. 8056-8061, 2015.

[2] N. Loubet et al., "Stacked Nanosheet Gate-All-Around Transistor to Enable Scaling Beyond FinFET," in Proc. Symp. VLSI Tech. Dig., pp. 230-231, 2017.

[3] E. O. Kane, "Band structure of indium antimonide," J. Phys. Chem. Solids, vol. 1, pp. 249-261, 1957.

[4] W. R. Frensley, "Boundary conditions for open quantum systems driven far from equilibrium," Rev. Mod. Phys., vol. 62, no. 3, pp. 745-791, 1990.

[5] J. C. Slater and G. F. Koster, "Simplified LCAO Method for the Periodic Potential Problem," Phys. Rev., vol. 94, no. 6, pp. 1498-1524, June 1954.

[6] T. B. Boykin, "Strain-indluced, off-diagonal, same-atom parameters in empirical tight-binding theroy suitable for [110] uniaxial strain applied to a silicon parameterization," Phys. Rev. B 81, 125202, 2010.

[7] S. Jin, Y. J. Park, and H. S. Min, "A three-dimensional simulation of quantum transport in silicon nanowire transistor in the presence of electron-phonon interactions," J. Appl. Phys., vol. 99 (12), pp. 123719-123729, 2006.

[8] C. Jacoboni and L. Reggiani, "The Monte Carlo method for the solution of charge transport in semiconductors with applications to covalent materials," Rev. Mod. Phys. 55, pp. 645-705, 1983.

[9] M. V. Fischetti and S. E. Laux, "Band Structure, deformation potentials, and carrier mobility in strained Si, Ge, and SiGe alloys," J. Appl. Phys. 80, pp. 2234-2252, 1996.

[10] T. Ohashi, T. Tanaka, T. Takahashi, S. Oda, and K. Uchida, "Experimental Study on Deformation potential (Dac) in MOSFETs: Demonstration of Increased Dac at MOS Interfaces and Its Impact on Electron Mobility," IEEE J. Electron Devices Soc., vol. 4, pp. 278-285, 2016.

[11] A. Asenov et al., "Simulation of statistical variability in nano-CMOS transistors using drift-diffusion, Monte Carlo and non-equilibrium Green's function techniques," J Comput Electron, pp. 349-373, 2009.

[12] Y.-M. Niquet, V.-H. Nguyen, F. Triozon, I. Duchenmin, O. Nier, and D. Rideau, "Quantum calculations of the carrier mobility: Methodology, Matthiessen's rule, and comparison with semi-classical approaches," J. Appl. Phys. 118, 054512, 2014.

[13] M. G. Pala and D. Esseni, "Interface Traps in InAs Nanowire Tunnel-FETs and MOSFETs—Part I: Model Description and Single Trap Analysis in Tunnel-FETs," IEEE Trans. Elec. Dev. vol. 60, no. 9, pp. 2795-2801, 2013.

[14] M. Luisier, "Quantum Transport Beyond the Effective Mass Approximation," Ph.D. dissertation, ETH Zurich, 2007.

[15] Y. Li and H.-W. Cheng, "Random Work-Function-Induced Threshold Voltage Fluctuation in Metal-Gate MOS Devices by Monte Carlo Simulation," IEEE Trans. Semicond. Manuf., vol. 25, no. 2, pp. 266-271, 2012.

[16] R. Lake, G. Klimeck, R. C. Bowen, and D. Jovanovic, "Single and multiband modeling of quantum electron transport through layered semiconductor devices," J. Appl. Phys. 81 (12), pp. 7845-7869, 1997.

[17] H.-H. Park, W. Choi, M. A. Pourghaderi, J. Kim, U. Kwon, and D. S. Kim, "Toward more realistic NEGF simulations of vertically stacked multiple SiNW FETs," in Proc. SISPAD Conference, pp. 206-209, 2018.

[18] M. P. Lopez Sancho, J. M. Lopez Sancho, and J. Rubio, "Highly convergent schemes for the calculation of bulk and surface Green functions," J. Phys. F: Met. Phys. 15, pp. 851-858, 1985.

[19] M. Luisier, A. Schenk, and W. Fichtner, "Atomistic simulation of nanowires in the sp3d5s* tight-binding formalism: From boundary conditions to strain calculations," Phys. Rev. B 74, 205323, 2006.

978-1-7281-0941-1/19 $31.00 © 2019 IEEE   194

Modeling Silicon CMOS devices for quantum computing

Benjamin Venitucci
Univ. Grenoble Alpes, CEA, IRIG
MEM/L_Sim
Grenoble, France

Jing Li
Univ. Grenoble Alpes, CEA, IRIG
MEM/L_Sim
Grenoble, France

Léo Bourdet
Univ. Grenoble Alpes, CEA, IRIG
MEM/L_Sim
Grenoble, France

Yann-Michel Niquet
Univ. Grenoble Alpes, CEA, IRIG
MEM/L_Sim
Grenoble, France
yniquet@cea.fr

Abstract — **We review our recent results on the modeling of silicon spin qubits. We describe, in particular, the methodology we have set-up for the simulation of these devices, and give some illustrations on silicon-on-insulator (SOI) qubits. We discuss, in particular, the electrical manipulation of electron and hole spins.**

Keywords—Modeling, spin quantum bits, silicon, quantum information.

I. INTRODUCTION

A quantum bit (qubit) is a device where an information is stored as a coherent superposition $|\psi\rangle = \alpha|0\rangle + \beta|1\rangle$ of two quantum states $|0\rangle$ and $|1\rangle$ (for example, two spin states $|0\rangle \equiv |\downarrow\rangle, |1\rangle \equiv |\uparrow\rangle$). Such devices open new perspectives for information processing, thanks, e.g., to the intrinsic parallelism afforded by the superposition of inputs/outputs. Protecting this superposition from decoherence and relaxation (drifts in α, β due to noise and unwanted interactions) is, however, one tremendous challenge (among others) [1].

The spins of electrons confined in semiconductor quantum dots (QDs) actually make promising solid-state qubits with good prospects for large scale integration [2, 3]. High fidelity single and two qubits gates have been demonstrated in III-V materials as well as in silicon [4, 5, 6, 7]. Silicon is an attractive material for quantum information devices because the majority ^{28}Si isotope has no nuclear spin, effectively decoupling electron and hole spins from the lattice and allowing for longer coherence times [8]. Also, silicon benefits from the strong portfolio of technologies developed in conventional micro-electronics for the design of complex and scalable devices.

Many aspects of the physics of silicon qubits are, however, still poorly understood. It is, therefore, essential to complement the experimental activity with microscopic modeling able to give insights into the operation of these devices and provide guidelines for their optimization.

In this paper, we review the methodology we have set-up for the modeling of spin qubits in semiconductors. We then discuss some applications to electron and hole qubits in silicon-on-insulator (SOI) devices. We first give a short introduction to these devices (section II), then outline the modeling methodology (section III), and discuss the manipulation of spin qubits as an illustration (section IV).

II. SPIN QUBITS ON SOI

We focus on the SOI devices fabricated in Grenoble as an illustration (Fig. 1). The layout of these devices resembles conventional silicon nanowire Trigate MOSFETs [9, 10, 11]. Thanks to larger source/gate and gate/drain spacers, the devices go into the Coulomb blockade regime at low temperature: quantum dots form under the gates, in which the number of carriers can be controlled by the bias voltages. The information is then stored in the spin of the carrier(s) trapped under the gates. For that purpose, the $|\uparrow\rangle$ and $|\downarrow\rangle$ spin states are split by a static magnetic field **B**. This Zeeman splitting $\Delta E = g\mu_B B$ (where $g \simeq 2$ and μ_B is Bohr's magneton) is typically of the order of a few tens of μeV, which is one of the reasons why the devices must be operated at very low temperatures (< 100 mK).

The spin can then be manipulated by radio-frequency (RF) bursts on the gates, resonant with the Zeeman splitting between the up and down spins ($\Delta E \simeq 40$ μeV $\Rightarrow$ frequency $f_L \simeq 10$ GHz). Coherent photon absorption and stimulated emission indeed drive rotations of the spin called "Rabi oscillations". Starting from the ground-state $|\downarrow\rangle$, any superposition $|\psi\rangle = \alpha|\downarrow\rangle + \beta|\uparrow\rangle$ can in principle be reached that way. The number of spin rotations per unit of time is characterized by the Rabi frequency $f_R \ll f_L$ (typically 1 to 100 MHz). Note that the spin must be coupled to the orbital motion of the carrier under the gate by spin-orbit coupling (SOC) in order to be manipulated with a RF electric field. SOC is, however, known to be very weak in the conduction band of silicon. Alternatively, the spin can be manipulated with the RF magnetic field produced by a nearby current line.

The second dot on Fig. 1 is used to measure the spin in the qubit through "Pauli spin blockade": The source-drain current is indeed blocked when the spins in the two dots are parallel ("triplet" state) because the carriers can not tunnel from one dot to the other ("two carriers with same spin can not occupy the same level"). More complex one and two-dimensional layouts (bearing less resemblance to traditional MOSFETs) have been designed in order to allow for more versatile interactions between qubits (two qubit gates). These interactions are controlled by inter-dot tunneling and Coulomb repulsion, which translate into an effective "exchange" interaction between spins.

978-1-7281-0941-1/19 $31.00 © 2019 IEEE

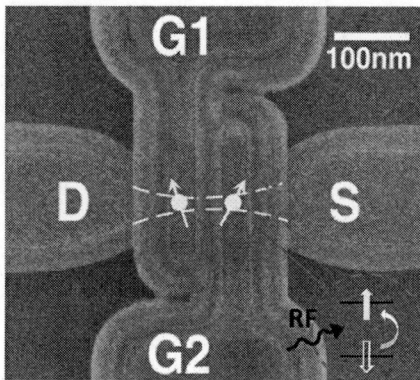

Fig. 1: SEM image of a SOI device. The gates G1 and G2 control two quantum dots along a [110] Si nanowire (outlined by the dashed white lines). The dot under G1 is used as a filter to measure the spin in the qubit under G2 through Pauli spin blockade of the source-drain (S-D) current (current blocked when spins parallel). The electronic structure of the qubit is depicted in the inset. The information is encoded as a superposition of the up and down spin states split by a static magnetic field. The qubit is manipulated by radio-frequency pulses on G2, resonant with the splitting between the two spin levels. Adapted from [10].

III. METHODS

We have developed a specific methodology for the microscopic modeling of spin qubit devices (Fig. 2).

The potential landscape in the devices is first computed in the empty dots with a finite volumes Poisson solver. Screening by source/drain reservoirs can be taken into account at this stage with semi-classical approximations for the electron/hole carrier density in the contacts.

Then, N single-particle states $|\varphi_i\rangle$ are calculated in this potential using either a multi-bands $\mathbf{k.p}$ or a tight-binding (TB) model [11, 12, 13]. TB captures the multi-valley character of the conduction band of silicon and SOC at the atomistic level. It can also describe atomic scale features such as surface roughness and impurities. It is, therefore, ideally suited to the microscopic modeling of qubits. Its numerical cost scales, however, at least linearly with the number of atoms. $\mathbf{k.p}$ calculations are, therefore, more suitable for large scale systems and hole qubits where the description of SOC is simpler (using either four or six bands $\mathbf{k.p}$ models). There also exists a four bands $\mathbf{k.p}$ model accounting for SOC in the conduction band [14]. The set of differential equations for the $\mathbf{k.p}$ envelope functions is discretized over a finite differences mesh. The matrices of relevant observables (such as spin, gate potentials, etc...) are then computed within the basis of the N $|\varphi_i\rangle$'s. These matrices can later be used to build effective, low-energy Hamiltonians for the qubits.

As an illustration, the 3D model of an electron spin qubit in a silicon nanowire controlled by a partially overlapping gate is shown in Fig. 3. The location of the electron trapped under this gate is outlined in orange; the squared TB wave function of the ground-state is plotted in a transverse cross section in the inset. The electrons (and holes) tend to localize in the corner(s) covered by the gate [15], where the electric field is maximum. These devices, therefore, form low symmetry-dots, which tends to enhance SOC [11, 16].

The many-electron states relevant for the modeling of multiply-charged dots, of Pauli spin blockade readout

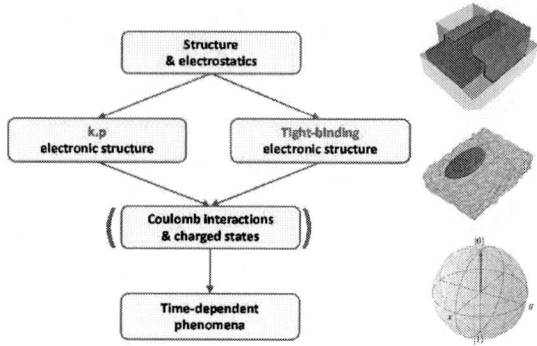

Fig. 2: Computational methodology developed for the qubits.

(singlet/triplet states), and of multi-qubit gates can then be computed with either mean-field approximations (Hartree/Hartree-Fock, usually performed within the reduced basis set of the N precomputed φ_i's), or with a configuration interaction (CI) method [17]. The latter captures the correlations among electrons and is most often the method of choice for the description of the many-electron physics. The principle of CI is to expand the many-particle wave function in a basis of Slater determinants built from a subset of M φ_i's. The basic inputs of CI are the Coulomb integrals:

$$U_{ijkl} = \iint d^3\mathbf{r}\, d^3\mathbf{r}'\, \varphi_i\,(\mathbf{r})\varphi_j^*(\mathbf{r})\, V(\mathbf{r},\mathbf{r}')\varphi_k^*(\mathbf{r}')\varphi_l\,(\mathbf{r}') \quad (1)$$

where $V(\mathbf{r},\mathbf{r}')$ is the potential created at point $\mathbf{r}'$ by a charge at point $\mathbf{r}$. These integrals can formally be written as:

$$U_{ijkl} = \int d^3\mathbf{r}\, \rho_{ij}^*(\mathbf{r})\, V_{kl}(\mathbf{r}) \quad (2)$$

where $\rho_{ij}(\mathbf{r}) = \varphi_i^*(\mathbf{r})\varphi_j\,(\mathbf{r})$ is a joint density and $V_{kl}(\mathbf{r})$ is the potential created by the joint density $\rho_{kl}(\mathbf{r})$. The latter is obtained from the finite volumes Poisson solver using this (possibly complex) joint density as input. The calculation of these integrals may be the most time-consuming part of the simulation. It can, however, be efficiently distributed over a large number of cores on a high-performance parallel cluster. The solution of Poisson and $\mathbf{k.p}$ equations as well as the

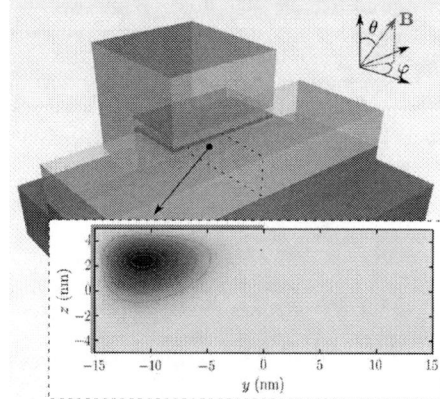

Figure 3: 3D model of a SOI device with a Si nanowire in yellow (cross section 30 nm × 10 nm), SiO₂ in dark blue, and a partly overlapping, 30 nm long gate in light blue. The location of the electron trapped under the gate is sketched in orange. A map of the squared TB wave function is plotted in the cross-section outlined by the dashed black lines. The orientation of the magnetic field $\mathbf{B}$ (see Fig. 6) is characterized by the angles θ and φ.

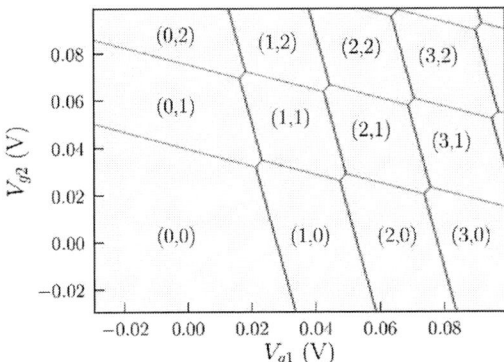

Figure 4: Stability diagram computed for a "double dot" system similar to Fig. 1. The number of electrons (n_1, n_2) in dots 1 and 2 is given as a function of the gates voltages V_{g1} and V_{g2}. Such maps can be reconstructed from the charging energies computed with CI.

calculation of the U_{ijkl}'s are actually parallelized within a mixed OpenMP+MPI scheme. The $M \varphi_i$'s used to build the CI determinants may be pre-optimized with Hatree-Fock and the basis set of CI determinants filtered out in order to further speed-up the calculation.

As an illustration, the stability diagram computed for a "double dot" system similar to Fig. 1 is plotted in Fig. 4. This map gives the number of electrons (n_1, n_2) in dots 1 and 2 as a function of the gates voltages V_{g1} and V_{g2}. It is used to find the relevant operating points for the qubits (dot filling, control of tunneling, etc...). The stability diagram can be constructed from the intra- and inter-dot charging energies extracted from the CI calculations.

Finally, the response of the system to various control signals and perturbations can be analyzed with a time-dependent solver. For that purpose, the time-dependent Schrödinger equation $[H_0 + V(t)]|\psi(t)\rangle = i\hbar\, \partial |\psi(t)\rangle / \partial t$ is solved in a basis set of eigenstates of the "static" single- or many-particle Hamiltonian H_0. The evolution operator between two time steps t and $t+dt$ is expanded as a Chebyshev polynomial [16, 18] of the Hamiltonian $H(t) = H_0 + V(t)$. Single/two-qubit gate operations and decoherence can be completely simulated in this framework starting from a microscopic (atomic scale) geometry, possibly including disorder (e.g., roughness and charged traps [16, 19]).

Such microscopic simulations can also provide valuable input parameters for the modeling of large scale arrays of qubits using effective Hamiltonians [20].

IV. APPLICATIONS

We now discuss two applications dealing with the electrical manipulation of electron and hole spin qubits.

A. Electron spin qubits.

SOC is known to be small at the conduction band edge of silicon, partly due to the indirect nature of the bandgap. On the one hand, this effectively decouples the electron spins from electrical and charge noise as well as phonons, hence enhances spin lifetimes. On the other hand, this prevents the direct

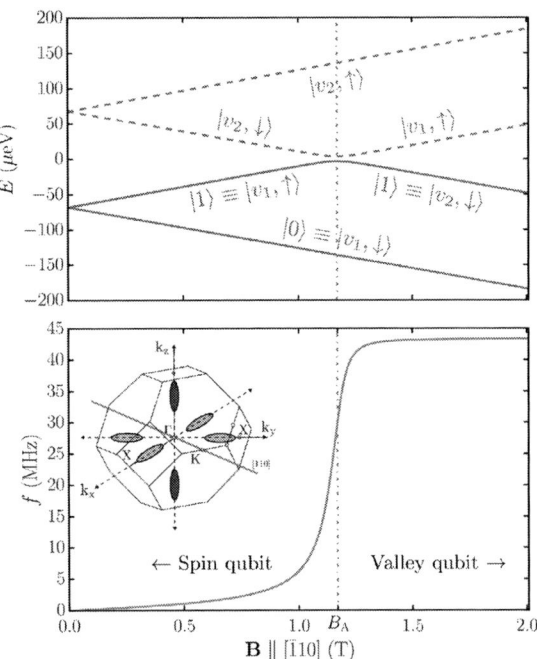

Figure 5: (top panel) Single-particle energy levels in the conduction band of a thin Si quantum dot (see Fig. 3) as a function of the magnetic field. At zero field, the degeneracy between the $\pm Z$ valleys (inset of bottom panel) is lifted by steep confinement at the Si/SiO$_2$ interface; the resulting v_1 and v_2 states are further split by the Zeeman interaction at finite magnetic field. The $|v_1, \uparrow\rangle$ state anti-crosses the $|v_2, \downarrow\rangle$ state around $B = B_A$ due to SOC; This enables electrically-driven Rabi oscillations between the lowest two states $|0\rangle$ and $|1\rangle$ as the electric dipole matrix element between $|v_1, \downarrow\rangle$ and $|v_2, \downarrow\rangle$ is finite. The calculated Rabi frequency is plotted in the lower panel (amplitude of the RF signal on the gate $V_{RF} = 1$mV). Adapted from [16].

manipulation of the spin by RF electric fields, and hinders the electrical tuning of the gyromagnetic factor g and Zeeman splitting ΔE that may be used to put spins in or out of resonance with a global RF magnetic field in order to address a particular qubit [21].

Yet recent experiments have shown that SOC can be sizable in electron qubits. Ref. [11] for example has demonstrated clear fingerprints of electric dipole spin resonance (EDSR), that is of electrically driven spin rotations, in a device similar to Fig. 1. Detailed microscopic modeling unveiled the mechanisms at play in this device [11].

The EDSR actually results from the interplay between spin and valley physics. Indeed, the X/Y/Z conduction band valleys are sixfold degenerate in bulk silicon [22]. Weak confinement in an anisotropic quantum dot such as those of Figs. 1 and 3 rises the X and Y valleys with respect to the ground-state Z valleys. Strong confinement at the steep Si/SiO$_2$ interface further couples the +Z and −Z valleys, leaving two valley states v_1 and v_2 at low energy split by a valley splitting energy Δ, which can range from a few tens to a few hundreds of μeV depending on the vertical electric field.

Under a finite magnetic field, the $|v_1, \downarrow\rangle$ and $|v_2, \downarrow\rangle$ states go down in energy, while $|v_1, \uparrow\rangle$ and $|v_2, \uparrow\rangle$ go up (Fig. 5). At some critical field B_A, $|v_1, \uparrow\rangle$ and $|v_2, \downarrow\rangle$ cross and get mixed by SOC. Near the anti-crossing point, the $|v_1, \uparrow\rangle$ state admixes a significant fraction of $|v_2, \downarrow\rangle$. Since $|v_1, \uparrow\rangle$ and $|v_2, \downarrow\rangle$ can be coupled by an electric field, this allows for electrically

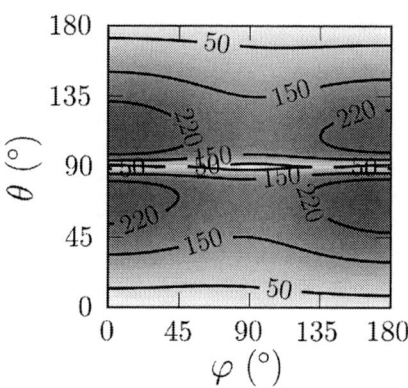

Figure 6: Map of the Rabi frequency (number of spin rotations per second, in MHz) of a heavy-hole spin qubit as a function of the orientation of the magnetic field, characterized by the angles θ and φ (see definition and device on Fig. 3) [13, 23]. The amplitude of the RF signal driving the rotations is $V_{RF} = 1 \text{mV}$ and the magnitude of the magnetic field is $B = 1 \text{T}$ (the Rabi frequency being proportional to V_{RF} and B).

driven rotations between the lowest two qubit states. The resulting Rabi frequency is plotted as a function of the magnetic field in the lower panel of Fig. 5. The Rabi frequency saturates beyond the anti-crossing point as the device becomes a "valley" qubit where the information is encoded into the valley states $|v_1, \downarrow\rangle$ and $|v_2, \downarrow\rangle$. Such a valley qubit is, however, very sensitive to charge noise.

SOC is enhanced by the very low symmetry of the "corner dots" formed in SOI devices (Fig. 3). We have demonstrated that the valley splitting and the SOC can actually be tailored by front and back-gate electric fields; and have proposed a manipulation scheme where an almost pure spin qubit (well protected from charge noise but hardly controllable electrically) can be transformed back and forth into a spin-valley qubit (or even a pure valley qubit) for fast electrical manipulation [16].

B. Hole spin qubits.

The physics of holes is very different yet extremely rich [10, 13]. There is no valley degree of freedom, but a strong interplay between heavy- and light-holes instead. The mixing between heavy- and light-hole envelopes in the quantum dot is actually a pre-requisite for the electrical manipulation of hole "spins". Since SOC is much stronger in the valence than in the conduction band, electrical driving of hole spins can be very efficient, with Rabi frequencies reaching a few tens of MHz [9, 10].

The linear response of hole spins to magnetic and electric fields can be described in a unified framework based on a gyromagnetic g-matrix and on its derivative with respect to the gate voltage [10, 13]. This g-matrix formalism can be used to analyze experimental data and to model spin qubits at a very low computational cost (Fig. 6). The figures of merit of holes, such as the Zeeman splittings and Rabi frequencies, show a much stronger dependence on the orientation of the fields than those of electrons − a fingerprint of the above-mentioned interplay between the heavy- and light-hole components. This can provide additional control knobs on the hole qubits, but may also enhance variability, and therefore calls for a careful design of the devices.

We have shown that silicon provides excellent opportunities for fast hole spin manipulation owing to its very anisotropic valence band that favors heavy- and light-hole mixing [23, 24]. We have also analyzed the response of hole spin qubits to strains [13, 23]. Semiconductor spin qubits in general tend to be very responsive to strains due to the small energy scales involved in these devices [25, 26].

V. CONCLUSIONS

We have discussed the modeling of silicon spin qubits derived from MOS technologies. While CMOS modeling usually focuses on transport at room temperature in the high carrier density regime, spin qubits are operated in the opposite, few electrons and very low temperature limit. Therefore, the methods and tools needed to address qubit devices are different from those used for conventional CMOS; yet they can leverage on the knowledge brought by CMOS modeling on, e.g., disorder and scattering in MOS devices.

We have illustrated the relevance of modeling on a few examples dealing with the electrical manipulation of electron and hole spins. Modeling of decoherence, spin readout, and two-qubit gates will provide in the near future a more complete picture of the expected strengths and weaknesses of silicon qubits with respect to competing technologies. Modeling will also certainly play a leading role in assessing the variability of silicon qubit devices [16, 19].

ACKNOWLEDGMENT

This work was supported by the French ANR project MAQSi and the EU H2020 project MOSQUITO.

REFERENCES

[1] T. D. Ladd et al., Nature **464**, 45 (2010).
[2] D. Loss and D. P. DiVincenzo, Physical Review A **57**, 120 (1998).
[3] R. Hanson and D. D. Awschalom, Nature **453**, 1043 (2008).
[4] J. J. L. Morton et al., Nature **479**, 345 (2011).
[5] J. Yoneda et al., Nature Nanotechnology **13**, 102 (2018).
[6] T. F. Watson et al., Nature **555**, 633 (2018).
[7] W. Huang et al., Nature **569**, 532 (2019).
[8] A. M. Tyryshkin et al., Nature Materials **11**, 143 (2012).
[9] R. Maurand et al., Nature Communications **7**, 13575 (2016).
[10] A. Crippa et al., Physical Review Letters **120**, 137702 (2018).
[11] A. Corna et al., npj Quantum Information **4**, 6 (2018).
[12] Y.-M. Niquet et al., Physical Review B **79**, 245201 (2009).
[13] B. Venitucci et al., Physical Review B **98**, 155319 (2018).
[14] D. Osintsev et al., Solid-State Electronics **90**, 34 (2013).
[15] B. Voisin et al., Nano Letters **14**, 2094 (2014).
[16] L. Bourdet and Y.-M. Niquet, Physical Review B **97**, 155433 (2018).
[17] C. D. Sherrill and H. F. Schaefer III,
    Advances in Quantum Chemistry **34**, 143 (1999).
[18] S. Roche and D. Mayou, Physical Review Letters **79**, 2518 (1997).
[19] D. J. Ibberson et al., Applied Physics Letters **113**, 053104 (2018).
[20] R. Li et al., Science Advances **4**, eaar3960 (2018).
[21] M. Veldhorst et al., Nature Nanotechnology **9**, 981 (2014).
[22] F. A. Zwanenburg et al., Review of Modern Physics **85**, 961 (2013).
[23] B. Venitucci and Y.-M. Niquet,
    Physical Review B **99**, 115317 (2019).
[24] C. Kloeffel, M. J. Rančić and D. Loss,
    Physical Review B **97**, 235422 (2018).
[25] J. Mansir et al., Physical Review Letters **120**, 167701 (2018).
[26] J. J. Pla et al., Physical Review Applied **9**, 044014 (2018).

ATOMOS: An ATOmistic MOdelling Solver for dissipative DFT transport in ultra-scaled HfS₂ and Black phosphorus MOSFETs

Aryan Afzalian
Imec
Kapeldreef 75, 3001 Leuven, Belgium
aryan.afzalian@imec.be

Geoffrey Pourtois
Imec
Kapeldreef 75, 3001 Leuven, Belgium
geoffrey.pourtois@imec.be

Abstract—A state-of-the-art DFT-NEGF based ATOmistic - MOdelling Solver (ATOMOS) was developed and used to assess the physics and fundamental-performance potential of various scaled mono-layer transition-metal-dichalcogenides and black-phosphorus (BP) MOSFETs down to a gate length of 5 nm, including the effect of electron-phonon scattering. Our study highlights the good scalability and drive-current potential of HfS2 and the impact of optical-phonon scattering for BP.

Keywords— Semiconductor Physics, 2D-material, Quantum transport, DFT NEGF, CMOS

I. Introduction

Transition metal dichalcogenides (TMDs) or black phosphorus (BP), are widely investigated by the scientific community nowadays [1-3]. Their large variety of bandgaps, effective masses, and their excellent electrostatic properties related to their 2-D nature hold promise to find in their midst the candidate for ultra-scaled CMOS or post-CMOS applications. Full-band transport simulations including electron-phonon (e-ph) scattering have been shown crucial to consider intricate band-structure and transport effects [1] and assess the performance of these devices. Because there are many materials on which little is known, an *ab-initio* based quantum-transport method, such as DFT NEGF, is ideal for such novel device/material exploration.

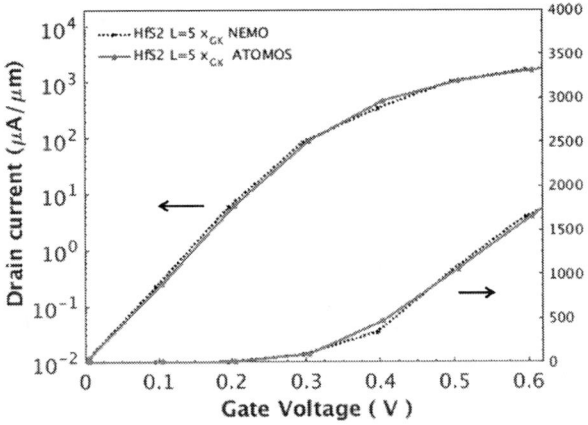

Fig. 1. $I_D(V_G)$ characteristics of a ballistic L = 5 nm HfS₂ nMOSFET (see schematic of Fig. 3) in the ΓK channel orientation simulated with DFT-NEGF using ATOMOS and NEMO5. V_{DD} = 0.5 V. I_{OFF} = 10 nA/µm.

Here, we report the development of a new DFT-NEGF based ATOmistic-MOdelling Solver (ATOMOS) and use it to assess the physics and fundamental-performance potential of various scaled mono-layers (1ML) TMD's and BP MOSFETs down to a gate length (L) of 5 nm, including the effect of e-ph scattering. Our results predict that the less-studied HfS₂

MOSFET has good scalability down to L = 5 nm with a promising high on-current (I_{ON}) level when oriented in the ΓM or ΓK directions. The 1ML BP-MOSFET I_{ON} is severely degraded by the intrinsic strong optical-phonon (OP) coupling that has been shown to appear in 1 or a few ML of free-standing BP [3]. It may, however, be possible to find a substrate that would attenuate the detrimental OP impact.

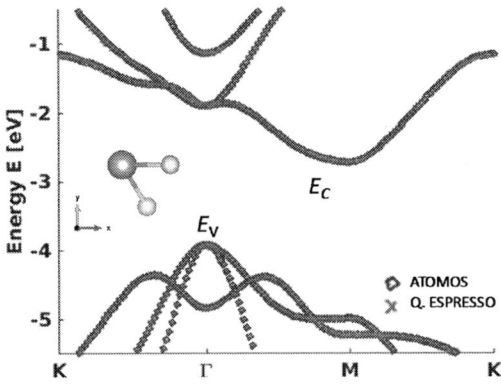

Fig. 2. Monolayer HfS₂ (1T-phase) band structure computed with QUANTUM ESSPRESSO using plane-wave DFT and with ATOMOS using the Wannierized Hamiltonian.

II. Solver Development

The core routines of ATOMOS were written for high-performance computing and computationally-demanding DFT Hamiltonians (H). They are based on C++ and multi-threaded MPI with various levels of parallelization.

Our Real-Space NEGF solver is based on the recursive-Green's function (RGF) algorithm [4]. A specific sparse block-matrix class, tailored for the RGF method, was developed to efficiently store H and other system matrices. Ultimately, the operation on the (slab) dense sub-block matrices are performed using BLAS and LAPACK. A dynamic scheduler, based on the master-slave approach, efficiently distribute the various energy-momentum (E-K) points between the different (MPI) ranks and ensure optimal load-balancing. A recursive adaptive-grid algorithm with global-error estimator is used for optimally generating the energy points. The contact-self energies are computed using the Sancho-Rubio method [5]. Electron-phonon scattering was considered using the self-consistent Born approximation [6]. For TMDs we used the DFT-computed e-ph parameters from [2], for BP those from [3]. To expedite the self-consistent Poisson-NEGF convergence, a predictor-corrector method [7] is used. To predict the carrier changes with respect to potential variation, various electron and hole functions, e.g., Fermi-Dirac integrals of order 0.5, 0, -0.5 [7], exponential, or their

978-1-7281-0941-1/19 $31.00 © 2019 IEEE

linearized versions have been implemented. A further adaptive-damping strategy for the charge or the potential can also be employed if necessary. We have used a DFT Hamiltonian that was expressed in a localized-orbital basis using the Wannierization technique [1]. ATOMOS finally embed Python to communicate with the outside world and get the user's inputs.

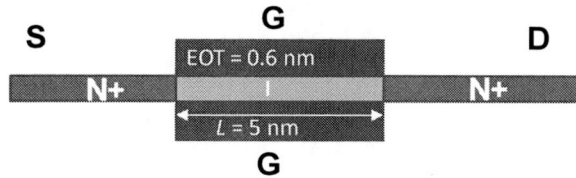

Fig. 3. Schematic view of an optimized 1ML DG nMOSFET.

The combination of state-of-the-art and fine-tuned algorithms with high-performance parallel computing leads to very fast and scalable computations. For the 2D devices simulated here, using a DFT-Hamiltonian including longer-range interactions and dissipative transport, the typical time for solving a single NEGF E-K point is typically ranging from a fraction of a second to a few seconds. On 200 cores, using the latest generation Intel Xeon CPU, the time to solve a single NEGF-Poisson loop bias point is of the order of tens of seconds to maximum a few minutes. A full IV curve is then typically achieved within about an hour to maximum a few hours on 200 cores.

The accuracy of our self-consistent NEGF simulator was also validated by comparing the drain current – gate voltage $I_D(V_G)$ characteristics of various devices, including a HfS$_2$ and BP MOSFETs, computed with ATOMOS and with the quantum-transport simulator NEMO5 [8] that we also augmented to use Wannierized H. Good agreement was achieved in all cases, as shown in Fig. 1 for the HfS$_2$ transistor.

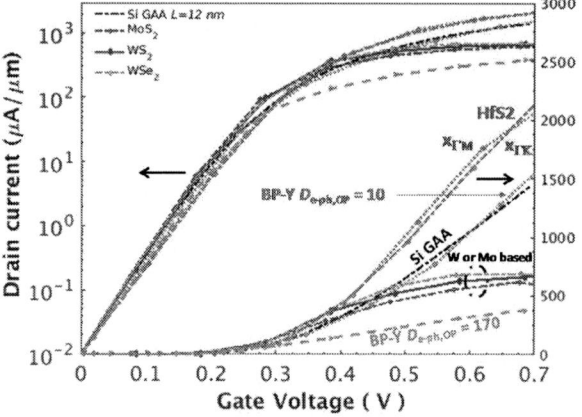

Fig. 4. $I_D(V_G)$ characteristics of the optimized L = 5 nm TMDs and BP-Y nMOSFETs , and the L = 12 nm Si GAA nMOS [12]. V_D = 0.5 V. I_{OFF} = 10 nA/μm. e-ph scattering is included. $D_{e-ph,OP}$ [eV/nm].

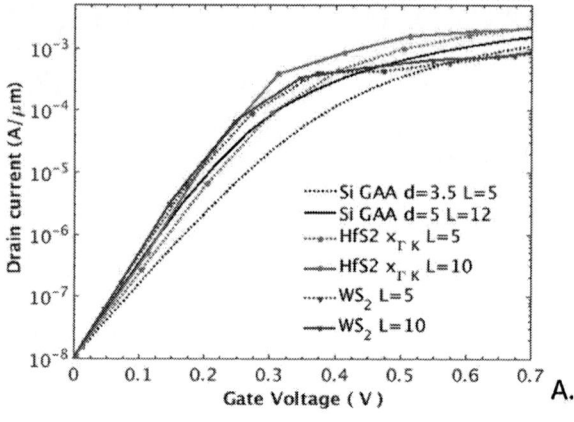

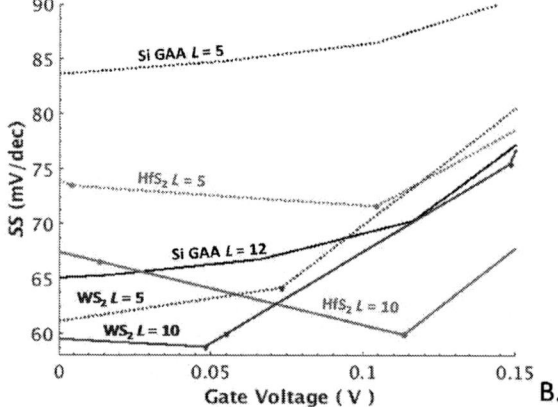

Fig. 5. $I_D(V_G)$ (A) and $SS(V_G)$ (B) characteristics of the WS$_2$ and HfS$_2$ nMOSFETs, as well as those of the Si GAA nMOS for different gate lengths as indicated in the legend (L is in nm). V_D = 0.5 V. I_{OFF} = 10 nA/μm. For the Si GAA, the diameter, d, (also indicated in nm in the legend) was optimized for each L.

III. DFT-HAMILTONIAN COMPUTATION

To model the electronic states in the various TMD and BP monolayers, we used the density-functional theory (DFT)-based *ab-initio* tool QUANTUM ESSPRESSO [9]. The exchange-correlation functional OPTB86B [10] was used both in the geometry relaxation and in the computation of the electronic structure. The plane-wave cutoff energy and Monkhorst-Pack k-point grid values that were used in the relaxation and bandstructure calculations (without spin-orbit coupling) were chosen so that the total energy was well converged. The convergence criteria are set to less than 10^{-3} eV/Å forces acting on each ion and a total energy difference smaller than 10^{-3} eV between two subsequent iterations. To cut off the periodic image along the out-of-plane z-direction (Fig. 2), a vacuum layer of 20 Å was employed in the DFT simulations.

The Bloch wavefunctions are then transformed into maximally-localized Wannier functions (MLWF) typically centered on the ions using the wannier90 package [11]. Fig. 2 demonstrates the validity of our MLWF representation for the case of HfS$_2$. The resulting supercell information, including atoms and MLWF positions, lattice vectors, as well as the localized "tight-binding-like" Hamiltonian matrix elements, are then loaded into ATOMOS and used as building blocks to create the full-device atomic structure and Hamiltonian matrix. The device geometry can be arbitrary rotated to a preferential channel orientation within the 2D layer. The device slab supercells typically encompass several conventional unit cells in the transport (x-axis) direction to keep, in the device Hamiltonian, the required Wannier Hamiltonian longer-range interactions (typically 12 to 15 Å). Periodic boundary conditions are assumed in the width (y-axis) direction and modeled with 16 k_y points. The longer-range interactions that extend to remote conventional cells in that direction are directly included in the periodic conditions.

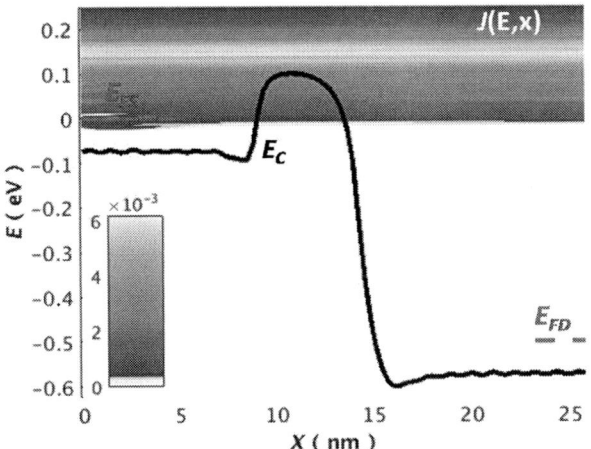

Fig. 6. Current spectrum $J(E, x)$ (surface plot), as well as conduction-band (solid) edges along the channel direction, x, in off-state, for the $L = 5$ nm HfS$_2$ nMOSFET in the ΓK channel orientation with electron-phonon scattering. Source and drain Fermi-level positions are also indicated (red dash).

IV. RESULTS

The schematic of the studied 1ML double-gated (DG) nMOSFETs is shown in Fig. 3. Fig. 4 benchmarks the $I_D(V_G)$ characteristics of various $L = 5$ nm TMD and BP MOSFETs at a typical off-state leakage $I_{OFF} = 10$ nA/µm against that of an optimized Si gate-all-around nanowire (GAA) [12], but at relaxed gate length ($L = 12$ nm). For the Si GAA, scaling L below 10 nm typically results in subthreshold slope (SS) and I_{ON} degradation (Fig. 5), e.g., due to electrostatic control losses, quantum confinement and source-to-drain tunneling (SDT). On the contrary, for all the 2D materials shown on Fig. 4, we observed marginal I_{ON} degradation when scaling L down to 5 nm. However, as also reported in other studies [13], the most commonly studied TMDs, i.e., those having a W or Mo chalcogen metal, do not provide enough drive current for high-performance applications [14], and other materials with current level similar or higher than that of the $L = 12$ nm Si GAA are desired.

Fig. 5 shows with more details the impact of gate length scaling on $I_D(V_G)$ and $SS(V_G)$ characteristics for the WS$_2$ and HfS$_2$ MOSFETs as well as for the Si device. For the Si GAA, we optimized (scaled down to achieve best I_{ON}) the nanowire diameter from 5 to 3.5 nm when scaling L. Still, the atomistically-thin 2D transistors suffer less degradation than the GAA MOSFET. HfS$_2$ features more SS degradation (but better I_{ON}) than WS$_2$ when scaling L down to 5 nm. This is related to a mild but stronger SDT effect (Fig. 6) related to its better transport properties.

Overall, our results predict (Fig. 4) that the less-studied HfS$_2$ (in the octahedral (1T) phase) MOSFET has good scalability down to $L = 5$ nm and features a promisingly high on-current (I_{ON}) level, especially when oriented in the ΓM or ΓK directions. Fig. 6 gives more details about the impact of crystal orientation and electron-phonon (e-ph) scattering on the performance of the device. As expected by the short gate length and low values of the electron-phonon coupling in HfS$_2$, e-ph has only a limited impact on I_{ON}.

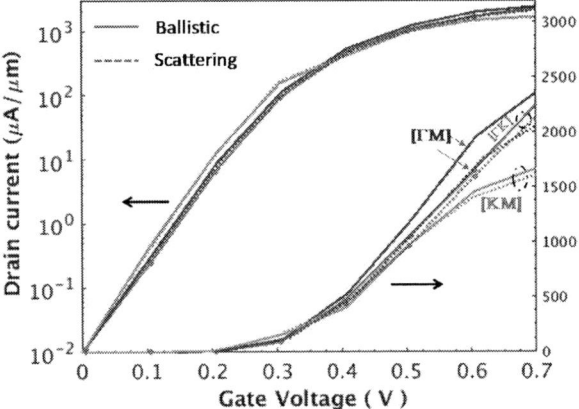

Fig. 7. $I_D(V_G)$ characteristics of the $L = 5$ nm HfS$_2$ nMOSFET for various channel orientations (ΓM, ΓK and KM) and scattering conditions (ballistic, with e-ph). $V_D = 0.5$ V. $I_{OFF} = 10$ nA/µm. HfS$_2$ with e-ph [2]: $D_{e\text{-}ph,AC} = 1.31$ eV. $\hbar\omega_{OP} = 42$ meV, $D_{e\text{-}ph,OP} = 9.9$ eV/nm.

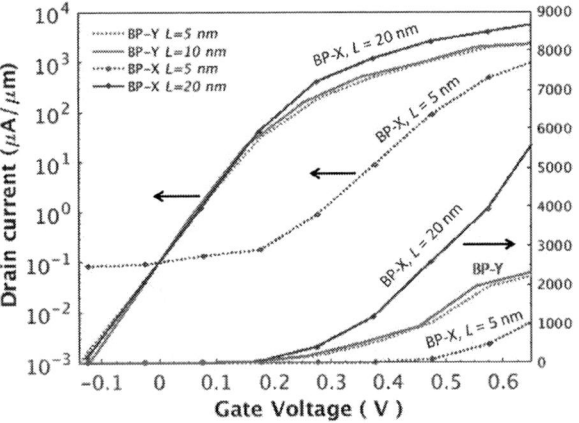

Fig. 8. $I_D(V_G)$ characteristics under ballistic approximation of the $L = 5$ and 10 nm BP-Y and $L = 5$ and 20 nm BP-X nMOSFETs. $V_D = 0.5$ V. $I_{OFF} = 100$ nA/µm.

Concerning, the 1ML BP-MOSFET, the ΓY channel orientation (BP-Y), i.e., the direction with the highest transport effective mass, was chosen as the $L = 5$ nm ΓX-oriented BP MOSFET characteristics were strongly degraded by SDT (Fig. 8 and Fig. 9). Owing to its very low transport effective mass, the BP-X device achieves the highest ballistic I_{ON} at relaxed gate length, but suffers more from SDT than BP-Y at scaled L. The cross-over point where I_{ON} BP-Y exceed BP-X is roughly around $L = 10$ nm.

Most of the transport studies on scaled BP devices with 1ML or a few layers BP to date have been dealing either with ballistic performance or have neglected the optical-phonon coupling [15]. The 1ML BP-MOSFET I_{ON} is severely degraded by the intrinsic strong optical-phonon (OP) coupling that is present in 1 or a few ML of BP ($D_{e\text{-ph,OP}} = 170$ eV/nm for a free standing single monolayer) [3] (Fig. 5). As BP has a strong interlayer coupling, it may, however, be possible to find a substrate that would attenuate the detrimental OP impact. Our results show, however, that a very strong reduction of the optical coupling constant ($D_{e\text{-ph,OP}} < 20$ eV/ nm) would be required to match the Si GAA I_{ON}. The existence of such a substrate is presently unclear.

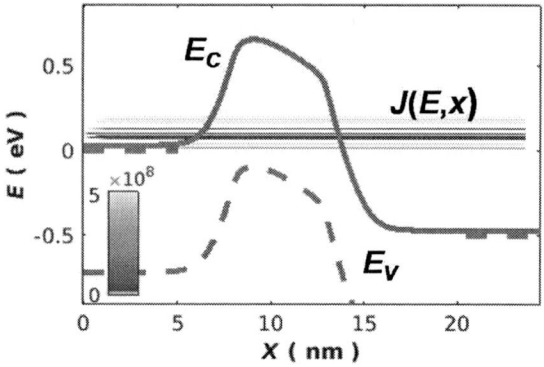

Fig. 9. Current spectrum $J(E, x)$ (surface plot), as well as conduction-band (solid) and valence-band (dashed) edges along the channel direction, x, in off-state, for the $L = 5$ nm ballistic BP-X nMOSFET. Source and drain Fermi-level positions are also indicated (red dash).

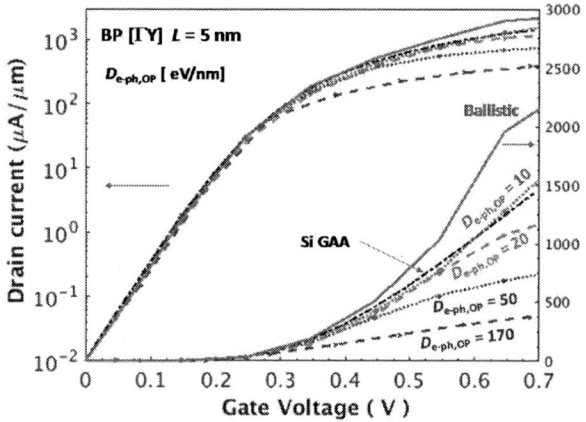

Fig. 10. $I_D(V_G)$ characteristics of the $L = 5$ nm BP-Y nMOSFET for different scattering strengths and the $L = 12$ nm Si GAA nMOS. $V_D = 0.5$ V. $I_{\text{OFF}} = 10$ nA/μm. BP-Y with e-ph [3]: $D_{e\text{-ph,AC}} = 7.11$ eV. •$\omega_{\text{OP}} = 32$ meV, $D_{e\text{-ph,OP}}$ as indicated in the figure ($D_{e\text{-ph,OP}} = 170$ eV/nm for a free standing 1ML).

V. CONCLUSIONS

A state-of-the-art DFT-NEGF based ATOmistic-MOdelling Solver (ATOMOS) was developed and used to assess the physics and fundamental-performance potential of various scaled mono-layers TMD's and BP MOSFETs down to a gate length of 5 nm, including the effect of e-ph scattering. Our results predict that the less-studied HfS₂ MOSFET has good scalability down to $L = 5$ nm with a promising high on-current (I_{ON}) level when oriented in the ΓM or ΓK directions. The 1ML BP-MOSFET I_{ON} is severely degraded by the intrinsic strong optical-phonon (OP) coupling that appears in

1 or a few ML of free-standing BP. It may, however, be possible to find a substrate that would attenuate the detrimental OP impact.

REFERENCES

[1] Á. Szabó, R. Rhyner, and M. Luisier, "Ab initio simulation of single- and few-layer MoS2 transistors: Effect of electron-phonon scattering", Phys. Rev. B 92, 035435, 2015. Doi: 10.1103/PhysRevB.92.035435.

[2] Z. Huang, W. Zhang and W. Zhang, "Computational Search for Two-Dimensional MX2 Semiconductors with Possible High Electron Mobility at Room Temperature", Materials, vol. 9, no.9, p. 716, 2016. Doi: 10.3390/ma9090716.

[3] G. Gaddemane, W. G. Vandenberghe, M. L. Van de Put, S. Chen, S. Tiwari, E. Chen, M. V. Fischetti, "Theoretical studies of electronic transport in mono- and bi-layer phosphorene: A critical overview", Phys. Rev. B, vol. 98, p. 115416, 2018. DOI:10.1103/PhysRevB.98.115416.

[4] A. Svizhenko, M. Anantram, T. Govindan, R. Biegel, R. Venugopal, "Two-dimensional quantum mechanical modeling of nanotransistors", J. Appl. Phys. 91, 2343–2354 (2002).

[5] M. Sancho, J. Sancho, J. Sancho, and J. Rubio, "Highly Convergent Schemes for the Calculation of Bulk and Surface Green Functions", J. Phys. F: Met. Phys., vol. 15, pp. 851–858, 1985.

[6] A. Afzalian, "Computationally Efficient self-consistent Born approximation treatments of phonon scattering for Coupled-Mode Space Non-Equilibrium Green's Functions", J. Appl. Phys., vol. 110, p. 094517, 2011. Doi: 10.1063/1.3658809.

[7] A. Trellakis and A.T. Galick, "Iteration scheme for the solution of the two-dimensional Schrödinger-Poisson equations in quantum structures", J. of Applied Physics, vol. 81, p. 7880, 1997. Doi: 10.1063/1.365396.

[8] S. Steiger, M. Povolotskyi, H.-H. Park, T. Kubis and G. Klimeck, "NEMO5: A Parallel Multiscale Nanoelectronics Modeling Tool", IEEE TNANO, vol. 10, p. 1464, 2011. Doi: 10.1109/TNANO.2011.2166164.

[9] Paolo Giannozzi; Stefano Baroni; Nicola Bonini; Matteo Calandra; Roberto Car; Carlo Cavazzoni; Davide Ceresoli; Guido L Chiarotti; Matteo Cococcioni; Ismaila Dabo; Andrea Dal Corso; Stefano de Gironcoli; Stefano Fabris; Guido Fratesi; Ralph Gebauer; Uwe Gerstmann; Christos Gougoussis; Anton Kokalj; Michele Lazzeri; Layla Martin-Samos; Nicola Marzari; Francesco Mauri; Riccardo Mazzarello; Stefano Paolini; Alfredo Pasquarello; Lorenzo Paulatto; Carlo Sbraccia; Sandro Scandolo; Gabriele Sclauzero; Ari P Seitsonen; Alexander Smogunov; Paolo Umari & Renata M Wentzcovitch (2009). "QUANTUM ESPRESSO: a modular and open-source software project for quantum simulations of materials". Journal of Physics: Condensed Matter. 21 (39): 395502. berg. pp. 155–178. Doi:10.1007/978-3-642-61478-1_10.

[10] J. Klimeš, D. R. Bowler, and A. Michaelides, "Chemical accuracy for the van der Waals density functional", J. Phys.: Cond. Matt. 22, 2, 022201 (2010). Doi: 10.1088/0953-8984/22/2/022201.

[11] A.A. Mostofi, J.R. Yates, G. Pizzi, Y.S. Lee, I. Souza, D. Vanderbilt, N. Marzari, "An updated version of wannier90: A tool for obtaining maximally-localised Wannier functions", Comput. Phys. Commun. 185, 2309 (2014). Doi:10.1016/j.cpc.2014.05.003.

[12] A. Afzalian, M. Passlack and Y.-C. Yeo, "Scaling perspective for III-V broken gap nanowire TFETs: An atomistic study using a fast tight-binding mode-space NEGF model", in Proc. IEEE Int. Electron Device Meeting (IEDM), San-Francisco, CA, USA, 2016, pp. 30.1.1-30.1.4. Doi: 10.1109/IEDM.2016.7838510.

[13] M. Luisier, A. Szabo, C. Stieger, C. Klinkert, S. Brück, A. Jain, L. Novotny, "First-principles simulations of 2-D semiconductor devices: Mobility, I-V characteristics, and contact resistance", in Proc. IEEE Int. Electron Device Meeting (IEDM), San-Francisco, CA, USA, 2016, pp. 5.4.1-5.4.4. DOI: 10.1109/IEDM.2016.7838353.

[14] https://irds.ieee.org/roadmap-2017.

[15] K.-T. Lam, S. Luo, B. Wang, C.-H. Hsu, A. Bansil, H, Lin, G. Liang, "Effects of interlayer interaction in van der Waals layered black phosphorus for sub-10 nm FET", in Proc. IEEE Int. Electron Device Meeting (IEDM), San-Francisco, CA, USA, 2015, pp. 12.2.1-12.2.4. DOI: 10.1109/IEDM.2015.7409681.

Atomistic modeling of nanoscale ferroelectric capacitors using a density functional theory and non-equilibrium Green's-function method

Daniele Stradi*, Ulrik G. Vej-Hansen*, Petr A. Khomyakov*, Maeng-Eun Lee*, Gabriele Penazzi*, Anders Blom*, Jess Wellendorff*, Søren Smidstrup*, and Kurt Stokbro*
*Synopsys Denmark ApS
Fruebjergvej 3, 2100 Copenhagen, Denmark
Email: stradi@synopsys.com

Abstract—We propose a first-principles atomistic method based on density functional theory and the non-equilibrium Green's-function method to investigate the electronic and structural response of metal-insulator-metal capacitors under applied bias voltages. We validate our method by showing its usefulness in two paradigmatic cases where including finite-bias structural relaxation effects is critical to describe the device behavior: formation of dielectric dead layers in a paraelectric SRO|STO|SRO capacitor due to an applied bias voltage, and the switching behavior of a ferroelectric SRO|BTO|SRO capacitor due to an external electric field.

I. Introduction

With the continuous downscaling of device components, understanding the interplay between electronic and structural response in the presence of an applied electrical bias (V_{bias}) becomes central to determine their functionality. For metal-insulator-metal (MIM) capacitors, the importance of structural effects due to V_{bias} has long been recognized. Atomistic simulations have shown that a significant portion of the capacitance decrease observed in thin paraelectric (PE) perovskite films comes from low-permittivity regions at the metal-insulator interface ("dead layers") [1], which develop as a result of structural distortions in the insulating layer due to V_{bias} [2]. The behavior of such dead layers beyond the linear-response regime is still poorly understood, possibly due to the lack of atomistic methods that are able to describe the structure of the capacitor at finite V_{bias} [3]. Elucidating the role of V_{bias} in the switching behavior of ferroelectric (FE) materials has also recently become a subject of attention, due to the application of FE-based structures in negative-capacitance field-effect transistors (NC-FETs) [4]. However, similarly to the case of paraelectric MIM capacitors, the atomistic-level description of the switching process is still mostly limited to classical simulations using parametrized Hamiltonians [5] or first-principles methods where other variables than V_{bias} are used to impose an electric field [6]–[8], thereby lacking a direct connection to experimental measurements.

Here, we propose an efficient *ab initio* approach to describe atomistically the electronic and structural response of nanoscale capacitors at finite V_{bias}. Our method accounts for the exact boundary conditions of the capacitor at $V_{bias} \neq 0$, as well as for the ballistic current between the capacitor plates. These could not be accounted for in previous approaches used to describe finite-bias effects in nanoscale capacitors [8], [9]. We validate our approach by studying the role of V_{bias} in the formation of dead layers in a PE $SrRuO_3|SrTiO_3|SrRuO_3$ (SRO|STO|SRO) capacitor, and in the FE switching of a $SrRuO_3|BaTiO_3|SrRuO_3$ (SRO|BTO|SRO) capacitor.

II. Methodology

All calculations were performed using density functional theory (DFT) and the non-equilibrium Green's-function (NEGF) method, [11] as implemented in the QuantumATK program developed by Synopsys [12], [13]. We used the linear combination of atomic orbitals (LCAO) method and normconserving pseudopotentials (PPs), see [13] for further details.

The SRO|STO|SRO capacitor was based on a thin film made of 7 STO unit cells (UCs), sandwiched between two semi-infinite SRO electrodes. The PBE exchange-correlation functional was used [14], together with PseudoDojo PPs and a PseudoDojo-Medium basis set [10]. The mesh cutoff was 750 Ha. Monkhorst–Pack (MP) k-point grids [15] of 13×13 and $13 \times 13 \times 401$ were used for the device and for the electrodes, respectively. In the MIM structure, the lattice constant in the xy-plane was set to the calculated lattice constant for STO, $a_{STO} = 3.95$ Å to simulate pseudomorphic growth on STO substrates. The zero-bias structure was relaxed along the z-direction using a 2-step procedure. First, a SRO|STO|SRO slab was constructed, with 6 SRO UCs on each side of the STO film. The atomic positions of the central part of the slab were optimized along the z-direction, while keeping fixed the 4 SRO UCs closest to the left surface, and treating the 4 SRO UCs closest to the right surface as a rigid body. A device configuration was then created based on the optimized slab geometry. This configuration was further optimized by

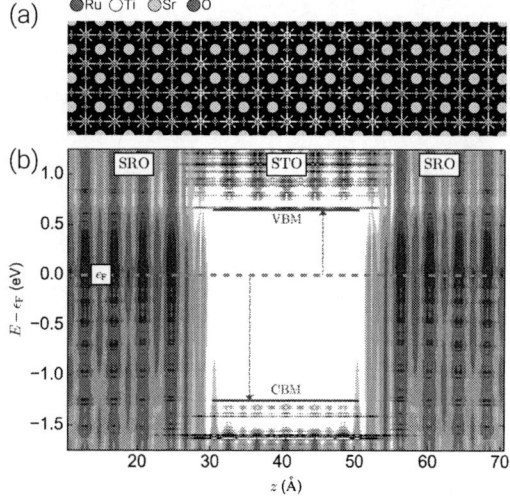

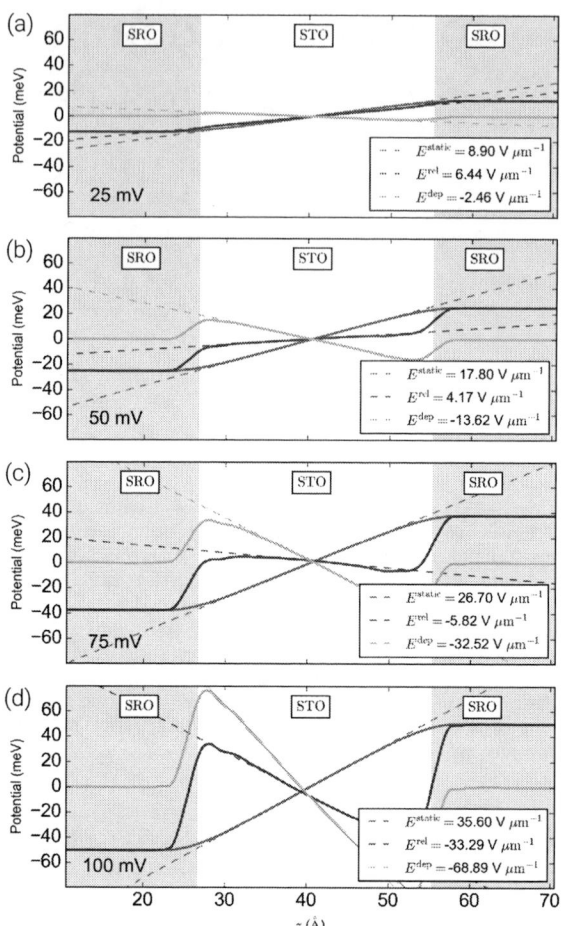

Fig. 1. (a): Structure of the SRO|STO|SRO capacitor. For sake of clarity, the structure has been repeated 4 times in the y direction. (b): PLDOS of the capacitor in (a). Darker (lighter) colors indicate regions of high (low) PLDOS. The shaded blue areas indicate the position of the SRO regions. The solid purple lines mark the position of the CBM and VBM. The orange dashed line marks the position of the Fermi level ϵ_F.

allowing the STO layers and the two SRO layers closest to each surface to relax along z for applied biases in the range $0\ \mathrm{mV} \leq V_\mathrm{bias} \leq 100\ \mathrm{mV}$. The threshold for force minimization was 5 meV/Å for all the structural relaxations.

The SRO|BTO|SRO structure was based on a FE thin film made of 9 BTO UCs. The LDA exchange-correlation functional was used, together with FHI PPs and a DZP basis set [10], and a 180 Ha mesh cutoff. 6×6 and $6 \times 6 \times 251$ MP k-point grids were used to sample the BZ of the device and of the electrodes, respectively. The structure at zero bias was optimized using the 2-step procedure described above, with a threshold criterion for the forces of 10 meV/Å. Starting from this zero-bias device configuration, we performed a sweep of V_bias to inverstigate the FE behavior of the capacitor. First, we gradually increased V_bias up to 300 mV, and optimized the structure of the BTO UCs and of the 2 SRO UCs closest to each interface along z at each intermediate value of V_bias. The procedure was then repeated backwards, starting from the structure optimized at 300 mV bias. For each optimized structure, the ballistic current I was calculated by means of the Landauer formula, using a 51×51 MP k-point sampling.

The macroscopic averages [16] of the induced potentials were calculated from the Hartree difference potential, δV_H [13], as $\langle \Delta V_\mathrm{H} \rangle = \langle \delta V_\mathrm{H}(V_\mathrm{bias} \neq 0\ \mathrm{mV}) - \delta V_\mathrm{H}(V_\mathrm{bias} = 0\ \mathrm{mV}) \rangle$. We calculated the potential drop before ($\langle \Delta V_\mathrm{H}^\mathrm{static} \rangle$) and after ($\langle \Delta V_\mathrm{H}^\mathrm{rel} \rangle$) structural relaxation. The depolarization potential was calculated as $\langle \Delta V_\mathrm{H}^\mathrm{dep} \rangle = \langle \Delta V_\mathrm{H}^\mathrm{rel} \rangle - \langle \Delta V_\mathrm{H}^\mathrm{static} \rangle$. The corresponding electric fields in the center of the capacitor were calculated from linear fits to those potentials in the center of the insulating region. For the PE capacitor, the profile

Fig. 2. (a): Macroscopic average of the induced potentials in the SRO|STO|SRO capacitor at $V_\mathrm{bias} = 25$ mV. The profiles of the potential drop before ($\langle \Delta V_\mathrm{H}^\mathrm{static} \rangle$) and after ($\langle \Delta V_\mathrm{H}^\mathrm{rel} \rangle$) structural optimization are shown as grey and purple solid lines, respectively. The depolarization potential ($\langle \Delta V_\mathrm{H}^\mathrm{dep} \rangle$) is shown as a green solid line. Dashed lines indicate the linear fits in the center of the STO region used to extract the fields associated with each potential. The values of the resulting field are shown in the legend. (b-d): same as (a), but for $V_\mathrm{bias} = 50$ mV, 75 mV, 100 mV.

of the inverse relative permittivity, $1/\varepsilon_r$, was calculated as $1/\varepsilon_r = -1/E_\mathrm{ext} \times \partial \Delta V_\mathrm{H}^\mathrm{rel}/\partial z$, where $E_\mathrm{ext} = \varepsilon E_\mathrm{int}$, and $\varepsilon = 239$ and E_int are the dielectric constant calculated for bulk STO and the internal field in the insulating layer, respectively. The projected local density of states (PLDOS) across the device at zero bias was calculated on a 25×25 MP k-point grid.

III. RESULTS

A. SRO|STO|SRO paraelectric capacitor

The PLDOS of the SRO|STO|SRO PE capacitor in Fig. 1b shows that the SRO electrodes are metallic, while the band gap in the center of the STO film is roughly 2 eV. Close to both SRO|STO interfaces, a finite DOS is present within the

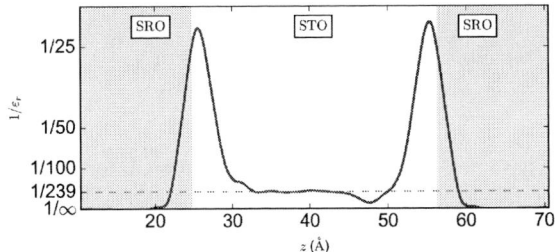

Fig. 3. Profile of the inverse permittivity $1/\varepsilon_r$ along the z Cartesian direction of the SRO|STO|SRO capacitor, obtained from $\langle \Delta V_H^{rel} \rangle$ in Fig. 2b. The dashed grey line marks the value of inverse permittivity $1/239$ calculated for bulk STO.

STO gap, which we identify as metal-induced gap states. The gap states pin the bands of the STO film to the Fermi energy (ϵ_F) of SRO, which results in the STO valence band maximum (VBM) lying about 1.25 eV below ϵ_F, in good agreement with literature results reported for a similar capacitor structure [2].

At $V_{bias} > 0$, the Ti (O) sublattice of STO translates along (opposite to) the direction of the applied field, indicating a structural distortion along the zone-center optical modes. At $V_{bias} = 25$ mV, the geometry is only slightly deformed, and the profile of $\langle \Delta V_H^{rel} \rangle$ closely matches that of $\langle \Delta V_H^{static} \rangle$, see Fig. 2a. As a result, $\langle \Delta V_H^{dep} \rangle$ is almost flat, and the depolarization field $E^{dep} \sim 0$ V/μm, indicating that the dead layers are not fully developed at this bias voltage. These results differ from those obtained in [2] using a linear-response method, since in that case the formation of dead layers was observed at comparable values of V_{bias}. Increasing further V_{bias} results in a highly nonlinear behavior of the profile of $\langle \Delta V_H^{rel} \rangle$, which points at the formation of dead layers at the SRO|STO interfaces. At $V_{bias} = 50$ mV, the profiles of $\langle \Delta V_H^{rel} \rangle$ and $\langle \Delta V_H^{dep} \rangle$, shown in Fig. 2b, are qualitatively similar to those reported in [2]. Note that $\langle \Delta V_H^{rel} \rangle$ is flat in the SRO regions away from the interface, indicating that the field induced by V_{bias} has been fully screened. The presence of dielectric dead layers is even more evident from the inverse permittivity profile in Fig. 3, which shows two peaks at the SRO|STO interfaces, indicating regions of reduced permittivity. The $1/\varepsilon_r$ profile is only qualitatively similar to that reported in [2]. In particular, the low-permittivity regions calculated here are considerably sharper, and $1/\varepsilon_r$ tends to the correct limit $1/\infty$ close to the metallic electrodes. Furthermore, additional features are resolved, in particular the asymmetry in the $1/\varepsilon_r$ profile between the dead layers, which is due to the formation of a dipole caused by the field-induced distortion of the STO lattice. The main reason for these improved results is that the DFT-NEGF method applied here imposes the physically correct boundary conditions on the device. In more quantitative terms, the values obtained here for the fields in the center of the insulating film are roughly twice as large as those obtained at 27.8 mV bias voltage in [2], which supports the numerical accuracy of our DFT-NEGF method.

For $V_{bias} > 50$ mV, the nonlinear behavior of $\langle \Delta V_H^{rel} \rangle$ becomes even more pronounced, see Fig. 2c,d. We may assume that the capacitor can be described as three effective capacitors in series, $C = (C_{int}^{-1} + C_0^{-1} + C_{int}^{-1})^{-1}$. With $C_0 \sim Q/E^{rel}d$, and since Fig. 2 shows that E^{rel} becomes negative at $V_{bias} \geq 75$ mV, our simulations suggest that the SRO|STO|SRO central region acquires a negative effective capacitance ($C_0 < 0$) as the bias voltage is ramped up.

B. SRO|BTO|SRO ferroelectric capacitor

The SRO|BTO|SRO capacitor exhibits a finite polarization $\vec{P}$ in the BTO film, which affects its electronic and electrical properties. Due to the finite $\vec{P}$, the capacitor behaves as a FE switch, as verified here by electronic-structure analysis and simulated I-V_{bias} characteristics. As shown in Fig. 4, the latter exhibits a clear hysteretic behavior characterized by a low- (LR) and a high-resistance (HR) state, which is a key feature of a FE switch.

The PLDOS shown in Fig. 5a for the LR state at $V_{bias} = 100$ mV is representative of its electronic structure at 0 mV $\leq V_{bias} < 300$ mV. In this bias range, the direction of $\vec{P}$ is along z, i.e., $P_z > 0$. This affects the CBM and VBM throughout the BTO film, which are bent in the opposite direction with respect to $\vec{P}$, and results in an asymmetric potential barrier, which can be approximated by a trapezoidal shape. For each V_{bias}, we estimated the trapezoidal barrier by performing a linear fit to $\langle V_H \rangle$, and subsequently aligned it with the CBM in the center of the BTO insulator at zero bias. Increasing V_{bias} results in a monotonic increase in the current I, but only in minor changes to the shape of the barrier. The profile of $\langle \Delta V_H^{rel} \rangle$ in Fig. 5b shows that dead layers appear at the SRO|BTO interfaces, resulting in a sizeable "depolarizing" field in the center of the capacitor. This behavior is fully consistent with that observed above for the PE capacitor. However, in the FE case, $P \neq 0$ at zero bias, and the effect of the dead layers is to inhibit changes in $\vec{P}$ with V_{bias} inside the BTO

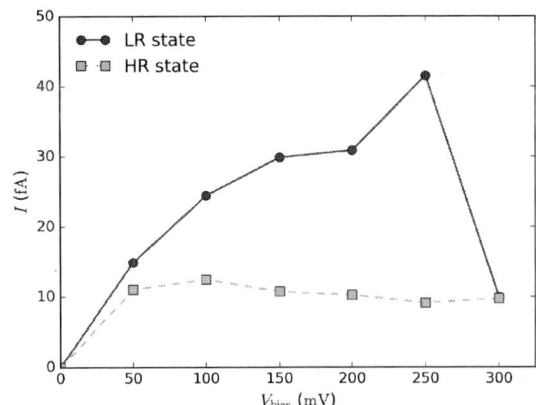

Fig. 4. I-V_{bias} characteristics calculated for the ferroelectric SRO|BTO|SRO capacitor. Solid purple line (circles): low-resistance (LR) state. Green dashed line (squares): high-resistance (HR) state.

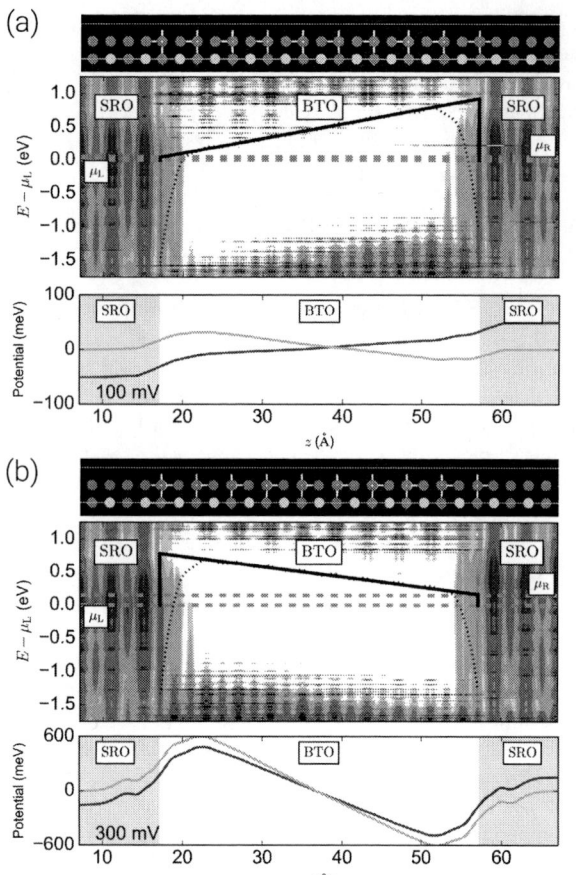

Fig. 5. (a): Low-resistance state of the SRO|BTO|SRO capacitor at $V_{bias} = 100$ mV. Top panel: structure of the capacitor. Central panel: PLDOS of the capacitor structure in the top panel. The color scheme is the same as in Fig. 1, with the difference that the orange dashes lines indicate the position of the left (μ_L) and right (μ_R) chemical potentials. The dotted line indicates the macroscopic average of the Hartree difference potential $\langle V_H \rangle$. The solid black line indicate the shape of the trapezoidal potential barrier. Bottom panel: profiles of the potential drop $\langle \Delta V_H^{rel} \rangle$ and of the depolarization potential $\langle \Delta V_H^{dep} \rangle$. (b): Same as (a) but for the high-resistance state at $V_{bias} = 300$ mV.

film, resulting in a small variation of the barrier shape with V_{bias}. The increase in the current I is therefore mostly due to the increasingly larger number of states that become available for electron transmission inside the bias window as V_{bias} is increased.

At $V_{bias} = 300$ mV the polarization of the BTO thin film switches spontaneously to $P_z < 0$. This change in polarization is associated with a sudden drop of I, as seen in Fig. 4. The PLDOS of the resulting HR state in Fig. 5b is representative of its electronic structure at 0 mV $\leq V_{bias} \leq 300$ mV. The CBM and VBM are bent in the opposite direction compared to the LR state, resulting in a trapezoidal barrier shape that is now inverted. This suggests that the capacitor behaves as a FE switch, which is irreversibly switched between a LR and a HR state when $V_{bias} \geq V_C$, $V_C = 300$ mV being the coercive field

required for polarization switching [17]. Indeed, following the LR→HR switching event, $I^{HR} < I^{LR}$ for each bias voltage as it is decreased towards zero, which confirms the irreversibly of the switching event.

IV. Conclusions

We have proposed a computational method based on density functional theory and non-equilibrium Green's functions to account for both electronic structure and geometry relaxation effects in MIM capacitors induced by a finite bias voltage. We have shown that the method allows for a straightforward interpretation of complex effects due to field-induced structural relaxation, such as formation of dielectric dead layers and ferroelectric switching. We expect that the method will find broad applicability for complex and technologically relevant capacitor structures beyond those considered in this work, and could be useful to extract fundamental parameters to be used as input for higher-level TCAD tools.

References

[1] C. Zhou and D. M. Newns, "Intrinsic dead layer effect and the performance of ferroelectric thin film capacitors" J. Appl. Phys., vol. 82, p. 3031, 1997.

[2] M. Stengel and N. A. Spaldin, "Origin of the dielectric dead layer in nanoscale capacitors" Nature, vol. 443, pp. 679–682, 2009.

[3] D. I. Bilc, F. D. Novaes, J. Íñiguez, P. Ordejón, and P. Ghosez, "Electroresistance effect in ferroelectric tunnel junctions with symmetric electrodes" ACS Nano, vol. 6, pp. 1473–1478, 2012.

[4] S. Salahuddin and S. Datta, "Use of negative capacitance to provide voltage amplification for low power nanoscale devices" Nano Lett., vol. 8, pp. 405–410, 2008.

[5] T. Nishimatsu, U. V. Waghmare, Y. Kawazoe, and D. Vanderbilt, "Fast molecular-dynamics simulation for ferroelectric thin-film capacitors using a first-principles effective Hamiltonian" Phys. Rev. B, vol. 78, p. 104104, 2008.

[6] S. Kasamatsu, S. Watanabe, and S. Han, "First-principles calculation of charged capacitors under open-circuit conditions using the orbital-separation approach" Phys. Rev. B, vol. 92, 115124, 2015.

[7] S. Kasamatsu, S. Watanabe, C. S. Hwang, and S. Han, "Emergence of negative capacitance in multidomain ferroelectric-paraelectric nanocapacitors at finite bias" Adv. Mater., vol. 28, pp. 335–340, 2016.

[8] M. Stengel, D. Vanderbilt, and N. A. Spaldin, "First-principles modeling of ferroelectric capacitors via constrained displacement field calculations" Phys. Rev. B, vol. 80, p. 224110, 2009.

[9] M. Stengel and N. A. Spaldin, "Ab initio theory of metal-insulator interfaces in a finite electric field" Phys. Rev. B, vol. 75, p. 205121, 2007.

[10] S. Kasamatsu, S. Watanabe, and S. Han, "Orbital-separation approach for consideration of finite electric bias within density-functional total-energy formalism" Phys. Rev. B, vol. 84, p. 085120, 2011.

[11] D. Stradi, U. Martinez, A. Blom, M. Brandbyge, and K. Stokbro, "General atomistic approach for modeling metal-semiconductor interfaces using density functional theory and nonequilibrium Green's function" Phys. Rev. B, vol. 93, p. 155302, 2016.

[12] https://www.synopsys.com/silicon/quantumatk.html

[13] S. Smidstrup et al., "QuantumATK: An integrated platform of electronic and atomic-scale modelling tools" arXiv:1905.02794v1.

[14] J. P. Perdew, K. Burke, and M. Ernzerhof, "Generalized gradient approximation made simple" Phys. Rev. Lett., vol. 77, p. 3865, 1996.

[15] H. J. Monkhorst and J. D. Pack, "Special points for Brillouin-zone integrations" Phys. Rev. B, vol. 13, p. 5188, 1976.

[16] A. Baldereschi, S. Baroni, and R. Resta, "Band offsets in lattice-matched heterojunctions: A model and first-principles calculations for GaAs/AlAs" Phys. Rev. Lett., vol. 61, p. 734, 1988.

[17] H. Kohlstedt, N. A. Pertsev, J. Rodriguez Contreras, and R. Waser, "Theoretical current-voltage characteristics of ferroelectric tunnel junctions" Phys. Rev B, vol. 72, p. 125341, 2005.

978-1-7281-0941-1/19 $31.00 © 2019 IEEE

Trigonal Tellurium Nanostructure Formation Energy and Band gap

Aaron Kramer[1], Maarten L. Van de Put[2], Christopher L. Hinkle[3], and William G. Vandenberghe[2]

[1]Department of Physics and [2]Department of Materials Science and Engineering,
University of Texas at Dallas, 800 W. Campbell Rd., Richardson, Texas 75080, USA
[3]Department of Electrical Engineering, University of Notre Dame, Notre Dame, Indiana 46556, USA

Abstract—**Trigonal-Tellurium (t-Te), a van der Waals material, recently garnered interest to the nanoelectronics community because a high hole mobility, a high bandgap, and low temperature growth have all been observed in nanostructures. We analyze various t-Te nanostructures (nanowires and layers) using first principles simulations. We compare bandgap variation and relative stability among different shapes and sizes of Te nanostructures. We determine that nanowires host higher bandgaps and are preferentially grown, rather than layers of t-Te. We also propose a simplified model using the number of van der Waals interactions in explaining relative stability among t-Te nanostructures. Finally, we study uniquely shaped (auxiliary) t-Te nanostructures and verify that their stability obeys the same simplified model.**

I. INTRODUCTION

As semiconductor devices shrink in size, common materials such as silicon have their mobilities reduced significantly [1]. Graphene offers a solution to the low mobility problem, but unfortunately does not offer a bandgap [2]. An alternative material of interest to the nanoelectronics community is trigonal tellurium (t-Te), which admits a high hole mobility even in nanostructures [3]. While t-Te, in contrast to graphene, does offer a nearly direct bulk bandgap, the bandgap is small (0.33 eV) [4]. However, at scaled dimensions, quantum confinement results in an increasing bandgap and this may reduce leakage current and improve tellurium's prospects for use in future transistors.

Structurally, t-Te comprises of one-dimensional (1D) helical chains with covalently bonded Te atoms. These Te-helices manifests into a trigonal lattice though a mixture of van der Waals (vdW) interactions and covalent bonding. The overall trigonal structure, vdW interactions, and the covalent

bonding are all consequences of Tellurium containing six valence electrons [5]. Historically t-Te was used in infrared detectors [6], thermoelectric [7], piezoelectric [8] and photoconductive devices [9].

Low-dimensional Tellurium allotropes, such as monolayer Tellurene have been investigated using first principles density functional theory (DFT). Notable allotropes are described in Zhu et. al. (α, β, and γ) [10], Liu, Lin and Tomanek (δ and η) [11], and Xian et. al. (square Tellurene) [12]. It is worth mentioning that only the β allotrope bears resemblance to bulk t-Te. For monolayers, the α, δ, and η allotropes are predicted to be more stable than the β-allotropes with the η allotrope being the most stable. Due to the approximations made in the DFT-treatment of the vdW interaction, determining the most stable t-Te nanostructure is challenging. Notably, different vdW models can change the order of the most stable monolayer structure.

Here we employ first principles calculations with DFT to compare and analyze the stability of various t-Te nanostructures (layers of t-Te and nanowires). We calculate formation energies, study surface-to-volume ratios, and develop a simplified model to determine the stability of 1D nanostructure using the total energy associated with a Te-helix with m helical-neighbors. We find that 1D hexagonal shaped nanowires are the most favored thermodynamically. We also include bandgaps and find that 1D triangle shaped nanowires feature the highest bandgaps (up to 1-2 eV) making t-Te an eye-catching channel material for extremely scaled field-effect transistors.

Section II contains our results and discussion, section III describes our methodology and section IV is our conclusion.

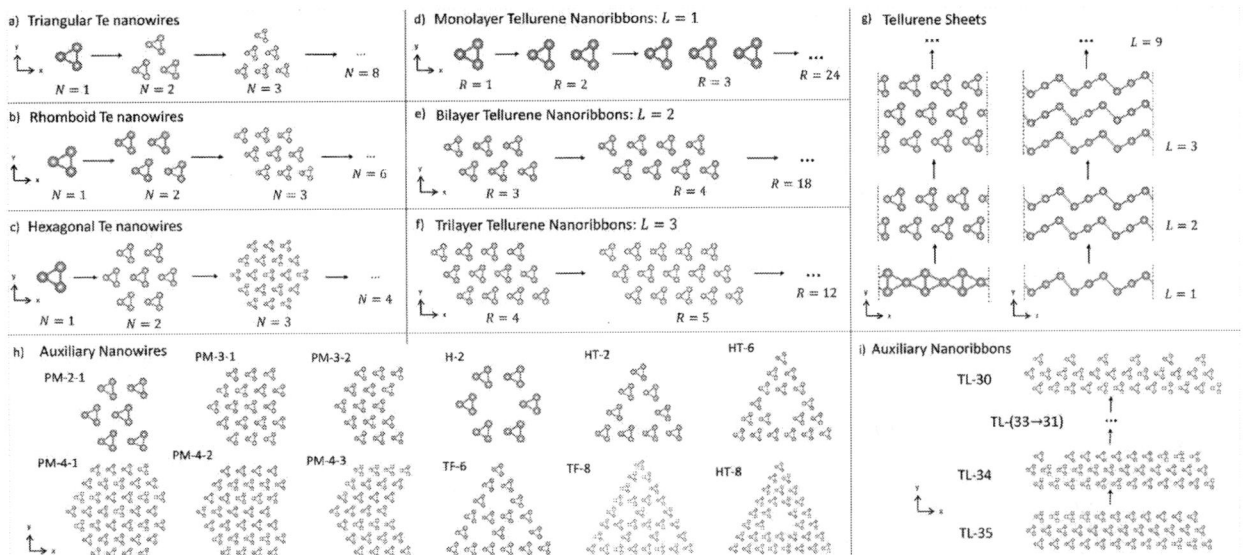

Figure 1. Illustration of all our computed t-Te nanostructures. Cross-sectional area of triangular (a), rhomboid (b), hexagonal (c) nanowires, and monolayer (d), bilayer (e), and trilayer (f) Tellurene nanoribbons, sheets of Tellurene (g), and auxiliary nanowires (h), and (i) nanoribbons. N is the number of helices per nanowire side. R is the number of helices on the longer nanoribbon side. L is number of layers for Tellurene sheets.

978-1-7281-0941-1/19 $31.00 © 2019 IEEE

II. RESULTS AND DISCUSSION

Fig. 1 illustrates all t-Te nanostructures under investigation. We have three classes of nanowires (triangular, rhomboid, and hexagonal), three classes of nanoribbons (monolayer, bilayer, and trilayer), and monolayer to multilayer sheets of Tellurene. Additionally, we create some "auxiliary" nanowires and nanoribbons to supplement our main structures.

A. Formation Energy

Fig. 2 shows the calculated formation energies for all nanowires and nanoribbons. The nanoribbons have a higher formation energy than the nanowires for the same number of Te-helices. The order going from highest to lowest formation energy for the same number of Te helices is: monolayer (1L), bilayer (2L), and trilayer (3L) nanoribbons, then Triangular, Rhomboid, and Hexagonal nanowires.

Rhomboid nanowires would be the most difficult Te nanowire to fabricate because hexagonal or triangular nanowires would be preferentially formed instead, depending on the environment. This is in line with earlier experimental findings [13].

Finally, Fig. 2 shows that the monolayer (1L) ribbon has an exceedingly high formation energy and will be very difficult to fabricate unless substrate interactions drastically alter formation energies. The 2L and 3L ribbons have formation energies much closer to those of the wires and may form in the presence of favorable substrate interaction. Note that the monolayer, shown in Fig 1g, has a fundamentally different structure. 1L nanoribbons approach this limit as they get wider, while the helices in 2L and 3L ribbons remain structurally similar to bulk t-Te.

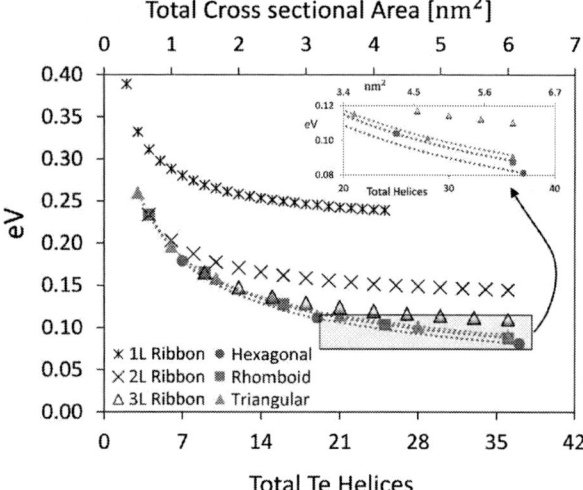

Figure 2. Formation energies for all 1D structures (excluding auxiliary structures) as a function of the total number of Te helices or cross-sectional area. The solid or open points are computed values. The dotted lines are power fits to guide the eye.

Table 1 shows the formation energy for the auxiliary structures and Tellurene sheets shown in Fig. 1h and 1g. Two auxiliary nanowires (e.g., PM-2-1 and H-2) can have the same number of helices (e.g., 6 helices) but vastly different formation energies (e.g., 0.20 eV/atom and 0.27 eV/atom). These structures do not follow the simple trend of nanowire formation energy in Fig. 2, where formation energy variation

is small across the triangular, rhomboid, and hexagonal nanowires as the number of helices is varied.

The first three rows in Table 1b represent the lowest formation energies that monolayer, bilayer and trilayer nanoribbons can obtain. The formation energy quickly decreases with increasing number of layers. At around 7 layers the formation energy decrease starts to taper as the layers more closely resemble bulk.

(a) Auxiliary Structures

Name	# of Helices	Formation Energy [eV / atom]	
PM-2-1	6		0.20
PM-3-1	18		0.12
PM-3-2	15	Wires	0.14
H-2	6		0.27
HT-2	9		0.21
HT-6	18		0.17
PM-4-1	36		0.08
PM-4-2	34		0.09
PM-4-3	28		0.10
TF-6	18		0.18
TF-8	30		0.16
HT-8	33		0.12
TL-35	35		0.11
TL-34	34	Ribbons	0.12
TL-33	33		0.12
TL-32	32		0.13
TL-31	31		0.14
TL-30	30		0.14

(b) Tellurene Sheets

Name	Formation Energy [eV / atom]
1-Layer	0.22
2-Layer	0.13
3-Layer	0.09
4-Layer	0.07
5-Layer	0.06
6-Layer	0.05
7-Layer	0.04
8-Layer	0.04
9-Layer	0.03

Table 1. a) The Formation energy of the auxiliary structures with the number of Helices present and b) 2D sheets of Tellurene.

To support our knowledge of the relative formation energies among nanostructures in Fig. 2, we calculate the surface-to-volume ratio, for all nanowires and the bilayer and trilayer nanoribbons (the auxiliary structures are excluded). The calculated surface-to-volume ratios are shown in Fig 3. Hexagonal nanowires have the smallest surface-to-volume ratio for the same number of helices. This agrees with Fig. 2 where hexagonal nanowires also have the lowest formation energy for the same number of helices. Unfortunately, surface-to-volume ratio does not illustrate all formation energy differences. For instance, the bilayer and trilayer nanoribbons have a lower surface-to-volume ratio compared to the triangular nanowires when the nanoribbons have more than 14 and 27 helices respectively. Since the nanowires in Fig. 2 always exhibit a lower formation energy compared to the nanoribbons, surface-to-volume ratio does not tell the full story.

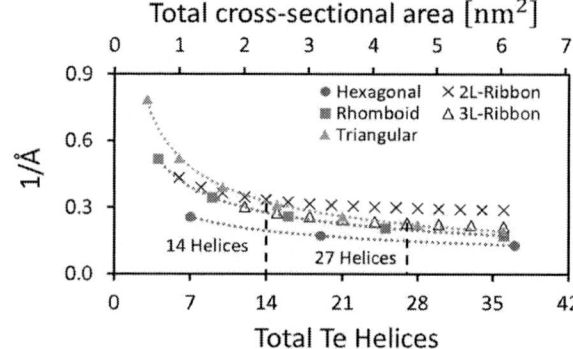

Figure 3. Surface-to-volume ratio for the nanowires and the nanoribbons. Solid points are calculated values and dotted lines are power fits to guide the eye.

An alternative explanation is in the differing number of "happy" helices that have six nearest helical-neighbors. Nanowires will always have more "happy" helices compared to nanoribbons for the same total number of helices. To determine whether the number of neighboring helices is a good metric, we determine an energy penalty ϵ_m associated with each helix that has a given number (m) helices.

Table 2 shows the resulting energy penalties determined using an ordinary least squares (OLS) fit as explained in the methods section. Significant energy penalties ranging from 1.17 eV to 0.39 eV are observed for helices with six to one missing helical-neighbors respectively. The small value $\epsilon_6 = 0.02$ eV indicates that the energy obtained from the OLS for helices with six helical-neighbors, is remarkably close to that of the bulk t-Te. To rule out large systematic errors between nanowire types, we verify that the order of largest to lowest formation energies observed in Fig. 2 is maintained using the energy penalties. We find that the formation energies of the auxiliary structures in Table 1a, the ones not used in the OLS fit, can be estimated within 3% error using the energy penalties.

Keeping the missing neighboring helix picture in mind, we inspect the results in Fig. 2 and Table 1. Our triangular, rhomboid, and hexagonal nanowires have Te-helices with fewer missing helical neighbors compared to layers of Tellurene. Our auxiliary structures have a large number of Te helices with many missing helical neighbors. Therefore, the missing neighboring helix picture, in contrast to surface-to-volume ratio picture, can explain why nanowires have lower formation energy than layers for the same number of Te-helices. As noted before, this agrees with experimental growth, which also found that wires rather than layers are preferred [3].

	Energy Penalty [eV /Unit-Cell]
ϵ_0	1.17
ϵ_1	0.89
ϵ_2	0.88
ϵ_3	0.55
ϵ_4	0.40
ϵ_5	0.39
ϵ_6	0.02

Table 2. The energy penalties for having m helical-neighbors per unit cell determined using the OLS fit on all our structures. The energy penalty is relative to bulk t-Te.

B. Electronic Properties

The computed bandgaps in Fig. 4 ranges from 0.34 eV to 1.46 eV. When interpreting these bandgaps, the tendency for DFT to underestimate the bandgaps should be taken in mind. For example, DFT predicts a vanishing t-Te bulk bandgap. The bandgap for the N=2 Hexagonal nanowire (Fig 1c N=2, 7 Helices) with hybrid HSE06 functionals [14] went from 0.50 eV to 0.99 eV, that is about a factor of 2.

Trends in nanowire bandgaps in Fig. 4 show that hexagonal, rhomboid, and triangular nanowires have lowest, intermediate, and the highest bandgaps respectively. Bandgap values start to decrease quickly at nanowire sizes around 7 helices ($\sim$1.3 nm^2).

The computed bandgaps for the auxiliary nanowires and sheets of Tellurene are given in Table 3. The auxiliary

nanowires and sheets of Tellurene have their bandgap range from 0.37 eV to 0.94 eV and 0.17 eV to 1.01 eV respectively. Unlike the nanowires in Fig. 4 where bandgaps start to decrease quickly at a size of 7 Te helices ($\sim$1.3 nm^2), the auxiliary nanowires do not exactly follow that same decreasing trend at 7 Te helices. For instance, the TF-8 nanowire has 30 Te helices ($\sim$4.8 nm^2) but exhibits a high bandgap of 0.74 eV.

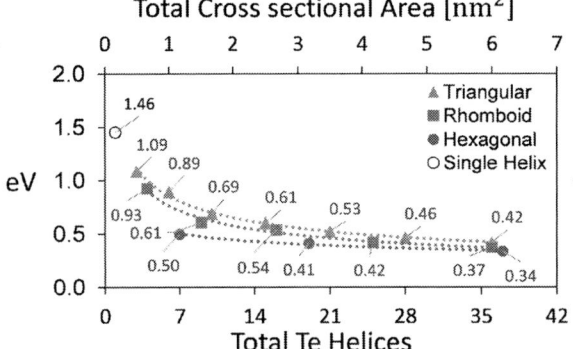

Figure 4. Bandgap values for our Triangular, rhomboid and hexagonal nanowires as a function of the total number of Te helices or total nanowire cross sectional area. Solid and open points are calculated points while the dotted lines are power and logarithmic fits to guide the eye.

(a) Auxiliary Nanowires			(b) Tellurene Sheets	
Name	# of Helices	Bandgaps [eV]	Name	Bandgaps [eV]
PM-2-1	6	0.64	1-Layer	1.01
PM-3-1	18	0.45	2-Layer	0.76
PM-3-2	15	0.50	3-Layer	0.58
H-2	6	0.57	4-Layer	0.44
HT-2	9	0.94	5-Layer	0.35
HT-6	18	0.79	6-Layer	0.28
PM-4-1	36	0.37	7-Layer	0.24
PM-4-2	33	0.45	8-Layer	0.20
PM-4-3	28	0.41	9-Layer	0.17
TF-6	18	0.85		
TF-8	30	0.74		
HT-8	33	0.58		

Table 3. The computed bandgaps for a) auxiliary nanowires and b) sheets of Tellurene.

III. Methodology

A. Computational Details

We employ DFT as implemented in the Vienna Ab initio Simulation Package (VASP) [15], using the generalized gradient approximated PBE functional [16], DFT-D3 vdW corrections [17], and a 200 eV kinetic energy cutoff for the plane wave basis. For charge density calculations: all nanowires, nanoribbons, sheets of Tellurene, and the bulk t-Te use a 1x1x4, 1x1x4, 6x1x4, and 6x6x4 Monkhorst-Pack k-point sampling, respectively [18].

We relax the atomic positions of the bulk t-Te until all forces are lower than 0.005 eV/Å. From the bulk t-Te atomic coordinates and lattice parameters we construct and relax three classes of nanowires (Triangular, Rhomboid, and Hexagonal), three classes of nanoribbons (monolayer, bilayer, and trilayer), and monolayer to multilayer sheets of Tellurene. All auxiliary structures also created in the same manner with the bulk t-Te atomic coordinates and lattice parameters.

The formation energy is $E_F = E_{tot}/N_{tot} - \epsilon_{bulk}$. Where E_{tot} is the total ground state energy of a nanostructure, N_{tot} is the total number of atom per supercell of a nanostructure, and ϵ_{bulk} is the cohesive energy of the bulk t-Te.

For bandgap calculations we sampled the first Brillouin zone with no less than 16 points between each symmetry point. Due to the relatively high mass of Tellurium we included spin-orbit coupling using methods developed by reference [19].

B. Surface-to-Volume Ratio

For surface-to-volume-ratio calculations, we use lattice constants (c) plus a "quasi-lattice-constant" ($\tilde{a}$) in the non-periodic directions.

To calculate $\tilde{a}$ we separate all Te atoms into three planes normal to the z-axis. Within each plane, we calculate and then average out all the nearest neighbor distances $d_{p,i}$ for each atom in the plane. Where i is the nearest neighbor atom (in terms of distance) index and p is the plane index. The average nearest neighbor distances across all three planes is the quasi lattice constant. As a closed form equation, the quasi-lattice-constant is:

$$\tilde{a} = \frac{1}{3N_i} \sum_{p=1}^{3} \sum_{i=1}^{N_i} d_{p,i} \qquad (1)$$

N_i equals the total number of nearest neighbor distance per plane. Figure 2 illustrates the methodology for the N=3 rhomboid t-Te nanowire.

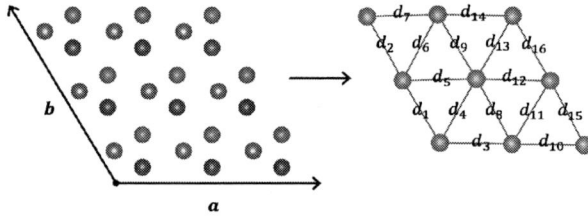

Figure 5. Calculation of $\tilde{a}$ for the N=3 Rhomboid nanowire. All green, blue, and red atoms are in separate planes. There are 16 distances in plane containing red atoms. There are 48 distances to average when calculating $\tilde{a}$ for the N=3 Rhomboid nanowire example when considering all three planes.

The surface-to-volume ratio for all nanowires and nanoribbons is A_L/V, where A_L is the lateral surface area and V is the volume occupied by the structure. The volume for the nanowires and nanoribbons is $V = A_B c$ where A_B is the base area and c is the lattice constant in the z-direction. The nanowire base area is either a hexagon ($A_B = 3\sqrt{3}(N\tilde{a})^2/2$), rhombus ($A_B = \sqrt{3}(N\tilde{a})^2/2$), or an equilateral triangle ($A_B = \sqrt{3}(N\tilde{a})^2/4$). The nanoribbon base areas are parallelograms ($A_B = LRa_{bulk}/2$), where a_{bulk} is the bulk lattice constant which we calculate as 4.40 Å. Monolayer ribbons lack a consistent definition of a quasi-lattice constant, so we do not determine their surface-to-volume ratio.

To calculate the lateral surface areas, we use six (N_R=6), four (N_R=4), and three (N_R=3) rectangles for the hexagonal, rhomboid, and triangular nanowire supercells, respectively.

The total lateral surface area is $A_L = N_R N \tilde{a} c$ where N_R is the number of rectangles and N is the number of Te-helices per nanowire side. The lateral surface for the nanoribbons is computed in the same manner using rectangles. Their total lateral surface area is approximated as $A_L \approx 2N_L a_{bulk} c + 2R a_{bulk} c$.

C. Formation Energy of a Te-Helix

We develop a simpler model of the formation energy by decomposing the formation energy E_F of a 1D Te nanostructure as an energy penalty per Te-helix, based on its number of helical neighbors. If the energy penalty of a Te-helix with m-helical neighbors is given by ϵ_m, the total formation energy is approximated by

$$E_{tot} \approx \sum_{m=1}^{6} \frac{n_m \epsilon_m}{N_{tot}} \qquad (2)$$

where n_m is the number of Te helices with m nearest helical neighbors in the structure.

We obtain the energy parameters ϵ_m by performing an ordinary least squares (OLS) fit on Eq. (2) for $m > 1$. We fit the parameters on the calculated formation energies for all 1D structures in Fig. 1, excluding the auxiliary nanoribbons. The energy penalty ϵ_0 is simply the total energy difference between a single Te helix and a helix in the bulk t-Te.

IV. CONCLUSION

Our calculations have shown that hexagonally shaped t-Te nanowires are the most thermodynamically stable, while triangular shaped nanowires can host high bandgaps while still being thermodynamically stable. Layers of t-Te are not favorable to be grown compared to the nanowires. Nanowires are more stable than layers of t-Te because nanowires allow more vdW interactions per Te-helix. Based on the number of neighbors of a Te-helix, we create a simple model to predict the formation energy of any nanostructure of t-Te. Computing the bandgap, we show that small nanowires can have a bandgap of more than 1 eV for a cross section lower than around 1 nm², making Te a promising material for future nanoelectronic applications.

REFERENCES

[1] F. Gamiz and M. V. Fischetti, *J. Appl. Phys*, **89**, 5478 (2001)
[2] A. H. C. Neto, et al., *Rev. Mod. Phys.*, **81**, 109 (2009)
[3] G. Zhou, et al., *Adv. Mater.* **30**, 1803109 (2018)
[4] V. B. Anzin, et al., *Phys. Status solidi*, **42**, 385 (1977)
[5] J. D. Joannopoulos, et al., *Phys. Rev. B*, **11**, 2186 (1975)
[6] H. Zoog, et al., *Semicond. Sci. Technol.*, **6**, C36 (1991)
[7] H. Peng, N. Kioussis, and G. J. Snyder, *Phys. Rev. B.*, **89**, 195206 (2014)
[8] T. I. Lee, et al., *Adv. Mater.*, **25**, 2920 (2013)
[9] W. Xu, et al., *Small* **4**, 888 (2008)
[10] Z. Zhu, et al., *Phy. Rev. Lett.* **119**, 106101 (2017)
[11] D. Liu, et al., *Nano Letters*, **18**, 4908 (2018)
[12] L. Xian, et al., *2D Mater*, **4**, 041003 (2017)
[13] B. Mayers and Y. Xia, *J. Mater. Chem.*, **12**, 1875 (2002)
[14] J. Heyd, et al., J. Chem. Phys. **118**, 8207 (2003)
[15] G. Kresse and J. Furthumuller, *Phys. Rev. B*, **54**, 11169 (1996)
[16] J. P. Perdew, et al., *Phys. Rev. Lett.* **77**, 3865 (1996)
[17] S. Grimme, et al, *J. Chem. Phys.* **132**, 154104 (2010)
[18] H. J. Monkhorst, and J. D. Pack, *Phys. Rev. B*, **13**, 5188 (1976)
[19] S. Steiner et al., *Phys. Rev. B*, **93**, 224425 (2016)

Defect creation and Diffusion under electric fields from first-principles: the prototypical case of silicon dioxide

N. Salles,[1,3] L. Martin-Samos,[1] S. de Gironcoli,[2] L. Giacomazzi,[1,3] M. Valant,[3] A. Hemeryck,[4] P. Blaise,[5] B. Sklenard,[5] N. Richard[6]

[1]CNR-IOM/Democritos National Simulation Center, Istituto Officina dei Materiali, c/o SISSA, via Bonomea 265, IT-34136 Trieste, Italy. [2]SISSA, via Bonomea 265, IT-34136 Trieste, Italy. [3]Materials Research Laboratory, University of Nova Gorica, Vipavska 11c 5270-Ajdovščina, Slovenija. [4]LAAS-CNRS, Université de Toulouse, CNRS, Toulouse, France [5]Univ. Grenoble Alpes, CEA, LETI, F-38000 Grenoble, France [6]CEA, DAM, DIF, F-91297 Arpajon, France.

email: nsalles33@gmail.com

Abstract—**In this paper we study the effect on the electric fields on the formation of bulk Frenkal Pairs and on the migration of oxygen interstitials, I_O, and oxygen vacancies, V_O, within the framework of Density Functional Theory and Modern Theory of Polarization. At typical OXRRAM field conditions, We show that a significant effect of the electric field is observed only for charged defect. Analyzing the polarization work, we found anomalously high polarization work, for the case of I_O^{-2}, with respect to the classical picture of the electric work of an isolated point charge. This large difference has to be ascribed to collective contributions coming from the environment.**

I. Introduction

Resistive switching phenomena taking place under electric fields on oxide-based memories have attracted much attention in the last decade. Under the modelling view point, a number of investigations have exploited empirical kinetic Monte Carlo (kMC) approaches [1], [2], [3], [4], [5] to model ionic transport. In all these approaches, the **effect of the electric field on the migration energy change is accounted by means of the classical electric work of a point charge, Q in vacuum, first proposed by Cabrera and Mott [6]**, as part of their Model of Oxidation of Metals. In other words, the migration enthalpy variation (the polarization work, W, at the saddle point of the minimal energy path of a given migration mechanism) is approximated by:

$$W = \frac{Qd}{2} \cdot \mathcal{E} \tag{1}$$

The saddle point is assumed to lie half way ($d/2$) between the initial and final configuration and the polarization variation ($Qd/2$) is implicitly assumed to be co-linear with the applied electric field and the jumping coordinate.

In this work we will present a study on the effect of electric fields in the formation of Frenkel Pairs (VOIO) and in the migration of oxygen vacancies (VO) and interstitials (IO) in SiO2 from first-principles. We show that, **at typical working field conditions in OXRRAM, the polarization work is negligible for neutral defects**, i.e. for V_O^0, I_O^0 and the Frenkel pair $V_O^0 I_O^0$. For the same conditions, **the contribution to the switching capacity can not be ascribe to the formation of bulk Frenkel pairs (CFP) composed by a V_O^{+2} and a I_O^{-2}.** An order of magnitude higher field is necessary to trigger a significant contribution from CFP formation. In addition, we find an **anomalously high polarization work** if compared to the classical electric work of a point charge in vacuum. We demonstrate that this difference has to be ascribed to **collective contributions coming from the environment**: the movement of the Migrating Atom (MA) is accompanied by the recoil of neighboring atoms such increasing the dipole moment. The proposed new point of view, provides a theoretically grounded framework to the classical "electric work approximation ", Equation 1, in which its success, and **the success of similar simplified classical single-moving charge pictures can be traced back to subtle system-dependent and symmetry-related compensations** between different effects.

II. Computational details

In our work we use Density Functional Theory (DFT) as in implemented in Quantum Espresso [7], [8]. A Perdew-Zunger[9] exchange-correlation functional together with a Norm-conserving pseudo-potential from the original QE library[1] have been used. The valence wave functions are expanded into plane waves basis up to an energy cutoff of 80 Ry. The Brillouin zone is sampled only at Γ k-point. For each studied defect, the migration/reaction path is decomposed into a series of intermediate configuration frames. The SiO2, Quartz model is a supercell of 216 atoms with group symmetry P3₂21. The model has been fully relaxed with a force and stress convergence threshold of 7.10^{-3} eV/Å. Initial and final defect configurations have been relaxed with the same force convergence threshold. This chain-of-frames is, then, optimized (Climbing Image Nudged Elastic Band [10], [11] (NEB) algorithm) to find the minimal energy path. The frame with the highest energy is the saddle point and this energy corresponds to the migration/reaction energy (see Figure 3 for

---

[1]https://www.quantum-espresso.org

a summary of some of the migration/formation mechanisms considered in this work). The Modern Theory of Polarization (MTP) [12], [13] is used to compute the electric enthalpy, macroscopic polarization changes, Born Effective Charges and the polarization work.

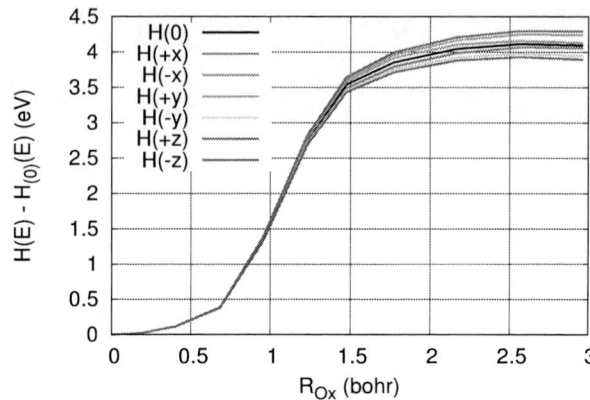

Fig. 1: Formation path of a charged Frenkel pair (CFP), $V_O^{+2}I_O^{-2}$, from a -Si-O-Si-O- quartz structure. Different colors address the enthalpy variation along the path when an electric field of 3.6 MV.cm^{-1} is applied along different directions ($\pm$ orthogonal axis of the box).

III. RESULTS

A. Structure and Formation

SiO$_2$ polymorphs (excluding high-pressure phases) are made of connected SiO$_4$ tetraedra. The building blocks of oxygen Frenkel pairs in silica are the oxygen vacancy (V_O) and the oxygen interstitial (I_O). Neutral Frenkel Pairs, NFP, can be formed either by the formation of the neutral defects pairs (V_O^0 and I_O^0) either by the formation of their respective charged counterparts (V_O^{+2} and I_O^{-2}), CFP. I_O^0s have the structure of a peroxide bridge. V_O^0s are formed by the covalent bonding of two silicon atoms belonging to two different tetraedra (Si-Si bond). Oxygen interstitials with charge -2 have the structure of a twofold Si-O-Si-O ring with pentacoordination for the silicon atoms involved in the ring. For the case of the charge +2 oxygen vacancy, two separated units can be identified, both made from a back-projected positively charged silicon atom weakly bonded to a three-coordinated back oxygen.

Figure 1 and Figure 2 shows the reaction path as a function of the electric field direction, for the formation of first-neighbour Frenkel Pair defects from a chemically ideal -Si-O-Si-. Because of the local symmetry, three non-equivalent FP can be formed. That is, there are three different first-neighbour positions for the vacancy and the interstital, with, therefore, different local environments. The dispersion between formation energies is of the order of few tens of electron volts, see Table I, in both the NFP and the CFP. The formation of a NFP requires about 7 eV, while the formation of a CFP only 4 eV. The difference between this two quantities is the coulomb interaction energy, that is attractive, between the negative and positive charges of the CFP. As can be seen by comparing Figure 1 with Figure 2, the electric field is only affecting the formation of CFP, with enthalpy variations of few tens of eV. Therefore, at typical working conditions in OXRRAMs, the field intensity is not sufficient to significantly trigger the formation of bulk FP.

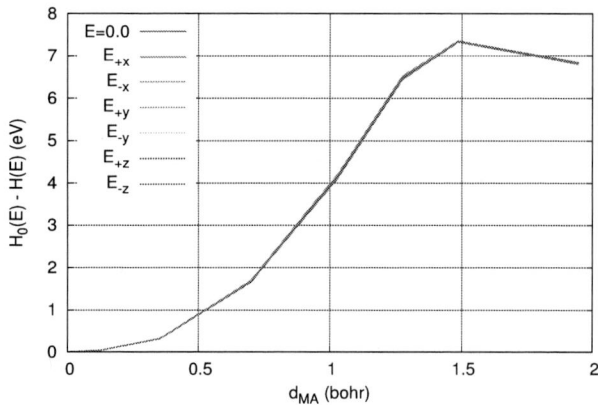

Fig. 2: Formation path of a neutral Frenkel pair (FP), V_OI_O, from a -Si-O-Si-O- quartz structure. Different colors address the enthalpy variation along the path when an electric field of 3.6 MV.cm^{-1} is applied along different directions ($\pm$ orthogonal axis of the box).

	I_O^q		V_O^q		FP	
q (e)	0	-2	0	+2	neutral	Charged
					6.88	4.86
E$_f$ (eV)	1.64	11.6	5.51	-3.31	7.3	4.98
					7.1	4.09

TABLE I: Formation energy of V_O^{+2}, I_O^{-2} and FP for the three non-equivalent first-neighbour positions. The formation energy of isolated defects is computed assuming equilibrium with an O$_2$ reservoir, and an electron reservoir with a Fermi energy the the DFT midgap.

For infinitely far FP defects, the formation energy of the FP is given by the sum of isolated defetcs, column one and two of Table I. The difference between the first-neighbour value and the aforementioned sum of isolate defect is the FP binding energy. The FP binding energy contains both, elastic and electrostatic interactions. For the NFP, the sum of isolated defect formation energies gives 7.15 eV, very close to the first-neighbour NFP. As the monopolar electrostatic term is zero, the small differences are a measure of the weak elastic interaction between V_O^0 and I_O^0. For the CFP the sum of isolated defect formation energy is 8.35 eV, significantly larger than the values of 4.86, 4.98 and 4.09 eV computed for first-neighbour FP. The large binding energies, between -4.26 and -3.37, arise mainly from the large attractive Coulomb between V_O^{+2} and I_O^{-2}.

B. Migration of charged vacancies and interstitials

For the case of the V_O^{+2} migration, the movement of one of the two building units involves the displacement of the back-projected silicon that gets attached to a different neighboring oxygen atom, i.e. the Moving Atom (MA) is a silicon, see Figure 3 a). The twofold ring, that characterize the I_O^{-2}, moves by exchanging one of its silicon atom with one of its neighboring silicon through the switching of one of its oxygen atoms, i.e the Migrating Atom (MA) is an oxygen. At the saddle point the aforementioned oxygen presents a dangling bond, see Figure 3 b).

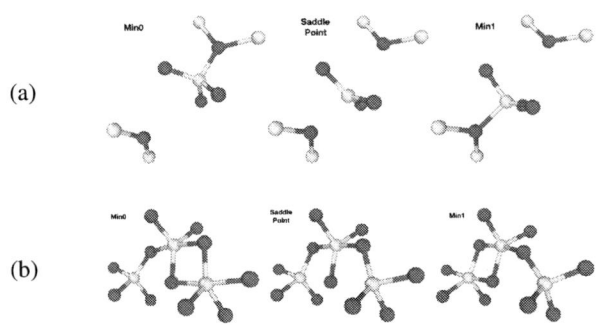

(a)

(b)

Fig. 3: (a) migration of charged +2 Oxygen Vacancy (V_O^{+2}), (b) migration of a charged -2 Oxygen Interstitial (I_O^{-2}).

Along the migration path, the Enthalpy vary proportionally to the polarization work, i.e. between two points, A and B, of a reaction path, the work can be written as:

$$W_{AB} = \int_A^B \Omega d\vec{P} \cdot \vec{\mathcal{E}} = \Omega \left(\vec{P^B} - \vec{P^A} \right) \cdot \vec{\mathcal{E}}. \quad (2)$$

Figures 5 and 4 show the effect of the polarization work on the migration of V_O^{+2} and I_O^{-2}, respectively.

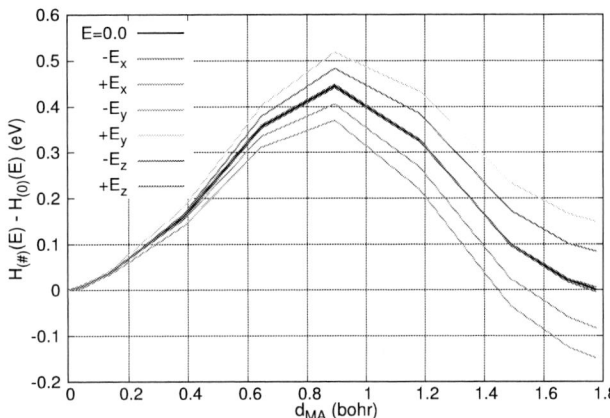

Fig. 4: Energetic path of diffusion of I_O^{-2} under different electric field orientation. Different colors address the enthalpy variation along the path when an electric field of 3.6 MV.cm^{-1} is applied along different directions ($\pm$ orthogonal axis of the box).

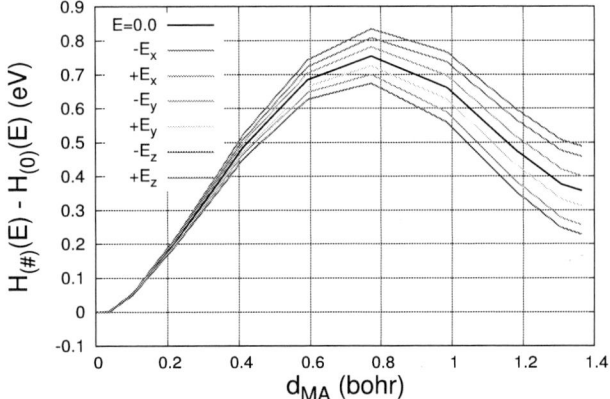

Fig. 5: Energetic path of diffusion of V_O^{+2} under different electric field orientation. Different colors address the enthalpy variation along the path when an electric field of 3.6 MV.cm^{-1} is applied along different directions ($\pm$ orthogonal axis of the box).

From initial to Saddle point we find a polarization work for the V_O^{+2} consistent with an equivalent ionic charge of about +3.4 eV, for a jump distance of half 1.36 bohr (see Table II). Such a charge value agrees very well with the Born Effective Charge (BEC) of silicon in SiO$_2$ and with the BEC of a back-projected positively charged silicon atom weakly bonded to a three-coordinated back oxygen. For the I_O^{-2}, the Polarization Work Equivalent Charge (PWEC) exceed by more than twice, -4.9, the BEC of oxygen in SiO$_2$ or the BEC of oxygens in a charged -2 twofold ring (-2.2), see Table II. With some paperwork, it can be demonstrated that differences between the ionic charge and the first-principles PWEC have to be ascribed **to collective contributions coming from the environment, i.e. the MA is accompanied by the recoil of environment atoms. Equation 1 can be generalized to explicitly include such collective terms.**

Defect	d_{MA}	$|\Delta\vec{P}_{\parallel}|$	Q ionic	Q_{eff}
V_O^{+2}	1.46	5.19	3.4	3.4
I_O^{-2}	0.89	4.43	-2.2	-4.9
CFP	1.98	17.99	-	8.45

TABLE II: Distance (in Bohr) traveled by the MA (an oxygen atom for I_O^{-2} and a silicon atom in V_O^{+2}) up to the saddle point, the macroscopic polarization ($\Delta\vec{P}_{\parallel}$, in units of $e\cdot$Bohr) at the saddle point, and Born effective charges and effective polarization work charge (in units of e) for V_O^{+2}, I_O^{-2}, and for the CFP formation.

Table III summarizes he key quantities that impact on the defect kinetics. The rate of each event (defect diffusion or defect creation) is computed from Arrhenius law. The magnitude of the effect of polarization work on the rate is relative to the rate without the field. We obtain a factor 4 for the V_O^{+2} and 3 for the I_O^{-2} while a factor hundred for the CFP.

Defect	E_{ac} (eV)	W (eV)	γ_0	$\gamma(\pm\mathcal{E})$
V_O^{+2}	0.75	-0.026 $+0.026$	0.25	0.28 0.23
I_O^{-2}	0.44	-0.03 $+0.03$	$4\cdot10^4$	$1.3\cdot10^5$ $1.3\cdot10^4$
CFP	4.1	-0.08 $+0.08$	10^{-57}	10^{-56} 10^{-59}

TABLE III: Activation energy, polarization work (W) at saddle point for an electric field of 1.8MV.cm^{-1}, reaction rate (in s^{-1}) with $\gamma(\pm\mathcal{E})$ and without, γ_0, the electric field for V_O^{+2}, I_O^{-2}), and CFP formation. Reaction rates are calculated at room temperature and with an attack frequency of $\nu_0 = 10^{12}s^{-1}$.

IV. CONCLUSION

To summarize, we have performed fisrt-principle calculations, including Modern Theory of Polarization, for the formation of oxygen Frenkel Pairs and for the migration of oxygen vacancies and interstitials. Due to electrostatic interactions the CFP has a formation energy of about 3 eV below the NFP. **At typical working field conditions in OXRRAM, the polarization work is negligible for neutral defects**, i.e. for V_O^0, I_O^0 and the Frenkel pair $V_O^0 I_O^0$. For the same conditions, **the contribution to the switching capacity can not be ascribe to the formation of bulk Frenkel pairs (CFP) composed by a V_O^{+2} and a I_O^{-2}**. In addition, we find an **anomalously high polarization work** resulting in a very high polarization work equivalent charge for I_O^{-2}, more than twice the ionic charge of oxygen or its BEC. This difference has to be ascribed to **collective contributions coming from the environment**: the movement of the Migrating Atom (MA) is accompanied by the recoil of neighboring atoms.

ACKNOWLEDGEMENTS

High Performance Computing ressources have been provided by the company Arctur (https://hpc.arctur.si) and by the super computing center CINECA (www.cineca.it, grant number ISCRA C HP10CQGOFM). Authors are grateful to stefano Baroni for useful discussions.

REFERENCES

[1] A. Apolinário, P. Quitério, C. T. Sousa, J. Ventura, J. B. Sousa, L. Andrade, A. M. Mendes, and J. P. Araújo, "Modeling the growth kinetics of anodic TiO$_2$ nanotubes," *The Journal of Physical Chemistry Letters*, vol. 6, no. 5, pp. 845–851, 2015. PMID: 26262661.

[2] A. J. Choksi, R. Lal, and A. N. Chandorkar, "Growth kinetics of silicon dioxide on silicon in an inductively coupled rf plasma at constant anodization currents," *Journal of Applied Physics*, vol. 72, no. 4, pp. 1550–1557, 1992.

[3] K. Sasikumar, B. Narayanan, M. Cherukara, A. Kinaci, F. G. Sen, S. K. Gray, M. K. Y. Chan, and S. K. R. S. Sankaranarayanan, "Evolutionary optimization of a charge transfer ionic potential model for Ta/Ta-Oxide heterointerfaces," *Chemistry of Materials*, vol. 29, no. 8, pp. 3603–3614, 2017.

[4] K. S. Martirosyan and M. Zyskin, "Reactive self-heating model of aluminum spherical nanoparticles," *Applied Physics Letters*, vol. 102, no. 5, p. 053112, 2013.

[5] F. Raffone and G. Cicero, "Unveiling the fundamental role of temperature in RRAM Switching Mechanism by multiscale simulations," *ACS Applied Materials & Interfaces*, vol. 10, no. 8, pp. 7512–7519, 2018. PMID: 29388424.

[6] N. Cabrera and N. F. Mott, "Theory of the oxidation of metals," *Reports on Progress in Physics*, vol. 12, no. 1, p. 163, 1949.

[7] P. Giannozzi, S. Baroni, N. Bonini, M. Calandra, R. Car, C. Cavazzoni, D. Ceresoli, G. L. Chiarotti, M. Cococcioni, I. Dabo, A. D. Corso, S. de Gironcoli, S. Fabris, G. Fratesi, R. Gebauer, U. Gerstmann, C. Gougoussis, A. Kokalj, M. Lazzeri, L. Martin-Samos, N. Marzari, F. Mauri, R. Mazzarello, S. Paolini, A. Pasquarello, L. Paulatto, C. Sbraccia, S. Scandolo, G. Sclauzero, A. P. Seitsonen, A. Smogunov, P. Umari, and R. M. Wentzcovitch, "QUANTUM ESPRESSO: a modular and open-source software project for quantum simulations of materials," *Journal of Physics: Condensed Matter*, vol. 21, no. 39, p. 395502, 2009.

[8] P. Giannozzi, O. Andreussi, T. Brumme, O. Bunau, M. B. Nardelli, M. Calandra, R. Car, C. Cavazzoni, D. Ceresoli, M. Cococcioni, N. Colonna, I. Carnimeo, A. D. Corso, S. de Gironcoli, P. Delugas, R. A. D. Jr, A. Ferretti, A. Floris, G. Fratesi, G. Fugallo, R. Gebauer, U. Gerstmann, F. Giustino, T. Gorni, J. Jia, M. Kawamura, H.-Y. Ko, A. Kokalj, E. Küçükbenli, M. Lazzeri, M. Marsili, N. Marzari, F. Mauri, N. L. Nguyen, H.-V. Nguyen, A. O. de-la Roza, L. Paulatto, S. Poncé, D. Rocca, R. Sabatini, B. Santra, M. Schlipf, A. P. Seitsonen, A. Smogunov, I. Timrov, T. Thonhauser, P. Umari, N. Vast, X. Wu, and S. Baroni, "Advanced capabilities for materials modelling with Quantum ESPRESSO," *Journal of Physics: Condensed Matter*, vol. 29, no. 46, p. 465901, 2017.

[9] J. P. Perdew and A. Zunger, "Self-interaction correction to density-functional approximations for many-electron systems," *Phys. Rev. B*, vol. 23, pp. 5048–5079, May 1981.

[10] G. Henkelman and H. Jónsson, "Improved tangent estimate in the nudged elastic band method for finding minimum energy paths and saddle points," *The Journal of Chemical Physics*, vol. 113, no. 22, pp. 9978–9985, 2000.

[11] G. Henkelman, B. P. Uberuaga, and H. Jónsson, "A climbing image nudged elastic band method for finding saddle points and minimum energy paths," *The Journal of Chemical Physics*, vol. 113, no. 22, pp. 9901–9904, 2000.

[12] R. Resta, "Macroscopic polarization in crystalline dielectrics: the geometric phase approach," *Rev. Mod. Phys.*, vol. 66, pp. 899–915, Jul 1994.

[13] P. Umari and A. Pasquarello, "Ab initio molecular dynamics in a finite homogeneous electric field," *Phys. Rev. Lett.*, vol. 89, p. 157602, Sep 2002.

Effective work-function tuning of TiN/HfO$_2$/SiO$_2$ gate-stack; a density functional tight binding study

Hesameddin Ilatikhameneh[1*], Hong-Hyun Park[1], Zhengping Jiang[1], Woosung Choi[1],
Mohammad Ali Pourghaderi[2], Jongchol Kim[2], Uihui Kwon[2], and Dae Sin Kim[2]

1 Device lab, Samsung Semiconductor Inc., San Jose, California, USA
2 Semiconductor Research and Development Center, Samsung Electronics, Hwasung-si, Korea
* hesam.ilati@samsung.com

Abstract—In this work, density functional tight binding (DFTB) calculations are used to study the characteristics of full gate stack TiN/HfO$_2$/SiO$_2$/Si and possible effective work-function (EWF) tuning options. First, the DFTB parameterization method to produce both electronic and repulsion information for all atom pairs is introduced briefly. Since the simulated gate-stack structure has thousands of atoms, conventional relaxation methods are computationally intensive. Hence a method to relax and passivate the material interfaces is introduced. Next, the impact of aluminum substitution is studied. It is shown that the change in EWF strongly depends on the atom which is substituting Aluminum; e.g. Aluminum substitutions of Hf and Ti show opposite impact on EWF. Finally, the origin of this different behavior is discussed.

Index Terms—Gate stack, density functional tight binding, DFTB, work-function

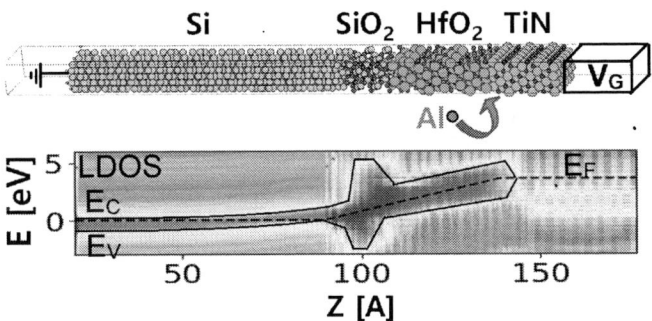

Fig. 1. Atomistic structure of gate stack including about 2000 atoms and corresponding LDOS under applied gate bias.

I. INTRODUCTION

Shrinking transistor dimensions has pushed electronic industry toward transistor structures with better gate control like FinFETs and gate all around (GAA) FETs [1]. The 3D feature of these devices requires very tight control on the height of gate stack. On the other hand, designers need multi-Vth options for target performance. Additionally, reliability, leakage, and process compatibility requirements drastically limit the material choice of gate stack [2, 3, 4]. Taking all these considerations into account, the metal-barrier system of Al and TiN has been matured and optimized for successive generation of high-k/metal gate [2, 3, 5, 6]. To understand EWF engineering of the gate-stacks in atomistic level and under applied bias (e.g. non-equilibrium), density functional tight binding (DFTB) method is one of the best candidates; DFTB has proper transferability for material interfaces and computational efficiency for large structure simulation and allows fine-tuning of material properties like band gap and dielectric constant towards experimental values. Moreover, DFTB is based on LCAO (linear combination of atomic orbitals) and hence suitable for non-equilibrium simulations.

Here, we investigate different configurations for Aluminum dopants and their impact using DFTB. The simulation structure is shown in Fig. 1 including 2000 atoms. Such simulation requires DFTB parameter generation and structure relaxation, explained in first and second sections. The impact of Aluminum dopants on EWF is discussed in the last section.

II. DFTB PARAMETERIZATION

The computational efficiency of DFTB arises from the fact that all coupling and overlap integrals are pre-calculated and stored in files. Hence, at runtime, program only needs to read the corresponding DFTB file to find the coupling and overlap matrix elements. To produce the Slater-Koster tables, the Kohn-Sham equations should be solved for each atom type. Since, wave-function of single atom can be spatially extended, a confining potential is used usually to truncate the wave-functions. The shape of this confining potential can be used to fine tune the electronic properties toward reference system. We use an in-house DFTB parameterzation tool which takes a set of materials as target environment and tunes the confining potential for each atom such that the resulting DFTB band-structure fits the target bandstructure from experimental results or DFT calculations. To use DFTB for geometry relaxation, one needs to calculate repulsion potential. The repulsion

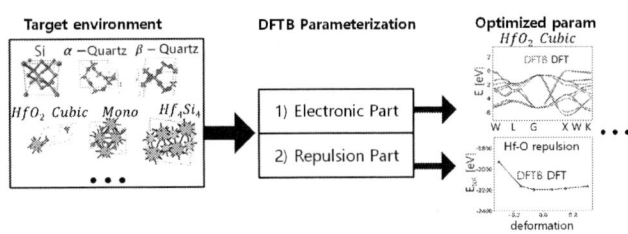

Fig. 2. Parameterization procedure for a target environment.

978-1-7281-0941-1/19 $31.00 © 2019 IEEE

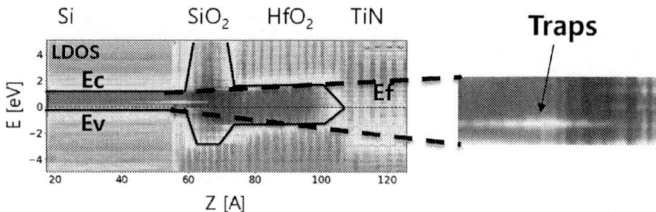

Fig. 3. Appearance of interface trap states without the passivation-relaxation procedure in Fig. 4.

potential is optimized such that the total energy curves for different deformations of materials in environment match the reference values.

The target environment is composed of bulk materials which include atom pairs in different configurations: Al, AlN, Al$_2$O$_3$, Hf (hcp), HfO$_2$ (monoclinic and cubic) Si, SiO$_2$ (alpha and beta), Ti, TiN, HfN, HfSi, Si$_3$N$_4$.

A. Poisson-DFTB Method

Usually DFTB II is used for self-consistent DFTB calculations [7]. However, the boundary conditions for gate contact cannot be applied in DFTB II. To overcome this problem, DFTB I method is solved self-consistently with Poisson equation. It is easy to prove that in the case of Neumann boundary condition, these two methods are equivalent.

$$(H_0 + V^M S)\psi = ES\psi \tag{1}$$

$$\nabla^2 V = \frac{\rho}{\epsilon_0} \tag{2}$$

Notice that H_0 and S are constructed using the Slater-Koster tables obtained from DFTB parameterization. V^M is a matrix obtained from potential V with matrix elements $V_{i,j}^M = 0.5(V_i + V_j)$ where V_i is potential at atom i. Moreover, since all valence electrons are considered in this method, the polarization is induced by electron transfer and hence ϵ_0 is used for everywhere.

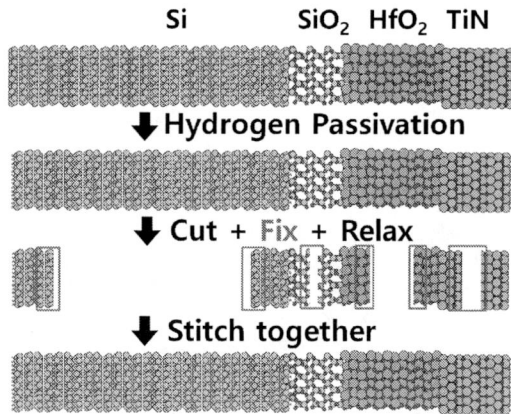

Fig. 4. Steps of efficient passivation-relaxation procedure.

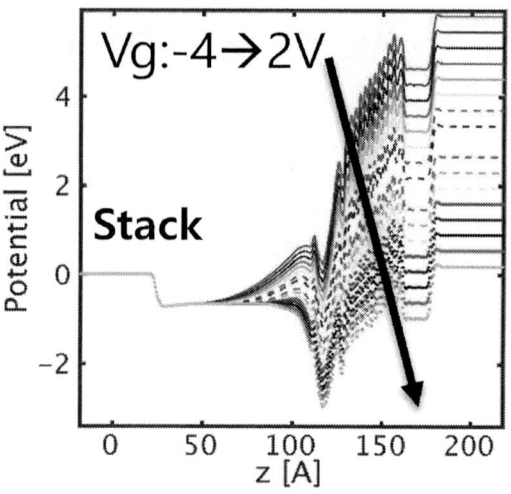

Fig. 5. Potential profiles for gate voltage sweep. The difference between potential of gate stack with Aluminium dopants and without them.

B. Relaxation Method

The simulated gate structure contains about 2000 atoms. Relaxation of complete structure is computationally intensive. Here, a method is developed to relax structures efficiently without losing too much accuracy. The method is based on the fact that the chemical bonds and forces are local in nature. Hence, far from the interfaces, atoms do not distort significantly. Without correct passivation and relaxation, the interface states appears and may affect the results. Such issue is shown in Fig. 3 which results in screening electric field. The interface states study is important, but out of scope of this work. The procedure is shown in Fig. 4 and has following steps: 1) Initial structure is generated such that the strain is in acceptable range, 2) dangling bonds at interface are passivated with Hydrogen atoms, 3) Atoms far from interfaces are removed, boundary atoms are fixed and the rest are relaxed using DFTB or DFT, 4) stitch back all together.

C. Results

Fig. 5 shows the potential profile for different gate voltages of the gate stack without Al atoms. The depletion and inversion regimes are clear from the depth of electric field penetration into the silicon channel. To investigate the impact of Al dopants, we replace Hf or Ti atoms at the HfO$_2$/TiN interface with Aluminum atoms and repeat the procedure to relax the structure. The relaxed structure is simulated under same gate bias conditions.

The potential and charge outputs of Al substitution of Hf atoms are shown in Fig. 6. Notice that the potential difference between gate stack with and without Al atoms are shown, hence there is no atomic level variation of potential. The induced dipole due to Al substitution of Hf atoms moves the band diagram in silicon towards p-type. Moreover, increasing the concentration of Al dopants results into linear increase of induced dipole.

978-1-7281-0941-1/19 $31.00 © 2019 IEEE   216

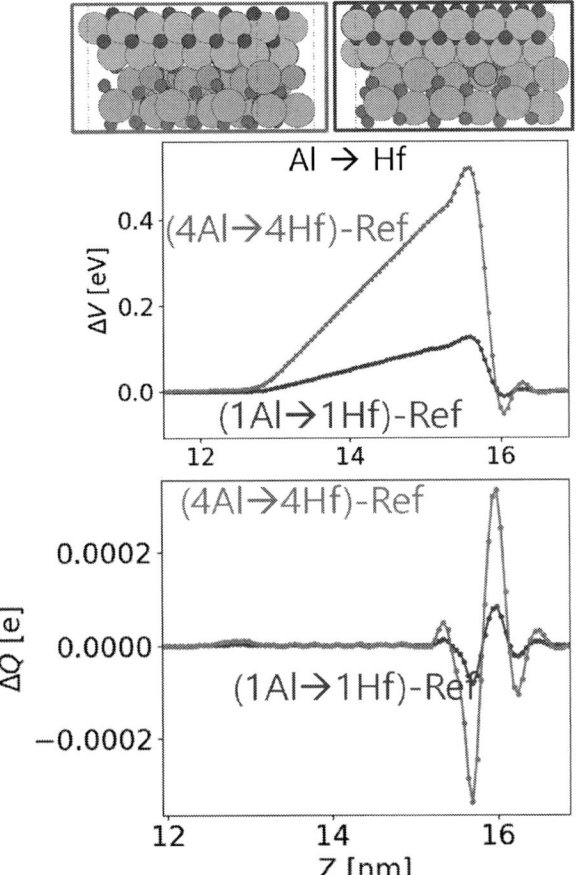

Fig. 6. The potential and charge outputs of Al substitution of Hf atoms. Potential of gate stack with 4 and 1 Al replacement is compared to the one without any Al atom.

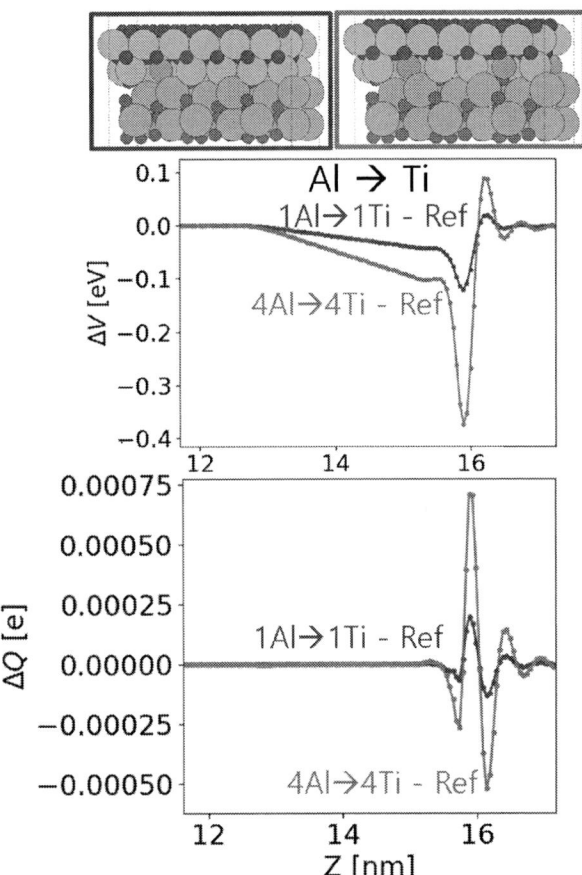

Fig. 7. The potential and charge outputs of Al substitution of Ti atoms. Potential of gate stack with 4 and 1 Al replacement is compared to the one without any Al atom.

Fig. 7 shows the results for Al substitution of Ti atoms at the TiN/HfO$_2$ interface. Although Ti and Hf are from the same group in periodic table, the Al substitution with these two atoms results to a very different behavior. Al substitution of Ti atoms moves the band diagram in Silicon toward n-type. The origin of this behavior is explained in the next section. Another difference between Hf and Ti atom is that in the latter case, the increase of Al concentration results into a sublinear increase of dipole. Additionally, the magnitude of dipole for Ti substitution is smaller.

Fig. 8 depicts the polarization field as a function of number of dopant atoms within the super-cell. The linear dependence of polarization field on Al substitution of Hf is clear. The main question is that despite the fact that Ti and Hf are from the same group in periodic table, the Al substitution with these two atoms results into a very different behavior. To answer these questions qualitatively, we doped super-cells of HfN and TiN with Al and investigate their work-function change.

The main question which Fig. 8 arises is that what is the cause for difference in Al substitution of Ti and Hf; is this difference related to atom properties (i.e. electronegativity) or material itself. To answer these questions, we have Al doped super cell super cell of TiN and HfO$_2$. Fig. 9 shows the energy resolved density of states for TiN and HfO$_2$. Adding Al dopants to these materials results into 2 effects: 1) Reducing the number of valence electrons since Al has one valence electron less than Ti and Hf. 2) Reducing and shifting the density of states. These two effects are competing against each

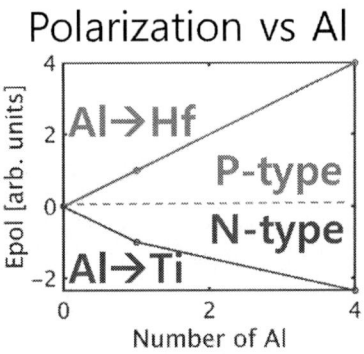

Fig. 8. The polarization electric of field as a function of number of Al substitutions of Ti and Hf atoms at TiN/HfO$_2$ interface.

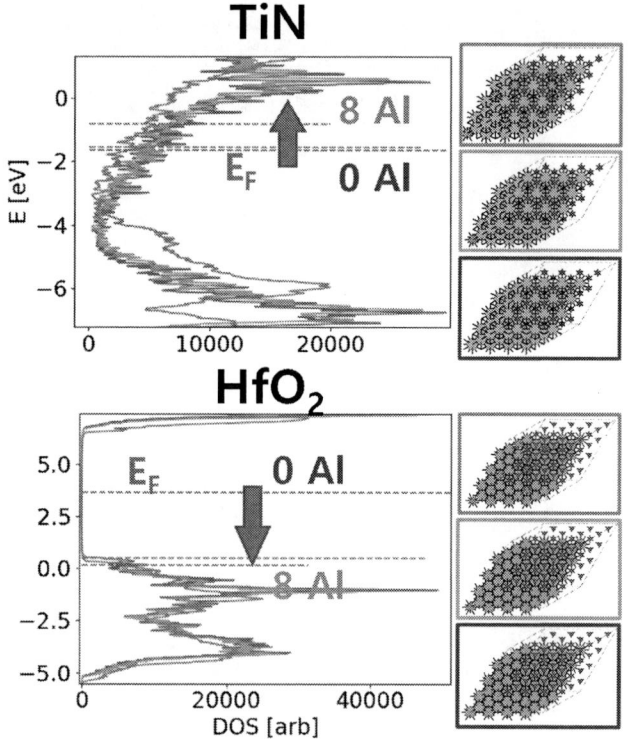

Fig. 9. Energy resolved density of states of TiN and HfO$_2$ for different amount of Al doping.

the impact of aluminum substitution of Ti and Hf atoms at TiN/HfO2 interface and the origin of its difference is studied.

REFERENCES

[1] M. Salmani-Jelodar et al., "Design Guidelines for Sub-12 nm Nanowire MOSFETs," in IEEE Transactions on Nanotechnology, vol. 14, no. 2, pp. 210-213, March 2015. doi: 10.1109/TNANO.2015.2395441

[2] A. Veloso et al., "Process control and integration options of RMG technology for aggressively scaled device," in Proc. Symp. VLSI Tech. Dig., pp. 33-34. 2012.

[3] A. Veloso et al., "Effective Work Function Engineering for Aggressively Scaled Planar and Multi-Gate Fin Field-Effect Transistor-Based Devices with High-k Last Replacement Metal Gate Technology," Jpn. J. Appl. Phys., Park 1 52, 04CA02, 2013.

[4] M. Salmani-Jelodar et al., "Optimum high-k oxide for the best performance of ultra-scaled double-gate MOSFETs" IEEE Transactions on Nanotechnology, vol. 15, no. 6, pp. 904-910, Nov. 2016. doi: 10.1109/TNANO.2016.2583411

[5] Chen, Y. W., et al. "Effective Work Function Modulation by Aluminum Ion Implantation on Hf-Based High-k/Metal Gate pMOSFET." IEEE EDL 31.11 (2010): 1290-1292.

[6] Singanamalla, R., et al. "Effective Work-Function Modulation by Aluminum-Ion Implantation for Metal-Gate Technology (Poly-Si/TiN/SiO$_2$)." IEEE EDL 28.12 (2007)

[7] Elstner, Marcus, et al. "Self-consistent-charge density-functional tight-binding method for simulations of complex materials properties." PRB 58.11 (1998): 7260.

other. If DOS does not change significantly, then the reduced number of electrons push material toward p-type. However, if the shift in DOS is significant, then material becomes more n-type.

To make sure if the origin is material related or atom related, we have also doped HfN and TiO$_2$. Interestingly, Al doping of both HfN and TiN shifts work-function toward n-type, whereas Al doping of HfO$_2$ and TiO$_2$ makes work-function p-type. It is apparent that the origin of shift in work-function is more related to material. For metals, such as TiN and HfN, the shift in DOS is much more significant than the reduced number of electrons. Hence, work-function of the metal and Aluminium matters most. However, in the oxides, the change in DOS is negligible and the change in number of electrons determines the shift in work-function.

D. Conclusion

In summary, a computationally efficient approach to study gate stack engineering at atomistic level is proposed. The density functional tight binding (DFTB) parameters for the full gate stack TiN/HfO$_2$/SiO$_2$/Si are obtained from an in-house tool based on band structure and total energies of reference systems (Al, AlN, Al$_2$O$_3$, Hf (hcp), HfO$_2$ (monoclinic and cubic) Si, SiO$_2$ (alpha and beta), Ti, TiN, HfN, HfSi, Si$_3$N$_4$). Moreover, an efficient method to relax and passivate the material interfaces is introduced. Using this infrastructure,

978-1-7281-0941-1/19 $31.00 © 2019 IEEE

Single Event Transient Compact Model for FDSOI MOSFETs Taking Bipolar Amplification and Circuit Level Arbitrary Generation Into Account

Neil Rostand
DPHY
ONERA
Toulouse, France
neil.rostand@onera.fr

Sébastien Martinie
DCOS
CEA-LETI
Grenoble, France
sebastien.martinie@cea.fr

Joris Lacord
DCOS
CEA-LETI
Grenoble, France
joris.lacord@cea.fr

Olivier Rozeau
DCOS
CEA-LETI
Grenoble, France
olivier.rozeau@cea.fr

Thierry Poiroux
DCOS
CEA-LETI ·
Grenoble, France
thierry.poiroux@cea.fr

Guillaume Hubert
DPHY
ONERA
Toulouse, France
guillaume.hubert@onera.fr

Abstract— Single Event Transients (SET) are ionizing particles induced current pulses which are able to generate soft errors in CMOS circuits. In Silicon-on-Insulator (SOI) technologies, bipolar amplification phenomena is more significant due to presence of the Burried Oxide (BOX), which is detrimental to soft errors sensitivity. State of the art FDSOI SET models account for bipolar amplification through a dynamic pre-factor. This approach is mainly empirical and not compact. In this work, we propose a SET compact model for FDSOI MOSFETs including a physical modeling of bipolar amplification. Results are validated through TCAD simulations. A circuit level approach is proposed considering arbitrary generation within functional SRAM cell. This approach allows more realistic Single Event Upset (SEU) prediction and we show how circuit level generation can influence SEU prediction.

Keywords— SET, Bipolar Amplification, SOI MOSFET, Compact Model, SPICE, TCAD.

I. INTRODUCTION

Soft errors in circuits are generally due to parasitic current induced by ionizing particles within MOSFETs, called Single Event Transients (SETs) [1]. In recent technologies based on SOI technology, bipolar amplification phenomena is more significant due the presence of the BOX and increase circuit sensitivity to single events [2-7]. It consists in parasitic source – drain current induced by storage of generated holes (in NMOS) or electrons (in PMOS) in the Silicon film. Another issue is the morphology of the particle induced charge deposit at the circuit level which matters for high level of integration as its spatial extension can overlap with many transistors volumes. Both circuit level charge deposit and bipolar amplification have to be taken into account in SET models in order to perform reliable soft errors risk assessment.

In the literature, the approach to model bipolar amplification in SOI technologies relies on consideration of equivalent access resistance to determine triggering of bipolar amplification. Classical SET current model (i.e without bipolar amplification; called 1st discharge in [2]) is then multiplied by an empirical pre-factor [8]. This

approach is not suitable for compact modeling (or SPICE modeling) point of view [9].

In this paper, we propose a compact model of SET taking both bipolar amplification and circuit level arbitrary charge deposit into account suitable for Fully Depleted SOI (FDSOI) structures (exposed for NMOSFETs). In Section. II, we evidence bipolar amplification through TCAD simulations of FDSOI MOSFET. In Section. III, we develop our bipolar amplification model and the Verilog-A implementation method. In Section. IV, we show the resulting SET compact model considering, 1st discharge, bipolar amplification, and circuit level arbitrary charge deposit. The relevance of accounting for circuit level charge deposit is highlighted through SPICE simulations of Single Event Upsets (SEUs).

II. EVIDENCE OF BIPOLAR AMPLIFICATION

In this section, bipolar amplification is evidenced performing transient TCAD simulations [10] of a 2D FDSOI NMOSFET ($L_{ch} = 0.1\mu m$, $T_{si} = 10nm$, $V_{ds} = 1$, $V_{gs} = 0V$) with heavy ion strike considering simulation setup explained in [1]. After particle strikes the transistor, generated holes remain in the body (due to SOI structure) as illustrated in Fig. 1.a. This involves barrier lowering (see Fig.1.b) which allows electrons to flow from source to drain if $V_{ds}>0$, this extra current is the bipolar amplification. As illustrated in Fig. 1.c, after prompt 1st discharges, we see these pulses exhibit relaxation tails which decay very slowly involving an higher collected charge at the drain $Q(t) = \int I_d dt$ than deposited charge Q_{dep} which is the main feature of bipolar amplification. Fig 1.d illustrates the hole and electron quantity in the body (respectively denoted P and N) for different *LET* versus time, and highlights that holes are stored inside the body after a short time interval corresponding to quick decay of P (for high Linear Energy Transfer *LET*). We can actually show that this decay is due to quick recombination in source area. We also see electrons are injected in the body (still N remains lower than P) in order to contribute to bipolar amplification current.

978-1-7281-0941-1/19 $31.00 © 2019 IEEE

III. MODELIING OF BIPOLAR AMPLIFICATION

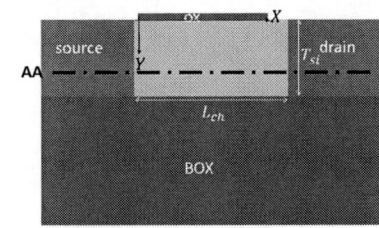

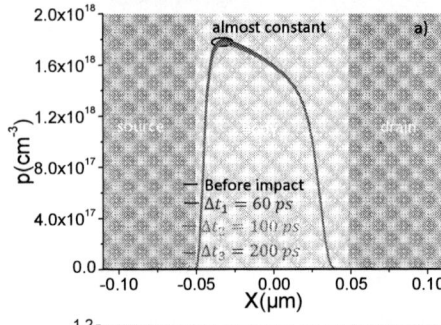

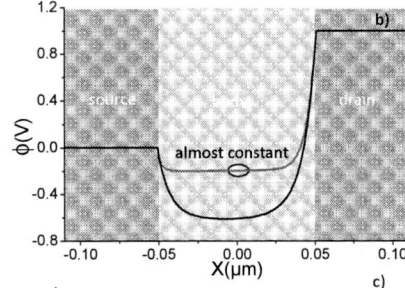

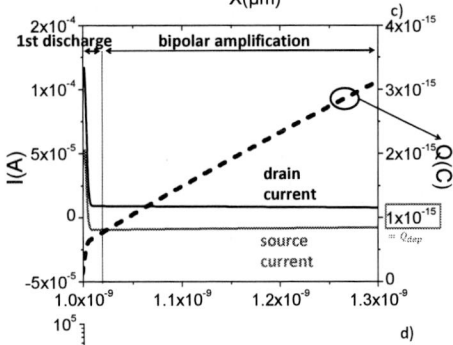

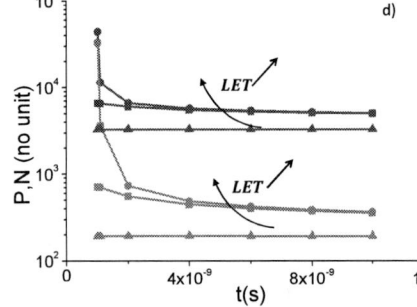

Fig. 1: TCAD evidence of bipolar amplification for *LET=0.1 pC/µm* . a,b): plots of hole density p and of the electrical potential φ in AA cutline for different times after particle impact Δt. c): SET pulses at drain (black full line) and source (blue full line) electrodes and collected charge at the drain electrode (black dash line). d): quantity of holes (blue) and of electrons (red) relative to time after particle impact in the body. Circle, square, and triangle symbols correspond respectively to *LET={1,0.1,0.05} pC/µm*.

A. Explicit bipolar amplification current expression

In this part, we consider the particle generated $P_{dep} = LET.T_{si}/q$ electron/hole pairs in the body during impact at time t_i. We assume electron current density $\boldsymbol{J_n}$ is conservative within the body, as evidenced by TCAD ($\boldsymbol{\nabla.J_n} \approx 0$). In a 1D problem along X, it means $\boldsymbol{J_n}$ is uniform. Integrating $\boldsymbol{J_n}$ in drift-diffusion formalism along $\boldsymbol{X}$, we obtained:

$$j_n = -q.\frac{\mu_n}{L_{ch}}.\int_0^{L_{ch}} n.\frac{\partial \phi_n}{\partial x}.dx \qquad (1)$$

In (1), ϕ_n is the electron quasi-Fermi potential, μ_n the electron mobility in the body and n the electron density. To capture V_{ds} dependence, we consider simplified case of linear ϕ_n often called "long channel approximation". We can then derive (2), corresponding to source to drain bipolar amplification current:

$$I_{ba}(t) = \frac{-q.\mu_n}{L_{ch}^2}.N(t).V_{ds} \qquad (2)$$

At this point we introduce the electroneutrality factor defined by $X(t) = N(t)/P(t)$. As the relation between time and P is a bijection, we can redefine X so that it is $P(t)$ dependant. Then:

$$I_{ba}(P(t)) = \frac{-q.\mu_n}{L_{ch}^2}.X(P(t)).P(t).V_{ds} \qquad (3)$$

An empirical function is chosen for the electroneutrality factor $X(P(t))$ (which is actually *LET* dependant). We now need to determine $P(t)$.

B. Non-linear differential equation for $P(t)$

Fig.2 shows the hole quasi-Fermi potential ϕ_p in X direction for different times after particle impact. We clearly see that ϕ_p is almost uniform in the body. Writing $d\phi_p/dx = 0$ in the body, we obtain the 1st order partial differential equation:

$$\frac{\partial \Delta p}{\partial X} = -\frac{\partial \phi}{\partial X}.\frac{1}{V_t}.\Delta p \qquad (4)$$

In (4), Δp is the excess hole density (generated by the particle), ϕ the electrical potential, and V_t the thermic voltage. Considering Δp property $\int_0^{L_{ch}} \Delta p.dX.W.T_{si} = P(t)$ (W being the body width), the solution of this equation can be written as follow:

$$\Delta p(X,t) = \frac{P(t)}{W.T_{si}.I_\phi(t)}.e^{-\frac{\phi(X,t)}{V_t}} \qquad (5)$$

Where $I_\phi(t) = \int_0^{L_{ch}} e^{-\frac{\phi(X,t)}{V_t}}.dX$. The next step is to express the hole conservation law in the body, considering zero hole current at the source – body and body – drain junctions. However, surface recombination currents at these jonctions have to be taken into account because of high doping in source/drain areas. Note that no volume recombination occurs in the body for this time scale due to very low doping level $N_a = 10^{15}cm^{-3}$. Denoting V_{rec} for the recombination speed at the PN junctions, the resulting conservation law can be written as follow:

$$\frac{dP}{dt} = -\frac{1}{T(P)}.P \qquad (6.a)$$

978-1-7281-0941-1/19 $31.00 © 2019 IEEE

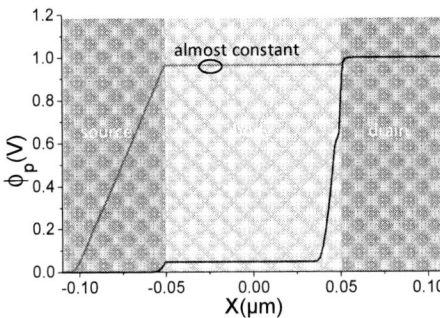

Fig. 2: plot of hole quasi Fermi potential ϕ_p in AA cutline for different times after particle impact Δt.

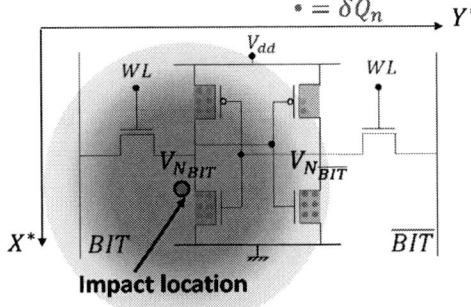

Fig. 3: 6T-SRAM cell submitted to particle strike. Illustration of discretization of circuit level charge deposit morphology induced by the particle in vertical incidence (orthogonal to the SRAM cell plane).

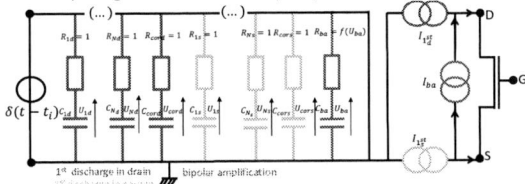

Fig. 4: equivalent electrical circuit of SET model in case of FDSOI. Modified RC circuit related to bipolar amplification completes initial 1st discharge RC circuits, see [1,9].

$$T(P) = \frac{I_\phi(P)}{V_{rec}\left(e^{-\frac{V_s}{V_t}} + e^{-\frac{V_d}{V_t}}\right)} \quad (6.b)$$

As the time constant (6.b) is not analytical, we prefer extract a simpler expression from TCAD, plotting the quantity $-P^{-1}.dP/dt = \tau(P)^{-1}$. We obtain the following function:

$$T_{ex}(P) = \tau_r.\left(1 + \left(\frac{P_c}{P}\right)^\gamma\right), \forall P > 0, \gamma > 0, P_c > 0 \quad (7)$$

In (7), τ_r is the recombination time in source/drain while γ and P_c are parameters depending on geometry parameters (in particular L_{ch}, T_{si} and W) but this dependence has not been modeled. The set of implicit equations represented by (6.a) and (7) can be solved by the SPICE simulator, assigning P/P_{dep} to the voltage drop U_{ba} of a capacitance being part of a modified RC circuit, where R value depends on U_{ba}, this circuit being submitted to 1V voltage pulse at impact time of the particle. This equivalent circuit is implemented in Verilog-A.

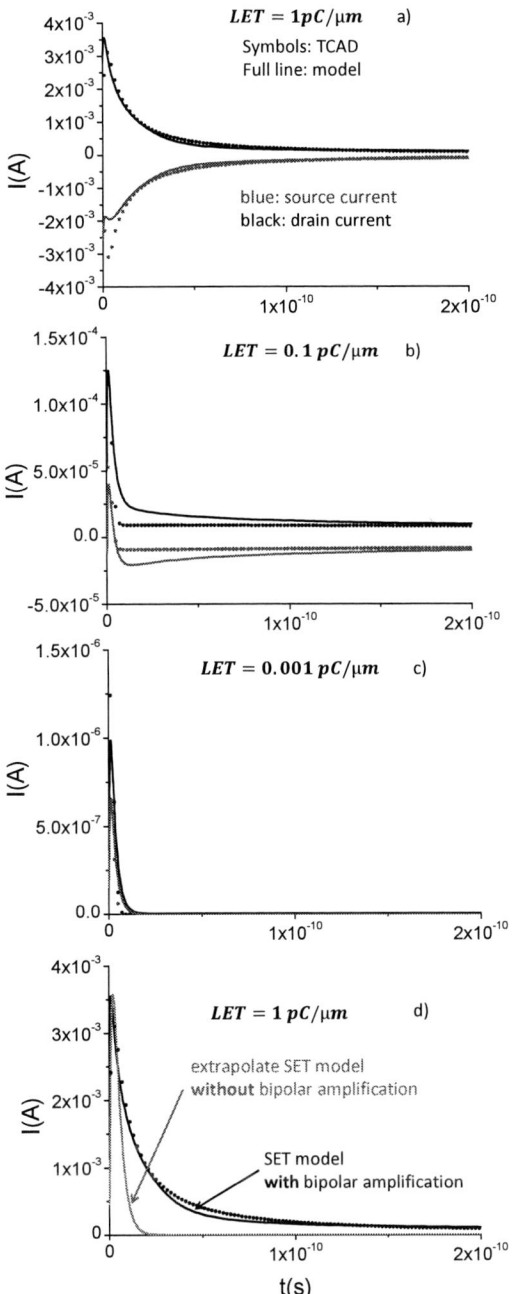

Fig. 5: a-c): comparison between TCAD and proposed SET model at source and drain electrodes for different *LET* values. Such a good agreement is obtained setting model parameters to consistent values: $D = 2.7.10^{-4} m^2.s^{-1}, v_X = 4000\ m.s^{-1}, D_n = 2.25.10^{-4}\ m^2.s^{-1}, P_c = 3.10^3 C, \gamma = 3$. d): comparison between SET model with and without bipolar amplification and TCAD pulse at drain electrode after adjustment of model parameters for $LET = 1\ pC/\mu m$: we cannot describe the tail of the pulse.

IV. SET COMPACT MODEL FOR CIRCUIT LEVEL ARBITRARY CHARGE DEPOSIT: SEU PREDICTION IN SRAMS CELLS

As bipolar amplification occurs after 1st discharge, we can assume that 1st discharge and bipolar amplification are independent. For 1st discharge modeling, we improved former work [1,9] assuming the particle induced many punctual charge deposits δQ_i at different locations of the circuit (made of many transistors) with coordinate X_i^* along X^*, see Fig.3. We also consider uniform drift velocity v_X in source – drain direction in the body of each transistor. For one given transistor of the circuit, the resulting 1st discharge model at the drain for example is then:

$$I_{1^{st}}(t) = \sum_{k=1}^{\infty} G_k . e^{-\frac{t-t_i}{\tau_k}} . H(t - t_i) \tag{8.a}$$

$$G_k = -q . D_n . T_{si} . W . \frac{\partial A_k}{\partial X}(L_{ch}) \tag{8.b}$$

$$A_k = \frac{2.\left(\sum_{i \in body} \delta Q_i . \sin\left(\frac{k.\pi}{L_{ch}}.X_i\right) . e^{\frac{v_X(X-X_i)}{2.D}}\right) . \sin\left(\frac{k.\pi}{L_{ch}}.X\right)}{W . T_{si} . L_{ch} . q} \tag{8.c}$$

$$\tau_k = \left(\frac{\pi^2 . k^2 . D}{L_{ch}^2} + \frac{v_X^2}{4.D}\right)^{-1} \tag{8.d}$$

In (8.a-d), H is the Heaviside function, D_n the electron diffusivity in the body, $X_i = X_i^* - X_s^*$ (X_s^* being the source – body junction along X^*), $X = X^* - X_s^*$, and D is the ambipolar diffusivity.

Resulting SET model at the drain of one given transistor is then the sum of $I_{1^{st}}$ and I_{ba}. Similar model can be obtained for the source SET current. The corresponding equivalent electrical circuit is shown in Fig. 4. Note that at circuit level, the number of deposited holes in the body are expressed as $P_{dep} = \sum_{i \in body} \delta Q_i / q$. SET Model has been validated through extraction of source and drain TCAD pulses for one generation point in the middle of the channel. We obtain good agreement between TCAD and model, after realistic calibration of 1st discharge parameters D, D_n, v_X and bipolar amplification parameters τ_r, P_c, γ, as shown in Fig.5.a-c. Fig.5.d evidences the relevance of the bipolar amplification modeling work, comparing model with and without bipolar amplification.

We then apply the proposed SET model to SEU prediction performing transient SPICE simulations of standard 14nm FDSOI SRAM [11]. In Fig.6.a, we consider 2 different cases of charge deposit at $t = 0$: case 1 corresponds to deposit localized in non-sensitive PMOS for the considered initial bit state and case 2 to more realistic charge sharing with sensitive NMOS latch. We see that circuit level generation can influence SEU prediction as shown in Fig.6.b as a slight charge generation in sensitive NMOS latch can overcome large generation in non-sensitive PMOS (the latter actually reinforcing the initial state) and trigger the bit state flipping.

V. CONCLUSION

In this paper, we proposed a compact model of SET taking bipolar amplification and circuit level arbitrary charge deposit into account. Such a model is dedicated to FDSOI structures (FDSOI MOSFET, FinFET …) and is suitable for performing realistic soft error risk assessment. TCAD simulations supported model development. More extensive work will be dedicated to improve predictability of the model focusing on electroneutrality function and description of the triggering of bipolar amplification around impact time.

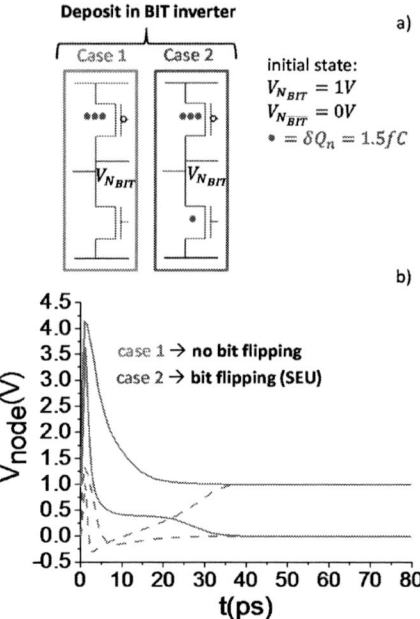

Fig. 6:a): illustration of 2 different cases of particle induced charge deposit at initial time within the BIT inverter b): dynamics of V_{BIT} and $V_{\overline{BIT}}$ after particle strike for these cases: SEU is able to occur even if most of the charge is deposited in non-sensitive PMOS because of some charge deposited in sensitive NMOS Latch.

REFERENCES

[1] N. Rostand, S. Martinie, J. Lacord, O. Rozeau, J. Barbe and G. Hubert, "Single event transient in bulk MOSFETs: Original modelling for SPICE application," 2017 International Conference on Simulation of Semiconductor Processes and Devices (SISPAD), Kamakura, 2017, pp. 89-92.

[2] D. Kobayashi, M. Aimi, H. Saito and K. Hirose, "Time-Domain Component Analysis of Heavy-Ion-Induced Transient Currents in Fully-Depleted SOI MOSFETs," in IEEE Transactions on Nuclear Science, vol. 53, no. 6, pp. 3372-3378, Dec. 2006.

[3] M. Gaillardin et al., "Transient Radiation Response of Single- and Multiple-Gate FD SOI Transistors," in IEEE Transactions on Nuclear Science, vol. 54, no. 6, pp. 2355-2362, Dec. 2007.

[4] J. L. Autran and D. Munteanu – "Soft errors: from particle to circuits" CRC press Taylor & Francis book 2015

[5] V. Ferlet-Cavrois et al., "Characterization of the parasitic bipolar amplification in SOI technologies submitted to transient irradiation," in IEEE Transactions on Nuclear Science, vol. 49, no. 3, pp. 1456-1461, June 2002.

[6] V. Ferlet-Cavrois et al., "Charge enhancement effect in NMOS bulk transistors induced by heavy ion Irradiation-comparison with SOI," in IEEE Transactions on Nuclear Science, vol. 51, no. 6, pp. 3255-3262, Dec. 2004.

[7] V. Ferlet-Cavrois et al., "Direct measurement of transient pulses induced by laser and heavy ion irradiation in deca-nanometer devices," in IEEE Transactions on Nuclear Science, vol. 52, no. 6, pp. 2104-2113, Dec. 2005.

[8] L. Artola, G. Hubert and R. D. Schrimpf, "Modeling of radiation-induced single event transients in SOI FinFETS," 2013 IEEE International Reliability Physics Symposium (IRPS), Anaheim, CA, 2013, pp. SE.1.1-SE.1.6.

[9] N. Rostand et al., "Compact Modelling of Single Event Transient in Bulk MOSFET for SPICE: Application to Elementary Circuit," 2018 International Conference on Simulation of Semiconductor Processes and Devices (SISPAD), Austin, TX, 2018, pp. 364-368.

[10] TCAD Sentaurus Device Manual, Synopsys, Inc.: J-2014.09

[11] R. Berthelon et al., "Investigation of SiGe channel introduction in FDSOI SRAM cell pFET and assessment of the Complementary-SRAM," 2018 Joint International EUROSOI Workshop and International Conference on Ultimate Integration on Silicon (EUROSOI-ULIS), Granada, 2018, pp. 1-4.

From devices to circuits: modelling the performance of 5nm nanosheets

Andrew R. Brown
Synopsys Northern Europe Ltd.
Glasgow, G3 8HB, UK
Andrew.Brown@synopsys.com

Liping Wang
Synopsys Northern Europe Ltd.
Glasgow, G3 8HB, UK
Liping.Wang@synopsys.com

Plamen Asenov
Synopsys Northern Europe Ltd.
Glasgow, G3 8HB, UK
Plamen.Asenov@synopsys.com

Fabian J. Klüpfel
Fraunhofer IISB
91058 Erlangen, Germany
Fabian.Kluepfel@iisb.fraunhofer.de

Binjie Cheng
Synopsys Northern Europe Ltd.
Glasgow, G3 8HB, UK
Binjie.Cheng@synopsys.com

Sébastien Martinie
CEA Leti
38054 Grenoble, France
Sebastien.Martinie@cea.fr

Olivier Rozeau
CEA Leti
38054 Grenoble, France
Olivier.Rozeau@cea.fr

Sylvain Barraud
CEA Leti
38054 Grenoble, France
Sylvain.Barraud@cea.fr

Jean-Charles Barbé
CEA Leti
38054 Grenoble, France
Jean-Charles.Barbe@cea.fr

Campbell Millar
Synopsys Northern Europe Ltd.
Glasgow, G3 8HB, UK
Campbell.Millar@synopsys.com

Jürgen K. Lorenz
Fraunhofer IISB
91058 Erlangen, Germany
Juergen.Lorenz@iisb.fraunhofer.de

Abstract—A simulation flow for design-technology co-optimisation using 5nm stacked nanowires is presented. The effect of variation in key process parameters on the behaviour of benchmark circuits is examined through the use of variability-aware compact models, accounting for both global and local variability.

Keywords—DTCO, TCAD, variability, SPICE, compact models, circuits, nanowires

I. INTRODUCTION

Design-Technology Co-Optimisation (DTCO) has developed into a key methodology to reduce technology development costs and speed up time to market. Starting from technological specifications and customer requests, Technology Computer Aided Design (TCAD) is used to simulate the performance of transistors and interconnects. This is followed by compact model and resistor-capacitor (RC) extraction for SPICE simulation, and the assessment of key technology performance metrics [1].

Here we present a TCAD-to-SPICE simulation flow that demonstrates the evaluation of transistor performance using circuit-level metrics. The work presented here was done as part of the European research project SUPERAID7*.

II. TCAD PROCESS AND DEVICE SIMULATION

The benchmark device used here is a 5nm stacked nanowire transistor. The fabrication of nanowire transistors has been demonstrated by CEA-Leti [2], and detailed information about their fabrication process was provided in the form of technical specifications as well as transmission electron micrographs (Fig. 1). Process simulation is used not only to generate the nominal dimensions, dopant and stress distribution of the transistor to be investigated, but also to predict the changes of these quantities in the case of systematic variations of the processes occurring. While Synopsys has a complete set of TCAD tools including lithography (Sentaurus Lithography) and topography (Sentaurus Topography), for the process flow simulation in the SUPERAID7 project Fraunhofer IISB used a combination of Sentaurus Process [3] and their in-house simulation tools ANETCH, BNDEDIT, DEP3D and Dr.LiTHO to produce the simulation structure shown in Fig. 2.

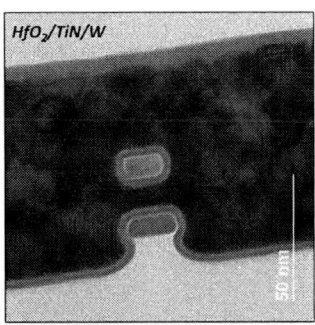

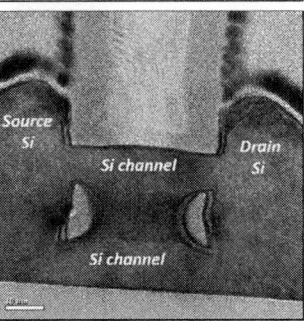

Fig. 1. Example of nanowire transistors fabricated by CEA-Leti and considered in this work: Cross-sectional TEM images of stacked nanowires (top), with SiN inner spacers (bottom).

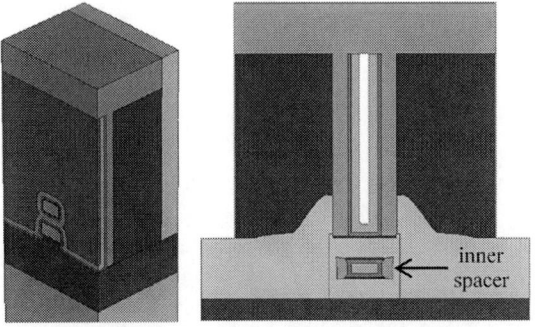

Fig. 2. Simulated nanowire transistor: (left) half of the structure cut across the nanowires, and (right) cross section along source-drain direction.

---

* The research leading to these results received funding from the European Union's Horizon 2020 research and innovation programme under grant agreement No. 688101 SUPERAID7.

Using simulations of transfer characteristics in Sentaurus Device [4], a sensitivity analysis of key process parameters was performed (Fig. 3) and three complementary parameters covering different processes (etching, deposition, and lithography) were chosen to form a design of experiment (DoE) matrix, where each parameter is varied by a +/- amount. The three process parameters chosen were the thickness of the deposited nitride layer (d_{SADP}) in the self-aligned double patterning process employed for the fin, the defocus of the gate lithography setup (F_{gate}) and the Germanium content of the sacrificial layer (x_{Ge}), which influences position and shape of the inner spacers. The nominal values and the size of

variation applied to form the DoE are given in TABLE I. This is used to study the systematic global variability (GV) due to process variations.

To capture the effects of the process variations across the DoE, Sentaurus Device simulations of the required I_D-V_D and I_D-V_G characteristics are run at each point in the DoE, which will be sufficient for the extraction of process-variability-aware compact models. Fig. 4 presents an analysis of the effects of varying each individual process parameter on key figures of merit (FoM). Each line has three data points with the centre point representing the nominal parameter value, while the left and right points respectively represent the negative (-ve) and positive (+ve) variation. Of note here is the non-monotonic response to variation in F_{gate}, which will need to be captured in the process-variability-aware compact model. F_{gate} has a large effect on Vt_{sat} and also impacts Ion_{sat}. The largest variation in Ion_{sat} comes from d_{SADP}.

To capture the statistical variability effects, the dedicated local variability (LV) simulator Garand VE [5] is used. For the 5nm technology being considered here, it is expected that metal gate granularity will no longer be a significant source of LV, as gate-last processing will lead to a mostly amorphous gate metal. Therefore, the sources of LV considered here are random discrete dopants (RDD) and line edge roughness (LER) in the gate edge. The dominant source of LV in these devices will be RDD and an example of RDD within the benchmark nanowire structure is shown in Fig. 5, with resultant I_D-V_G characteristics for the nominal DoE point shown in Fig. 6, for an ensemble of 500 devices.

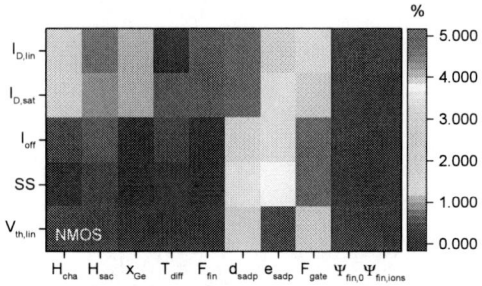

Fig. 3. Sensitivity of electrical FET properties (left axis) to variations in process parameters (bottom axis).

TABLE I. PROCESS PARAMETERS CHOSEN TO FORM THE DoE, NOMINAL VALUES AND THE MAGNITUDE OF THE VARIATIONS USED.

Process Parameter	Symbol	Nominal	Variation
SiGe mole fraction	x_{Ge}	0.3	± 0.03
Gate litho defocus	F_{gate}	0.0	± 0.04 µm
Fin SADP deposition factor	d_{sadp}	1.0	± 0.1

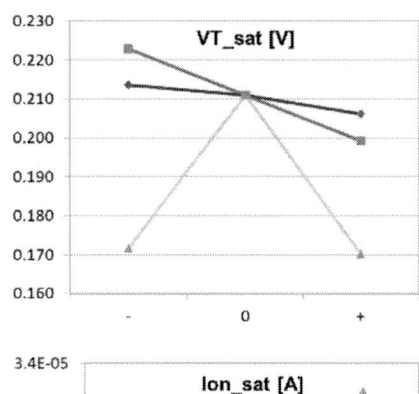

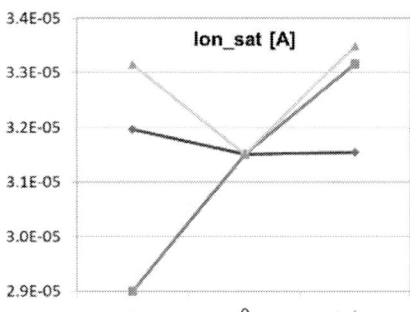

Fig. 4. Variation of key figures of merit with the -ve, nominal and +ve variation of each process parameter for the nMOS nanowire. Key: blue - x_{Ge}; orange - d_{SADP}; yellow - F_{gate}.

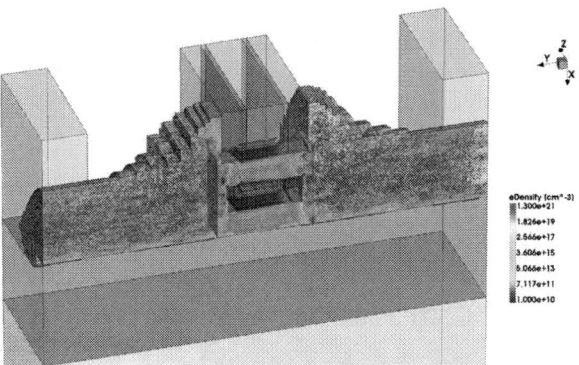

Fig. 5. Benchmark nanosheet device simulated in Synopsys Garand with random discrete dopants.

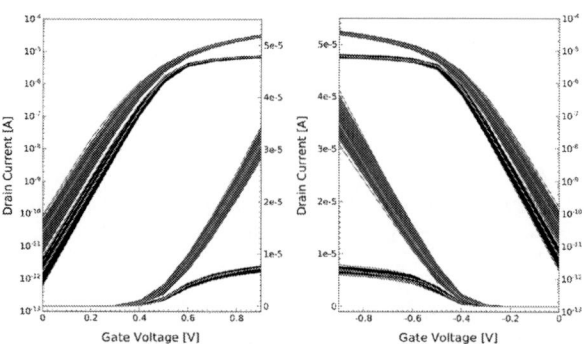

Fig. 6. I_D-V_G characteristics for the nMOS and pMOS devices at the nominal point in the DoE, with a statistical ensemble of 500 devices. Black: V_D=0.05V and blue: V_D=0.9V.

978-1-7281-0941-1/19 $31.00 © 2019 IEEE

III. SPICE MODEL EXTRACTION

The L-NSP model [6] is a surface-potential-based model dedicated to advanced 3D CMOS device architectures. The model was developed with special emphasis on supporting vertically stacked nanowire/nanosheet Gate-All-Around (GAA) CMOS technologies, and fits naturally with the stacked nanowire architecture that is investigated here.

L-NSP is constructed in a hierarchical way with two levels of parameter sets: a global-mode parameter set, and a local-mode parameter set. In the Synopsys TCAD-to-SPICE flow [7], instead of using the global mode to describe devices with different geometries, the Response Surface Model (RSM) approach is applied. RSM can not only capture the device geometry-dependent effects, but also the impact of non-geometry parameters, such as implantation energy, annealing condition, lithography defocus, etc., on device electrical characteristics. To use the RSM approach, only the local mode is required for SPICE modelling of individual devices.

The SPICE compact model extraction is divided into three stages: nominal device extraction stage (provides the base SPICE model for the target device at the nominal process design point), the response surface model extraction stage (provides SPICE models that cover device global process variation in the DoE space), and, finally, the statistical model extraction stage (provides SPICE models that cover device local statistical variability across the DoE).

To reduce the large number of statistical device simulations required, the LV simulations and statistical compact model extraction were not done at every DoE point. The benefit of the compact model approach adopted here is that the distributions and correlations of figures of merit due to LV are maintained across the DoE, even at points for which they are not extracted. This is demonstrated in Fig. 7, which compares the correlations of key FoM generated by TCAD simulations (black) and coming from the compact model interpolated by the RSM (red) at a point in the DoE at which the TCAD local variability simulations were not included in the RSM.

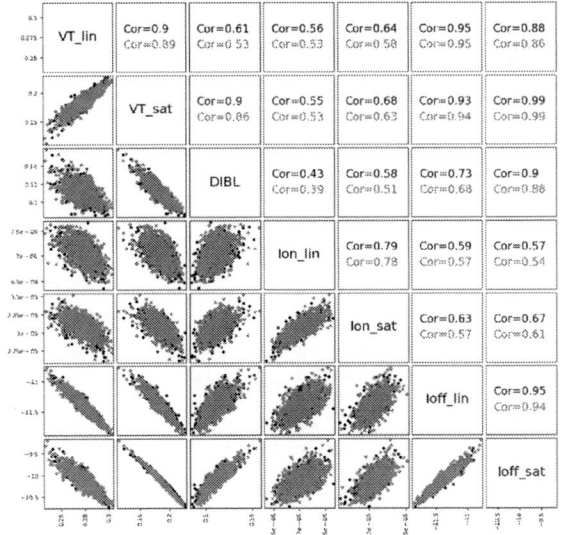

Fig. 7. Statistical compact model fitting results for nMOS nanowire devices (Black: TCAD results; red: compact model results) at x_{Ge} =0.3, d_{SADP} =0.9 and F_{gate} = -0.04 µm.

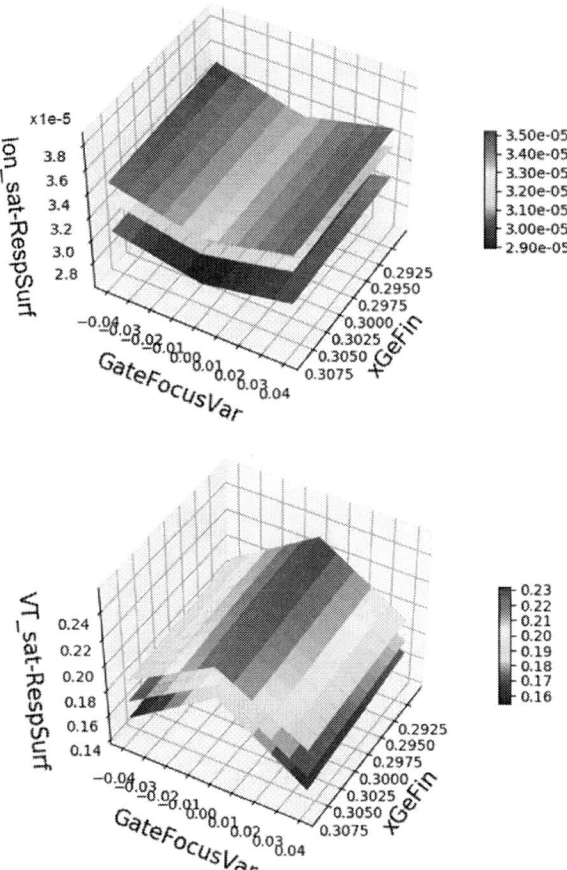

Fig. 8. Response surface compact model, regenerated by RandomSpice, covering DoE of nMOS nanowire devices. Different surfaces correspond to the three different DoE values of d_{sadp}.

An unlimited number of compact models can be generated for any point in the DoE following the extracted response surface (Fig. 8), allowing large-scale statistical circuit simulations. Note that the observed non-monotonic dependence on F_{gate} is captured in the compact model.

When generating random instances of devices within the DoE we will assume a Gaussian variation of the three process parameters about the nominal value. In the case of F_{gate}, the "V"-shaped response observed in Fig. 4, when applied to a Gaussian distribution, will lead to distinctly skewed FoM. The highest Vt_{sat} occurs at the nominal gate focus factor (peak of the Gaussian distribution), therefore there will be a large number of devices with close-to-maximum Vt_{sat}. At the same time, the minimum Vt_{sat} occurs at both +ve and -ve tails of the distribution in gate focus factor and therefore has significantly lower probability of occurring. Conversely, the Ion_{sat} distribution will be oppositely skewed.

IV. CIRCUIT SIMULATION

The circuit used as a demonstrator is an AND-OR-Inverter (AOI), with a realistic 5nm backend structure produced in Process Explorer [8] (Fig. 9), based on an AOI221 standard cell layout in GDSII format. Once a 3D structure has been generated using process emulation, this can be used to extract a netlist of the RC equivalent parasitic elements for the BEOL, which in combination with front-end compact models can be used to construct a complete SPICE netlist for circuit

978-1-7281-0941-1/19 $31.00 © 2019 IEEE

Fig. 9. 3D emulated process structure of an AOI221 standard cell for the 5nm technological node.

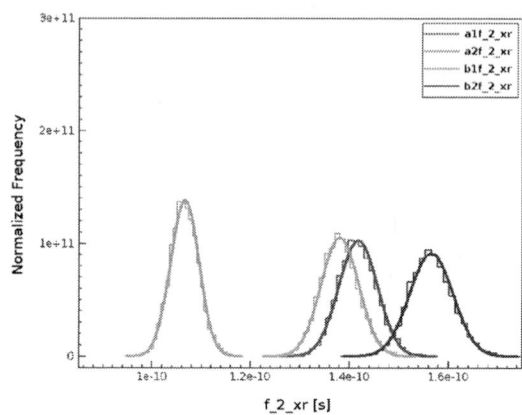

Fig. 10. Signal timing histograms with fitted Gaussian distributions for AOI propagation delays in the presence of combined GV and LV.

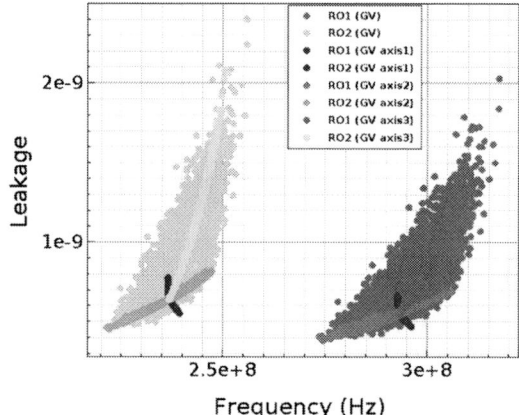

Fig. 11. Variations introduced by different process parameters for the leakage of two extreme case ROs (axis1: x_{Ge}; axis2: d_{sadp}; axis3: F_{gate}).

simulation of the AOI standard cell. The equivalent RC netlist for the interconnects is extracted using Raphael [9].

We have performed large-scale Monte Carlo circuit simulation to investigate the effects of GV and LV on delay times for different state transitions. For example, in Fig. 10 we see that transition $b2f\_2\_xr$ (delay from input $b2$ falling to output x rising) is usually the critical delay path, but due to a combination of global and local variability it is possible to have a circuit in which $a1f\_2\_xr$ becomes the longer delay. As a result, it is critical to consider the combined impact of GV and LV when developing cell characterisation flows which feed into place and route, and static timing analysis (STA) for advanced technologies.

We have also studied the performance of ring oscillators (ROs) by wiring the AOI to mimic digital logic delay chains. The leakage and frequency variations are dominated by GV, and we can isolate the effects of the individual process parameters (Fig. 11). As shown, d_{SADP} variation mainly impacts the RO frequency, while the variation of F_{gate} mainly impacts on the RO leakage. x_{Ge} variation has a small impact on RO performance, however, it contributes to the decorrelation between frequency and leakage.

V. CONCLUSIONS

We have presented a fully integrated process-to-circuit simulation flow. This can be used to identify the relative impact of both global and local variations, and can help to direct process improvement efforts to the aspects of variation where the biggest gains in circuit performance can be achieved, thus closing the feedback loop on design-technology co-optimisation.

[1] A. Asenov, B. Cheng, X. Wang, A. R. Brown, C. Millar, C. Alexander, S. M. Amoroso, J. B. Kuang and S. R. Nassif, "Variability Aware Simulation Based Design- Technology Cooptimization (DTCO) Flow in 14 nm FinFET/SRAM Cooptimization," *IEEE Transactions on Electron Devices,* vol. 62, no. 6, pp. 1682-1690, 2015.

[2] S. Barraud, V. Lapras, M. P. Samson, L. Gaben, L. Grenouillet, V. Maffini-Alvaro, Y. Morand, J. Daranlot, N. Rambal, B. Previtalli, S. Reboh, C. Tabone, R. Coquand, E. Augendre, O. Rozeau, J. M. Hartmann, C. Vizioz, C. Arvet, P. Pimenta-Barros and N. Posseme, "Vertically stacked-NanoWires MOSFETs in a replacement metal gate process with inner spacer and SiGe source/drain," in *IEEE International Electron Devices Meeting,* San Francisco, 2016.

[3] Synopsys, *Sentaurus Process User Guide,* Mountain View, CA, USA, O-2018.06.

[4] Synopsys, *Sentaurus Device User Guide,* Mountain View, CA, USA, O-2018.06.

[5] Synopsys, *Garand User Guide,* Mountain View, CA, USA, O-2018.06.

[6] O. Rozeau, S. Martinie, T. Poiroux, F. Triozon, S. Barraud, J. Lacord, Y. M. Niquet, C. Tabone, R. Coquand, E. Augendre, M. Vinet, O. Faynot and J.-C. Barbé, "NSP: Physical compact model for stacked-planar and vertical Gate-All-Around MOSFETs," in *IEEE International Electron Devices Meeting (IEDM),* San Francisco, 2016.

[7] A. Asenov, K. E. Sayed, R. Borges, P. Asenov, C. Millar and T. Ma, "TCAD based Design-Technology Co-Optimisations in advanced technology nodes," in *International Symposium on VLSI Design, Automation and Test (VLSI-DAT),* Hsinchu, Taiwan, 2017.

[8] Synopsys, *Sentaurus Process Explorer User Guide,* Mountain View, CA, USA, O-2018.06.

[9] Synopsys, *Raphael Reference Manual,* Mountain View, CA, USA, O-2018.06.

Compact Modelling of Resistive Switching Devices based on the Valence Change Mechanism

Camilla La Torre
Instutitut für Werkstoffe der
Elektrotechnik II, RWTH Aachen
University & JARA-FIT
Aachen, Germany
latorre@iwe.rwth-aachen.de

Alexander F. Zurhelle
Instutitut für Werkstoffe der
Elektrotechnik II, RWTH Aachen
University & JARA-FIT
Aachen, Germany
a.zurhelle@iwe.rwth-aachen.de

Stephan Menzel
Peter Grünberg Institut (PGI-7)
Forschungszentrum Juelich GmbH &
JARA-FIT
Juelich, Germany
st.menzel@fz-juelich.de

Abstract—**In this paper, a compact model for filamentary, resistive switching devices based on the valence change mechanism is proposed. It is based on the motion of ionic defects in a filamentary region. In contrast to previous model, it uses two state variables representing the ionic defect concentration close to the two opposing electrodes. This enables the modelling of ionic diffusion and, hence, drift–diffusion equilibria. In addition, the model can be used to simulate complementary switching in addition to the standard bipolar switching behavior.**

Keywords—ReRAM, complementary switching, modelling, valence change mechanism

I. INTRODUCTION

Nonvolatile resistive switching devices based on the valence change mechanism (VCM) are a promising candidate for data storage, neuromorphic computing or computation-in-memory applications [1, 2]. By the application of appropriate voltage stimuli, VCM devices can be switched between at least two resistance states: a high resistive state (HRS) and a low resistive state (LRS). The switching mechanism relies on the non-isothermal drift and diffusion of ionic defects, typically oxygen vacancies, within a filamentary region of the oxide [3]. The change of the atomic configuration modulates the electrostatic barriers at the metal/oxide interfaces leading to a change in the local conductivity [3, 4]. To model this complex physical behavior, Marchewka and co-workers developed a physical continuum model. The simulation results show, for example, that the gradual RESET transition originates from the drift and diffusion of the ionic defects approaching a dynamic equilibrium [5, 6]. In addition, the transition from bipolar to complementary switching is related to the symmetrization of the electrostatic barriers at the opposing metal/oxide interfaces [7].

The rather long simulation times of continuum models, however, are not suited for circuit simulations. Thus, compact models need to be developed that allow for fast computation while still reproducing the experimental behavior. The compact models proposed in literature often use one state variable and only consider ionic drift [8-10]. These models can reproduce, for example, the nonlinear switching kinetics of VCM devices, but cannot account for a RESET equilibrium or complementary switching.

In this paper, we present a compact model that is capable of simulating the RESET equilibrium as well as complementary switching. The compact model is consistent with the results of our VCM continuum model [5, 6].

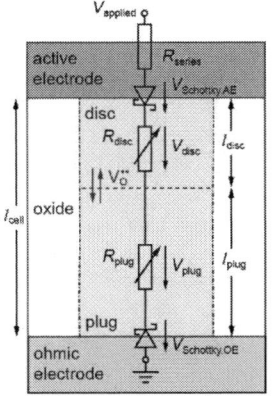

Fig. 1. Equivalent circuit diagram representing the electrical model of an asymmetric ReRAM device. The voltage is applied to the active electrode (AE), whereas the ohmic electrode (OE) is grounded.

The compact model is a part of our Juelich-Aachen Resistive Switching Tools (JART) and is called JART VCM v2.

II. SIMULATION MODEL

In the JART VCM v2 model, a filamentary region within the oxide matrix is considered as shown in Fig. 1a. The filamentary region of length l_{cell} is divided into two regions: the disc region and the plug region of length l_{disc} and l_{plug}, respectively. To each of these regions, one state variable is assigned, N_{disc} and N_{plug}, which represents the ionic defect concentrations (typically oxygen vacancies) in the disc and the plug, respectively. For each of these state variables one ordinary differential equation (ODE) is solved resulting in the coupled ODEs

$$\frac{dN_{\text{disc}}}{dt} = -\frac{1}{z_{\text{Vo}}\, e\, A\, l_{\text{disc}}} \cdot I_{\text{ion}} \qquad (1)$$

and

$$\frac{dN_{\text{plug}}}{dt} = \frac{1}{z_{\text{Vo}}\, e\, A\, l_{\text{plug}}} \cdot I_{\text{ion}}, \qquad (2)$$

where A is the cross-section of the filament, e is the elementary charge, and z_{Vo} is the charge number of the ionic defects. The change of the defect concentrations is proportional to the ionic current I_{ion}. As the increase/decrease of the disc and the plug concentration are interchanged in this model, the signs in eqs. (1) and (2) are opposite to each other. Thus, the total number of ionic defects in the whole filament stays constant during switching. The two state variables allow

TABLE I
SIMULATION PARAMETERS

Symbol	Value	Symbol	Value
l_{cell}	5 nm	$e\phi_{Bn0,AE}$	0.5 eV
l_{disc}	1.5 nm	$e\phi_{Bn0,OE}$	0.1 eV
l_{plug}	$l_{cell} - l_{disc}$	A^*	$6.01 \cdot 10^5$ A/(m²K²)
r_{fil}	35 nm	μ_{n0}	$5 \cdot 10^{-6}$ m²/(Vs)
z_{Vo}	2	ΔE_{ac}	0.05 eV
a	0.4 nm	N_{max}	$6 \cdot 10^{27}$ m⁻³
v_0	$8 \cdot 10^{12}$ Hz	N_{min}	$1/(A \cdot l_{disc})$
ΔW_A	0.9 eV	$R_{series,0}$	1200 Ω
ε	$17 \cdot \varepsilon_0$	$R_{th,eff}$	$1.6 \cdot 10^6$ K/W
ε_{ϕ_B}	$5.5 \cdot \varepsilon_0$	T_0	293 K

us to model the ionic diffusion current density $J_{ion,diffusion}$ between these two regions, in addition to the ionic drift current density $J_{ion,drift}$ according to

$$I_{ion} = A\left(J_{ion,drift} + J_{ion,diffusion} \right)$$
$$= A\left(CN \sinh\left(\frac{a\, z_{Vo} eE}{2\, k_B T} \right) \cdot F_{limit} - C\frac{a}{2}\frac{dN}{dx}\cosh\left(\frac{a\, z_{Vo} eE}{2\, k_B T} \right) \right), \quad (3)$$

with

$$C = 2 z_{Vo} e a\, v_0 \exp\left(-\frac{\Delta W_A}{k_B T}\left[\sqrt{1-\gamma^2} + \gamma \arcsin \gamma \right] \right), \quad (4)$$

and

$$\gamma = \frac{a\, z_{Vo} eE}{\pi \Delta W_A}. \quad (5)$$

In eqs. (3)-(5), a is the hopping distance, k_B is the Boltzmann constant, ΔW_A is the migration barrier, and v_0 is the attempt frequency for ion hopping. The function F_{limit} is introduced to prevent unreasonable defect concentrations and keep them between a maximum N_{max} and minimum concentration N_{min} [11]. The concentration gradient in eq. (3) is defined according to

$$\frac{dN}{dx} = \frac{N_{plug} - N_{disc}}{0.5\, l_{cell}}. \quad (6)$$

Furthermore, the mean concentration N in eq. (3) is calculated using

$$N = \sqrt{N_{disc} \cdot N_{plug}}. \quad (7)$$

The ionic current depends in a nonlinear manner on the electric field E. In principle, the electric field varies along the filament due to different work functions and space charge effects. This complex space-dependency needs to be approximated for compact modelling. If the work functions of the two metals are identical, the length of disc and plug are considered to be identical. In this symmetric case, the electric field is defined as

$$E_{symmetric} = \frac{V_{disc} + V_{plug}}{l_{cell}}, \quad (8)$$

where V_{disc} (V_{plug}) defines the voltage drop over the disc (plug) region. For differing work functions and lengths, the band bending under forward and reverse bias of the disc/metal interface is different. Thus, the definition of the electric field becomes polarity-dependent and

$$E_{asymmetric} = \begin{cases} \dfrac{V_{disc} + V_{plug}}{l_{cell}} & V_{applied} > 0 \text{ (RESET)} \\[2mm] \dfrac{V_{disc}}{l_{disc}} & V_{applied} < 0 \text{ (SET)} \end{cases} \quad (9)$$

is used. Here, it is always assumed that the disc is close to the electrode with the higher work function, which is called active electrode (AE). The electrode close to the plug region is called ohmic electrode (OE) as it shows a very small interface barrier.

The ionic current depends on the filament temperature T, which can increase from the ambient temperature T_0 due to local Joule heating. Thus, the temperature becomes a function of the dissipated electrical power and it is approximated by

$$T = (V_{disc} + V_{plug}) \cdot I \cdot R_{th,eff} + T_0. \quad (10)$$

In eq. (10), $R_{th,eff}$ is the equivalent thermal resistance of the device. It represents the heat dissipation via the electrodes and the surrounding material. The total current flowing through the cell (the filament) is determined using the equivalent circuit diagram in Fig. 1. The two "Schottky" contacts are modelled by assuming thermionic field emission I_{TFE} and thermionic emission I_{TE} in reverse and forward direction of the contact, respectively. As the two "Schottky" contacts have opposite polarities, a signum function needs to be used in the calculation of the current as

$$I = \begin{cases} \text{sign}\left(V_{applied}\right) \cdot I_{TE} & V_{Schottky} > 0 \text{ (forward)} \\[1mm] \text{sign}\left(V_{applied}\right) \cdot I_{TFE} & V_{Schottky} > 0 \text{ (reverse)} \end{cases} \quad (11)$$

Thus, while one Schottky contact is biased in forward direction the other one is biased in reverse direction. The currents are strongly nonlinear dependent on the ionic defect concentration close to the respective contacts. If the defect concentration increases, the current increases, too. The disc (plug) resistance R_{disc} (R_{plug}) is also a function of the disc (plug) concentration and is calculated according to

$$R_{disc/plug} = \frac{l_{disc/plug}}{A \cdot z_{Vo} e \mu_{n0} N_{disc/plug}} \exp\left(\frac{\Delta E_{ac}}{k_B T} \right). \quad (12)$$

A band conduction mechanism with mobility μ_{n0} is assumed, in which the defect concentration defines the numbers of electrons in the conduction band. A slight temperature activation ΔE_{ac} of the transport is assumed as well. Further details of the model, like the used equations for F_{limit}, I_{TFE}, and I_{TE}, can be found in [11]. The model has been implemented in MATLAB and Verilog-A. Both model implementations of the JART VCM v2 model can be downloaded from our web site [12]. The used simulation parameters are given in Table I.

III. SIMULATION RESULTS

In a first simulation, constant voltage pulses with different amplitudes are used to study the RESET transition. Here, we assume an asymmetric cell. Fig. 2a shows the simulated current transients for three different RESET voltage amplitudes. The transition starts first for the highest voltage amplitude and then the current decreases gradually. This current transition is caused by the change of the disc concentration due to the applied voltage (Fig. 2b). Eventually, the current does not change anymore. In the constant current regime, the disc concentration is constant, too. Thus, the disc concentration does not change anymore while the voltage is still applied.

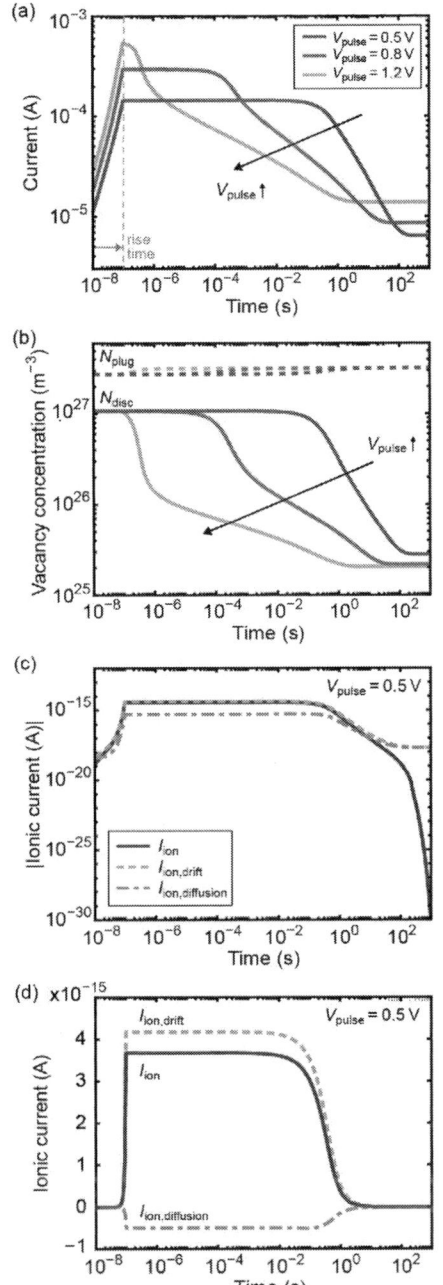

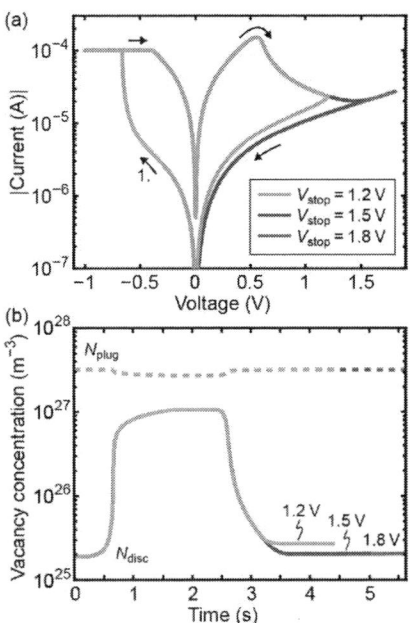

Fig. 3. Influence of V_{stop} on the HRS during the RESET and its limit. All sweeps are simulated with a sweep rate of 1 V/s. The initial HRS is defined by $N_{disc,initial} = 1.9 \cdot 10^{25}$ m^{-3} and $N_{plug,initial} = 3.2 \cdot 10^{27}$ m^{-3}. (a) I–V curves with a maximum positive sweep voltage of 1.2 V, 1.5 V, and 1.8 V, respectively. (b) Oxygen vacancy concentration in the plug and the disc for the three stop voltages.

V_{pulse}. The higher the pulse amplitude, the smaller is the final disc concentration. Thus, the final resistance is higher for a higher RESET voltage used. These results are equivalent to the results of our continuum model and consistent with experimental data [6]. In a second simulation study, the RESET equilibrium is investigated for voltage sweeps with varying RESET stop voltage V_{stop}. As shown in Fig. 3a, the increase of V_{stop} leads to different HRS, but at very high V_{stop}, the HRS stays constant. Similar to the constant voltage operation a drift–diffusion equilibrium evolves, which results in constant values for the disc and plug concentration (Fig. 3b). The tuning of the HRS with the RESET stop voltage has been reported for various VCM cells in literature [5, 13-16]. The occurrence of an HRS saturation has been reported as well [17, 18].

Besides the simulation of the RESET equilibrium, the JART VCM v2 model allows us to simulate complementary switching behavior. For the simulation results shown in Figs. 2 and 3, an asymmetric device structure is assumed as discussed before. The electrostatic barrier at the active electrode is a lot higher than the one at the ohmic electrode. In this case, N_{disc} determines the overall current through the device and thus the switching is bipolar. If a symmetric stack is considered, the lower one of the two concentrations will determine the overall current. In Fig. 4a, the concentration in the left half of the cell is lower. As long as the concentration on the left side is limited to lower values than on the right side due to the (chosen) current compliance, bipolar switching is obtained. If no current compliance is used, the ionic defects can move completely from the left region to the right region and vice versa. This leads to complementary switching behavior as shown in Fig. 4b. Figure 4c shows a bipolar switching mode, which results from the change of the concentration in the right region. This behavior is consistent with experimental data [19, 20] and our continuum model [7].

Fig. 2. For three RESET pulses (rise time 100 ns) with $V_{pulse} = 0.5$ V, $V_{pulse} = 0.8$ V, and $V_{pulse} = 1.2$ V, the current transients (a) and the oxygen vacancy concentrations in the plug and the disc (b) are compared. For the initial LRS, $N_{disc,initial} = 1.07 \cdot 10^{27}$ m^{-3} and $N_{plug,initial} = 2.75 \cdot 10^{27}$ m^{-3} are used. For $V_{pulse} = 0.5$ V, the total ionic current and its drift and diffusion component are plotted on (c) a logarithmic and (d) a linear scale.

As shown in Fig. 2c and d, the constant disc concentration originates from the approached equilibrium of drift and diffusion. During the RESET transition, the ionic drift moves the positively charged oxygen vacancies from the disc region to the plug region. Thus, a concentration gradient builds up, which gives rise to an ionic diffusion current counteracting the drift current. Eventually, these two currents reach an equilibrium. The simulation results also show that the final disc concentration is a function of the applied pulse voltage

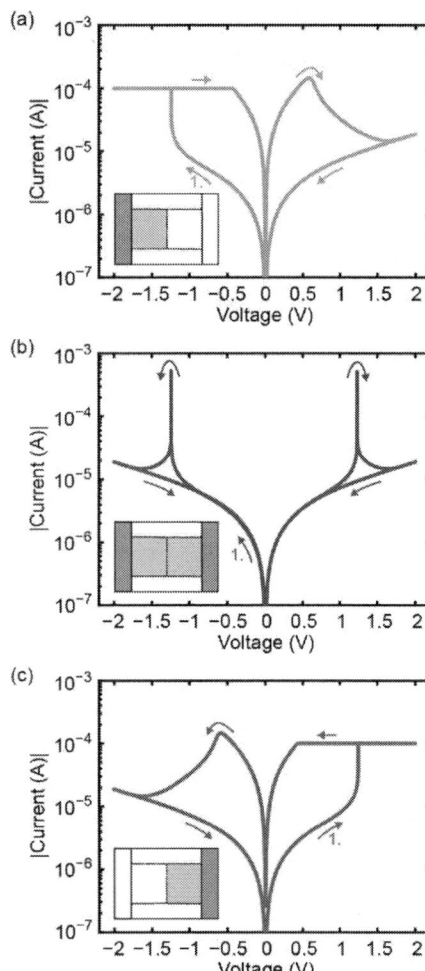

Fig. 4. Transition of bipolar and complementary switching. (a) Bipolar switching at one electrode by applying a current compliance for negative voltages. (b) Complementary switching arises in a symmetric model if no current compliance is active. (c) Bipolar switching at the other electrode by applying a current compliance for positive voltages. The initial values for the defect concentrations in both regions are similar to the simulation in Fig. 3. The device stack is symmetric with $l_{disc} = l_{plug} = 2.5$ ns and $e\phi_{Bn0,AE} = e\phi_{Bn0,OE} = 0.3$ eV.

IV. CONCLUSIONS

In this paper, we present a compact model that can simulate the RESET equilibrium and complementary switching behavior. To this end, a second state variable is introduced, which enables us to implement ionic diffusion in addition to ionic drift. The simulation results are consistent with our previous physical continuum model, but significantly reduce the simulation times.

ACKNOWLEDGMENT

This work was supported in parts by the Deutsche Forschungsgemeinschaft (SFB917).

REFERENCES

[1] J. J. Yang, D. B. Strukov and D. R. Stewart, "Memristive Devices for Computing," *Nat. Nanotechnol.*, vol. 8, pp. 13-24, 2013.

[2] D. Ielmini and H. P. Wong, "In-memory computing with resistive switching devices," *Nature Electronics*, vol. 1, pp. 333-343, 2018.

[3] R. Waser, R. Dittmann, G. Staikov and K. Szot, "Redox-Based Resistive Switching Memories - Nanoionic Mechanisms, Prospects, and Challenges," *Adv. Mater.*, vol. 21, pp. 2632-2663, 2009.

[4] J. J. Yang, M. D. Pickett, X. Li, D. A. A. Ohlberg, D. R. Stewart and R. S. Williams, "Memristive switching mechanism for metal/oxide/metal nanodevices," *Nat. Nanotechnol.*, vol. 3, pp. 429-433, 2008.

[5] A. Marchewka, B. Roesgen, K. Skaja, H. Du, C. L. Jia, J. Mayer, V. Rana, R. Waser and S. Menzel, "Nanoionic Resistive Switching Memories: On the Physical Nature of the Dynamic Reset Process," *Adv. Electron. Mater.*, vol. 2, pp. 1500233/1-13, 2016.

[6] A. Marchewka, R. Waser and S. Menzel, "A 2D Axisymmetric Dynamic Drift-Diffusion Model for Numerical Simulation of Resistive Switching Phenomena in Metal Oxides," *2016 International Conference On Simulation of Semiconductor Processes and Devices (SISPAD), Nuremberg, Germany, September 6-8, 2016*, 2016, pp. 145-148.

[7] A. Marchewka, R. Waser and S. Menzel, "Physical Simulation of Dynamic Resistive Switching in Metal Oxides Using a Schottky Contact Barrier Model," *2015 International Conference On Simulation of Semiconductor Processes and Devices (SISPAD), 9-11 September, Washington D.C, USA*, 2015, pp. 297-300.

[8] Z. Jiang, Y. Wu, S. Yu, L. Yang, K. Song, Z. Karim and H.-P. Wong, "A Compact Model for Metal-Oxide Resistive Random Access Memory With Experiment Verification," *IEEE Trans. Electron Devices*, vol. 63, pp. 1884-1892, 2016.

[9] A. Hardtdegen, C. La Torre, F. Cüppers, S. Menzel, R. Waser and S. Hoffmann-Eifert, "Improved Switching Stability and the Effect of an Internal Series Resistor in HfO2/TiOx Bilayer ReRAM Cells," *IEEE Trans. Electron Devices*, vol. 65, pp. 3229-3236, 2018.

[10] S. Ambrogio, S. Balatti, D. C. Gilmer and D. Ielmini, "Analytical Modeling of Oxide-Based Bipolar Resistive Memories and Complementary Resistive Switches," *IEEE Trans. Electron Devices*, vol. 61, pp. 2378-2386, 2014.

[11] C. La Torre, A. F. Zurhelle, T. Breuer, R. Waser and S. Menzel, "Compact Modeling of Complementary Switching in Oxide-Based ReRAM Devices," *IEEE Trans. Electron Devices*, vol. 66, pp. 1268-1275, 2019.

[12] JART, "Juelich Aachen Resistive Switching Tools (JART)," www.emrl.de/JART.html, 2019.

[13] W. Hu, L. Zou, C. Gao, Y. Guo and D. Bao, "High speed and multi-level resistive switching capability of Ta2O5 thin films for nonvolatile memory application," *J. Alloy. Compd.*, vol. 676, pp. 356-360, 2016.

[14] S. Yu, Y. Wu, R. Jeyasingh, D. Kuzum and H. P. Wong, "An Electronic Synapse Device Based on Metal Oxide Resistive Switching Memory for Neuromorphic Computation," *IEEE Trans. Electron Devices*, vol. 58, pp. 2729-2737, 2011.

[15] H. K. Li, T. P. Chen, S. G. Hu, S. Liu, Y. Liu, P. S. Lee, X. P. Wang, H. Y. Li and G. Q. Lo, "Study of Multilevel High-Resistance States in HfOx-Based Resistive Switching Random Access Memory by Impedance Spectroscopy," *IEEE Trans. Electron Devices*, vol. 62, pp. 2684-2688, 2015.

[16] F. Yuan, Z. Zhang, L. Pan and J. Xu, "A Combined Modulation of Set Current With Reset Voltage to Achieve 2-bit/cell Performance for Filament-Based RRAM," *IEEE J. Electron Devices Soc.*, vol. 2, pp. 154-157, 2014.

[17] J. Frascaroli, S. Brivio, E. Covi and S. Spiga, "Evidence of soft bound behaviour in analogue memristive devices for neuromorphic computing," *Sci Rep*, vol. 8, pp. 7178/1-12, 2018.

[18] L. Larcher, F. M. Puglisi, P. Pavan, A. Padovani, L. Vandelli and G. Bersuker, "A Compact Model of Program Window in HfOx RRAM Devices for Conductive Filament Characteristics Analysis," *IEEE Trans. Electron Devices*, vol. 61, pp. 2668-2673, 2014.

[19] F. Nardi, S. Balatti, S. Larentis, D.C. Gilmer and D. Ielmini, "Complementary Switching in Oxide-Based Bipolar Resistive-Switching Random Memory," *IEEE Trans. Electron Devices*, vol. 60, pp. 70-77, 2013.

[20] S. Balatti, S. Larentis, D. C. Gilmer and D. Ielmini, "Multiple Memory States in Resistive Switching Devices Through Controlled Size and Orientation of the Conductive Filament," *Advanced Materials*, vol. 25, pp. 1474-1478, 2013.

Effect of Stacking Faults on the Thermoelectric Figure of Merit of Si Nanowires

Kantawong Vuttivorakulchai*, Mathieu Luisier, and Andreas Schenk

Integrated Systems Laboratory
ETH Zürich
Gloriastrasse 35, CH-8092 Zürich, Switzerland
*email: kvuttivo@iis.ee.ethz.ch

Abstract—The efficiency of converting waste heat to electricity requires a large value of the thermoelectric figure of merit (ZT). This can be achieved by patterning bulk material into nanostructures like nanowires (NWs). Further improvement results from an increased surface roughness (SR) of such NWs [1]. In this work, Si NWs with stacking faults (SFs) are studied. It is shown that SFs can significantly reduce the lattice thermal conductivity as compared to ideal NWs [2]. A recent derivation of the phonon relaxation time for SF scattering [3] is adapted to the electronic case. It turns out that in most cases the thermoelectric power factor (PF) decreases to a lesser extent than the thermal conductivity. This can double ZT provided that SR scattering of electrons is negligible.

Keywords—thermoelectricity, Si nanowire, stacking faults, Scattering, device modeling

I. INTRODUCTION

It has been shown experimentally that the presence of SFs strongly increases the ZT value of InAs NWs with a diameter of 20 nm [4]. To the best of our knowledge, such experiments have not been conducted yet with Si NWs, where SFs arise as the interfaces between alternating regions having either diamond (DM) or wurtzite (WZ) structure (see Fig. 1). For the first time, the simulation of such a Si NW with 70 nm in diameter will be presented.

II. THEORY AND APPROACH

First-principle calculations based on projector augmented-wave (PAW) pseudopotentials with hybrid functional of Heyd, Scuseria, and Ernzerhof (HSE06) [5] as implemented in the Vienna ab initio simulation package (VASP) were performed [6,7]. The electronic band structures are calculated for the cubic unit cells of DM oriented in ⟨100⟩ and ⟨111⟩ direction, respectively, and WZ oriented in ⟨0001⟩ direction along the x-axis. These band structures are the input for our in-house linearized Boltzmann transport equation (BTE) solver.

A. Transport Coefficients from BTE

The theoretical background of the BTE will not be outlined here as it can be found elsewhere, e.g. in Ref. [8]. The transport properties of real materials can be explained by this theory. In the presence of an electrical field (**E**) and a temperature gradient (∇T), the electrical current density **J** and the heat current (or energy flux density) J_Q can be written, respectively, as

$$J = \sigma \left(\mathbf{E} - S\nabla\mathrm{T} \right), \tag{1}$$

$$J_Q = \mathrm{T}\sigma S\mathbf{E} - K\nabla\mathrm{T}, \tag{2}$$

where σ, S, and K are rank-two tensors which reduce to scalars for isotropic media.

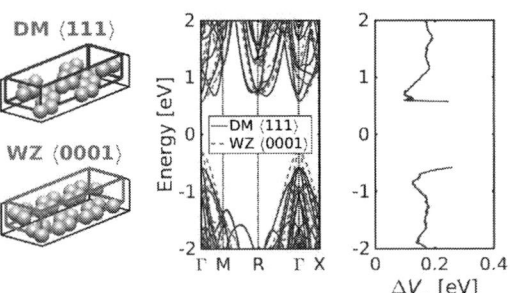

Figure 1: Left panel: Schematic view of the two (three) smallest cubic unit cells of DM (WZ) in ⟨111⟩ (⟨0001⟩) direction. Right panel: The corresponding electronic band structures calculated by density functional theory (DFT) and the energy difference between DM and WZ structures. A **k**-point sampling of "401×101×101" is used.

The electronic part of the thermal conductivity is defined as (minus) the heat current per unit of temperature gradient in open-circuit conditions (i.e., **J** = 0). It is given by

$$k_e = K - \mathrm{T}\sigma S^2. \tag{3}$$

By solving the BTE under relaxation time (RT) approximation and the assumption that the system is in steady state with a distribution function slightly different from its equilibrium form, the transport coefficients read

$$\left[\sigma \right]_{ij} \left(E_F, \mathrm{T} \right) = e^2 \int_{-\infty}^{\infty} dE \left(-\frac{\partial f \left(E, E_F, \mathrm{T} \right)}{\partial E} \right) \Sigma_{ij} \left(E \right), \tag{4}$$

$$\left[\sigma S \right]_{ij} \left(E_F, \mathrm{T} \right) = \frac{e}{\mathrm{T}} \int_{-\infty}^{\infty} dE \left(-\frac{\partial f \left(E, E_F, \mathrm{T} \right)}{\partial E} \right) \left(E - E_F \right) \Sigma_{ij} \left(E \right), \tag{5}$$

$$\left[K \right]_{ij} \left(E_F, \mathrm{T} \right) = \frac{1}{\mathrm{T}} \int_{-\infty}^{\infty} dE \left(-\frac{\partial f \left(E, E_F, \mathrm{T} \right)}{\partial E} \right) \left(E - E_F \right)^2 \Sigma_{ij} \left(E \right), \tag{6}$$

where i and j are Cartesian indices, $\partial f / \partial E$ is the derivative of the Fermi-Dirac distribution function with respect to the energy (E), E_F is the Fermi level or the electro-chemical potential, σ, S, and k_e are the electrical conductivity, Seebeck coefficient, and electron thermal conductivity, respectively. The transport distribution function (TDF) is defined as

$$\Sigma_{ij} \left(E \right) = \frac{1}{V} \sum_{n,k} v_i \left(n, k \right) v_j \left(n, k \right) \tau_{n,k} \delta \left(E - E_{n,k} \right), \tag{7}$$

where the summation is over all bands n and the entire Brillouin zone (BZ). v_i is the i-component of the group velocity at $(n, \mathbf{k})$. The electron lifetime $\tau_{n,k}$ is a function of both n and $\mathbf{k}$. The electron and hole density are given by

$$n_e\left(E_F, \mathrm{T}\right) = \int_{E_c}^{\infty} dE f\left(E, E_F, \mathrm{T}\right) DOS\left(E\right), \qquad (8)$$

$$n_h\left(E_F, \mathrm{T}\right) = \int_{-\infty}^{E_v} dE \left[1 - f\left(E, E_F, \mathrm{T}\right)\right] DOS\left(E\right), \qquad (9)$$

respectively, where DOS is the electron density of states, E_C the conduction band edge, and E_V the valence band edge. Finally, the electron and hole mobilities are, respectively, defined as

$$\mu_e = \sigma / n_e e, \quad \mu_h = \sigma / n_h e. \qquad (10)$$

The electron scattering rate is obtained by summing the inverse partial RTs of all involved scattering processes.

B. Scattering of Electrons at Stacking Faults

The SF scattering model developed for phonons in Ref. [3] is modified for electron scattering at SFs. The only difference is that the phonon wave function is replaced by the electron wave function ($\mathbf{q}$ replaced by $\mathbf{k}$). Only an exponential distribution of SFs is considered here. In this case the relaxation time takes the form

$$\frac{1}{\tau_{sf}(E)} = \frac{l_{sf} \Delta V^2}{\hbar^2 v_x\left(E\right)\left[1 + 4k_x^2 l_{sf}^2\right]}. \qquad (11)$$

The average distance between SFs (l_{sf}) along the NW is assumed to be 2.5 nm. The computation of the electronic relaxation time for SF scattering requires the knowledge of the energy difference between DM in $\langle 111 \rangle$ direction and WZ in $\langle 0001 \rangle$ direction (see Fig. 1). The number of atoms must be the same in both lattice configurations, hence two (three) smallest unit cells of DM (WZ) are used. For a dense $\mathbf{k}$-point sampling in the BTE calculation, the first Brillouin zone is discretized by 401 points in x-direction and 101 points in the other directions.

C. Electron-phonon Scattering on Electrons

Apart from SF scattering, the electronic relaxation time is determined by electron-phonon scattering as shown in Fig. 2. Here, the semi-empirical treatment is used to derive the electron-phonon coupling matrix. The relaxation time approximation from the linearized BTE is defined by the relation

$$\frac{1}{\tau_{nk}} = \frac{2\pi}{\hbar} \sum_{mp} \int \frac{d\mathbf{q}}{\Omega_{BZ}} \left|g_{mnp}\left(\mathbf{k}, \mathbf{q}\right)\right|^2$$
$$\times \left[\left(n_{qp} + f_{mk+q}\right) \delta\left(E_{mk+q} - \left(E_{nk} + \hbar\omega_{qp}\right)\right)\right. \qquad (12)$$
$$\left. + \left(1 + n_{qp} - f_{mk+q}\right) \delta\left(E_{mk+q} - \left(E_{nk} - \hbar\omega_{qp}\right)\right)\right]$$

The right-hand side presents the modification of the distribution function arising from electron-phonon scattering in and out of the state $|n\mathbf{k}\rangle$, by emission or absorption of phonons with frequency ω_{qp}, and branch index p. n_{qp} is the Bose-Einstein distribution function. The matrix elements $g_{mnp}(\mathbf{k}, \mathbf{q})$ are the probability amplitude for scattering from an initial electronic state $|n\mathbf{k}\rangle$ into a final state $|n\mathbf{k} + \mathbf{q}\rangle$ by a phonon $|\mathbf{q}p\rangle$.

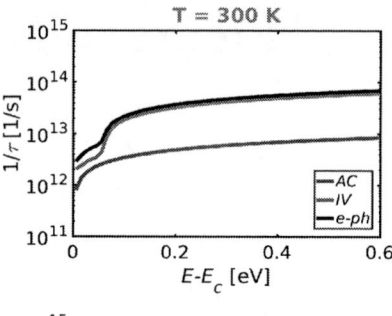

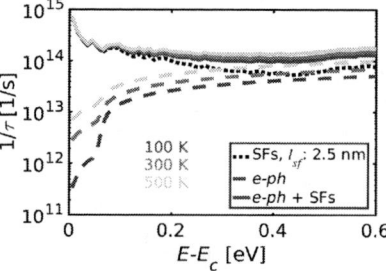

Figure 2: Energetic scattering rate of electrons from different scattering mechanisms. Top panel: The contributions of the electron-phonon (e-ph) scattering from acoustic phonon (AC) and intervalley (IV) scatterings at room temperature. Bottom panel: e-ph scattering and SF scattering with l_{sf} equal to 2.5 nm, at different temperatures. The energy zero is the conduction band edge (E_c).

The coupling of carriers with the lattice vibrations is described by the deformation potential interaction. The deformation potential involves only the short-range interaction between electrons and long-wavelength phonons. All deformation potential parameters for Si can be found in Ref. [9]. The electron-phonon coupling constants are given by the expressions

$$g_p\left(\mathbf{q}\right) = \begin{cases} \sqrt{\dfrac{\hbar}{2\Omega\rho\omega_{pq}}} D_A |\mathbf{q}|, & AC \\[3mm] \sqrt{\dfrac{\hbar}{2\Omega\rho\omega_{pq}}} D_O, & OP \end{cases}, \qquad (13)$$

where D_A is the acoustic (AC) deformation potential, D_O is the constant zero-order optical (OP) deformation potential and ρ is the mass density of the material. By inserting Eq. (13) into Eq. (12) with the use of a simple parabolic band model, the rates for scattering at acoustic and optical phonons, respectively, read

$$\frac{1}{\tau_{AC}} = \frac{2^{1/2} D_A^2 m_{eff}^{3/2} k_B T E_k^{1/2}}{\pi \hbar^4 v_{AC}}, \qquad (14)$$

$$\frac{1}{\tau_{OP}} = \frac{D_O^2 m_{eff}^{3/2}}{2^{1/2} \pi \hbar^3 \omega_{OP} \rho} \left[\left(n_q + \frac{1}{2} \mp \frac{1}{2}\right) \cdot \left(E_k \pm \hbar\omega_{OP}\right)^{1/2}\right], \qquad (15)$$

where v_{AC} is the acoustic phonon group velocity (sound velocity) and m_{eff} is the density-of-states effective mass which is equal to $(m_t^2 m_l)^{1/3}$ (m_t is the transverse effective mass and m_l is the longitudinal effective mass).

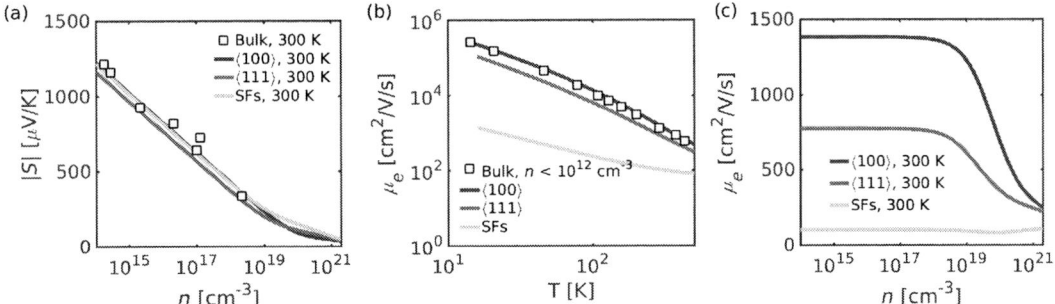

Figure 3: (a) Absolute value of Seebeck coefficient as function of electron concentration (n) at room temperature. The blue (red) line corresponds to a cubic unit cell of DM oriented in $\langle 100 \rangle$ ($\langle 111 \rangle$) direction, respectively. Symbols represent the experimental data of Ref. [11]. (b) Electron mobility as function of temperature at an electron concentration of 10^{12} cm^{-3}. Symbols are experimental data from Ref. [12]. (c) Electron mobility as function of electron concentration at 300 K.

The equi-energy surfaces of Si have several valleys. Therefore, scattering between valleys (intervalley scattering) can occur. For Si, the valleys are equivalent near the zone boundary along $\langle 100 \rangle$ directions. There are two types of intervalley scattering. (i) The g-type processes scatter a carrier from a valley to the opposite one. (ii) The f-type processes scatter a carrier into one of the remaining valleys. The scattering rate of intervalley scattering can be written as

$$\frac{1}{\tau_{IV}} = \frac{D_{if}^2 m_{eff}^{3/2} Z_f}{2^{1/2} \pi \hbar^3 \omega_{if} \rho} \left[\left(n_q + \frac{1}{2} \mp \frac{1}{2} \right) \cdot \left(E_k \pm \hbar \omega_{if} - \Delta E_{if} \right)^{1/2} \right].$$
(16)

where D_{if} is the intervalley deformation potential which characterizes the strength of the scattering from the initial valley i to the final valley f, Z_f is the number of final valleys, ΔE_{if} is the difference between the bottom of the conduction bands in the final and the initial valleys.

SR scattering and all kinds of Coulomb scattering are ignored. The electron density is treated as a parameter. The possible impact of SR scattering would depend on the surface field that emerges as the electrostatic consequence of doping, interface charges, and fixed oxide charges. A doping level of $\sim 1 \times 10^{20}$ cm^{-3} as chosen in Ref. [10] generates flat-band conditions and makes SR scattering a second-order effect in bulk-like NWs.

III. RESULTS

Near the conduction band edge, the energetic rate of SF scattering strongly dominates over the rate of electron-phonon scattering (Fig. 2). The latter becomes stronger when the energy increases. The figure of merit ZT is the product of PF and average temperature between two contacts, divided by the sum of lattice and electron thermal conductivities. The PF is given by the square of the Seebeck coefficient (S) multiplied by the electrical conductivity. As shown in Fig. 3(a), the calculated S of the DM structure perfectly fits the experimental data. As $|S|$ is the conductance-averaged energy difference $|E - E_F|$, it decreases with increasing density. The electron mobility of the DM structure oriented in $\langle 100 \rangle$-direction matches the measured bulk electron mobility for negligible doping concentration ($< 10^{12}$ cm^{-3}) as can be seen in Fig. 3(b). However, the mobility of the DM structure oriented in $\langle 111 \rangle$-direction has smaller values as function of both temperature and electron concentration (see Fig. 3(c)). The structure with SFs exhibits a clear reduction of the electron mobility compared to the two DM structures.

Calculations of the lattice thermal conductivity of bulk Si and Si NWs are reproduced from Ref. [2] in Fig. 4. As in the case of InAs NWs [3], comparable reductions of the thermal conductivity of Si NWs can be obtained with SFs instead of SR. Figure 4 also presents the electronic part of the thermal conductivity as function of electron density. Its values increase with the electron concentration. When SFs are introduced into the material, the electron thermal conductivity is reduced significantly from the perfect crystal. Figure 5 shows the PF as function of electron density at 300 K and 500 K. SFs lower the PF significantly over the entire concentration range. The percentage reduction of the PF of Si NWs as the consequence of SFs reduces when the temperature increases. The ZT as function of electron concentration is presented in Fig. 5. As expected, engineering NWs increases the ZT value

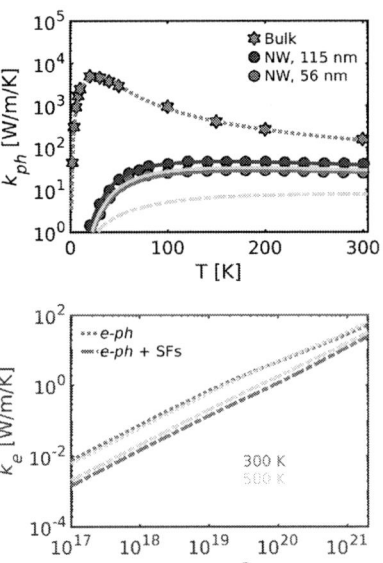

Figure 4: Top panel: Lattice thermal conductivity as function of temperature. Symbols denote the measurements from Refs. [1,13-14]. Lines are simulated thermal conductivities. The dotted line is for the bulk case, whereas solid lines are for the ideal NW case. The dash-dotted line is for a NW with SFs. Blue (green, red) lines are for 120 nm (70 nm, 56 nm) diameter, respectively. Bottom panel: Electron thermal conductivity as function of electron density with temperatures equal to 300 K and 500 K.

compared to the bulk case. The benefit of SFs to the improvement of ZT is clearly observed in the high-density range and at high temperatures. A doubling of the ZT value compared to that of an ideal NW could be achievable.

IV. CONCLUSION

An in-house linearized BTE solver was used to derive the ZT of bulk Si, ideal Si NWs, and Si NWs with SFs based on DFT band structure calculations for electrons and phonons. NWs with SFs have a lower electron mobility and a smaller PF. At high electron concentration, this suppression is reduced by the increasing role of electron-phonon scattering, whereas the lattice thermal conductivity remains the same. This leads to an improved ZT. Our simulations show the possibility of engineering the ZT of Si NWs by the introduction of SFs. This could encourage experimentalists to explore the benefit of such NWs for thermoelectric converters.

ACKNOWLEDGMENT

We acknowledge funding from the Swiss National Science Foundation through SNF under project 149454 (TORNAD).

REFERENCES

[1] A. I. Hochbaum et al., "Enhanced thermoelectric performance of rough silicon nanowires," Nature, vol. 451, no. 7175, pp. 163-167, January 2008.

[2] K. Vuttivorakulchai, M. Luisier, and A. Schenk, "Effect of stacking faults and surface roughness on the thermal conductivity of Si nanowires," IWCN, pp. 23-24, May 2019.

[3] K. Vuttivorakulchai, M. Luisier, and A. Schenk, "Effect of stacking faults and surface roughness on the thermal conductivity of InAs nanowires," J. Appl. Phys., vol. 124, no. 20, pp. 205101, November 2018.

[4] P. Mensch, S. Karg, V. Schmidt, B. Gotsmann, H. Schmid, and H. Riel, "One-dimensional behavior and high thermoelectric power factor in thin indium arsenide nanowires," Appl. Phys. Lett., vol. 106, no. 9, pp. 093101, March 2015.

[5] J. Heyd, G. E. Scuseria, and M. Ernzerhof, "Hybrid functionals based on a screened Coulomb potential," J. Chem. Phys., vol. 118, no. 18, pp. 8207-8215, April 2003.

[6] G. Kresse and J. Furthmuller, "Efficient iterative schemes for ab initio total-energy calculations using a plane-wave basis set," Phys. Rev. B, vol. 54, no. 16, pp. 11169-11186, October 1996.

[7] J. P. Perdew et al., "Restoring the density-gradient expansion for exchange in solids and surfaces," Phys. Rev. Lett., vol. 100, no. 13, pp. 136406, April 2008.

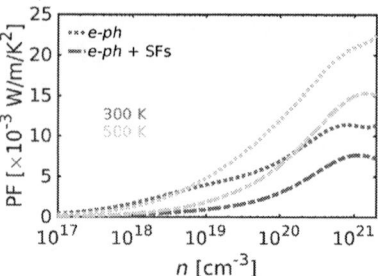

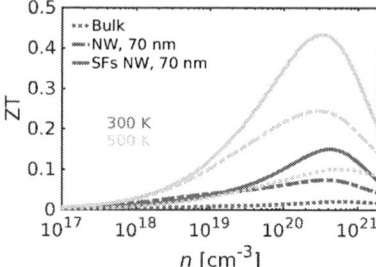

Figure 5: Top panel: Power factor as function of electron density with temperatures equal to 300 K and 500 K. Bottom panel: Thermoelectric figure of merit as function of electron density. The NW diameter is 70 nm.

[8] J. M. Ziman, "Electrons and phonons: The theory of transport phenomena in solids," ser. Oxford Classic Texts in the Physical Sciences. Oxford, New York: Oxford University Press, February 2001.

[9] C. Jacoboni and L. Reggiani, "The Monte Carlo method for the solution of charge transport in semiconductors with applications to covalent materials," Rev. Mod. Phys., vol. 55, no. 3, pp. 645-705, July 1983.

[10] A. I. Boukai et al., "Silicon nanowires as efficient thermoelectric materials," Nature, vol. 451, no. 7175, pp. 168-171, January 2008.

[11] Z. Wang et al., "Thermoelectric transport properties of silicon: Toward an ab initio approach," Phys. Rev. B, vol. 83, no. 20, pp. 205208, May 2011.

[12] C. Jacoboni, C. Canali, G. Ottaviani, and A. Alberigi Quaranta, "A review of some charge transport properties of silicon," Solid State Electron., vol. 20, no. 2, pp. 77-89, February 1977.

[13] C. J. Glassbrenner and G. A. Slack, "Thermal conductivity of silicon and germanium from 3 K to the melting point," Phys. Rev., vol. 134, no. 4A, pp. A1058-A1069, May 1964.

[14] M. G. Holland, "Analysis of lattice thermal conductivity," Phys. Rev., vol. 132, no. 6 pp. 2461-2471, December 1963.

978-1-7281-0941-1/19 $31.00 © 2019 IEEE

Modeling of Temperature-Dependent MOSFET Aging

Fernando Ávila Herrera
HiSIM Research Center
Hiroshima University
Higashihiroshima, Japan
herrera@hiroshima-u.ac.jp

Mitiko Miura-Mattausch
HiSIM Research Center
Hiroshima University
Higashihiroshima, Japan
mmm@hiroshima-u.ac.jp

Hideyuki Kikuchihara
HiSIM Research Center
Hiroshima University
Higashihiroshima, Japan
kikuchihara532@
hiroshima-u.ac.jp

Takahiro Iizuka
HiSIM Research Center
Hiroshima University
Higashihiroshima, Japan
iizuka@hiroshima-u.ac.jp

Hans Jürgen Mattausch
HiSIM Research Center
Hiroshima University
Higashihiroshima, Japan
hjm@hiroshima-u.ac.jp

Hirotaka Takatsuka
Technology Development
Division
Mie Fujitsu Semiconductor
Limited
Yokoyama, Japan
takatuka@jp.fujitsu.com

Abstract—**We have modeled MOSFET-device aging based on the trap-density increase, which is included in the Poisson equation to consider aging explicitly and physically correct. To preserve consistency, the Poisson equation is solved iteratively. Measured temperature dependence of aged *I-V* characteristics are well reproduced with implementation of this aging model into the industry-standard model HiSIM. The extracted physical device quantities with the developed model from measurements have been investigated to characterize the aging features. It is observed that the activation energy E_a as a function of V_{gs} is nearly identical for non-aged and aged devices. This concludes that the temperature dependence of aging originates mostly from the temperature-dependent electrostatic potential, resulting in negligible temperature dependency of extracted trap density N_{trap}. To generalize the conclusion, 2D-device simulation is investigated for a double-gate (DG) MOSFET with increased stress-induced trap density. The same results as obtained from measurements are achieved, namely the activation energy is nearly identical for either non-aged or aged cases. This concludes that the temperature dependence of device aging can be accurately predicted using the temperature-dependent *I-V* characteristics of non-aged device.**

Keywords—*MOSFET aging, Temperature Dependence, Trap-density increase, bulk MOSFET, Double-Gate MOSFET*

I. INTRODUCTION

Stress applied to devices during circuit operation induces a trap-density N_{trap} increase at the insulator/substrate interface, which is the origin of MOSFET aging [1]. Our investigation focuses on nMOSFET aging, where hot carriers are responsible for the N_{trap} increase. Particular focus is given on the temperature-dependent device aging, to determine whether the N_{trap} increases as the temperature increases. It has been shown that the trap-time constant is temperature dependent, reducing with increased temperature [2]. The emission-time constant shows the same characteristic. This is important for circuit simulation. However, investigations for aging of real device after stress duration with different temperatures have not been done yet. Our purpose here is to analyze the measured temperature dependence of aged *I-V* characteristics on the basis of the trap density. The temperature dependence of the trap density will be analyzed,

to determine whether the temperature dependency can be written only by the temperature-dependent physical quantity of the thermal voltage. In this work aged *I-V* characteristics of a leading-edge bulk MOSFET are studied with use of the developed aging model, which considers the trap density explicitly within the Poisson equation. The investigation is further extended to a multi-gate technology to verify whether the obtained results are still valid for this advanced generation.

II. ANALISYS OF TEMPERATURE-DEPENDENT AGING

Fig. 1 shows measured I_{ds}-V_{gs} characteristics at low drain voltage (V_{ds}) after enhanced stress duration at different temperatures of 25, 85 and 125ºC. A leading-edge 50nm CMOS technology is applied for the device fabrication. Figs. 1a and b show the transfer characteristics at room temperature in semi-log and linear scale, respectively. Enhanced stress conditions are also applied with different duration time at different temperatures. The device aging is clearly observed in the subthreshold region, while the device characteristics are mostly governed by the carrier scattering in the high field, induced by V_{gs} for the strong inversion condition. Additionally, it is seen that the aging is enhanced for elevated temperatures, as shown in Figs. 1c and d.

The measured I_{ds}-V_{gs} characteristics are studied with the compact model HiSIM, where the trap density is explicitly considered in the Poisson equation as [3, 4]

$$\nabla\phi^2 = -q/\varepsilon_S \cdot \left(p - n + N_D^+ - N_A^- - N_{trap}^- \right) \quad (1)$$

$$N_{trap} = N_0 \exp\left[(E_F - E_C)/E_S \right] \quad (2)$$

$$N_0 = g_C \cdot E_S \cdot kT/E_S \big/ \sin(kT/E_S) \quad (3)$$

$$E_F - E_C = -\left[qV_{ds} - q\phi_S + (E_C - E_V) - E_V - kT\ln(N_V/N_A) \right] \quad (4)$$

where only the acceptor-type trap N_{trap} is considered. Above equations are solved iteratively to obtain a consistent solution of the static potential distribution ϕ. N_{trap} includes the thermal energy and the bandgap, which induces a temperature dependence.

The parameter extraction is performed with HiSIM including the N_{trap} increase due to the applied stress conditions [3]. The stress is modeled as a function of the integrated substrate current, reflecting the stress condition, during the stress duration. In the HiSIM model the temperature dependence is modeled by physical quantities, as listed in Table 1, which are identical with the 2D-device simulation treatment. All measured temperature dependences of the unstressed-device performances are well reproduced, as demonstrated in Fig. 2a for two V_{ds} values. It is further seen that good temperature-dependence reproduction in the subthreshold region is nearly exclusively due to the contributions from thermal voltage and band-gap, while the phonon scattering is responsible for data reproduction under the strong-inversion condition. As verified in Fig. 2 the aged measurements can also be reproduced with the same temperature dependence of the trap density, which is modeled only by the thermal-energy and band-gap differences. Fig. 3 compares the measured transfer characteristics after stress at high temperature for low applied V_{ds} to the model-calculation results. A good agreement for all aged I-V characteristics verifies that the temperature dependence of the trap density can be written only by the thermal energy and band-gap dependences as it has been expected. This simplifies the model development for circuit simulation as well as increases the reliability of the circuit simulation.

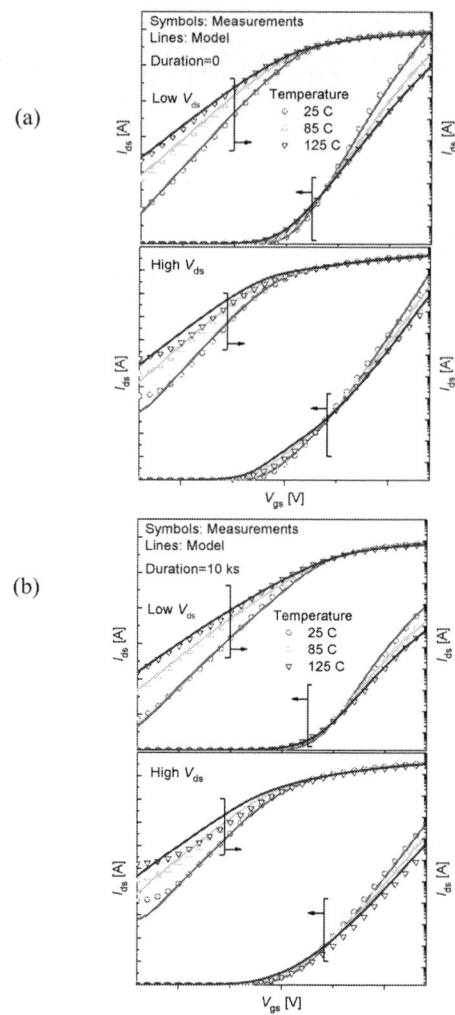

Fig. 2. Comparison of measurements and modeled results for different temperatures of (a) an unstressed device and of (b) a device with degradation time of 10ks at low and high V_{ds}.

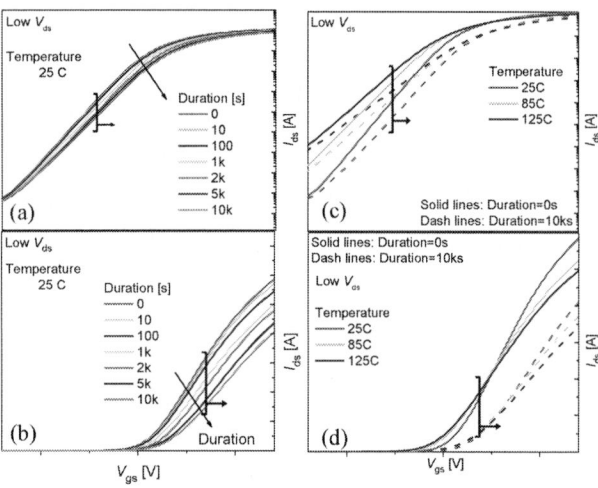

Fig. 1. Measured drain current (I_{ds}) characteristics of the studied device with aging are shown in (a) semilog and (b) linear scale at low V_{ds}. Aged and not aged I_{ds}-V_{gs} measurements for temperatures of 25, 85 and 125C are shown in (c) semilog and (d) linear scale at low V_{ds}.

TABLE I.   TEMPERATURE DEPENDENT PHYSICAL QUANTITIES

Physical Quantity	Value[a]
Thermal voltage	$\beta^{-1} = kT/q$
Band-gap	$E_g = E_{g0} + \textbf{BGTMP1} \cdot X1 + \textbf{BGTMP2} \cdot X2$ $X1 = T - \text{TNOM} ; X2 = T^2 - \text{TNOM}^2$
Intrinsic Carrier concentration	$n_i = \sqrt{N_C \cdot N_V} \cdot \exp(-\beta E_g / 2)$
Velocity saturation	$v_{\text{sat}} = \dfrac{\textbf{VMAX}}{1.8 + 0.4X1 + 0.1X1^2 - \textbf{VTMP}(1 - X1)}$
Phonon Mobility	$\mu_{\text{ph}} = \mu_{\text{ph0}} \cdot X1^{\textbf{MUETMP}}$
Substrate current	$I_{\text{sub}} = I_{\text{sub0}} \cdot (1 + \textbf{SUBTMP} \cdot X1)$

[a.] TNOM is the nominal temperature, N_C and N_V are the densities of states in the conduction and valence band, respectively. E_{g0} is the band-gap value at 0K. μ_{ph0} and I_{sub0} are the phonon mobility and substrate current at nominal temperature, respectively. BGTMP1, BGTMP2, VMAX, VTMP, MUETMP and SUBTMP are model parameters

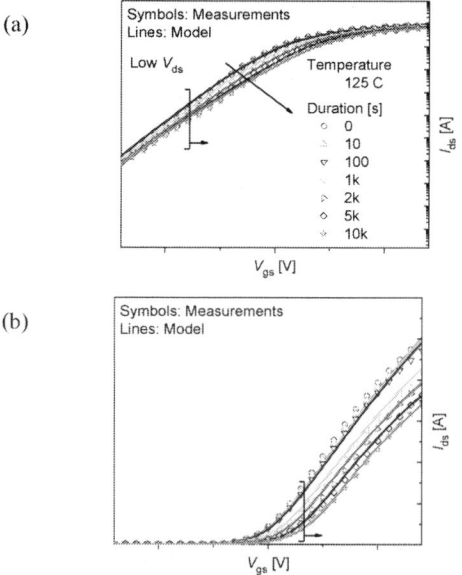

Fig. 3. Comparison of measurements and the best fit of modeled aging results for several degradation times at temperature of 125C for low V_{ds} in (a) semilog and (b) linear scale.

978-1-7281-0941-1/19 $31.00 © 2019 IEEE

III. TEMPERATURE-DEPENDENT TRAP DENSITY

The extracted trap-density increase with measured *I-V* characteristics is investigated in detail here. Fig. 4a shows extracted density of state (DoS) distributions within the bandgap for different stress durations under a constant stress bias at different temperatures. Here N_{trap} is the multiplication of the intercept with conduction-band edge E_C and the inverse of the gradient of the logarithmic DoS plot, as can be seen in Eq. 2 [5, 6]. The gradients of the DoS-curves in Fig. 4a also effectively determine the V_{gs} dependence of N_{trap}. It can be seen that the N_{trap} increase starts to saturate for long stress durations. In spite of the quite drastic DoS increase as a function of the stress magnitude, the stress-temperature dependency remains at a practically negligible level, as can be seen in zoomed Fig. 4b. This temperature dependence is caused by the thermal-voltage and band-gap changes in the extracted results for all stress conditions. The stress estimation, used for modeling the trap-density increase, is done with the substrate current, which is strongly dependent on the temperature. For the DoS extraction this dependency must be ignored to reproduce the measurements. The reason could be that the hot electrons suffer from the very high scattering probability as observed in the drain current. Thus not all of the hot carriers can contribute to the trapping but only lucky carriers. A trap/detrap-time-constant reduction could be also the reason. Anyway, our result concludes that the trapping probability does not increase, in spite of the hot carrier increase.

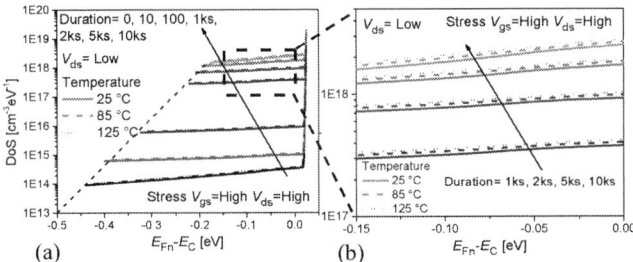

Fig. 4. (a) Density-of State (DoS) increase *vs.* energy difference between the Fermi level and the conduction band, $E_{Fn} - E_C$, for several degradation times and different temperatures at low V_{ds}. (b) Zoom of the selected upper-right range of Fig. 4a.

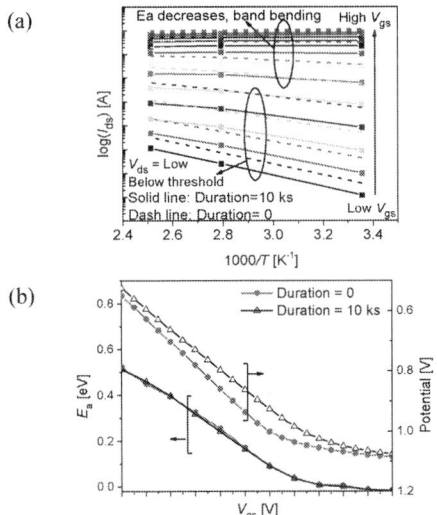

Fig. 5. (a) Temperature dependence of drain current with/without stress for a bulk device. (b) Activation energy as a function of V_{gs} for bulk device and source potential for degradation times of 0 and 10ks.

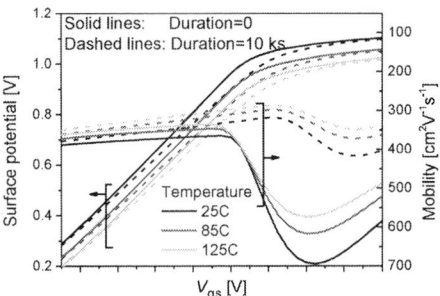

Fig. 6. Surface potential and mobility degradation for different temperatures and for non-aged and aged measurements of a bulk MOS device.

The temperature dependence of the measured drain current was investigated. Fig. 5a compares the Arrhenius plots of non-aged and aged measurements. Only results for the longest stress-duration time are depicted. Fig. 5b compares the extracted activation energies E_a as a function of V_{gs}, showing a reduction of E_a according to the V_{gs} increase. For comparison, the electrostatic surface potentials ϕ_s are depicted together. The drastic reduction of E_a in the subthreshold region is due to the temperature dependency of the carrier density, where the thermal voltage and the band-gap are the origins, since their values determine the charge densities within the Poisson equation, resulting in the temperature dependence of ϕ_s. Above the threshold voltage, E_a decreases smoothly to zero, referring negligible temperature dependence. The reason is that the potential starts to saturate, resulting also in a weaker temperature dependence. On the contrary the phonon scattering of carriers increases, which cause a mobility reduction as shown in Fig. 6. The comparison of activation energies with/without aging in Fig. 5b verifies that E_a is nearly independent of device aging, even though the surface potential characteristic is degraded by aging. Therefore, it can be concluded that the device features observed from the non-aged measurements are preserved in spite of the N_{trap} increase. This result coincides with that obtained by investigating the compact model HiSIM, namely the reduction of the temperature dependence to that induced by the thermal voltage and the band-gap.

IV. DISCUSSION

To confirm the proposed model for temperature dependence of the trap density, 2D-device simulation is investigated for a DG-MOSFET [7]. The studied structure is depicted in Fig. 7. The simulation includes the mobility degradation, the velocity saturation as well as the interface trap density [6]. The DoS distribution within the band-gap is ignored and treated as homogeneous (see Fig. 8) as can be observed in Fig. 4 after aging. The value of the interface state density D_{it} is varied to imitate the different stress conditions. As the compact model for the investigation HiSIM_MG is applied, where the only difference from HiSIM for bulk MOSFETs is that the Poisson equation has to consider the two gate controls. 2D-device simulation results of the I_{ds}-V_{gs} characteristics at low and high V_{ds} are depicted in Figs. 9a and b, respectively. Calculation results with HiSIM-MG are depicted together. These results confirm that the modeling of the trap density is still valid for such an advanced MOSFET generation, where the carriers are controlled by multiple fields within the device.

Fig. 10 shows the E_a characteristics as a function of V_{gs} for non-aged and aged cases with surface trap-densities of 5×10^{11} and 10^{12} cm^{-2}eV^{-1}, where D_{it} of 10^{11} cm^{-2}eV^{-1} is considered as the appropriate value for a fresh device. The same results as obtained for the measurements (see Fig. 5b) are obtained. This verifies again that the temperature-dependent trap density is correctly modeled only by the contributions from thermal energy and band-gap. Further, it confirms that the conventional temperature-dependent models are well applicable for aging investigations.

V. CONCLUSION

The temperature dependence of an aged device can be modeled simply with the temperature dependence of physical quantities as it is usually done. Even though the substrate current and stress-duration time are controlling the induced trap-density, the device characteristics after aging at higher temperature show only a very small temperature effect on the trap-density. The trap densities generated through stress at room temperature give enough accuracy in the calculation of the aged drain-current characteristics. Further the trap-density models for temperature dependence of unstressed devices are reliable for temperature-dependence simulations of the stressed devices. It is also observed that in bulk devices the temperature dependency of the N_{trap} increase is not detectable in the E_a characteristics. Therefore, device aging mainly originates from the aging of the electrostatic device parameters. The flexibility of the developed trap-density-based aging model has been corroborated by extending the model to multi-gate devices, such as double-gate MOSFETs.

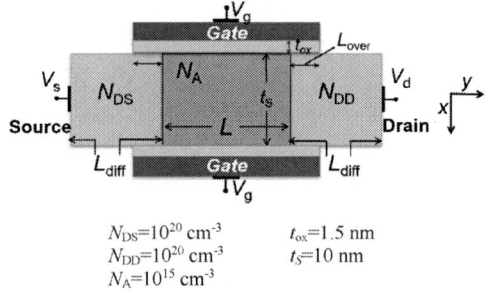

$N_{DS}=10^{20}$ cm^{-3} $t_{ox}=1.5$ nm
$N_{DD}=10^{20}$ cm^{-3} $t_S=10$ nm
$N_A=10^{15}$ cm^{-3}

Fig. 7. Schematic of the studied double-gate MOS structure and the device parameters values.

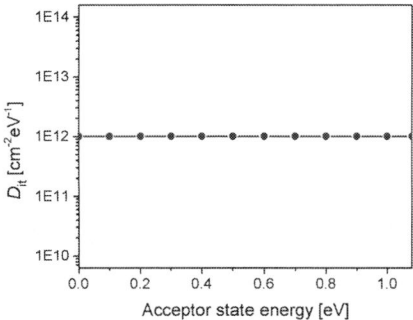

Fig. 8. Assumed homogeneous interface DoS distribution.

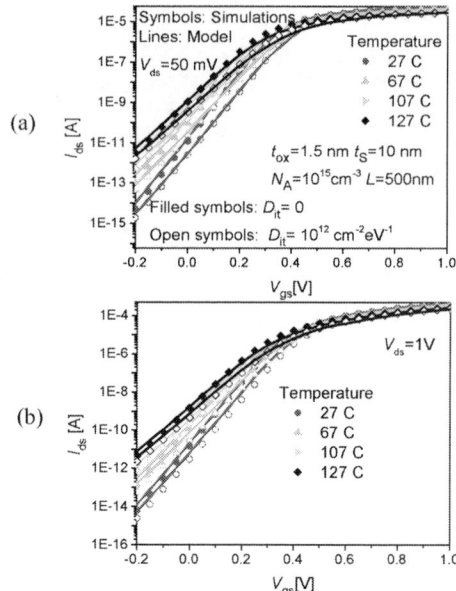

Fig. 9. Drain current characteristics for a DG-MOS considering the trap density distribution, $D_{tA}=10^{12}$cm^{-2}eV^{-1}, for different temperatures at (a) $V_{ds}=50$mV and (b) $V_{ds}=1$V.

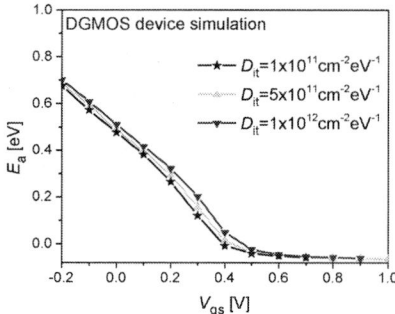

Fig. 10. Activation energy for different trap-density distributions as a function of V_{gs}.

REFERENCES

[1] V. Huard, C. Parthasarathy, A. Bravaix, C. Guerin and E. Pion, "CMOS device design-in reliability approach in advanced nodes," *2009 IEEE International Reliability Physics Symposium*, Montreal, QC, 2009, pp. 624-633, doi: 10.1109/IRPS.2009.5173321.

[2] T. Grasser, P.-J. Wagner, H. Reisinger, Th. Aichinger, G. Pobegen, M. Nelhiebel and B. Kaczer, "Analytic modeling of the bias temperature instability using capture/emission time maps," *2011 International Electron Devices Meeting*, Washington, DC, 2011, pp. 27.4.1-27.4.4, doi: 10.1109/IEDM.2011.6131624.

[3] *HiSIM2 3.1 User's Manual*, 2018.

[4] T. Leroux, "Static and dynamic analysis of amorphous-silicon field-effect transistors," *Solid-State Electronics*, vol. 29, no. 1, pp. 47-58, Jan. 1986, doi: 10.1016/0038-1101(86)90197-8.

[5] H. Tanoue, A. Tanaka, Y. Oodate, T. Nakahagi, D. Sugiyama, C. Ma, H. J. Mattausch and M. Miura-Mattausch, "Compact Modeling of Dynamic MOSFET Degradation Due to Hot-Electrons," *IEEE Trans. Device and Materials Reliability*, vol. 17, no. 1, pp. 52-58, March 2017, doi: 10.1109/TDMR.2017.2655519.

[6] M. Miura-Mattausch, H. Miyamoto, H. Kikuchihara, T. K. Maiti, N. Rohbani, D. Navarro and H. J. Mattausch, "Compact modeling of dynamic trap density for predicting circuit-performance aging," *Solid-State Electronics*, vol. 80, pp. 164-175, Jan. 2018, doi: 10.1016/j.microrel.2017.12.003.

[7] *ATLAS User's Manual*, Silvaco, Inc., Santa Clara, CA, USA, Apr. 2018.

TCAD analysis of FinFET temperature-dependent variability for analog applications

S. Donati Guerrieri[1], F. Bonani[1], G. Ghione[1]

[1]Dipartimento di Elettronica e Telecomunicazioni, Politecnico di Torino
Corso Duca degli Abruzzi, 24, I-10129 Torino, ITALY

Abstract—The Green's Function based TCAD device variability analysis is extended to allow for temperature-dependent variability, with negligible overhead in terms of simulation time with respect to fixed temperature simulations. We provide temperature and bias-dependent 3D variability analysis of the DC current for a FinFET structure from the 22 nm node, showing how to predict and mitigate the effects of poor thermal management. Based on the quasi-stationary assumption, preliminary analysis of self-heating effects of a FinFET medium power amplifier is also presented.

Index Terms—FinFET, Variability, Temperature modeling

I. Introduction

FinFET technology is nowadays well established for digital applications, and its interest for analog RF applications is growing [1]. From the thermal standpoint, though, FinFETs are known to pose significant concern and require accurate modeling, leading to deep investigations of the heat dissipation mechanisms [2]. The inefficient dissipation through the substrate suggests FinFETs can be better described as single fin floating devices, embedded into thermal networks accounting for the heat dissipation through nearby fins and/or metallizations [3]. Modeling electrical FinFET behavior, including variability, as a function of the fin lattice temperature is therefore needed. Variability aware temperature-dependent models are relevant for compact modeling, used e.g. for the concurrent dynamic control of chip temperature, frequency and voltage in digital circuits [4]; and for the optimization of the digital circuit performance under concurrent process, temperature and voltage variations (PVT [5], [6]), where modeling self-heating along with electrical variability is relevant, e.g. for the development of medium power RF stages [1].

In this work we exploit a novel and computationally efficient method to incorporate temperature dependency into Green Function-based (GF) variability analysis [7]–[9], also referred to as the Impedance Field Method (IFM) [10], with negligible numerical overhead with respect to fixed-temperature simulations. The new technique was first introduced in [11], limited to simplified two dimensional double-gate structures. Here we exploit the new approach on a 22 nm FiNFET to further validate the technique and to investigate the sensitivity of the RF performance of a medium power amplifier. We demonstrate that limited temperature dependency is present on a class A-AB bias point, while the knee current is extremely sensitive to self-heating. High bias currents, despite leading to higher gain and better Q factors, need to be carefully evaluated in

terms of variations and self-heating, since they also introduce a temperature dependency in the bias point, which is fully correlated to the maximum AC swing (output power).

II. Linearized Approach to Temperature-Dependent Variability

An effective approach to device variability through TCAD physics-based simulations can be carried out via a linearized approach, exploiting the Green's Functions (GFs) of the lienarized physical model, as described in [8], [9]. The same approach is also implemented in Sentaurus Synopsys, improperly referenced to as the Impedance Field Method (IFM), although limited to the variability analysis of the DC device performances only. In particular, the so called *statistical IMF* allows for statistical analysis, while for deterministic variations, the `ParameterVariation` command is explicitly dedicated to user-defined parameter variations. In this work, we propose to exploit this Sentaurus feature to investigate deterministic variations of the lattice temperature T_L (details can be found in [11]) in conjunction with the statistical technological variability, finally allowing for a *temperature-dependent variability analysis*. Based on the treatment of [8], when the device is subject to the variation of any parameter P with respect to a nominal value P_0, the device current I_k at each terminal k will undergo a variation which can be expressed as

$$\Delta I_k^P = \int_\Omega G_k(x_0, P_0)\, S_P(x_0, P)\, \mathrm{d}V$$

where the integral extends over the device volume Ω, x_0 collectively denotes the physical model solution with nominal parameters, G_k is the GF related to terminal k and the source term $S_P(x_0, P)$ accounts for the parameter variation ΔP. It can be calculated by the derivative of the physical model equations $F(x)$ with respect to P or in an approximated way by the same physical model calculated with the nominal solution but *varied parameter*, i.e.

$$S_P(x_0, P) = F(x_0, P) - F(x_0, P_0) = F(x_0, P)$$

since $F(x_0, P_0)$ is null by definition, being the residual of the physical model calculated in the nominal solution. Notice that the above source is accurate only for small variation of P and it is a linear (first order) perturbation in ΔP. We now add the lattice temperature dependency, extending the above formulas as

$$\Delta I_k^P = \int_\Omega G_k(x_0, P_0, T)\, S_P(x_0, P, T)\, \mathrm{d}V \qquad (1)$$

We refer to this approach as Method 1 (M1). M1 requires the solution of the physical model and GFs for each temperature, with a linear increase of simulation time with the number of temperatures for which we require the analysis. The computational overhead for the source term $S_P(x_0, P, T)$ is instead negligible. Finally the overall current reads

$$I_k = I_k(T, P_0) + \Delta I_k^P$$

To further reduce the simulation time, though, we follow [11] to linearize the device current with respect to the temperature,

$$I_k(T, P_0) = I_k(T_0, P_0) + \Delta I_k^T$$

where

$$\Delta I_k^T = \int_\Omega G_k(x_0, P_0, T_0) \, S_T(x_0, P_0, T) \, \mathrm{d}V$$

where the source term is

$$S_T(x_0, P_0, T) = F(x_0, P_0, T) - F(x_0, P_0, T_0) = F(x_0, P_0, T)$$

Here the required solution of the physical model is only one (at nominal temperature and nominal parameter). Collecting all the variations

$$I_k(T, P) = I_k(T_0, P_0) + \Delta I_k^T + \Delta I_k^P$$

still requires the solution for ΔI_k^P at multiple temperatures. At first order with respect to the concurrent variations of P and T, though, we approximate (1) as:

$$\Delta I_k^P = \int_\Omega G_k(x_0, P_0, T_0) \, S_P(x_0, P, T_0) \, \mathrm{d}V \qquad (2)$$

hereafter denoted by Method 2 (M2) and sketched in Fig. 1, again requiring the solution only with nominal parameters and at a single temperature. To assess the consistency of this *double linearization* and to which extent neglecting second order terms limits the model accuracy, one should verify whether both the GFs (G_k) and the source term S_P show a weak T dependency. Notice, however, that the source term S_P is not directly available in Synopsys, while the GFs can be extracted. To this aim, we consider a FinFET device from the 22 nm technology node, including all the typical technology features, such as the raised source and drain extensions, a high-k dielectric and the fin oxide side-walls. The structure is taken from the Synopsys Library [12] and the device geometry is presented in Fig. 2. The device is simulated taking into account accurate physical models, including strain, interface traps and quantum effects through the Density Gradient approach (see [12] for details). Fig. 2 shows the behaviour of the drain terminal GF at room temperature $T_0 = 300$ K. To assess the T dependency, the same was calculated at $T = 320$ K: Figs. 3 and 4 show the nominal value and the variations of the GFs for the Poisson and electron current continuity equations as a function of the position along the channel, in the mid-fin point and at a different depth in the fin. Is evident that the GFs variations are extremely limited, always at maximum equal to 5% of the nominal value, thus validating the approximation in (2). The same trend was observed moving the cut-line C1 away from

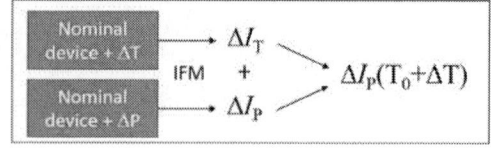

Fig. 1. Schematic representation of the proposed approach.

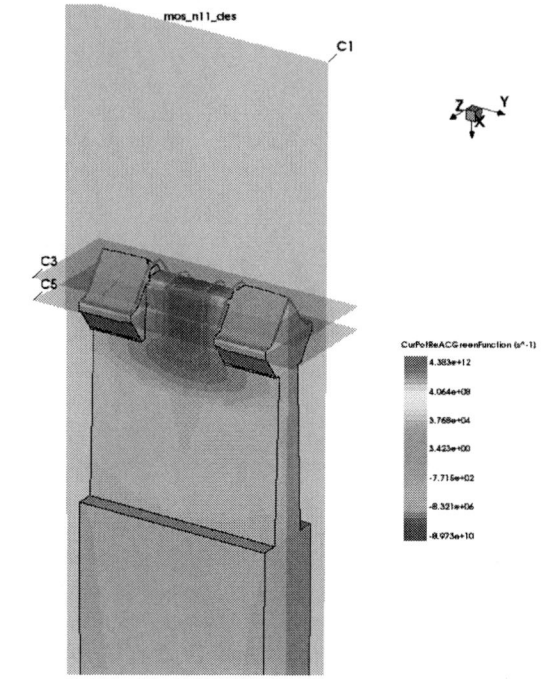

Fig. 2. 3D structure of the simulated 22 nm FinFET (after [12]). Oxide layers are not included in the figure, to allow for internal quantity inspection. In particular the plot shows the drain Green's Function of the Poisson equation at room temperature and the cut-lines used for the T dependent analysis.

the fin mid-point, i.e. towards the side gates. Concerning the source S_P, albeit it can't be directly inspected, notice that, even when the approximation of (2) is not verified, it may still be evaluated without approximations at the varied temperature T, still with negligible numerical overhead, but it would require *ad hoc* tools from Synopsys or in-house tools like [8].

III. 22 NM FINFET MEDIUM POWER AMPLIFIER VARIABILITY

To fix the ideas, we refer to the standard design of a tuned load power amplifier (PA), whose schematic is shown in Fig. 5: the bias point and the load line are selected in order to find the best compromise in terms of output power, efficiency, gain and linearity. Therefore the bias point can be chosen at higher or lower gate voltages (drain currents) for a selected drain bias in saturation, while the maximum output power will be determined by the amount of current at the knee voltage. It is therefore significant to select these two operating conditions of the device (linear at the knee or saturation) and investigate how variability and self-heating affect the overall PA performance.

978-1-7281-0941-1/19 $31.00 © 2019 IEEE

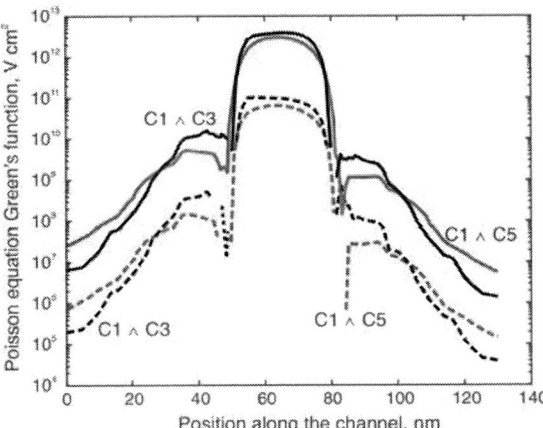

Fig. 3. Poisson drain GF at 300 K and its variation for $T = 320$ K. The cutlines are at the intersection of C1 and C3 and C1 and C5 shown in Fig. 2.

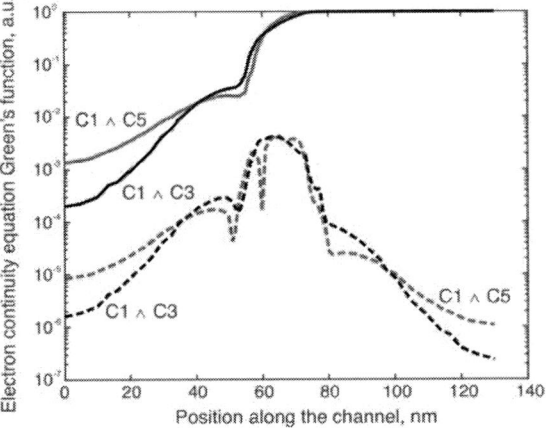

Fig. 4. Electron continuity equation drain GF at 300 K and its variation for $T = 320$ K. The cutlines are the same as Fig. 3.

We apply this concept to the analysis of a possible power amplifier (PA) built on the 22 nm FinFET analysed in the previous section. In this device, we limit the supply voltage to 1.1 V. Figs. 6–9 show the statistical distribution of the FinFET DC drain current above threshold for various values of the gate voltage. In particular, Figs. 6 and 7 show the linear region ($V_D = 0.1$ V), to investigate the effect of the PA knee voltage. On the other hand, the saturation region (here, $V_D = 0.6$ V), shown in Figs. 8 and 9, is important for the bias condition, usually at lower/intermediate gate voltages, e.g. $V_G = 0.6 \div 0.8$ V (Fig. 5). The variability analysis addresses concurrent deterministic lattice temperature variations and random gate workfunction variations (WFV, average metal grain size 5 nm) and random doping fluctuations (RDF). These can be regarded as the main sources of variability [12]. The proposed approach M2 based on the *double linearization*, was compared to M1 showing always a good agreement, up to 50 K of heating, with a remarkable reduction of simulation time (about 80% less). Turning to the detailed analysis of

variations, in all operating conditions the drain current temperature dependency has opposite trends as a function of the gate voltage: for gate bias below 0.6 V, where the carrier exponential temperature dependency dominates, the current is increasing with temperature while above 0.6 V, where mobility degradation with T dominates, it decreases. Notice also that the RDF spread is always higher than WFV, at all temperatures. In the linear region (Figs. 6 and 7), for lower gate voltage (just above the threshold, here 0.25 V), the overall spread is lower and the T-dependency very weak. At intermediate gate bias (around $V_G = 0.6$ V), the spread is larger (maximum gaussian variance), but nearly insensitive to T (i.e. device self-heating). For large gate bias (amplifier maximum current and power), the current becomes more T dependent, even if the technological spread is reduced: in this case T severely affects the device knee even for a moderate temperature increase. Summarizing, the temperature sensitivity is increasing with V_G, while the gate workfunction and RDF variability is decreasing with V_G: since the knee voltage is usually exploited in a power amplifier for larger current values (see Fig. 5), we conclude that the knee voltage variations are usually dominated by self-heating. Turning to Figs. 8 and 9, at intermediate gate voltages, usually exploited for the class A PA design, the temperature sensitivity is nearly null, but the overall technological spread is always significant, unless the operating condition is chosen very close to the threshold voltage (class B). On the contrary, even higher gate voltages could be exploited for gain and bandwidth enhancement but would introduce T variability in the bias current, which would add to the knee variability, resulting in even more perturbations of the overall AC voltage swing.

A reasonable estimate of self-heating for this PA can be made supposing for the PA periphery 10 fingers of 30 fins with a thermal resistance $R_{TH} = 3$ K/mW [13]. When $V_G = 0.76$ V, we obtain 15 mA overall current (~ 0.05 mA/fin), and a DC power consumption $P_{DC} \approx 11$ mW. This is completely dissipated to the thermal sink if the PA is not delivering any power (or in back-off), causing a temperature increase of 33 K, while at peak power and 50% efficiency this amount is halved. The predicted self-heating is therefore well within the temperature range for which the analysis has been carried out.

IV. CONCLUSIONS

We have presented a numerically efficient approach for the temperature dependent variability analysis of electron devices, carefully exploiting TCAD simulations and the Green's Functions of the linearized physical model. This approach is seamlessly integrated with the well-known Impedance Field Methos used e.g. in Synopsys Sentaurus to account for device technological variations. This approach can be successfully applied in a variety of circuits where the device self-heating and variability impact the overall circuit performance. In particular, a preliminary analysis of a power amplifier based on the 22 nm FinFET technology as a function of the WF and RDF variations was successfully carried out.

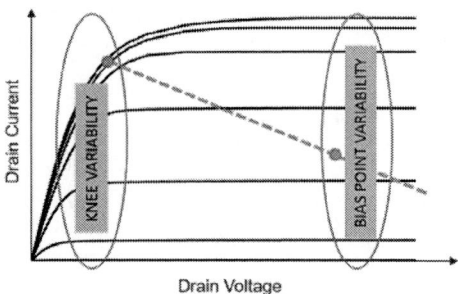

Fig. 5. Investigation of power amplifier variations as a function of current variations at the knee and saturation operating conditions.

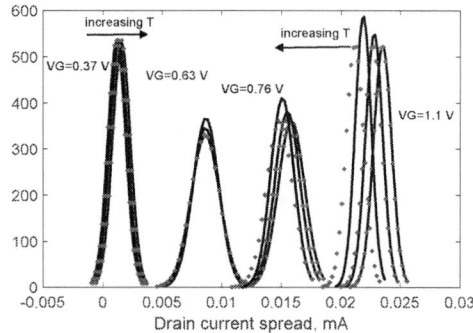

Fig. 6. WFV probability density function of the drain current in linear bias ($V_D = 0.1$ V). T=300, 320 and 350 K. Black lines: M1; Red symbols: M2.

REFERENCES

[1] S. Callender,, et. al., "FinFET for mm Wave - Technology and Circuit Design Challenges", *Proc. BCICTS 2018*, pp. 168–173, 15-17 Oct. 2018, San Diego, CA, USA.

[2] J. Jeon, et. al., "Investigation of Electrothermal Behaviors of 5-nm Bulk FinFET", *IEEE Trans. El. Dev.*, Volume: 64, No: 12, pp. 5284–5287, Dec. 2017.

[3] B. Swahn, et. al., "Electro-Thermal Analysis of Multi-Fin Devices", *IEEE Trans. VLSI*, Volume: 16, No: 7, pp. 816–829, July 2008.

[4] W. Lee, et. al., "Dynamic Thermal Management for FinFET-Based Circuits Exploiting the Temperature Effect Inversion Phenomenon", *Proc. ISLPED 2014*, Aug. 2014, La Jolla, CA, USA.

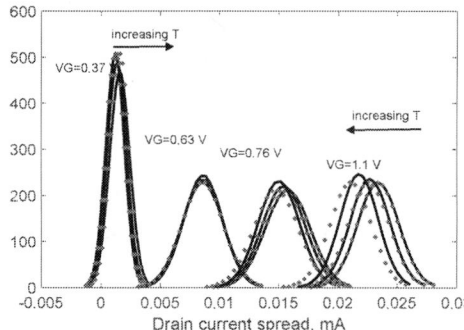

Fig. 7. RDF probability density function of the drain current in linear bias ($V_D = 0.1$ V). T=300, 320 and 350 K. Black lines: M1; Red symbols: M2.

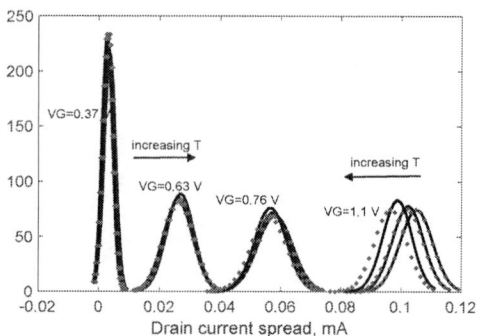

Fig. 8. WFV probability density function of the drain current in saturation ($V_D = 0.6$ V, possible power amplifier bias). Temperatures of T=300, 320 and 350 K. Black lines: M1; Red symbols: M2.

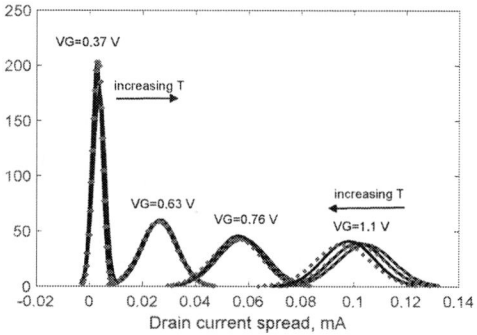

Fig. 9. RDF probability density function of the drain current in saturation ($V_D = 0.6$ V, possible power amplifier bias). Temperatures of T=300, 320 and 350 K. Black lines: M1; Red symbols: M2.

[5] Y. Yang, et. al., "FinPrin: FinFET Logic Circuit Analysis and Optimization Under PVT Variations", *IEEE Trans. VLSI*, Vol. 22, No: 12, pp. 2462–2475, Dec. 2014.

[6] J. H. Choi, et. al., "The effect of process variation on device temperature in finFET circuits", *Proc. ICCAD 2007*, pp. 747–751, Nov. 2007, San Jose, CA, USA.

[7] F. Bertazzi et. al., "Physics-based SS and SSLS variability assessment of microwave devices through efficient sensitivity analysis", *Proc. INMMIC 2012*, 3-4 Sept. 2012, Dublin, Ireland

[8] S. Donati Guerrieri, et. al., "A Unified Approach to the Sensitivity and Variability Physics-Based Modeling of Semiconductor Devices Operated in Dynamic Conditions. Part I: Large-signal sensitivity", *IEEE Trans. El. Dev.*, Vol. ED-63, No: 3, pp. 1195–1201, March 2016.

[9] S. Donati Guerrieri, et. al., "A Unified Approach to the Sensitivity and Variability Physics-Based Modeling of Semiconductor Devices Operated in Dynamic Conditions. Part II—Small-Signal and Conversion Matrix Sensitivity", *IEEE Trans. El. Dev.*, Vol. ED-63, No: 3, pp. 1202–1208, March 2016.

[10] see online: https://www.synopsys.com/silicon/tcad/device-simulation/sentaurus-device.html

[11] S. Donati Guerrieri, F. Bonani, G. Ghione, "A novel TCAD approach to temperature dependent DC FinFET variability analysis", *2018 European Microwave Integrated Circuits Conference (EuMIC)*, 23-25 Sept. 2018, Madrid, Spain.

[12] Available from TCAD Sentaurus Version O-2018 installation, go to *Applications_Library/Variability/FinFET_Variability_sIFM*.

[13] B. González, et. al., "DC self-heating effects modelling in SOI and bulk FinFETs", *Microelectronics Journal*, Volume 46, Issue 4, April 2015, Pages 320-326

978-1-7281-0941-1/19 $31.00 © 2019 IEEE

Numerical Investigation of the Leakage Current and Blocking Capabilities of High-Power Diodes with Doped DLC Passivation Layers

Luigi Balestra
ARCES and DEI, University of Bologna
Bologna, Italy.
luigi.balestra5@unibo.it

Susanna Reggiani
ARCES and DEI, University of Bologna
Bologna, Italy.
susanna.reggiani@unibo.it

Antonio Gnudi
ARCES and DEI, University of Bologna
Bologna, Italy.
antonio.gnudi@unibo.it

Elena Gnani
ARCES and DEI, University of Bologna
Bologna, Italy.
elena.gnani@unibo.it

Giorgio Baccarani
ARCES and DEI, University of Bologna
Bologna, Italy.
giorgio.baccarani@unibo.it

Jagoda Dobrzynska
ABB Switzerland Ltd. Semiconductors
Lenzburg, Switzerland.
jagoda.dobrzynska@ch.abb.com

Jan Vobecký
ABB Switzerland Ltd. Semiconductors
Lenzburg, Switzerland.
jan.vobecky@ch.abb.com

Abstract— **Diamond-like carbon (DLC) is a very attractive material for Microelectronics, as it can be used to create robust passivation layers in semiconductor devices. In this work, the modelling of DLC in a TCAD framework is addressed, with special attention to the role played as the bevel coating of large-area high-voltage diodes. The TCAD simulations are nicely compared with experiments, giving rise to a detailed explanation of the role played by the DLC conductivity on the diode performance.**

Keywords—TCAD modeling, Diamond-Like Carbon simulation, bevel termination, large-area diode.

I. INTRODUCTION

Every high-voltage device requires an optimized junction termination to reach a stable blocking with minimal consumption of wafer periphery area. To this purpose, state-of-the-art beveling is applied to discrete devices [1]. The blocking stability is then dependent on the electrical strength and conductivity of the surface passivation material, as it can be exposed to a significant leakage current and electric field. Understanding of basic physical principles of the DLC, including the ways to control the surface electric field and hereby the breakdown voltage, has been recently improved thanks to TCAD simulation [2]. As the DLC is in a direct contact with the silicon bevel, understanding the complex DLC transport through the Si/DLC interface supports an optimal design of junction termination and deserves further improvement as presented in this paper.

II. MODELING OF THE DLC MATERIAL IN THE TCAD SETUP

The metal-DLC-Si structure reported in Fig. 1 has been used to study the DLC transport properties. In [3], a TCAD setup for DLC layers was proposed identifying the most relevant physical effects: DLC is an amorphous material which can be modelled by using the drift-diffusion (DD) transport equation with a Poole-Frenkel-like hopping mobility and the first-order Debye equation of the ferroelectric model giving the polarization effect. As far as

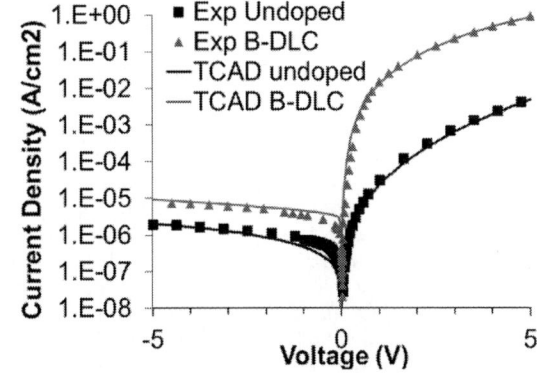

Fig. 1. Schematic view of the cross-section of a Metal-DLC-Si device. TCAD simulations were carried out in cylindrical coordinates in order to predict the charge spreading in the Si substrate. Dot-dashed line: symmetry axis. Variations of the DLC properties are expected at the Si interface, thus an interlayer (DLC1) and a top-side passivation (DLC2) have been realized in the TCAD setup. The structure is not in scale.

Fig. 2. Simulated current density characteristics of n-type MIS structures with undoped and Boron-doped DLC compared with experiments. Symbols: experiments. Solid lines: TCAD results.

This work was supported by ABB Semiconductors, Switzerland.

978-1-7281-0941-1/19 $31.00 © 2019 IEEE

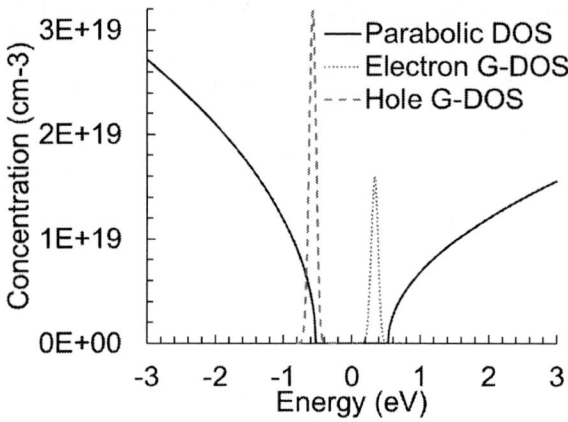

Fig. 3. Asymmetric band structure (solid line) as for undoped DLC. The disorder due to Boron doping was accounted for with two Gaussian Density of States (G-DOS) (green and red dashed lines).

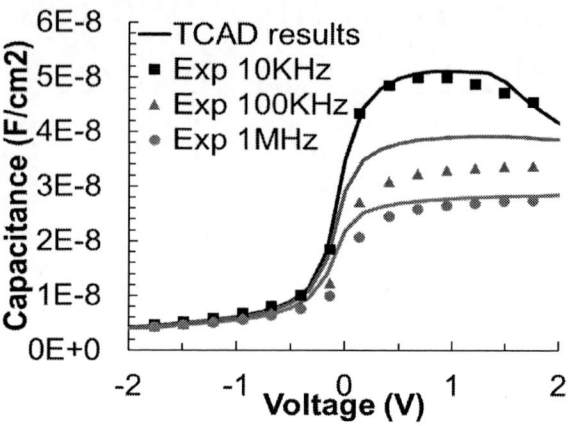

Fig. 5. Simulated C-V curves of the n-type MIS device with undoped DLC on top compared with experiments at different frequencies.

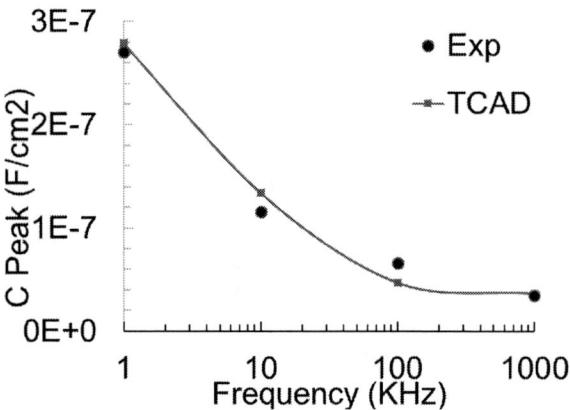

Fig. 4 Simulated capacitance peak value extracted from the C-V curves of a n-type MIS device with Boron-doped DLC on top compared with experiments. The polarization model nicely reproduces the frequency dependence. Solid line: TCAD results. Symbols: experiments.

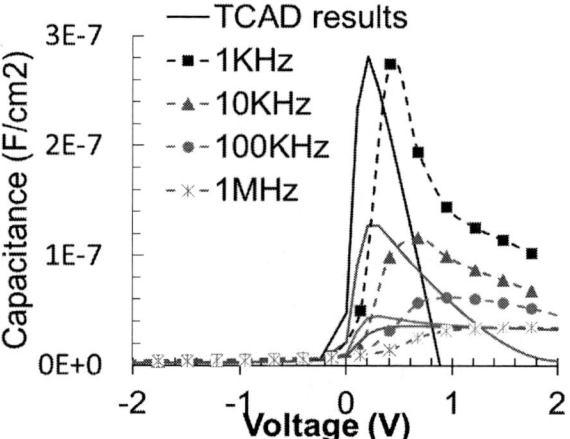

Fig. 6. Simulated C-V curves of the n-type MIS device with Boron-doped DLC on top compared with experiments at different frequencies. TCAD nicely reproduces experiments except for a small shift of threshold voltage.

the band structure adopted for undoped DLC in [3], it is asymmetric and was fitted against DFT calculations [4] and analytical formulations of the diamond-like σ and graphene-like π bands used to describe the optical response of similar DLC layers [5]. The role of Nitrogen doping in the DLC was addressed in [2], where an amount of traps in the bandgap was used to model it. In this work, a Boron doping is addressed, which is known to give a larger conductivity but with no clear evidence of a p-type behavior: no threshold-voltage shift is experimentally observed in the current characteristics (Fig. 2). The same band structure as for the undoped and N-DLC can be used, as all DLC layers showed a quite constant optical band gap of about 1 eV given by the presence of a π band (Fig. 3). In agreement with the theoretical explanations in [6], we assumed that the acceptor level lies within the modified π band and gives rise to modified band edges. In order to correctly model it, the Gaussian density-of-states (G-DOS) model for disordered organic semiconductors available in the TCAD tool has been used [7].

The dielectric response of complex materials is always dependent on the frequency of the applied signal. Polar molecules tend to reorient under the influence of an external electric field, thus contributing to the polarization. Free charges inside a heterogeneous material can be blocked by interfaces inside the material, also causing dielectric relaxation. Mixtures of molecules with different properties, as those giving rise to the σ and π bands, can be a cause of such effects. In accordance with the latter consideration, the dielectric polarization vector P induced by the DLC disorder has been calibrated against C-V experiments carried out at different frequencies. To this purpose, the single-time-constant response given by the first-order Debye equation, available in the ferroelectric model of the TCAD tool, has been used [7]:

$$\frac{dP}{dt} = \frac{P_0 - P}{\tau},$$

where P is the polarization vector, τ the relaxation time and P_0 the polarization vector induced by the electric field through the dielectric relaxation strength. In particular, $P_0 = (\varepsilon_s - \varepsilon_\infty) E$, with ε_s the static dielectric constant, ε_∞ the dielectric constant at high frequencies and E the electric field. The TCAD parameters (τ and P_0) have been fitted

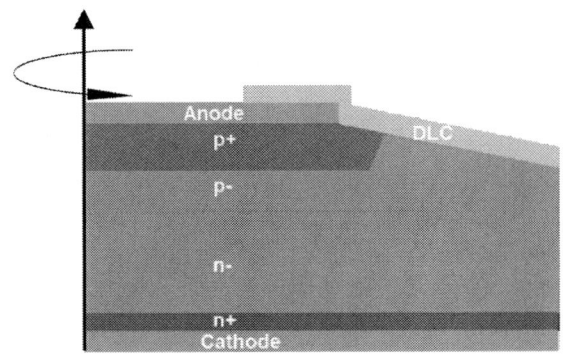

Fig. 7. Schematic view of the cross-section of the investigated large-area 4.5 kV diode with cylindrical symmetry. The structure is not in scale.

against the peak values of C vs. frequency (Fig. 4). A larger P_0 has been fixed for the Boron-doped DLC with respect to the undoped one in order to fit the larger capacitance peak values: the latter difference can be ascribed to the expected larger disorder induced by the doping effect. No relevant different among the different DLCs has been found on the value of the relaxation time.

The TCAD model has been finally validated against the C-V curves at different frequencies (Figs. 5 and 6). The reported curves nicely predict the depletion condition of the Si substrate in reverse regime and the significant dependence on frequency in the forward regime. A slightly anticipated threshold voltage was found in the C-V curves of the doped-DLC case, which is probably due to the proposed energetic distribution of the Gaussian DOS and its interaction with the Schottky contact at the top metal. A finer tuning of the energetic G-DOS distribution should be necessary to improve the fitting.

As far as the DLC characteristics at the Silicon interface, relevant variations of the transport properties are expected in the first interface layer with respect to the bulk DLC. A difference of about a factor 100 was found between the conductivity extracted from the J-V measurements on MIS devices and the data measured from lateral structures (with contacts on top of the DLC), independently of the doping content. In order to consistently model it in the TCAD setup, the DLC was simulated as a two-layer system as schematically illustrated in Fig. 1. A similar model was proposed in [8] for the interpretation of the optical properties of Boron-doped DLCs, which led to the extraction of different interlayer and doped DLC thicknesses. In the proposed TCAD setup, a first interface layer of 70 nm DLC close to the silicon substrate was modelled with a limited conductivity (labelled as "DLC1" in Fig. 1), followed by a thicker DLC bulk with total thicknesses of 190-300 nm depending on the deposition times ("DLC2").

III. SIMULATIONS OF THE HIGH-POWER DIODES

The schematic view of the simulated diode is reported in Fig. 7. The structure is cylindrical; thus transport equations are solved under cylindrical coordinates to obtain full 3D results. The total diameter of the device is about 9 cm. Doping profiles were measured from the spreading resistance profiling. A high carrier lifetime was fitted against the reverse leakage current curves at different temperatures. The

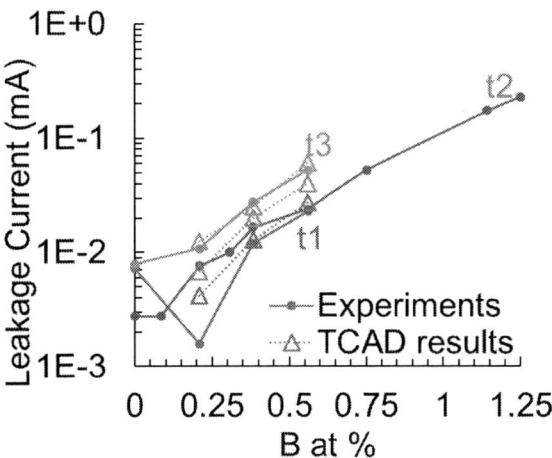

Fig. 8. Simulated leakage currents versus Boron atomic concentration for three different thicknesses of the DLC passivation layer compared with experiments.

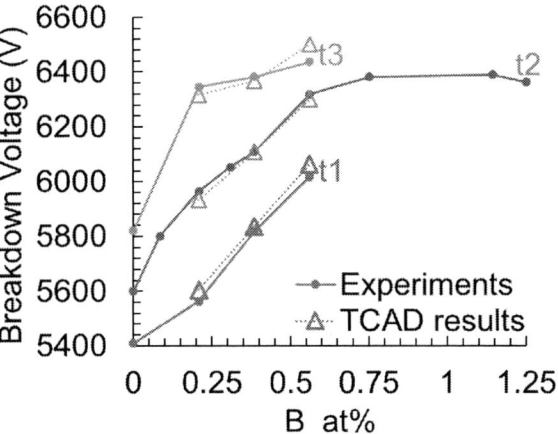

Fig. 9. Simulated leakage currents versus Boron atomic concentration for three different thicknesses of the DLC passivation layer compared with experiments.

Van Overstraeten model for the impact-ionization generation has been used with default parameters. The termination region is realized with a negative bevel passivated by a DLC layer. The DLC structure has been modelled using the same two-layer structure as in Fig. 1 and the TCAD setup used for the metal-DLC-Si devices as described in the previous Section. An additional encapsulation material is realized on top of the diode periphery to set required creepage distance between anode and cathode: it has been realized in the TCAD setup as an ideal insulating material (not drawn in Fig. 7).

Figs. 8 and 9 summarize the simulation results on the diodes, showing I_{OFF} at 4 kV and V_{BD} at 15 mA as functions of the DLC doping for different passivation thicknesses. The DLC thickness has been changed by adopting three different deposition times, with time $t_2 = 2t_1$ and $t_3 = 3t_1$. Thus, thicknesses equal to t_{DLC}, $2t_{DLC}$, $3t_{DLC}$ are obtained, with t_{DLC} the reference thickness of the layer ranging from about 190 to 300 nm. The latter values depend on the doping

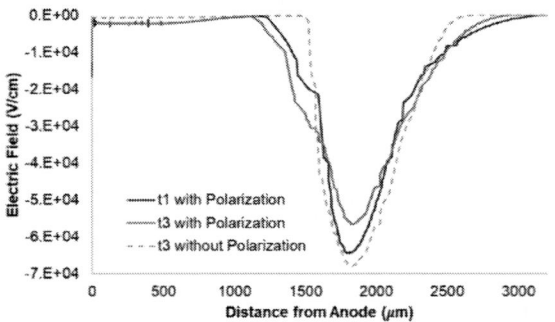

Fig. 10. Simulated electric field in Silicon along the bevel vs. distance from anode edge. A Boron doping concentration of 0.4at% is used. Black line: DLC thickness fixed to t_1. Red line: DLC thickness fixed to t_3. Dashed line: DLC thickness fixed to t_3 without polarization.

concentration as different chemical reactions are activated by the flow of dopants during the deposition leading to a lower growth rate for doped DLCs.

Measurements and TCAD simulations have been carried out at room temperature. In this specific case, we compared the DLC experiments with numerical simulations carried out by assuming that different doping doses simply affect the peak of the Gaussian DOS. The predicted I_{OFF} is in nice agreement with experiments, clearly showing the role of the DLC layer on the leakage current at the periphery of the diode. A clear significant dependence of I_{OFF} on the doping was found, while similar I_{OFF} values are measured with different DLC thicknesses. The latter effect might be mostly due to the coupling of the silicon interface with the DLC layer, leading to a significant variation of the depleted region at the bevel and consequent increase of the leakage current. The V_{BD} TCAD results are in good quantitative agreement with experiments as well, showing also a clear dependence on the adopted thickness of the DLC: the onset of avalanche in silicon takes place below the surface along the bevel, thus the conductivity of DLC provides an electrostatic effect similar to the SIPOS field-plate structures, improving V_{BD} when thicker layers and larger doping doses are used. This is clearly visible in Fig. 10, where the electric field profile within Silicon along the bevel is reported at a fixed bias of 4000V, showing the role of the DLC thickness on the critical surface electric field. Due to the presence of the DLC1, the thinner DLC passivation shows a limited charge spreading on top of the bevel, with a less effective reduction of the peak electric field. The effect of a larger thickness is much more significant in spreading the electric field and reducing its peak, thus improving breakdown.

Finally, the effect of polarization on the electrostatics of the diode was checked by simulations. In Fig. 10, the TCAD curve obtained by switching off the P effect is reported, clearly showing an increase of the peak electric field, which causes a reduction of the diode breakdown voltage of about 600 V in the analyzed case.

IV. CONCLUSIONS

In this work, the current-voltage and capacitance-voltage-frequency characteristics of metal-DLC-Si devices have been simulated in the TCAD framework for Boron-doped DLC layers. The Gaussian densities of states, along with the Poole-Frenkel conduction and the frequency dependent polarization, showed to nicely explain the features of the Boron-doped DLC. The TCAD model was finally demonstrated to effectively reproduce the performance of large-area diodes including the passivation layer.

REFERENCES

[1] C.-Y. Wu, Y. Wang and C.-C. Zhu, "Effect of equivalent surface charge density on electrical field of positively beveled p-n junction", J. Shanghai Univ., n. 12, pp. 43-46, 2008.

[2] S. Reggiani, L. Balestra, A. Gnudi, E. Gnani, G. Baccarani, J. Dobrzynska, J. Vobecký and C.Tosi, "TCAD study of DLC coatings for large-area high-power diodes", Microelectronics Reliability 88-90, pp. 1094-1097, 2018.

[3] S. Reggiani, C. Giordano, A. Gnudi, E. Gnani, G. Baccarani, J. Dobrzynska, J. Vobecky , M. Bellini, "TCAD-based investigation on transport properties of Diamond-like carbon coatings for HV-ICs", IEEE IEDM 2016, p. 36.7.1.

[4] M. A. Caro, R. Zoubkoff, O. Lopez-Acevedo, T. Laurila, "Atomic and electronic structure of tetrahedral amorphous carbon surfaces from density functional theory: Properties and simulation strategies", Carbon 77, pp. 1168-1182, 2014.

[5] D. Franta, D. Nečas, L. Zajíčková, V. Buršíková, "Limitations and possible improvements of DLC dielectric response model based on parameterization of density of states", Diamond Relat. Mater. 18, pp. 413–418, 2009.

[6] P. K. Sitch, Th. Kühler, G. Jungnickel, D. Porezag and Th. Frauenheim, "Theoretical Study of Boron and Nitrogen Doping in Tetrahedral Amorphous Carbon", Solid State Communications, Vol. 100, pp. 549-553, 1996.

[7] Synopsys Inc., "Sentaurus Device User Guide, Version M-2016.12", December 2016.

[8] A. A. Ahmad, "Optical and electrical properties of synthesized reactive rf sputter deposited boron-rich and boron-doped diamond-like carbon thin films", J. Mater. Sci.: Mater. Electron. 28, p. 1695, 2017.

978-1-7281-0941-1/19 $31.00 © 2019 IEEE

Influence of Accurate Electron Drift Velocity Modelling on the Electrical Characteristics in GaN-on-Si HEMTs

Korbinian Reiser[*‡], John Twynam[†], Christian Eckl[†], Helmut Brech[†] and Robert Weigel[‡]

[*]Infineon Technologies AG, Regensburg, Germany
Email: Korbinian.Reiser@infineon.com
[†]Infineon Technologies AG, Regensburg, Germany
[‡]Institute for Electronics Engineering
Friedrich-Alexander-Universitaet Erlangen-Nuernberg, Erlangen-Nuernberg, Germany

Abstract—The influence of an accurate electron velocity-field relationship modelling on pulsed IV and small-signal RF characteristics in GaN-on-Si HEMTs is discussed and compared to measurements. We show by technology computer-aided design (TCAD) simulation and measurements that not only the low-field mobility and saturation velocity are of great importance, but also the transition behaviour in between has to be modelled accurately. Experimentally, we extract the velocity-field relationship using device simulation and measured data with ultra short pulse lengths. To the best of our knowledge, this is the first study on the velocity-field relationship in GaN-on-Si devices.

Index Terms—Gallium nitride, saturation velocity, velocity field curve, GaN-on-Si, TCAD

I. INTRODUCTION

Gallium nitride based devices gained a lot of attention in the last years for power as well as RF-power applications due to the inherent advantages of the III-nitride material system. In combination with silicon as substrate, the advantages of the III-nitrides can be offered for cost sensitive applications. However, optimizing GaN High-Electron-Mobility-Transistors (HEMTs) remains challenging. Virtual prototyping based on technology computer aided design (TCAD) is hence a valuable guidance for device design. Compared to already successfully established silicon device simulations, TCAD simulations of III-nitrides are so far not as predictive. A major reason is the lack of well-calibrated empirical models supporting the simulations. Special focus should be put on the correct modelling of the electron velocity-field relationship, since high electric fields are commonly present in GaN based devices. Several studies extracting the velocity-field relationship for GaN-on-Sapphire and GaN-on-SiC are available in literature [1]–[6]. However, the increased density of dislocations in GaN-on-Si devices [7] compared to GaN-on-Sapphire or GaN-on-SiC affects the coulomb scattering mechanisms [8] and thus is expected to influence also the velocity-field curve [9]. Furthermore, studies extracting the full velocity-field relationship based on an analytical model that can be used for TCAD based simulations are barely available [10]. Hence, common practice

in TCAD studies is to model low-field mobility and saturation velocity only, whereas different models for the full velocity-field curve are used [11]–[14]. In this study, we extract the full GaN-on-Si velocity-field relationship including an analytical representation that can be easily used in TCAD simulations. We show that in particular the transition regime has a big impact on the modelled characteristics and needs to be taken into account carefully. To our best knowledges, this is the first report discussing the full velocity-field relationship of GaN-on-Si devices.

II. EXPERIMENTAL METHODS AND SIMULATION SETUP

A. GaN-on-Si Technology and Fabrication

AlGaN/GaN heterostructures grown on high resistivity (4000 Ωcm) silicon (111) substrate by metal-organic chemical vapour deposition (MOCVD) are investigated. The epitaxial layers consist of a nucleation layer, a stress mitigation buffer, a GaN channel and a 21 nm thick $Al_{0.20}Ga_{0.80}N$ barrier layer. Test structures for velocity-field relationship measurements and AlGaN/GaN HEMTs with 450 nm gate length as well as gate and source connected field-plates were fabricated and passivated by SiN. The specially designed test structure for the velocity-field relationship characterization is shown in Fig. 1. The ohmic contact length l was chosen to be much bigger than the contact transfer length $l > 2L_T$.

B. Electrical Characterization

For characterizing the velocity field relationship and pulsed transistor curves we use a four point Kelvin TLP method [15]. The force and sense pads were connected with GGB picoprobes each to eliminate possible errors from non-zero contact resistance of the probes at high currents. An ultra short pulse width of 5.0 ns and a rise time of 0.3 ns were used to avoid unintentional self-heating during the characterization of the velocity field relationship. Similarly, the transfer and output characteristics of the fabricated GaN HEMTs were measured with short pulses. Small-signal RF measurements were performed on an Agilent N5230A vector network analyzer (VNA) by measuring the s-parameters. For calibration an off-chip SOLT standard as well as short and open on-wafer de-embedding was used.

C. TCAD Simulation

Two-dimensional TCAD drift and diffusion simulations based on Fermi statistics are performed using the commercial tool Synopsis Sentaurus. The geometric dimension of the test and device structure were obtained from layout as well as scanning electron microscope (SEM) pictures. The buffer above the substrate is simplified by using AlGaN layers with various concentrations for the carbon doping. The doped layers are modelled by an auto compensation model using shallow donors and deep acceptors [16]. The 2DEG density of $7 \cdot 10^{12}$ cm^{-2} as well as the low-field mobility were fitted according to Hall-measurements. For the low-field mobility the constant-mobility model was used after [17]. The ohmic contact resistances were carefully determined with TLM measurements [18] and accordingly used during simulation. For the surface physics a simplified Fermi-level pinning model based on surface donors is used [19]. The piezoelectric charge is taken into account according to the built in model of the simulator following [20]. For high-field saturation the Farahmand-Model [1] is used and fitted according to own measurements as discussed below.

III. RESULTS

A. Extraction of Velocity-Field Relation

To characterize the velocity-field relationship experimentally, we measure the pulsed I-V characteristics of the specially designed test structure and extract the parameters for the Farahmand model by comparing to simulations of the same structure. This procedure has the advantage of taking non-

account as well as not using the approximation $v_d = I/qn_S$. The fitted Farahmand model, given by Eq. 1

$$v_d = \frac{\mu_0 E + v_{sat} \left(\frac{E}{E_0} \right)^{n_1}}{1 + \alpha \left(\frac{E}{E_0} \right)^{n_2} + \left(\frac{E}{E_0} \right)^{n_1}},\tag{1}$$

together with the measured data is shown in in Fig. 1. The low field mobility is denoted by μ_0 and the saturation velocity by v_{sat}. The parameters used are given in Tab. I.

TABLE I: Extracted parameters for the fitted Farahmand velocity-field model.

Extracted parameter	Value
v_{sat} $[cm/s]$	$1.50 \; 10^7$
E_0 $[V/cm]$	$2.4550 \; 10^4$
α	6.187
n_1	2.912
n_2	1.025

B. Impact of Accurate Velocity-Field Relation Modelling

The importance of accurately modelling the full velocity-field relationship is discussed in the following. Two velocity-field models are plotted in Fig. 1, were we compare our fitted model to the often used Caughey-Thomas model [21] given by $v_d = \mu_0 E / \left(1 + (\mu_0 E/v_{sat})^\beta \right)^{1/\beta}$. The major difference for the two models can bee seen for electric fields between 10^3 V/cm and 10^5 V/cm. Our fitted model exhibits much lower drift velocities in the transition regime compared to the Caughey-Thomas model. Comparing to device simulations of the fabricated GaN-on-Si HEMTs, this difference directly translates into a much lower saturation current and transconductance as shown in Fig. 2a and Fig. 2b. Only by using an accurate modelling of the transition regime results in a good agreement with measurements.

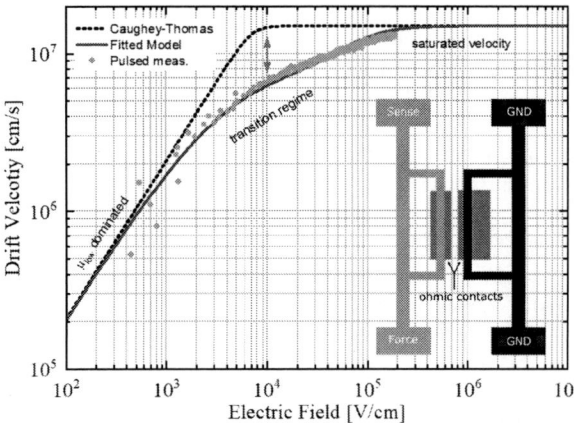

Fig. 1: Velocity-field relationship comparison between the measured values, the fitted Farahmand and the Caughey-Thomas model.

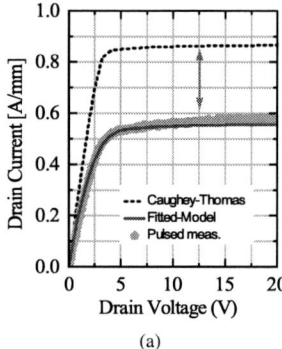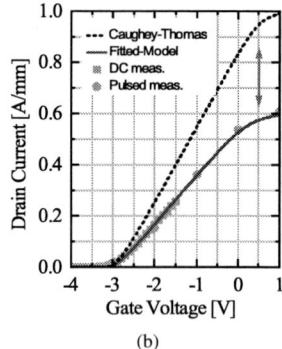

Fig. 2: Comparison of simulated and measured transistor characteristics under pulsed and DC conditions. In (a) the IdVd curve for $V_{gs} = 0$ V, and in (b) the IdVg curve for $V_{ds} = 5$ V is shown.

To further discuss this difference between the two models used we compare the simulated drift velocity and electric field

uniform carrier concentration and electric field variations into parallel to the channel between the drain side edge of the gate and the drain contact as shown in Fig. 3. The electric

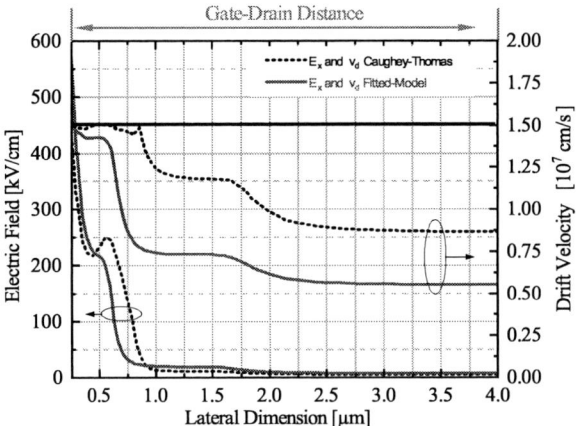

Fig. 3: Drift velocity (dashed lines) and electric field along the channel (solid lines) for the two different velocity-field relationships at $V_{gs} = 0\ V$ and $V_{ds} = 20\ V$. The Caughey-Thomas velocity-field relationship overestimates the drift velocity inside the device.

fields along the channel are qualitatively comparable for both models with a peak electric field at the gate edge and a second lower field peak at the gate field plate corner. However, for the Caughey-Thomas model the region where the drift velocity remains at the saturated value is much bigger and shows an overall higher value throughout the whole gate-drain region. This difference in drift velocity between the Caughey-Thomas model and the fitted model explains the significant difference in the simulated characteristics.

Finally, we compare the simulated h_{21} RF characteristic to measurements, as shown in Fig. 4. There, the current gain h_{21} for a fixed drain current and drain voltage of 20 V is shown. The higher drift velocity for the Caughey-Thomas model leads to an increased h_{21} and to an overestimated f_t compared to the measured values, whereas the simulations based on the fitted velocity-field relationship matches the measurements well.

IV. CONCLUSION

We have experimentally characterized the velocity-field relationship in GaN-on-Si devices at room temperature and presented an analytical model that can be easily used in TCAD simulations. We discussed the influence of the velocity-field relationship on the device characteristics and concluded that not only modelling the low-field mobility and saturation velocity is of great importance, but also the full velocity-field relationship has to be modelled accurately in order to match experimental obtain DC and RF characteristics of GaN-on-Si transistors. To the best of our knowledge this is the first report discussing the velocity-field relationship in GaN-on-Si based transistors.

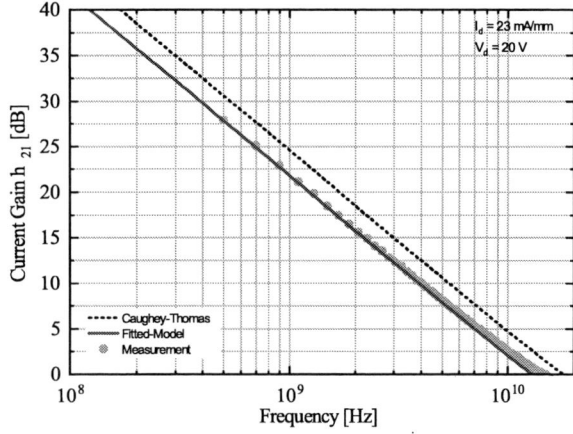

Fig. 4: Current gain h_{21} for fixed drain current and voltage. The measurements are compared with two different velocity-field relations, where only the fitted model matches the measured results.

REFERENCES

[1] M. Farahmand, C. Garetto, E. Bellotti, K. F. Brennan, M. Goano, E. Ghillino, G. Ghione, J. D. Albrecht, and P. P. Ruden, "Monte carlo simulation of electron transport in the III-nitride wurtzite phase materials system: binaries and ternaries," *IEEE Transactions on Electron Devices*, vol. 48, no. 3, pp. 535–542, March 2001.

[2] L. Ardaravičius, A. Matulionis, J. Liberis, O. Kiprijanovic, M. Ramonas, L. F. Eastman, J. R. Shealy, and A. Vertiatchikh, "Electron drift velocity in AlGaN/GaN channel at high electric fields," *Applied Physics Letters*, vol. 83, no. 19, pp. 4038–4040, 2003.

[3] L. Ardaravičius, M. Ramonas, O. Kiprijanovic, J. Liberis, A. Matulionis, L. F. Eastman, J. R. Shealy, X. Chen, and Y. J. Sun, "Comparative analysis of hot-electron transport in AlGaN/GaN and AlGaN/AlN/GaN 2DEG channels," *physica status solidi (a)*, vol. 202, no. 5, pp. 808–811, 2005.

[4] J. M. Barker, D. K. Ferry, D. D. Koleske, and R. J. Shul, "Bulk GaN and AlGaN/GaN heterostructure drift velocity measurements and comparison to theoretical models," *Journal of Applied Physics*, vol. 97, no. 6, p. 063705, 2005.

[5] S. Bajaj, O. F. Shoron, P. S. Park, S. Krishnamoorthy, F. Akyol, T.-H. Hung, S. Reza, E. M. Chumbes, J. Khurgin, and S. Rajan, "Density-dependent electron transport and precise modeling of GaN high electron mobility transistors," *Applied Physics Letters*, vol. 107, no. 15, p. 153504, 2015.

[6] L. Guo, X. Yang, Z. Feng, Y. Lv, J. Cheng, L. Sang, F. Xu, N. Tang, X. Wang, W. Ge, and B. Shen, "Effects of light illumination on electron velocity of AlGaN/GaN heterostructures under high electric field," *Applied Physics Letters*, vol. 105, no. 24, p. 242104, 2014.

[7] S. Ghosh, A. Bag, P. Mukhapadhay, S. M. Dinara, S. K. Jana, S. Kabi, and D. Biswas, "Threading dislocations in GaN HEMTs on silicon: Origin of large time constant transients?" in *Proc. CS MANTECH Conf.*, 2014, pp. 349–352.

[8] F. A. Marino, N. Faralli, T. Palacios, D. K. Ferry, S. M. Goodnick, and M. Saraniti, "Effects of threading dislocations on AlGaN/GaN high-electron mobility transistors," *IEEE Transactions on Electron Devices*, vol. 57, no. 1, pp. 353–360, Jan 2010.

[9] S. Sridharan, A. Christensen, A. Venkatachalam, S. Graham, and P. D. Yoder, "Temperature- and doping-dependent anisotropic stationary electron velocity in wurtzite GaN," *IEEE Electron Device Letters*, vol. 32, no. 11, pp. 1522–1524, Nov 2011.

[10] G. Atmaca, P. Narin, E. Kutlu, T. V. Malin, V. G. Mansurov, K. S. Zhuravlev, S. B. Lişesivdin, and E. Özbay, "Negative differential resistance observation and a new fitting model for electron drift velocity in GaN-based heterostructures," *IEEE Transactions on Electron Devices*, vol. 65, no. 3, pp. 950–956, March 2018.

978-1-7281-0941-1/19 $31.00 © 2019 IEEE

[11] N. K. Subramani, A. K. Sahoo, J. Nallatamby, R. Sommet, N. Rolland, F. Medjdoub, and R. Quéré, "Characterization of parasitic resistances of AlN/GaN/AlGaN HEMTs through TCAD-based device simulations and on-wafer measurements," *IEEE Transactions on Microwave Theory and Techniques*, vol. 64, no. 5, pp. 1351–1358, May 2016.

[12] M. G. Ancona, J. P. Calame, D. J. Meyer, and S. Rajan, "Device modeling of graded III-N HEMTs for improved linearity," in *2018 International Conference on Simulation of Semiconductor Processes and Devices (SISPAD)*, Sep. 2018, pp. 154–158.

[13] K. Ahmeda, B. Ubochi, K. Kalna, B. Benbakhti, S. J. Duffy, W. Zhang, and A. Soltani, "Self-heating and polarization effects in AlGaN/AlN/GaN/AlGaN based devices," in *2017 12th European Microwave Integrated Circuits Conference (EuMIC)*, Oct 2017, pp. 37–40.

[14] V. Joshi, A. Soni, S. P. Tiwari, and M. Shrivastava, "A comprehensive computational modeling approach for AlGaN/GaN HEMTs," *IEEE Transactions on Nanotechnology*, vol. 15, no. 6, pp. 947–955, Nov 2016.

[15] W. Simbürger, D. Johnsson, and M. Stecher, "High current TLP characterisation: An effective tool for the development of semiconductor devices and ESD protection solutions," *ARMMS RF & microwave society*, 2012.

[16] G. Verzellesi, L. Morassi, G. Meneghesso, M. Meneghini, E. Zanoni, G. Pozzovivo, S. Lavanga, T. Detzel, O. Häberlen, and G. Curatola, "Influence of buffer carbon doping on pulse and ac behavior of insulated-gate field-plated power AlGaN/GaN HEMTs," *IEEE Electron Device Letters*, vol. 35, no. 4, pp. 443–445, April 2014.

[17] C. Lombardi, S. Manzini, A. Saporito, and M. Vanzi, "A physically based mobility model for numerical simulation of nonplanar devices," *IEEE Transactions on Computer-Aided Design of Integrated Circuits and Systems*, vol. 7, no. 11, pp. 1164–1171, Nov 1988.

[18] G. K. Reeves and H. B. Harrison, "Obtaining the specific contact resistance from transmission line model measurements," *IEEE Electron Device Letters*, vol. 3, no. 5, pp. 111–113, May 1982.

[19] J. P. Ibbetson, P. T. Fini, K. D. Ness, S. P. DenBaars, J. S. Speck, and U. K. Mishra, "Polarization effects, surface states, and the source of electrons in AlGaN/GaN heterostructure field effect transistors," *Applied Physics Letters*, vol. 77, no. 2, pp. 250–252, 2000. [Online]. Available: https://doi.org/10.1063/1.126940

[20] O. Ambacher, B. Foutz, J. Smart, J. R. Shealy, N. G. Weimann, K. Chu, M. Murphy, A. J. Sierakowski, W. J. Schaff, L. F. Eastman, R. Dimitrov, A. Mitchell, and M. Stutzmann, "Two dimensional electron gases induced by spontaneous and piezoelectric polarization in undoped and doped AlGaN/GaN heterostructures," *Journal of Applied Physics*, vol. 87, no. 1, pp. 334–344, 2000. [Online]. Available: https://doi.org/10.1063/1.371866

[21] D. M. Caughey and R. E. Thomas, "Carrier mobilities in silicon empirically related to doping and field," *Proceedings of the IEEE*, vol. 55, no. 12, pp. 2192–2193, Dec 1967.

TCAD Simulations Combined with Free Carrier Absorption Experiments Revealing the Physical Nature of Hydrogen-Related Donors in IGBTs

Andreas Korzenietz
Chair for Physics of Electrotechnology,
Technical University of Munich
Munich, Germany
andreas.korzenietz@tum.de

Christian Sandow
Infineon Technologies AG
Neubiberg, Germany

Frank Hille
Infineon Technologies AG
Neubiberg, Germany

Gerhard Wachutka
Chair for Physics of Electrotechnology,
Technical University of Munich
Munich, Germany

Franz-Josef Niedernostheide
Infineon Technologies AG
Neubiberg, Germany

Gabriele Schrag
Chair for Physics of Electrotechnology,
Technical University of Munich
Munich, Germany

Abstract— **Hydrogen-related donors can be advantageously used in IGBTs and power diodes with a view to creating field-stop layers and to optimising the electrical performance. In this work, the influence of hydrogen-related donors on the on-state plasma profile in field-stop IGBTs is analysed by means of free-carrier absorption measurements. For these investigations, dedicated IGBT test structures were used, which had been adapted to the specific properties of the employed measurement set-up. Two different hydrogen-related donor profiles were implanted into these IGBT samples and, subsequently, measurements with different current densities were compared to 2D TCAD numerical simulations. In the next step, the simulation models were adjusted, with respect to carrier lifetime and mobility to reflect the impact of a possible variation of these properties.**

Keywords— *IGBT, Free Carrier Absorption Measurements, Proton Induced Donors, Carrier Lifetime, Carrier Mobility*

I. INTRODUCTION

The profound knowledge and understanding of the internal electronic behaviour of semiconductor power devices under specific operating conditions constitutes an indispensable prerequisite for their proper design and efficient device optimisation. Usually, such devices are characterised by electrical measurements resulting in the terminal behaviour of the device, while their local internal behaviour is not easily accessible by measurements. In order to enhance the accuracy and prediction of the applied simulation models, numerical analysis has to be supported by appropriate characterisation and model parameter extraction. To this end, a dedicated free-carrier absorption (FCA) measurement technique for the experimental analysis of the internal spatial- and time-resolved excess charge-carrier density distribution in vertical semiconductor power devices is employed by exploiting the plasma-optical effect [1]. In the present work, this technique is used for investigating the impact of hydrogen-related donors (HD) on the stationary excess charge-carrier density profile in IGBTs. HDs, generated in silicon devices by proton implantation, exhibit two major advantages: the low thermal budget [2] necessary for their creation, and the broad range of penetration depth which is beneficial for realising donor profiles acting as field stop (FS) layers in semiconductor power devices [3]. However, the quantitative analysis of microscopic physical properties such as carrier lifetime and mobility, which are

necessarily needed to predict the electronic transport properties of these devices, is still problematic. Combining the data extracted by FCA measurements with a dedicated simulation and parameter extraction scheme enables the proper and accurate calibration of the transport models and, hence, the reliable design and efficient optimisation of HD-based devices by predictive TCAD simulations. Three IGBT structures, one reference sample without and two samples with implanted HD profiles, have been investigated by optical FCA as well as by standard electrical measurements (stationary transfer characteristics, dynamic turn-off) at room temperature and at 150°C. In order to assure an optimum interaction of the probing beam with the charge carriers inside the DUT and to ascertain that the disturbance of the probing beam by reflection from the contact edges is minimal, the peaks of the HD profile have been placed in the middle of the device (Fig.1).

A 2D simulation model (Fig.2) serves as basis for the interpretation of the measured data following an iterative inverse modelling procedure, which, in a first calibration step, is applied to the reference sample in order to minimise parasitic effects of the measurement set-up (e.g., parasitic capacitances) and to extract device-specific effects like oxide charges and contact resistances. The calibrated simulation model is applied for investigating the influence of HDs on the FS DUTs and the assessment of the related transport coefficients.

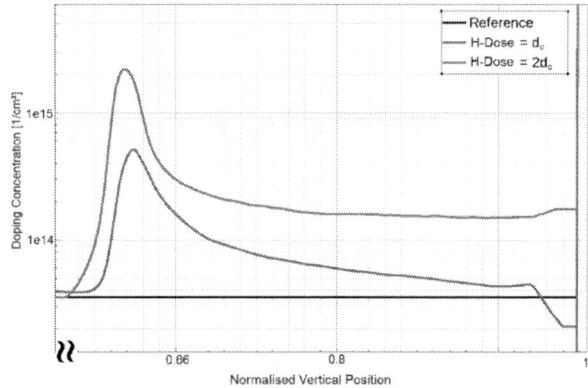

Fig. 1: Implanted hydrogen donor profiles and background doping of the reference device obtained from spreading resistance measurements.

978-1-7281-0941-1/19 $31.00 © 2019 IEEE

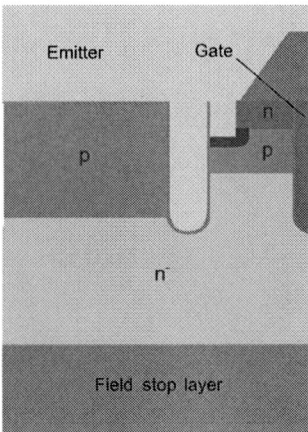

Fig. 2: Two-dimensional TCAD model of the IGBT trench cell under test.

II. FREE-CARRIER ABSORPTION MEASUREMENT TECHNIQUE

The applied FCA measurement technique exploits the plasma-optical effect [1], which describes the attenuation and deflection of an electromagnetic wave propagating through the interior of the DUT. The experimental set-up comprises a nearly incoherent light beam generated by an amplified-spontaneous-emission light source which is focused onto the centre of the DUT along the direction of the probing beam, while the device is subjected to periodic electric current pulses. The light beam traverses the DUT in lateral direction at a vertical position x through the drift region (Fig.3). In order to avoid free electron-hole pair generation by photons, a wavelength of $\lambda = 1550nm$ is used for the probing beam to ensure that the energy of the impinging photons is lower than the width of the silicon band gap ($E_{photon} < E_{gap}$). The nearly negligible optical transitions of electrons from the valence band to the conduction band do not significantly contribute to the total carrier concentration, also in consequence of the high density of electron states near the band edges [4]. The injection of charge-carriers in the active region of the DUT causes an attenuation of the propagating light beam, resulting in a reduction of the intensity of the transmitted light, which in turn is detected by a photo-diode. A basic sketch illustrating the orientation and location of the propagating light beam with respect to the DUT is shown in Figure 3.

The dominant processes that affect the optical absorption inside a semiconductor are band-to-band absorption (interband transition) and free-carrier absorption (intraband transition) [5]. Since the energy of the photons in the light beam is lower than the band-gap of silicon, the modulation of the light absorption coefficient α originates primarily from the rise of the free-carrier absorption caused by high injection of charge-carriers into the intrinsic region of the forward-biased IGBT. The high injection leads to a quasi-neutral electron-hole plasma ($n(x) \approx p(x)$) and, hence, α becomes proportional to the sum of the electron and hole density [1]. Therefore, the resulting modulation of the electromagnetic infrared wave gives an image of the local time- and position-dependent excess charge-carrier density $\Delta C(x,z,t)$, which can be extracted from the measured transmitted light intensity profile by exploiting the Beer-Lambert law:

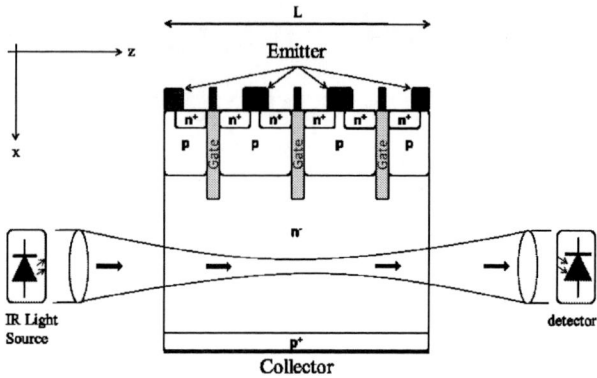

Fig. 3: Schematic drawing of the optical beam path in free-carrier absorption experiments.

$$I_{on}(x,t) = I_{off}(x) \exp\left[-\frac{\partial \alpha}{\partial C} \int_0^L \Delta C(x,z,t)dz\right] \quad (1)$$

Here I_{on} and I_{off} are the light intensities during the on-state and the off-state of the current pulse, respectively, and the z-integral extends in lateral direction along the optical path with interaction length L.

III. MEASUREMENT & NUMERICAL RESULTS

In the TCAD simulations, the field-stop profiles shown in Fig. 1 were used. For both the donor profile extracted from spreading resistance profiling (SRP) and the donor parameters in the TCAD model the well-known electrical properties of phosphorous doping were assumed.

Assuming quasi-stationary conditions during the on-state and during the off-state of the IGBT ($\partial/\partial t=0$) and quasi-1D current flow in vertical (x-)direction with negligible carrier recombination, the carrier balance equations for electrons and holes simplify to

$$0 = \text{div } \vec{j}_\alpha = \frac{d}{dx}j_\alpha(x) \approx -\frac{d}{dx}\left(D_\alpha \frac{d}{dx}c_\alpha(x)\right) \text{ for } \alpha=n,p \quad (2)$$

where $c_n(x) = n(x)$ and $c_p(x) = p(x)$, respectively, and D_α denote the carrier diffusion coefficients. Consequently, we find

$$\frac{dn}{dx} = const., \qquad \frac{dp}{dx} = const. \qquad (3)$$

in the intrinsic region flooded by the electron-hole plasma and, hence, we obtain linear carrier density profiles as a result. This explains the shape of the measured and simulated carrier profiles inside the reference DUT (Fig. 4) for nominal current and one tenth of it at room temperature and at 150°C, respectively.

Simulations of the transfer characteristics (Fig.5) provided the basis for adjusting the density of oxide charges and the contact resistance. While the oxide charges affect the threshold voltage of the IGBT, the adjustment of the contact resistance has an impact on the slope of the transfer characteristics. In this way, the charge carrier density at the emitter side is controlled and can be adjusted in such a way that it corresponds to the data extracted from the FCA measurements.

978-1-7281-0941-1/19 $31.00 © 2019 IEEE

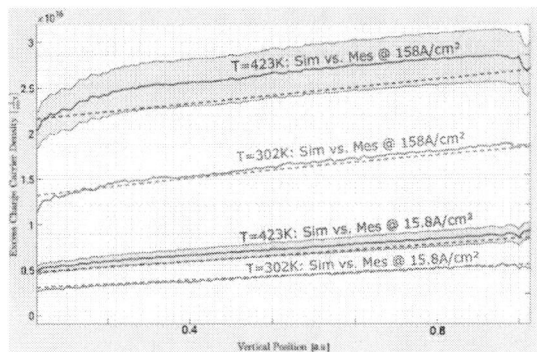

Fig. 4: Measured (solid) and simulated (dashed) excess charge carrier profiles in the reference DUT at room temperature (green) and 150°C (red).

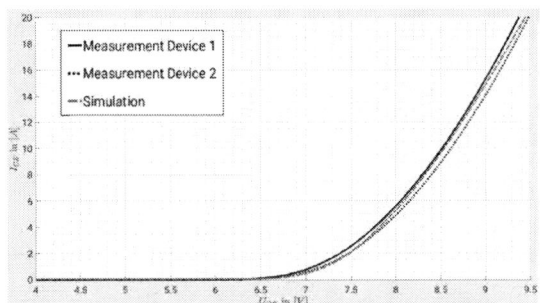

Fig. 5: Measured and calibrated simulated transfer characteristic as obtained after adjusting the density of oxide charges and the contact resistance.

The calibrated excess charge carrier profiles shown in Fig.4 have been used as initial values for simulating the turn-off transients of a double pulse test. The gradient of the calculated voltage transient dU/dt as well as the current transient dI/dt are in very good agreement with the measured turn-off transients, for room temperature (Fig.6) and for 150°C (Fig.7). This corroborates that the measured excess charge carrier gradient and the absolute values have been extracted very accurately by the FCA experiments.

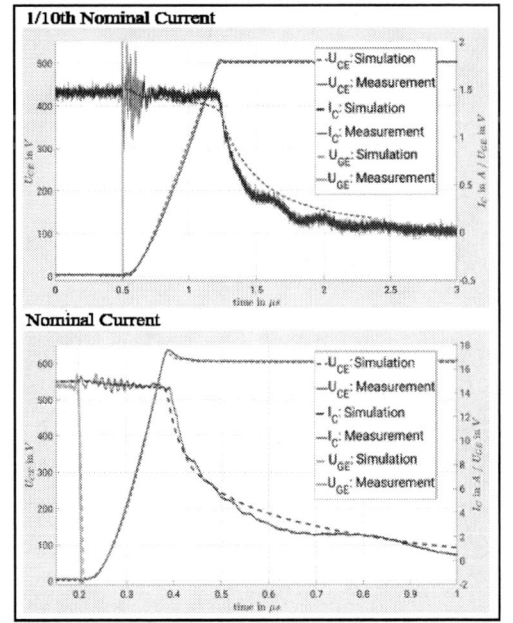

Fig. 6: Comparison of simulated and measured turn-off transients of the reference DUT at room temperature.

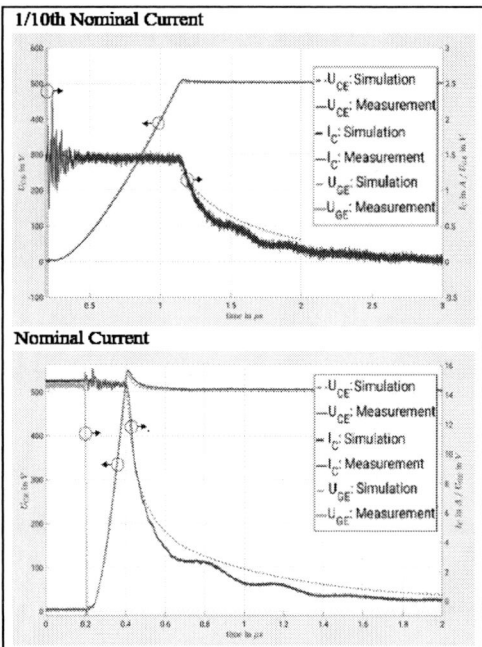

Fig. 7: Comparison of simulated and measured turn-off transients of the reference DUT at T=150°C.

The optically measured excess charge-carrier density profiles of the two IGBTs with FS layers (Fig.8) differ significantly from the linear profiles of the reference DUT without any FS profile (Fig.4) for the same terminal currents The excess charge carrier gradient along the FS profile is steepening with increasing proton dose. Inverse simulations reveal that the gradient of the excess charge carrier profile can be attributed either to a lifetime reduction in vicinity to the proton peak concentration or to a local reduction of the carrier mobility along the implantation path.

In order to model the reduction of carrier lifetime around the doping concentration peak of the HDs, the SRP doping profile is translated into a charge carrier lifetime profile $\tau(x)$ (Fig.9) with the relation:

$$\tau = 1/(c_{fact} N_{D,SRP}) \qquad (6)$$

where the factor c_{fact} is a fit parameter. Consequently, the carrier lifetime will locally degrade very steeply due to the high donor concentration N_D and rises again, beyond the donor peak, up to several µs. This δ-pulse formed lifetime profile produces the observed charge carrier gradient, which can be controlled by the factor c_{fact} in such a way that the

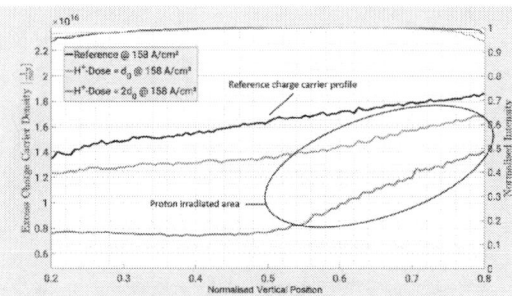

Fig. 8: FCA measurements at room temperature on the FS DUTs reveal that the charge carrier gradient along the implantation path becomes steeper with increasing hydrogen implantation dose.

978-1-7281-0941-1/19 $31.00 © 2019 IEEE   253

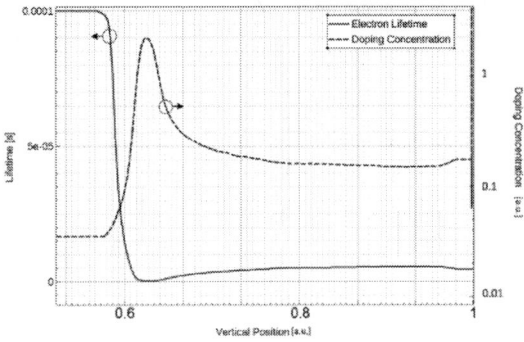

Fig. 9: Carrier lifetime profile (blue) as used for simulation. it has been generated on the basis of spreading resistance measurements of the doping concentration and the relation $\tau = 1/(c_{fact} \cdot N_{D,SRP})$.

simulated excess charge carrier gradient will match the simulated one. In the case of the DUT with the lower HD dose, the lifetime in the doping concentration peak decreases to 0.5µs.

It has been shown in [6] that the excess charge carrier gradient may also be affected by a reduction of the hole mobility μ_h. In order to generate the measured charge carrier gradient of the DUT with the lower HD dose, a reduction of the hole mobility by 5% along the implantation path is required (Fig.10).

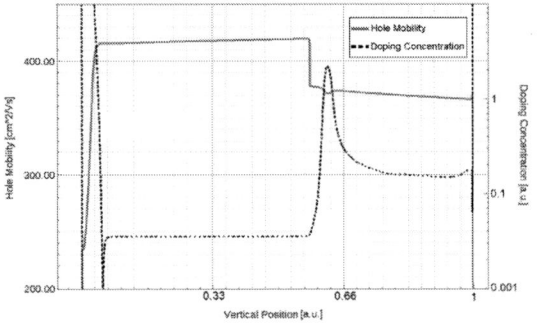

Fig. 10: Hole mobility profile (red) as used for simuation. The mobility along the implantation path is reduced by 5%.

Implementing either the lifetime profile or the local reduction of the hole mobility in the simulation allows us to adjust the carrier transport model in such a way that the measured excess charge carrier profile (Fig.11) is correctly reproduced. The relative deviation of the calibrated profiles and the measured ones is below 10% along the whole profile.

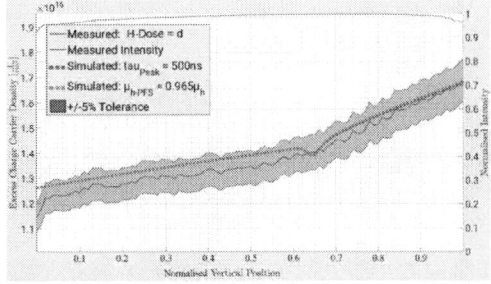

Fig. 11: Measured excess charge carrier density compared to simulations with adjusted model parameter for room temperature.

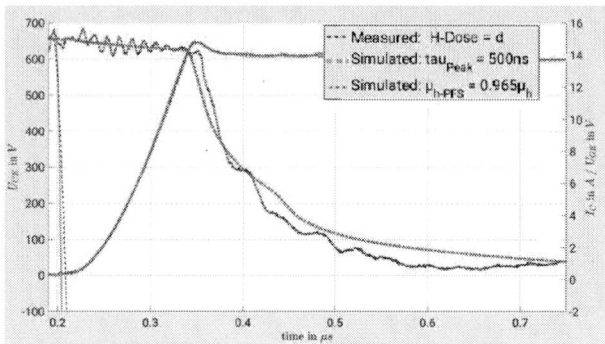

Fig. 12: Comparison of the two simulated turn-pff transients based on adjusted empirical models and the measured turn-off transients of the FS devices at room temperature.

The calibrated excess charge carrier profiles obtained from the FS samples (Fig.11) are used again as initial values for simulating the turn-off transients. The comparison of the data acquired from the double pulse test and with each of the two alternative simulation models shows that the hole mobility reduction as well as the reduced lifetime profile yield turn-off transients that conform very well with the measured data as long as the calibrated simulated excess charge carrier profiles concur to the measured ones (Fig.12).

IV. CONCLUSION

We have shown that the FCA measurement method is well suited for extracting excess charge carrier profiles in IGBTs. Furthermore, it allows for the investigation of additional FS layers which are created by proton implantation. Combining the results extracted from FCA analysis with an inverse modelling technique enables the calibration of the transport models for field stop devices. Our study reveals that a local reduction of the lifetime around the peak of the HD concentration as well as a reduction of the hole mobility along the implantation path can explain the experimental findings. Each of the two effects may be accounted for the good agreement between the experimental and simulated data, as long as the simulated excess charge carrier profiles are calibrated in such a way that they concur with the profiles extracted by the applied FCA experiments.

[1] R. Thalhammer: "Internal Laser Probing Techniques for Power Devices: Analysis, Modeling and Simulation" in "Selected Topics of Electronics and Micromechatronics", Volume 1, Eds: G. Wachutka and D. Schmitt-Landsiedel, SHAKER Verlag, Aachen, 2001.

[2] F.-J. Niedernostheide et. al: „Tailoring of Field-stop Layers in Power Devices by Hydrogen-related Donor Formation", Proc. of ISPSD 2016, Prague, Czech Republic, pp. 351-354.

[3] J. Laven: „Protonendotierung von Silizium.", Universität Erlangen, Dissertation, 2013.

[4] R. A. Soref and B. R. Bennett, "Electro-Optical Effects in Silicon", IEEE J. Quantum Electronics, vol. 23, pp. 123, 1987.

[5] J. C. Sturm and C. M. Reaves, "Silicon Temperature Measurement by Infrared Absorption: Fundamental Processes and Doping Effects", IEEE Trans. On Electron Devices, vol. 39, pp. 81, 1992.

[6] A. Korzenietz et. al: "Free-Carrier Absorption Experiments for the Investigation of the physical device properties in IGBTs with Hydrogen-Related Donors" Proc. of ISPSD 2017, Sapporo, Japan, pp. 163-166

TCAD investigation of zero-cost high voltage transistor architectures for logic memory circuits

Jordan Locati,
STMicroelectronics
13106 Rousset, France
jordan.locati@st.com

Christian Rivero
STMicroelectronics
13106 Rousset, France
christian.rivero@st.com

Julien Delalleau
STMicroelectronics
13106 Rousset, France
julien.delalleau@st.com

Vincenzo Della Marca
Aix-Marseille University
CNRS, IM2NP UMR 7334
F-13397 Marseille, France
vincenzo.della-marca@im2np.fr

Karine Coulié
Aix-Marseille University
CNRS, IM2NP UMR 7334
F-13397 Marseille, France
karine.coulie@im2np.fr

Jordan Innocenti
STMicroelectronics
13106 Rousset, France
jordan.innocenti@st.com

Olivier Paulet
STMicroelectronics
13106 Rousset, France
olivier.paulet@st.com

Arnaud Regnier
STMicroelectronics
13106 Rousset, France
arnaud.regnier@st.com

Stephan Niel
STMicroelectronics
13106 Rousset, France
stephan.niel@st.com

Abstract— **In this paper, a new device architecture has been studied by TCAD process simulations in order to provide the improvements on the electrical characteristics. We focus mainly on the drain-bulk junction breakdown voltage, of a double 130 nm poly gate transistor for Non-Volatile Memory technology. It is used as a word line select transistor, handling the drain voltage up to 13 V. The proposed structure has been implemented on silicon and the electrical measurements demonstrate the good predictability given by simulations. Finally, a new zero-cost added process asymmetric architecture is also studied to propose further improvements in terms of footprint or electrical characteristics.**

Keywords—MOSFET architecture, junction engineering, junction breakdown

I. INTRODUCTION

Nowadays, as the memory density is increasing, the maintaining of low energy consumption is a major challenge for the logic circuits. The scaling of peripheral components, such as high voltage transistors (HV), remains complicated, even if it is possible to reduce the size of memory point. This is a crucial point for the memory array operation in charge storage technologies. In order to guarantee a good product functionality, the HV transistors must assure a drain-bulk junction breakdown voltage (BV) greater than 13 V with a low leakage current. The "more than Moore" [1] approach leads, in the case of HV transistors to a drastic drop in break down voltage or a strong increase of static leakage. To prevent these problems, it is necessary to explore new HV architectures.

Therefore, in this work, we investigate a new double gate MOS transistor to improve performances without adding extra photolithography operations and then process cost. Two-dimensional simulations are performed using Synopsys® [2] [3] Technology Computer Aided Design (TCAD) simulator to predict and compare the fabrication process and electrical characteristics of proposed device with respect to the conventional one. First, TCAD process and electrical calibration of the standard architecture is needed to be predictive for the development of new architectures. Then, after obtaining promising simulation results on the proposed architecture, it has been manufactured in 130 nm Non-Volatile Memory (NVM) technology node. The electrical characterizations demonstrate the good predictability given by simulations. Finally, to deeply push the study on TCAD simulations, we propose an asymmetrical HV transistor for scaled technologies.

II. DEVICES ARCHITECTURES

Fig. 1a shows the assembled TCAD/TEM (Transmission Electron Microscopy) view of the conventional 13V double gate nMOS architecture having a self-aligned (SA) gate length in a range of 0.6-0.9 μm. As the fabrication is integrated in a charge storage memory process flow, it is free to obtain a select transistor with a double-gate. The first poly1 layer, result from the floating gate deposition step, and the second layer (poly2) is a consequence of a control gate fabrication. These are shorted to obtain a single gate terminal. Both polysilicon gates are etched by a single operation called Self Aligned Etch (SAE). TCAD simulation in the Fig. 1b shows that the leakage current, when a 13 V drain voltage is applied, mainly comes from the drain-bulk junction under the gate overlap region. This is due to the doping junction profile. The aim of our study is to propose a new architecture that solves this weakness by varying the length of the poly2 layer with respect to the poly1.

In Fig. 2a a well-known architecture of drift-MOS [4] is also represented. It enables to reach a better BV. The electric field across the drain-bulk junction can be mitigated using low doped implants [5]. Moreover, the increase of the distance between the drain contact and the junction allows to obtain a

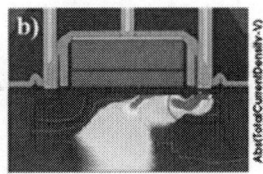

Fig. 1. Cross-section of the conventionnal double gate nMOS a) Two-dimensional Sprocess simulation and Transmission Electron Microscopy (TEM) are assembled. b) Leakage current observed during a high voltage biasing (13 V) on drain terminal using Sdevice simulation. The source, gate and bulk are grounded.

978-1-7281-0941-1/19 $31.00 © 2019 IEEE

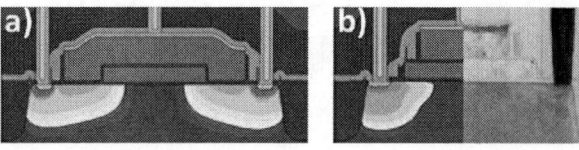

Fig. 2. Cross-section of double gate architecture: a) drift-nMOS and b) nMOS-T assembled with a Transmission Electron Microscopy (TEM) picture.

high drain resistance and thus to enhance the breakdown voltage limit [6]. After the poly1 gate etching, three consecutives boron implantation are made in order to control the punch-through current leakage [7] and the threshold voltage (Vth). Then, a phosphorus lightly doped drain (LDD) implantation is made to create source and drain regions. Oxide-nitride-oxide stack (ONO) is processed for the NVM and the second polysilicon layer is thus deposited by Low Pressure Chemical Vapor Deposition (LPCVD) at 620 °C, and etched with an overflow on both sides, followed by a heavily N+ arsenic implantation close to the surface to finally obtain the source and drain regions. The main drawback of this architecture is the footprint. The transistor length increased by 40% compared to the conventional architecture.

A new architecture studied in this paper is proposed in Fig. 2b[8]. We called it "MOS-T" [8]. The TEM picture is assembled with the TCAD sketch. This device configuration allows to use a portion of poly1 gate to implant the LDD regions saving the transistor size. Contrary to the SA architecture, the double-gate is etched by two different steps. First, the poly1 is immediately etched after the deposition. Then, the poly2 is etched after deposition shorter to an identical dimension on both sides than the poly1 gate. In this case, the drain-bulk junction under the gate overlap region has a smoothed profile to achieve an increased junction BV [9]. Fig. 3a shows the main standard process operations to develop the SA architecture. The LDD source/drain implantations are made after the SAE operation and before the poly2 etch (not used for the SA transistor). However, for the MOS-T transistor, the LDD implantations must be shifted after the poly2 etching (Fig. 3b). This is mandatory to obtain the source and drain regions, because of the higher thickness of poly2 that is two times ticker than the poly1. The shift of these operations does not impact the functioning of the other components because they are specific to the SA transistor.

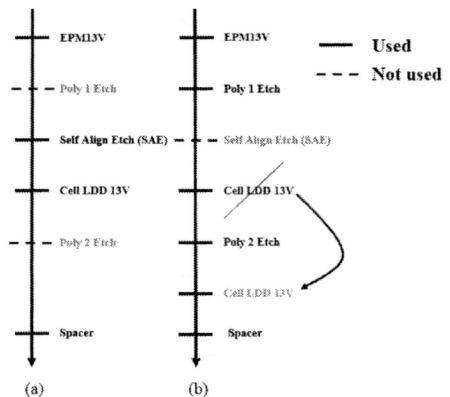

Fig. 3. Main process operations of (a) Standard SA architecture (b) MOS-T architecture.

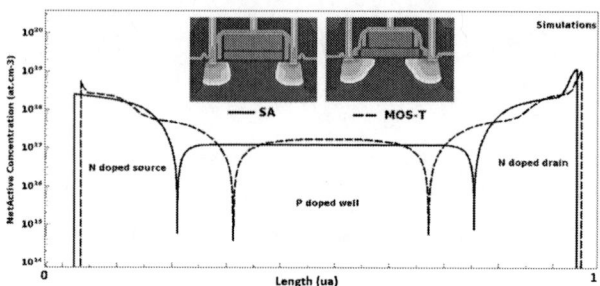

Fig. 4. Simulated NetActive Concentration in channel device versus channel length for conventional SA and MOS-T transistors.

The conventional SA and the MOS-T architectures use an identical well (EPM13V) and source/drain implants. For these last, the phosphorus implantations are made in two steps in order to achieve a smooth junction profile. In Fig. 4, the simulation of the effective channel length (L_{eff}) for the MOS-T and SA transistors is plotted. Because of the LDD implantations through the polysilicon gate, we can observe a L_{eff} reduction of 35 %, for the MOS-T. This impacts the threshold voltage with a decrease of 200 mV. Thanks to TCAD simulations, the dose of the lowest boron energy implantation localized in the channel region has been increased by 40 % with respect to the SA transistor. The implantation energy was also investigated through TCAD without significant improvements. Changing the order of the process steps and avoiding extra lithography operations, the fabrication cost is thus saved.

III. SA AND MOS-T COMPARISON

A. Electrical calibration for TCAD prediction

In order to evaluate the performances of the new architecture, the simulations were optimized on the already existing SA transistor to be predictive for the proposed one. In preamble, the device morphologic calibration such as gate oxide thickness or poly gate doping, was done before the MOS architecture investigation. For example, the boron segregation parameter [10] has been modified from $3 \cdot 10^{18}$ cm^{-3} to $5 \cdot 10^{17}$ cm^{-3} to adjust the boron concentration near to the silicon oxide interface. Poisson and drift diffusion models are used for the simulation of carrier transport associated to an avalanche model to simulate the junction comportment at high drain voltages. We are thus able to fit experimental measurements to be predictive in the simulation of the MOS-T. Higher width gate than the real device dimensions has been used for the calibration to avoid narrow channel effect [11]. The Fig. 5a and Fig. 5b show the simulated and measured electrical characteristics of drain current (Id) versus drain (Vd) and gate (Vg) voltage (Fig. 5a and Fig. 5b respectively). In 2D simulations, the back end of line (BEOL) is not considered. A 1.3 kΩ source/drain access resistance has been used in order to be realistic in the ON state operation (Vg>Vth).

978-1-7281-0941-1/19 $31.00 © 2019 IEEE

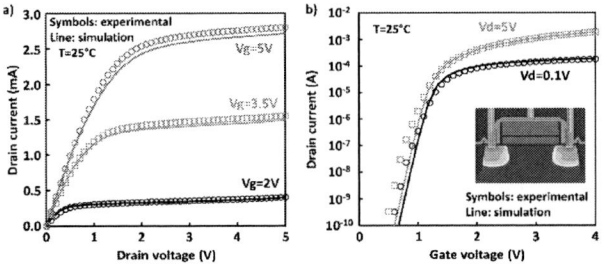

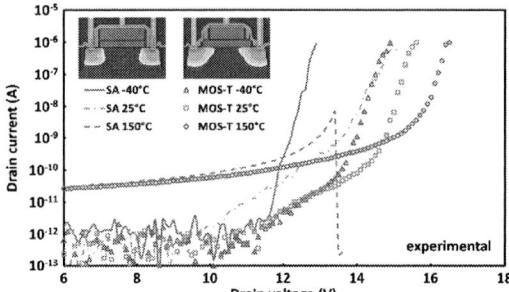

Fig. 5. Simulated and measured electrical characteristics of conventional transistor. a) Drain current versus drain voltage for different gate biasing. b) Drain current versus gate voltage for different drain bias conditions.

Fig. 7. Drain current measured of MOS-T and SA transistor as a function of drain voltage for different temperatures.

The MOS-T simulations have been carried out using the same models implemented for the SA transistor. The drain current as a function of the drain voltage characteristics were simulated to qualitatively verify the improvement on BV parameter. The comparison between the conventional SA and MOS-T architectures is shown in Fig. 6, no significant variation of leakage current is observed while an improvement of the BV up to 1.5 V is reached. Even if the L_{eff} of MOST transistor is decreased, the leackage current is limited by the triple Boron channel implantation and a double LDD step. The achieving of a smoother drain/bulk juction enables an electric field decreasing with a consequent BV increasing. This, together with the maintaining of the footprint, makes the MOS-T architecture suitable for the non-volatile memory logic circuit design.

B. Electrical characterization

Based on the good results obtained by TCAD simulations, MOS-T has been manufactured on 200 mm wafers.

The Fig. 7 shows the experimental Id-Vd characteristics of the SA and MOS-T transistors in a wide range of temperatures in the OFF-state. During these tests the gate, source and bulk terminals were grounded. The simulation prediction was accurate at room temperature. At low and room temperature, the electrical characteristics show two different slopes while the drain voltage is increased. At the beginning, the leakage current starts to increase slowly following the typical impact ionization contribution [12]. While the avalanche contribution becomes dominant for the drain voltage increasing. We can see that for the highest temperature (150 °C) the leakage current is suddenly around 30pA for both devices due to the punch through current, but

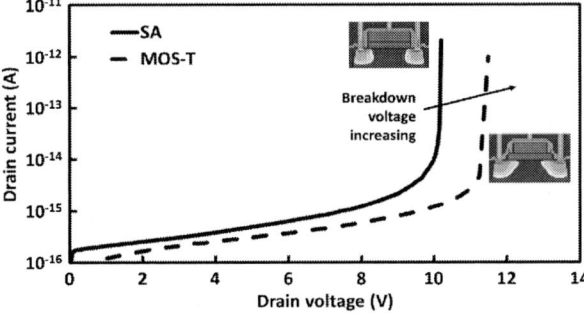

Fig. 6. Simulated drain current versus drain voltage for conventional SA and MOS-T architecture.

the MOS-T architecture improves the breakdown voltage up to 2.5 V. Hence, with the same footprint, both leakage current and breakdown voltage constraints have been respected. Concerning the operation in the ON-state, experimental comparison of Id-Vg characteristics is shown in Fig. 8. We can see there is no variation in the threshold voltage as desired and a difference in the conduction current (Vg>Vth) around 5 % between the SA and MOS-T explained by the higher resistance of the source and drain region induced by the LDD implants.

The promising results obtained by simulations have been confirmed by electrical characterizations. Taking advantage of implanting through the poly1 gate allows to gain up to 2.5 V on BV, 1 V more than expected by simulation.

In order to explain this improvement, we took a look at different simulation results such as the electric field, the impact ionization and total current density generated during a 10 V stress applied on the drain. As we can see in Fig. 9a, the electric field is clearly mitigated by the LDD until the P-N junction for the MOS-T while it remains important and concentrated in the junction space charge region for the SA transistor.

This, if the electric field is strong enough, can results in the creation of free electron-hole pairs induced by the impact ionization mechanism (Fig. 9b) for the SA transistor thus inducing quickly the avalanche breakdown. Meanwhile, the electric field across the junction for the MOS-T is weak to improve the avalanche mechanism and it needs higher voltage on the drain. The consequences are shown in Fig. 9c, where the junction leakage current density is plotted. The current induced by the avalanche phenomenon is minimized for the MOS-T.

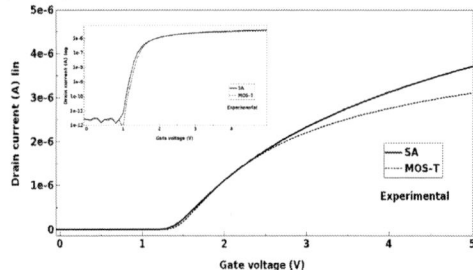

Fig. 8. Drain current measured of MOS-T and SA transistor as a function of gate voltage for a 0.1 V drain bias applicated.

978-1-7281-0941-1/19 $31.00 © 2019 IEEE   257

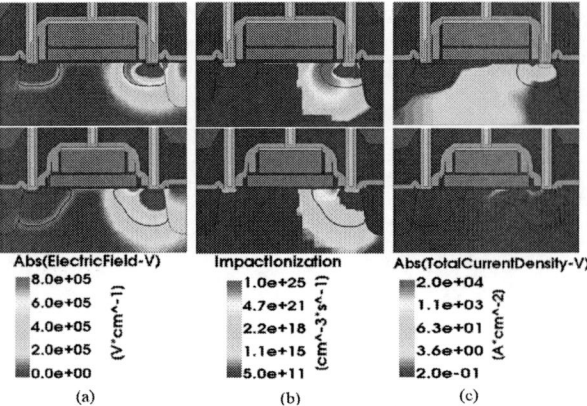

Fig. 9. Two-dimensional electrical simulation of SA and MOS-T transistor. a) electric field, b) impact ionization and c) total current density generated during 10 V stress applied on drain terminal using Sdevice simulation. The source, gate and bulk are grounded.

IV. ASYMMETRICAL MOS-T

In order to go further in the MOS-T study, we have evaluated the impact to the poly2 gate alignment on the threshold and breakdown voltage. Several simulations were done from the initial poly2 gate position until the case where the poly1 is aligned to the poly2 gate on the source side, in order to extend the LDD implant for the drain region. No additional process steps are required to evaluate this architecture.

The consequences on the BV are shown in the Fig. 10. We can expect an increase up to 1 V taking advantage of asymmetrical architecture. No significant Vth variations have been observed (<60 mV) since there is no variation in the effective length. Furthermore, this option potentially allows to find a tradeoff between the transistor length and the BV to address different applications.

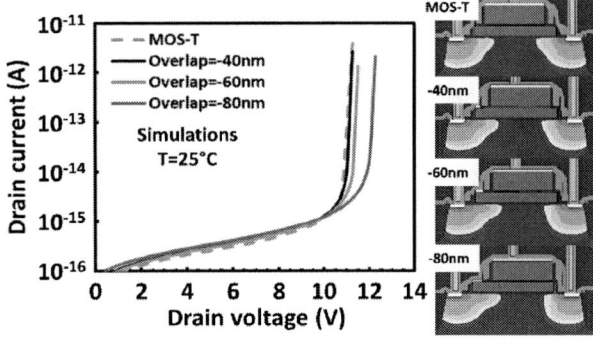

Fig. 10. Simulated drain current as a function of drain voltage for asymmetrical MOS-T.

V. CONCLUSION

This paper presents a new double gate MOS architecture. The simulation results are in agreement with the experimental electrical characterizations, demonstrating the TCAD modeling predictability. With the combination of a source and drain implantation and a shorter control gate, we were able to create LDD regions for free and so to provide with the MOS-T architecture a better breakdown voltage in regards of the SA architecture without additional lithography operations. Finally, we demonstrate that it is possible to implement an asymmetrical gate to extend the LDD region for the drain side increasing the BV allowing to explore footprint reduction.

REFERENCES

[1] Gordon E. Moore, "Cramming more components onto integrated circuits," Electronics, vol. 38, NO. 8, pp. 114, April 1965

[2] Synopsys®, "Sentaurus Process," Version N-2017-09-SP1

[3] Synopsys®, "Sentaurus Device," Version N-2017-09-SP1

[4] H. Brech et al, "Record Efficiency and Gain at 2.1GHz of High Power RF Transistors for Cellular and 3G Base Stations," IEEE International Electron Devices Meeting 2003

[5] R. Marjorie, G. PA, L. Kishore K, "Analysis Breakdown in New Step gate Structures with Graded LDD," IJCSNS International Journal of Computer Science and Network Security, vol. 17, NO. 4, April 2017

[6] J. Vinson, J. Bernier, G. Croft, J. Liou, "ESD Design and Analysis Handbook," Springer Science & Business Media, December 2012

[7] R. Dennard, F. Gaensslen, H. Yu, V. Rideout, E. Bassous, A. Leblanc, "Design of ion-implanted MOSFET's with very small physical dimensions", Proceeding of EEE, vol. 87, NO. 4, April 1999.

[8] C. Rivero, J. Delalleau, "Double-gate MOS transistor with increased breakdown voltage", US 2019/0027566 A1, January 2019

[9] Y. Yamaguchi et al, "Source to Drain Breakdown Voltage Improvement in Ultrathin-Film SOI MOSFET's Using a Gate Overlapped LDD Structure", IEEE Transactions on Electron Devices, vol. 41, NO. 7, July 1994

[10] C. Machala, R. Wise, D. Mercer, A. Chatterjee, "The Role of Boron Segregation Parameter and Transient Enhanced Diffusion on Reverse Short Channel Effect", SISPAD'97. 1997 International Conference on Simulation of Semiconductor Processes and Devices

[11] P. Wang, "Device Characteristics of Short-Channel and Narrow-Width MOSFET's", IEEE Transactions on Electron Devices, vol. ED-25, NO. 7, July 1978

[12] S.M. Sze, "Physics of Semiconductor Devices", A. John Wiley & Sons, p. 45, 1981

978-1-7281-0941-1/19 $31.00 © 2019 IEEE

Barrier Engineering of Lattice Matched AlInGaN/ GaN Heterostructure Toward High Performance E-mode Operation

Niraj Man Shrestha
Department of Electrical and Computer Engineering and Center for mmWave Smart Radar System and Technologies, National Chiao Tung University, Hsinchu 30010, Taiwan
Email: nirajnctu@gmail.com

Chao-Hsuan Chen
Institute of Communications Engineering, National Chiao Tung University, Hsinchu 30010, Taiwan

Zuo-Min Tsai
Institute of Communications Engineering, Department of Electrical and Computer Engineering, and Center for mmWave Smart Radar System and Technologies, National Chiao Tung University, Hsinchu 30010, Taiwan

Yiming Li
Institute of Communications Engineering, Department of Electrical and Computer Engineering, and Center for mmWave Smart Radar System and Technologies, National Chiao Tung University, Hsinchu 30010, Taiwan
Email: ymli@faculty.nctu.edu.tw

Jenn-Hawn Tarng
Institute of Communications Engineering, Department of Electrical and Computer Engineering, and Center for mmWave Smart Radar System and Technologies, National Chiao Tung University, Hsinchu 30010, Taiwan

Seiji Samukawa
Center for mmWave Smart Radar System and Technologies, National Chiao Tung University, Hsinchu 30010, Taiwan and Institute of Fluid Science, Tohoku University, Sendai 980-8557, Japan

Abstract— Electrical characteristics of lattice matched AlInGaN/GaN high electron mobility transistors with different barrier engineering was studied theoretically by solving drift diffusion equation. The results of the study thoroughly disclose the mitigation of induced polarization charge on lowering Al and In content in barrier resulting in a positive shift of threshold voltage with huge deduction on drain current. The newly designed lattice match double $Al_{0.54}In_{0.12}Ga_{0.34}N/$ $Al_{0.18}In_{0.04}Ga_{0.78}N$ barrier recess gate HEMT helps to boost the drain current by reducing the access resistance and enhancing the polarization charge density. The proposed HEMT exalted current density and transconductance by two times with significant shift of threshold voltage in positive axis than that of single barrier structure. Conclusively, the high performance novel double barrier recess gate E-mode HEMT will be key for real and efficient high power switching application.

Keywords—AlInGaN/GaN HEMT, Double barrier, recess gate, simulation, Physical models, Drift diffusion, lattice matched.

I. INTRODUCTION

AlGaN/GaN High Electron Mobility Transistor (HEMT) breathtaking performance has been made in the field of high-power and high-frequency electronic applications due to the existence of high-mobility two-dimensional electron gas (2DEG) at the AlGaN/GaN interface, large conduction band offset, and strong piezoelectric and spontaneous polarization effects. Performance of the device can be improved by increasing the Al content of the AlGaN barrier layer. However, significant drop in electron mobility (μ_e) for high aluminum composition than 30% in the barrier layer due to the onset of AlGaN relaxation has detrimental effect on performance of AlGaN/GaN HEMT [1]. In order to get rid of this problem, the AlGaN barrier is replaced by quaternary AlInGaN because the quaternary has wide range

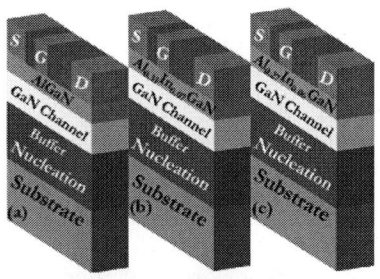

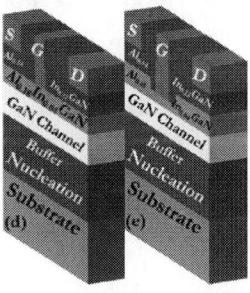

Fig. 1. The cross-sectional views of (a) conventional AlGaN/GaN HEMT, (b) $Al_{0.31}In_{0.07}Ga_{0.62}N$/GaN HEMT, (c) $Al_{0.27}In_{0.06}Ga_{0.67}N$/GaN HEMT, (d) $Al_{0.54}In_{0.12}Ga_{0.34}N/$ $Al_{0.18}In_{0.04}Ga_{0.78}N$/GaN HEMT, and (e) Recess gate $Al_{0.54}In_{0.12}Ga_{0.34}N/$ $Al_{0.18}In_{0.04}Ga_{0.78}N$/GaN HEMT.

of adjustments of the bandgap and lattice constant and sufficiently large spontaneous polarization [2]. Researchers shift their interest to lattice-matched quaternary barrier layer grown to GaN due to spontaneous polarization-induced high-density 2DEG at AlInGaN/GaN heterojunction interface [1, 3-4] which facilitates to reduce strain-related defects on adjusting the different bandgap. Furthermore, the stronger polarization effect and higher mobility of AlInGaN

978-1-7281-0941-1/19 $31.00 © 2019 IEEE

TABLE I ADOPTED DEVICE PARAMETERS

Sample Parameter	Cal. Sample	I	II	III	IV
Channel Thickness (μm)	2	2	2	2	2
Recess Gate Depth (nm)	-	-	-	5	8
Al comp.(1st barrier) %	29	31	27	18	18
Al comp. (2nd barrier) %	-	-	-	54	54
Gate Length (μm)	3	0.16	0.16	0.16	0.16
Barrier Thickness (nm)	15.5	8.5	8.5	8.5	8.5
2nd Barrier Thickness (nm)	-	-	-	5	5
Indium Comp. (1st barrier) %	-	7	6	4	4
Indium Comp. (2nd barrier) %	-	-	-	12	12

for the same lattice strain as in AlGaN/GaN, help to boost the performance of the devices beyond the limit of ternary barrier layers. Moreover, use of AlInGaN as the barrier layer is beneficial to reduce the gate leakage current and to increase carrier mobility with increase of 2DEG density [5].

However, a higher 2DEG density induced at such heterointerface results the device operated in depletion mode [6]. Due to cost effective and safety issue, enhancement-mode (E-mode) devices are more desirable in the practical applications [7]. Several approaches have been popular to realize E-mode operation such as recess gate [8], fluoride implantation, p-type Gate [7] etc. However, these technologies include complicated fabrication procedures. Ketteniss et al. reported quaternary AlInGaN/GaN E-mode HEMT by using low Al-content in quaternary AlInGaN barrier [9]. Unfortunately, drastic reduction of current density limits the performance of the HEMT.

In this report, we have studied the effect of AlInGaN barrier engineering on threshold voltage (V_{th}) and reported lattice matched double barrier AlInGaN/GaN e-mode HEMT to boost the performance of the device.

II. DEVICE STRUCTURE & METHODOLOGY

To verify and validate the physical and transport model, simulation results were calibrated with experimental results from AlGaN/GaN HEMT device. The schematic cross-section of the AlGaN/GaN HEMT used for calibration is shown in Figure 1(a). The details of the HEMT which was used to examine the accuracy of simulation data are found in our previous report [8]. After calibration, new HEMTs structure are proposed and the proposed device are theoretically studied by using calibrated models. AlInGaN/GaN HEMT with the (Sample I), $Al_{0.27}In_{0.06}Ga_{0.67}N$ barrier (Sample II), $Al_{0.54}In_{0.12}Ga_{0.34}N$ /$Al_{0.18}In_{0.04}Ga_{0.78}N$ barriers with 5 nm recess gate (Sample III) and $Al_{0.54}In_{0.12}Ga_{0.34}N$/ $Al_{0.18}In_{0.04}Ga_{0.78}N$ double barrier with 8 nm recess (Sample IV), as shown in Figs. 1(b)-1(e), are the proposed structures for theoretical study of dc and transfer characteristics. The adopted parameters of studied device are listed in Table I.

The device characteristics are simulated by numerically solving 2D drift-diffusion (DD) transport model together with the strain generated polarization model [10], the high field saturation and the interface fixed charge at interface.

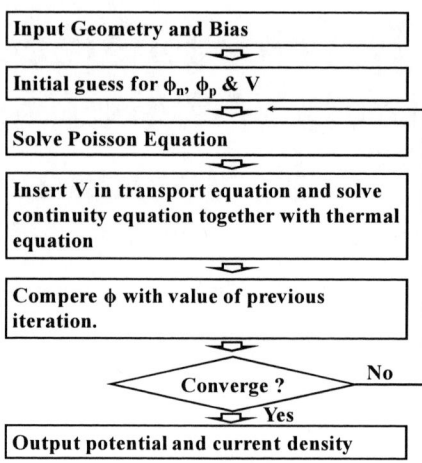

Fig. 2 Flowchart to solve the DD system

Additionally, mobility degradation due to impurity scattering are also properly modeled. To study HEMTs with recess gate structure, acoustic phonon scattering and surface roughness scattering are also taken into the account. Furthermore, the interface fixed charge at the interfaces between the AlGaN barrier layer and nitride layer and the acceptor buffer trap at GaN buffer are activated. The simulation flow of the device is shown in Fig. 2.

III. RESULTS AND DISCUSSION

It is noted that AlInGaN barriers used in this study were latticed matched with GaN. The calculated Al, In and Ga composition for latticed matched AlInGaN with GaN are plotted in Fig. 3(a).

The perfectly overlapping simulated and measured transfer (at $V_D = 5$ V) of AlGaN/GaN HEMT with nearly equal V_{th} and maximum transconductance ($G_{m,max}$) as shown in Fig. 3(b) reveal that the transport and physical models solved during simulation exactly defined the physical phenomenon. The accuracy of the physical and transport models is further verified by perfectly matched simulated dc characteristics with measured outcomes for different gate voltages as shown in Fig. 3(c).

The transfer characteristics and G_m of the lattice matched AlInGaN/GaN HEMT for different barrier engineering as shown in Figs. 3(d) and 3(e) reveal that both current density ($I_{D,max}$) and $G_{m,max}$ of device shrink with positive shift of V_{th} on decreasing Al composition in the AlInGaN barrier from 0.31 to 0.27. Furthermore, it is observed that the HEMT with double $Al_{0.54}In_{0.12}Ga_{0.34}N$/ $Al_{0.18}In_{0.04}Ga_{0.78}N$ barriers with 5 nm recess (Sample III) in gate region offer $I_{D,max}$ and G_m of 768 mA/mm and 493 mS/mm respectively which are lower in comparison of sample I and sample II. Moreover, sample III was redesigned with increasing recess depth 8 nm, $I_{D,max}$ was reduced to 688 mA/mm and G_m is increased to 558 mS/mm with significant increase of V_{th} toward the positive direction. Strong polarization induced interface charge at heterointerface is the key factor behind the excellent transfer properties of the proposed device with double AlInGaN barriers. It is found that the polarization induced interface charge (σ_{int}) is strongly depend on Al and In composition and is expressed as [11],

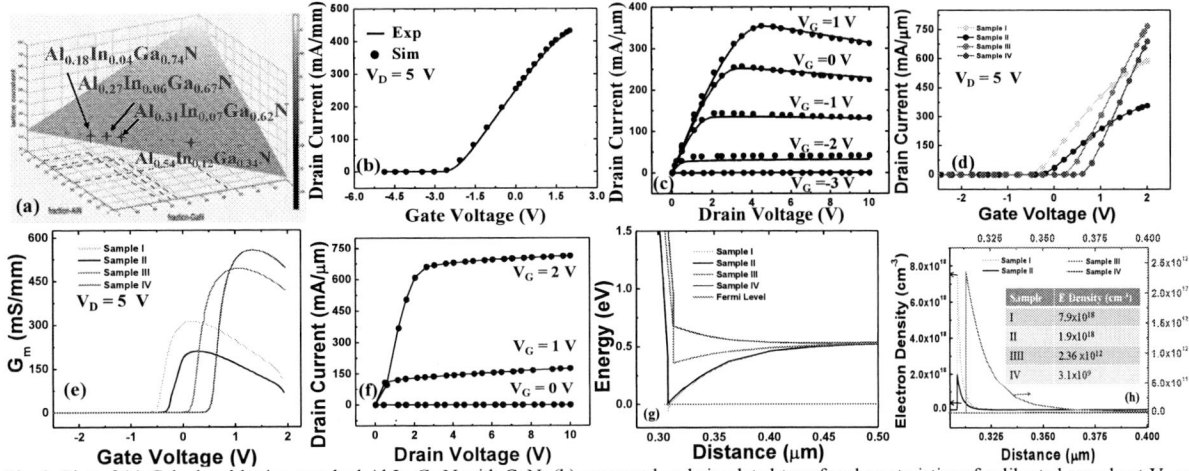

Fig. 3. Plot of (a) Calculated lattice matched $Al_xIn_yGa_zN$ with GaN, (b) measured and simulated transfer characteristics of calibrated sample at $V_D = 5$ V, (c) measured and simulated dc characteristics of calibrated sample, (d) I_D-V_G characteristics of AlInGaN/GaN HEMTs, (e) G_m of AlInGaN/GaN HEMTs, (f) transfer characteristics of sample IV, (g) Conduction band profile, and (h) n_s distribution, in gate area of AlInGaN/ GaN HEMT under zero bias condition (cut vertically in gate region).

$$\sigma_{int} = P_{GaN} - P_{InAlGaN} = P_{sp}^{GaN} - x \cdot P_{sp}^{AlN}, \quad (1)$$
$$- y \cdot P_{sp}^{InN} - z \cdot P_{sp}^{GaN} - b_{AlGaN} x \cdot z$$
$$- b_{InGaN} \cdot y \cdot z - b_{InAlN} \cdot x \cdot y - P_{pz}^{InAlGaN}$$

where P_{sp} and P_{pz} are spontaneous and piezoelectric polarization of respective binary compound and b_{AlGaN}, b_{InGaN} and b_{AlInN} are bowing parameters for ternary nitrides and are calculated by using following expression [12],

$$b_{AlGaN} = 4 P_{sp}^{AlGaN} - 2 \left(P_{sp}^{AlN} + P_{sp}^{GaN} \right), \quad (2)$$

$$b_{InGaN} = 4 P_{sp}^{InGaN} - 2 \left(P_{sp}^{InN} + P_{sp}^{GaN} \right), \quad (3)$$

$$b_{AlInN} = 4 P_{sp}^{AlInN} - 2 \left(P_{sp}^{AlN} + P_{sp}^{InN} \right), \quad (4)$$

Spontaneous polarization of ternary compound can be calculated by [12]

$$P_{sp}^{AlGaN} = -0.09x - 0.034(1-x) + 0.021x(1-x), \quad (5)$$

$$P_{sp}^{InGaN} = -0.042x - 0.034(1-x) + 0.037x(1-x), \quad (6)$$

$$P_{sp}^{AlInN} = -0.09x - 0.042(1-x) + 0.07x(1-x), \quad (7)$$

Since AlInGaN barrier layer are lattice matched with GaN, strain induced piezoelectric polarization was neglected in simulation. Therefore, effective polarization induced interface charge (σ_{int}) is expressed as,

$$\sigma_{int} = P_{GaN} - P_{InAlGaN} = P_{sp}^{GaN} - x \cdot P_{sp}^{AlN}$$
$$- y \cdot P_{sp}^{InN} - z \cdot P_{sp}^{GaN} - b_{AlGaN} x \cdot z \quad (8)$$
$$- b_{InGaN} \cdot y \cdot z - b_{InAlN} \cdot x \cdot y,$$

And, similarly, 2DEG density (n_s) is a function of interface charge and is given by [11]

$$n_s = \frac{\sigma_{int}}{q} - \frac{\varepsilon_0 \varepsilon_r}{q^2 d} \left(q\phi_b - \Delta - \Delta E_c \right), \quad (9)$$

where d is the barrier thickness, ΔE_c is the conduction band offset and Δ is the penetration of the conduction band below the Fermi level at the heterointerface. Conduction band

energy profile at zero biased condition as shown in Fig. 3(g) suggests that comparatively smaller ΔE_c is observed for low Al composition barrier and further decrease on increasing recess depth. Due to combined effect of σ_{int} and small ΔE_c, sample II has smaller n_s than that of sample I as listed in Table II. Notably, high Al composition of lattice matched second barrier of sample III and sample IV result high induced spontaneous polarization charge at the interface [12] which results in improvement of n_s significantly (Table II). Therefore, larger current density observed for Sample III than that of Sample I and Sample II whereas large positive threshold voltage is due to recess of second AlInGaN barrier beneath the gate region. Nonetheless, recess depth leads highly reduced n_s in gate region in sample III and sample IV as shown in Fig. 3 (h). Therefore, these samples offer large positive threshold voltages along with large current density. Notably, high effective potential to the channel, increased parasitic resistances, and deep level traps in barrier due to recess depth, small current density is observed in sample III and sample IV [8]. The dc characteristics of sample IV is shown in Fig. 3(f). The current flow across 2DEG in conventional HEMT devices (Sample I and Sample II) shown in Fig. 4(a) revels that there is single current flowing path in normal HEMT structure. However, there are two paths for current flow in proposed double barrier HEMT (Sample IV); one is along $Al_{0.18}In_{0.04}Ga_{0.78}N$/GaN interface and the other is along $Al_{0.54}In_{0.12}Ga_{0.34}N$/$Al_{0.18}In_{0.04}Ga_{0.78}N$ as shown in Fig 4 (b) which is similar to the resistors combined parallel where effective resistance is decreased [14]. The upper channel is cut by the recess gate and the electrons don't have enough energy to overcome the potential barrier and so they pass from the top channel. The decrease in effective resistance of the channel helps to enhance the current density of the device. The second channel contributes to the total current. The equivalent circuit for double-channel HEMT [15] is shown in Fig. 4(c). The lattice matched $Al_{0.54}In_{0.12}Ga_{0.34}N$ /$Al_{0.18}In_{0.04}Ga_{0.78}N$ /GaN recess gate HEMT shows unexpectedly better output performance with significantly large $I_{D,max}$ and G_m, and the performance with significantly large $I_{D,max}$ and G_m, and

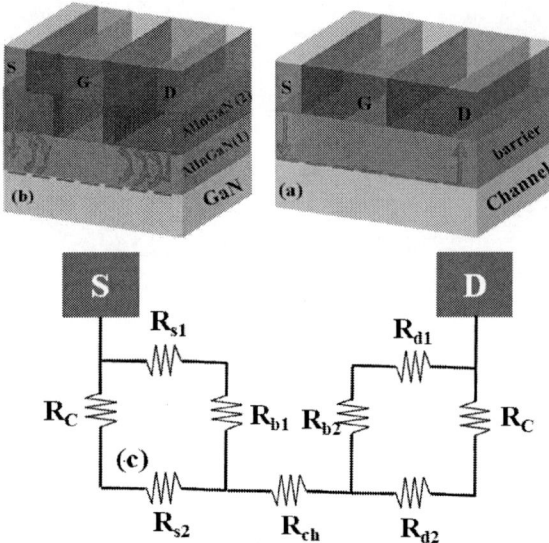

Fig.4. Schematic of the current flow in the (a) Single barrier (b) double-barrier devices, and (c) Equivalent circuit of the double-barrier HEMTs.

TABLE II COMPARISON OF OUR RESULTS WITH SIMILAR REPORTS

Sample	$I_{D,max}$ (mA/mm)	$G_{m,max}$ (mS/mm)	n_s (cm^{-3}) ($\times 10^{19}$)	V_{th} (V)
I	586	312	1.89	-0.5
II	356	210	1.02	-0.2
III	768	493	4.15	0.4
IV	688	558	4.15	0.7
Ref [16]	**338**	**165**	-	**0.2**

surprisingly large and positive V_{th} than previous report for e-mode AlGaInN/GaN HEMT as shown in Table II.

IV. CONCLUSION

Electrical properties of the different lattice matched AlInGaN/GaN HEMT were studied by using experimentally calibrated transport and physical models. Results observed that e-mode HEMT with low Al and In composition in AlInGaN barrier offer significant deduction of current density. The proposed novel double barrier recess gate HEMT gives surprisingly large current density with large positive V_{th} due to high polarization charge and low channel resistance. Finally, the high performance proposed e-mode will be advantageous for future high performance real high power high frequency application.

ACKNOWLEDGMENT

This work was supported in part by the Ministry of Science and Technology, Taiwan, under Grant MOST 106-2221-E-009-149, Grant MOST 106-2622-8-009-013-TM, Grant MOST 107-2622-8-009-011-TM, Grant MOST 107-2221-E-009-094, and Grant MOST 107-3017-F-009-001, in part by the "Center for mmWave Smart Radar Systems and Technologies" under the Featured Areas Research Center Program within the framework of the Higher Education Sprout Project by the Ministry of Education in Taiwan.

REFERENCES

[1] M. Gonschorek, J.-F. Carlin, E. Feltin, M. A. Py, and N. Grandjean, "High electron mobility lattice-matched AlInN/GaN field-effect transistor heterostructures," Appl. Phys. Lett. vol. 89, pp. 062106, August 2006.

[2] Y. Liu, H. Jiang, S. Arulkumaran, T. Egawa, B. Zhang, and H. Ishikawa, "Demonstration of undoped quaternary AlInGaN/GaN heterostructure field-effect transistor on sapphire substrate," Appl. Phys. Lett, vol. 86, pp 223510, May 2005.

[3] B. Reuters, A. Wille, B. Hollander, E. Saklauskas, N. Ketteniss, C. Mauder, R. Goldhahn, M. Heuken, H. Kalisch, and A. Vescan, "Growth Studies on Quaternary AlInGaN Layers for HEMT Application," J. Electron. Mater., vol. 41, pp. 905-909, March 2012.

[4] Y. Liu, T. Egawa, H. Jiang, B. Zhang, H. Ishikawa, and M. Hao, "Near-ideal Schottky contact on quaternary AlInGaN epilayer lattice-matched with GaN," Appl. Phys. Lett, vol.85, pp.6030, October 2004.

[5] N Ketteniss, L-R Khoshroo, M Eickelkamp1, M Heuken, H Kalisch, R H Jansen and A Vescan, "Study on quaternary AlInGaN/GaN HFETs grown on sapphire substrates," Semicond. Sci. Technol. Vol. 25, pp. 075013, May 2010.

[6] H. Hahn, B. Reuters, A Wille, N. Ketteniss, F. Benkhelifa, O. Ambacher, H. Kalisch, and A. Vescan, "First polarization-engineered compressively strained AlInGaN barrier enhancement-mode MISHFET," Semicond. Sci. Technol. vol. 27, pp. 055004, March 2012.

[7] N. M. Shrestha, Y. Li, and E. Y. Chang, "Step buffer layer of $Al_{0.25}Ga_{0.75}N/Al_{0.08}Ga_{0.92}N$ on P-InAlN gate normally-off high electron mobility transistors," Semicond. Sci. Technol., vol. 3, pp. 075006, June 2016.

[8] N. M. Shrestha, Y. Li, T. Suemitsu, and S. Samukawa, "Electrical Characteristic of AlGaN/GaN High-Electron-Mobility Transistors With Recess Gate Structure," IEEE Trans. Electron Devices, vol. 66, pp. 1694-1698, April 2019.

[9] N. Ketteniss, A. Askar, B. Reuters, A. Noculak, B. Hollander, H Kalisch, and A. Vescan, "Polarization-reduced quaternary InAlGaN/GaN HFET and MISHFET," Semicond. Sci. Technol., vol. 27, pp. 055012, April 2012.

[10] O. Ambacher, B. Foutz, J. Smart, J. R. Shealy, N. G. Weimann, K. Chu, M. Murphy, A. J. Sierakowski, W. J. Schaff, L. F. Eastman, R. Dimitrov, A. Mitchell, and M. Stutzmann, "Two dimensional electron gases induced by spontaneous and piezoelectric polarization in undoped and doped AlGaN/GaN heterostructures," J. Appl. Phys., vol. 87, no. 1, pp. 334–344, December 2000.

[11] N. Ketteniss, L. R. Khoshroo, M. Eickelkamp, M. Heuken, H. Kalisch, R. H. Jansen and A. Vescan, "Study on quaternary AlInGaN/GaN HFETs grown on sapphire substrates," Semicond. Sci. Technol., vol. 2, pp. 075013, June 2010.

[12] D. Godwinraj, H. Pardeshi, S. K. Pati, N. Mohankumar, and C. K. Sarkar, "Polarization based charge density drain current and small-signal model for nano-scale AlInGaN/AlN/GaN HEMT devices," Superlattices Microstruct, vol. 54, pp. 188–203, December 2013.

[13] N. M. Shrestha, Y. Li, E. Y. Chang, "Optimal design of the multiple-apertures-GaN-based vertical HEMTs with SiO2 current blocking layer," J Comput Electron, vol. 15, pp. 154-162, March 2016.

[14] N. M. Shrestha, Y. Li, E. Y. Chang, "Optimal design of the multiple-apertures-GaN-based vertical HEMTs with SiO2 current blocking layer," J Comput Electron, vol. 15, pp. 154-162, March 2016.

[15] T. Palacios, A. Chini, D. Buttari, S. Heikman, A. Chakraborty, S. Keller, S. P. DenBaars, and U. K. Mishra, "Use of Double-Channel Heterostructures to Improve the Access Resistance and Linearity in GaN-Based HEMTs," IEEE Trans. Electron Devices, vol. 53, pp. 562-565, March 2006.

[16] B. Reuters, A. Wille, N. Ketteniss, H. Hahn, B. Hollander, M. Heuken, H. Kalisch, and A. Vescan, "Polarization-Engineered Enhancement-Mode High-Electron-Mobility Transistors Using Quaternary AlInGaN Barrier Layers," J. Electron. Mater., Vol. 42, pp. 826-832, February 2013.

RF performance improvement on 22FDX® platform and beyond

Tom Herrmann
GLOBALFOUNDRIES Dresden
Dresden, Germany
tom.herrmann@globalfoundries.com

Alban Zaka
GLOBALFOUNDRIES Dresden
Dresden, Germany
alban.zaka@globalfoundries.com

Nandha Kumar Subramani
GLOBALFOUNDRIES Dresden
Dresden, Germany
nandhakumar.subramani
@globalfoundries.com

Zhixing Zhao
GLOBALFOUNDRIES Dresden
Dresden, Germany
zhixing.zhao @globalfoundries.com

Steffen Lehmann
GLOBALFOUNDRIES Dresden
Dresden, Germany
steffen.lehmann1
@globalfoundries.com

Yogadissen Andee
GLOBALFOUNDRIES Dresden
Dresden, Germany
yogadissen.andee
@globalfoundries.com

Abstract—The paper describes manufacturing process and layout optimizations to improve RF performance of 22FDX® N/PFET devices, based on a comprehensive calibration of DC and RF figures of merit. Process and Device simulations of the individual and combined elements show ft/fmax improvement up to about 1.13/1.1x (NFET) and about 1.32/1.24x (PFET) over standard devices mainly driven by mechanical stress and parasitic R/C elements.

Keywords—FDSOI, ft, fmax, 22FDX®, TCAD, RF, mmWave

I. INTRODUCTION

The benefits of 22FDX® in terms of ultra low leakage and power applications as well as RF features have been extensively analyzed in several publications [1,2,3]. This paper is now focusing on the TCAD simulation of further manufacturing processes and layout options to enhance RF performance for mmWave devices. Section II discusses the calibration of DC and RF Figures-of-Merit (FoMs) for both NFET and PFET devices. The calibration procedure in mixed-mode includes intrinsic transistor behavior as well as parasitic RC components, extracted after a layout analysis. Section III of the paper analyzes RF performance elements such as poly-poly pitch, gate length, different raised Source/Drain options and additional PFET Middle-of-Line stressors in detail. Significant increase in ft and fmax can be achieved by merging the single elements together.

II. PROCESS FLOW AND TCAD CALIBRATION

A. Process flow

The manufacturing processes relevant for the RF analysis of single devices consists of Front-End-of-Line until Metal1, since de-embedding procedure removes Back-End-of-Line parasitic elements from the measurement data. 22FDX® FEOL features are Si/SiGe channel as well as in-situ doped Si/SiGe raised source/drain for N/PFET, a high-k metal gate process with tensile strained liner to improve NFET performance with minimal impact on PFET devices. Flip-well (Super Low Vth) architecture allows further forward back biasing and exceptional low noise behavior due to low channel doping [1,2].

B. TCAD calibration

Process and device simulations have been performed with SENTAURUS TCAD including quantum-drift-diffusion framework, thin layer mobility, mechanical stress modulation of mobility and band structure. By implementing relevant process steps and doing careful structural matching to inline and TEM data (Fig.1) a solid DC matching to median values of the electrical test parameters can be achieved as shown in Fig.2 for NFET and Fig.3 for PFET devices. Parasitic resistances and capacitances from salicidation and contact process are included according to the layout of the measured devices and by considering additional measurements from specific test structures like contact chains.

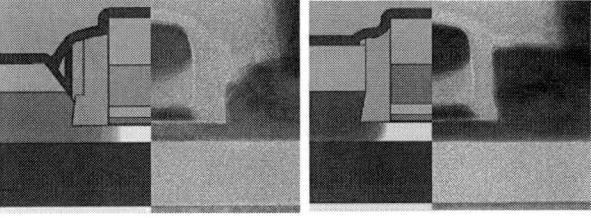

Fig. 1. TCAD and TEM cross section of a) NFET and b)PFET, along the channel

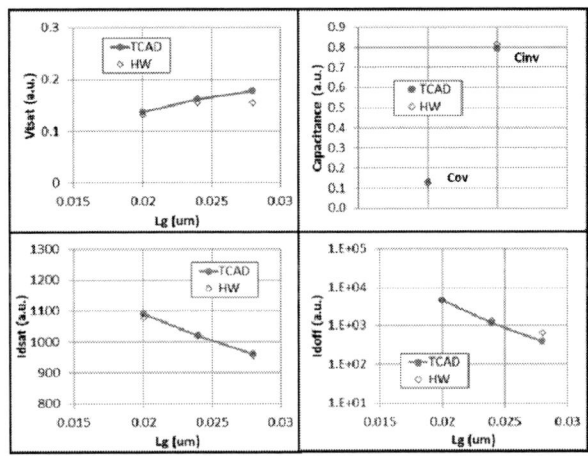

Fig. 2. NFET DC calibration across gate length (TCAD - line / HW - symbols)

Additionally, simulation time reduction could be achieved by simulating only one finger of the typical multi-finger RF devices by taking into account the active length effect on mechanical stress and the change in effective contact resistance

The resulting IdVg and GmVg characteristics in saturation (Vd=Vg=0.8V) of a typical device are shown in Fig.4 (NFET) and Fig.5 (PFET). These comparisons illustrate that TCAD simulations well reproduce the device

978-1-7281-0941-1/19 $31.00 © 2019 IEEE

DC behavior in the sub-threshold regime, at low as well as in high inversion regimes, with a particular focus on the transconductance behavior over gate bias, which is an important parameter for the RF FoMs. In order to match ft and fmax a careful analysis of the test structure (Fig.6) is essential to include the necessary parasitic components into the mixed mode simulation. Therefore additional separate 3D simulations have been performed to quantify the gate overhang and substrate capacitances as well as gate to contact and Metal1 capacitance (Fig.7). After taking into account all the above mentioned elements in a mixed-mode simulation, a good ft matching over gate bias as well as over drain current can be achieved as shown in Fig.8 (NFET) and Fig.9 (PFET).

Salicide-Contact interfaces and the MetalGate, Polysilicon, Salicide and Contact itself. Putting all of them together with the parasitic capacitances in a network of lumped elements, fmax will follow ft (Fig.8-NFET / Fig.9-PFET). The calculated gate resistance out of simulated S-Parameter matrix can be compared to the measured values for reference.

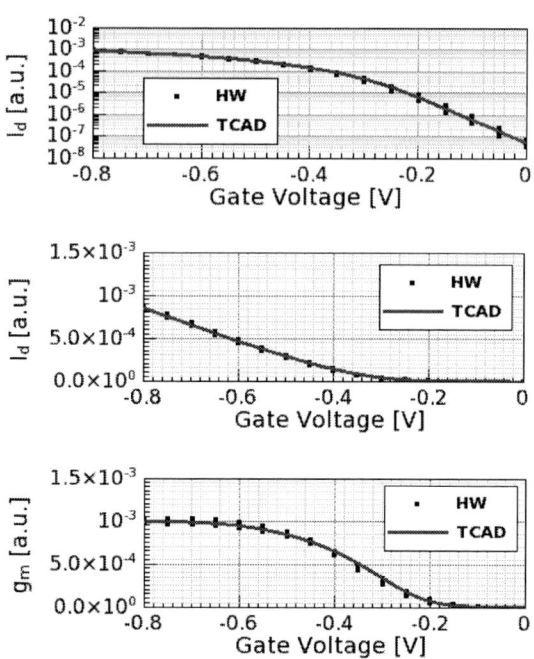

Fig. 5. Comparison of TCAD (line) against HW (symbols) in terms of IdVg and GmVg characteristics for a typical PFET device.

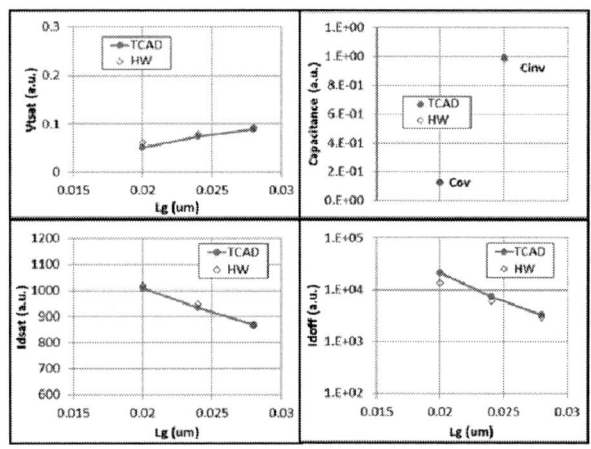

Fig. 3. PFET DC calibration across gate length (TCAD - line / HW - symbols)

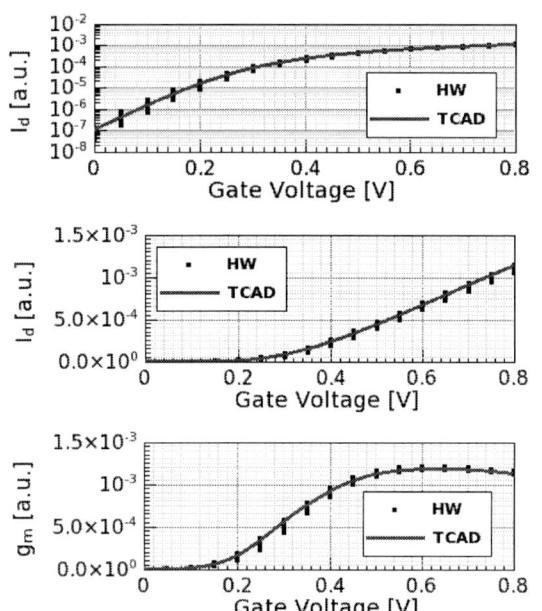

Fig. 4. Comparison of TCAD (line) against HW (symbols) in terms of IdVg and gmVg characteristcs for a typical NFET device.

Having ft calibrated, fmax is mainly impacted by gate resistance. Here the high-k metal gate first technology includes a couple of parallel and series resistances, namely the MetalGate-PolySilicon, PolySilicon-Salicide and

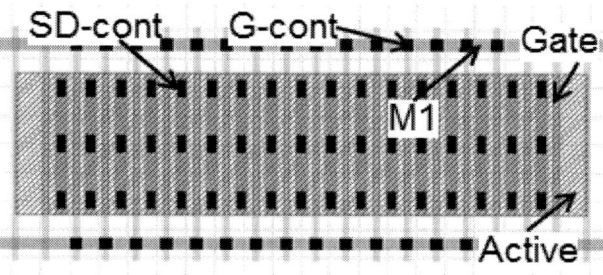

Fig. 6. Layout of RF single device test structure using Multi-Finger architecture and top/bottom gate contacts

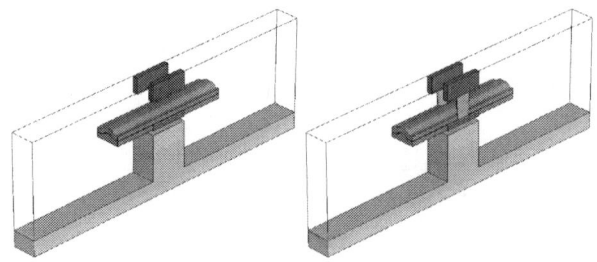

Fig. 7. 3D simulation structure to quantify parasitics of a) poly overhang to S/D/Substrate and b) poly to contact and Metal1 layer

978-1-7281-0941-1/19 $31.00 © 2019 IEEE

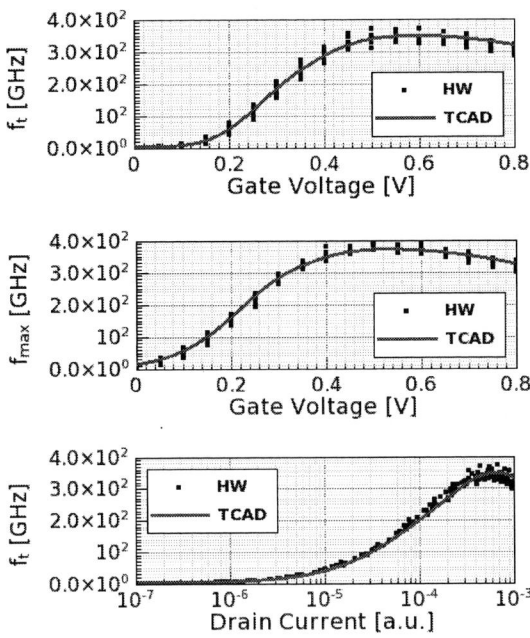

Fig. 8. ft/fmax calibration for NFET (TCAD - solid line / HW - squares)

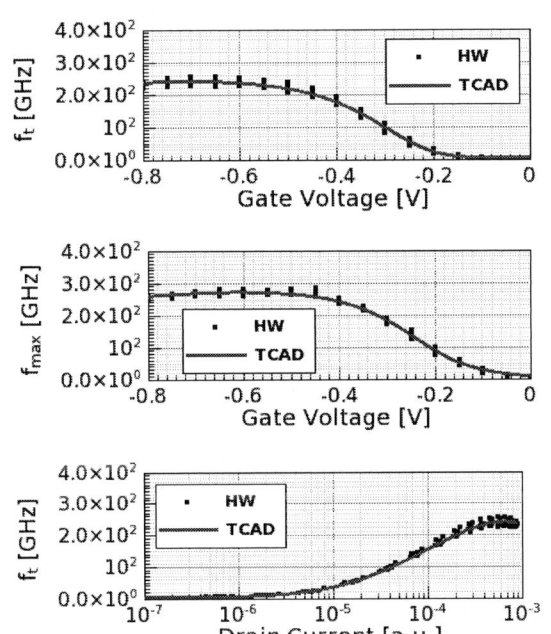

Fig. 9. ft/fmax calibration for PFET (TCAD - solid line / HW - squares)

III. RF OPTIMIZATION KNOBS

A. Individual elements analysis

Improvement of the RF FoMs ft and fmax can be done in two ways: by manufacturing process and by layout optimization. The considered elements in this study are summarized in Fig.10 and their independent relative impact on Ft, Fmax, gm are given in Fig.11 (NFET) and Fig.12 (PFET).

The process optimizations investigated here are 2 different raised S/D options and the introduction of a Middle-of-Line stressor element for PFET devices. With careful implementation, raised S/D modifications are mainly reducing the parasitic capacitance with only marginal impact on transconductance. The 2 raised S/D options consist of an epitaxy height reduction and the implementation of a partial facet epitaxy [4]. Adding the Middle-of-Line stressor element to the PFET device boosts uniaxial strain and improves gm significantly, hence increases ft and fmax.

An option combining manufacturing process and layout optimization is the gate length reduction. This is enabled by the intrinsic superior electrostatic control of FDSOI architecture. In addition, the increased Idoff leakage due to shorter Lgate is not detrimental for many RF applications. Thus, a lower effective inversion capacitance as well as higher transconductance as a consequence of shorter channel length is beneficial for RF FoMs gm and ft. Interestingly, fmax remains stable since the Lgate reduction brings about an Rgate incease, which eventually compensates the capacitance and transconductance benefits.,Relaxation of poly-poly pitch is investigated as a layout driven performance element [5]. The improvement of the RF FoMs can be mainly attributed to mechanical strain increase due to larger volume of raised S/D SiGe for PFET devices. NFET devices show better mechanical stress transfer from the MoL stressor into the channel region due to the larger open area between the polysilicon fingers. In addition, the larger raised S/D area allows for double row contact placement and has longer Silicon-Salicide interface region, which reduces the source/drain parasitic resistance for N- and PFET simultaneously.

Single splits	Description	
- #	NFET	PFET
1	Standard Process/Layout	
2	epi option 1	
3	epi option 2	
4	Gate length reduction	
5	Poly-Poly pitch relaxation	
6	--	MoL stressor

Fig. 10. Split table of the individual elements for RF improvement

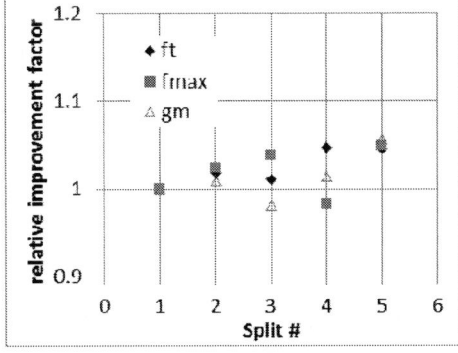

Fig. 11. Relative ft/fmax/gm improvement for single elements (listed in Fig.10) for NFET device.

978-1-7281-0941-1/19 $31.00 © 2019 IEEE

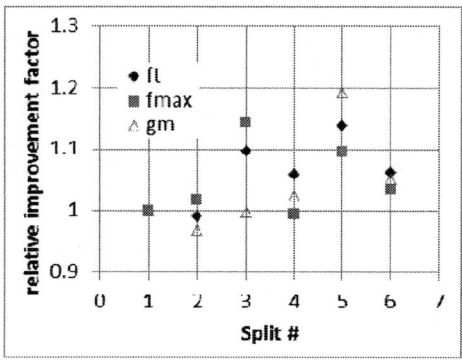

Fig. 12. Relative ft/fmax/gm improvement for single elements (listed in Fig.10) for PFET device

B. Combined elements analysis

The step-by-step combination of the elements is described in Fig.13. The corresponding mid channel stress for NFET and PFET devices is shown in Fig. 14. Here epi option 1 acts slightly and the Poly-Poly pitch relaxation stronger on the NFET channel stress, which leads to a gm and hence also ft/fmax improvement (Fig. 15). PFET devices show similar behavior of increasing channel stress using MoL stressor and Poly-Poly pitch relaxation resulting in gm/ft/fmax improvement (Fig.16). In contrast, the gate length reduction is increasing gm and ft, but slightly reduces fmax due to higher gate resistance. As stated in III.A the epi options are mainly reducing parasitic capacitances with a minor impact on gm, but improvement of ft and fmax. Combining all elements a ft/fmax improvement of ~1.13/1.11 for NFET and ~1.32/1.24 for PFET can be achieved with respect to current process and standard layout.

Combined splits - #		Description	
NFET	PFET		
1	1	Standard Process/Layout	
	2	--	+ MoL stressor
2	3	+ epi option 1	
3	4	+ epi option 2	
4	5	+ Gate length reduction	
5	6	+ Poly-Poly pitch relaxation	

Fig. 13. Split table of the combined elements for RF improvement

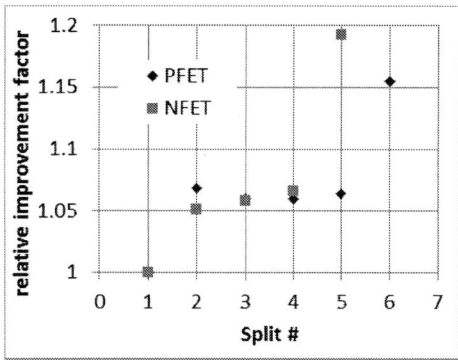

Fig. 14. Mechanical stress improvement factor of NFET and PFET device at mid channel position.

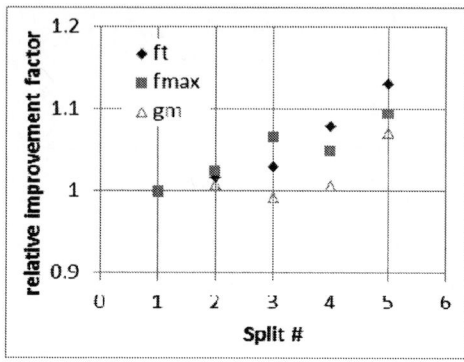

Fig. 15. Relative ft/fmax/gm improvement for NFET with the combined elements (shown in Fig.13)

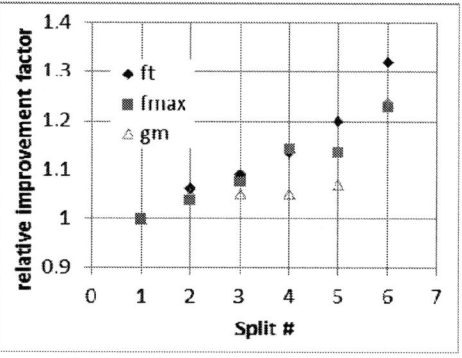

Fig. 16. Relative ft/fmax/gm improvement for PFET with the combined elements (shown in Fig.13).

IV. CONCLUSION

The potential improvement of 22FDX® technology with regard to gm/ft/fmax could be shown by combining different process and layout elements in TCAD simulations. Based on careful process and device calibration the different individual elements have been investigated and a combination of those elements shows a ft/fmax improvement of ~1.13/1.11 for NFET and ~1.32/1.24 for PFET. Main drivers are channel mechanical stress and parasitic resistance/capacitance.

ACKNOWLEDGMENT

This work is funded by the German Bundesministerium für Wirtschaft und Energie (BMWI) and by the State of Saxony in the frame of the "Important Project of Common European Interest" (IPCEI).

REFERENCES

[1] R. Carter et. al., "22nm FDSOI Technology for Emerging Mobile, Internet-of-Things, and RF Applications", IEDM, 2016, pp.27-30

[2] E.M. Bazizi et. al., "Versatile technology modeling for 22FDX platform development", SISPAD, 2017

[3] S.N. Ong et. al., "A 22nm FDSOI Technology Optimized for RF/mmWave Applications", RFICS, 2018, pp. 72-75

[4] Ö.I. Aydin et al.: "Advantages of Faceted P-Raised Source/Drain in Fully Depleted Silicon on Insulator Technology", ECS Trans. 2018 86(7), pp. 199-206

[5] Z. Zhao et al.: ""22FDX® fMAX Optimization Through Parasitics Reduction and GM Boost", ESSDERC 2019, to be published

Leakage Performance Improvement in Multi-Bridge-Channel Field Effect Transistor (MBCFET) by Adding Core Insulator Layer

Saehoon Joung[1,2], *Student Member, IEEE* and SoYoung Kim[2], *Senior Member, IEEE*

[1]Samsung Electronics Co. Foundry Division, Yield Enhancement, Process Integration Engineering Group, Ltd Kiheung, Republic of Korea

[2]College of Information and Communication Engineering,Sungkyunkwan University, Suwon,Gyeounggi-do,Republic of Korea

Email: ksyoung@skku.edu

Abstract—Altering from existing planar devices to FinFETs has revolutionized device performance, but demands of leakage and gate controllability are increasing relentlessly. Gate all around field effect transistor (GAAFET) is expected to be the next-generation device that meets these needs. This paper suggests a way to improve the gate electrostatic characteristics by adding an oxidation process to the conventional multi-bridge-channel field effect transistor (MBCFET) process. The main advantage of the proposed method is that a device with ultimate electrostatic properties can be implemented without changing the complex and expensive photo-patterning. In the proposed device, the immunity of short channel effects is enhanced in a single transistor. And the performance of ring oscillator (RO) and SRAM was confirmed to be improved by Sentaurus technology computer aided design (TCAD) mixed-mode simulation.

Index Terms—Gate-all-around FET, Transistor leakage, Insulator, Gate controllability, Electrostatic potential, MBCFET.

I. INTRODUCTION

Although changes from planar 2D devices to FinFET have revolutionized device performance, demands of leakage and gate controllability are growing more rapidly. So recently, FinFET faces many challenges, such as high cost of scaling, performance limitation, process variation immunity, process difficulty for steep fins [1-3]. While operating voltage should continue to decrease for longer battery times, nevertheless the performance of the product should be superior to that of the existing product. Almost all of the FinFET problems are expected to be dramatically solved with gate all around field effect transistor (GAAFET) [4-5]. However, as FinFET did in the past, GAAFET will face limitations that cannot satisfy the demands of better properties. Therefore, it is still mandatory to improve leakage and electrostatic characteristics. This paper proposes a method to improve the leakage and gate electrostatic characteristics of the device by adding an oxidation layer to multi bridge-channel-field effect transistor (MBCFET) which is the most promising next-generation device. The proposed method in this study can be applied to existing MBCFET processes with less effort and can improve

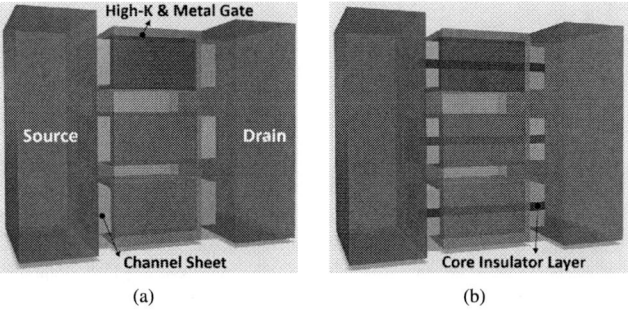

Fig. 1: Structure comparison. (a) Conventional MBCFET, (b) Core insulator MBCFET.

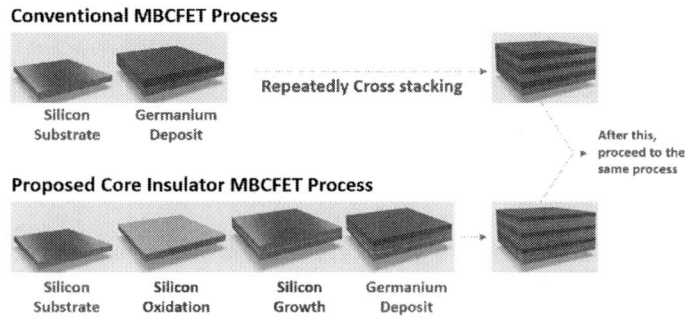

Fig. 2: Process flow comparison.

the electrostatic properties without tuning the complex and expensive photo-patterning process.

II. DEVICE STRUCTURE AND SIMULATION APPROACH

Fig. 1 (a) shows the conventional MBCFET and fig. 1 (b) shows the proposed core insulator MBCFET. Compared to

978-1-7281-0941-1/19 $31.00 © 2019 IEEE

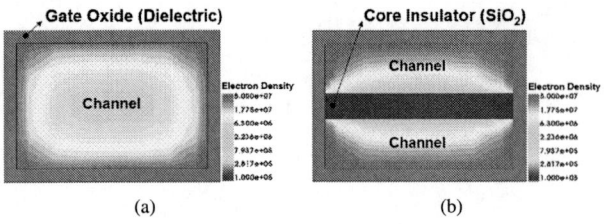

Fig. 3: Channel cross-sectional view of electron current density. (a) Conventional MBCFET. (b) Core insulator MBCFET.

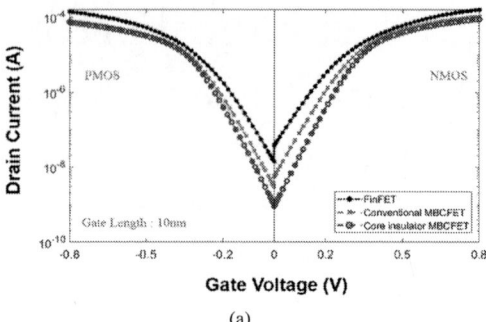

Fig. 4: Comparing the I_d-V_g curves of NMOS and PMOS.

the conventional MBCFET, the proposed device has a thin oxidation layer in the core of the channel. And equally, the HfO_2 gate oxide and metal gate surround the channel on four sides, including the core oxide layer [6]. Fig. 2 is a process sequence for implementing this MBCFET structure. The conventional MBCFET is created through cross-stack silicon and silicon-germanium (SiGe) and implements channel by removing only the SiGe layer using the selective etching ratio between two layers [7-8]. Then a thermal oxidation step is added in the middle of this process to create an insulating layer between silicon and silicon. Increasing the distance from the gate electrode reduces the effect of gate bias on the channel, which degrades gate controllability and leakage characteristics. Fig. 3 (a), cross-sectional view of electron density in the conventional MBCFET, shows that the closer the gate electrodes, the higher the electron density, and the farther away the less the effect of the gate biases. Fig. 3 (b) is the same view of the proposed core insulator MBCFET. The core insulator MBCFET improves electrostatic properties by effectively removing poor gate control area. Creating the channel very thin for device characteristics increases device variation with the damage of silicon when SiGe is etched [9]. As the process continues to scale, it is becoming much more difficult to form a very thin channel. By inserting the insulator in the core of the channel, it can have the effect that a very thin channel is implemented. This improved method is difficult to use in the FinFET process. The reason is that it is quite difficult to etch a very thin fin core that has already been formed by the shallow trench process [10]. On the other hand, in the MBCFET process, which forms channels later through selective etch after cross-stacking of silicon and SiGe, it is easy to insert another layer in the middle.

Based on this idea, technology computer-aided design (TCAD) simulation was conducted with MBCFET, which will be used in of 3nm process and lower. TCAD simulator was used with various physics models to electrically analyze the characteristics of transistors ; Trap-assisted Auger recombination model was applied to analyze the electrical characteristics of highly doped devices and the Shockley-Read-Hall (SRH) model was applied to estimate the carrier generation and recombination mechanism for estimating doping-dependent device characteristics [11]. The physical and electrical specifications of the device used for the simulation are given in Table 1. In order to analyze to short channel effects immunity, it is used as a characteristic indicator of drain induced barrier lowering (DIBL) and subthreshold slope (S_{slope}).

III. SINGLE TRANSISTOR SIMULATION RESULTS

Fig. 4 shows the correlation between the gate bias (V_g) and the drain current (I_d). The left side shows the characteristics of the PMOS and the right side shows the that of the NMOS. The steeper the curve of the I_d-V_g, the better electrostatic properties. Analysis of the data shows that the core insulator MBCFET has the best switching characteristics. In order to analyze short channel effects immunity in detail, we compared DIBL and S_{slope} characteristics according to gate length as shown in fig. 5. Fig. 5 (a) shows the DIBL according to the gate length and fig. 5 (b) is that of the S_{slope}. The proposed core insulator MBCFET has 29.5% better DIBL compared to the conventional MBCFET, 55.5% improvement compared with the FinFET and S_{slope} improvement is 5.9% compared to the conventional MBCFET and more than 40% compared to FinFET at 8nm gate length. The immunity of short channel effects, represented by the DIBL and S_{slope}, improved at the proposed device because it effectively removed the poor gate control area by using an insulating. Because short channel effects increase exponentially as the gate length is shortened, the advantages of the proposed device are increased significantly. It means that even if the gate length is shortened, the device can have robust and mass-productive properties [12].

As shown in fig. 6, leakage characteristics also be improved by inserting a core insulator. Fig. 6 (a) shows the relationship

TABLE I: Parameters of transistors

	FinFET	Normal MBCET	Core Insulator MBCFET
Gate Length	5-90nm	5-90nm	5-90nm
Channel Height (Fin Width)	10nm	10nm	10nm
Channel Width (Fin Height)	45nm	15nm	15nm
Gate Oxide Thickness	2nm	2nm	2nm
Operating Volatage	0.8V	0.8V	0.8V
Channel Doping Concentration	5e17	5e17	5e17
S/D Doping Concentration	2e20	2e20	2e20
Workfunction of NMOS	4.45eV	4.45eV	4.45eV
Workfunction of PMOS	4.85eV	4.85eV	4.85eV

978-1-7281-0941-1/19 $31.00 © 2019 IEEE

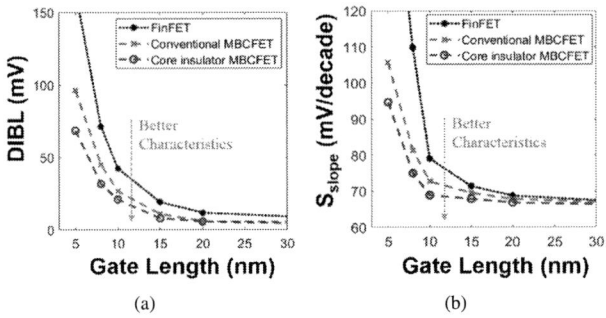

(a)   (b)

Fig. 5: Gate length dependency of short channel effects characteristics. (a) DIBL. (b) S_{slope}.

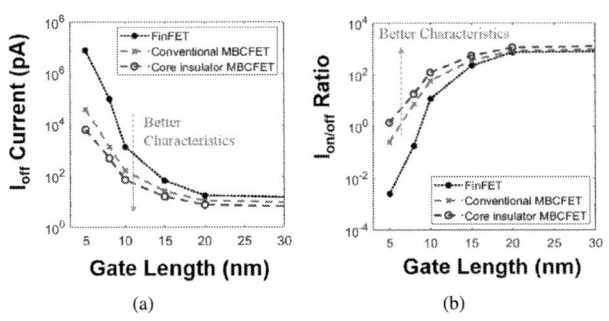

(a)   (b)

Fig. 6: Leakage characteristics. (a) I_{off}-L_g dependency. (b) $I_{on/off}$-L_g ratio dependency.

between gate length and leakage current (I_{off}), and fig. 6 (b) shows the ratio on drive and leakage current ($I_{on/off}$) for checking the risk of I_{on} reduction. In fig. 6 (a), although leakage current (I_{off}) increased exponentially at short gate lengths in the three devices in common, I_{off} of the proposed device increases less extremely than that of other devices. Also, as shown in fig. 6 (b), the $I_{on/off}$ characteristics of the proposed device were superior to that of other transistors. The main components of the transistor leakage are subthreshold leakage due to minority carrier diffusion, gate leakage from dielectric tunneling and junction leakage due to reverse bias. The reason for the good result of the $I_{on/off}$ in the proposed device is that it reduces the area of the channel with the insulator to lower the subthreshold leakage. At the same time, it increases the sensitivity to the gate bias, thus suppressing the reduction of I_{on} as much as possible.

IV. MIXED-MODE SIMULATION RESULTS

A. Ring Oscillator Simulation Results

In this subsection, we analyzed the improvement of logic circuit composed of NMOS and PMOS using TCAD mixed-mode simulation. First, we compared the performance of ring oscillator (RO) which is the basis of design in a lot of applications. The experiment consisted of a 3 stage RO

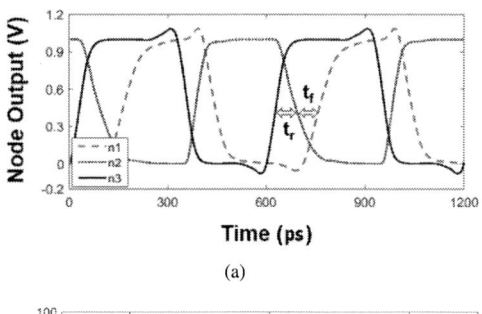

(a)

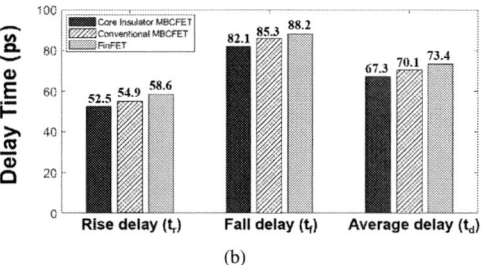

(b)

Fig. 7: RO simulation results. (a) Time domain output waveform. (b) Delay comparisons.

consisting of three transistors, respectively, to analyze the change in delay. Fig. 7 (a) is a graph of the output waveform of RO configured with the proposed device. The rise delay (t_r) and fall delay (t_f) were measured based on 0.5 * VDD and the average delay (t_d) was calculated as $t_d = (t_r + t_f) / 2$ [13]. Fig. 7 (b) compares the delay times between the three devices. The average delay of the core insulator RO is improved by 4.0% compared with the conventional MBCFET RO and 8.3% improved compared with the FinFET RO. The reason for this improvement is that the leakage of off-state transistors improves the overall switching performance of the inverter.

B. SRAM Simulation Results

Static random access memory (SRAM) is a logic circuit that is as important as an RO in an actual logic integrated circuit chip. Since the SRAM carries out its role as a cache memory in the application processor requiring a high-speed operation, the speed of the memory itself is very important for the performance of the entire product. SRAM is generally implemented with four NMOS and two PMOS devices as shown in fig. 8 (a). When the word line (WL) enable signal is applied to read or write data, the current flows from the precharged bit line (BL) to the SRAM bit cell [14]. If the current from this BL cannot be sent to the ground sufficiently, the existing cell data can be flipped. This failure mechanism is called disturb fail. And SRAM noise margin is a indicator that tests and extracts at this environment. SRAM noise margin is measured by a butterfly curve resulting from sweeping V_{in} and V_{out} across the half-cells of the SRAM. The higher the

978-1-7281-0941-1/19 $31.00 © 2019 IEEE   269

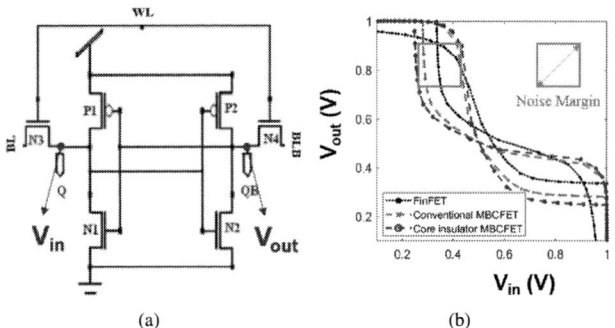

(a)   (b)

Fig. 8: SRAM simulations. (a) Circuit schematic of SRAM. (b) Comparison of SRAM noise margin.

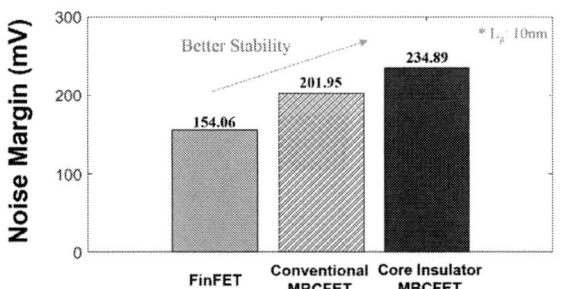

Fig. 9: Noise margin comparison in SRAM.

noise margin value, the more stable the cell can store data. Fig. 8 (b) shows the superposition of the butterfly curves of SRAMs made with three devices. In order to obtain a superior noise margin, the switching characteristics should be improved, and the leakage current in the off state have to be suppressed. Fig. 8 (b) shows that SRAM with the proposed device has better noise margin. For a more quantitative analysis, fig. 9 shows a 16.3% improvement in the SRAM implemented using the core insulator compared with conventional MBCFET SRAM and more than 40% improved compared with the FinFET SRAM. Higher tolerance to the disturb failure can significantly improve the overall memory speed because fewer redundancy repair circuits are used in the event of a memory failure.

V. CONCLUSIONS

The electrostatic and leakage properties of the GAAFET is expected to be further improved, but the relentless require-ment demand will exceed the performance limit of general GAAFET device. So, the needs for research on better electro-static performance are never reduced. Most applications are designed with a combination of devices with a high drive current or good leakage control. Generally, these types of devices are implemented by diversifying the gate length by tunning the photo-patterning process. Tuning the patterning involves increasing process complexity and costs. In this paper, it was confirmed that the leakage and electrostatic characteris-

tics were improved through simple oxidation process addition. The most attractive point is that it can achieve better device performance through a simple oxidation without changing a difficult photo-patterning. Furthermore, as the device continues to scale, the proposed device can get higher resistance to process variation than photo-patterning method in terms of controlling the short channel effects. It is also meaningful to verify progress in composite devices such as RO and SRAM. In this study, we concluded that the proposed method can suppress leakage current without significantly reducing the driving current through the core insulator.

VI. ACKNOWLEDGEMENT

This work was supported by the National Research Foun-dation of Korea (NRF) grant funded by the Korea government (MSIP) (No. NRF-2017R1A2B2003240). The TCAD tools were supported by the IC Design Edu-cation Center (IDEC).

REFERENCES

[1] M. Garcia Bardon et al., "Dimensioning for power and performance under 10nm: The limits of FinFETs scaling," International Conference on IC Design & Technology (ICICDT)., pp. 1-4, June. 2015.
[2] B. S. Kumar et al., "On the design challenges of drain extended FinFETs for advance SoC integration," International Conference on Simulation of Semiconductor Processes and Devices (SISPAD)., pp. 189- 192, Sep. 2017
[3] Gaurav Saini, Ashwani K Rana , "Physical Scaling Limits of FinFET Structure: A Simulation Study," International Journal of VLSI design & Communication Systems (VLSICS)., Vol.2, pp. 26-35, Mar. 2011
[4] Doyoung Jang et al., "Device Exploration of NanoSheet Transistors for Sub-7-nm Technology Node," IEEE Transactions on Electron Devices. ,vol. 64 ,no. 6, pp. 2707-2713, May. 2017
[5] N. Loubet et al., "Stacked nanosheet gate-all-around transistor to enable scaling beyond FinFET," Symposium on VLSI Technology. , pp. 230-231, June. 2017
[6] S. Lee et al., "Sub-25 nm single-metal gate CMOS multi-bridge-channel MOSFET (MBCFET) for high performance and low power application," VLSI Symp. Tech. Dig., pp. 154-155, June. 2005
[7] Sung-Young Lee et al., "A novel multibridge-channel MOSFET (MBCFET): fabrication technologies and characteristics," IEEE Trans-actions on Nanotechnology.,vol.2, no.4, pp. 253-257, Dec. 2003
[8] Emilie Bernard et al., "Multi-Channel Field-Effect Transistor (MCFET)—Part I: Electrical Performance and Current Gain Analysis," IEEE Transactions on Electron Devices., vol. 56, no. 6, pp. 1243-1251, June. 2009
[9] H. Kawasaki et al., "Challenges and Solutions of FinFET Integration in an SRAM Cell and a Logic Circuit for 22nm node and beyond," IEEE International Electron Devices Meeting (IEDM)., pp. 1-5, Dec. 2009
[10] Suk-Kang Sung et al., "Fully Integrated SONOS Flash Memory Cell Array With BT (Body Tied)-FinFET Structure," IEEE Transactions on Nanotechnology., vol. 5, no. 3, pp. 174-179, May. 2006
[11] Sentaurus Device User Guide, K-2015.06-SP1, Synopsys, Mountain View, CA, USA, 2015.
[12] E. M. Bazizi et al., "USJ engineering impacts on FinFETs and RDF investigation using full 3D process/device simulation," International Conference on Simulation of Semiconductor Processes and Devices (SISPAD)., pp. 25-28, Sep. 2014
[13] SeongSik Choe et al., "Performance Analysis of Tri-gate FinFET for Different Fin Shape and Source/Drain Structures," Journal of The Institute of Electronics and Information Engineers., vol. 51 , no. 7, pp. 1497-1507, Dec. 2009
[14] X. Wang et al., "Process informed accurate compact modelling of 14-nm FinFET variability and application to statistical 6T-SRAM simulations," International Conference on Simulation of Semiconductor Processes and Devices (SISPAD)., pp. 303-306, July. 2014

978-1-7281-0941-1/19 $31.00 © 2019 IEEE

Impact of MOL/BEOL Air-Spacer on Parasitic Capacitance and Circuit Performance at 3 nm Node

Ashish Pal, Sushant Mittal, El Mehdi Bazizi, Angada Sachid, Mehdi Saremi, Benjamin Colombeau, Gaurav Thareja, Samuel Lin, Blessy Alexander, Sanjay Natarajan and Buvna Ayyagari

Applied Materials, Santa Clara, USA

Abstract— Impact of air-spacer at MOL and BEOL on circuit performance at 3nm technology node is studied. Our modeling results show that by introducing air-spacer at MOL and BEOL, parasitic capacitance can be reduced by 18% and circuit performance as simulated on a 31-stage ring oscillator can be improved by 6%. Other advanced parasitic improvement technologies, such as Ruthenium, also show similar performance improvement. Finally, we show that best circuit performance is achieved when these 2 technologies are combined, yielding to a circuit performance boost of 16%.

Keywords—Air-Spacer, FinFET, ring oscillator, middle-of-line, back-end-of-line

I. INTRODUCTION

Air-spacer, also known as air-gap is an attractive option for logic technology to reduce parasitic capacitance (C_P) and increase circuit performance. Various simulation and experimental research have demonstrated air-spacers in the front-end-of-line (FEOL) transistor to improve circuit performance [1]. However, because of the presence of numerous materials, and the complexity of etch selectivity between these materials at FEOL level, semiconductor industry still has not been able to adopt the air-spacer technology at FEOL level. To utilize the concept of air-spacer, while still maintaining the integrity of other materials around it, one approach can be to introduce it at MOL (middle-of-line) and BEOL (back-end-of-line) level, where the size of the materials-set exposed to the air-spacer etch is comparatively small. Intel has already introduced air-spacer at M4 and M6 level in their 14nm technology node [2]. In this paper, we proceed one step further, introducing the air-spacer at MOL, M0 and M1 level and by studying its impact on circuit performance.

II. DESCRIPTION OF APPROACH

31-stage ring oscillator (RO) is used as a representative circuit to investigate the impact of air-spacer at MOL/BEOL for 3 nm technology node. The RO simulation framework is described in figure 1. First, TCAD simulations are performed for individual n-channel and p-channel FEOL FinFET devices (shown in figure 2) using typical 3nm node parameters (shown in figure 3). Then BSIM-CMG compact model is calibrated with TCAD generated current-voltage and capacitance-voltage characteristics, as shown in figure 4.

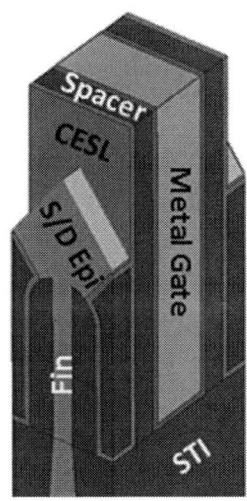

Fig. 2. FEOL FinFET structure for individual NMOS and PMOS TCAD device simulations

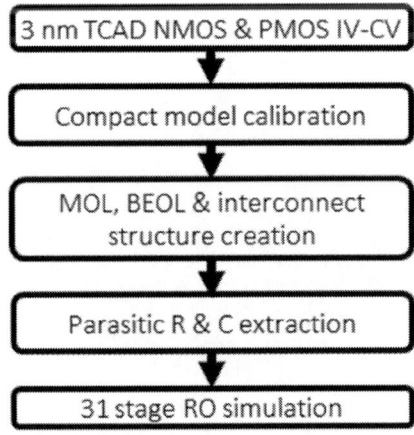

Fig. 1. Framework for 31 stage ring oscillator simulation

Parameter	Value	
Gate Pitch	45 nm	
Fin Pitch	26 nm	
Fin Width	5 nm	
Fin Height	60 nm	
Gate Length	15 nm	
Parameter	NMOS	PMOS
I_{ON}(mA/µm)	2.1	1.9
SS(mV/dec)	70.7	71.6
DIBL(mV/V)	47.4	47.9

Fig. 3. Typical FEOL dimensions used for 3 nm node FinFET, together with short-channel characteristics and on-current (at 10 nA/µm off-current).

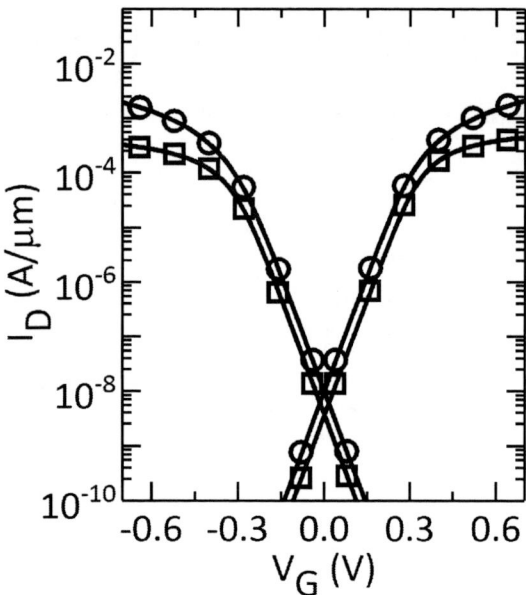

Fig. 4. Transfer characteristics of NMOS and PMOS devices from TCAD and BSIM-CMG compact model, showing the goodness of fit between those.

To model the interconnect of an inverter, a 3D structure comprising sub-contact SC or contact-plug, MOL (contact-trench CT & contact-gate CG and surrounding dielectrics) and BEOL (via-0, M0, via-1, M1 and surrounding dielectrics) is constructed, using a 3 nm inverter layout (shown in figure 5a), FinFET process flow (shown in figure 5b) and currently established material system for foundry 7nm node. The industry standard FinFET process flow is modified by inserting air-spacer modules at different levels of MOL & BEOL.

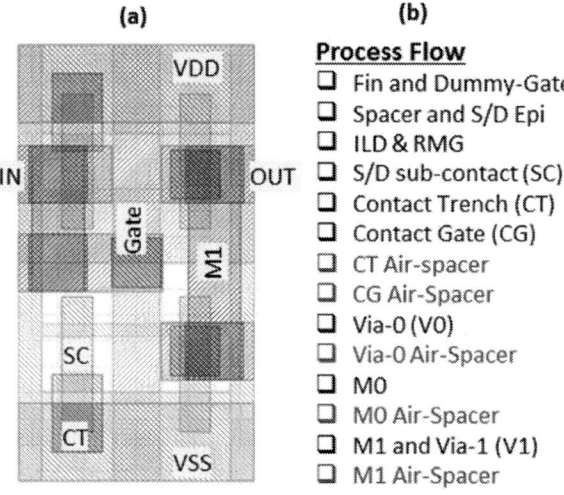

Fig. 5. (a) Inverter layout created with extrapolated 3 nm design rules. 2(b) FinFET process flow used to create 3D structures. Different air-spacer modules shown in blue font are optional.

The baseline 3D-structure (without air-spacer) is shown in figure 6. The 3D structure is used to extract interconnect parasitic resistance (R_P) and parasitic capacitance (C_P) of an inverter. For RO simulations, similar 3D structures are formed separately to model the interconnect between inverters, whose length is assumed to be 90 nm (2 gate pitch). Finally, transient simulations of RO are performed by combining 31 connected inverters as active elements, the calibrated BSIM-CMG compact model and interconnect R_P & C_P components. Stage delay of the ring oscillator is extracted and used as performance metric.

Figure 7 is a pie-chart, showing RO delay contribution of FEOL, sub-contact C_P, combined C_P and R_P of BEOL and MOL. Figure 7 does show that after FEOL device, the sub-contact C_P has maximum contribution to circuit performance, therefore justifying the effort of research community on investigating air-spacer at FEOL device.

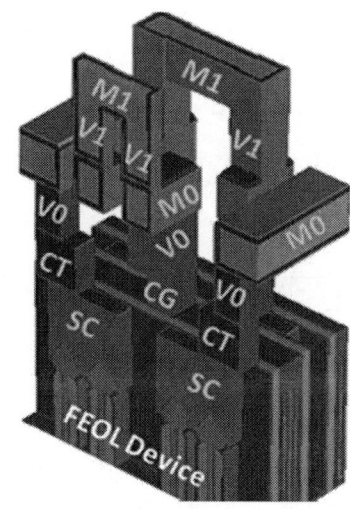

Fig. 6. Example of a 3D-structure with different layers annotated. The 3D-structure is built using the layout in and the process flow as described above. BEOL and MOL parasitic resistance and capacitances are extracted from this 3D-structure

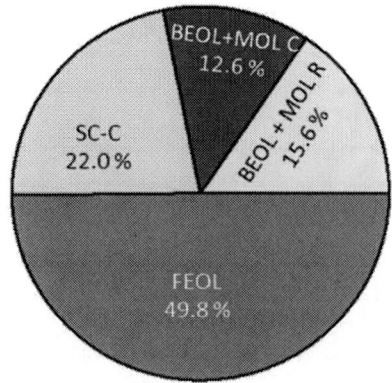

Fig. 7. Pie-chart distribution showing contribution of different elements to 31-stage ring-oscillator delay: FEOL device, sub-contact parasitic capacitance, BEOL & rest of MOL parasitic capacitance and BEOL & MOL parasitic capacitance.

978-1-7281-0941-1/19 $31.00 © 2019 IEEE

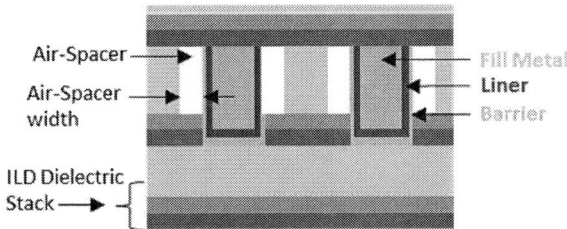

Fig. 8. Simplified air-spacer structure used in the simulations.

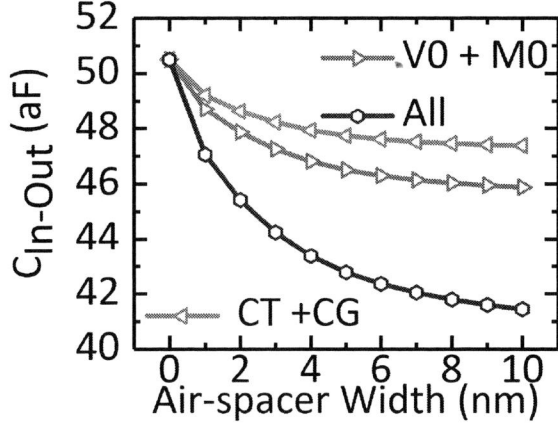

Fig. 10. Parasitic capacitance between input and output port of an inverter as a function of air-spacer width, when it is introduced at multiple locations

III. DISCUSSIONS AND RESULTS

Currently significant effort is dedicated to reduce the R_P of MOL & BEOL by introducing new conductors - liners, barriers and fill metals. However, the pie-chart in figure 7 shows that MOL/BEOL C_P contributes almost equally to circuit performance, when compared to MOL/BEOL R_P. This indicates that air-spacer at MOL/BEOL can bring similar performance benefit as of advanced MOL/BEOL conductor related innovations.

For illustration, an air-spacer structure of constant spacer width is assumed, as shown in figure 8. In real fabrication process, the air-spacer might be of larger width at top and tapered at bottom. Typically, an ILD stack would consist of 3 layers, where the topmost layer, also called ULK (ultra-low K) layer is the thickest and with lowest dielectric constant. In our process emulation of air-spacer, we etch this ULK layer, selectively to the 2nd layer of the ILD stack. To understand impact of air-spacer on CP at different locations of MOL & BEOL, air-spacer is introduced separately at different stages (CG, CT, V0, M0 and M1) of the process flow, as shown in figure 5b. For each case, the air-spacer width is varied between 0 to 10nm and the parasitic capacitance between inverter's input and output port (C_{In-Out}) is extracted. Introduction of air-spacer at V0 has the largest benefit, reducing C_{In-Out} by 6% (figure 9). Air-spacer at CG and CT can reduce C_{In-Out} by 5%. On the other hand, air-spacer at M0 and M1 has relatively lower impact on C_{In-Out}, reducing it by 3%. By inserting air-spacer at multiple locations, the benefits can also be combined as shown in figure 10. Implementation of air-spacer at both V0 and M0 can reduce C_{In-Out} by 9%, whereas insertion of air-spacer at both CG and CT yields a

reduction of 6%. When deployed at all MOL and BEOL level, C_{In-Out} can be reduced by about 18%, as shown in figure 10. Figure 11 shows the impact of air-spacer on RO performance, when air-spacer is introduced separately at different MOL & BEOL locations. Similar trend in RO stage delay is found as of C_{In-Out}. With air gap only at V0, RO delay can be improved by about 1.9%, whereas air-spacer at CG, CT and M0 can reduce RO delay each by 1.3%. Similar to C_{In-Out}, RO performance benefit can be further boosted by inserting air-spacer at multiple locations as shown in figure 12. With air-spacer inserted at all locations, RO delay can be reduced by about 6%.

Finally, to emphasize the performance benefit of air-spacer at BEOL technology, we compare it with other MOL/BEOL parasitic reduction technology. Among different conductors, ruthenium has shown most promises because of it lower resistivity at lower critical dimensions. Our RO simulation results (figure 13) show that when Ruthenium (with 2nm barrier) is introduced at all MOL and BEOL levels, the RO delay reduces by about 5.6%, a similar performance improvement figure as of air-spacer. Furthermore, we show

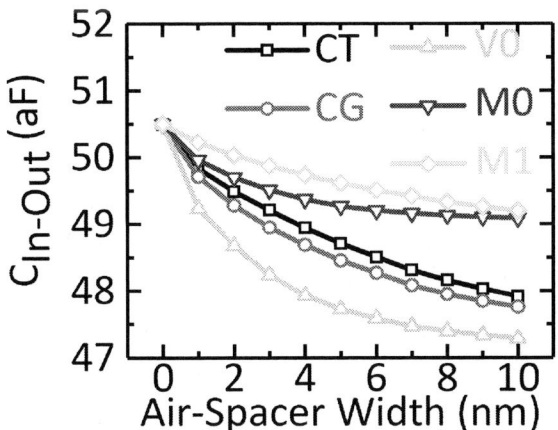

Fig. 9. Parasitic capacitance between input and output port of an inverter as a function of air-spacer width, when it is introduced at only at one location

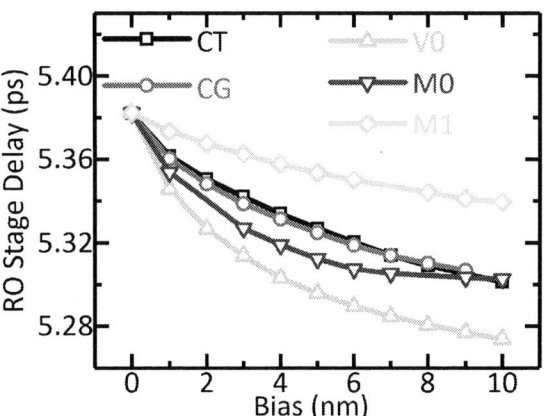

Fig. 11. Ring-Oscillator delay as a function of air-spacer width when air-spacer is introduced only at one location.

978-1-7281-0941-1/19 $31.00 © 2019 IEEE   273

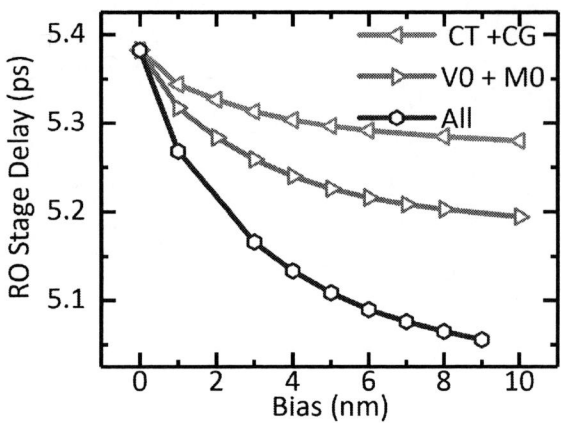

Fig. 12. Ring-Oscillator delay as a function of air-spacer width when at multiple locations of the FinFET process flow

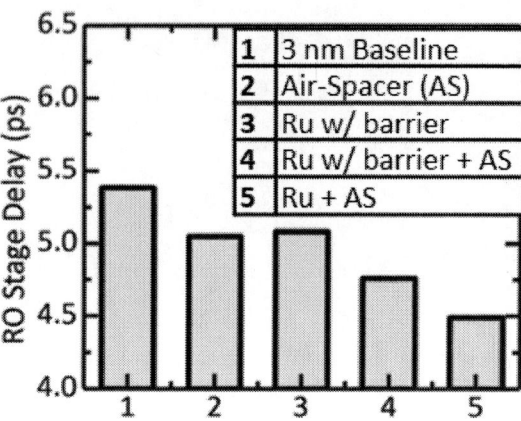

Fig. 13. Comparison of RO performance with air-spacer (AS) technology and Ruthenium (Ru)-based systems. Ru and air-spacer technology, when combined give best circuit-level performance. An air-spacer width of 10nm is used for these simulations.

that Ru and air-spacer technologies can be combined to reduce both MOL/BEOL R_P and C_P, achieving a circuit performance benefit of about 11%. The circuit performance can be further boosted by using air-spacer and eliminating the barrier for Ru, as demonstrated recently [3], since because of the air-spacer, Ru cannot diffuse in surrounding ILD dielectric. Combining barrier-less Ru and air-spacer technology, a maximum circuit performance benefit of about 16% can be achieved, which is difficult to obtain just by using Ruthenium technology alone.

REFERENCES

[1] K. Cheng et al, "Air spacer for 10nm FinFET CMOS and beyond", in IEEE International Electron Devices Meeting (IEDM), pp. 17.1.1 – 17.1.4, Dec 2016

[2] S. Natarajan et al, "A 14nm Logic Technology Featuring 2nd-Generation FinFET Transistors, Air-Gapped Interconnects, Self-Aligned Double Patterning and a 0.0588 μm² SRAM cell size," in IEEE Internaional Electron Device Meeting (IEDM), pp. 3.7.1 – 3.7.3, Dec 2014

[3] K. Croes et al, "Interconnect metals beyond copper: reliability challenges and opportunities," in IEEE Internaional Electron Device Meeting (IEDM), pp. 5.3.1 – 5.3.4, Dec 2018

Scaling-aware TCAD Parameter Extraction Methodology for Mobility Prediction in Tri-gate Nanowire Transistors

Cristina Medina-Bailon
School of Engineering
University of Glasgow
Glasgow, United Kingdom
Cristina.MedinaBailon@glasgow.ac.uk

Tapas Dutta
School of Engineering
University of Glasgow
Glasgow, United Kingdom
Tapas.Dutta@glasgow.ac.uk

Fabian Klüpfel
Fraunhofer Institut für Integrierte
Systeme und Bauelementetechnologie
Erlangen, Germany
Fabian.Kluepfel@iisb.fraunhofer.de

Sylvain Barraud
CEA, LETI, MINATEC campus
and Universit Grenoble Alpes
Grenoble, France
Sylvain.Barraud@cea.fr

Vihar Georgiev
School of Engineering
University of Glasgow
Glasgow, United Kingdom
Vihar.Georgiev@glasgow.ac.uk

Jürgen Lorenz
Fraunhofer Institut für Integrierte
Systeme und Bauelementetechnologie
Erlangen, Germany
Juergen.Lorenz@iisb.fraunhofer.de

Asen Asenov
School of Engineering
University of Glasgow
Glasgow, United Kingdom
Asen.Asenov@glasgow.ac.uk

Abstract—In the simulation framework for the study of aggressively scaled CMOS transistors, it is mandatory to capture the dependence of the model parameters on the physical structure of the devices in order to perform predictive device simulations. TCAD models typically have tunable parameters to characterize physical phenomena that ultimately determine different measurable electrical quantities. In this work, we extract the density gradient quantum correction parameters and Monte Carlo scattering parameters in order to fit the C-V characteristics and the low field mobility to experimental data in the case of Tri-gate nanowire transistors, which are of high importance for the semiconductor industry. Once the relevant parameters are calibrated, we have obtained a good agreement between the experimentally measured mobility and the predictions from the Monte Carlo module of the Synopsys TCAD tool Garand.

Index Terms—Density Gradient; Monte Carlo; Remote Coulomb Scattering; Phonon Scattering; Surface Roughness Scattering; Capacitance - Voltage characteristics; Mobility; Tri-gate MOSFETs

I. Introduction

Physical models that determine the macroscopic behaviour of the field-effect transistors (FETs), such as the carrier mobility, in TCAD simulations do not scale properly with device geometry in the nanometer regime. Therefore, it is

This work was supported by the European Unions Horizon 2020 research and innovation programme under grant agreement No 688101 SUPERAID7. The authors would like to thank Dr. Ewan Towie for useful discussions.

critical to capture the scaling dependence of TCAD model parameters that allow for predictive device simulations [1], [2]. After the appropriate calibration, simulation models become indispensable tools to explain the underlying physics behind the impact of scaling on FETs [3].

In this work, we present a methodology to obtain the correct mobility in Monte Carlo (MC) TCAD simulations by capturing the evolution of relevant parameters that affect the electrostatics and carrier scattering with scaling. The strategy considered combines: *(i)* the accurate device structure generation following the fabrication process; *(ii)* the density gradient (DG) based quantum correction model in three dimensional (3D) MC simulations to capture the Capacitance - Voltage (C-V) characteristics; *(iii)* the impact of the most important scattering mechanisms in fitting the experimental electron mobility. Tri-gate MOSFETs are considered in this study, as they are expected to become the transistor architecture of choice for the forthcoming CMOS technology generations [4].

The aim of this work is to show the methodology and the importance of the parameter calibration in order to capture the experimental mobility of tri-gate MOSFETs in the nanometer regime. General discussion of the device generation methodology is provided in Section II. Section III summarizes the calibration of the DG parameters to replicate the gate voltage dependence of measured inversion charge. The mobility calibration approach together with the details of the relevant scattering mechanisms implemented in the 3D MC module of Garand are given in Section IV. The comparison between the experimental and the simulation results for the C-V characteristics and the mobility are reported in Section III and IV, respectively, including the parameter dependence on the tri-gate MOSFETs width and orientation. Finally, conclusions are drawn in Section IV.

II. METHODOLOGY

In this work, we have used the DG parameters and the MC scattering parameters for fitting the experimental results of n-type Tri-Gate nanowire transistors (NWTs) provided by CEA-LETI (Fig. 1) [5]. Initially, we have focused on the impact of the NW width (W) ranging from 8nm to 38nm, while the height (H) and the channel orientation are fixed to 11nm and [110], respectively. The rest of the technological parameters remains constant. Then, the impact of the device orientation on the electron mobility has also been studied.

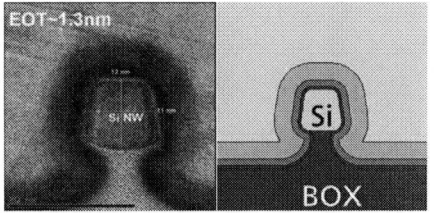

Fig. 1. Cross-sections of experimental (left) and simulated (right) Tri-gate nanowire transistor. The process simulation sequence was calibrated against TEM images provided by CEA-LETI.

Fig. 2 shows the different steps involved in this study as well as the simulation tools used at each stage. Device structures have been generated with Sentaurus Process [6] following the experimental process sequence described by Coquand et al. for Tri-gate NWT [7]. The simulated technology uses SOI substrates and features high-k/metal gates and epitaxially-grown raised S/D contacts. The process simulation sequence has been calibrated against TEM images provided by CEA-LETI, as shown exemplary in Fig. 1. Then, we have performed the C-V and mobility simulations with Garand [8] which is part of the TCAD to SPICE tool chain from Synopsys. In order to connect the device structure generated and the simulation tool, it is necessary to translate the mesh of the device to a rectilinear grid suitable for Garand through the use of the Synopsys tool SNMesh. We have curve fitted all extracted parameters to interpolate their values for any arbitrary cross-sectional width. Since the experimental data has been obtained from devices with 10 μm gate length, the long channel approximation has been adopted in the C-V and mobility simulations. It allows us to simulate only a single slice in the center of the device with periodic boundary conditions in the transport direction.

III. CAPACITANCE VOLTAGE SIMULATIONS

The main objective of this simulation has been to replicate the experimentally measured gate voltage dependence of the inversion charge. For this task, we have used drift-diffusion (DD) simulations with DG quantum corrections [9] with

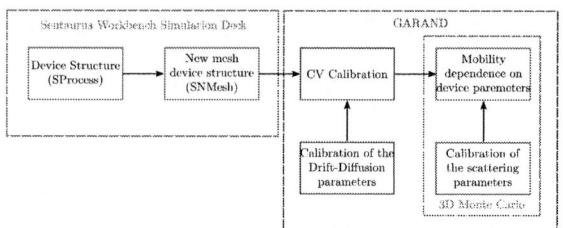

Fig. 2. Flowchart of the steps involved in the comparison between the experimental Tri-gate devices and TCAD simulations: the device structure is generated with Sentaurus Process [6]; then, the C-V and mobility simulations are performed with Garand [8] which is part of the TCAD to SPICE tool chain from Synopsys.

scaling dependent DG parameters that change with the device cross-section. The gate workfunction has been calibrated for the widest device, whereas the two density gradient masses in the confinement plane (Fig. 3(a)), and gate insulator (hafnia) permittivity (Fig. 3(b)) have been calibrated as a function of the device width. The fitting curves for each parameter as well as the equation that follows the curve are also shown. Fig. 4 illustrates the results of the C-V calibration for the devices with different cross-sectional widths showing an excellent match between the experimental data and TCAD simulations.

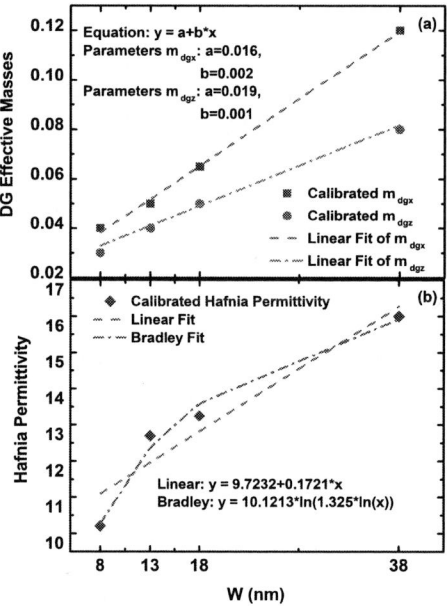

Fig. 3. Calibrated parameters for the CV simulation as a function of the Tri-gate device width (W) with [110] channel orientation: (a) density gradient effective masses, and (b) hafnia permittivity. m_{dgx} and m_{dgz} correspond to DG effective masses along the the NW height and width directions, respectively.

IV. MOBILITY CALCULATIONS

In the long channel simulations [10], the channel is assumed to be infinitely long and the electric field in the transport

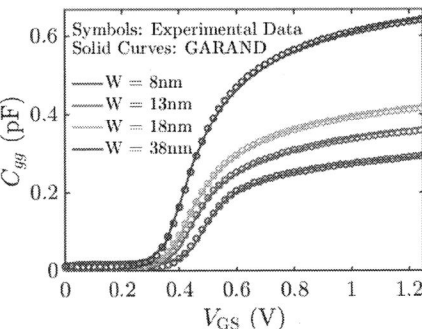

Fig. 4. Comparison of C-V curves obtained from experiments and TCAD simulations with [110] channel orientation.

direction is fixed to a low value (1kV/cm in this work). Long channel simulations provide a convenient framework for assessing low-field electron mobilities in devices with strong confinement effects, such as NWTs or FinFETs. The mobility is computed from the average velocity of the particles obtained from the MC simulations and the corresponding electric field.

We have calibrated each scattering mechanism for the particular inversion charge region in which it dominates. The following scattering mechanisms have been considered in the simulations (their detailed models as well as their expressions can be found in [8]):

- *Remote coulomb (RC) scattering* has been calibrated at low inversion charge density. This model is a general treatment of the Coulomb scattering mechanism described in [11], which assumes a metal/high-k/oxide gate stack. It simplifies the models by considering only trapped charges at the material interfaces instead of their distribution throughout the gate dielectrics. In this mechanism, the charge centroid depth (Z_D) and the trap density at the Si-SiO$_2$ have been calibrated as a function of the width. The trap density at the SiO$_2$/HfO$_2$ interface remains fixed at 1.4×10^{13} cm^{-2}.

- *Acoustic (Ac) and the optical (Op) phonon scattering* limit the maximum mobility. The Ac model treats this scattering within the elastic approximation and intra-valley transitions considering a single phonon branch [12]. In this case, we have calibrated the longitudinal and transverse Ac deformation potential ($D_{AC,Long}$=12.5eV and $D_{AC,Trans}$=9.5eV, respectively) for the widest Tri-gate device considered in the 3D MC simulations (W=38nm) and the same channel orientation as for the CV calibration [110]. We have considered Garand's default values for the 3 f-type and the 3 g-type processes for the inter-valley Op phonon parameters.

- *Surface roughness (SR) scattering* is responsible for the mobility decrease at high inversion charge density. This model is treated elastically and in the plane of the rough surface. In this mechanism, the correlation length (C_L) has been calibrated at the Si-SiO$_2$ interface as a function

of the device width and orientation, whereas the root mean square (Δ_{RMS}) has been fixed to 0.43nm.

For the studied Tri-gate transistors, we have considered that the scattering mechanisms at the top surface (001) are not affected by the reduction of the width, whereas the parameters at the sidewall interface (-110) have stronger impact with the width reduction. In particular, Fig. 5 shows the scattering parameters obtained as a result of the calibration as a function of the device width as well as the fitting curves for each parameter and the equation that follows the curve.

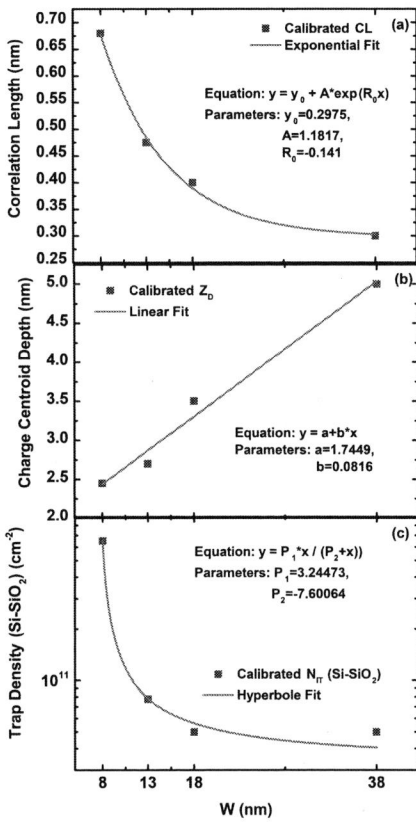

Fig. 5. Calibrated parameters for the mobility simulation as a function of the Tri-gate device width (W) for the sidewall interface and with [110] channel orientation: (a) correlation length (C_L) included in surface roughness scattering, (b) charge centroid depth (Z_D) and (c) trap density (N_{IT}) at the Si-SiO$_2$ interface included in remote coulomb scattering mechanism.

The comparison between the electron mobility as a function of the inversion charge (N_{inv}) obtained from the simulations and the experimental data for the four considered widths is depicted in Fig. 6. The inversion charge per unit area is normalized with respect to the perimeter W_{eff} of the SiNWs: W_{eff}=W+2H.

Finally, the dependence of the electron mobility on the device orientation has been investigated. In the 3D MC simulator Garand, the quantum confinement model by default takes only into account a change in the electrostatics (DG model). There is no valley splitting, and as a result, the density of

978-1-7281-0941-1/19 $31.00 © 2019 IEEE   277

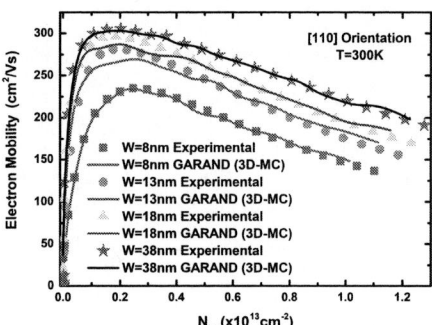

Fig. 6. Electron mobility from simulation and experiment as a function of the inversion charge (N_{inv}) for Tri-gate devices with [110] channel orientation.

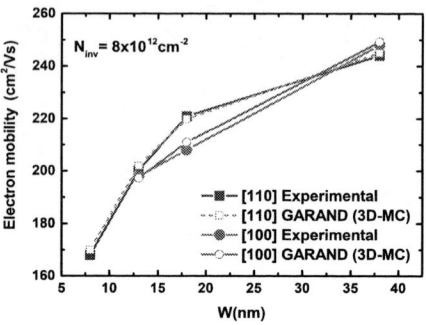

Fig. 8. Electron mobility from simulation and experiment as a function of the Tri-gate device width (W) at N_{inv}=8x10^{12}cm^{-2} for two channel orientations: [110] and [100].

states applied to the scattering mechanisms is unaffected by the change in orientation. Therefore, the scattering models by default are not modified in any way by a change in orientation.

Nevertheless, as the change in the crystal orientation has direct impact on the interface roughness, we have studied the modification of the C_L parameter (Fig. 7) when the channel orientation changes from [110] to [100]. Despite the identical top interface for both channel orientations, the orientation of the sidewall interface is now aligned with the principal axis changing from [-110] to [010], respectively. The rest of the scattering parameters have remained constant during this comparison.

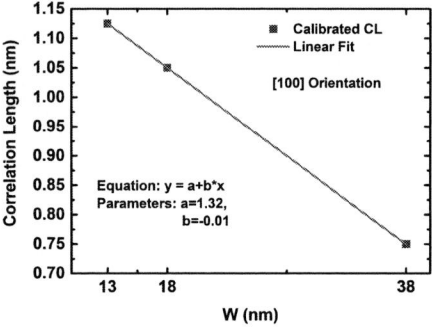

Fig. 7. Calibrated correlation length (CL) parameter included in surface roughness scattering mechanism for the mobility simulation as a function of the Tri-gate device width (W) for the sidewall interface with [100] channel orientation.

Fig. 8 shows the electron mobility enhancement for the [100] channel orientation in comparison to [110] orientation. Moreover, the mobility curve has an exponential behavior for the [110] channel orientation, and a linear behavior for the [100] orientation. This change in the calibrated C_L as a function of the channel orientation is also shown in Fig. 5(a) and Fig. 7, respectively.

V. CONCLUSIONS

In this work, based on comprehensive calibration, we have obtained width scaling dependent empirical expressions for effective gate insulator permittivity, density gradient effective masses, and Monte-Carlo surface roughness and remote coulomb scattering parameters which accurately match the measured C-V characteristics and carrier mobility. The expressions can be used to predict these parameters and hence carrier mobilities in scaled Tri-gate devices with arbitrary widths and for different substrate orientations.

REFERENCES

[1] H. Park, M. Bafleur, L. Borucki, C. Sughama, T. Zirkle, and A. Wild, "Systematic calibration of process simulators for predictive TCAD," in *1997 International Conference on Simulation of Semiconductor Processes and Devices (SISPAD)*. IEEE, 1997, pp. 273–275, DOI: 10.1109/SISPAD.1997.621390.

[2] A. Biswas, S. S. Dan, C. Le Royer, W. Grabinski, and A. M. Ionescu, "TCAD simulation of SOI TFETs and calibration of non-local band-to-band tunneling model," *Microelectronic Engineering*, vol. 98, pp. 334–337, 2012, DOI: 10.1109/SISPAD.1997.621390.

[3] S. Sahay and M. J. Kumar, *Simulation of JLFETS Using Sentaurus TCAD*. Wiley-IEEE Press, 2019, DOI: 10.1002/9781119523543.ch9.

[4] J. Cartwright, "Intel enters the third dimension," *Nature News*, 2011, DOI: 10.1038/news.2011.274.

[5] Z. Zeng, F. Triozon, S. Barraud, and Y.-M. Niquet, "A Simple interpolation model for the carrier mobility in trigate and gate-all-around silicon NWFETs," *IEEE Transactions on Electron Devices*, vol. 64, no. 6, pp. 2485–2491, 2017, DOI: 10.1109/TED.2017.2691406.

[6] Synopsys, *Sentaurus Process User Guide Version O-2018.06*, Mountain View, CA, USA, 2018, no. June.

[7] R. Coquand, S. Barraud, M. Cassé, P. Leroux, C. Vizioz, C. Comboroure, P. Perreau, E. Ernst, M.-P. Samson, V. Maffini-Alvaro, C. Tabone, S. Barnola, D. Munteanu, G. Ghibaudo, S. Monfray, F. Boeuf, and T. Poiroux, "Scaling of high-κ/metal-gate TriGate SOI nanowire transistors down to 10 nm width," *Solid-State Electronics*, vol. 88, pp. 32–36, 2013, DOI: 10.1016/j.sse.2013.04.006.

[8] Synopsys, *Garand User Guide, O-2018.06*, Mountain View, CA, USA, 2018, no. June.

[9] A. R. Brown, J. R. Watling, G. Roy, C. Riddet, C. L. Alexander, U. Kovac, A. Martinez, and A. Asenov, "Use of density gradient quantum corrections in the simulation of statistical variability in MOSFETs," *Journal of computational electronics*, vol. 9, no. 3-4, pp. 187–196, 2010, DOI: 10.1007/s10825-010-0314-y.

[10] E. B. Ramayya, D. Vasileska, S. M. Goodnick, and I. Knezevic, "Electron transport in silicon nanowires: The role of acoustic phonon confinement and surface roughness scattering," *Journal of Applied Physics*, vol. 104, no. 6, p. 063711, 2008, DOI: 10.1063/1.2977758.

[11] D. Esseni, P. Palestri, and L. Selmi, *Nanoscale MOS Transistors: Semiclassical Transport And Applications*. New York, USA: Cambridge University Press, 2011.

[12] P. L. C. Jacoboni, *The Monte Carlo method for semiconductor device simulation*. Vienna: Springer-Verlag, 1989.

978-1-7281-0941-1/19 $31.00 © 2019 IEEE

DFT study of graphene doping due to metal contacts

P. Khakbaz*, F. Driussi*, A. Gambi*,
P. Giannozzi[†], S. Venica*, D. Esseni*
*DPIA, University of Udine, Udine, Italy
[†]DMIF, University of Udine, Udine, Italy
Email: francesco.driussi@uniud.it

A. Gahoi[‡], S. Kataria[‡], M.C. Lemme[‡§]
[‡]RWTH Aachen University, Aachen, Germany
[§]AMO GmbH, Advanced Microelectronic Center, Aachen, Germany

Abstract—The experimental results of Metal–graphene (M–G) contact resistance (R_C) have been investigated in–depth by means of Density Functional Theory (DFT). The simulations allowed us to build a consistent picture explaining the R_C dependence on the metal contact materials employed in this work and on the applied back–gate voltage. In this respect, the M–G distance is paramount in determining the R_C behavior.

Index Terms—Graphene, Contacts, DFT

I. INTRODUCTION

In recent years, graphene raised great interest for many electronic applications [1], [2]. However, metal–graphene (M–G) contacts still severely degrade the electrical performance of graphene based devices because of the large contact resistance (R_C). For instance, R_C largely degrades the output conductance and the maximum oscillation frequency of graphene-FETs (GFETs) [1], [3]. Thus, to boost the graphene technology, design strategies to optimize R_C are urgently required to make graphene a viable solution for high performance electronic devices [4].

In this respect, for a proper contact engineering, the physics underlying the conduction through M–G contacts needs to be completely understood. Therefore, in this work we made use of Density Functional Theory (DFT) simulations to interpret the experimental R_C values obtained for different metal materials [5] and to gain an insight into the M–G contact physics.

II. EXPERIMENTAL BEHAVIOR OF M–G CONTACT RESISTANCE

Figure 1 reports the contact resistance values measured on back–gated TLM structures with contacts to graphene fabricated with nickel, coppper and gold. Details concerning the measured devices and the exploited characterization technique can be found in [6], [7].

As it can be seen, R_C largely depends on the applied back–gate voltage V_{BG}. Furthermore, despite the similar graphene quality between the different samples (not shown) [8], the R_C values do depend on the metal contact, with larger R_C values for Ni and smaller R_C values for Au. This indicates that the M–G interactions influence the electrical properties of the M–G contact, which is not a surprising result [9].

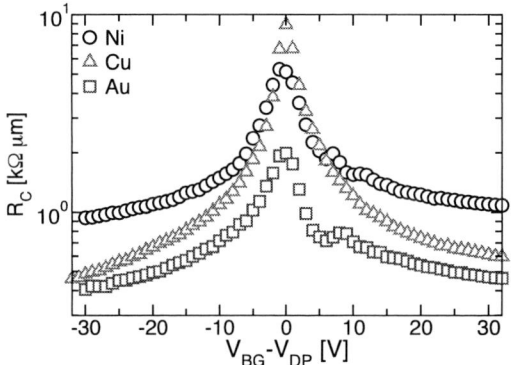

Fig. 1: Experimental contact resistance R_C as a function of V_{BG} for samples with Ni, Cu and Au contacts. V_{DP} is the V_{BG} corresponding to the Dirac Point, at which the graphene resistance is maximum.

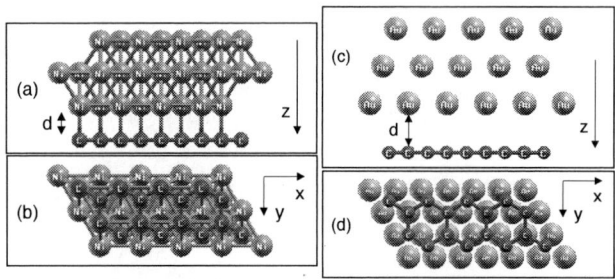

Fig. 2: Lateral (a) and bottom (b) view of the simulated Ni–G structure. A 1×1 graphene cell with a lattice constant of $a = 0.246$ nm is considered. For the Cu–G stack, we used the same cell. Lateral (c) and bottom (d) view of the simulated Au–G structure. The Au lattice is matched with a 2×2 graphene supercell [10].

III. DENSITY FUNCTIONAL THEORY SIMULATIONS

To interpret the experimental results in Fig. 1, we performed DFT simulations with the Quantum ESPRESSO suite [11]. We built 3–layer–thick clusters for Ni, Cu and Au contacts and we matched the 111 metal surface with the graphene (G) lattice, which primitive cell as an in-plane lattice constant of $a = 0.246$ nm [10] (Fig. 2). We verified that the three metal layers are sufficient to develop a band structure consistent with the bulk metallic material (not shown). In the supercell, we

978-1-7281-0941-1/19 $31.00 © 2019 IEEE

also included a vacuum region of about 2 nm and we applied dipole correction in order to avoid spurious interactions between periodic images of the slab [11].

We used local–spin density approximation, plane–wave basis sets and gradient–corrected exchange–correlation functionals (Perdew-Burke-Ernzerhof) for Ni, Cu, Au and C atoms. Van der Waals forces employing the Grimme–D2 empirical method were also accounted for. Energy minimisation allowed us to obtain the equilibrium geometry and the minimum energy distance (MED) between graphene and Au ($d = 0.31$ nm), Cu ($d = 0.31$ nm) or Ni ($d = 0.21$ nm) contacts, in agreement with [10].

Figure 3 reports the energy bands of the Cu–G (a) and Au–G (b) stacks at MED: it is worth to notice that these bands are simply the superimposition of metal and graphene bands (not shown) and no hybridization of the metal and graphene orbitals is expected [10]. Here, the typical Dirac cone of graphene at the K point is clearly visible. Furthermore, opposite spin states are degenerate, because Cu, Au and graphene are diamagnetic.

Figure 4 shows instead the energy bands of the Ni–G structure at two Ni–G distances: MED $d = 0.21$ nm (a) and $d = 0.3$ nm (b). Spin degeneracy is here lifted, because Ni is ferromagnetic. The Ni–G interaction is very strong and the hybridization of graphene and Ni orbitals changes the bands compared to isolated materials. Indeed, at the K point, it is

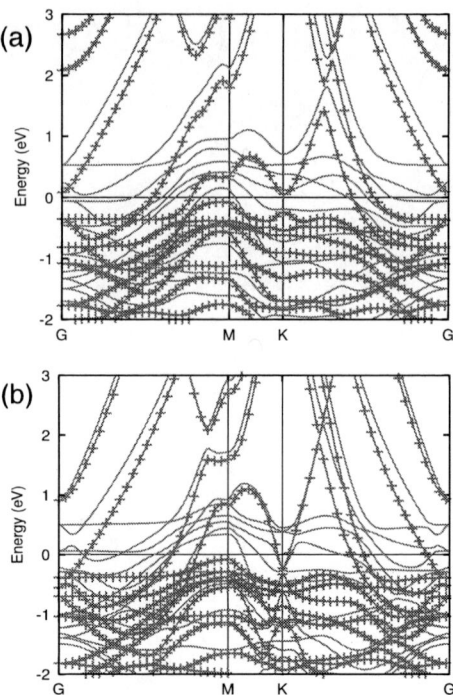

Fig. 4: Energy bands of Ni–G stack for two d values: (a) MED, $d = 0.21$ nm; (b) $d = 0.3$ nm. Ni induces gaps in graphene and the Dirac cone at the K point vanishes. Blue crosses/lines are for spin up, while red lines are for spin down.

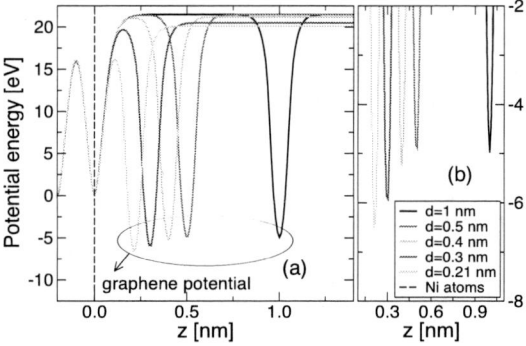

Fig. 5: (a) Potential energy along z (perpendicular to graphene) of the Ni–G stack for different d values. The energy reference has been taken at the Ni atoms position. (b) Zoom in the graphene region. The graphene relative energy with respect to Ni atoms changes when the layers are approaching, indicating the charging of the two materials.

no longer possible to identify the Dirac cone of graphene for any of the distances in Fig. 4 [10].

IV. METAL–INDUCED DOPING OF GRAPHENE

Figure 5(a) shows the potential energy calculated along the z direction of the Ni–G stack (perpendicular to graphene, Fig. 2) and averaged over the x–y plane. The energy profiles are reported for different Ni–G distances (d) and they describe the variation of the energy of the two materials when they are

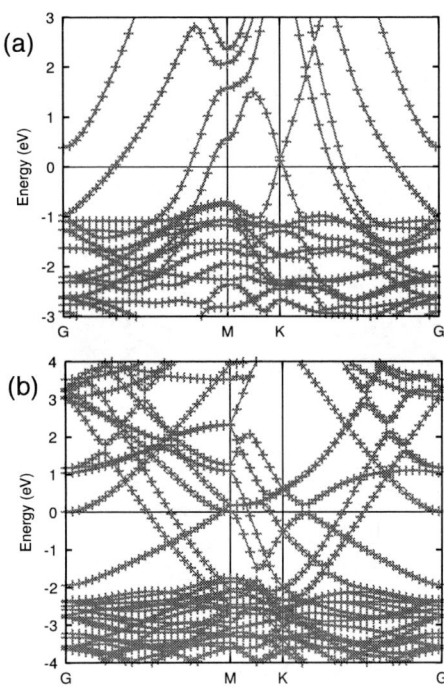

Fig. 3: Energy bands of Cu–G (a) and Au–G (b) stacks at minimum energy distance ($d = 0.31$ nm). For both cases bands are the superimposition of metal and graphene bands and the Dirac cone is visible at the K point. Note that the Au–G supercell is based on a 2×2 graphene cell, resulting in the folding of graphene bands. Blue crosses/lines are for spin up, red lines are for spin down.

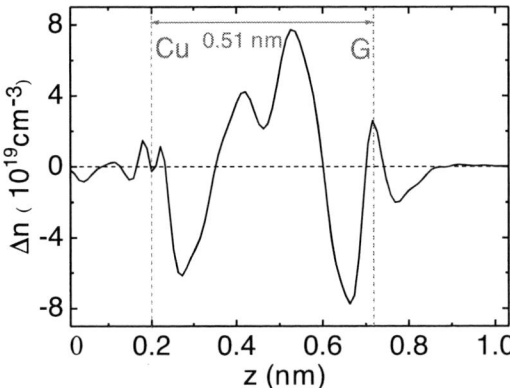

Fig. 6: Induced net charge along z (perpendicular to graphene) for the Cu–G stack at $d = 0.51$ nm. Vertical dashed lines are the positions of graphene and of the closest Cu layer.

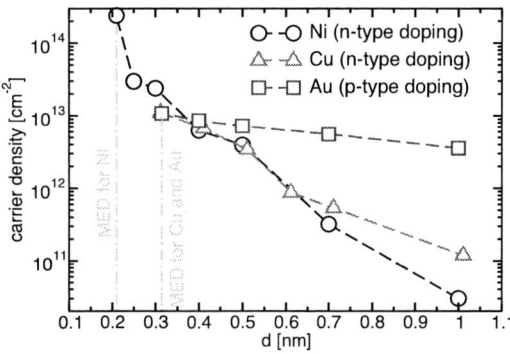

Fig. 7: Graphene doping as a function of the distance d, calculated through Bader analysis. Au–G contact shows long range interaction, while Ni–G and Cu–G stacks show shorter range interactions. Minimum energy distance (MED) values are highlighted for the three M–G contacts.

approaching. The lower peaks are located at the atom positions and, by reducing d, the graphene peak lowers with respect to the Ni peaks (see Fig. 5(b)), indicating the charging of the two materials.

We studied the charge transfer due to the chemical interaction between metal and graphene by calculating the difference (Δn) between the valence electron density of the M–G stacks and of the isolated metal and graphene layers [12], [13]. Fig. 6 shows a typical example of these calculations averaged over the graphene plane for the Cu–G stack (at $d = 0.51$ nm). The interaction between the metal and graphene induces a non negligible charge transfer, with the building up of dipoles along the stack ($\Delta n > 0$ corresponds to negative charge, $\Delta n < 0$ indicates positive charge). In this respect, R_C values are expected to be largely affected by the charge in the graphene underneath the contact [10]. Thus, there is the need of estimating the graphene doping due to the metal proximity. However, from this plot it is not possible to evaluate the metal-induced G doping, since it is not straightforward to distinguish between the charge belonging either to graphene or to the Cu contact [8].

A. Graphene doping extraction through Bader analysis

In order to extract a dependable value for the graphene doping, also avoiding spurious contributions by the charge redistribution due to Pauli repulsion [13], we made used of the Bader analysis to calculate the charge variation of each C atom induced by the presence of the metal contact [14]. Then, we converted this charge into an areal doping density of the graphene sheet.

Figure 7 shows the calculated doping for the different M–G contacts and for several d values down to the MED of each structure. Cu and Ni induce n–type doping in graphene, while, for the Au contact, graphene becomes p–type [13]. It is evident that, at small d, the metals largely dope the graphene, with a huge doping value in the case of the Ni–G stack at MED. However, under these circumstances, it becomes difficult to interpret the Ni–G stack in terms of a Ni and a graphene

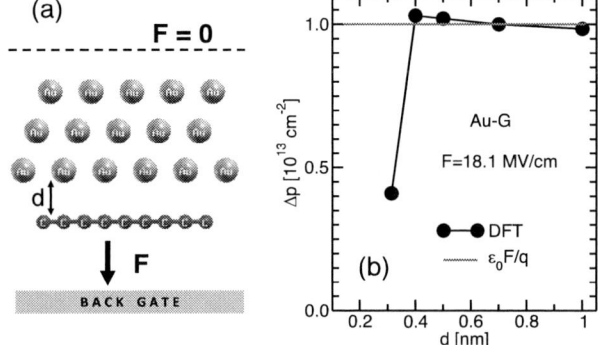

Fig. 8: Back gated Au-G stack: (a) Sketch of the applied electric field F inducing a total charge $\varepsilon_0 F/q$ in the stack; (b) Graphene hole density variation Δp determined via Bader analysis and compared to the total charge $\varepsilon_0 F/q$.

sub-systems, because of the strong hybridization of the two materials already discussed in Fig. 4.

Figure 7 demonstrates the very different dependence on d of the induced graphene doping for the different metals, with a very long range interaction between Au and graphene. For Ni and Cu, instead, the interaction with graphene has a much shorter range.

This trend for distances larger than the MED is consistent with the experimental results of Fig. 1. Indeed, the Au-G stack shows the lowest R_C values suggesting a significant doping of graphene, while the Ni–G system has the worst R_C, thus indicating a weak Ni–G interaction. The comparison between the experimental R_C behavior and the simulations in Fig. 7 suggests that the actual M–G distance in the fabricated contacts is larger than 0.5 nm, thus leading to an inefficient doping of graphene by the Ni (with the worst R_C). Au instead is still effective in doping the underlying graphene (with the best R_C), even at a non optimal Au–G distance.

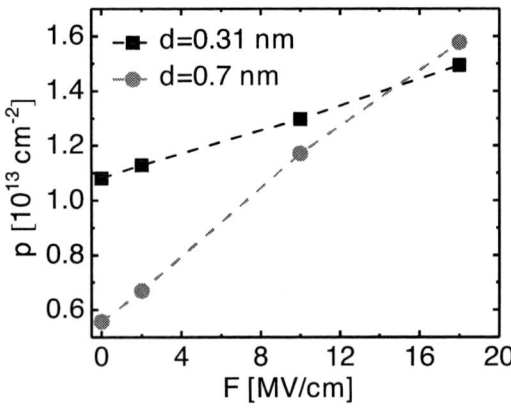

Fig. 9: Back gated Au-G stack: Graphene doping density p determined via Bader analysis at different back–gate electric fields F for two Au–G distances. At MED, the gold contact is more effective in defining the doping of the underlying graphene, also reducing the p dependence on F.

B. R_C dependence on the back–gate voltage

Finally, in order to interpret the experimentally observed R_C dependence on V_{BG}, we used DFT simulations and Bader analysis to study also the influence on the graphene doping of a back-gate induced electric field F. In particular, as sketched in Fig. 8(a), we repeated the analysis in Fig. 7 on M–G stacks undergoing the electrostatic effect of gate plate, inducing F along the z direction perpendicular the graphene sheet. The dipole correction option in the Quantum Espresso suite allowed us to apply a non null F only at the graphene side of the M–G stack, thus reproducing an electrostatic configuration similar to that in the experiments [8].

Figure 8(b) reports the F induced variation of the graphene doping (Δp), calculated as the difference between the charge configuration in the Au–G stack subject to the field F and that for the unperturbed stack (with $F = 0$, thus without the gate plate). The results show that, for $d \geq 0.4$ nm, the Δp accounts for the entire charge $\varepsilon_0 F/q$ induced in the Au–G stack, whereas at the MED $d = 0.31$ nm only about the 40% of $\varepsilon_0 F/q$ is induced in the graphene.

Figure 9 compares the dependence of the graphene doping p on the applied F value, for Au–G stacks at the MED and at $d = 0.7$ nm. At MED, the gold contact is more effective in defining the doping of the underlying graphene, also reducing the p dependence on F, that is more pronounced for $d = 0.7$ nm. Therefore, the large R_C dependence on the V_{BG} bias in the experiments of Fig. 1 is consistent with a distance $d \geq 0.4$ nm, such that V_{BG} can still modulate the graphene charge to a large extent.

V. CONCLUSIONS

In summary, our simulations show that Ni and Cu largely dope graphene at MED, but at larger distance they fail to dope graphene. In this respect, the experiments for Ni–G and Cu–G contacts report a large R_C (especially at $V_{BG} = 0$), thus suggesting $d > 0.5$ nm. This increased distance may be the

result of wrinkles, roughness and/or impurities impacting the quality of the M–G interface.

For the Au–G contact, a longer range interaction and significant G doping are predicted for d up to 1 nm, explaining the lower experimental R_C compared to Ni–G and Cu–G. Furthermore, the simulations predict a large R_C dependence on V_{BG} at $d > 0.4$ nm. This result seems to grasp the experimental behavior, thus supporting the picture of a M–G distance larger than the minimum distance.

ACKNOWLEDGEMENTS

Work was funded by the EU through the FP7 STREP Project GRADE (317839) and an ERC grant (InteGraDe, 307311). The German Research Foundation is acknowledged.

REFERENCES

[1] Y. Wu, X. Zou, M. Sun, Z. Cao, X. Wang, S. Huo, J. Zhou, Y. Yang, X. Yu, Y. Kong, G. Yu, L. Liao, and T. Chen, "'200 GHz maximum oscillation frequency in CVD graphene radio frequency transistors'," *ACS Applied Materials and Interfaces*, vol. 8, pp. 25 645–25 649, 2016.

[2] C. Chen and J. Hone, "Graphene nanoelectromechanical systems," *Proceedings of the IEEE*, vol. 101, no. 7, pp. 1766–1779, 2013.

[3] S. Venica, F. Driussi, P. Palestri, D. Esseni, S. Vaziri, and L. Selmi, "On the adequacy of the transmission line model to describe the graphene–metal contact resistance," *IEEE Trans. on Electron Devices*, vol. 61, no. 7, pp. 2570–2576, 2014.

[4] A. Meersha, H. B. Variar, K. Bhardwaj, A. Mishra, S. Raghavan, N. Bhat, and M. Shrivastava, "Record low metal – (CVD) graphene contact resistance using atomic orbital overlap engineering," in *IEEE IEDM Technical Digest*, 2016, p. 5.3.15.3.4.

[5] S. Venica, F. Driussi, A. Gahoi, S. Kataria, P. Palestri, M. C. Lemme, and L. Selmi, "Reliability analysis of the metal–graphene contact resistance extracted by the transfer length method," *Proceedings of ICMTS*, pp. 57–62, 2018.

[6] A. Gahoi, S. Wagner, A. Bablich, S. Kataria, V. Passi, and M. C. Lemme, "Contact resistance study of various metal electrodes with CVD graphene," *Solid State Electronics*, vol. 125, pp. 234–239, 2016.

[7] S. Venica, F. Driussi, A. Gahoi, V. Passi, P. Palestri, M. C. Lemme, and L. Selmi, "Detailed characterization and critical discussion of series resistance in graphene–metal contacts," *Proceedings of ICMTS*, pp. 27–31, 2017.

[8] F. Driussi, S. Venica, A. Gahoi, A. Gambi, P. Giannozzi, S. Kataria, M. C. Lemme, P. Palestri, and D. Esseni, "Improved understanding of metal–graphene contacts," *Microelectronic Engineering*, vol. 216, p. 111035, 2019.

[9] S. Venica, F. Driussi, A. Gahoi, P. Palestri, M. C. Lemme, and L. Selmi, "On the adequacy of the transmission line model to describe the graphene–metal contact resistance," *IEEE Trans. on Electron Devices*, vol. 65, no. 4, pp. 1589–1596, 2018.

[10] P. A. Khomyakov, G. Giovannetti, P. C. Rusu, G. Brocks, J. van den Brink, , and P. J. Kelly, "First-principles study of the interaction and charge transfer between graphene and metals," *Physical Review B*, vol. 79, p. 195425, 2009.

[11] P. Giannozzi, O. Andreussi, T. Brumme, O. Bunau, M. B. Nardelli, M. Calandra, R. Car, C. Cavazzoni, D. Ceresoli, and M. Cococcioni, "Advanced capabilities for materials modelling with quantum espresso," *Journal of Physics: Condensed Matter*, vol. 29, no. 46, p. 465901, 2017.

[12] C. Gong, G. Lee, B. Shan, E. M. Vogel, R. M. Wallace, , and K. Cho, "First-principles study of metalgraphene interfaces," *Journal of Applied Physics*, vol. 108, no. 12, p. 123711, 2010.

[13] J. Gebhardt, F. Viñes, and A. Görling, "Influence of the surface dipole layer and Pauli repulsion on band energies and doping in graphene adsorbed on metal surfaces," *Phys. Rev. B*, vol. 86, p. 195431, 2012.

[14] G. Henkelman, A. Arnaldsson, and H. Jonsson, "A fast and robust algorithm for bader decomposition of charge density," *Computational Materials Science*, vol. 36, pp. 354–360, 2006.

Device and Circuit Level Gate Configuration Optimization for 2D Material Field-Effect Transistors

Devin Verreck[*†], Goutham Arutchelvan[*†], Marc M. Heyns[*†], Iuliana P. Radu[*],
[*]imec, 3001 Leuven, Belgium, email: devin.verreck@imec.be
[†] Department of Materials Engineering, KU Leuven, 3001 Leuven, Belgium

Abstract—**A tied double gate structure has been shown to deliver optimal device-level performance in few-layer MoS$_2$ field-effect transistors. However, the enlarged gate capacitance from the added gate increases circuit-level power consumption and negatively affects minimum obtainable delay. Here, we therefore use a calibrated design-technology co-optimization approach that includes the interconnect load to evaluate back gate size reduction strategies in terms of power and delay. We consider the impact of a spacer region and varying interconnect length. We find that power consumption can be decreased by almost 20% by reducing the back gate overlap with the source-drain contacts without negatively affecting delay, as the carrier injection is occurring dominantly at the contact edges. We also show that opening the back gate underneath the channel provides additional benefit for locally interconnected devices.**

I. INTRODUCTION

Two-dimensional (2D) semiconductor channel materials open a path for the scaling of field-effect-transistors (FETs) to the 5nm node (N5) and beyond, through the promise of excellent gate control and limited short-channel effects [1], [2]. Of the many candidate 2D materials, MoS$_2$ is among the most studied, because it is stable in air, can be grown in large areas and has a high theoretical carrier mobility [3], [4]. So far, though, experimental MoS$_2$ FETs have not been able to realize their full potential, as they are hampered by non-idealities such as Schottky barriers at the source/drain (S/D) contacts, traps at the oxide interfaces and reduced mobility due to poor grown material quality [5]. While research is ongoing to continuously improve material growth, a parallel avenue to improve 2D FET performance is through gate configuration engineering. Previous simulation research which took into account the mentioned non-idealities showed that a tied double gate is preferable over a single gate as it improves gate control over both the channel and the Schottky contacts, while providing electrostatic doping of the access regions [6]. However, an extra gate means a larger gate capacitance, which increases power consumption (P) and limits the obtainable minimal delay (τ_D). In this work, we therefore optimize the back gate size to reduce P without negatively impacting τ_D. In doing so, we investigate the origin of the double gate improvement and determine which parts of the channel are most critical to be overlapped by both gates. The evaluation is carried out with calibrated Sentaurus Device TCAD simulations on

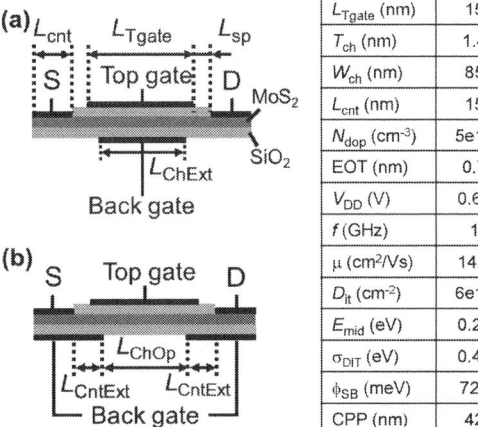

L_{Tgate} (nm)	15
T_{ch} (nm)	1.4
W_{ch} (nm)	85
L_{cnt} (nm)	15
N_{dop} (cm^{-3})	5e18
EOT (nm)	0.7
V_{DD} (V)	0.65
f (GHz)	1
μ (cm^2/Vs)	14.5
D_{it} (cm^{-2})	6e12
E_{mid} (eV)	0.25
σ_{DIT} (eV)	0.48
ϕ_{SB} (meV)	720
CPP (nm)	42

Fig. 1: Simulated bilayer MoS$_2$ FETs and parameters. Mobility (μ), density and distribution of acceptor-type MoS$_2$-oxide traps (D_{it}, E_{mid}, σ_{mid}) and Schottky barrier height at the S/D contacts (ϕ_{SB}) were calibrated to experiment [5]. Dimensions and bias conditions align with imec's N5 process assumptions.

the device level, which feed into a simple interconnect-aware circuit model.

II. MODELING AND CONFIGURATION SET-UP

We consider two different tied double gate bilayer MoS$_2$ configurations to assess in which part of the channel the added back gate control is most critical. In the first configuration (shown in Fig. 1(a)) the back gate covers the center of the channel and is progressively extended symmetrically towards the S/D contacts by increasing the parameter L_{ChExt}. In the second configuration (shown in Fig. 1(b)), the back gate is split and covers the S/D contacts. The two parts are progressively extended symmetrically towards the channel center when L_{CntExt} is increased. Both 2D FET configurations have top contacted S/D contacts and allow for the presence of a spacer region (L_{sp}) between S/D contacts and the top gate. Default parameter values are shown in the table of Fig. 1.

Our modeling approach relies on a Synopys SDevice set-up modified with a 2D density of states, which was verified with atomistic non-equilibrium Green's function simulations

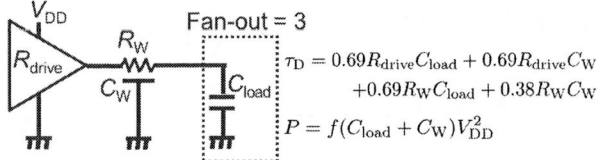

Fig. 2: Circuit model for the calculation of delay (τ_D) and power (P). R_{drive} and C_{load} are extracted from the 2D FET SDevice simulations. R_W and C_W represent wire resistance and capacitance and are extracted from a calibrated resistivity model for a three-level Cu-TaN/Ru interconnect scheme and Sentaurus Raphael simulations respectively [6], [7].

and for which non-idealities were calibrated to experimental bilayer devices [5]. Considered non-idealities are traps at the MoS$_2$-oxide interfaces, Schottky barriers at the S/D contacts and a limited mobility. The calibrated parameter values are shown in the table of Fig. 1. For the circuit-level evaluation, the SDevice model provides input to a simple circuit model that includes the interconnect load (shown in Fig. 2) [6]. Delay and power are calculated in the Elmore approximation.

III. DEVICE-LEVEL EVALUATION

From the device-level transfer characteristics in Fig. 3, it is clear that the overlap of the back gate with the S/D contacts can be reduced without a significant performance penalty. Performance in terms of on-current (I_{ON}) and subthreshold swing (SS) improves as L_{ChExt} increases, with the improvement quickly saturating as the back gate extends underneath the S/D contacts. Even for the large, calibrated Schottky barrier (ϕ_{SB}) of 720 meV, there is only a small benefit to extending the back gate beyond alignment with the S/D contacts (L_{ChExt}=15 nm, grey dashed line in Fig. 3), as the carrier injection is occurring mainly at the contact edges [8], [9]. For a lower ϕ_{SB} of 200 meV, the improvement with increasing L_{ChExt} only comes from better channel control, so even for zero overlap with the S/D contacts the same performance as a full double gate is obtained.

Fig. 4, in which the back gate is extended from the contact edges, additionally shows that the overlap of the back gate with the center of the channel can be removed with limited impact on performance. This is particularly true for the large ϕ_{SB}, where the case of 5 nm L_{CntExt} (corresponding to 5 nm channel opening) almost coincides with the full back gate curve. This again confirms that for this case, control over the Schottky barrier through proximity to the S/D contact edges accounts for the largest share of the improvement delivered by the back gate. In the case of the low ϕ_{SB}, however, the current is channel limited and the overlap contributes significantly to performance, which means performance is progressively degraded as the back gate is opened underneath the channel.

From the device level evaluation, two strategies therefore emerge to reduce back gate size with limited impact on I_{ON} or SS: decreasing the S/D contact overlap and splitting the back gate, creating an opening underneath the channel.

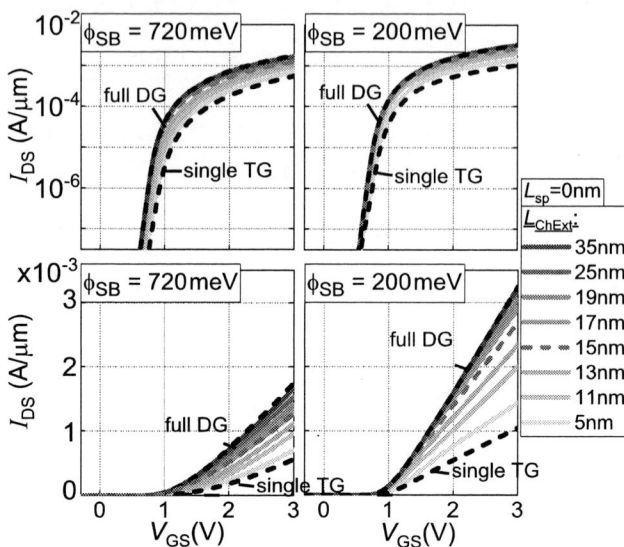

Fig. 3: Simulated transfer characteristics of the configuration in Fig. 1(a) for varying back gate extension underneath the channel and for two values of ϕ_{SB}. Top plots are in logarithmic scale, bottom in linear. The dashed lines correspond to a full double gate (DG) and single top gate (TG).

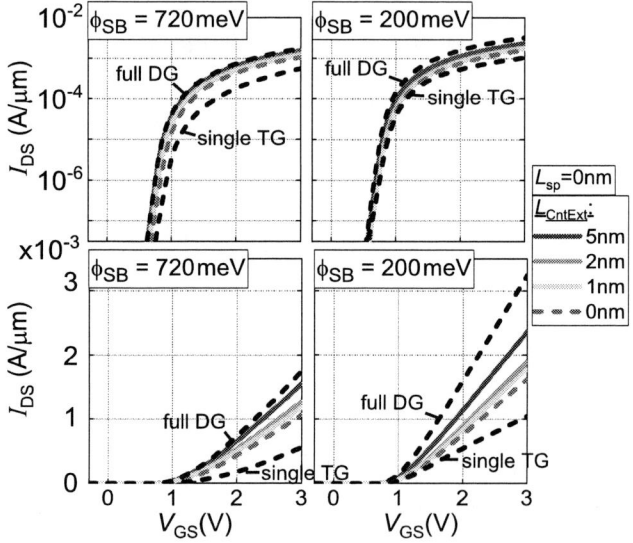

Fig. 4: Simulated transfer characteristics of the configuration in Fig. 1(b) with varying extension of the back gate from the contacts and for two values of ϕ_{SB}. Top plots are in logarithmic scale, bottom in linear. The dashed lines correspond to a full double gate (DG) and single top gate (TG).

978-1-7281-0941-1/19 $31.00 © 2019 IEEE

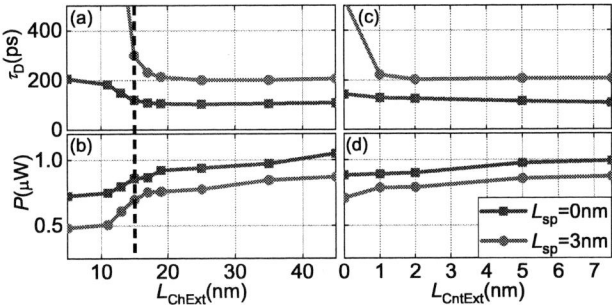

Fig. 5: Simulated delay at optimal wire dimensions and corresponding power consumption for the 2D FET in Fig. 1 for varying extension of the back gate underneath the channel and from the contacts. L_w is 300CPP. The dashed line indicates the boundary between the S/D contacts and the channel.

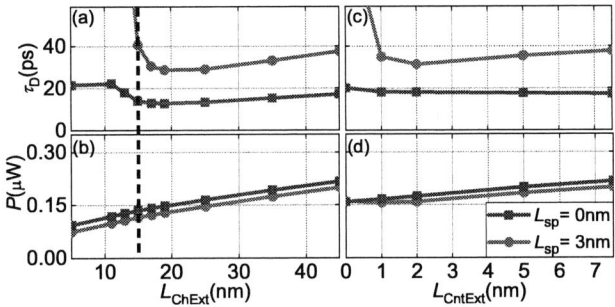

Fig. 6: Simulated delay at optimal wire dimensions and corresponding power consumption for the 2D FET in Fig. 1 for varying extension of the back gate underneath the channel and from the contacts. L_w is 3CPP. The dashed line indicates the boundary between the S/D contacts and the channel.

IV. CIRCUIT-LEVEL EVALUATION

We now evaluate these strategies in terms of circuit-level τ_D and P in Figs. 5 and 6, first in the case without a spacer. Fig. 5(a-b) shows for a 300CPP wire length (L_w) that τ_D is almost unaffected from a reduction of the S/D overlap from 15 nm to 1 nm, while P drops with 18% to 0.87 μW. Reducing L_{ChExt} more decreases P further, but at the expense of τ_D. Fig. 5(c-d) shows that opening the back gate underneath the channel at the most reduces P with 12%, at the expense of a 30% τ_D increase. For constant τ_D, a 5 nm opening only delivers 2% P improvement, which might not be worth the added process complexity. For shorter wires (L_w=3CPP, Fig. 6), the wire capacitance is less dominant, allowing for greater benefits from the back gate reduction: a S/D overlap reduction to 1 nm results in a 35% reduction in P to 0.14 μW, accompanied with a 25% decrease in τ_D. Even a 5 nm channel opening underneath the channel now reduces P with 8%.

Next, we check if the same trade-offs hold when a spacer is present for the top gate. Such spacer is often unavoidable due to process limitations, but can also intentionally be introduced

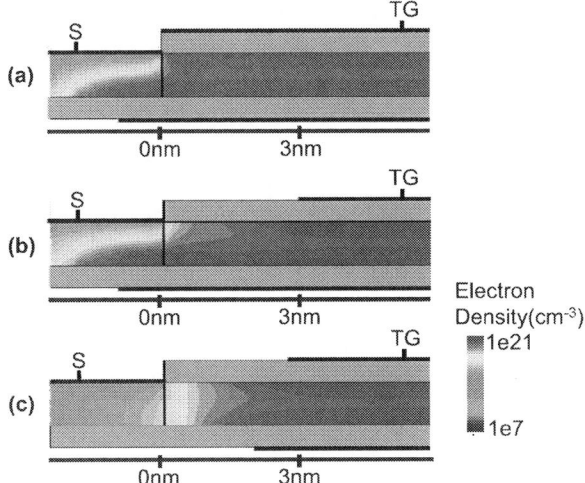

Fig. 7: Electron density profile at the source contact for (a) L_{sp}=0 nm, L_{ChExt}=17 nm, (b) L_{sp}=3 nm, L_{ChExt}=17 nm and (c) L_{sp}=3 nm, L_{ChExt}=11 nm. V_{GS} is 3 V. Other configuration details are listed in Fig. 1.

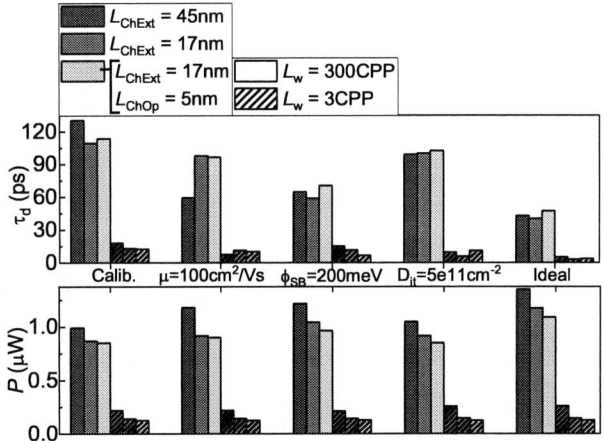

Fig. 8: Delay at optimal wire dimensions and corresponding power consumption in various projected scenarios for a full gate, compared to a configuration with reduced S/D contact overlap and a configuration with an additional channel opening. The ideal case combines all projected improvements. L_{sp} is 0nm. Other configuration details are listed in Fig. 1.

to reduce parasitic capacitance [10]. Indeed, Figs. 5 and 6 confirm the same trends for L_{sp}=3 nm, although a larger penalty exists for reducing the overlap of the back gate with the spacer region, because of a sharply increasing access resistance. Fig. 7 shows a 1 nm S/D overlap is sufficient to induce an electron density in the spacer region similar to the no spacer case (Fig. 7(a) and (b)). If the overlap with the spacer region is reduced, the electron density drops (Fig. 7(c)), resulting in a larger resistivity.

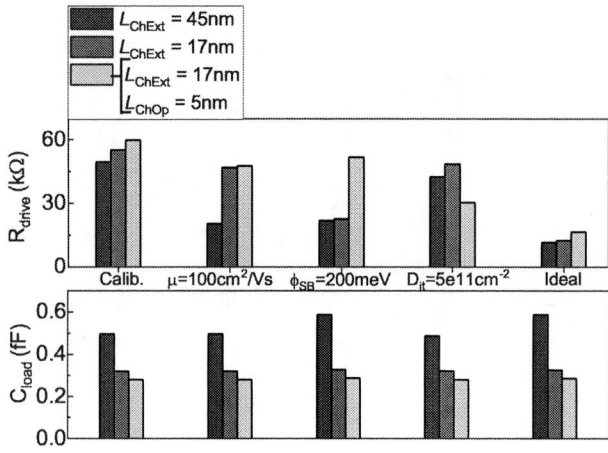

Fig. 9: Drive resistance and load capacitance in various projected scenarios for a full gate, compared to a configuration with reduced S/D contact overlap and a configuration with an additional channel opening. The ideal case combines all projected improvements. L_{sp} is 0nm. Other configuration details are listed in Fig. 1.

Finally, we project whether the τ_D and P advantages of reducing the back gate hold in a number of improvement scenarios that represent a future maturing of the technology. Fig. 8 shows that the two back gate reduction scenarios, one with the S/D overlap reduced to 1 nm and one with an additional channel opening of 5 nm, can be seen to retain the P advantage at similar τ_D to the full back gate configuration. This is generally true for both wire lengths, but for the longer wires ($L_w = 300$CPP), adding the channel opening on top of the S/D overlap reduction can negatively affect τ_D. Fig. 9 shows that it is indeed the consistent drop in C_{load} with the decrease in back gate size that causes the lower P, even when R_{drive} tends to increase.

V. Conclusion

We showed using interconnect-aware calibrated TCAD simulations that back gate reduction is a viable strategy in double gate devices to avoid power consumption penalties from the increased gate capacitance while retaining the delay advantage over a single gate device. Although this delay advantage comes both from an improved control over the channel and the Schottky contacts, the overlap with the latter can be strongly reduced as a result of edge injection of the carriers. We showed that the advantages of back gate reduction increase for locally interconnected devices. This study provides a guideline in the trade-off between delay-power performance and process complexity.

Acknowledgments

This work was supported by imec's Industrial Affiliation Program.

References

[1] B. Liu, A. Abbas, and C. Zhou, "Two-Dimensional Semiconductors: From Materials Preparation to Electronic Applications," *Advanced Electronic Materials*, 2017.

[2] Y. Yoon, K. Ganapathi, and S. Salahuddin, "How good can monolayer MoS2 transistors be?" *Nano letters*, vol. 11, no. 9, pp. 3768–3773, 2011.

[3] Y.-H. Lee, X.-Q. Zhang, W. Zhang, M.-T. Chang, C.-T. Lin, K.-D. Chang, Y.-C. Yu, J. T.-W. Wang, C.-S. Chang, L.-J. Li *et al.*, "Synthesis of large-area MoS2 atomic layers with chemical vapor deposition," *Advanced Materials*, vol. 24, no. 17, pp. 2320–2325, 2012.

[4] C. D. English, G. Shine, V. E. Dorgan, K. C. Saraswat, and E. Pop, "Improved contacts to MoS2 transistors by ultra-high vacuum metal deposition," *Nano letters*, vol. 16, no. 6, pp. 3824–3830, 2016.

[5] D. Verreck, G. Arutchelvan, C. J. Lockhart De La Rosa, A. Leonhardt, D. Chiappe, A. K. A. Lu, G. Pourtois, P. Matagne, M. M. Heyns, S. De Gendt, A. Mocuta, and I. P. Radu, "The Role of Nonidealities in the Scaling of MoS2 FETs," *IEEE Transactions on Electron Devices*, vol. 65, no. 10, pp. 4635–4640, Oct 2018.

[6] D. Verreck, G. Arutchelvan, I. Ciofi, M. M. Heyns, and I. P. Radu, "Interconnect-device co-optimization for field-effect transistors with two-dimensional materials," in *2018 IEEE International Interconnect Technology Conference (IITC)*, June 2018, pp. 73–75.

[7] I. Ciofi, A. Contino, P. J. Roussel, R. Baert, V.-H. Vega-Gonzalez, K. Croes, M. Badaroglu, C. J. Wilson, P. Raghavan, A. Mercha *et al.*, "Impact of wire geometry on interconnect RC and circuit delay," *IEEE Trans. Elec. Dev.*, vol. 63, no. 6, pp. 2488–2496, 2016.

[8] G. Arutchelvan, C. J. L. de la Rosa, P. Matagne, S. Sutar, I. Radu, C. Huyghebaert, S. De Gendt, and M. Heyns, "From the metal to the channel: a study of carrier injection through the metal/2D MoS2 interface," *Nanoscale*, vol. 9, no. 30, pp. 10 869–10 879, 2017.

[9] A. Prakash, H. Ilatikhameneh, P. Wu, and J. Appenzeller, "Understanding contact gating in Schottky barrier transistors from 2D channels," *Scientific Reports*, vol. 7, p. 12596, July 2017.

[10] C. D. English, K. K. Smithe, R. L. Xu, and E. Pop, "Approaching ballistic transport in monolayer MoS2 transistors with self-aligned 10 nm top gates," in *Electron Devices Meeting (IEDM), 2016 IEEE International*. IEEE, 2016, pp. 5–6.

978-1-7281-0941-1/19 $31.00 © 2019 IEEE

Accurate and Efficient Dynamic Simulations of Ferroelectric Based Electron Devices

T. Rollo[1], L. Daniel[2], D. Esseni[1] Email: rollo.tommaso@spes.uniud.it

[1] DPIA, University of Udine, Italy; [2] Electrical Engineering and Computer Science Depart., MIT, Cambridge, MA, USA.

I. INTRODUCTION

In recent years electron devices based on ferroelectric materials have attracted a lot of interest well beyond FeRAM memories. Negative capacitance transistors (NC-FETs) have been investigated as steep slope transistors [1], [2], and Ferroelectric FETs (Fe-FETs) are under intense scrutiny also as synaptic devices for neuromorphc computing, where the minor loops in ferroelectrics can allow to achieve multiple values of conductance in read mode [3], [4], [5]. Furthermore, the persistence of ferroelectricity in ultra-thin ferroelectric layers paved the way to ferroelectric tunnelling junctions [6], where a polarization dependent tunneling current can be exploited to realize high impedance memristors, amenable for ultra power-efficient and thus massive parallel computation.

In all the above devices the dynamics of ferroelectric domains must be solved self-consistently with the device electrostatics. Furthermore, in MOS transistors having a semiconductor channel, the semiconductor introduces a strong non linearity in the electrostatics, and consequently in the dynamic equations describing the ferroelectric device evolution. The defects at the interfaces also play an intriguing role in ferroelectric FETs [7], in contrast to the well established and detrimental effects in conventional FETs [8]. Moreover, the presence of different trap levels imply a large range of charging and discharging time constants, possibly very different compared to the ferroelectric time constants.

In this paper we compare different numerical integration methods to achieve an accurate and effective simulation of NC-FETs, where the dynamics is governed by possibly very different time constants for either the ferroelectric or interface traps.

II. MODEL DESCRIPTION AND NUMERICAL ALGORITHMS

In the multi-domain, time-dependent Landau-Khalatnikov Equations (LKE) the description of ferroelectric domains requires the numerical solution of a set of differential equations for the polarization P_i of the i-th domain that read [9], [2]

$$\rho \frac{dP_i}{dt} = -(a_i P_i + b_i P_i^3 + c_i P_i^5) + \frac{V_{fe,i}}{T_{Fe}} + k \sum_j (P_j - P_i) \quad (1)$$

where ρ is a resistivity associated to domain switching, k is a coupling factor between nearest neighbor domains and $V_{fe,i}$ is the ferroelectric voltage drop for the i-the domain. The ferroelectric parameters were calibrated by comparing to experiments in [10], and the resulting set is [2] a=-9.5×10^8 m/F, b=2.01×10^{10} $m^5/F/C^2$, c=5.11×10^{10} $m^9/F/C^4$. When the ferroelectric is operated at small P (e.g. in NC-FETs), the linear term in Eq. (1)

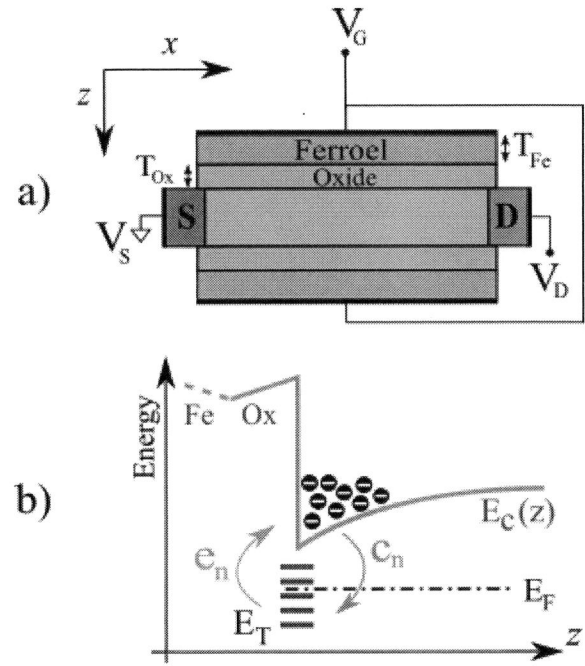

Fig. 1: a) Sketch of the double-gate ferroelectric NC-FET used in our simulations. b) Representation of the capture, c_n, and emission, e_n, processes for electrons at the semiconductor-oxide interface. E_T denotes the energy of an individual trap, E_C is the conduction band and E_F is the local Fermi-level.

is dominant and a time constant $\tau_{Fe}=\rho/|a|$ is readily identified, which is a property of the ferroelectric material.

Our analysis will be focused on an n-type, double-gate ultra-thin body (DG-UTB), nanoscale NC-FET (see Fig.1(a)), and a single domain analysis is used because the channel length is comparable to the size of ferroelectric domains [11]. Current is calculated with a simple ballistic top-of-the-barrier (ToB) model [12], [13], where electrons at the ToB with positive and negative velocity are taken to be in equilibrium with respectively the source, $E_{f,S}$, and drain Fermi level $E_{f,D}=(E_{fS}-qV_{DS})$ [12].

Quantization in the semiconductor is described with a 1D, parabolic effective mass Schrödinger solver, with valley multiplicities, and effective masses corresponding to a [100] silicon interface [14]. The link between the semiconductor and dielectrics is given by continuity conditions for the electric displacement at the interfaces; more modelling details may be found in [15].

We included in our analysis acceptor-type traps in the upper half of the silicon energy gap (see Fig. 1(b)), which exchange electrons with the conduction band with an emission, e_n, and capture rate c_n. The continuity equation for the carrier density n_T in traps with energy E_T can be written as [16]

$$\rho \frac{\partial n_T}{\partial t} = c_n (N_T - n_T) - e_n n_T \qquad (2)$$

where N_T is the trap density at energy E_T, and

$$e_n = \sigma \, v_{th} \, N_C \qquad (E_T - E_C)/(K_B T) \qquad (3a)$$

$$c_n = \sigma \, v_{th} \, N_C \qquad (E_f - E_C)/(K_B T) \qquad (3b)$$

with E_C and E_f being the conduction band edge and local Fermi level, and σ, v_{th}, N_C denoting respectively the trap cross-section, thermal velocity and conduction band effective density of states. For any trap energy E_T, we solve Eq. (2) self-consistently with the ferroelectric dynamics governed by Eq. (1), in fact the charge in the traps $Q_{it} = -q \sum_{E_T} n_T(E_T)$ influences the overall electrostatics in the gate stack and thus across the ferroelectric.

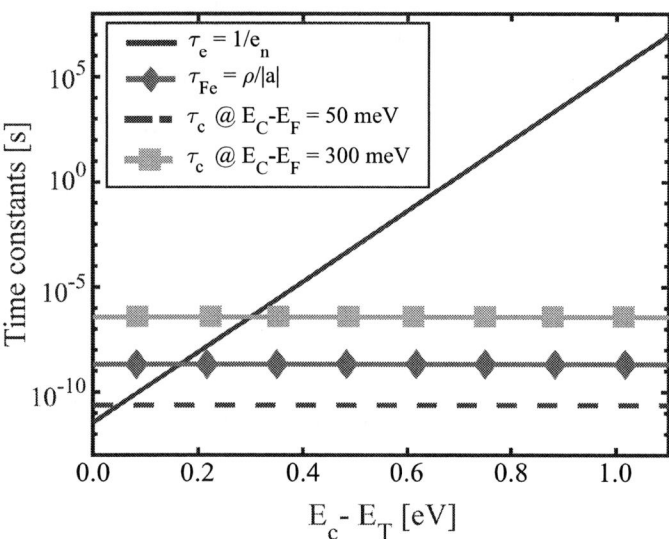

Fig. 2: Estimated time constants for the ferroelectric dynamic $\tau_{Fe} = \rho/|a|$, and for the emission, $\tau_e = 1/e_n$, and capture process, $\tau_c = 1/c_n$, at the silicon oxide interface. In this work we use $\sigma = 10^{-15} \text{cm}^2$, $v_{th} = 2.3 \cdot 10^7$ cm/s, $N_C = 3.2 \cdot 10^{19}$ cm^{-3}, which are the values experimentally extracted values for a Si-SiO2 interface [17]. Moreover, for the ferroelectric we calibrated the model against data for large area metal-ferroelectric-metal structures [10][15].

Fig. 2 shows the time constant of the ferroelectric, whose parameters were extracted in [15] by comparison to experiments in [10]. Fig. 2 further compares those time constants to those of the interface traps, that depend on the external bias through the alignment between E_C and E_f. The figure confirms that a wide range of time constants are present in the problem at study. In other words the problem is numerically extremely stiff. In the numerical simulation community it is well known that, for such a problem, using standard explicit integrators would force very small time steps in order to avoid numerical instability. Fortunately an efficient implementation, not requiring small time steps and guaranteeing numerical stability, can still be obtained by employing implicit time domain integrators [18], [19], such as the trapezoidal method.

III. RESULTS AND DISCUSSION

Our results were obtained for an n-type, DG-UTB NC-FET illustrated in Fig.1(a), having a 7 nm silicon film thickness, and a ferroelectric and interfacial SiO2 layer of respectively $T_{Fe} = 20$nm

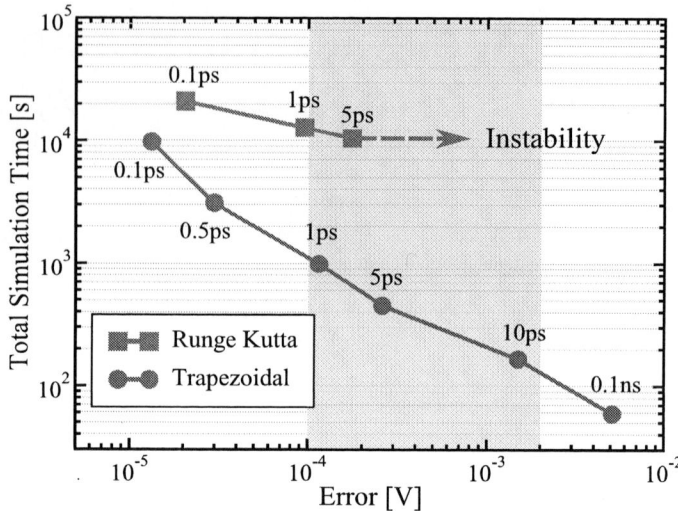

Fig. 3: Total simulation time (for one period of the input signal and for f=5MHz) versus maximum surface-potential error (i.e. potential at the silicon-SiO$_2$ interface) for an explicit Runge Kutta and the implicit trapezoidal methods. The shaded area indicates the typical range of acceptable maximum errors for the surface potential, namely between 10^{-4} and 2×10^{-3}V at room temperature. Note that the largest time step before the explicit method becomes unstable corresponds to a very small 0.15mV error.

and T_{ox}=0.5nm. An energetically uniform distribution of acceptor traps is assumed in the upper half of the energy-gap, with a concentration $D_{it} = 10^{13}$ cm^{-2}eV^{-1}.

In Fig.3 we compare the performance of an explicit integrator (Runge Kutta) with an implicit integrator (Trapezoidal). The figure shows that the total simulation time decreases in both methods for increasing values of the error, computed as the infinity norm of the semiconductor surface potential φ at different time-steps δ_t:

$$Error = ||\varphi_{\delta_t} - \varphi_{10fs}||_\infty := max \left(|\phi_{\delta_t} - \phi_{10fst}| \right) \quad (4)$$

where the potential calculated for δ_t=10fs was used used as the reference. However Fig.3 also shows that the explicit integrator becomes unstable for time steps larger than those needed to produce an error of 0.2mV. In other words, the explicit method is forced to continue using very small time steps and producing a small error even when larger errors would be acceptable. On the other hand the implicit integrator does not have stability problems, not even when using larger time steps when targeting larger values of the error in exchange for faster simulation times. For instance, if an error of 1mV is considered acceptable by the user, the implicit integrator could solve the problem in just 3 minutes while the explicit integrator would still be forced to require more than 4 hours to solve the problem to avoid numerical instability.

Fig.4 reports the simulated I_{DS}-V_G curves for two frequencies of the triangular gate voltage waveform. A few periods of the V_G input waveform were simulated and we verified that the I_{DS} becomes periodic after the first two or three periods, so that the I_{DS}-V_G curve plot can be obtained by taking the corresponding I_{DS} and V_G values in the last period of the V_G waveform. The results obtained with explicit and implicit methods are compared at two time steps, demonstrating a remarkable speed-up of the

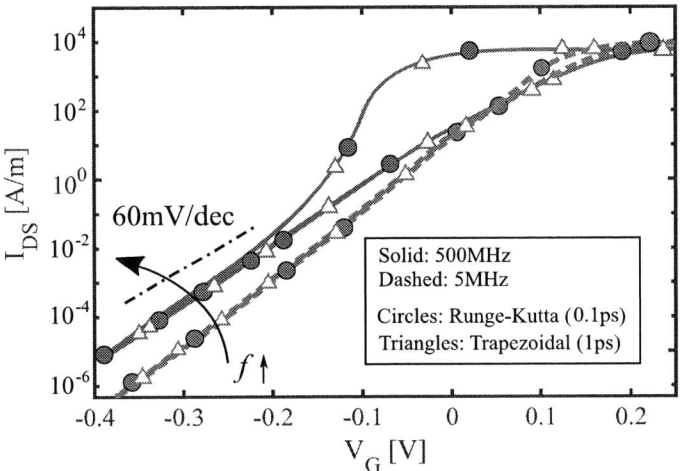

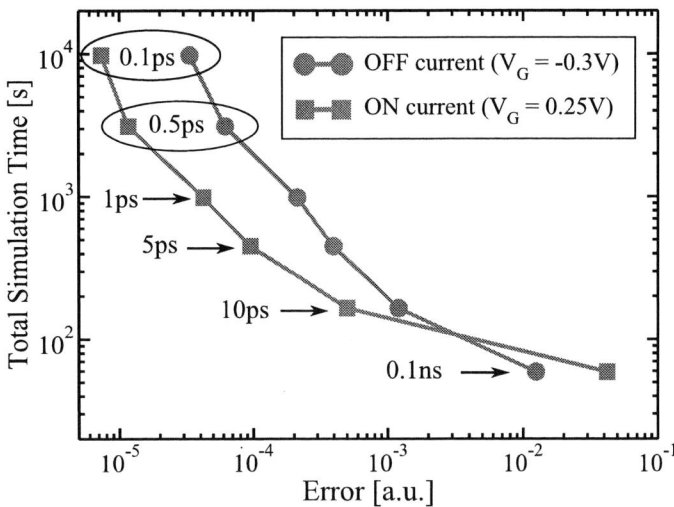

Fig. 4: Comparison of I_{DS}-V_G characteristics at different frequencies and for two integration methods . The curves are obtained with a uniform trap density $D_{it}=10^{13}$cm^{-2}/eV, the LKE coefficients reported in fig.2 and for a ρ=0.5Ωm. The curves obtained with the different integration methods are almost indistinguishable, but the speedups are consistent with the ones reported in Fig.3.

Fig. 6: Total simulation time (for one period of the input signal and for f=5MHz) versus the relative error on the current I_{DS} for the implicit integrator; the reference used to calculate the error is the simulated current at $delta_t$=10fs. We here label as on- and off-state current the I_{DS} at respectively V_G=0.25 V and V_G=−0.3 V (see also the I_{DS} versus V_G characteristics in Fig.4.

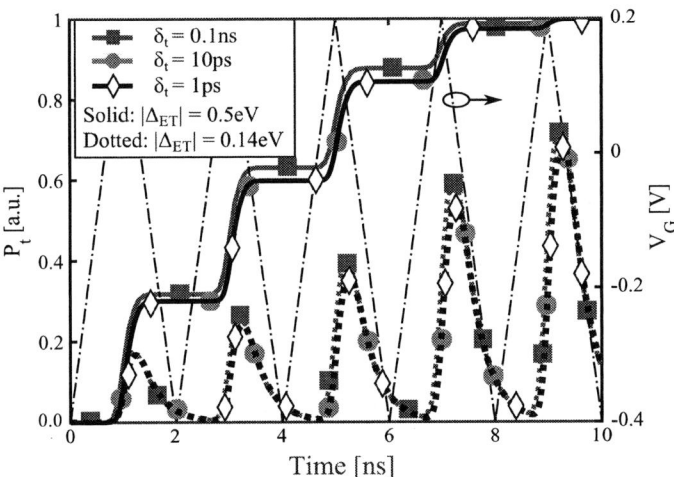

Fig. 5: Occupation probability $P_t=n_t/N_t$ for two trap levels (in midgap position and closer to the conduction band) corresponding to the f=500MHz curve in Fig.4 and for three integration time-steps δ_t; the V_G input waveform is also shown (right y axis). $\Delta_{ET}=E_C$-E_T is the distance between the energy level of the trap and the conduction band at the semiconductor-dielectric interface.

simulation time, enabled by a drastic reduction of the time step for a given accuracy. From a device perspective, Fig.4 also confirms how the presence of acceptor type interface traps can help reduce the subthreshold swing below 60mV/dec in NC-FETs (in contrast to the well-known detrimental effect it has in conventional MOSFETs), essentially because the traps improve the capacitance matching between the ferroelectric and the SiO$_2$-semiconductor stack [15]. However such a benefit tends to vanish at higher frequencies because the occupation of the traps cannot follow the gate voltage waveform, so that the subthreshold steepness of the I_{DS}-V_G curve degrades and the amplitude of the hysteresis enlarges by increasing the frequency.

To highlight the sensitivity of simulation outcomes to integration-step variations, in Fig.5 we have reported the occupa-

tion probabilities P_t for two energy levels E_T at three different time-step δ_t values, only for the trapezoidal integrator. At the frequency of 500MHz the deepest trap, placed 500meV below the conduction band, is not responding to the input waveform V_G, meaning that the emission mechanism is not fast enough to discharge the level E_T: this is again consistent with the behavior of the I_{DS}-V_G characteristics reported in Fig.4 of reference [15]. Fig.5 also shows that for δ_t smaller than 10ps the curves for different δ_t values are essentially overlapping, whereas for δ_t=0.1ns the difference in the P_t waveforms is sizeable. Still these differences are small and cannot result in an appreciable difference, for example, in the I_{DS}-V_G characteristic of the transistor, such as the curves reported in Fig.4. This is also confirmed by the analysis in Fig.6, where we have reported the relative error for the on- and off-state I_{DS}. As it can be seen for δ_t<0.1 ns the I_{DS} difference with respect to the reference value (i.e. the current calculated for δ_t=10fs) is smaller than 1%, which is sufficiently small for most TCAD applications.

IV. CONCLUSIONS

In summary, in this work we systematically demonstrated the advantages of an implicit trapezoidal integrator with respect to a explicit Runge-Kutta method for the simulation of NC-FETs having a wide range of time constants set either by the ferroelectric or by the interface traps dynamics. Advantages are observed in terms of robustness of convergence and in terms of simulation time at fixed accuracy. Our results are expected to be useful in the development of robust TCAD tools for ferroelectric based devices, that are important for the design and optimization of ferroelectric FETs and ferroelectric tunnelling junctions. The growing interest for negative capacitance, steep slope FETs and ferroelectric based synaptic devices for neuromorphc computing will make the modeling and simulation of ferroelectric devices a topic of increasing technological relevance in the near future.

V. ACKNOWLEDGMENTS

The project has received financial support from the MIT International Science and Technology Initiatives (MISTI) Global Seed Funds, within the MIT-FVG Project (University of Udine, Trieste and SISSA).

REFERENCES

[1] S. Salahuddin and S. Datta, "Use of Negative Capacitance to Provide Voltage Amplification for Low Power Nanoscale Devices," *Nano Letters*, vol. 8, no. 2, 2008.

[2] T. Rollo and D. Esseni, "New Design Perspective for Ferroelectric NC-FETs," *IEEE Electron Device Letters*, vol. 39, no. 4, pp. 603–606, April 2018.

[3] M. Jerry, P.-Y. Chen, J. Zhang, P. Sharma, K. Ni, S. Yu, and S. Datta, "Ferroelectric FET Analog Synapse for Acceleration of Deep Neural Network Training," in *IEEE IEDM Technical Digest*, Dec 2017, pp. 139–142.

[4] B. Obradovic, T. Rakshit, R. Hatcher, J. Kittl, R. Sengupta, J. G. Hong, and M. S. Rodder, "A multi-bit neuromorphic weight cell using ferroelectric fets, suitable for soc integration," *IEEE Journal of the Electron Devices Society*, vol. 6, pp. 438–448, 2018.

[5] H. Mulaosmanovic, J. Ocker, S. Mller, M. Noack, J. Mller, P. Polakowski, T. Mikolajick, and S. Slesazeck, "Novel ferroelectric fet based synapse for neuromorphic systems," in *2017 Symposium on VLSI Technology*, June 2017, pp. T176–T177.

[6] B. Max, M. Hoffmann, S. Slesazeck, and T. Mikolajick, "Ferroelectric Tunnel Junctions based on Ferroelectric-Dielectric Hf0.5Zr0.5O2/Al2O3 Capacitor Stacks," in *Proc. European Solid State Device Res. Conf.*, 2018, pp. 142–145.

[7] T. Rollo and D. Esseni, "Influence of Interface Traps on Ferroelectric NC-FETs," *IEEE Electron Device Letters*, vol. 39, no. 7, pp. 1100–1103, 2018.

[8] E. H. Nicollian and J. R. Brews, *MOS (metal oxide semiconductor) Physics and Technology*. Wiley Interscience, 1982.

[9] Z. C. Yuan, S. Rizwan, M. Wong, K. Holland, S. Anderson, T. B. Hook, D. Kienle, S. Gadelrab, P. S. Gudem, and M. Vaidyanathan, "Switching-Speed Limitations of Ferroelectric Negative-Capacitance FETs," *IEEE Transactions on Electron Devices*, vol. 63, no. 10, pp. 4046–4052, Oct 2016.

[10] D. Zhou, Y. Guan, M. M. Vopson, J. Xu, H. Liang, F. Cao, X. Dong, J. Mueller, and T. S. U. Schroeder, "Electric field and temperature scaling of polarization reversal in silicon doped hafnium oxide ferroelectric thin films," *Acta Materialia*, vol. 99, pp. 240–246, 2015.

[11] A. Roelofs, T. Schneller, K. Szot, and R. Waser, "Towards the limit of ferroelectric nanosized grains," *Nanotechnology*, vol. 14, pp. 250–253, 2003.

[12] A. Rahman, J. Guo, S. Datta, and M.S. Lundstrom, "Theory of Ballistic Nanotransistors," *IEEE Trans. on Electron Devices*, vol. 50, no. 9, pp. 1853–1863, 2003.

[13] S. Rakheja, M.S. Lundstrom, and D. A. Antoniadis, "An Improved Virtual-Source-Based Transport Model for Quasi-Ballistic TransistorsPart I: Capturing Effects of Carrier Degeneracy, Drain-Bias Dependence of Gate Capacitance, and Nonlinear Channel-Access Resistance," *IEEE Trans. on Electron Devices*, vol. 62, no. 9, pp. 2786–2793, 2015.

[14] D. Esseni, P. Palestri, and L. Selmi, *"Nanoscale MOS Transistors - Semi-Classical Transport and Applications"*, 1st ed. Cambridge University Press., 2011.

[15] T. Rollo, H. Wang, G. Han, and D. Esseni, "A simulation based study of NC-FETs design: off-state versus on-state perspective," in *IEEE IEDM Technical Digest*, Dec 2018, pp. 213–216.

[16] M.Rudan, *Physics of Semiconductor Devices*. Springer International Publishing, 2018.

[17] G. Brammertz, K. Martens, S. Sioncke, A. Delabie, M. Caymax, M. Meuris, and M. Heyns, "Characteristic trapping lifetime and capacitance-voltage measurements of GaAs metal-oxide-semiconductor structures," *Applied Physics Letters*, vol. 91, p. 133510, 2007.

[18] G. Dahlquist, "A Special Stability Problem for Linear Multistep Methods," *BIT Numerical Mathematics*, vol. 3, no. 1, pp. 27–43, Mar 1963.

[19] G. Dahlquist and B. Lindberg, "On some implicit one-step methods for stiff differential equations," Dept. of Information Processing, Royal Inst. of Tech., Stockholm, Tech. Rep., 1973.

Precise Transient Mechanism of Steep Subthreshold Slope PN-Body-Tied SOI-FET and Proposal of a New Structure for Reducing Leakage Current upon Turn-off

Takayuki Mori
Kanazawa Inst. of Tech.
Nonoichi, Japan
t_mori@neptune.kanazawa-it.ac.jp

Jiro Ida
Kanazawa Inst. of Tech.
Nonoichi, Japan
ida@neptune.kanazawa-it.ac.jp

Hiroki Endo
Kanazawa Inst. of Tech.
Nonoichi, Japan
.

Yasuo Arai
High Energy Accelerator Research Org., KEK
Tsukuba, Japan

Abstract—**In this study, the precise transient mechanism of the super-steep subthreshold slope PN-body-tied (PNBT) silicon on insulator field-effect transistor (SOI-FET) is clarified by using technology computer-aided design. We found out that the operation mechanism differs between the turn-on and turn-off. Additionally, a new PNBT SOI-FET structure with a second gate for the high-speed operation is proposed and we showed that the new structure can reduce the leakage current upon the turn-off.**

Keywords—body tied, feedback, floating body, steep subthreshold slope

I. INTRODUCTION

Ultralow power devices for the Internet of Things and artificial intelligence systems are needed to achieve a next-generation society such as Industry 4.0 [1] and Society 5.0 [2]. We proposed PN-body-tied (PNBT) silicon on insulator field-effect transistors (SOI-FETs), which have a super-steep subthreshold slope (SS) (< 1 mV/dec) with a low drain voltage ($V_d = 0.1$ V) [3][4]. A PNBT SOI-FET has the possibility to be applied to ultralow power consumption LSIs, sensors, and neuromorphic chips. We report the DC and transient characteristics of the PNBT SOI-FET; however, an understanding of the operation principle is insufficient, and the device shows a long time leakage current upon a turn-off [5][6]. In this study, we clarify the precise transient mechanism on a PNBT SOI-FET using technology computer-aided design and propose a new structure PNBT SOI-FET with a second gate to reduce the leakage current.

II. DEVICE STRUCTURE AND SIMULATION CONDITION

Fig. 1 shows the structure of the PNBT SOI-FET. The actual device was fabricated using a LAPIS Semiconductor 0.2 μm SOI CMOS process, and device simulations were

applied using HyENEXSS [7]. The device has a 50-nm thick SOI (T_{Si}), a 200-nm box (T_{Box}), a 4.4-nm thick gate oxide (T_{ox}) (in the simulation, $T_{ox} = 5$ nm), a 0.2-μm gate length (L_g), a 1-μm gate width (W_g), and a 1.2-μm base width (W_b). In the simulation, we used the Shockley–Read–Hall (SRH) recombination, auger recombination, and band-to-band tunneling model. The carrier lifetime parameters in the SRH model were modified (default lifetime × 0.02) to fit the measurement results [8].

III. RESULTS AND DISCUSSIONS

Fig. 2 shows the measured [5] and simulated [6] DC and transient (turn-on) characteristics of the PNBT SOI-FET. The PNBT SOI-FETs have a super-steep SS and turn-on delay time in both measurements and simulations. However, the turn-on delay time decreases when the gate voltage V_g slightly increases, as shown in Fig. 3. Therefore, the turn-on delay time is expected to not be a critical issue for a high-speed operation.

In order to clarify the more precise mechanism of transient effect, we focused on the time dependence of the potential (conduction band energy) shift of the body terminal direction to clarify the precise transient mechanism. The

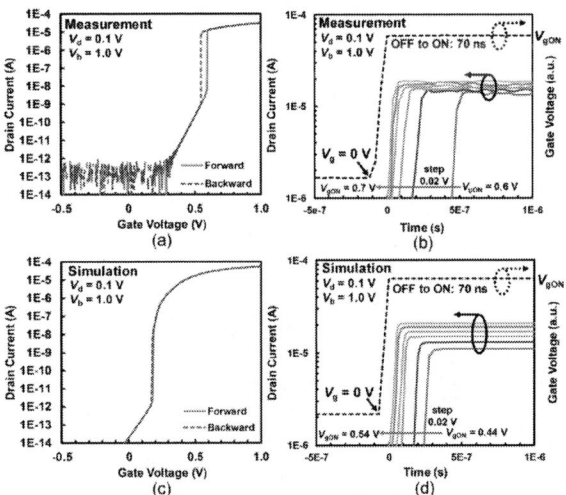

Fig. 2. DC and transient (turn-on) characteristics of PNBT SOI-FET: (a) measured I_d–V_g, (b) measured transient characteristics, (c) simulated I_d–V_g, and (d) simulated transient characteristics.

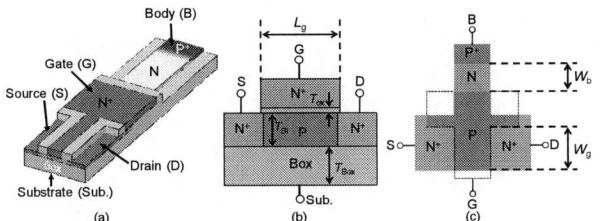

Fig. 1. Device structure of PNBT SOI-FET: (a) bird's eye view, (b) front view, and (c) top view.

978-1-7281-0941-1/19 $31.00 © 2019 IEEE

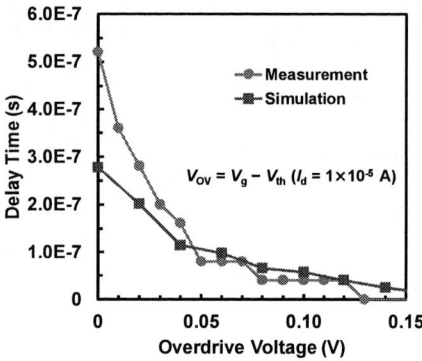

Fig. 3. Overdrive voltage dependence on turn-on delay time.

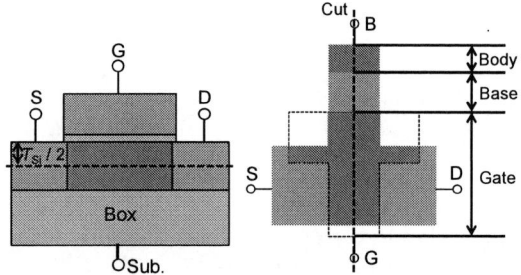

Fig. 4. Cutting direction used to extract potential.

cutting direction used to extract the potential is shown in Fig. 4. Fig. 5 shows the time-shift of the potential upon the turn-on. The potential varies as the time proceeds; however, the potential in the gate region and that in the base region behave as different modes of change. Fig. 6 shows the time dependence on the potential in the gate and the base region extracted from Fig. 5. In the base region, the potential gradually decreases (the energy increases). In contrast, the potential in the gate region remains constant and then abruptly varies at (3) in Figs. 5 and 6. We consider the transient mechanism as follows. First, electrons flow from the source to the body just after a turn-on. The electrons modulate the potential in the base region. Second, holes are provided from the body and flow toward the base region. However, the holes recombine in the base region. Therefore, this phenomenon requires a certain amount of time, and the potential gradually modulates. Finally, the holes reach the gate region and modulate the potential of the gate region. This induces a strong

positive feedback, and thus the potentials jump to (3) in Figs. 5 and 6. Therefore, Large I_d abrupt flow as shown in Fig. 5. This behavior is similar to that of other devices using a feedback mechanism, such as Z^2-FET [9].

Fig. 7 shows the measured [5] and simulated [6] transient characteristics upon a turn-off. The leakage current is

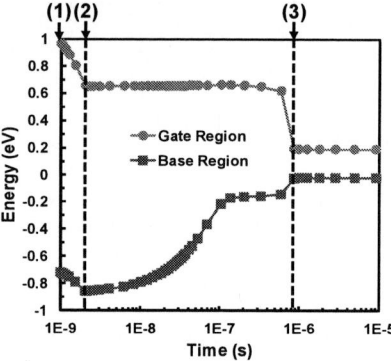

Fig. 6. Time dependence on potential in gate and base region extracted from Fig. 5.

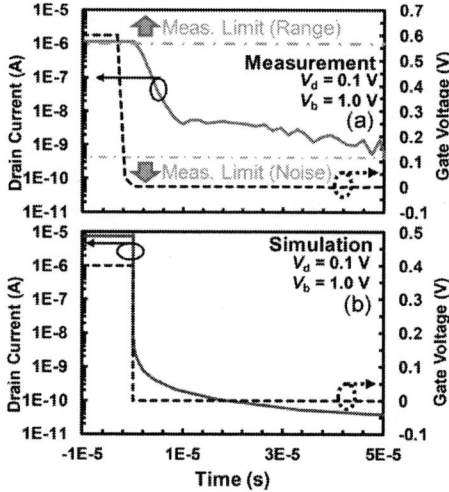

Fig. 7. Transient (turn-off) characteristics of PNBT SOI-FET: (a) measurement and (b) simulation results.

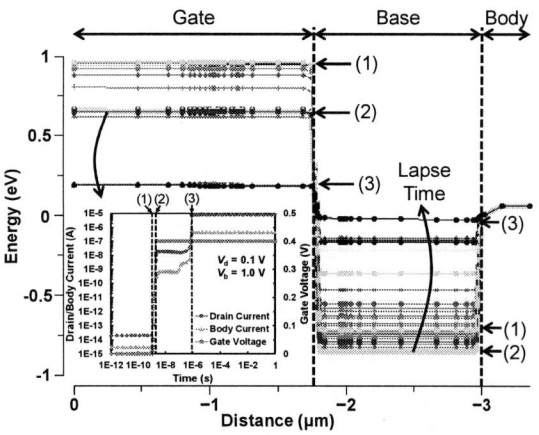

Fig. 5. Time-shift of potential (conduction band energy) upon a turn-on.

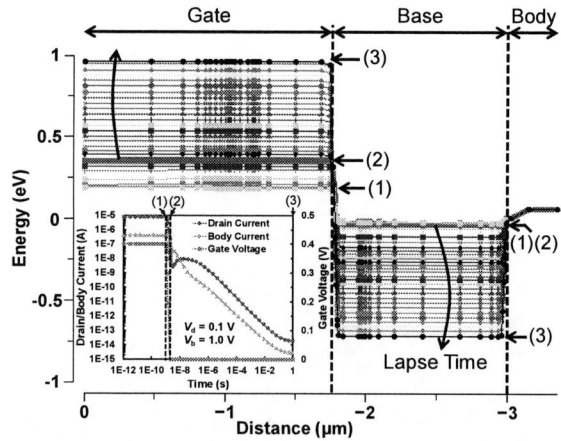

Fig. 8. Time-shift of potential upon a turn-off.

978-1-7281-0941-1/19 $31.00 © 2019 IEEE   292

sustained for a few seconds. This is a significant issue of PNBT SOI-FET. Figs. 8 and 9 show the details of the time dependence on the potential shift in the gate and base region upon a turn-off. The potentials of both the gate region and the base region gradually vary as contrasted with the turn-on. For the turn-off, the accumulation carriers in both the gate region and the base region are ejected through a recombination, which unfortunately is a slow operation.

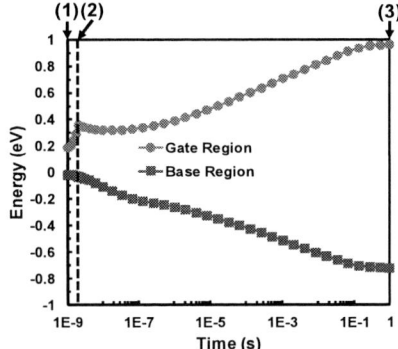

Fig. 9. Time dependence on potential in gate and base region extracted from Fig. 8.

To reduce the leakage current upon a turn-off, we propose a new structure, PNBT SOI-FET with a second gate, as shown in Fig. 10. The objective of the new PNBT SOI-FET structure is to control carrier injection and ejection by using the second gate. The second gate width set to be 1.2 μm and the width between the first gate and second gate is 50 nm. Fig. 11 shows the simulated DC characteristics of the new PNBT SOI-FET structure. The device also has steep SS characteristics. Additionally, the body current I_b is smaller than I_d as shown in

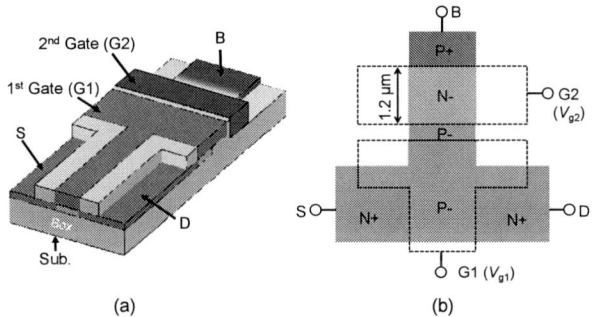

Fig. 10. New PNBT SOI-FET structure with second gate: (a) bird's eye view and (b) top view.

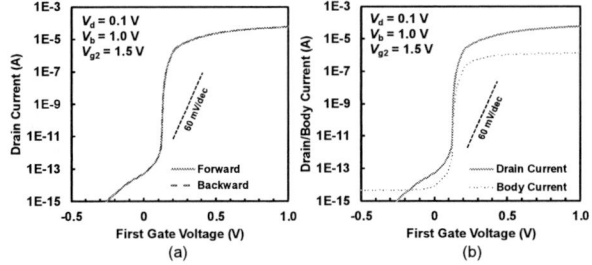

Fig. 11. Simulated DC characteristics of new PNBT SOI-FET structure with second gate. (a) Double sweep I_d–V_g characteristics and (b) I_d and I_b–V_g characteristics.

Fig. 11(b). Therefore, it is possible that the power consumption of the body lower than the power consumption of the drain. 12 shows the simulated transient (turn-off) characteristics of the new PNBT SOI-FET. When the first gate voltage V_{g1} only turns off, the leakage current flow is the same as in Fig. 8. However, for V_{g1}, body voltage V_b, and the V_{g2} turn off under the same timing, the leakage current decreases faster than the conventional PNBT SOI-FET. This phenomenon does not occur when V_{g1} and V_b turn off under the same timing (V_{g2} does not turn off). Switching the second gate means that the operation mode changes from PNBT mode to a common body-tied mode. Therefore, the turn-off delay decreases because the accumulation carriers smoothly eject to the body terminal.

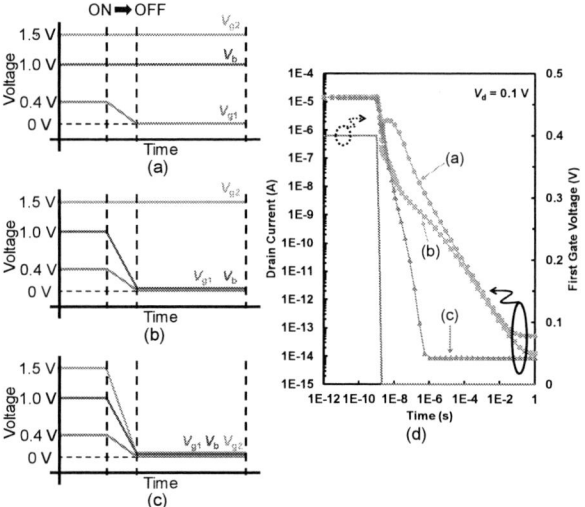

Fig. 12. Waveform of (a) V_{g1} only turn-off, (b) V_{g1} and V_b turn-off, and (c) V_{g1}, V_b, and V_{g2} turn-off. (d) Simulated transient (turn-off) characteristics of new PNBT SOI-FET structure with second gate.

IV. CONCLUSIONS

We clarified the precise transient mechanism of the PNBT SOI-FET and proposed a new PNBT SOI-FET structure with a second gate. We found for the first time that the operation mechanism of the PNBT SOI-FET differs between the turn-on and turn-off from the time dependence of the potential shift. Additionally, the leakage current on the turn-off can be decreased by V_b and V_{g2} control on the new PNBT SOI-FET structure. The new PNBT SOI-FET structure has the possibility of a high-speed operation.

ACKNOWLEDGMENT

This work is the result of collaborations with the High Energy Accelerator Research Organization (KEK) and LAPIS Semiconductor Co., Ltd. This work is supported by VLSI Design and Education Center (VDEC), the University of Tokyo in collaboration with Cadence Corporation and Mentor Graphics Corporation.

This work was supported in part by JST-CREST Grant Number JPMJCR16Q1, and in part by MEXT KAKENHI Grant Number 25109002.

REFERENCES

[1] (2013) Recommendations for implementing the strategic initiative INDUSTRIE 4.0. [Online]. Available: https://www.acatech.de/wp-content/uploads/2018/03/Final_report__Industrie_4.0_accessible.pdf

[2] Society 5.0 [Online]. Available: https://www.gov-online.go.jp/cam/s5/eng/

[3] J. Ida *et al*, "Super steep subthreshold slope PN-body tied SOI FET with ultra low drain voltage down to 0.1V," in *IEDM Tech. Dig.*, Dec. 2015, pp. 624–627.

[4] T. Mori and J. Ida, "P-Channel and N-Channel super-steep subthreshold slope PN-body tied SOI-FET for ultralow power CMOS," *IEEE J. Electron Devices Soc.*, vol. 6, pp. 1213-1219, Oct. 2018.

[5] H. Endo, J. Ida, T. Mori, K. Ishibashi, and Y. Arai, "First experimental confirmation of transient effect on super steep SS "PN-body tied SOI FET" with pulse measurements," in *EDTM Proc. of Tech. Papers*, Mar. 2019, pp. 91-93.

[6] T. Mori, J. Ida, and H. Endo, "Analysis of transient effect on super-steep SS PN-body tied SOI-FET," in *2019 International Symposium on VLSI Technology, Systems and Application (VLSI-TSA)*, Hsinchu, Apr. 2019, pp. 1-2.

[7] HyENEXSS™ ver. 5.5, Selete, 2011.

[8] T. Mori and J. Ida, "Analysis and optimization of device parameters on super-steep subthreshold slope PN-body tied SOI-FET," in *proceeding of International Conference on Solid State Devices and Materials (SSDM)*, Sep. 2018. pp. 839–840.

[9] C. Navarro *et al.*, "Extended analysis of the Z2-FET: Operation as capacitorless eDRAM," *IEEE Transactions on Electron Devices*, vol. 64, no. 11, pp. 4486-4491, Nov. 2017.

978-1-7281-0941-1/19 $31.00 © 2019 IEEE

Negative Capacitance Field-Effect Transistor Based on a Two-Dimensional Ferroelectric

M. Soleimani, N. Asoudegi, P. Khakbaz, M. Pourfath*

School of Electrical and Computer Engineering
University College of Engineering
University of Tehran
Tehran, Iran
pourfath@ut.ac.ir

Abstract—Negative capacitance field effect transistors (NCFETs) based on ferroelectric materials have been the focus of intensive research activities because of their relatively small sub-threshold swing. This work proposes and presents a comprehensive study of a NCFET based on few-layer α-In$_2$Se$_3$ as the ferroelectric in order to reduce the sub-threshold swing through voltage amplification effect. By employing first principles electronic structure calculations, the Landau constants of mono and few-layer α-In$_2$Se$_3$ are extracted which were utilized for analyzing the characteristics of a NCFET with a monolayer MoS$_2$ as the channel material. Sub-threshold swings in the range of ~27-59 mV/dec were achieved for few-layer α-In$_2$Se$_3$ that can be further improved by increasing the thickness of the ferroelectric layer and by using a thinner or high-κ insulate layer.

Index Terms—Negative capacitance, Transistor, Sub-threshold swing, Ferroelectric, In$_2$Se$_3$

I. Introduction

The scaling of metal-oxide-semiconductor field-effect transistors (MOS-FETs) is being limited by the inability to remove the heat generated during the switching process that originates from the poor scaling of the operating voltage (V_{DD}). A key factor that hinders the scaling of V_{DD} is the subthreshold swing (SS) which is thermally limited to ~60 mV/dec at room temperature because of the Boltzmann statistic. The (SS)of a FET is generally given by [1]:

$$SS = \frac{\partial V_{\text{G}}}{\partial(\log_{10} I_{\text{D}})} = \frac{\partial V_{\text{G}}}{\partial \psi_{\text{S}}} \frac{\partial \psi_{\text{S}}}{\partial(\log_{10} I_{\text{D}})}$$
$$= \left(1 + \frac{C_{\text{S}}}{C_{\text{Ins}}}\right)\left(\frac{k_{\text{B}}T}{q}\ln 10\right), \qquad (1)$$

where V_{G} is the gate voltage, I_{D} is the drain current, ψ_{S} is the surface potential of the semiconducting channel, C_{S} is the substrate capacitance, C_{Ins} is the gate insulator capacitance and k$_{\text{B}}$ is the Boltzmann constant, T is the temperature, and q is the electron unit charge. At room temperature, a sub-threshold swing of approximately 60 mV/dec can be obtained if the body factor becomes close to 1. This happens if a relatively thin high-κ insulator material is used ($C_{\text{S}}/C_{\text{Ins}} \ll 1$). A sub-60 mV/dec for SS can be achieved if the device body factor becomes

smaller than 1, where C_{Ins} can be negative while the C_{S} remains positive [2], [3]. This can be reached by using the recently proposed negative capacitance field effect transistor (NCFET), in which a ferroelectric material is added to the gate stack of conventional MOSFETs [3], [4]. Because of a negative voltage drop through the ferroelectric layer, which is due to the combination of the external electric field and polarization in the ferroelectric material, the gate voltage is amplified which in turn reduces the SS.

Two main structures are commonly studied for NCFETs: the metal-ferroelectric-metal-insulator-semiconductor (MFMIS) [5] and metal-ferroelectric-insulator-semiconductor (MFIS) [6]. It is more straightforward to realize MFMIS structures, where the inserted metal layer provides a uniform surface potential for simultaneous polarization and makes it easy to detect the internal voltage in the experiment. To obtain stabilized NC hysteretic behavior, which is the characteristic for ferroelectric materials, should be eliminated. The requirement of a hysteresis-free NC is related to the maximum critical ferroelectric thickness which can be reduced by the background permittivity.

The application of conventional ferroelectric materials for 2D channels is not optimal; therefore, this work investigates the potential of 2D ferroelectrics in NCFETs. A few 2D van der Waals (vdW) materials exhibit ferroelectricity, where for NC those with out-of-plane polarizability are of interest. An appropriate selection of a 2D ferroelectric insulator for integrating with 2D semiconductors is essential for achieving high performance NCFETs. Ferroelectric polarization switching and hysteresis loop have been recently observed down to the bilayer and monolayer (1L) α-In$_2$Se$_3$ [7] which demonstrates that the thinnest layered ferroelectrics can be realized with 2D materials.

This work proposes and presents a comprehensive study on a NCFET based on the MFMIS structure, where a monolayer MoS$_2$ serves as the channel material and few-layer α-In$_2$Se$_3$ acts as the ferroelectric material. As shown in Fig.1(a), compared to the conventional MOSFETs, the NCFETs have an additional ferroelectric layer deposited on the metal gate. Recent experiments indicate that in the employed structure the metal gate provides the same gate charge density to both the internal gate and ferroelectric surfaces [5], [8]. The NCFETs can be described by the capacitor divider model of Fig.1(b), where C_{Fe}, C_{Ins} and C_{S} are the capacitors due to the ferroelectric layer, the dielectric layer and the semiconductor, respectively. The total gate capacitance of a NCFET (C_{G}) consists of a series combination of C_{Ins} and C_{Fe}.

In this paper, we focus on the α-In$_2$Se$_3$ as the ferroelectric material shown in Fig.1(c). First principle calculations based on density functional theory (DFT) are applied to systematically study the electronic properties of α-In$_2$Se$_3$ through layer-dependent cleavage energy, Gibbs free energy, polarization, averaged electrostatic potential energy and bandgap. Using the electronic structure calculations we determine the Landau constants of α-In$_2$Se$_3$ and quantitatively investigate the device properties of the NCFET.

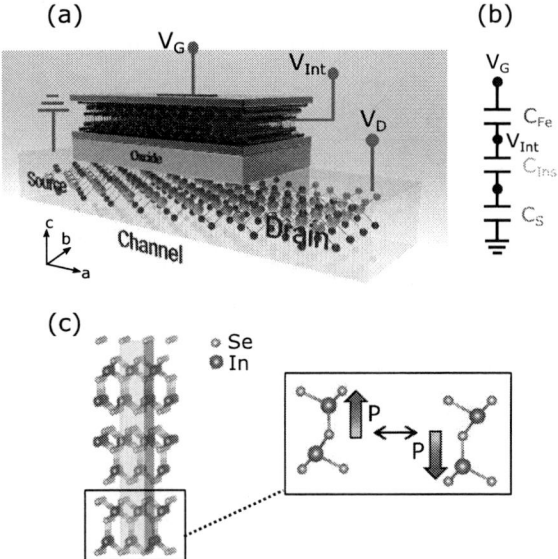

(a)

(b)

(c)

Fig. 1: The schematic of a NCFET with monolayer MoS$_2$ as the channel material and α-In$_2$Se$_3$ as the ferroelectric. (b) The equivalent capacitor model for the combination of the gate, ferroelectric and insulator in the MFMIS structure. (c) The atomistic structure of layered α-In$_2$Se$_3$ ferroelectric. α-In$_2$Se$_3$ switching from the upward polarization (top) to the downward polarization (bottom).

II. Approach

To model the current-voltage characteristic of nanoscale transistors, the top-of-the barrier model proposed in Ref. [9] has been utilized. The drain-source current can be obtained by:

$$I_{\text{DS}} = \int_{-\infty}^{+\infty} J(E - U_{\text{scf}})[f_1(E) - f_2(E)]dE, \quad (2)$$

where f_1, f_2 are the Fermi-Dirac distribution functions of the contacts and J is the current density, and U_{scf}

is the self-consistent potential at the top of the barrier along the channel. To model the dynamics of ferroelectric polarization, the Landau-Khalatnikov (LK) equation can be used [3]:

$$\rho \frac{d\vec{P}}{dt} + \nabla_{\vec{P}} G = 0, \quad (3)$$

where ρ and P are the resistivity and polarization, respectively. Here, F represents the Gibbs free energy of the ferroelectric material and is a function of a series expansion of the polarization as:

$$F = aP^2 + bP^4 + cP^6 - \vec{E}_{\text{ext}}.\vec{P}, \quad (4)$$

where a, b and c are Landau coefficients and $\vec{E}_{\text{ext}}$ is the external electric field. The gate voltage of the NCFET can be written as

$$V_{\text{G}} = V_{\text{Int}} + \left[2aQ_{\text{G}} + 4bQ_{\text{G}}^3\right] t_{\text{Fe}}, \quad (5)$$

where V_{Int} is the internal voltage the second term is the voltage drop across the ferroelectric. Also, t_{Fe} and Q_{G} are the ferroelectric thickness and gate charge density, respectively.

DFT calculations have been carried out within the framework of the projector augmented-wave (PAW) formalism [10], [11] as implemented in the Vienna Ab initio Simulation Package (VASP). Structural relaxations and calculation of the electronic properties were carried using the generalized gradient approximation (GGA) as parameterized by Perdew-Burke-Ernzerhof functional for solids (PBEsol) [12] with a convergence criterion that forces on the atoms become smaller than 0.001 eV/Å. For the bandgap calculations, we also used the hybrid functional of Hyed-Scuseria-Ernzerhof (HSE06) [13] The plane wave cutoff energy is 500 eV and a 12×12×1 grid points were used in k-space to sample the Brillouin zone. The dispersion interaction was accounted for layered α-In$_2$Se$_3$ by using the Grimme's DFT-D3 method [14]. The Gibbs free energy under mild conditions of temperature, T=300 K can be expressed as

$$F = H^0 + TS + \text{ZPE}, \quad (6)$$

where H^0, S and ZPE are the enthalpy at absolute zero, the entropy and intrinsic zero point energy.

III. Results and Discussion

Cleavage energies of α-In$_2$Se$_3$ 1L and 2L from a five-layer (5L) slab were calculated to confirm feasibility of getting a 1L and 2L α-In$_2$Se$_3$ by exfoliation. As shown in Fig.2(a), the cleavage energies required to slice the 1L and 2L α-In$_2$Se$_3$ along c plane are 0.60 J/m^2 and 0.66 J/m^2, respectively. It should be noted that the cleavage energy of slicing bulk α-In$_2$Se$_3$ into 1L and 2L is close to that of slicing graphite into graphene (0.36 J/m^2) [15]. This indicates that slicing the α-In$_2$Se$_3$ into 2D monolayers along a specific plane is highly feasible.

Fig.2(b) shows the Gibbs free energy for 1L to 5L α-In$_2$Se$_3$ at room temperature. It is found that, thicker α-

978-1-7281-0941-1/19 $31.00 © 2019 IEEE

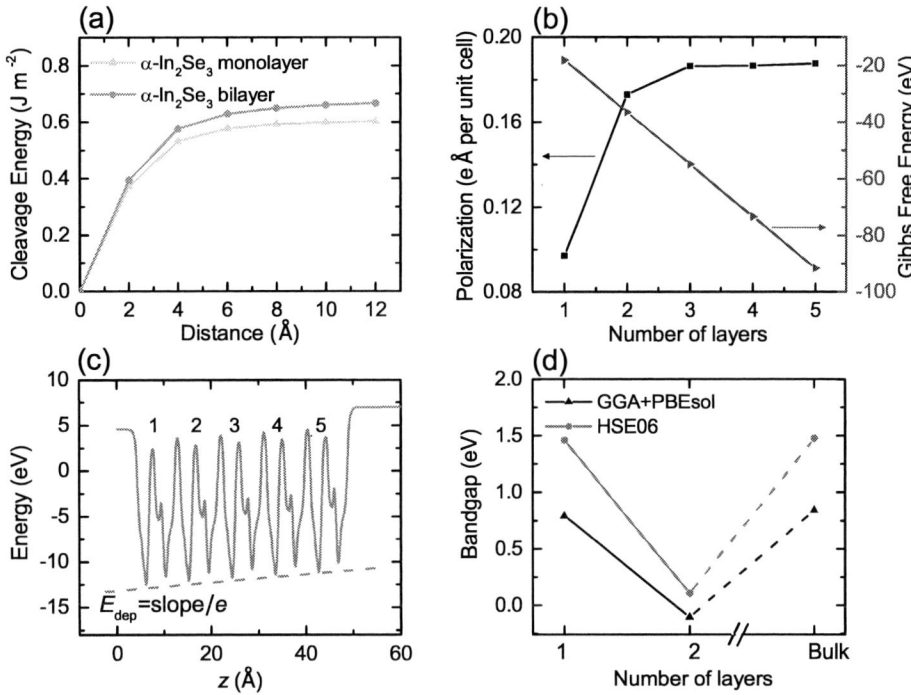

Fig. 2: The cleavage energies estimation regarding the slicing of α-In$_2$Se$_3$ along c plane the structure of α-In$_2$Se$_3$, (b) the polarization and Gibbs free energies at T=300 K, (c) the planar-averaged electrostatic potential energy along c plane for 5L α-In$_2$Se$_3$. (d) Evolution of bandgap as a function of the number of layers with the GGA-PBEsol and HSE06 methods, respectively.

In$_2$Se$_3$ is more stable than monolayer of this materials. The difference in stability between monolayer and few-layer α-In$_2$Se$_3$ is related to the vdW coupling between the α-In$_2$Se$_3$ layers. In addition, Fig.2(b) indicates that the average out-of-plane polarization of the 2L is over two times larger than that of the monolayer and the magnitude of the electric dipole saturates beyond two layers. The absence of the energy band gap and overlaying of the conduction and valence bands above 2L lead to charge transfer induced by the depolarization field on both surfaces of the film which in turn hinders further increase of the polarization.

The average electrostatic potential energy in the c plane for 5L α-In$_2$Se$_3$ is shown in Fig.2(c). Also, the depolarization field E_{dep} was calculated from the electrostatic potential energy. The depolarization fields of the 5L are not completely compensated, and the values decrease as the number of layers increases and will disappear in thick films. Fig.2(d) indicates the bandgap evolution as a function of the number of layers. Due to the existence of bandgap underestimation in GGA-PBEsol functional, we also employed the hybrid functional approximation using HSE06 exchange-correlation term. It was obvious that there was a bandgap shortening from the monolayer to bilayer by both methods. Bandgaps of α-In$_2$Se$_3$ are extracted as 1.46 eV (1L), 0.11 eV (2L) and 1.48 eV (bulk).

To analyze α-In$_2$Se$_3$ as the ferroelectric in NCFET, the

Landau constants are extracted (Table.I) from the Gibbs free energy at T=300 K versus polarization during the phase transition.

TABLE I: Landau constants extracted by fitting parameters in Eq. (4).

α-In$_2$Se$_3$	a (V/eÅ^2)	b (V/e$^3\text{Å}^4$)
1L	-28.65	14.575e+2
2L	-6.96	116.27
Bulk	-2.75	48.40

In an NCFET, the external gate voltage V_{G} induces polarization within the ferroelectric. The net charge density inside the ferroelectric capacitor, defined as the difference between the polarization-induced surface charge and the screening charge density, determines the voltage drop across a ferroelectric V_{Fe}. The negative capacitance appears appears when the polarized charges are not completely screened during the switching of polarization [4]. This implies that the voltage drop across the ferroelectric is a negative quantity and $V_{\mathrm{FET}} > V_{\mathrm{G}}$ (from $V_{\mathrm{G}} = V_{\mathrm{FET}} + V_{\mathrm{Fe}}$), where V_{FET} is the voltage across the MOS-FET. Therefore, the same charge density can be achieved in the channel with a smaller gate voltage which leads to a SS steeper than 60 mV/dec [16].

As shown in Fig.3(a) SS=67 mV/dec for MoS$_2$-FET, while the NCFET with a 5 nm (5L) α-In$_2$Se$_3$ achieved the SS of 59 mV/dec that is reduced to 27 mV/dec for 25

978-1-7281-0941-1/19 \$31.00 © 2019 IEEE

nm. By increasing the thickness of the ferroelectric layer, the performance enhancement increases because of the reduction of C_{Fe}, ($|C_{\text{Fe}}| \sim C_{\text{FET}}$), which in turn results in larger amplification of the gate voltage. In addition a more direct way to improve the device performance is the increasing the capacitance of the insulator layer. The dependence of the C_{Ins} on the EOT suggests that thinner gate insulators result in smaller SS Also, the SS of NCFET can be further improved by using high-κ insulate layer. In Fig.3(b) different EOT and gate insulator layer materials are applied in the NCFET with a 5 nm ferroelectric layer. It can be seen that the SS can reach smaller than 60 mV/dec by using thinner or high-κ insulate layer.

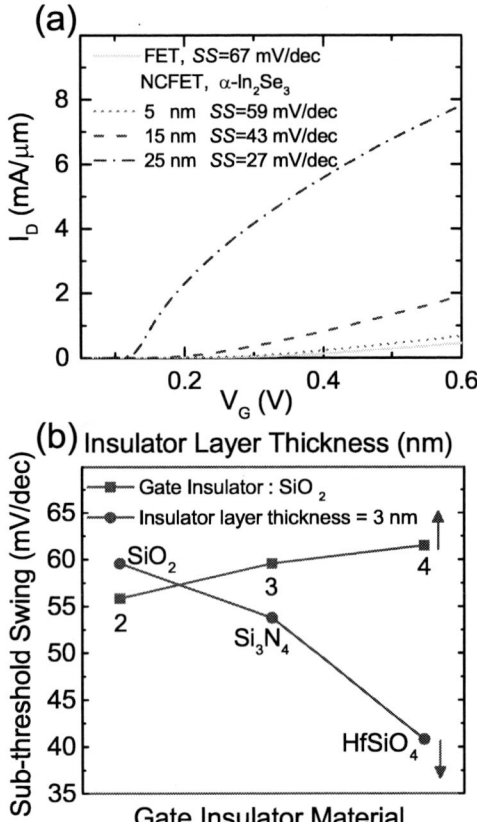

Fig. 3: (a) I_{D}-V_{G} for a MoS$_2$-based FET and NCFETs at V_{D}=0.6 V with a 3 nm SiO$_2$ insulator layer for various thicknesses of α-In$_2$Se$_3$. (b) The SS at various thicknesses of insulator layer and gate insulator materials for NCFET with 5 nm α-In$_2$Se$_3$.

IV. Conclusion

This work presents a comprehensive study of NCFETs with monolayer MoS$_2$ as the channel material and few-layer α-In$_2$Se$_3$ as the ferroelectric material. First principles calculations were employed for determining the electronic properties and extracting Landau constants of the α-

In$_2$Se$_3$. Monolayer α-In$_2$Se$_3$ as the thinnest layered ferroelectric reveals the out-of-plane polarization. The SS in the range of $\sim$27-59 mV/dec were achieved for 25 nm to 5 nm of α-In$_2$Se$_3$ which can be further reduced by using thicker α-In$_2$Se$_3$ and thinner or high-κ insulate layer. The presented results pave the way for realizing high performance 2D NCFETs.

Acknowledgements

This work was partly supported by Iran National Science Foundation (INSF).

References

[1] R. J. Van Overstraeten, G. J. Declerck, and P. A. Muls, "Theory of the mos transistor in weak inversion-new method to determine the number of surface states," IEEE T-ED, vol. 22, pp. 282–288, May 1975.

[2] A. M. Ionescu and H. Riel, "Tunnel field-effect transistors as energy-efficient electronic switches," nature, vol. 479, p. 329, November 2011.

[3] S. Salahuddin and S. Datta, "Use of negative capacitance to provide voltage amplification for low power nanoscale devices," Nano lett., vol. 8, pp. 405–410, March 2008.

[4] V. V. Zhirnov and R. K. Cavin, "Nanoelectronics: Negative capacitance to the rescue?" Nat. Nanotechnol., vol. 3, p. 77, February 2008.

[5] J. Jo, W. Y. Choi, J.-D. Park, J. W. Shim, H.-Y. Yu, and C. Shin, "Negative capacitance in organic/ferroelectric capacitor to implement steep switching mos devices," Nano lett., vol. 15, pp. 4553–4556, June 2015.

[6] X. Wang, Y. Chen, G. Wu, D. Li, L. Tu, S. Sun, H. Shen, T. Lin, Y. Xiao, M. Tang et al., "Two-dimensional negative capacitance transistor with polyvinylidene fluoride-based ferroelectric polymer gating," NPJ 2D Mater. Appl., vol. 1, p. 38, November 2017.

[7] F. Xue, W. Hu, K.-C. Lee, L.-S. Lu, J. Zhang, H.-L. Tang, A. Han, W.-T. Hsu, S. Tu, W.-H. Chang et al., "Room-temperature ferroelectricity in hexagonally layered α-in2se3 nanoflakes down to the monolayer limit," Adv. Funct. Mater., vol. 28, p. 1803738, October 2018.

[8] G. Pahwa, T. Dutta, A. Agarwal, and Y. S. Chauhan, "Physical insights on negative capacitance transistors in nonhysteresis and hysteresis regimes: Mfmis versus mfis structures," IEEE T-ED, vol. 65, pp. 867–873, January 2018.

[9] A. Rahman, J. Guo, S. Datta, and M. S. Lundstrom, "Theory of ballistic nanotransistors," IEEE T-ED, vol. 50, pp. 1853–1864, September 2003.

[10] P. E. Blöchl, "Projector augmented-wave method," Phys. Rev. B, vol. 50, p. 17953, December 1994.

[11] G. Kresse and D. Joubert, "From ultrasoft pseudopotentials to the projector augmented-wave method," Phys. Rev. B, vol. 59, p. 1758, January 1999.

[12] J. P. Perdew, A. Ruzsinszky, G. I. Csonka, O. A. Vydrov, G. E. Scuseria, L. A. Constantin, X. Zhou, and K. Burke, "Restoring the density-gradient expansion for exchange in solids and surfaces," Phys. Rev. Lett., vol. 100, p. 136406, April 2008.

[13] J. Heyd, J. E. Peralta, G. E. Scuseria, and R. L. Martin, "Energy band gaps and lattice parameters evaluated with the heyd-scuseria-ernzerhof screened hybrid functional," J. Chem. Phys., vol. 123, p. 174101, October 2005.

[14] S. Grimme, J. Antony, S. Ehrlich, and H. Krieg, "A consistent and accurate ab initio parametrization of density functional dispersion correction (dft-d) for the 94 elements h-pu," J. Chem. Phys., vol. 132, p. 154104, March 2010.

[15] R. Zacharia, H. Ulbricht, and T. Hertel, "Interlayer cohesive energy of graphite from thermal desorption of polyaromatic hydrocarbons," Phys. Rev. B, vol. 69, p. 155406, April 2004.

[16] K. Ng, S. J. Hillenius, and A. Gruverman, "Transient nature of negative capacitance in ferroelectric field-effect transistors," Solid State Commun., vol. 265, pp. 12–14, October 2017.

Progress in dislocation stress field model and its appications

Uihui Kwon*, Jeong-Guk Min, Seon-Young Lee, Alexander Schmidt, Dae Sin Kim
Semiconductor R&D Center
Samsung Electronics Corp. Ltd.
Hwasung-si, Gyeonggi-do, Korea
*e-mail: uihui.kwon@samsung.com

Yasuyuki Kayama, Yutaka Nishizawa, Kiyoshi Ishikawa
Device Solution Center
Samsung R&D Institute Japan
Tshurumi-ku, Yokohama, Japan

Abstract— TCAD prediction of the stress field generated by dislocation is crucial for the optimization of stressors for next generation logic devices. In this paper, we present a new hybrid approach for dislocation stress field calculation and its application to strained Si devices. New methodology combines an analytic stress field model for dislocation cores and consecutive FEM stress solving to get mechanical equilibrium. It was applied to the design optimization of dislocation stress memorization technique (D-SMT), its local layout effect (LLE) modeling, and the relaxation of lattice mismatch strain at Si/SiGe interface which degrades eSiGe stress. All the simulation results were verified with experimental results.

Keywords—dislocation, stress memorization technique, SMT, strained silicon, stress engineering, Peierls-Nabarro model

I. INTRODUCTION

Strained Si technology itself is a rather old topic in the history of semiconductor technology, which traces back to the early studies of deformation potential [1] and piezo-resistivity [2]. However, it caught the attention of semiconductor industry after the first commercialization of strained Si device by Intel in 2002. Since then, many stressors have been proposed and introduced in logic technology; tensile contact etch stopper layer (CESL) for NMOS and embedded SiGe (eSiGe) for PMOS in 90nm node, poly-Si stress memorization technique (SMT) for NMOS in 65nm node, tensile trench contact (TTC) for NMOS and sigma eSiGe for PMOS in 45nm, dislocation SMT (D-SMT) for NMOS in 32nm. Even in the era of FinFET beyond 22nm, stressors like eSiGe & stress-relaxed buffer (SRB) are still playing an important role as a performance booster. All these stressors can be categorized into 6 types depending on its deformation mechanism, which are listed in Table.1. All the stressors except high dimensional defect can be solved numerically in conventional TCAD framework only if we apply proper initial stress, mechanical properties & boundary condition. However, high dimensional defects like dislocation (1D defect), stacking fault/twin boundary (2D defect), and grain boundary (3D defect), are still difficult to simulate in conventional TCAD framework due to its phenomenological complexity.

Table 1 Categorization of stressors

Deformation mechanism		Examples
Elastic	Residual film stress	CESL
	Lattice mismatch stress	SiGe/SiC
	Thermal inclusion stress	TTC, TSV
Inelastic	Volume change by shrinkage / phase change	Oxide, Silicide
	Plastic deformation (yield)	SMT
	High dimensional defect	D-SMT, SRB

In Poly-Si/SiON (PSiON) gate first process used until 45~32nm planar transistor, SMT was a major stressor for NMOS, whose main effect comes from the plastic strain memorized within poly Si gate. However, in HK/MG gate last process introduced beyond 45~32nm, PSiON gate stack is replaced by HK/MG stack, which means the plastic strain memorized within Poly-Si gate disappears. To overcome this limitation, a new SMT process is proposed [3], which forms high dimensional defects within S/D region, what we call dislocation SMT (D-SMT). It deploys the mechanical stress exerted by the stacking fault generated during solid phase epitaxial regrowth (SPER) in SMT annealing step as shown in Fig.1.

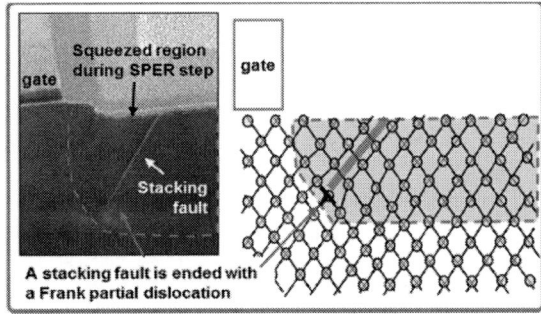

Fig. 1 A schematic illustration of the stacking fault generated during SPER.

Many researchers reported their experimental and modeling results [4] [5] describing the generation mechanism of stacking faults during SPER in SMT process. To estimate the channel stress enhancement by D-SMT, we should calculate the misfit strain exerted by a stacking fault, which means the conventional plastic strain model [6] is no more valid. To calculate the misfit strain exerted by a stacking fault rigorously, we needs a sophisticated ab-initio calculation [7] requiring very high computing power. However, the problem becomes extremely simply, if we assume the misfit strain of a stacking fault mostly comes from the partial dislocations forming the boundary of stacking fault. In fact, the strain field of stacking fault itself is negligible compared to that of surrounding partial dislocations. The equilibrium between the strain energy by surrounding partial dislocations and the stacking fault energy determines the area of stacking fault. So the simplest but effective way to estimate the stress field by stacking fault is to calculate the stress field from its surrounding partial dislocations. Dislocation also plays an important role in stress relaxation at the interface where lattice mismatch stress is applied. There are many types of relaxation by dislocations working in strained Si technology; misfit dislocations in Si/SiGe interface, threading dislocations in stress relaxed buffer (SRB). So many researchers analyzed dislocation problems in inhomogeneous media and the behavior of dislocations near material interface boundary.

Up to now, three different approaches have been made to calculate the stress field generated by dislocation core. The first is based on analytic approach represented by venerable Peierls–Nabarro model [8][9] inspired by Eshelby's inclusion problems [10]. The second is based on slab insertion type finite element method (FEM) [11], which requires very careful mesh handling at dislocation core to avoid the numerical instability at the singularity point. The third is a direct atomistic simulation, which is considerably more accurate but computationally expensive [12] [13]. Since the first application of analytic approach based on P-N model to the logic gate last devices [14], the flaw of analytic approach compared to FEM approach is pointed out by some group [15] ; i.e. the interaction between dislocation and material interfaces & free surface. A new approach remedying the drawback of conventional analytic approach is proposed in this paper.

II. SIMULATION METHODOLOGY

To overcome the shortcomings of conventional analytic approach when the dislocation is located at near free surface or hetero-interfaces, we solved mechanical equilibrium with appropriate boundary conditions after setting displacement field based on analytic dislocation field model. The overall calculation flow is as shown in Fig.2.

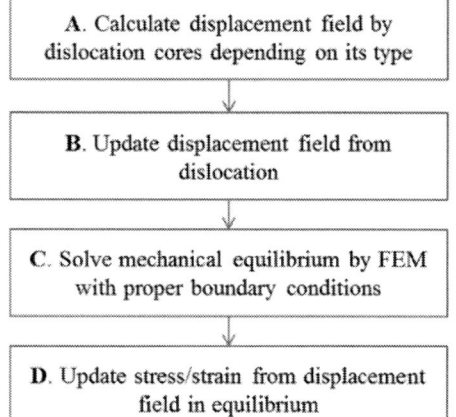

Fig. 2 Schematics chart of simulation flow

A. *Calculate displacement vector $\{u^{disl}\}$ by a dislocation core depending on dislocation type;*

(For edge dislocation)

$$u_1 = \frac{-b}{2\pi}\left[\tan^{-1}\left(\frac{x_2}{x_1}\right) + \frac{1}{2(1-v)}\frac{x_1 x_2}{(x_1^2 + x_2^2)}\right] \quad (1)$$

$$u_2 = \frac{b}{8\pi(1-v)}\left[(1-2v)\ln(x_1^2 + x_2^2) + \frac{(x_1^2 - x_2^2)}{(x_1^2 + x_2^2)}\right]$$

$$u_3 = 0$$

(For screw dislocation)

$$u_1 = u_2 = 0$$

$$u_3 = \frac{b}{2\pi}\tan^{-1}(x_2 / x_1) \quad (2)$$

Where v is Poisson ratio and b is Burgers vector, which depends on dislocation types as below;

Dislocation Types	Burgers Vector
Perfect dislocation	$a_0/2$ [110]
Shockley partial	$a_0/6$ [112]
Frank partial	$a_0/3$ [111]

B. *Update displacement field from multiple dislocations;*

$$\{u\} = \{u^0\} + \Sigma\{u^{disl}\} \quad (3)$$

where $\{u^0\}$ is a displacement field vector from previous step by other stress sources like residual film stress, lattice match stress, thermal stress, and so on.

C. *Solve FEM to get mechanical equilibrium;*

Solve FEM to get displacement field $\{u\}$ in mechanical equilibrium with appropriate boundary conditions for hetro-interfaces within simulation domain

$$K\{u\} = \{F\} \quad (4)$$

where K is a stiffness matrix, $\{F\}$ is a load vector.

D. *Update strain and stress field*

Update strain and stress field from the updated displacement field after equilibrium solving

$$\{\varepsilon\} = B\{u\}, \{\sigma\} = E\{\varepsilon\} \quad (5)$$

where B is a matrix differential operator, E is elasticity matrix.

III. APPLICATION RESULTS AND DISCUSSION

New dislocation field model is implemented into our in-house process simulator and tested for various cases to show an improvement over conventional analytic / FEM approaches. And also, all the simulation results were verified with experimental results.

A. *An improvement over coventional analytic approach*

In conventional analytic approach, the stress field from analytic formulation, what we call P-N model, is just mapped on the whole node point of Si region as shown in Fig.3.

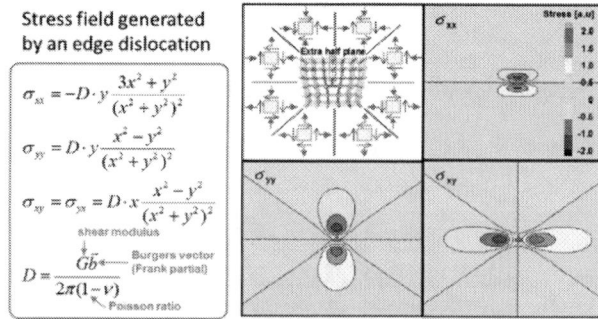

Fig. 3 Dislocation stress field mapping technique used in conventional analytic approach

This causes two drawbacks; Firstly, it cannot consider the stress relaxation at the surface or at the interface with softer material, which is the main drawback of analytic approach compared to FEM approach. Secondly, it is hard to combine with other consecutive stressors solving mechanical equilibrium with FEM, which has to manage historic displacement information for stress history calculation.

An increase of ~0.2% strain in the channel is observed when we remove the poly-Si gate in gate last process, which was successfully reproduced by new approach as shown in Fig.4. The stress contours by hybrid approach match well the corresponding FEM cases when the dislocation core comes close to the surfaces, which is verifying the new hybrid approach.

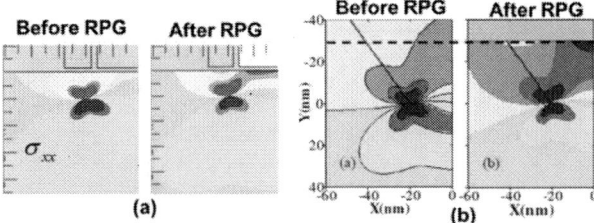

Fig. 4 Free surface effect in D-SMT for NMOS when dummy poly-Si gate is removed with (a) new hybrid approach and (b) a slab-insertion FEM approach [15].

B. An improvement over coventional FEM approach

Conventional FEM approach, whose concept is to remove an atom plane from perfect crystal lattice and to map displacement vector across the missing atom plane (slab), has several drawbacks; Firstly, it is a matter of cause that you have to get over the numerical instability at the singularity point of dislocation core through mesh refinement. Secondly, it requires extra geometry handling because you have to define the slab region (extra half plane) manually. The difficulty increases exponentially when you try to consider the nested dislocation stress field from multiple dislocation cores to estimate the impact of length of diffusion (LOD) or contacted poly pitch (CPP). Finally, it is hard to apply to mixed dislocations whose slab insertion shape is not a simple cuboid but complexly curved one.

(a) TEM image of SMT stacking faults

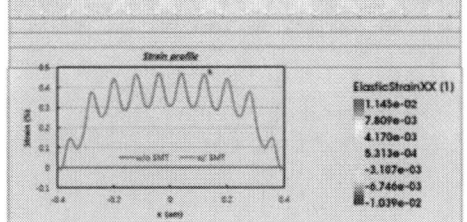

(b) Strain profile across active region

Fig. 5 (a) TEM image of stacking faults in LOD TEG [16] (b) Strain profile simulated by new hybrid approach

With new hybrid dislocation field approach, the LOD effect generated by the hetero-interface between rigid Si and soft STI oxide is successfully reproduced as shown in Fig.5.

And also, the channel stress increase by nested dislocation field with increasing transistor number and smaller contacted poly pitch (CPP) are successfully simulated as shown in Fig.6 and Fig.7. Different from other stressors likes embedded SiGe and CESL, the beneficial effect of D-SMT does not decrease with CPP scaling but rather increases according to our analysis.

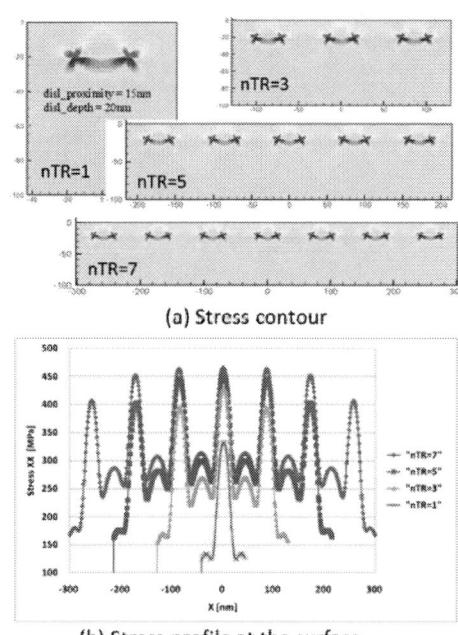

(a) Stress contour

(b) Stress profile at the surface

Fig. 6 Effect of nested dislocation field (a) 2D stress contour (b) stress profile at the channel surface.

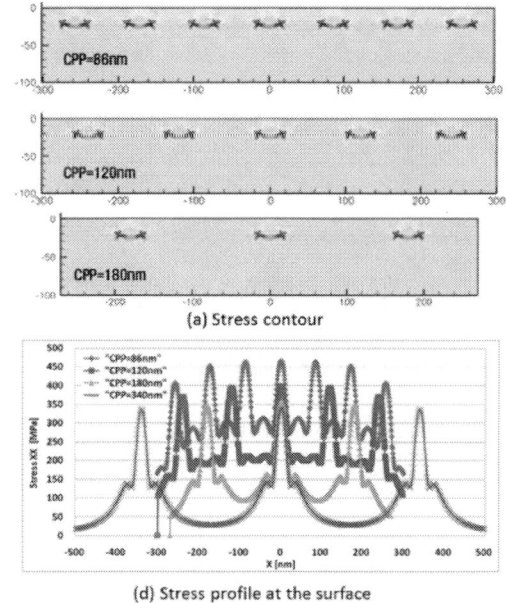

(a) Stress contour

(d) Stress profile at the surface

Fig. 7 The effect of CPP (a) 2D stress contour (b) stress profile at the channel surface.

C. Accuracy check with experimental measurement

The accuracy of new hybrid approach is confirmed by comparing to experimental strain measurement result with scanning transmission electron microscopy-geometrical phase analysis (STEM-GPA) & scanning moiré fringe (SMF) pattern [17] shown in Fig.8.

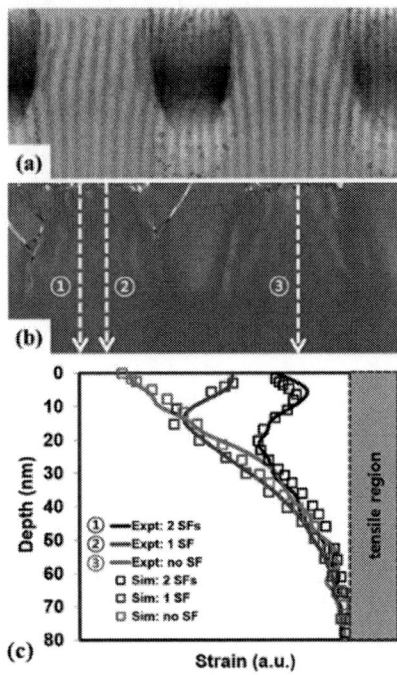

Fig. 8 Strain comparisons with experiments. (a) SMF image (b) Strain map obtained by STEM-GPA. (c) Simulation vs. experimental strain profile

With new approach, it is possible to find the optimal position & number of stacking faults for D-SMT to maximize NMOS performance. And also, you can quantitatively model the degradation of compressive strain in the channel, which causes DC saturation in state-of-the-art PMOS, when [Ge] gradient in the layered structure of eSiGe is too steep.

IV. CONCLUSION

TCAD prediction of the stress field generated by dislocations is crucial for the optimization of D-SMT for NMOS, its length of diffusion (LOD) dependency on local layout effect (LLE) model, and the degradation of lattice mismatch strain at the Si/SiGe interface. A new approach for dislocation field model, the combination of analytic dislocation field model and succeeding FEM solving to get mechanical equilibrium, is developed and successfully applied to state-of-the-art strained Si devices. New approach works well with multiple dislocations and arbitrary boundary shape so we could model the nested dislocation stress field effect fast with high accuracy. Moreover, it showed good match with experimental STEM-GPA/SMF strain measurements.

REFERENCES

[1] J. Bardeen and W. Shockley, "Deformation Potentials and Mobilities in non-polar crystals", Phys. Rev. 80, 72 (1950)

[2] C. S. Smith, "Piezoresistance effect in germanium and silicon", Phys. Rev. 94, p.42 (1954)

[3] A. Wei et al., "Multiple Stress Memorization In Advanced SOI CMOS Technologies", IEEE VLSI, p.216 (2007)

[4] N. G. Rudawski et al., "Effect of uniaxial stress on solid phase epitaxy in patterned Si wafers", APL 89, p.082107 (2006)

[5] S. Morarka et al., "Modeling of two-dimensional solid-phase epitaxial regrowth using level set methods", JAP 105, 053701 (2009)

[6] X. Wang et al., "Progress in modeling of SMT "stress memorization technique" and prediction of stress enhancement by a novel PMOS SMT process", SISPAD, p.117 (2008)

[7] H. V. Swygenhoven et al., "Stacking fault energies and slip in nanocrystalline metals", Nature Material 3, p.399 (2004)

[8] Peierls RE. "The size of a dislocation." Proceedings of the Physical Society; 52:34 –37 (1940)

[9] Nabarro FRN. "Dislocation in Solids." North-Holland: Amsterdam (1979)

[10] J. D. Eshelby, "The determination of the elastic field of an ellipsoidal inclusion, and related problems", Proceedings of the Royal Society, London, Series A, 241, 376–396. (1957)

[11] K. Sasaki, et al., "Stress analysis in continuous media with an edge dislocation by finite element dislocation model", Int. J. Numer. Meth. Engng; 54:671–683 (2002)

[12] Wei Cai, "Atomistic and Mesoscale Modeling of Dislocation Mobility", Ph.D Thesis, MIT (2001)

[13] P.M. Derlet et al., "Atomistic simulations of dislocations in confined volumes", MRS Bulletin, vol 34, Mar. (2009)

[14] K-Y. Lim et al., "Novel stress-memorization-technology (smt) for high electron mobility enhancement of gate last high-k/metal gate `devices," IEDM Tech. Dig., 229-232 (2010)

[15] C. E. Weber et. al. "Modeling of NMOS Performance Gains from Edge Dislocation Stress", IEDM (2011)

[16] F. Sato et al., "Process and Local Layout Effect interaction on a high performance planar 20nm CMOS", VLSI (2013)

[17] Jeong-Guk Min et al., "The Impact of Dislocation on Bulk -Si Fin FET Technologies: Physical Modeling of Strain Relaxation and Enhancement by Dislocation", Nanotech. Materials and Devices (2018)

Effect of Trap on Carrier Transport in InAs FET with Al$_2$O$_3$ Oxide: DFT-based NEGF simulations

Mincheol Shin
School of Electrical Engineering
Korea Advanced Institute of Science and Technology
Daejeon 34141, Republic of Korea
mshin@kaist.ac.kr

Yucheol Cho
School of Electrical Engineering
Korea Advanced Institute of Science and Technology
Daejeon 34141, Republic of Korea
yc_cho@kaist.ac.kr

Seonghyeok Jeon
School of Electrical Engineering
Korea Advanced Institute of Science and Technology
Daejeon 34141, Republic of Korea
jsh7700@kaist.ac.kr

Abstract— To accurately assess the effect of trap on the performance of field effect transistors (FETs), atom-level first-principles modeling of channel/oxide/trap and rigorous quantum mechanical transport calculations are necessary. In this work we have developed an innovative approach to solve the challenging problem efficiently. Non-equilibrium Green's function simulation of InAs FET with a trap in the channel/oxide interface that is atomically modeled by using the density functional theory is demonstrated.

Keywords— Non-equilibrium Green's function, Density functional theory, Defect/trap, III-V channel material, field effect transistor

I. INTRODUCTION

As the feature size of the field effect transistors (FETs) approaches a few nanometer scale, it has become increasingly important to simulate the devices at the atomistic level. While the empirical tight-binding method seems suited for atomistic modeling, it has many limitations, especially when heterostructure with interface and trap is involved. First-principles density functional theory (DFT) which does not rely on empirical parameters would give an ultimate solution. But the great hurdle in utilizing DFT Hamiltonians is that the size of the non-orthogonal Hamiltonian matrices becomes prohibitively large for transport simulations of realistically sized devices.

A scheme to overcome the computational hurdle [1,2] has been introduced recently and its applications to homojunction devices were demonstrated [3]. In the approach, the DFT and non-equilibrium Green's function method (NEGF) parts are separated, where the latter method is used for electron transport calculation in the device region. Namely, one calculates the electronic structure of the unit cell by the DFT method, repeats the unit cell to construct the device region with source and drain regions attached, and implements the NEGF transport calculations for the device using the imported DFT Hamiltonians of the unit cell. Then using the mode space transformation, the Hamiltonian is reduced in size without any practical loss in accuracy in terms of device simulation outputs. In this work, the approach is extended to include an interface trap in the channel region. An oxide trap which gives rise to a defect state within the band gap is particularly paid attention to because it is expected to provide a leakage path affecting subthreshold current.

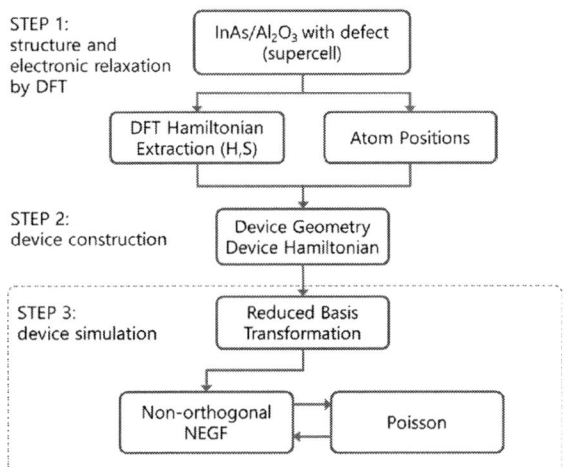

Fig. 1: Simulation flowchart. Simulations are performed in the three stages: density functional theory (DFT) relaxation of supercell, followed by the extraction of Hamiltonian and construction of the device region. After Hamiltonian matrices are effectively reduced, they are fed to the NEGF-Poisson self-consistent calculation. Upon convergence, current is calculated.

II. DFT MODELING

The overall simulation flow is shown in Fig. 1. Firstly, linear combination of atomic orbitals based DFT structural relaxations and band structure calculations were performed by using SIESTA tool [4]. PSML norm-conserving pseudopotentials [5] were used to obtain transferable results as practiced with the plane-wave based DFT. The generalized gradient approximation (GGA) and Perdew, Burke and Ernzerhof exchange (PBE) functional were employed. The band gap underestimation of GGA-PBE were adjusted by using the DFT-1/2 technique [6]. For the bulk InAs, the DFT-1/2 scheme opens the band gap of 0.42 eV which agrees well with the experimental result [7].

Atomic configuration of the defect-free InAs/Al$_2$O$_3$ structure is shown in Fig. 2 (a). The crystalline α-Al$_2$O$_3$ was considered for the oxide, which was strained to match the lattice constant of InAs. As-O bonds were formed at the interface to mimic the As-rich condition usually practiced in the experiments [8].

978-1-7281-0941-1/19 $31.00 © 2019 IEEE

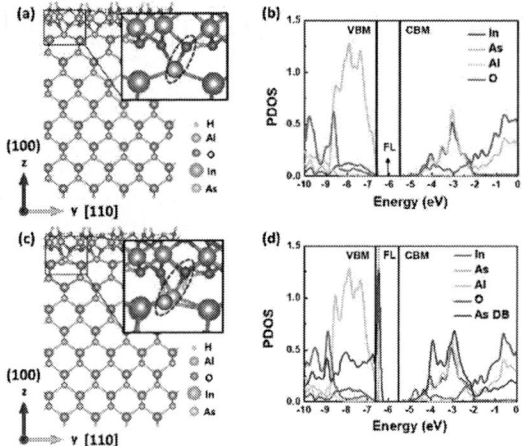

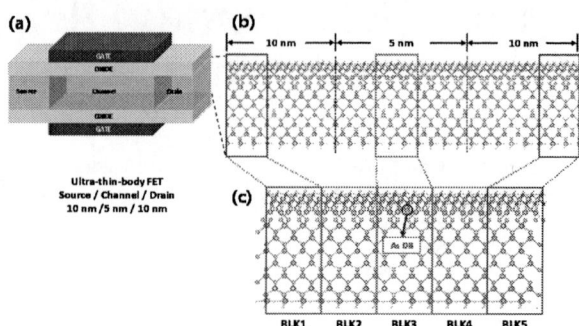

Fig. 2: (a) Atomic configuration of the defect-free InAs/Al₂O₃ cell with an As-O bond magnified. (b) Projected density of state (PDOS) of the supercell (a) where the Fermi level (FL) is located in the middle of the band gap. (c) Atomic configuration of the InAs/Al₂O₃ cell with a broken As-O bond resulting an As dangling bond. (d) PDOS of (c) where the defect energy state is shown as the sharp peak close to the valence band maximum (VBM).

Fig. 3: (a) Schematic diagram of ultrathin-body InAs simulated in this work. The source/channel/drain regions modelled atomically are shown in (b), which are constructed by using the blocks of the relaxed DFT supercell shown in (c). BLK3 contains an As DB while BLK1 and BLK5 are very close to standalone, pure InAs/Al₂O₃ unit cell. The latter two are used as the source and drain blocks and also as the blocks to fill up the rest of the device region besides the block containing the trap and its neighbor blocks.

A defect at the interface was created when an As-O bond was broken resulting an As dangling bond (DB) and hydrogen passivated oxygen. See Fig. 2 (c). The supercell containing the defect is shown in Fig. 3 (c), which has the thickness of 22.54 Å in the z direction and periodically repeated in the x and y directions. The As DB is positioned in the middle block (BLK3) which is sided by defect-free InAs/Al₂O₃ blocks (BLK2 and BLK4) to ensure no interaction between the defect and its periodic images. BLK1 and BLK5 are very close, in terms of the geometric and electronic structures, to the standalone, pure InAs/Al₂O₃ unit cell. Each block in the figure has the size of 3x4 primitive cells or 13.14 Å by 17.51 Å and contains the total of 360 atoms (361 atoms for BLK3). The block size was chosen to guarantee the nearest neighbor interaction condition between neighboring blocks (and between the blocks and their periodic images) as each block is to be used as an unit cell in the NEGF transport calculation in the followings.

Figs. 2 (b) and (d) show the projected density of state (PDOS) of the defect-free and the structure with an As DB, respectively. As shown in the figures, the As DB gives rise to a sharp defect energy state above and close to the valence band maximum (VBM) and as the consequence the Fermi level becomes located close to the defect energy state. Fig. 4 shows the band diagrams of the two structures, where the flat defect state is clearly seen in the right panel. The Fermi level indicated in the figure is the one of the entire supercell of Fig. 3 (c). Thus it may be estimated that the defect energy states lies about 0.06 eV above the VBM of the defect free structure. It is just an estimation because the band diagram of the right panel of Fig. 4 corresponds to the case where BLK3 is repeated periodically, which is obviously not the case for the actual BLK3 which is a part of the supercell shown in Fig. 3 (c).

III. NEGF TRANSPORT CALCULATION

After relaxation of the supercell, the nonorthogonal DFT Hamiltonian and overlap matrices of the supercell were

extracted and used to construct the device region as shown in Figs. 3 (a) and (b). Each block of the supercell of Fig. 3 (c) is treated as a unit cell in the transport calculation; for instance, BLK1 is repeated to construct the source region and BLK5 is repeated to construct the drain region, while the channel region is constructed by 4 blocks of BLK2, BLK3, BLK4, and BLK5. Then the As DB which acts like a trap is located about 2.0 nm from the beginning of the channel region.

The Hamiltonian size of each unit cell containing 360 (361) atoms is 5064x5064 (5069x5069), where the number between the parenthesis is for the block containing the defect. Using the scheme in Ref. [2], the Hamiltonian matrices were reduced to about 100, which is about 2% of its original size. The Hamiltonian corresponding to BLK1, BLK3, and BLK5 were reduced and their unitary transformation matrices were applied to the original device Hamiltonian. The reduced Hamiltonian size was bigger than that of homojunction systems due to the coupling of wider range of modes of hetero-cells.

For transport calculations, the Hamiltonian and overlap matrices were fed to an in-house non-orthogonal non-equilibrium Green's function (NEGF) simulator, where the hole density is given by

$$p = \int \frac{dE}{2\pi} [G^p S]_{diag}$$

where G^p is the hole Green's function matrix and S is the overlap matrix. The hole density was calculated because the p-type device was considered in this work. With the hole density, the three dimensional Poisson's equation is solved for the electrostatic potential

$$\nabla^2 \phi(\vec{r}) = -\frac{q_0}{\epsilon}(p(\vec{r}) - N_A)$$

where ϵ is the permittivity and N_A is the doping density. Upon the convergence of the self-consistent calculation between the hole density and the electrostatic potential, the current was calculated. Ballistic transport was assumed. To account for the discrete nature of atoms, the finite element method was used to solve the Poisson's equation.

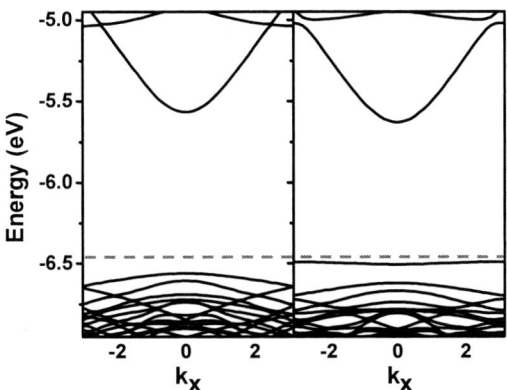

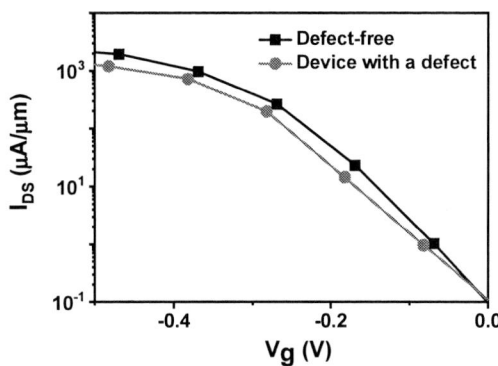

Fig. 4: (Left) The band diagram of the defect-free cell. (Right) The band diagram of the cell with an As DB as described in the text. The dashed line represents the Fermi level of the whole structure shown in Fig. 3 (c). The band diagrams are for $k_y = 0$ where k_y is the wave vector in the transverse (y) direction. Note that the unit cell structures assume the periodicity in the y direction, with the periodicity of $4u_y$ where u_y is the width of the primitive unit cell. As mentioned in the text, there are 3x4 primitive cells in the unit cells.

Fig. 6: Current-voltage characteristics of the simulated p-type UTB devices with 5.25 nm channel length and 2.25 nm channel width. The device with an As DB in the channel region (red line with circles) and the defect-free device (black line with squares) are compared. The OFF state current of 0.1 $\mu A/\mu m$ is assumed and the curves are adjusted accordingly. The drain voltage of -0.5 V is applied. The device with the defect shows the subthreshold swing of 85 mV/decade while the defect-free device shows 74 mV/decade. The ON currents are 2130 $\mu A/\mu m$ and 1280 $\mu A/\mu m$, respectively, for the defect-free and the device with defect.

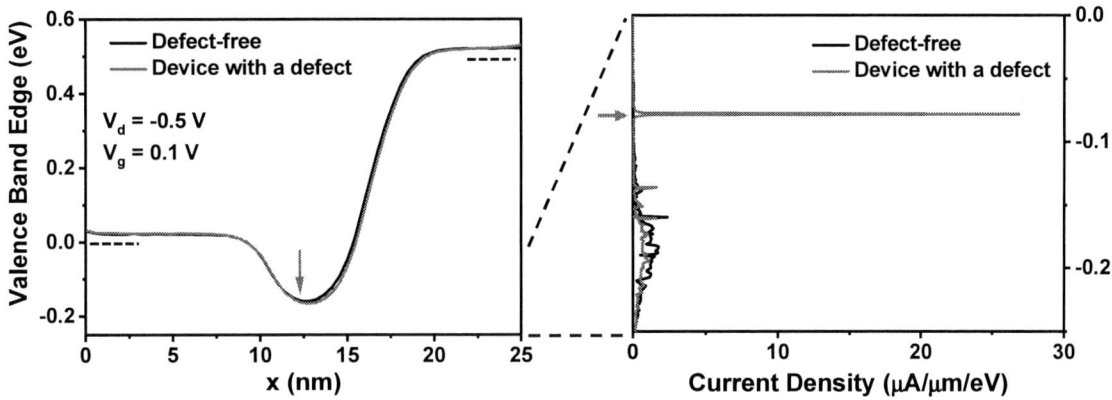

Fig. 5: The valence band edge profile for the p-type InAs FET shown in Fig. 3 and the corresponding energy-resolved current density. Black and red lines are for defect-free device and the device with an As DB in the channel (its location indicated as arrow in the left panel), respectively. In the current density graph, the peak due to tunneling current through the trap is indicated by the arrow.

IV. DEVICE SIMULATION

Ultra-thin-body (UTB) double-gate InAs FET is constructed as shown in Fig. 3 (b) using the blocks of Fig. 3 (c). InAs in the channel region is 2.25 nm thick, and 1 nm thick oxide consists of minimum layers of actual Al₂O₃ atoms to keep intact the band structure of InAs/Al₂O₃ and virtual atoms to fill up the rest of the oxide region. The bulk dielectric constants of 15 and 20 are assumed for channel and oxide, respectively. The source and channel regions are 10.5 nm long each and the channel length is 5.25 nm.

Since the defect state lies close to VBM, a p-type UTB transistor has been simulated to demonstrate the effect of the trap on the transport properties; the source and drain regions are assumed to be p-doped with the doping density of 10^{20} cm^{-3}. To account for the periodicity in the transverse (y) direction, 12 transverse modes (k_y) are selected for the

transport calculation, each of which acts as an independent channel. The drain voltage of -0.5 V is applied.

Fig. 5 shows the valence band profile of the simulated device. For comparison, the device with the same geometry but without the trap is also simulated. It is seen that, as the gate electric field is weakened due to the presence of the trap, whose location is pointed by an arrow in the figure, the potential becomes slightly higher (negatively) in the channel region. The energy resolved current density shown in Fig. 5 clearly shows the peak due to the tunneling through the trap state. The current density was calculated for the case of $k_y = 0$.

The leakage current through the defect state leads to deterioration of the subthreshold swing. Fig. 6 shows the current-voltage characteristics of the simulated p-type UTB devices where the defect-free device and the device with an

978-1-7281-0941-1/19 $31.00 © 2019 IEEE   305

As DB are compared. The curves are adjusted such that the OFF state current is 0.1 $\mu A/\mu m$. The subthreshold swing (SS) increases significantly: SS increases from 74 mV/decade for the defect-free device to 85 mV/decade for the device with defect in the channel. As a consequence, the ON current drops to 1280 $\mu A/\mu m$ from 2130 $\mu A/\mu m$.

V. Conclusions

We have performed a rigorous DFT based NEGF simulation on ultra-thin-body InAs double-gate FET with an interface trap modeled by an As dangling bond. We have demonstrated that the device performance becomes significantly degraded by the presence of defect in the channel, due to the trap-assisted leakage tunneling current that takes place in the subthreshold region.

Acknowledgment

This work has been supported by the project SRFC-TA1703-10 under Samsung Research Funding & Incubation Center for Future Technology.

References

[1] G. Mil'nikov et al, "Equivalent transport models in atomistic quantum wires," Phys. Rev. B, vol. 85, 035317, 2012.

[2] M. Shin et al, "Density functional theory based simulations of silicon nanowire field effect transistors," J. Appl. Phys, vol. 119, 154505, 2016.

[3] M. Shin et al, "First-principles based quantum transport simulations of nanoscale field effect transistors," 2017 IEEE International Electron Device Meeting, San Francisco, 2017.

[4] José M Soler et al, "The SIESTA method for ab initio order-N materials simulation," J. Phys.: Condens. Mat., vol. 14, pp. 2745-2779, 2002

[5] A. García, M. Verstraete, Y. Pouillon, J. Junquera, "The PSML format and library for norm-conserving pseudopotential data curation and interoperability" Computer Physics Communications 227, 51-71, 2018

[6] L. G. Ferreira et al, "Approximation to density functional theory for the calculation of band gaps of semiconductors," Phys. Rev. B, vol 78, 125116, 2008.

[7] I. Vurgaftman, J. R. Meyer, and L. R. Ram-Mohan," Band parameters for III–V compound semiconductors and their alloys", J. Appl. Phys. 89, 5815, 2001.

[8] J. Robertson et al, "Defect state passivation at III-V oxide interfaces for complementary metal–oxide–semiconductor devices," J. Appl. Phys. vol. 117, 112806, 2015.

Trap Dynamics based 3D Kinetic Monte Carlo Simulation for Reliability Evaluation of UTBB MOSFETs

Wangyong Chen, Linlin Cai, Xiaoyan Liu* and Gang Du+
Institute of Microelectronics
Peking University
Beijing 100871, China
+gangdu@pku.edu.cn, *xyliu@ime.pku.edu.cn

Abstract—**Trap dynamics based 3D Kinetic Monte Carlo (KMC) simulator is developed to offer physical insights into the electrical characteristics degradation and quantitative reliability evaluation for advanced MOSFETs. The physics-based 3D KMC simulation enables to reproduce the evolution of stress-induced charge distribution in the multi-layer dielectrics and identify the trap impact on the degradation of device performance. Simulation results of UTBB FDSOI MOSFETs reveal that assumption of the uniform charge distribution in the dielectrics induced by stress underestimates the statistical degradation and variability. It also shows that the higher intrinsic trap density of back-gate oxide leads to the larger degradation and its variability, especially for the increased back-gate bias case.**

Keywords—*trap dynamics, kinetic Monte Carlo, UTBB, performance degradation, variability*

I. INTRODUCTION

Trap induced time-dependent variability and reliability have emerged as one of the major challenging issues for advanced technology [1-3]. The trap dynamics during operation, including charging and discharging as well as the interactions with the ions or other traps, have a crucial impact on the device performance and reliability [4-5]. Therefore, the comprehensive physics-based simulation is highly desired for the understanding of trap behaviors and reliability identification.

In this work, the multiple trap-related physics processes in the multi-layer gate stacks are simulated by 3D Kinetic Monte Carlo (KMC) method, which is capable of reproducing the stress-dependent charge distribution and evaluating the degradation of the electrical characteristics for reliability optimization. Ultra-Thin Body and Box (UTBB) FDSOI device delivers superior benefits in the fields of mobile and IoT application [6-8]. The UTBB device reliability associated with the unique back gate modulation is investigated by the developed simulator. The impacts of back-oxide traps as well as remote Coulomb scattering (RCS) on the reliability degradation during different back-gate biases are clarified.

II. SIMULATION METHOD

The underlying stochastic trap behaviors during stress are responsible for the time-dependent variability and degradation [9-10]. Therefore, the 3D KMC method is employed to address the complex physical processes. The fully-coupled trap behaviors in the multi-layer gate stacks implemented in the 3D KMC simulator are depicted in Fig. 1. The processes can be categorized as the charging/discharging of intrinsic/newly-generated traps, trap generation and recombination with neighbor ions, as well as the interactions

with other traps [11]. The modeling equations for corresponding trap behaviors sketched in Fig. 1 are listed as follows.

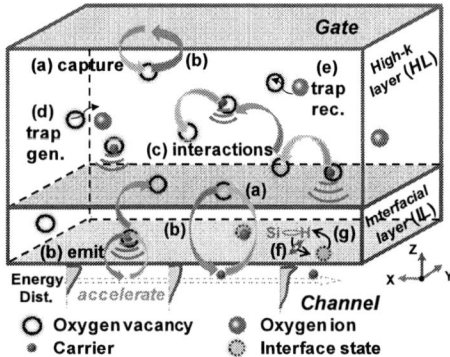

Fig. 1. Schematics of trap behaviors (a)-(g) in the gate dielectrics and the corresponding modeling equations: Eqs. (1)-(12) [9]. The mobility degradation due to remote Coulomb scattering (RCS) is described with Eqs. (13)-(15) [15].

Process (a)/(b) denotes the charge trapping/de-trapping from channel and gate as expressed in Eqs. (1)-(2).

$$P_{c,cg} = \delta_{c,cg} v_n n_{c,g} \exp(-\frac{u_c - \lambda(F/F_0)^\rho}{k_B T}) \quad (1)$$

$$P_{e,eg} = \delta_{e,eg} v_n N_{c,g} \exp(-\frac{u_e + \lambda(F/F_0)^\rho}{k_B T}) \quad (2)$$

where v_n is thermal velocity, $\delta_{c,cg}/\delta_{e,eg}$ is capture/emission coefficient, λ and ρ are enhancement factors of the electric field, $u_{c,e}$ are zero-field thermal barriers.

Process (c) represents the charge exchange between traps as described in Eq. (3).

$$P_{ij} = T_{ij} f_c \exp(-\frac{\varepsilon_{ij}}{k_B T}) \quad (3)$$

where T_{ij} is tunneling coefficient, ε_{ij} is the thermal barrier from the ith trap to jth trap.

Process (d)/(e) shows the generation/recombination of bulk traps, whose probability can be written as Eqs. (4)-(5).

$$P_g = f_g \exp(-\frac{E_a - \gamma F}{k_B T}) \quad (4)$$

$$P_r = f_r \exp(-\frac{E_r}{k_B T}) \quad (5)$$

where $f_{g,r}$ are lattice vibration frequencies, γ is bond polarization factor, $E_{a,r}$ are the zero-field activation energies of trap generation and recombination, respectively.

Process (f)/(g) illustrates the generation/recombination of interface states, which follows the truncated harmonic oscillator model as expressed in Eqs. (6)-(12). The bond breakage of Si-H bonds is contributed to the competitive

978-1-7281-0941-1/19 $31.00 © 2019 IEEE

mechanisms of antibonding (AB) and multi-vibrational excitation (MVE) [12-13].

$$P_{AB,n} = \int f(E)g(E)\sigma_0(E - E_b + E_t)^p v(E)dE \qquad (6)$$

$$P_{MVE} = \int f(E)g(E)\sigma_0(E - \hbar\omega)v(E)dE \qquad (7)$$

$$P_u = P_{MVE} + f_e \exp(-\frac{\hbar\omega}{k_B T}) \qquad (8)$$

$$P_d = P_{MVE} + f_e \qquad (9)$$

$$P_{ig,n} = P_{AB,n} + f_e \exp(-\frac{E_b - E_t}{k_B T}) \qquad (10)$$

$$P_{ig} = \frac{1}{k}\sum_n P_{ig,n}(\frac{P_u}{P_d})^n, \; k = \sum_n (\frac{P_u}{P_d})^n \qquad (11)$$

$$P_{ir} = f_e \exp(-\frac{E_{ir}}{k_B T}) \qquad (12)$$

where $E_{b,ir}$ are the dissociation and passivation energies of Si-H bond, respectively.

The detailed flowchart of trap dynamics simulation is illustrated in Fig. 2. The model parameters related to material properties containing band-breakage energy and thermal barriers for state transition (i.e. empty↔occupied) have been validated in [9]. The 3D electrostatics distributions obtained from device simulator are used for computing the event probabilities according to the above equations. The trap and charge distributions are updated along with the simulation time.

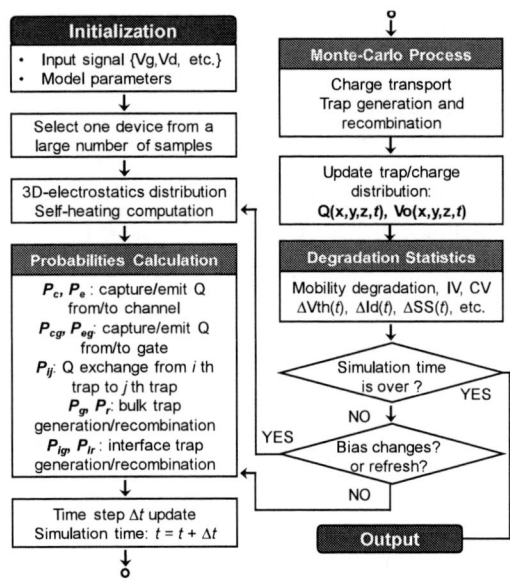

Fig. 2. The flowchart of 3D KMC implement of trap dynamics simulation for reliability evaluation.

The developed 3D KMC simulator incorporating the commercial TCAD tool e.g. Sentaurus [14] allows to evaluate the mobility degradation considering the effect of RCS [15] with the following equations (13)-(15), thereby the holistic time-dependent degradation of electrical characteristics can be acquired according to the time evolution of charge distribution. The variability is based on the statistical simulations of a large ensemble of devices which are unique in the trap configurations (e.g. initial locations, energy levels and thermal barriers).

$$\frac{1}{\mu} = \frac{1}{\mu_{other}} + \frac{D_{rcs} D_{rcs\_HL}}{\Delta\mu_{rcs}} \qquad (13)$$

$$\Delta\mu_{rcs} = \mu_{rcs} f(N_{A,D}, T, g_{screening}) / f(F_\perp) \qquad (14)$$

$$f(F_\perp) = 1 - \exp(-\xi F_\perp / N_{dep}) \qquad (15)$$

Here, the enhanced Lombardi model is applied to account for the mobility degradation due to RCS effect, where μ_{other} is the mobility contributions from other degradations. $F_\perp$ is the transverse field, D_{rcs}, D_{rcs_HL} are damping factors and g_{screen} is screening factor.

III. RESULTS AND DISCUSSION

The schematic of UTBB FDSOI device and the corresponding spatial distributions of traps in the front-gate stacks and back-oxide layer (Box) are illustrated in Fig. 3.

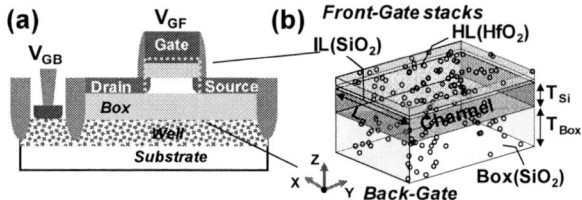

Fig. 3. (a) Schematic of the UTBB FDSOI device. (b) The traps (black circles) and charges (red circles) in the multi-layer gate stacks and back-oxide layer (Box).

The time evolutions of charge number (#) and distribution in the device during the typical BTI stress and recovery phases are shown in Fig. 4(a) and Fig.4 (b), respectively. It indicates that charges in the interfacial layer (IL) and Box are trapped/de-trapped in a short time due to the relatively small capture/emission time constants, which is mainly attributed to the traps with the shallow energy level. The charges in the high-k layer (HL) turn into the dominant role with increased time due to the high trap density in the HL and show the non-uniform distribution across the gate stacks owing to the differences of the trap locations and electrostatics.

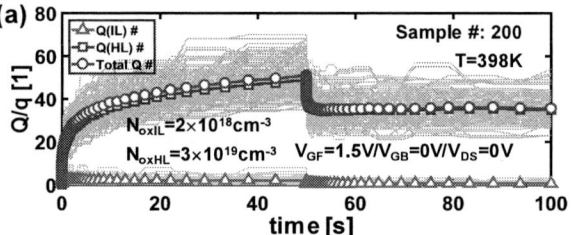

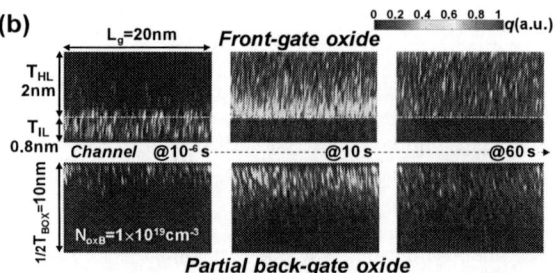

Fig. 4. Time evolutions collected from 200 samples: (a) charge # in IL and HL of the UTBB nFETs subjected to PBTI stress. (b) charge distributions in the front-gate oxide and the partial region of Box.

Fig. 5 shows the degradation difference between the assumed uniform charge distribution and simulated charge distributions. To make a fair comparison, the average charged

978-1-7281-0941-1/19 $31.00 © 2019 IEEE

traps number is the same for the two cases. It can be noted that the assumption of uniform charge distribution during stress underestimates ΔVth and ΔSS as well as the variabilities. The more serious Vth and SS degradation can be attributed to the fact that most of the charges generated by PBTI stress are trapped near the channel compared to the case of uniform charge distribution. To investigate the role of back-gate oxide in the reliability of the UTBB device, the statistical simulations of PBTI induced parameter degradation with and without the impact of back-gate oxide traps are compared. The results in Fig. 6 show that the degradation and variability increase with the consideration of trapped charges in the back-gate oxide. It can be seen from Fig. 7 that charged traps in the front-gate and back-gate oxides would induce the potential variations at both front- and back channel surfaces, thus deteriorating the electrical characteristics.

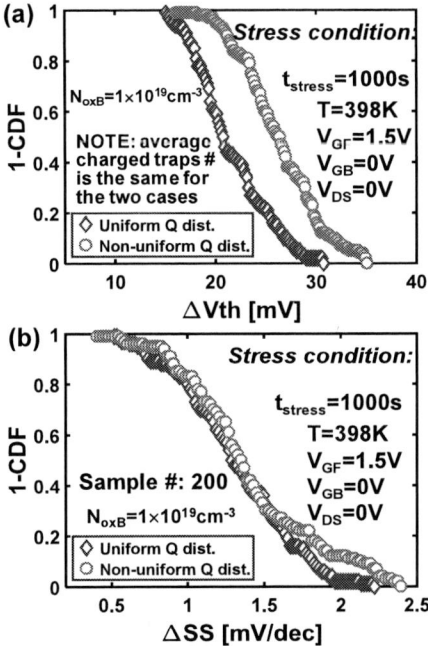

Fig. 5. Comparison of statistical (a) Vth and (b) SS degradations induced by the assumed uniform and simulated non-uniform charge distributions in the dielectrics.

To distinguish the degradation contribution by the traps located in the back-gate oxide, the energy band diagram with the illustration of trap distribution is shown in Fig. 8. The distributions of deep and shallow trap levels in HfO_2 and SiO_2 are taken into account [16]. The active regions in Fig. 8 denote that traps in the Box respond to back-gate bias, indicating the major contribution of shallow energy traps in the Box to the degradation of FDSOI nFET with p-Well during back-biasing.

The impact of different intrinsic trap densities in the Box on trapped charge # during stress is investigated. It can be found from Fig. 9 that the higher intrinsic trap density in the Box leads to more trapped charges and number variation during BTI stress, especially for the case of large back-gate bias. Therefore, improving the quality of back-gate oxide not only reduces the degradation but also mitigates its variability. Current degradation can be also acquired with the developed simulator. More Id degradation in the device with higher trap density in the Box can be observed from Fig. 10 when the back gate bias is enlarged. Moreover, it can be noted that ~1.4% Id

after 1000s stress is contributed by RCS under the case of V_{GB} = 3V and $N_{oxB}=1\times10^{19}cm^{-3}$.

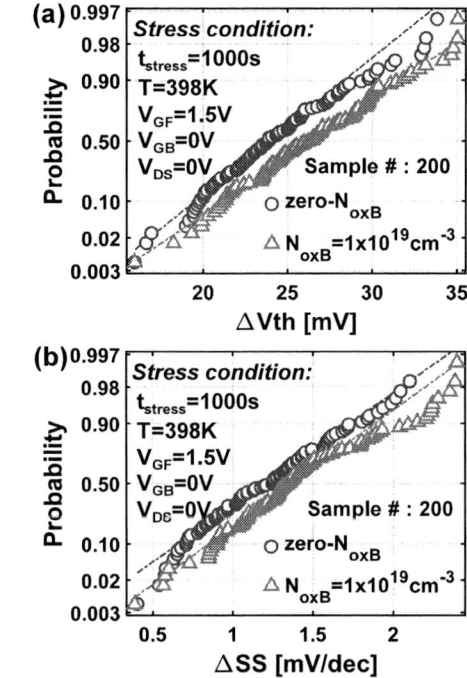

Fig. 6. Comparison of statistical (a) Vth and (b) SS degradation with/without the consideration of back-oxide traps.

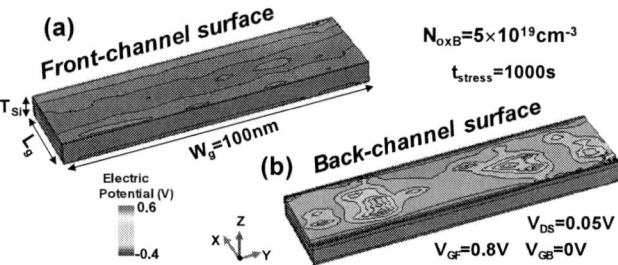

Fig. 7. Charged traps in the dielectrics induce the variations of the electric potential at the (a) front- and (b) back- channel surfaces, respectively.

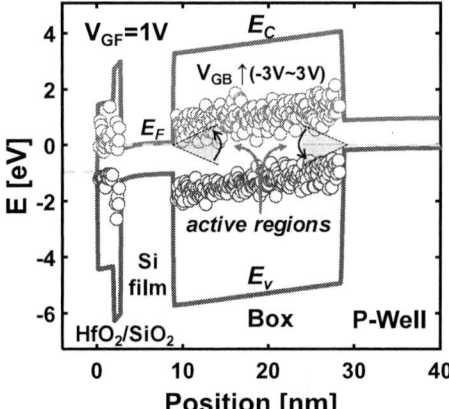

Fig. 8. Traps in the Box of the FDSOI nFET respond to the back-biasing from the energy band diagram perspective.

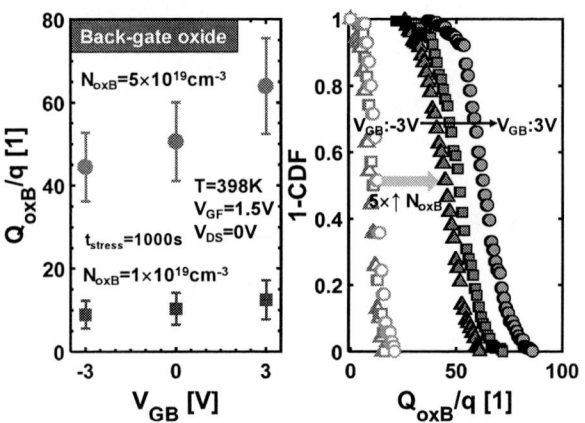

Fig. 9. (a) Charge statistics of the device with different intrinsic trap densities in the Box. (b) Complementary cumulative distribution plots.

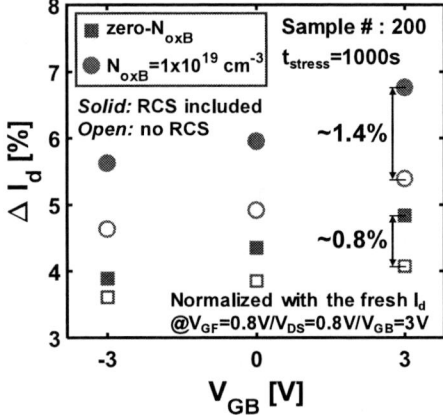

Fig. 10. The effects of RCS on statistical average ΔId of the device with/without traps in the Box during varying back-gate biases.

IV. CONCLUSION

The 3D KMC simulation approach of trap dynamics in the dielectrics is proposed to track time-dependent charge distribution and degradation. With the proposed method, simulation results show that the assumption of the uniform charge distribution in the dielectrics induced by stress underestimates the statistical degradation and variability. Moreover, the larger initial back-oxide trap density in the UTBB FDSOI brings more serious degradation and variability as the back-gate bias increases.

ACKNOWLEDGMENT

This work is supported by National Natural Science Foundation of China 2016YFA0202101, No. 61674008, and No. 61421005.

REFERENCES

[1] C. Liu, K. T. Lee, H. Lee, Y. Kim, S. Pae, and J. Park, "New observations on the random telegraph noise induced Vth variation in nano-scale MOSFETs," in *IEEE Int. Rel. Phys. Symposium*, pp. XT.17.1-XT.17.5, June 2014.

[2] J. Franco, B. Kaczer, N. Waldron, Ph. J. Roussel, A. Alian, M. A. Pourghaderi, Z. Ji, T. Grasser1, T. Kauerauf, S. Sioncke, N. Collaert, A. Thean, and G. Groeseneken, "RTN and PBTI-induced Time-Dependent Variability of Replacement Metal-Gate High-k InGaAs FinFETs, " in *IEDM Tech. Dig.*, pp. 20.2.1-20.2.4, December 2014.

[3] P. Weckx, B. Kaczer, P. Raghavan, J. Franco, M. Simicic, Ph. J. Roussel, D. Linten, A. Thean, D. Verkest, F. Catthoor, and G. Groeseneken, "Characterization and simulation methodology for time-dependent variability in advanced technologies," in *IEEE Int. Rel. Phys. Symposium*, pp. XT.17.1-XT.17.5, June 2015.

[4] L. Vandelli, L. Larcher, D. Veksler, A. Padovani, G. Bersuker, and K. Matthews, "A Charge-Trapping Model for the Fast Component of Positive Bias Temperature Instability (PBTI) in High-κ Gate-Stacks," *IEEE Trans. Electron Devices*, vol. 61, pp. 2287-2293, July 2014.

[5] B.Kaczer, J.Franco, P.Weckx, Ph.J.Roussel, V.Putcha, E.Bury, M.Simicic, A.Chasin, D.Linten, B.Parvais, F.Catthoor, G.Rzepa, M.Waltl, and T.Grasser, "A brief overview of gate oxide defect properties and their relation to MOSFET instabilities and device and circuit time-dependent variability," *Microelectronics Reliability*, vol. 81, pp186-194, February 2018.

[6] N. Xu, F. Andrieu, B. Ho, B-Y. Nguyen, O. Weber, C. Mazure, O. Faynot, T. Poiroux, and T-J. K. Liu, "Impact of back biasing on carrier transport in ultra-thin-body and BOX (UTBB) Fully Depleted SOI MOSFETs," in *Symposium on VLSI Tech. Dig.*, pp. 113-114, June 2012.

[7] E. Beigne, J-F. Christmann, A. Valentian, O. Billoint, E. Amat, and D. Morche, "UTBB FDSOI technology flexibility for ultra low power internet-of-things applications," in *ESSDERC*, pp. 164-167, September 2015.

[8] L. Grenouillet *et al.*, "UTBB FDSOI transistors with dual STI for a multi-Vt strategy at 20nm node and below," in *IEDM Tech. Dig.*, pp. 3.6.1-3.6.4, December 2012.

[9] W. Chen, Y. Li, L. Cai, P. Chang, G. Du, and X. Liu, "Entire Bias Space Statistical Reliability Simulation By 3D-KMC Method and Its Application to the Reliability Assessment of Nanosheet FETs based Circuits," in *IEDM Tech. Dig.*, pp. 33.5.1-33.5.4, December 2018.

[10] Y. Li, Z. Lun, P. Huang, Y. Wang, H. Jiang, G. Du, and X. Liu, "3D KMC Reliability Simulation of Nano-Scaled HKMG nMOSFETs with Multiple Traps Coupling," in *SISPAD*, pp. 148-151, September 2015.

[11] Y. Li, P. Huang, S. Di, X. Zhang, G. Du, and X. Liu, "Comprehensive Investigation of Multi-traps Induced Degradation in HfO2 Based nMOSFETs with Interfacial Layer by 3D-KMC Method," *IEEE Trans. Nanotechnology*, vol. 17, pp. 198-204, March 2018.

[12] C. Guerin, V. Huard, and A. Bravaix, "General framework about defect creation at the Si/SiO2 interface," *Journal of Applied Physics*, vol. 105, 114513, June 2009.

[13] S. Tyaginov, I. Starkov, H. Enichlmair, J. M. Park, C. Jungeman, and T. Grasser, "Physics-Based Hot-Carrier Degradation Models," *ECS Trans.*, pp. 321-352, 2011.

[14] *Sentaurus Device User Guide*, v-2013.06, Synopsys, Mountain View, CA, USA, 2013.

[15] H. Tanimoto, M. Kondo, T. Enda, N. Aoki, R. Iijimat, T. Watanabe, M. Takayanagi, and H. Ishiuchi, "Modeling of Electron Mobility Degradation for HfSiON MISFETs," in *SISPAD*, pp. 47-50, September 2006.

[16] G. Rzepa, J. Franco, B. O'Sullivan, A. Subirats, M. Simicic, G. Hellings, P. Weckx, M. Jech, T. Knobloch, M. Waltl, P.J. Roussel, D. Linten, B. Kaczer, and T. Grasser, "Comphy — A compact-physics framework for unified modeling of BTI," *Microelectronics Reliability*, vol. 85, pp49-65, April 2018.

Polarization Effect Induced by Discrete Impurity at Semiconductor/Oxide Interface in Si-FinFET

Katsuhisa Yoshida
Institute of Applied Physics,
University of Tsukuba
Ibaraki 305-8573, Japan
yoshida@bk.tsukuba.ac.jp

Kohei Tsukahara
Institute of Applied Physics,
University of Tsukuba
Ibaraki 305-8573, Japan
s1920328@s.tsukuba.ac.jp

Nobuyuki Sano
Institute of Applied Physics,
University of Tsukuba
Ibaraki 305-8573, Japan
sano.nobuyuki.gw@u.tsukuba.ac.jp

Abstract—The random dopant fluctuation (RDF) is a dominant source of statistical variability for nano-scale metal-oxide-semiconductor-field-effect-transistors (MOSFETs). We study RDF with the polarization effect induced by the discreteness of impurity and the dielectric mismatch at the Si/oxide interface by 3D drift-diffusion simulation. The charge distribution model employed in this study for the discrete impurity clarifies RDF dependence on the dielectric constant of oxide material. It is shown that explicit modeling of the polarization charge associated with discrete impurities is inevitable for reliable prediction of threshold voltage.

Keywords—*discrete impurity, random dopant fluctuation, drift-diffusion method, FinFET, MOSFET, polarization*

I. INTRODUCTION

The scaling of Si metal-oxide-field-effect-transistor (MOSFET) has been attained by adopting the multi-gate structure, such as FinFET [1], and high-κ oxide materials. However, nano-scaled devices are seriously suffered by statistical variability problems due to its small volume of semiconductor materials and an increase of the surface/volume ratio. The random dopant fluctuation (RDF) [2-8] is one of the crucial sources of the variability problems and unavoidable in any nano-scale devices employing a doping technique.

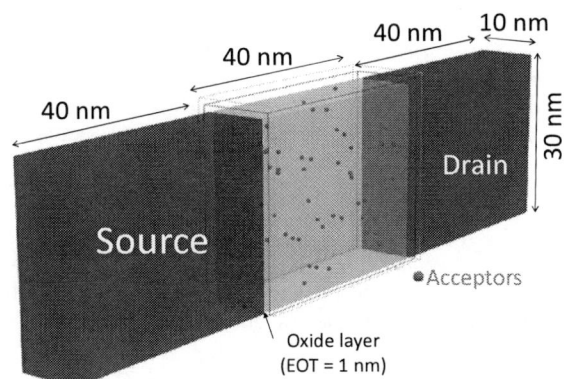

Fig. 1. Schematic of Si-FinFET structure employed in this study. In the channel region, 60 discrete acceptors (presented as red points) are randomly distributed. Source and drain regions are uniformly doped by donor concentration of 10^{20} cm^{-3} (blue shaded regions).

The physical origin of RDF is the potential fluctuation induced by the variabilities of the position and the number of dopants. In order to carry out predictive device simulation for RDF, it is inevitable to treat the Coulomb potential in the framework of the simulation [9]. In the device simulation, the Coulomb potential of discrete impurities is divided into the short-range and the long-range potentials, and those components are separately considered in the transport equation and the Poisson equation, respectively. It should be noted that the jelly dopant model corresponds to the long-wavelength limit of the long-range potential and, thus, there is no potential fluctuation.

For a bulk semiconductor, the short-range element is modeled as a screened potential by carriers in the substrate assuming charge neutral and thermal equilibrium. On the other hand, the long-range potential is the compensated part of the full Coulomb potential by carriers when we evaluate the short-range one. Therefore, once we know the distribution of screening carriers, we can regard it as a charge distribution model of the discrete impurity and obtain only the long-range potential as a solution of the Poisson equation without double-counting the short-range one. Additionally, in the nano-scale multi-gate devices, we should take care of a fact that the discreteness of impurities induces a polarization charge on the semiconductor/oxide interface due to a difference of dielectric constants.

Recently, we have reported a charge model for discrete impurities taken into account the polarization charge [10]. Since the polarization charge modifies the Coulomb potential compared to that without the interface and, thus, the charge distribution model for the long-range potential should be corrected. In this study, we carried out the drift-diffusion (DD) simulation incorporating the polarization effect at the interface associated with discrete impurities for the long-range Coulomb potential under a nano-scale FinFET operation.

II. NUMERICAL METHOD

We employ a Si-FinFET device structure as shown in Fig. 1 and carry out 3D-DD simulation. The Fin structure has 10 nm of the width and 30 nm of the height. The channel length is 40 nm. The source and drain regions are uniformly doped by donor concentration of 10^{20} cm^{-3}. In the channel region, 60 acceptors are randomly distributed, and 500 different acceptor configurations are considered. Therefore, we only introduce position variation of acceptors within the channel region as a

978-1-7281-0941-1/19 $31.00 © 2019 IEEE

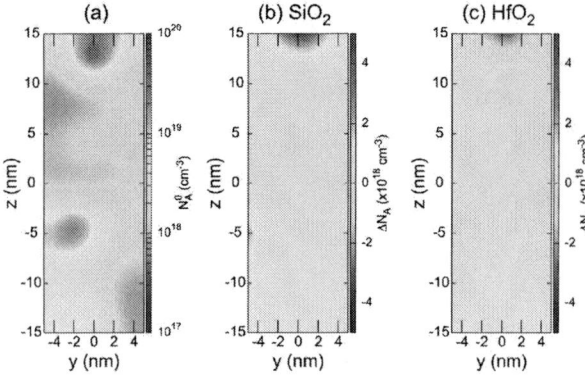

Fig. 2. Profiles of acceptor concentration in the middle corss-section of the channel region presented in Fig. 1 with the same acceptor configuration. (a) N_A^0 is acceptor concentration based on the Yukawa-like charge model for discrete impurities without polarization correction term induced by the interface. Polarization correction terms for acceptor concentration, ΔN_A, with (b) SiO$_2$ and (c) HfO$_2$ oxide layers. The acceptor concentration, N_A, used in the Poisson equation is a sum of N_A^0 and ΔN_A.

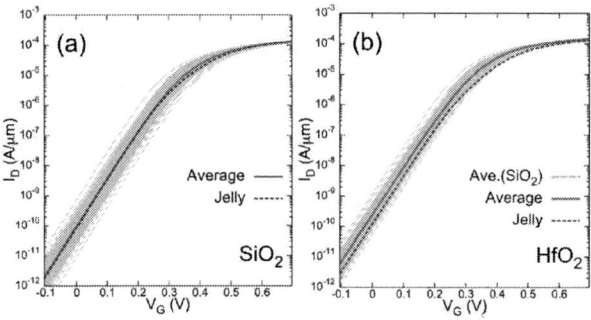

Fig. 3. I_D-V_G characteristics with 500 different acceptor configurations (gray lines) under $V_D = 0.05$ V. (a) SiO$_2$ and (b) HfO$_2$ oxide layers. Red solid lines are averaged value over 500 different configuration for each oxide material. The results with jelly model for each oxide material are plotted in blue dotted lines. The acceptor concentration employed in the jelly models is 5×10^{18} cm^{-3}. The green dashed line presented in (b) is the same curve named as "Average" in (a).

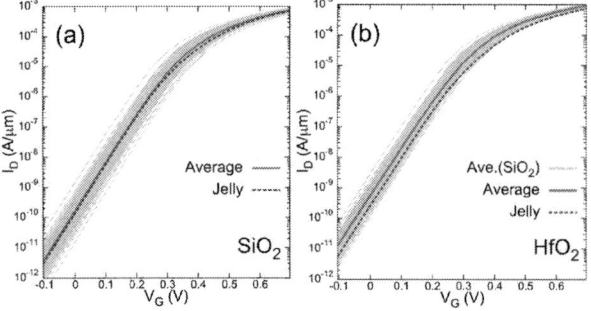

Fig. 4. I_D-V_G characteristics with 500 different acceptor configurations (gray lines) under $V_D = 0.8$ V. (a) SiO$_2$ and (b) HfO$_2$ oxide layers. Calculation conditions and lines are same employed in Fig. 3 except the drain voltage.

source of statistical variability in the present study. The charge distribution model for the discrete impurity and its long-range Coulomb potential is based on our recent work [10]. The acceptor concentration, N_A, used in the Poisson equation is described as,

$$N_A = \Sigma_i^{60}(N_{A,i}^0 + \Delta N_{A,i}), \qquad (1)$$

where $N_{A,i}^0$ is the Yukawa-like acceptor distribution for the i-th acceptor and given by

$$N_{A,i}^0(\mathbf{r}, \mathbf{r}_i) = q_C^2 \exp(-q_C^2|\mathbf{r} - \mathbf{r}_i|)/(4\pi|\mathbf{r} - \mathbf{r}_i|), \qquad (2)$$

where q_C is an inverse of the screening length, and $\mathbf{r}_i$ is a center of the i-th acceptor. The screening length is estimated by the Debye-Hückel or the Thomas-Fermi model depending on the concentration. $\Delta N_{A,i}$ is the polarization correction term and is functions of the difference of dielectric constants between Si and the oxide material and of the distance between the center of impurity and the interface. For convenience, we define two quantities as $N_A^0 = \Sigma_i^{60} N_{A,i}^0$ and $\Delta N_A = \Sigma_i^{60} \Delta N_{A,i}$. SiO$_2$ and HfO$_2$ are employed as oxide materials, and their thicknesses are 1.0 nm and 6.13 nm, respectively. The HfO$_2$ thickness is equivalent to 1.0 nm of the effective oxide thickness (EOT). The relative dielectric constants of Si, SiO$_2$, and HfO$_2$ are 11.8, 3.9, and 23.9, respectively. The acceptor concentration averaged in the channel volume is 5.0×10^{18} cm^{-3}, and the screening length is 1.63 nm at 300K. In the jelly model, the acceptor concentration in the channel region is 5.0×10^{18} cm^{-3}. The threshold voltage, V_{Th}, is defined when the drain current, I_D, exceeds 14.9 nA/μm.

III. Discussion

Fig. 2 (a) shows the acceptor concentration profile, N_A^0, without the polarization correction term in a cross-section at a middle plane of the channel region. Since we employ the Yukawa-like charge model for each acceptor, N_A^0 is widely distributed in the substrate, and its high concentration area - namely $N_A^0 > 2 \times 10^{19}$ cm^{-3} roughly spreads over a square of the screening length. This fact means that the charge of the discrete impurity is not a point-like charge but has a fine size to exclude the short-range potential.

Figs. 2 (b) and (c) are the distributions of the polarization correction term, ΔN_A, for SiO$_2$ and HfO$_2$, respectively, with the same acceptor configuration. The correction term ΔN_A is only distributed near the interface and rapidly decreases with distance from the oxide layer. Additionally, it indicates that a larger dielectric constant of the oxide layer than Si screens the Coulomb potential of a discrete impurity and, thus, reduces the screening carriers required to satisfy the charge neutral condition as pointed above. As a result, the correction term becomes negative to decrease acceptor concentration for the long-range potential. When the dielectric constant is smaller than Si, the opposite effect has to be taken into account and, thus, the result with SiO$_2$ has a positive correction value.

Figs. 3 and 4 show I_D-V_G characteristics associated with 500 different acceptor configurations under the drain voltages, V_D, of 0.05 V and 0.8 V, respectively. The results with the jelly acceptor model are also presented for each oxide material. The acceptor concentration of the jelly model is 5×10^{18} cm^{-3} within the channel region. I_D dependence on the acceptor configuration (plotted by thin gray lines) becomes small as V_G increases under both drain voltages. Additionally, the difference between the jelly and the discrete impurity models decreases. These properties indicate that the present charge distribution model for the discrete impurities is screened by the carriers when the carrier concentration becomes large, and, thus, the configuration dependence becomes small. On the other hand, when V_G is small, I_D-V_G curves shows a significant configuration dependence. Therefore, the potential fluctuation induced by discrete impurity is mainly caused by the long-range Coulomb potential since the short-range potential depends on the acceptor concentration and is independent of the carrier concentration induced by V_G.

Table I summarizes V_{Th} under various calculation conditions, and statistical distributions of V_{Th} are found in Fig. 5 for $V_D = 0.05$ V and 0.8 V. Extracted standard deviations with SiO$_2$ layer are 13.4 mV and 14.4 mV under $V_D = 0.05$ V and 0.8 V, respectively. Those with HfO$_2$ are 10.2 mV and 11.2 mV, respectively. By using V_{Th} average and the standard deviation for each condition, the Gaussian distribution fits well the calculated result.

In the present Fin structure, the short-channel-effect (SCE) especially drain-induced-barrier-lowering (DIBL) is not negligible as the drain-voltage dependence of V_{Th} found in Table I and also Fig. 5. In the jelly doping cases, V_{Th} differences between those under $V_D = 0.05$ V and 0.8 V are 14.1 mV and 19.4 mV for SiO$_2$ and HfO$_2$, respectively. This difference between the SiO$_2$ and HfO$_2$ could be caused by a slight difference of gate capacitances. In the discrete impurity cases, V_{Th} under $V_D = 0.8$ V is shift to 15.7 mV and 22.2 mV for devices with SiO$_2$ and HfO$_2$, respectively, compared to the value under $V_D = 0.05$ V. The discrete model shows a larger shift than the jelly model, and the discrete impurity models with HfO$_2$ is weak to SCE in the present structure. This reason could be explained as follows: The electron concentration becomes large as approaching to the interface from the center of Si substrate when V_G is applied. On the other hand, the polarization correction dominates around the interface as presented in Figs. 2 (b) and (c). Therefore, it is expected that V_{Th} significantly depends on the concentration of acceptors and electrons at the interface. In the case with HfO$_2$, the correction term reduces the acceptor concentration and, thus, effective acceptor concentration, where $N_A = N_A^0 + \Delta N_A$, in the channel becomes smaller than the jelly concentration. In the device with SiO$_2$, the correction term increases the net-acceptor concentration around the interface and can suppress DIBL effect. As a result, V_{Th} shift with HfO$_2$ becomes larger than that with SiO$_2$.

The standard deviation reflects the strength of V_{Th} dependence on the acceptor configurations. When we consider the dependence of standard deviation on the drain voltage, the number of accepters in the region where the gate can dominantly control the potential in the channel. As a result, the number fluctuation in such a sensitive region also contributes to the statistical variability and, thus, results in an increase of the standard deviation in both cases. In the case with SiO$_2$, the acceptor concentration, $N_A = N_A^0 + \Delta N_A$, around the interface strongly depends on the position of acceptors since the correction term increases the acceptor concentration and is only significant around the interface as presented in Fig. 2 (b). On the other hand, the polarization in HfO$_2$ reduces the concentration fluctuation induced by the

acceptor configurations as shown in Fig. 2 (c). It is because that the correction term reduces the net-acceptor concentration when an acceptor is close to the interface. As a result, the correction term weakens the acceptor configuration dependence. In summary, HfO$_2$ shows a smaller standard deviation than that with SiO$_2$.

IV. CONCLUSION

We have carried out 3D-DD simulation incorporated with discrete impurities modeled by Yukawa-like charge distributions. The discrete nature of impurities induces a polarization charge at the interface between the substrate and the oxide material depending on the difference of the dielectric constants. The present charge model produces the correct long-range Coulomb potential in the framework of DD device simulation. As a result, the screening of the long-range potential i.e., the potential fluctuation induced by the discrete impurity becomes dependent of the gate voltage and also carrier concentration in the channel region. The polarization correction term is able to distinguish the oxide material dependence of the potential fluctuation. It is inevitable for predictive device simulations to explicitly take into account the interface polarization induced by the discreteness of impurity.

TABLE I. TABLE OF THRESHOLD VOLTAGES

Oxide	SiO$_2$ (mV)		HfO$_2$ (mV)	
Model	Jelly	Discrete	Jelly	Discrete
$V_D = 0.05$ V	140.0	140.4	132.5	115.6
$V_D = 0.8$ V	125.9	124.7	113.1	93.4

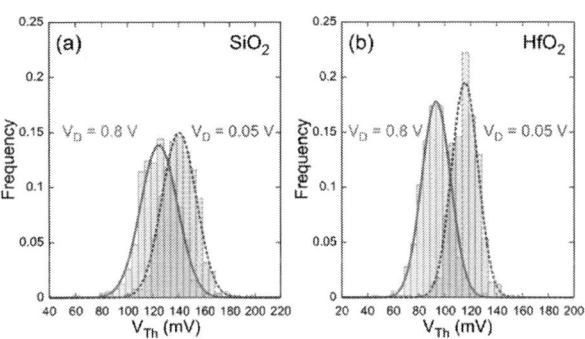

Fig. 5. Statistical distributions of threshold voltage for the devices with (a) SiO$_2$ and (b) HfO$_2$ gate oxides under $V_D = 0.05$ (orange) and 0.8 V (light-gray). Threshold voltage is defined when the drain current exceeds 14.3 nA/μm. The red solid and gray dotted curves are the Gaussian fits under $V_D = 0.05$ V and 0.8 V, respectively. The average of V_{Th} are summrized in Table I. For SiO$_2$ results, the standard deviations employed in Gaussian curves are 13.4 mV and 14.4 mV under $V_D = 0.05$ V and 0.8 V, respectively. Those values for HfO$_2$ are 10.2 mV and 11.2 mV under $V_D = 0.05$ V and 0.8 V, respectively.

ACKNOWLEDGMENT

This research was supported by MEXT as ''Priority Issue on Post-K computer'' (Creation of new functional Devices and high-performance Materials to Support next-generation Industries).

REFERENCES

[1] D. Hisamoto, W. C. Lee, J. Kedzierski, H. Takeuchi, and K. Asano, "FinFET-a self-aligned double-gate MOSFET scalable to 20 nm," IEEE Transactions on Electron Devices, vol. 47, no. 12, pp. 2320–2325, Dec. 2000, doi: 10.1109/16.887014.

[2] K. Nishinohara, N. Shigyo, and T. Wada, "Effects of Microscopic Fluctuations in Dopant Distributions on MOSFET Threshold Voltage," IEEE Transactions on Electron Devices, vol. 39, no. 3, pp. 634–639, Mar. 1992, doi: 10.1109/16.123489.

[3] H. S. Wong and Y. Taur, "Three-dimensional "Atomistic" Simu- lation of Discrete Random Dopant Distribution Effects in Sub-0.1 □/m MOSFET's," in IEDM Tech. Dig., 1993, pp. 705–708, doi: 10.1109/IEDM.1993.347215.

[4] Y. Li, C. H. Hwang, and T. Y. Li, "Random-Dopant-Induced Variability in Nano-CMOS Devices and Digital Circuits," IEEE Transactions on Electron Devices, vol. 56, no.8, pp. 1588–1597, Aug. 2009, doi: 10.1109/TED.2009.2022692.

[5] C. Shin, X. Sun, T. J. K. Liu, "Study of Random-Dopant-Fluctuation (RDF)Effects for the Trigate Bulk MOSFET," IEEE Transactions on Electron Devices, vol. 56, no. 7, pp. 1538–1542, Jul. 2009, doi: 10.1109/TED.2009.2020321.

[6] C. H. Hwang, Y. Li, and M. H. Han, "Statistical variability in FinFET de- vices with intrinsic parameter fluctuations," Microelectronics Reliability, vol. 50, pp. 635–638, Mar. 2010, doi: 10.1016/j.microrel.2010.01.041.

[7] L. Gerrer, A. R. Brown, C. Millar, R. Hussin, S. M. Amoroso, B. Chen, D. Reid, C. Alexander, D. Fried, M. Hargrove, K. Greiner, A. Asenov, "Accurate Simulation of Transistor-Level Variability for the Purposes of TCAD-Based Device-Technology Cooptimization," IEEE Transactions on Electron Devices, vol. 62, no. 6, pp. 1739–1745, Jun. 2015, doi: 10.1109/TED.2015.2402440.

[8] A. Asenov, G. Slavcheva, A. R. Brown, J. H. Davies, and S. Saini, "Increase in the Random Dopant Induced Threshold Fluctuations and Lowering in Sub-100 nm MOSFETs Due to Quantum Effects: A 3- D Density-Gradient Simulation Study," IEEE Transactions on Electron Devices, vol. 48, no. 4, pp. 722–729, Apr. 2001, doi:

[9] N. Sano, K. Matsuzawa, M. Mukai, and N. Nakayama, "On discrete random dopant modeling in drift-diffusion simulations: Physical meaning of 'atomistic' dopants," Microelectron. Rel., vol 42, no. 2, pp. 189–199, Feb. 2002, doi: 10.1016/S0026-2714(01)00138-X.

[10] N. Sano, K. Yoshida, C.-W. Yao, and H. Watanabe, "Physics of Discrete Impurities under the Framework of Device Simulations for Nanostructure Devices," Materials, vol. 11, pp. 2559, Dec. 2018, doi: 10.3390/ma11122559.

978-1-7281-0941-1/19 $31.00 © 2019 IEEE

Relationship between capacitance and conductance in MOS capacitors

E. Caruso*, J. Lin*, S. Monaghan*, K. Cherkaoui*, L. Floyd*, F. Gity*, P. Palestri¥, D. Esseni¥, L. Selmi#, P. K. Hurley*

*Tyndall National Institute, University College Cork, Cork, Ireland
¥ DPIA, University of Udine, Via delle Scienze 206, 33100, Udine, Italy
DIEF, University of Modena and Reggio Emilia, Via P. Vivarelli 10/1, 41125, Modena, Italy

Abstract—In this work, we describe how the frequency dependence of conductance (G) and capacitance (C) of a generic MOS capacitor results in peaks of the functions G/ω and -ωdC/dω. By means of TCAD simulations, we show that G/ω and -ωdC/dω peak at the same value and at the same frequency for every bias point from accumulation to inversion. We illustrate how the properties of the peaks change with the semiconductor doping (N_D), oxide capacitance (C_{OX}), minority carrier lifetime (τ_g), interface defect parameters (N_{IT}, σ) and majority carrier dielectric relaxation time (τ_r). Finally, we demonstrate how these insights on G/ω and -ωdC/dω can be used to extract C_{OX}, N_D and τ_g from InGaAs MOSCAP measurements

Keywords—Characterization, extraction technique, MOS, multi-frequency, C-V, G-V, minority carrier lifetime, oxide capacitance, doping

I. INTRODUCTION

Measuring and analyzing the impedance of the metal-oxide-semiconductor (MOS) system has played a central role in the development of MOS structures in electronic, photovoltaic and photocatalytic devices. The impedance/ admittance modulus (Z, Y) and phase angle (θ) of the MOS capacitor are typically measured over the frequency range from 20 Hz to 1 MHz, and resolved into capacitive (C) and conductive (G) elements. The behavior of C and G as a function of gate voltage, temperature and frequency are used to investigate and quantify physical parameters and defects of the MOS system [1].

It was recently shown using experimental results, simulations and mathematical analysis, that for any MOS capacitor in inversion, functions of the capacitance (-ωdC/dω) and conductance (G/ω) are related at all angular frequencies (ω=2πf) and that the two functions both exhibit a maximum value at the transition frequency (f_m). In addition, the values of G/ω and -ωdC/dω are equal at f_m [2].

The objective of this work is to extend on [2], to show that the functions G/ω and -ωdC/dω are related in all bias regions of the MOS C-V response and that this holds for ideal MOS systems and for non-ideal MOS structures with interface traps.

II. THE G/ω AND - ω dC/dω RELATIONSHIP: VALIDATION

To demonstrate the relationship between the C and G parameters, we consider the case of an InGaAs MOS capacitor (53% In). It is important to emphasize that the relationship is expected to hold for all MOS structures. InGaAs was selected as an example MOS system for the following reasons: (a) based on the energy gap (E_g=0.74 eV) and typical τ_g values, InGaAs MOS capacitors can exhibit an inversion response at room temperature [3], (b) MOS systems based on wider band gap semiconductors, (e.g., Si, GaN), and typical values t_g, exhibit peak frequency values (f_m) in inversion which are typically below 1 mHz, making difficult the measurements, (c) the C and G functions also have coincident peaks in the

GHz regime, of relevance to RF and mm wave applications, where In(Ga)As devices are often employed [4].

Figure 1(a) and 1(b) report the multi-frequency C-V and G-V responses (1 kHz to 1 MHz) of an $In_{0.53}Ga_{0.47}As$ MOS capacitor simulated with Sentaurus device simulator [5], including Fermi-Dirac statistics, multi-valley band structure with non-parabolic corrections and Shockley–Read–Hall (SRH) generation/recombination. Figure 2 plots G/ω and -ωdC/dω in inversion (V_G=-3 V) from 10^2 to 10^{14} Hz. The plot confirms that the two functions feature the same value at f = 2285 Hz, as described in [2], where the peak frequency is determined by the SRH time τ_g. The plot also shows that the two functions exhibit a second peak value at f = 8.53 THz. Figure 3 plots the magnitude of the peak values of G/ω and -ωdC/dω as the gate bias is varied. The plot demonstrates how the relationship holds at all bias points in the ideal CV/GV response.

Figure 4 shows how the two peak frequencies vary with the gate voltage. The high frequency peak is always present and relates to the majority carrier dielectric relation time, which is discussed later. The peak due to the supply of minority carriers to the inversion layer goes to zero, as expected, when the MOS structure moves from strong inversion to depletion. This is a general relationship and holds for any MOS system.

We next consider if the G/ω and -ωdC/dω relationships still hold for the non-ideal case, namely in the presence of interface states. Figure 5 illustrates an example for a Gaussian D_{IT} profile, described by the equation $D_{IT}(E) = N_{IT}\exp\left(-(E - E_{IT})^2/(2S_{IT}^2)\right)$. The traps are donor type, as proposed in previous publications [e.g., [6]]. This profile is introduced into the ideal InGaAs MOS structure considered in Figs. 1. The D_{IT} introduces a frequency dependent distortion into the CV and GV response, as reported in other publications [e.g., [6]]. Plotting the G/ω and -ωdC/dω peak values versus gate voltage (Figure 6(a)) and the frequency of the peak value versus bias (Figure 6(b)), indicates that the relationship still holds for interface states. The high frequency peaks (> 10^{12} Hz) are not influenced by interface states, as expected, and are removed for clarity.

The physical meaning of the peak values of G/ω and -ωdC/dω is illustrated in Figures. 7 to 10. Figure 7 shows that the peak frequency in inversion is directly related to $1/\tau_g$. At high values of the minority carrier lifetime, f_m eventually saturates, when the inversion layer carrier supply rate from minority carrier density in quasi neutral region (n_i^2/N_D) exceeds the SRH generation rate. In depletion region, where interface traps dominates the response, the peak value of G/ω linearly depends on N_{IT} (the peak value of D_{IT}) for G/ω<C_{OX}/2 (see Figure 8), and this is a well known result from the conductance method approach [1][7]. The peak frequency shows a linear dependence on the capture cross section (Figure 9). In accumulation, the high frequency peak is set by the majority carrier dielectric relaxation time (Figure 10)

978-1-7281-0941-1/19 $31.00 © 2019 IEEE

$\tau_r = \varepsilon_0 \cdot \varepsilon_r \cdot \rho$, with ε_0, ε_r and ρ being the free space permittivity and the semiconductor relative permittivity and resistivity [8].

The measured/simulated admittance is resolved in the equivalent circuit shown in Figure 11(c). This circuit representation in addition to the equivalent circuit topology of the device under test (Fig 11(b)) set the relationship between G/ω and $-\omega dC/d\omega$. By first writing C and G in terms of C_{OX}, G_S, C_S, and then by calculating the corresponding maximum/minimum, it is also possible to obtain analytic expressions for the peak value and position of the G/ω and $-\omega dC/d\omega$ functions:

$$f_m = \frac{G_S}{2\pi(C_S + C_{OX})} \tag{1.a}$$

$$\left(\frac{G}{\omega}\right)_{max} = \left(-\omega \frac{dC}{d\omega}\right)_{max} = \frac{C_{OX}^2}{2(C_S + C_{OX})} \tag{1.b}$$

where G_S and C_S are the conductance and capacitance of the semiconductor respectively. It is worth noting that at every bias point the semiconductor impedance can be modeled using a capacitor and a conductance, also in the presence of interface traps. The only assumption made in the topology reported in Figure 11(b) is that the oxide is free of traps. In fact, in the presence of border traps an R-C distributed network should be added in the circuit, as reported in [9].

III. THE G/ω AND $-\omega dC/d\omega$ RELATIONSHIP: APPLICATION

By performing a direct comparison between G/ω and $-\omega dC/d\omega$ from simulations and experimental results, MOS parameters can be extracted. An example is shown in Figure 11(a) based on experimental multi-frequency C-V and G-V measured from 1 kHz to 1 MHz for a Ni/Al$_2$O$_3$/n-In$_{0.53}$Ga$_{0.47}$As/InP structure [10]. The only region that can be analyzed is inversion, where the frequency peak of G/ω and $-\omega dC/d\omega$ is within the measured frequency range (Figure 4).

For non-silicon based MOS systems, defects in the oxide (border traps) can have a significant impact on the MOS/MOSFET characteristics [11][12]. However, in strong inversion the behavior of C and G is mainly given by the minority carriers response, so it is reasonable to neglect the effect of the border traps and to use the topology shown in Figure 11(b) also in this case.

Using C_{OX}=0.098 F/m^2, τ_g= 80 ps and N_D=4.6·10^{17} cm^{-3} we obtain an excellent agreement between simulations and experiments (Fig 11(a)). The extracted parameters are consistent with other extraction techniques [10]. In particular, N_D agrees with the mean value of the doping concentration extracted by ECV, while C_{OX} is compatible with the oxide thickness measured by TEM images (6 nm) and assuming a Al$_2$O$_3$ dielectric constant of 6.6, which is consistent with other experiments [13].

IV. CONCLUSIONS

Simulations have shown that the G/ω and $-\omega dC/d\omega$ versus frequency plots exhibit marked peaks. The value and the frequency of the peaks can be interpreted with the circuit elements of Figure 11(b) and thus eventually in terms of C_{OX}, N_D, τ_g, τ_r, D_{IT} and σ. In strong inversion, where the effect of border traps can be neglected for this analysis, it is possible to do a direct comparison of the G/ω and $-\omega dC/d\omega$ functions

between experimental and simulated results (without traps). The triplet (C_{OX}, N_D, τ_g) that minimizes the error between these functions identifies the extracted quantities. This technique is completely general and applicable to any MOS capacitor system, the information that can be extracted from the experiments depends however on the frequency range that canbe practically explored.

ACKNOWLEDGMENT

The research leading to these results has received funding from the European Commission H2020 INSIGHT project No 688784 and Science Foundation Ireland (15/IA/3131).

REFERENCES

[1] E. H. Nicollian and John R. Brews, *MOS Physics and Technology*, Wiley, 2002

[2] S. Monaghan, É. O'Connor, R. Rios, F. Ferdousi, L. Floyd, E. Ryan, K. Cherkaoui, I. M. Povey, K. J. Kuhn and P. K. Hurley, "Capacitance and Conductance for an MOS System in Inversion, with Oxide Capacitance and Minority Carrier Lifetime Extractions," in IEEE Transactions on Electron Devices, vol. 61, no. 12, pp. 4176-4185, 2014.

[3] É. O'Connor, K. Cherkaoui, S. Monaghan, B. Sheehan, I. M. Poey and P. K. Hurley "Inversion in the In$_{0.53}$Ga$_{0.47}$As metal-oxide-semiconductor system: Impact of the In$_{0.53}$Ga$_{0.47}$As doping concentration", Applied Physics Letters, vol. 110, no. 3, 032902, 2017

[4] O. Kilpi, J. Svensson and L. Wernersson, "Sub-100-nm gate-length scaling of vertical InAs/InGaAs nanowire MOSFETs on Si," Conf. Proc IEDM, pp. 17.3.1-17.3.4, 2017

[5] *Sentaurus Device Manual L-2016.03-SP2*, Synopsys Inc., 2016

[6] G. Brammertz, A. Alian, D. H. Lin, M. Meuris, M. Caymax and W. -. Wang, "A Combined Interface and Border Trap Model for High-Mobility Substrate Metal–Oxide–Semiconductor Devices Applied to In$_{0.53}$Ga$_{0.47}$As and InP Capacitors," in IEEE Transactions on Electron Devices, vol. 58, no. 11, pp. 3890-3897, 2011.

[7] K. Martens, G. Brammertz, B. De Jaeger,D. Kuzum, M. Meuris, M. M. Heyns, T. Krishnamohan, K. Saraswat, H. E. Maes, and G. Groeseneken, "On the Correct Extraction of Interface Trap Density of MOS Devices With High-Mobility Semiconductor Substrates," in IEEE Transactions on Electron Devices, vol. 55, no. 2, pp. 547-556, Feb. 2008

[8] T. Ohmi, M. Tsubota, T. Tsuneto, "Relaxation of Brinkman-Smith Mode in Superfluid 3He–B between Parallel Plates." Japanese Journal of Applied Physics vol. 26, no. S3-1, p. 169, 1987

[9] Y. Yuan, B. Yu, J. Ahn, P. C. McIntyre, P. M. Asbeck, M. J. W. Rodwell and Y. Taur, "A Distributed Bulk-Oxide Trap Model for Al$_2$O$_3$ InGaAs MOS Devices," in IEEE Transactions on Electron Devices, vol. 59, no. 8, pp. 2100-2106, 2012.

[10] E. Caruso, J. Lin, K. F. Burke, K. Cherkaoui, D. Esseni, F. Gity, S. Monaghan, P. Palestri, P. Hurley and L. Selmi, "Profiling border-traps by TCAD analysis of multifrequency CV-curves in Al$_2$O$_3$/InGaAs stacks," Conf. Proc. EUROSOI-ULIS, pp. 1-4, 2018

[11] J. Lin, Y. Y. Gomeniuk, S. Monaghan, I. M. Povey, K. Cherkaoui, E. O'Connor and P. K. Hurley, "An investigation of capacitance-voltage hysteresis in metal/high-k/ In$_{0.53}$Ga$_{0.47}$As metal-oxide-semiconductor capacitors", Journal of Applied Physics, vol. 114, no 14, p. 144105, 2013

[12] S. Johansson, M. Berg, K. Persson and E. Lind, "A High-Frequency Transconductance Method for Characterization of High-κ Border Traps in III-V MOSFETs," in IEEE Transactions on Electron Devices, vol. 60, no. 2, pp. 776-781, 2013

[13] R. Y. Khosa, E. B Thorsteinsson, M. Winters, N. Rorsman, R. Karhu, J. Hassan, and E. Ö. Sveinbjörnsson, "Electrical characterization of amorphous Al2O3 dielectric films on n-type 4H-SiC". AIP Advances, vol. 8, no. 2,p. 025304, 2018

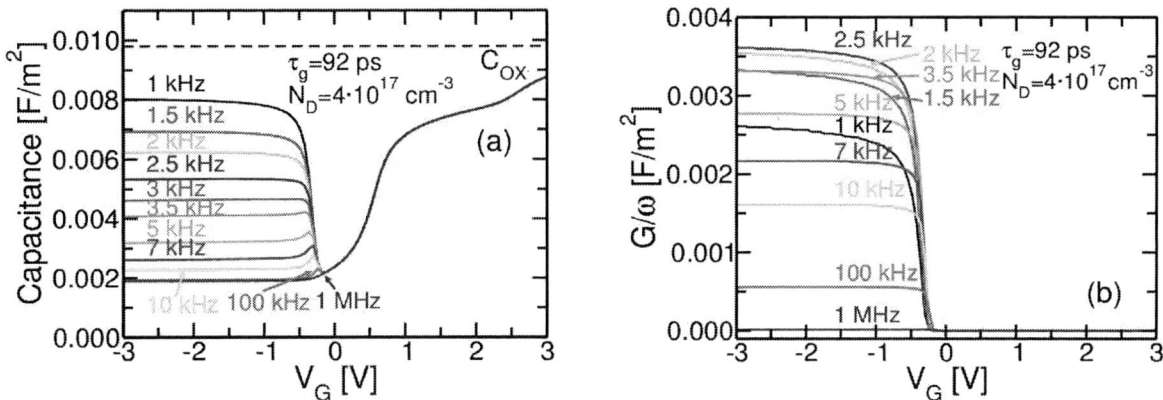

Figure 1. Simulated multi-frequency C-V (a) and G/ω -V (b) at 300 K w/o including traps. Simulations use $N_D=4\cdot10^{17}$ cm^{-3}, $C_{OX}=0.0098$ F/m^2 and $\tau_g= 92$ ps.

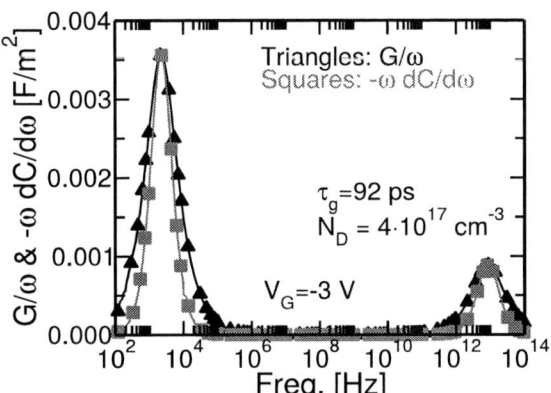

Figure 2. Simulated curves of G/ω (triangles) and -ω·dC/dω (squares) extracted from the MOS structure in Figure 1 and plotted as a function of frequency in inversion (V_G=-3 V).

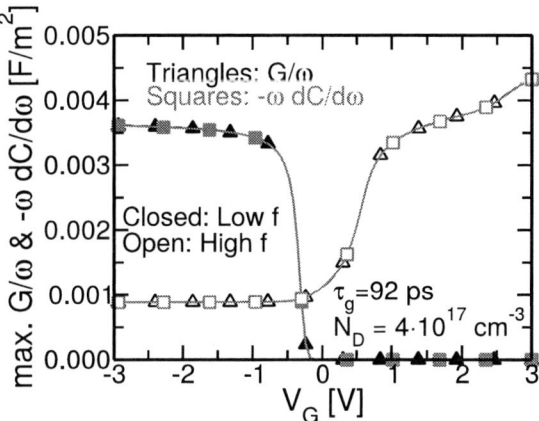

Figure 3. Peak value of G/ω (triangles) and -ω·dC/dω (squares) as a function of V_G for the simulations in Figure1

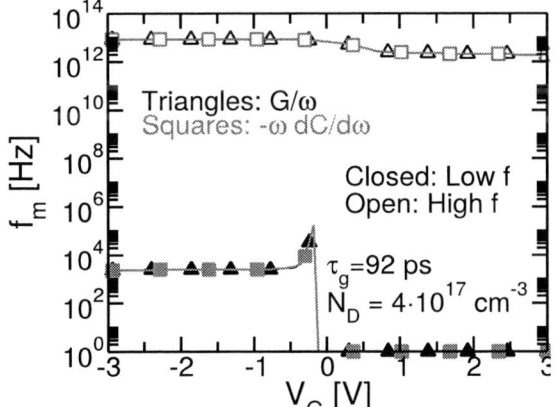

Figure 4. Frequency position of the peak value of G/ω (triangles) and -ω·dC/dω (squares) as a function of V_G for the simulations in Figure1.

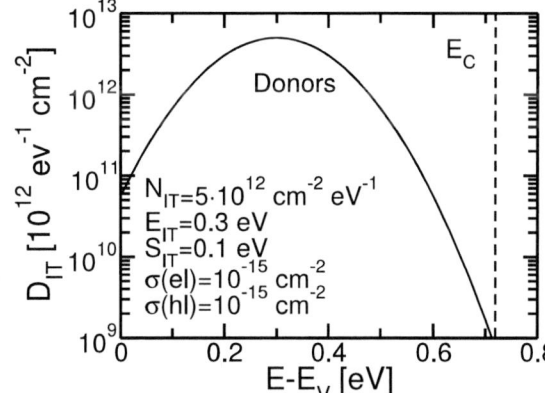

Figure 5. D_{IT} profile of donor traps included in the simulations of Figure 6. The energy is referred to the valence band of InGaAs. The figure reports also the parameter used for the Gaussian distribution and the values of the capture cross section for electron and holes.

978-1-7281-0941-1/19 $31.00 © 2019 IEEE   317

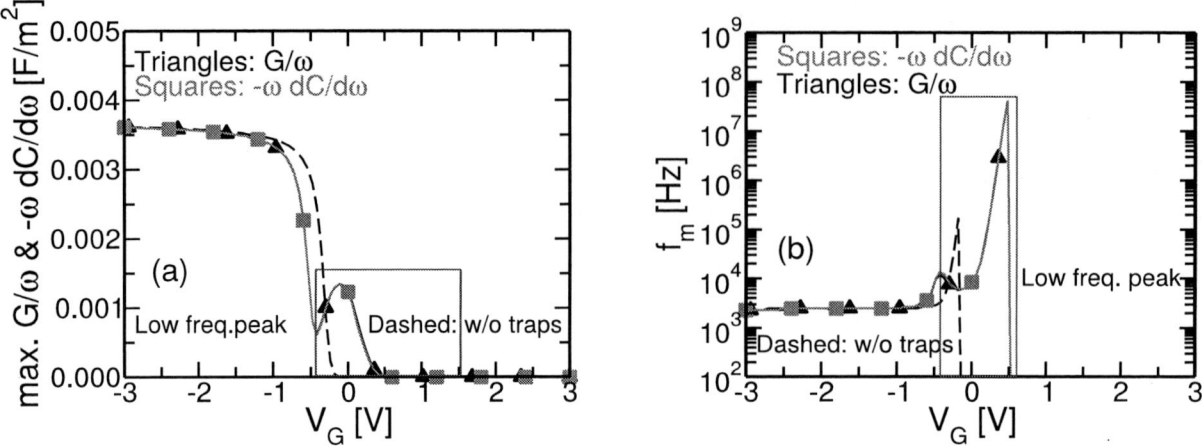

Figure 6. Simulations of the MOS structure in Figs. 1 including the Gaussian D_{IT} profile shown in Figure 5. The peak value of G/ω and -$\omega \cdot dC/d\omega$ versus V_G is shown in (a) and its frequency position in (b). The dashed lines are the simulations results without traps and the blue box highlights the region where the D_{IT} affects the properties of the peaks. The high frequency peaks ($> 10^{12}$ Hz) are not impacted by D_{IT}, as expected, and are removed for clarity.

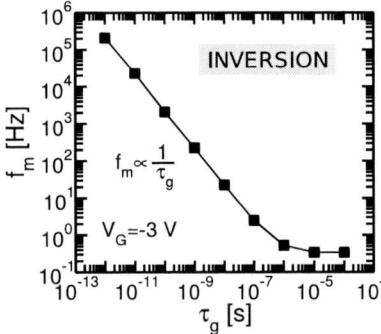

Figure. 7. Simulated f_m as a function of the minority carrier life time extracted from the MOS structure in Figure 1.

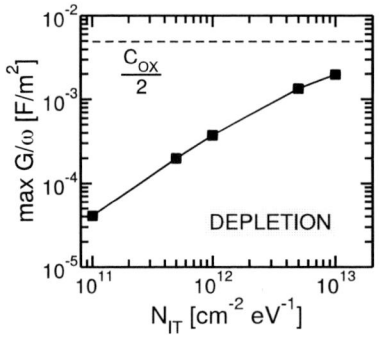

Figure. 8. Simulated maximum value of G/ω as a function of N_{IT} for the MOS structure in Figure 6. All the other parameter of D_{IT} are the ones reported in Figure 5. For high values of N_{IT}, the function tends to $C_{OX}/2$, which is shown with the dashed line.

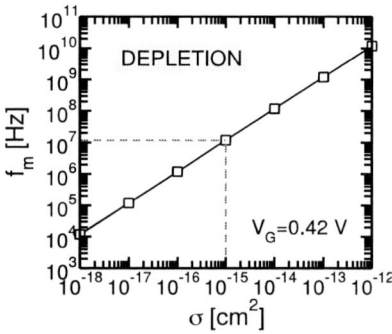

Figure 9. Simulated f_m associated to traps response of the D_{IT} in Figure 5 as a function of the capture cross section. Note that for $\sigma = 1 \cdot 10^{-15}$ cm^2, f_m is ~ 10 MHz, which is outside the typical measurement range (up to 1 MHz).

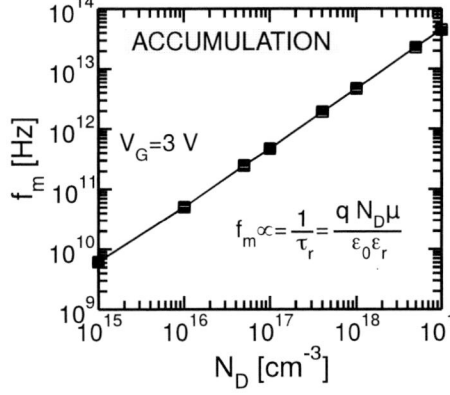

Figure 10. Simulated f_m of the majority carriers as a function of N_D from the MOS structure in Figure 1 (relevant for *mm* wave devices).

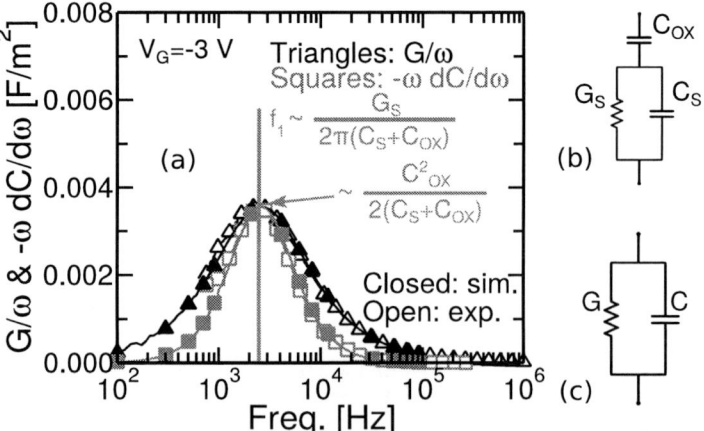

Figure. 11. (a) Simulated (closed) and experimental (open) curves at $V_G = -3$ V of G/ω (triangles) and -$\omega \cdot dC/d\omega$ (squares) plotted as a function of frequency. (b) Equivalent electrical circuit of a MOS structure interpreted using a parallel G-C circuit (c).

978-1-7281-0941-1/19 $31.00 © 2019 IEEE   318

Compact Modeling Perspetive – Bridge to Industrial Applications

Mitiko Miura-Mattausch
HiSIM Research Center
Hiroshima University
Higashi-Hiroshima 739−8530, Japan
mmm@hiroshima-u.ac.jp

Abstract—This paper summarizes briefly compact-model development history, which is characterized by the evolution into the role as a bridge between devices and circuits. It is demonstrated that the task of predicting circuitry performance accurately has been realized by considering the microscopic features of the device phenomena in the compact model, which had been previously treated only macroscopically.

Keywords—compact modeling, MOSFET, carrier dynamics, Poisson's equation, microscopic phenomena, circuit performance

I. INTRODUCTION

Diversity is growing in the world with respect to many aspects such as the internet of things (IoT). This became possible mostly due to the rapid technology development enabling and supporting many expectations growing in the society. It is clear that the semiconductor community has been playing an important role in this context. Owing to the difficulty of real-fabrication experiments with advanced technologies, simulation-based experiments have been intensively investigated to speed-up the achievement of targeted goals. The focus of this review is given on the compact-modeling technique, required for practical connection from new device developments to the key circuits of the final product. It is shown that a lot about the individual device characteristics can be further learned through the analysis of circuit performances in addition to the conventional device performances. Thus it can be said that compact models play the role of a bridge connecting devices and circuits. Due to such an important role, compact modeling is shifting from the simple threshold-voltage-based modeling approach, which has been studied for long time by the circuit simulation community, to modeling based on the potential distribution [1]. Progress achieved by this shift is summarized. Further, the compact-modeling improvements, following the evolution of applications, are overviewed and future aspects are also discussed.

II. COMPACT MODELING AND SIMULATION RESULTS

A. Compact Modeling

Compact models have been focusing on describing observed phenomena only on the basis of their essential origins in analytical forms for accurate prediction of circuit performances with less simulation time. The first compact model for circuit simulation was developed by Meyer in 1971, describing the MOSFET features with an equivalent circuit consisting of a current source, capacitances and resistors as shown in Fig. 1 [2]. The applied current equation is based on the simple drift approximation, which describes *I-V*

characteristics just as a function of applied biases with the threshold voltage V_{th} [3]. 2D-device simulation taught us that the saturation behavior of MOSFET is due to the loss of the gate control together with the increase of carrier scattering. These MOSFET features have been compact modeled afterwards. Increased needs for RF circuits and reduction of the non-linearity of device performances, causing cross talks in the communication systems, became serious problems to be solved. It was found that the carrier scatterings due to the field induced in the system are the main origin of the non-linearity, as demonstrated in Fig. 2 [4]. Thus the task for compact modeling entered into a new era, namely compact modeling has to consider the microscopic features of the carrier dynamics, which were modeled previously only phenomenologically. At the same time accurate prediction became an important task to complement actual measurements. Therefore, potential-based descriptions, capturing the origin of the device performances, became common and replaced the V_{th}-based modeling approach.

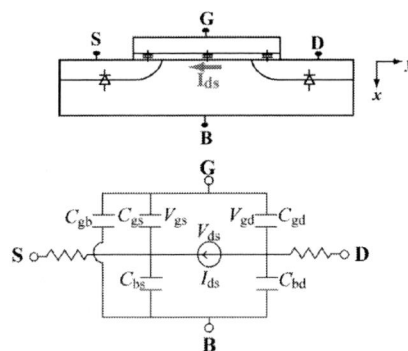

Fig. 1. Equivalent circuit (lower) developed by Meyer for describing a MOSFET (upper) [2].

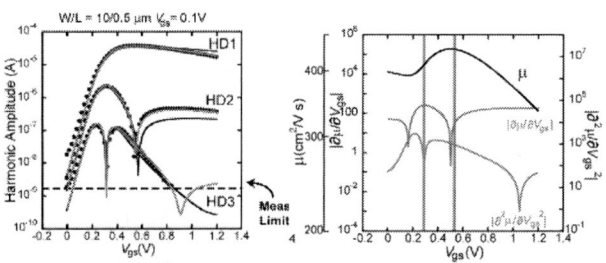

Fig. 2. Measured harmonic amplitudes (a) in comparison to the carrier mobility derivatives (b), showing that the main origin of the harmonics is the non-linearity of the carrier mobility.

978-1-7281-0941-1/19 $31.00 © 2019 IEEE

B. Relationship between Circuit and Device Performances

As depiced in Fig. 1, a simple MOSFET model consists of a current source, capacitances and resistances. Generally circuits include external elements such as loading capacitances and resistances to realize requested circuit performances. Fig. 3 shows a simple circuit applied for device-switching-performance measurements [5]. External elements form a low-path filter, which determines switching performance together with the internal device properties, as demonstrated in Fig. 4, where analytical cut-off-frequency descriptions are applied. Internal device properties dominate the switching performance with increased frequency. Therefore, it can be said that the device characteristics determine the RF circuit performances. The origin for this limitation is the carrier-transit delay, causing switching delay as depicted in Fig. 5. Thus the compact modeling has to be done by considering the transit delay explicitly [4].

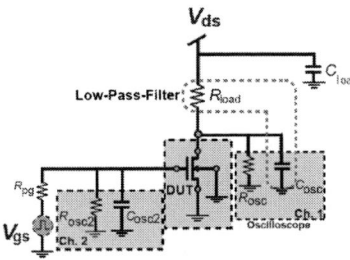

Fig. 3. A circuit applied for transient measurement.

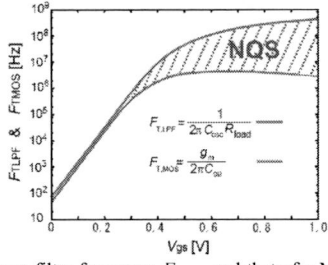

Fig. 4. The low-pass filter frequency $F_{\text{T,LPF}}$ and that of a MOSFET $F_{\text{T,MOS}}$, used as the DUT shown in Fig. 3, as a function of the input voltage V_{gs}.

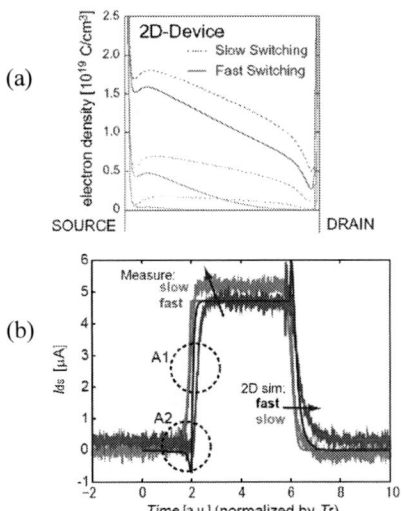

Fig. 5. (a) 2D-device simulation results of the electron-density distribution along the channel during switching-on condition, (b) measured transient currents. Comparisons are shown for two different switching speeds.

III. FUTURE ASPECTS

The device development is approaching size limits, causing non-ideal device performances such as increased short-channel effects as well as fabrication variations. Nevertheless, devices for ultra-low power applications are one of the forces driving the development forward. Spintronics is an example. Beyond the potential-distribution modeling approach, compact modeling requires to understand the origin of the microscopic features in addition to averaged macroscopic features. Such investigations might bring an idea whether a new device configuration leads to any advantage over the present status.

A. Compact Modeling for Individual Carrier Dynamics

One important circuit-performance degradation is caused by non-controllable carrier dynamics, which are observed as noise. To predict the noise characteristics accurately, it has been demonstrated that the carrier dynamics within the MOSFET channel is important [6]. Fig. 6 shows $1/f$ noise measurements for different devices and channel lengths. By approximating that trap sites are mostly located at the Fermi energy and the trap density is homogeneously distributed within the oxide with the attenuation coefficient, the $1/f$ feature can be analytically derived. Previously, modeling was done for the $1/f$ characteristics with averaged measurements (see Fig. 6a), to predict circuit performance accurately. However, Fig. 6b shows that averaging might not results in a $1/f$ characteristics for advanced technologies. The reason for deviating from the $1/f$ characteristics is explained by several specific trap states, which dominate the noise characteristics as depicted in Fig. 7a. An analytical description has been developed by considering the carrier-density distribution within the oxide explicitly (see Fig. 7b) [7].

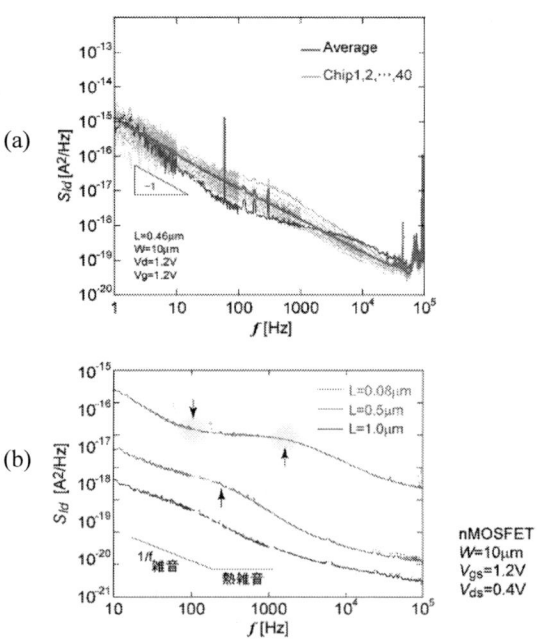

Figure 6. (a) Measured noise characteristics of 40 relatively long-channel MOSFETs and their average showing the 1/f noise characteristics. (b) Measured noise characteristics for different channel lengths.

978-1-7281-0941-1/19 $31.00 © 2019 IEEE   320

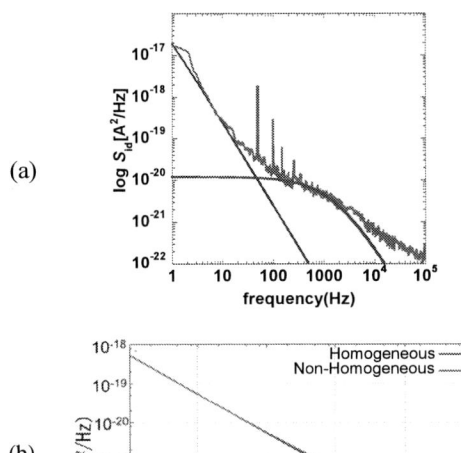

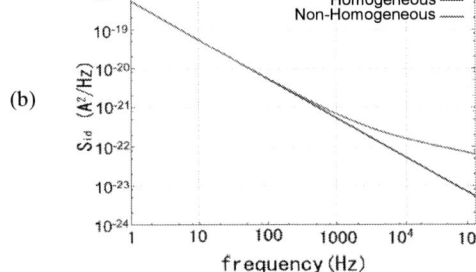

Figure 7. (a) Measured non 1/f noise intensity is attributed to the specific Lorentzian noise, whereas the tail for high frequency is explained by the non-homogeneous trap density within the insulator as shown in (b) [7].

B. Circuit Aging Prediction

Since the signals propagating within circuits are getting weaker and weaker due to the requirements for low-voltage application with extremely scaled-down device size, risks of circuit malfunction are increasing. Carrier trapping is one of the reasons for the malfunction. The trap-density increase in devices is exactly the origin of the circuit aging. Therefore, modeling the trap-density increase during circuit operation is the most essential task to be done. It is known that the carrier trapping is caused by hot carriers, where carriers must have sufficient energy to be trapped. The main difficulty in investigating circuitry aging is that the predicted trap density can never be verified experimentally. The best way to overcome this problem is to develop a sufficiently reliable methodology for the prediction.

The reliability of the simulation results is usually confirmed with accurate reproduction of DC measurements under enhanced stress condition. It is known that the trap-density increase is due to increased crystal defects induced by hot carriers. To extract the trap-density increase accurately, the $1/f$ noise measurement is applied, which can be reduced only to the trap-density increase without additional carrier characteristics as demonstrated in Fig. 8 [8]. Since the noise measurement is usually not included in the conventional measurement package for characterizing device features, the conventional I-V measurements, as shown in Fig. 8a, are applied for the trap-density extraction. The most dominant aging feature, namely the aging in the subthreshold slope, suggests that mainly the deep-level trap-density increase is enhanced. In this way, the conventional I-V measurements are sufficient to extract the trap-density increase, as shown in Fig. 8b. The extraction result concludes that the trap-density increase can be modeled by the substrate-current integration during stress, where the substrate current is a measure for the occurring amount of the hot carriers. Fig. 9 compares the $1/f$

noise measurements after stress application for two different devices. Obviously different aging features are observed. After enhanced-stress applications for characterizing the device-aging properties, it is seen that measured noise characteristics give rise to an additional feature similar to the Lorentzian noise shown in Fig. 7. More experimental investigation is required to clarify the origin of the different noise feature.

A circuit is an aggregation of different devices, and each transistor has microscopically its individual features, as can be expected from Fig. 6. Till now such a condition is usually treated in the circuit community by a worst/best-case analysis, using boundary definitions. Precise investigations of individual device-parameter variations, such as for the channel length, have been considered by applying the Monte-Carlo method. As shown in Fig. 1, a MOSFET is mainly modeled by a current and capacitances during the initial phase of the development. Circuit simulation is further proceeding to include the non-ideal features of the device characteristics. One investigation possibility is combining the newly detected non-ideality with the Monte Carlo method with different weights for different devices in circuit. For the next generation modeling era more microscopic variation such as the trap density must be considered. Here, another question arises, namely whether the device characteristics measured under the DC condition can predict the circuit performance accurately for e.g. such trapping event. It has been argued that the trap-density extraction under the DC condition cannot be directly applied for the switching investigation [9]. The reason is that a trapping event requires time to accomplish completion, namely the necessity of the trap-time constant. To overcome this problem, the switching measurement has to always accompany the DC measurement.

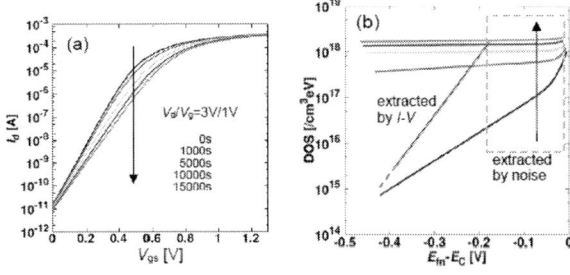

Fig. 8. (a) Measured I-V characteristics after different stress duration. (b) Extracted density of state (DOS) with measured noise intensity increase.

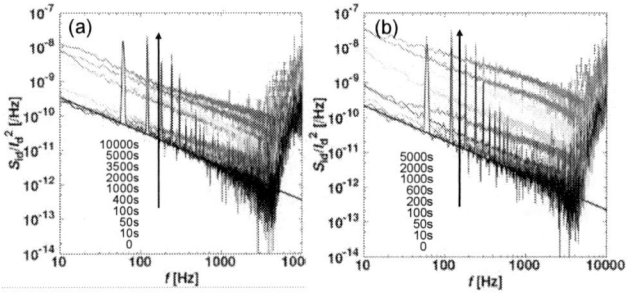

Fig. 9. Measured normalized noise intensity as a function of frequency under different stress duration, (a) with smaller stress condition and (b) with large stress condition.

The conventional circuit simulation for aging consists of two steps. First, the circuit simulation is performed without aging and is devoted only to estimate the stress during the circuit operation. With the estimated stress, aged model parameters are calculated. The second simulation then performs the circuit-aging simulation with the determined aged model parameters. We performed a so-called one-stop simulation for dynamic aging of the small circuit shown in Fig. 10a. The circuit iteration process is used for extracting the trap density accurately at the same time (see Fig. 10b). The resulting trap-density increase during circuit operation is depicted in Fig. 10c. However, this simulation method is possible only for relatively small circuits. If the circuit scale is large, such a simulation takes plenty of simulation time. Consequently, though this method removes the non-consistency of the conventional method, our one-stop method is hardly applicable for real circuit simulation. To overcome this problem, a simple and accurate description for aging of the device characteristics is required. However, the advantage of such macroscopic modeling approach is that the description is so simple that the existing simulation tool can be easily adopted to perform the simulation. For the purpose a reliable bridge between the microscopic feature and the macroscopic description is needed. Fig. 11 shows a measurement conventionally done by the circuit community as an example.

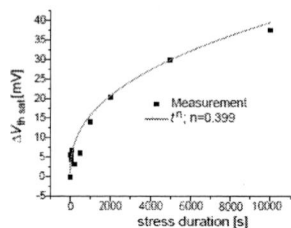

Fig. 11. Conventionally measured aging characteristics for characterizing aging features.

IV. CONCLUSION

The presented paper gives an overview of important aspects of the compact-modeling history. The future tasks in conjunction with microscopic device features are discussed. The main insight is the necessity to keep the Poisson equation being used. It has been demonstrated also that investigations of transient characteristics provide further information about the microscopic features of device characteristics.

ACKNOWLEDGMENT

The author would like to thank for continuous support though collaborations by STARC, ams, Broadcom, Infineon, Mie-Fujitsu, Panasonic, Ricoh, Renesas, TowerJazz, Toshiba, Sony, and all CMC member companies.

REFERENCES

[1] J. Watts, C. McAndrew, C. Enz, C. Galup-Montoro, G. Gildenblat, C. Hu, R. van Langevelde, M. Miura-Mattausch, R. Rios, C.-T. Sah, "Advanced compact models for MOSFETs," Modeling and Simulation of Microsystems, pp. 3-12, Anaheim, May 2005.

[2] J. E. Meyer, "MOS models and circuit simulation," RCA Rev., vol. 32, pp. 42-63, 1971.

[3] C.-T. Sah, "Characteristics of the metal-oxide-semiconductor transistors," IEEE Trans. Electron Devices, vol. 11, no. 7, pp. 324-345, 1964.

[4] M. Miura-Mattausch, N. Sadachika, D. Navarro, G. Suzuki, Y. Takeda, M. Miyake, T. Warabino, Y. Mizukane, R. Inagaki, T. Ezaki, H. J. Mattausch, T. Ohguro, T. Iizuka, M. Taguchi, S. Kumashiro, S. Miyamoto, "HiSIM2: Advanced MOSFET model valid for RF circuit simulation," IEEE Trans. Electron Devices, vol. 53, pp. 1994-2007, September 2006.

[5] D. Hori, Master Thesis, "For predicting ringoscillator oscillation frequency under low voltage operation," Hiroshima University, Higashi-Hiroshima, March 2010.

[6] S. Matsumoto, H. Ueno, S. Hosokawa, T. Kitamura, M. Miura-Mattausch, H. J. Mattausch, T. Ohguruo, S. Kumashiro, T. Yamaguchi, K. Yamashita, N. Nakayama, "1/f-noise characteristics in 100nm-MOSFETs and its modeling for circuit simulation," IEICE Trans. Electron., vol. E88-C, pp. 247-254, February 2005.

[7] T. Nakahagi, Master Thesis, "Analysis and modeling of low noise charactersitcs in Poly-Si TFT," Hiroshima University, Higashi-Hiroshima, March 2013.

[8] M. Miura-Mattausch, H. Miyamoto, H. Kikuchihara, T. K. Maiti, N. Rohbani, D. Navarro, H. J. Mattausch, "Compact modeling of dynamic trap density evolution for predicting circuit-performance aging," Microelectronics Reliability, vol. 80, pp. 164-175, 2018.

[9] Y. Tanimoto, A. Saito, K. Matsuura, H. Kikuchihara, H. J. Mattausch, M. Miura-Mattausch, and N. Kawano, "Power-loss prediction of high-voltage SiC-MOSFET circuits with compact model including carrier-trap influence," IEEE Trans. Power Electronics, vol. 31, pp. 4509-4516, June 2016.

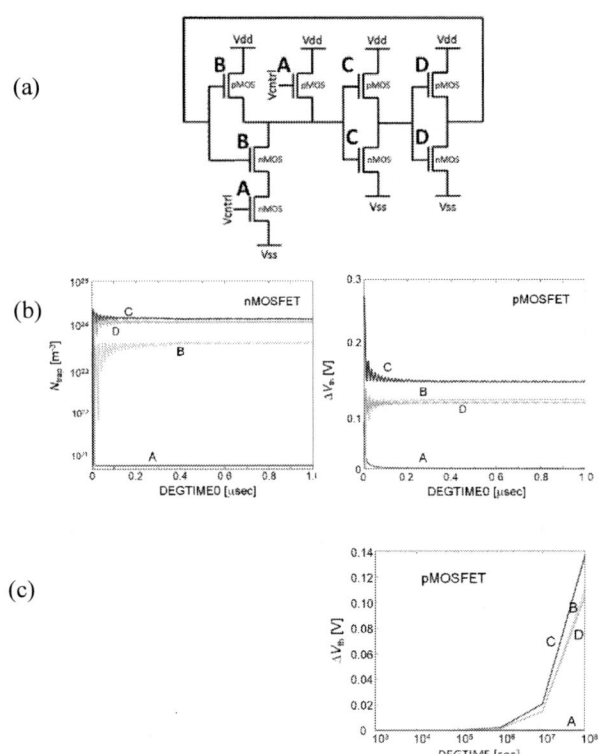

Fig. 10. (a) Studied ring-oscillator (upper left), (b) convergence features of the trap density N_{trap} in the left side the four nMOSFETs and ΔV_{th} in the right side the four pMOSFETs of the ring-oscillator as a function of circuit simulation time DEGTIME0 after the circuit operation of 10^8 sec, (c) N_{trap} increase of each MOSFET within the circuit as a function of circuit operation time, where pMOSFETs are modeled with the NBIT model shown [8].

Modeling and Simulation of Atomic Layer Deposition

Lado Filipovic

Institute for Microelectronics, Technische Universität Wien, Gußhausstraße 27-29/E360, 1040 Vienna, Austria

filipovic@iue.tuwien.ac.at

Abstract—**Two models for ALD of TiN and TiO$_2$ are incorporated in an in-house level set based topography simulator, ViennaTS. While the models are based on 1D surface kinetics, here they are extended to handle 2D and 3D geometries by applying single particle Monte Carlo ray tracing. The particle flux and sticking coefficients are used to calibrate the surface adsorption of precursors and ultimately to calculate the resulting surface velocity. The TiO$_2$ ALD model is based on the use of TTIP and H$_2$O precursors and includes all surface kinetics taking place during deposition. In contract, the model for the deposition of TiN is somewhat simplified by ignoring the purge steps which are introduced after surface exposure to either precursor. The simplified model is then applied to reproduce experimental results from plasma enhanced ALD process for TiN deposition from TDMAT and H$_2$-N$_2$ plasma precursors.**

Index Terms—**Atomic layer deposition, modeling and simulation, level set, TiN, TiO$_2$**

I. INTRODUCTION

TSMC has recently reported on vertically stacked lateral nanowire FETs where the high-k dielectric stack allows for a sub-1nm equivalent oxide thickness (EOT) and a sub-2nm physical thickness [1]. This was achieved using atomic layer deposition (ALD), the emergence of which was deemed essential to allow for the deposition of sub-2nm barrier layers [2]. ALD has been adopted aggressively at newer technology nodes and this trend is expected to continue well into future technology nodes at least until 2033, when a 0.5nm thin TiN barrier metal is foreseen to be used to limit copper (Cu) vacancy diffusion into the adjacent dielectric [3]. To ensure conformal deposition of thin TiN layers, alternatives to chemical vapor deposition (CVD) such as ALD are essential. In this manuscript, we describe two models for ALD and their implementation in the Level Set (LS) based feature-scale process and topography simulator ViennaTS [4].

ViennaTS is a Level Set framework which uses ray tracing and Monte Carlo methods to simulate the topography evolution of a surface after it is exposed to a given chemical reaction in a deposition/etching chamber. The simulator addresses the feature scale region of a wafer and for all models implemented thus far, the reactions on the surface were assumed to follow the Knudsen Law, meaning the diffusion length is smaller than the mean free path of the involved particles. Using ray tracing, the surface coverages for all involved particles

The research leading to these results has received funding from the European Union's Horizon 2020 research and innovation programme under grant agreement No 688101 SUPERAID7.

needs to be calculated, which directly influence the surface velocities. When simulating classical processes such as CVD the coverages are assumed to reach a steady state prior to applying a surface velocity. This cannot be assumed for ALD since the steady state will always be full coverage from a precursor. Therefore, in order to perform transient simulations of ALD, time discretization must be performed to track surface coverages during each step in the ALD process.

II. ATOMIC LAYER DEPOSITION

ALD is a cyclical process which proceeds with the sequential use of gas phase processes with two chemical precursors, as depicted in Fig. 1. In the first step, an atom or molecule A adsorbs on the surface while, during the second step, the chemical reaction between the adsorbed species and the gas atom or molecule B in the chamber results in the formation of one monolayer (ML) of a desired film AB. In the next cycle, species A adsorbs on the newly formed film AB in the first step, while in the subsequent step, species B once again reacts with the adsorbed A to form the second monolayer of film AB. This continues until the desired thickness or number of monolayers of AB is reached. In this manuscript, two models are addressed; the first one includes the simultaneous adsorption and desorption of precursors as well as pyrolytic and hyrolytic decomposition [5]. The second model is an alternative simplified implementation which ignores decomposition, thereby also avoiding the need to simulate the purge steps taking place after each gas phase during ALD [6].

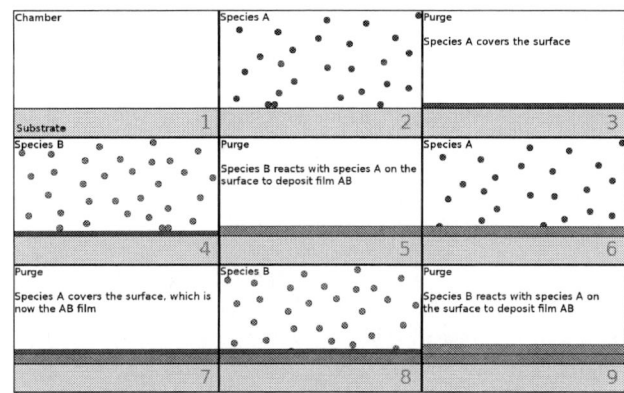

Fig. 1. ALD implemented with sequential use of gas phase processes with two chemical precursors (A and B) to form two monolayers of film AB.

978-1-7281-0941-1/19 $31.00 © 2019 IEEE

III. ALD MODELS

Two models for ALD have been implemented in an in-house topography simulator ViennaTS. The advantage of the models' implementation therein is its integration with other simulation steps in a full CMOS fabrication sequence and the ability to use them to analyze the influence of several parameters on two-dimensional (2D) and three-dimensional (3D) geometries. In the models, the surface coverages of the precursors are tracked in time and converted to a surface velocity at each step.

A. TiO_2-ALD using TTIP and H_2O

A model for the deposition of titanium dioxide (TiO_2) using cycles of precursor titanium tetraisopropoxide (TTIP) and water (H_2O) has been presented by [5] and includes the adsorption and desorption of precursors as well as pyrolytic and hydrolytic decomposition. TiO_2 has several potential applications due to its mixed ionic-electronic conducting properties [7] and is currently implemented in the memristor [8]. The surface kinetics taking place during the TTIP and H_2O precursor stages are depicted in Fig. 2. The adsorption of species depends on the impinging rates (J_X) and sticking coefficient (S_X). The desorption processes include hydrolytic decomposition at a rate of $k_{hydrolysis}$ and in the case of TTIP, an additional pyrolytic decomposition process takes place, with a rate of $k_{pyrolysis}$. In the above discussion X refers to $TTIP$ or H_2O, depending on which step in the cycle is being observed. During the purge steps, desorption, hydrolysis, and pyrolysis continues to take place [5].

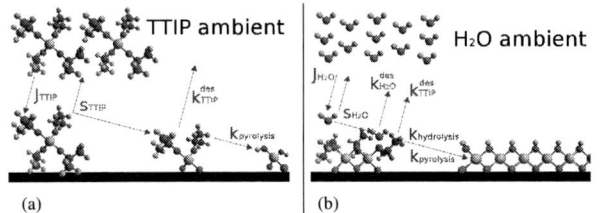

Fig. 2. Schematic representation of the surface kinetics during (a) the TTIP step and the (b) H_2O step.

The equations which describe the TTIP and H_2O coverages θ_X during their respective steps in the ALD cycle are given in equations (1) and (2), respectively. All reactions k_X are expressed by an Arrhenius law and for the exact parameter values which correspond to measurements, refer to [5]. The following equations shows the calculation of the change in TTIP and H_2O coverage with time during each respective cycle; θ_X^0 represent the total available surface coverage sites.
TTIP step:

$$\frac{d\theta_{TTIP}}{dt} = J_{TTIP} S_{TTIP} \left(1 - \frac{\theta_{TTIP}}{\theta_{TTIP}^0}\right) - k_{TTIP}^{des} \theta_{TTIP} \quad (1)$$
$$- k_{hydrol}\theta_{TTIP}\theta_{H_2O} - k_{pyrol}\theta_{TTIP}^2$$

H_2O step:

$$\frac{d\theta_{H_2O}}{dt} = J_{H_2O} S_{H_2O} \left(1 - \frac{\theta_{H_2O}}{\theta_{H_2O}^0}\right) - k_{H_2O}^{des} \theta_{H_2O} \quad (2)$$
$$- k_{hydrol}\theta_{TTIP}\theta_{H_2O}$$

Since the decomposition processes also take place during the purge steps, the surface coverages can change during this time. Also, surface deposition takes place only when both TTIP and H_2O are both covering a section of the surface and as the film is deposited, the TTIP coverage is reduced, as shown in the single-cycle simulation in Fig. 3. The growth rate applied in the topography simulator depends on the thickness of a single molecular layer of TiO_2 (ML_{TiO_2}) and follows

$$Rate = ML_{TiO_2} \left(\theta_{TTIP}\theta_{H_2O}\right). \quad (3)$$

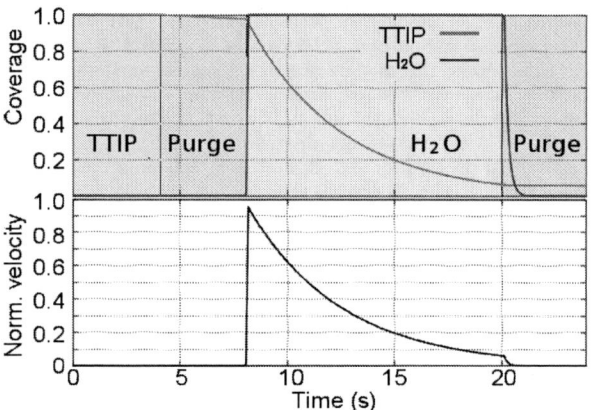

Fig. 3. Single cycle of the TiO_2 ALD process showing the coverages of TTIP and H_2O and the resulting surface velocity.

B. TiN-ALD using TDMAT and NH_3

In this section the ALD model for TiN deposition is described and a sample simulation is performed on a trench geometry. The purge steps are ignored since their influence on the surface coverages are minimal. The core of the model is the one-dimensional (1D) surface kinetics description of TiN-ALD using tetrakis(dimethylamido)titanium (TDMAT) and ammonia (NH_3) given in [6]. Since the base model is 1D, we improve on the implementation by extending it to reflect the influence of the sticking coefficient (s) of an atom or molecule on the surface coverage using Monte Carlo (MC) ray tracing in a LS framework. In this work we enable ALD simulation of a generic film AB from a cycle which includes adsorption of species A followed by species B. The universality of the model is demonstrated here by using it to simulate TiN plasma enhanced ALD (PE-ALD) by calibrating four adsorption parameters.

The TDMAT molecule, adsorbed on top of the substrate - or on the TiN film - in the first step of the ALD cycle reacts with the NH_3 from the second step, resulting in the deposition of TiN. Furthermore, during the second step, NH_3 can be adsorbed on the TiN, which is either the TiN formed by the reaction or the TiN which becomes an outer surface again after the adsorbed TDMAT has been removed by reaction with NH_3, shown in Fig. 4. These NH_3 molecules can serve to form TiN with the TDMAT from the first step in the next cycle. Therefore, it is possible that more than one monolayer (ML)

978-1-7281-0941-1/19 $31.00 © 2019 IEEE

of TiN is deposited in a single cycle [6]. In the simulation, this means that different coverages ($\theta_{X/Y}$) must be tracked consistently, as follows:

TDMAT step:

$$\frac{d\theta_{TDMAT/NH_3}}{dt} = C_{TDMAT/NH_3}$$
$$\times \left(\theta^{NH_3} - \theta_{TDMAT/NH_3}\right)^{n_{TDMAT}}$$

$$\frac{d\theta_{TDMAT/TiN}}{dt} = C_{TDMAT/TiN}$$
$$\times \left(\theta^{TiN} + \theta_{TDMAT/NH_3} - \theta_{TDMAT/TiN}\right)^{n_{TDMAT}} \quad (4)$$

NH$_3$ step:

$$\frac{d\theta_{NH_3/TDMAT}}{dt} = C_{NH_3/TDMAT}$$
$$\times \left(\theta^{NH_3} - \theta_{NH_3/TDMAT}\right)^{n_{NH_3}}$$

$$\frac{d\theta_{NH_3/TiN}}{dt} = C_{NH_3/TiN}$$
$$\times \left(\theta^{TiN} + \theta_{NH_3/TDMAT} - \theta_{NH_3/TiN}\right)^{n_{NH_3}} \quad (5)$$

In equations (4) and (5) above, $\theta_{X/Y}$ represents the adsorption coverage of species X on the film in the vicinity of adsorbed species Y and $\theta_{X/TiN}$ is the adsorption coverage of molecule X on the exposed TiN surface; θ^X is the coverage of species X at the end of the previous cycle; $C_{X/Y}$ is the adsorption constant for the adsorption of X onto Y; n_X is the adsorption order of the X molecule. Species X and Y can correspond to either the TDMAT or NH$_3$ molecule or, in a universal manner, to the molecule or atom of any reacting species A or B.

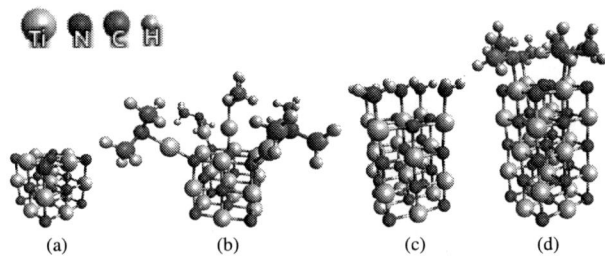

Fig. 4. TiN-ALD cycle with (a) the TiN unit cell, (b) after a single TDMAT step, (c) after a subsequent NH$_3$ step, and (d) after a 2nd TDMAT step.

1) Surface Velocity Calculation: While solving the coverage equations during CVD or plasma etching simulations, it is appropriate to assume that surface reactions are very fast, reaching a steady state before any deposition and etching takes place. However, this is not true of ALD, where coverages must be solved in time. Also, the set of coverage equations for both ALD steps, relating to the adsorption of molecules on the exposed TiN surface cannot be solved analytically and therefore approximations are used in the simulation. In order to account for this, time discretization for the calculation of surface coverages is introduced in ViennaTS; this requires a calibration between the time step used for the coverage calculation and the time step used for the LS surface evolution. Time discretization is implemented dynamically and the size of each step depends on the maximum allowed error, as set in the parameters file during runtime.

Another concern with this ALD process is that when TD-MAT reacts with adsorbed NH$_3$ to form TiN, there is a volume reduction taking place. The NH$_3$ molecule is much larger than a TiN molecule, when in the gas phase. Although the adsorbed reactants are not in the same chemical states as those in the gas phase, the ratio of the difference in molecular size is considered to evaluate a parameter m_X which represents the volume reduction and is used for the calculation of the maximum growth rate [6]. In the case of the TiN-ALD process, $m_{NH_3} = 4$ and $m_{TDMAT} = 1$. The TiN film thickness increase in a single cycle, in terms of ML/cycle, is therefore calculated using

$$Rate\,(ML/cycle) = \frac{\theta_{TDMAT/NH_3}}{m_{NH_3}} + \frac{\theta_{NH_3/TDMAT}}{m_{TDMAT}}, \quad (6)$$

where m_X refers to the reduction ratio of an occupied surface area when an adsorbed species X reacts with a gas molecule Y to form TiN. To determine the surface velocity, we note the thickness of a ML of TiN $ML_{TiN} = 0.4$ nm and that deposition occurs only when there is a change in coverage $\Delta\theta_{X/Y}$ during a time step Δt. The surface velocity is therefore determined by the change in the coverage per time step and is calculated using

$$V = \frac{ML_{TiN}}{\Delta t}\left(\frac{\Delta\theta_{TDMAT/NH_3}}{m_{NH_3}} + \frac{\Delta\theta_{NH_3/TDMAT}}{m_{TDMAT}}\right). \quad (7)$$

Note that the thickness per cycle $Rate(ML/cycle)$ calculated from equation (6) is obtained by integrating the surface velocity over the time of one full cycle. In Fig. 5 we plot the relationship between the surface velocity and respective coverages when using 1.5s and 3s step sizes for the TDMAT and NH$_3$ steps, respectively. It is evident that a surface velocity is only applied when the coverages are changing.

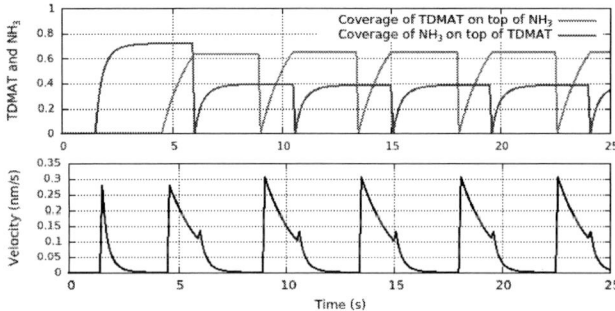

Fig. 5. Coverages (top) and the resulting surface velocity (bottom) when the TDMAT and NH$_3$ steps are set to 1.5s and 3s, respectively.

2) Model Extensions: In order to extend the above model to 2D and 3D geometries we introduce MC ray tracing, with a flux (J) and sticking coefficient (s) assigned to each involved precursor. This allows to further generalize the model to any precipitate and desired film. The effects of the surface flux are included by setting the adsorption constant relative to the distribution of the incoming MC particles using

$$C_{X/Y} = J_X \cdot s_X. \quad (8)$$

Based on the experimental results from [6] we find that s for TDMAT and NH_3 is in the order of approximately 10^{-5}, suggesting a highly conformal deposition. This means that the flux will see only a 13% drop at the bottom of a 3D cylindrical hole with aspect ratio (AR) of 100, when compared to the flux at the wafer top. However, this is not the case for all ALD processes, described in [9], where it is shown that, depending on the ALD precursors, s can vary by several orders of magnitude. In Fig. 6 we show a sample simulation on a trench with aspect ratio 2.5 and sticking coefficients $s_1 = 0.1$ and $s_2 = 0.2$, while varying the ALD cycle step times and the number of cycles. As an example, these values of s are in the range observed for $Al(CH_3)_3$ and O plasma precursors for the deposition of Al_2O_3 [9].

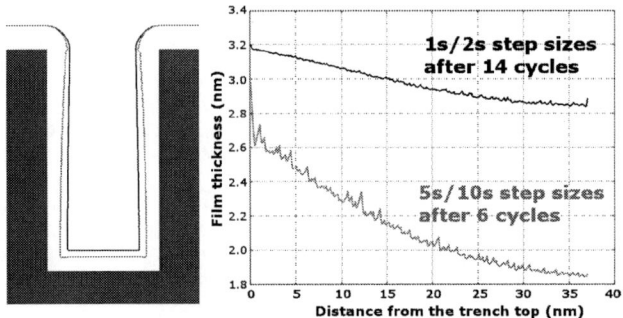

Fig. 6. The deposited film surface after a simulation of several cycles for a generic ALD model on a trench geometry with 2.5 AR using sticking probabilities 0.1 and 0.2 for species A and B, respectively.

C. Plasma Enhanced ALD

The TiN-ALD model described in the previous section and implemented in a level set framework was further applied to a low-temperature PE-ALD process to simulate the deposition of a thin TiN film using TDMAT and H_2-N_2 plasma as precursors using experiments from Caubet et al. [10]. The TiN film growth takes place almost exclusively during the plasma step with H_2-N_2 and the deposition rate is much faster than observed in the classical ALD. According to [10] no re-emission takes place; therefore, calculating the coverages of TDMAT (9) and H_2-N_2 (10) simplifies to:

TDMAT step:

$$\frac{d\theta_{TDMAT/H_2-N_2}}{dt} = C_{TDMAT/H_2-N_2} \times \left(\theta^{H_2/N_2} - \theta_{TDMAT/H_2-N_2}\right)^{n_{TDMAT}} \quad (9)$$

H_2-N_2 step:

$$\frac{d\theta_{H_2-N_2/TDMAT}}{dt} = C_{H_2-N_2/TDMAT} \times \left(\theta^{TDMAT} - \theta_{H_2-N_2/TDMAT}\right)^{n_{H_2-N_2}} \quad (10)$$

Since the deposition of TiN takes place exclusively during the H_2-N_2 plasma step, the TDMAT serves to prepare the surface with the appropriate coverages for the subsequent plasma deposition step. Furthermore, since N reacts directly

with exposed Ti and not with NH_3, no volume reduction needs consideration and m_X can be ignored. The surface rate is then

$$Rate\,(ML/cycle) = \theta_{H_2-N_2/TDMAT} \quad (11)$$

and the resulting surface velocity is calculated using

$$V = \frac{ML_{TiN}}{\Delta t}\left(\Delta\theta_{H_2-N_2/TDMAT}\right). \quad (12)$$

Using the above equations, the model the experimental results for a PE-ALD process described in [10] were replicated by adjusting the adsorption parameters, given on the right of Fig. 7, where the comparison between the simulated and experimental coverage is shown. This step determines how much TDMAT is adsorbed at the surface to prepare for the subsequent bonding with N atoms during the plasma step. Since the coverage during the plasma step is very fast, the TDMAT is the critical parameter for fast deposition.

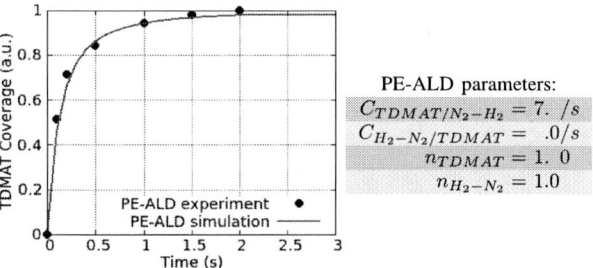

Fig. 7. Comparison between the experimental measurements from [10] and the PE-ALD model, with the adsorption parameters shown on the right.

IV. CONCLUSION

Two models for atomic layer deposition were presented and incorporated in a level set based topography simulator, ViennaTS. The models are for the deposition of TiO_2 and TiN from different precursors and are both based on one-dimensional surface kinetics. However, they have been extended to 2D and 3D geometries by introducing MC ray tracing and the influence of particle flux and sticking coefficients on the surface adsorption and resulting deposition velocity. The ALD-TiO_2 model is quite complex and includes adsorption and desorption of precursors as well as pyrolytic and hydrolytic decomposition. An alternative simplified universal model for TiN-ALD was also introduced in the level set environment. With these models we have reproduced experimental results from an ALD-TiO_2 and ALD-TiN processes as well as a PE-ALD process from TDMAT and H_2-N_2 plasma precursors.

REFERENCES

[1] M. van Dal et al. in *IEEE IEDM*, 2018, 492–495.
[2] "ITRS," IRC, Tech. Rep., 2013.
[3] "IRDS," IEEE, Tech. Rep., 2017.
[4] *ViennaTS User Guide*, Vienna, Austria: TU Wien, 2019.
[5] M. Reinke et al., *J. Phys. Chem. C*, 119(50), 27 965–27 971, 2015.
[6] J.-W. Lim et al., *J. Appl. Phys.*, 87(9), 4632–4634, 2000.
[7] H. Lim et al., *Nanotechnology*, 24(38), 384005(8pp), 2013.
[8] X. Yan et al., *Adv. Funct. Mater.*, 28(1), 1705320(9), 2018.
[9] H. Knoops et al., *J. Electrochem. Soc.*, 157(12), G241–G249, 2010.
[10] P. Caubet et al., *J. Electrochem. Soc.*, 155(8), H625–H632, 2008.

978-1-7281-0941-1/19 $31.00 © 2019 IEEE

Novel Numerical Dissipation Scheme for Level-Set Based Anisotropic Etching Simulations

Alexander Toifl
Christian Doppler Laboratory
for High Performance TCAD
Institute for Microelectronics, TU Wien
Wien, Austria
toifl@iue.tuwien.ac.at

Michael Quell
Christian Doppler Laboratory
for High Performance TCAD
Institute for Microelectronics, TU Wien
Wien, Austria
quell@iue.tuwien.ac.at

Andreas Hössinger
Silvaco Europe Ltd.
Cambridge, United Kingdom
andreas.hoessinger@silvaco.com

Artem Babayan
Silvaco Europe Ltd.
Cambridge, United Kingdom
artem.babayan@silvaco.com

Siegfried Selberherr
Institute for Microelectronics, TU Wien
Wien, Austria
selberherr@tuwien.ac.at

Josef Weinbub
Christian Doppler Laboratory
for High Performance TCAD
Institute for Microelectronics,TU Wien
Wien, Austria
weinbub@iue.tuwien.ac.at

Abstract—**We propose a novel dissipation scheme for level-set based wet etching simulations. The scheme enables modeling of the temporal evolution of the etch profile during anisotropic wet etching processes and is based on the local geometry and the crystallographic direction-dependent etch rate. We implemented the scheme into Silvaco's Victory Process simulator which is utilized in this work to simulate the fabrication of source/drain cavities for sub-28 nm strained metal-oxide-semiconductor field-effect transistors. Our results show excellent agreement with experimental data. In particular, the main cavity-related design variables are accurately predicted.**

I. INTRODUCTION

Anisotropic etching of silicon (Si) with wet etchants (e.g., potassium hydroxide (KOH) and tetramethylammonium hydroxide (TMAH)) is an important processing technique which utilizes the crystalline nature of the material. While KOH etching is mainly known for its application in the production of micro-electro-mechanical-systems (MEMS) [1], TMAH etching plays an important role for embedded silicon germanium (e-SiGe) in source/drain (S/D) engineered sub-28 nm node metal-oxide-semiconductor field-effect transistors (MOSFETs) [2], [3]. By employing a combination of dry and wet etching a S/D cavity formed by the characteristic {111}-planes can be produced. The exact geometry of the resulting sigma-shaped cavity determines the uniaxial strain in the MOSFET channel after epitaxial growth of SiGe [4]. Thus, it is very important to control the critical design variables, i.e., tip depth, channel-cavity distance, and cavity depth [2].

In this work we present a dissipation scheme for level-set based simulations of anisotropic wet etching, which allows to predict the temporal evolution during the etching step and the final profile with high accuracy. The proposed scheme was implemented into Silvaco's Victory Process simulator [5], which we utilize to simulate the process steps required for sigma-shaped cavities for S/D engineering.

II. ETCH RATE MODELING WITHIN THE LEVEL-SET FRAMEWORK

We employ the level-set method [6], [7] [8], where the wafer surface is described by the zero level-set of the function ϕ and the time evolution is determined by the level-set equation

$$\frac{\partial \phi}{\partial t} + H(\nabla \phi) = 0. \tag{1}$$

The level-set equation assumes the form of a Hamiltonian-Jacobi equation with the corresponding Hamiltonian $H = V(\nabla\phi)|\nabla\phi|$. The speed function V is constructed to reflect the highly anisotropic etch rates. We assume constant process temperature and neglect the influence of reactant transport on the etching kinetics. Under these conditions, the etch rate depends on the crystallographic planes which are exposed to the surface resulting in the self-limited etch profiles observed in experiments [1]. Consequently, we model V to assume the form

$$V = V(n^x, n^y, n^z) = V(\nabla\phi), \tag{2}$$

which is a function of the components n^l, $l \in \{x, y, z\}$ of the level-set normal vector $n = \nabla\phi/|\nabla\phi|$. The etch rates along the main crystal directions are well known from experiments. In order to define a continuous speed function, we use linear interpolation between a set of crystal directions, e.g., $\langle 100 \rangle$, $\langle 110 \rangle$, and $\langle 111 \rangle$ with their associated etch rates R_{100}, R_{110}, and R_{111}, respectively, while taking the cubic symmetry of silicon into account [9], [10]. Fig. 1 illustrates that the resulting speed function is spatially strongly varying. As a consequence, the resulting Hamiltonian is non-convex, which detrimentally impacts the stability of the numerical solution of the level-set equation [8].

978-1-7281-0941-1/19 $31.00 © 2019 IEEE

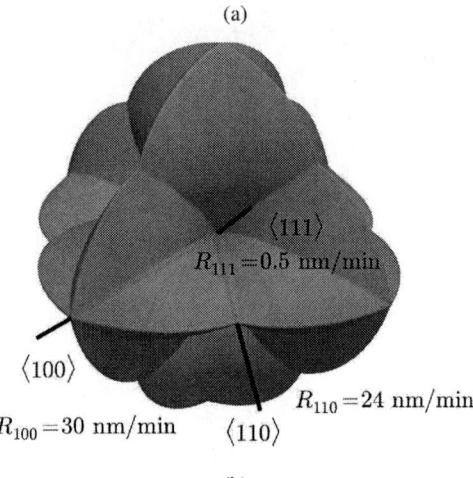

(a)

$R_{111} = 0.5$ nm/min

$\langle 111 \rangle$

$\langle 100 \rangle$

$R_{110} = 24$ nm/min

$R_{100} = 30$ nm/min

$\langle 110 \rangle$

(b)

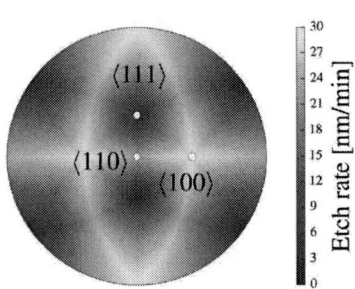

Fig. 1: The etch rate distribution function which originates from a linear interpolation between silicon's main crystallographic directions $\langle 100 \rangle$, $\langle 110 \rangle$, and $\langle 111 \rangle$ is visualized in a **(a)** three dimensional and **(b)** stereographic plot. The highly anisotropic etch rate distribution is typical for wet etchants [11] and is presented here for a TMAH solution (40 °C, 2.38 wt%) as given by *Qin et al.* [3].

A non-convex Hamiltonian is problematic for wet etching simulations, because the front is typically characterized by sharp corners, resulting in regions of high curvature. In particular, a non-convex speed function $V(n)$ (illustrated in the inset) assigns the level-set grid points along a high curvature front strongly varying speed values, as depicted in Fig. 2. The limited resolution invoked by the spatial discretization gives rise to the problem that the speed of the front between two grid points can only be estimated by some kind of combination of the values at exactly these two grid points, causing instable front propagation [12].

III. DISSIPATION SCHEME

In order to enable a stable and physically relevant numerical solution, the dissipative Lax-Friedrichs scheme [13], [15] is employed. The time integration is performed with the (first-

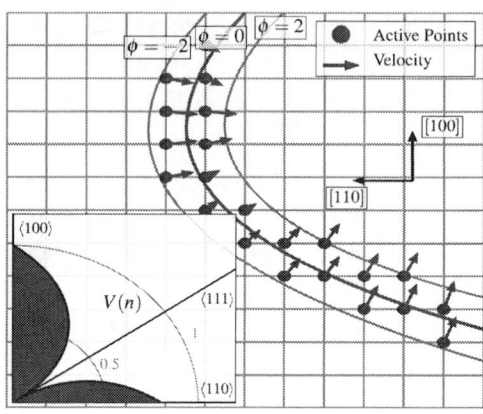

Fig. 2: The highly anisotropic etch rates in Fig. 1 give rise to spatially strongly varying velocities assigned to the level-set normals and a non-convex Hamiltonian. The level-set normals are shown for active grid points, which reside within the narrow band $|\phi| < 2.0$. The inset illustrates the speed function $V(n)$ in the depicted plane, which is characterized by a global minimum along $\langle 111 \rangle$ directions and two maxima.

order) Euler method. Associated with the Lax-Friedrichs-Scheme is the *numerical* Hamiltonian $\hat{H}$

$$\hat{H} = H\left(\frac{\phi_l^- + \phi_l^+}{2}\right) - \sum_l \alpha^l \left(\frac{\phi_l^+ - \phi_l^-}{2}\right), \qquad (3)$$

where ϕ_l^-, ϕ_l^+, $l \in \{x, y, z\}$ denote backward and forward differences of ϕ with respect to l. The Lax-Friedrichs scheme critically depends on the dissipation coefficients α^l which define the numerical dissipation (viscosity) $\sum_l \alpha^l \left(\phi_l^+ - \phi_l^-\right)/2$. The numerical dissipation has to be chosen appropriately to enable a stable and consistent surface evolution and to avoid artificially rounded corners. Consequently, for the particular case of wet etching it is essential to find a trade-off between accurate undercut rates and sharp corners formed by the slowly moving crystal planes.

In contrast to former approaches [14], [15] [16] we employ a novel *local* approach based on a stencil S (Fig. 3), which considers the processed and neighboring grid points P. In particular, the stencil integrates information about the local geometry and the *nature* of $V(n^x, n^y, n^z)$.

$$\alpha^l = \max_{P \in S} \Gamma \left| \frac{\partial V}{\partial n^l} \frac{\phi_p^2 + \phi_q^2}{|\nabla\phi|^2} - \frac{\partial V}{\partial n^p} \frac{\phi_p \phi_l}{|\nabla\phi|^2} - \frac{\partial V}{\partial n^q} \frac{\phi_q \phi_l}{|\nabla\phi|^2} + V n^l \right|$$

$$l, p, q \in \{x, y, z\}, \; l \neq p \neq q \qquad (4)$$

$\Gamma > 0$ denotes an etchant-specific prefactor which is treated as a calibration parameter. $\partial V/\partial n^l$ is numerically evaluated using central differences

$$\frac{\partial V}{\partial n^x} = \frac{V(n^x + \Delta N, n^y, n^z) - V(n^x - \Delta N, n^y, n^z)}{2\Delta N} \qquad (5)$$

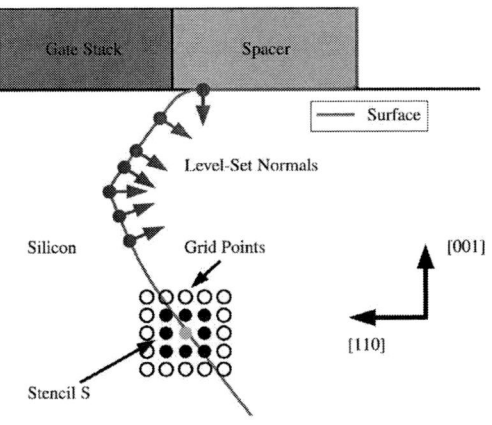

Fig. 3: The proposed dissipation scheme (4) is based on normals and the associated speed functions which are calculated for a stencil consisting of the central grid point and its immediate neighbors.

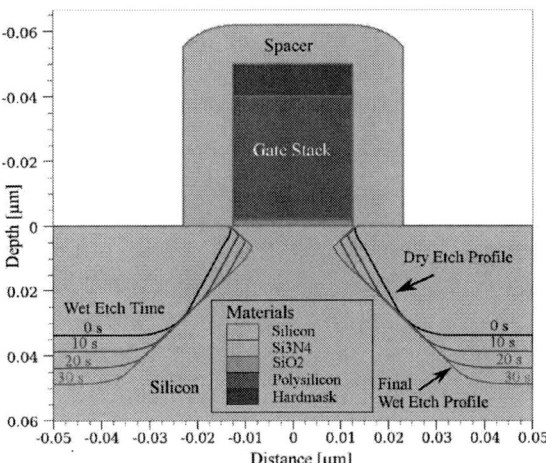

Fig. 4: Temporal evolution of the etch profile. During the self-limiting wet etching the slow-moving {111} planes form the sigma-shaped cavity.

and analogous expressions for the remaining spatial coordinates with $\Delta N = 10^{-6}$. The expression for the dissipation coefficients (4) originates from monotonicity considerations, which are based on the Lax-Friedrichs scheme being a monotone scheme [13]. Employing (4), the stencil Lax-Friedrichs scheme (3) introduces the appropriate amount of numerical dissipation at the mask-silicon interface and enables the prediction of the correct mask undercut rate.

IV. SIMULATION RESULTS AND DISCUSSION

The proposed dissipation scheme is applied to simulate the sigma-cavity process presented by Qin et al. [3], which is based on a two-step reactive ion etching (RIE) step and a sequential wet etching step. The RIE step is split into an anisotropic and isotropic sub-step in order to optimize the initial condition ('Dry Etch Profile' in Fig. 4) for a 30 s wet etch using a TMAH solution (40 °C, 2.38 wt%).

The fabrication of the sigma-cavities has been simulated with Silvaco's Victory Process simulator, including gate stack and spacer formation. In order to demonstrate the viability of the proposed dissipation scheme, we emulate a two-step RIE process (anisotropic and ideal isotropic step) to provide the initial profile depicted in Fig. 5. The proposed dissipation scheme has been utilized for the subsequent wet etching step with the speed function V (three-rate linear interpolation) defined by the rates along the high-symmetry planes $R_{100} = 30\,\mathrm{nm/min}$, $R_{110} = 24\,\mathrm{nm/min}$, and $R_{111} = 0.5\,\mathrm{nm/min}$ (Fig. 1).

The level-set equation is considered on a regular grid with a spatial resolution $\Delta x = \Delta y = 0.5\,\mathrm{nm}$. The dissipation coefficients (4) are calculated for all points residing in the narrow band, which is defined as the set of grid points with an associated level-set value satisfying $|\phi| < 2.0$. Within the narrow band all grid points hold a valid velocity value, which is strongly related to the surface velocity by extending the surface values to the respective grid points (velocity extension) [7]. Furthermore, we use $\Gamma = 3.0$ for the TMAH solution as a calibration parameter.

Fig. 4 visualizes the temporal evolution to the final profile, where the dry etch profile and the etch profile are shown after 10 s, 20 s, and 30 s. Due to the characteristically small etch rate of {111} planes, the final profile consists of two {111} planes which define a sharp corner at a certain position relative to the channel (sigma-cavity tip). The associated design parameters tip depth D and channel-cavity distance δ are depicted in Fig. 5, where the experimentally obtained etch profiles presented by Qin et al. [3] are compared to the simulation results. The wet etching simulations accurately reproduce D, δ, and the cavity depth. The dissipation scheme does not introduce non-physical rounding at the point of contact of the resulting two {111} planes.

Furthermore, we assess our dissipation scheme by performing the same process simulation steps with the elementary dissipation coefficients, which are computed purely based on the local speed function and level-set normal [7].

$$\alpha^l = \max_{P \in S} \left| V n^l \right|, \quad l \in \{x, y, z\} \qquad (6)$$

The elementary dissipation coefficients give rise to insufficient dissipation beneath the spacer and gate stack, which leads to an artificially high undercut rate and consequently to an incorrect sigma tip position. In particular, with respect to the experimental observations, the elementary scheme results in a displaced sigma tip position by 1.6 nm in x-direction and 4.2 nm in y-direction, which is significantly higher than the spatial resolution $\Delta x = 0.5\,\mathrm{nm}$ and thus not acceptable. In contrast, the proposed scheme (4) reproduces the physical undercut and is able to predict the geometry of the sigma-shaped cavity with high accuracy.

978-1-7281-0941-1/19 $31.00 © 2019 IEEE

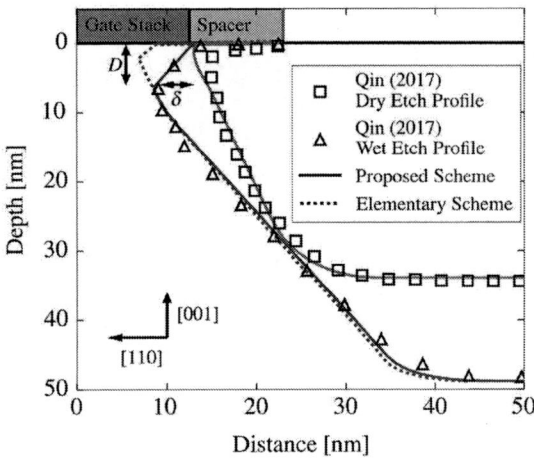

Fig. 5: The etch profiles presented by Qin *et al.* [3] are accurately reproduced using the proposed scheme (4). In contrast, the elementary scheme (6) starting from the same dry etch profile results in artificially high undercut.

V. SUMMARY

We have proposed a novel dissipation scheme for level-set based simulations of anisotropic wet etching processes. The scheme is based on the Lax-Friedrichs approach and is optimized for purely surface normal-dependent speed functions, which excellently model the etching kinetics of common etchants (e.g., TMAH and KOH). We have employed the dissipation scheme to accurately predict the sigma S/D cavity profiles for sub-28 nm node strained MOSFETs. In particular, the predicted cavity-related design parameters – tip depth, channel-cavity distance, and cavity depth – are shown to match with the experimental observations, while artificial smoothing of sharp corners is avoided.

ACKNOWLEDGMENT

The financial support by the *Austrian Federal Ministry for Digital and Economic Affairs* and the *National Foundation for Research, Technology and Development* is gratefully acknowledged.

REFERENCES

[1] H. Geng, *Semiconductor Manufacturing Handbook*. McGraw-Hill Education, 2017.

[2] H. Lo, J. Peng, E. Reis, B. Zhu, W. Ma, S. Y. Mun, S. Shintri, E. M. Bazizi, C. Gaire, Y. Qi, J. Chen, S. N. Ting, O. Hu, and S. Samavedam, "A Novel Approach to Control Source/Drain Cavity Profile for Device Performance Improvement," *IEEE Trans. Electron Devices*, vol. 65, no. 9, pp. 3640–3645, 2018.

[3] C. Qin, H. Yin, G. Wang, P. Hong, X. Ma, H. Cui, Y. Lu, L. Meng, H. Yin, H. Zhong, J. Yan, H. Zhu, Q. Xu, J. Li, C. Zhao, and H. H. Radamson, "Study of Sigma-Shaped Source/Drain Recesses for Embedded-SiGe pMOSFETs," *Microelec. Eng.*, vol. 181, pp. 22–28, 2017.

[4] Y.-Q. Sui, Q.-H. Han, H. Zhang, and K.-F. Lee, "A Study of Sigma-shaped Silicon Trench Formation," *ECS Trans.*, vol. 52, no. 1, pp. 331–335, 2013.

[5] "Silvaco Victory Process," https://www.silvaco.com/products/tcad.html, (accessed June 18, 2019).

[6] J. A. Sethian, "A Fast Marching Level Set Method for Monotonically Advancing Fronts," *Proc. Natl. Acad. Sci.*, vol. 93, pp. 1591–1595, 1996.

[7] S. Osher and R. Fedkiw, *Level Set Methods and Dynamic Implicit Surfaces*. Springer, 2003.

[8] J. A. Sethian, "Evolution, Implementation, and Application of Level Set and Fast Marching Methods for Advancing Fronts," *J. Comput. Phys.*, vol. 169, no. 2, pp. 503–555, 2001.

[9] T. J. Hubbard, "MEMS Design: The Geometry of Silicon Micromachining," Ph.D. dissertation, California Institute of Technology, 1994.

[10] B. Radjenović, J. K. Lee, and M. Radmilović-Radjenović, "Sparse Field Level Set Method for Non-Convex Hamiltonians in 3D Plasma Etching Profile Simulations," *Comput. Phys. Commun.*, vol. 174, no. 2, pp. 127–132, 2006.

[11] M. Gosálvez, P. Pal, and K. Sato, "Reconstructing the 3D Etch Rate Distribution of Silicon in Anisotropic Etchants Using Data From Vicinal {1 0 0}, {1 1 0} and {1 1 1} surfaces," *J. Micromech. Microeng.*, vol. 21, no. 10, p. 105018, 2011.

[12] J. A. Sethian and D. Adalsteinsson, "An Overview of Level Set Methods for Etching, Deposition, and Lithography Development," *IEEE Trans. Semicond. Manuf.*, vol. 10, no. 1, pp. 167–184, Feb 1997.

[13] B. M. G. Crandall and P. L. Lions, "Two Approximations of Solutions of Hamilton-Jacobi Equations," *Math. Comput.*, vol. 43, no. 167, pp. 1–19, 1984.

[14] C. Montoliu, N. Ferrando, M. A. Gosálvez, J. Cerdá, and R. J. Colom, "Implementation and Evaluation of the Level Set Method: Towards Efficient and Accurate Simulation of Wet Etching for Microengineering Applications," *Comput. Phys. Commun.*, vol. 184, no. 10, pp. 2299–2309, 2013.

[15] C. Montoliu, N. Ferrando, M. A. Gosálvez, J. Cerdá, and R. J. Colom, "Level Set Implementation for the Simulation of Anisotropic Etching: Application to Complex MEMS Micromachining," *J. Micromech. Microeng.*, vol. 23, no. 7, pp. 1–17, 2013.

[16] B. Radjenović and M. Radmilović-Radjenović, "3D Simulations of the Profile Evolution During Anisotropic Wet Etching of Silicon," *Thin Solid Films*, vol. 517, no. 14, pp. 4233–4237, 2009.

978-1-7281-0941-1/19 $31.00 © 2019 IEEE

Numerical simulations of nanosecond laser annealing of Si nanoparticles for plasmonic structures

A-S. Royet
CEA-LETI, MINATEC Campus
Université Grenoble Alpes
Grenoble, France
anne-sophie.royet@cea.fr

S. Kerdilès
CEA-LETI, MINATEC Campus
Université Grenoble Alpes
Grenoble, France
sebastien.kerdiles@cea.fr

P. Acosta Alba
CEA-LETI, MINATEC Campus
Université Grenoble Alpes
Grenoble, France
pablo.acostaalba@cea.fr

C. Bonafos
CEMES-CNRS
Université de Toulouse
Toulouse, France
caroline.bonafos@cemes.fr

V. Paillard
CEMES-CNRS
Université de Toulouse
Toulouse, France
vincent.paillard@cemes.fr

F.Cristiano
LAAS, CNRS
Université de Toulouse
Toulouse, France
cfuccio@laas.fr

B. Curvers
LASSE
SCREEN SPE
Gennevilliers, France
benoit.curvers@screen-lasse.com

K. Huet
LASSE
SCREEN SPE
Gennevilliers, France
karim.huet@screen-lasse.com

Abstract—**This paper reports numerical simulations of nanosecond laser thermal annealing of plasmonic structures based on Si-nanoparticles embedded in a SiO₂ matrix. From these simulations, we extracted guidelines for the structure design to be adopted. This study also investigates the expected laser annealing process window and the influence of nanoparticles coverage.**

Keywords—Laser annealing, Si nanoparticles, melt, plasmonic nanostructures.

I. INTRODUCTION

Noble metal nanoparticles exhibiting localized surface plasmon resonance (LSPR) show remarkable light scattering and absorption properties. The LSPR frequency is slightly tunable by the nanostructure size, geometry and local medium but is mainly controlled by the free electron density. Thus, a key advantage of using semiconductor nanoparticles for plasmonic structures is that their free carrier concentrations can be tuned by doping. Among plasmonic materials, heavily doped semiconductors such as Si Nanoparticles (Si-NPs) receive much attention thanks to their potential use in the infrared spectral range [1]. They find their application in the domains of biosensing, subwavelength microscopy or photovoltaics [2]. However, due to self-purification, doping Si-NPs is a challenge [3-5]. Doping levels and activation ratios become concepts to be reviewed. There are several different techniques compatible with CMOS technology to fabricate size controlled Si-NPs (2-10 nm) with narrow size distribution. A first one is leveraging Plasma Immersion Implantation (PIII) which allows the control of density and position of Si-NPs by tuning the plasma and the implantations conditions (energy and dose) [6]. The advantage of this technique is that it provides high throughput capabilities in the low energy regime and is well adapted for high dose implant applications. A second one uses the SiO/SiO₂ layer structure deposited by e-beam evaporation [7]. In that case, the doped Si-NPs embedded in a SiO₂ matrix are achieved thanks to the subsequent high temperature phase separation. Dopant introduction is made during the Si-rich layer deposition. This method is interesting for the formation of multilayers of small (2-5 nm) size controlled Si-NPs. For both processes, a subsequent high temperature annealing for dopant activation is mandatory. In this context, Ultra Violet Nanosecond Laser Annealing (UV-NLA) is a promising choice to achieve efficient dopant activation at the nanoscale. This technique allows working in near out-of-equilibrium conditions (nanosecond timescale at peak temperature near melting point) and fosters electrical dopant activation above the solid solubility limit [8]. This work presents preliminary simulation results carried out to optimize UV-NLA parameters. The aim of this study is to determine the UV-NLA energy density necessary to reach the Si-NPs melting point without melting the substrate which may generate detrimental surface roughness. We will also be able to determine the most favorable SiO₂ buffer layer thickness.

II. SI-NPS MODELING

The expected experimental structure is illustrated in Figure 1. In our 2D simulations, the model consists in a single layer of undoped Si-NPs with a quasi-spherical shape (3-7 nm diameter), embedded in a SiO₂ matrix below a 2 nm thin SiO₂ surface layer. Both cases amorphous and crystalline NPs must be considered, in particular for Si-NPs that will be synthesized by PIII process because initially crystalline NPs could be amorphised by further dopant ion implantation. Figure 1 illustrates the cross-section view of the corresponding test structure, used for 2D simulations, with 5 nm diameter Si-NPs with 5 nm spacing. These dimensions correspond to those given in [7]. The structure lays on a SiO₂ buffer layer (thickness varying from 10 to 50 nm) on top of a Si substrate. Based on this view, the ratio between the surface occupied by the NPs and the section area of the whole "SiO₂ embedded matrix" defines the coverage level of Si-NPs. In this example, the calculated coverage is about 33.7 % which can be related to a Si-NPs density of $1.7.10^{12}$ cm^{-2}. This value is in the same order of magnitude as in [7]. The first step was to consider extreme cases: maximum coverage (100 %) with a full sheet Si layer (amorphous and crystalline) and minimum coverage (0 %) by only keeping the SiO₂ matrix thickness. The resulting simplified structures can be investigated by 1D simulations (cf. Figure 2).

This work is supported by the ANR-18-CE09-0034 project (DONNA).

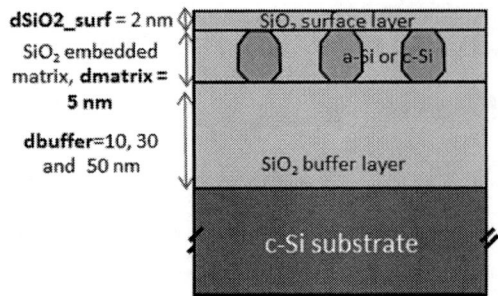

Fig. 1: Si-NPs modeling for 2D numerical UV-NLA simulations. 3 Si-NPs of 5 nm diameter spaced of 5nm, the coverage level is 33.7 %. This corresponds to a Si-NPs density of $1.7.10^{12}\,cm^{-2}$.

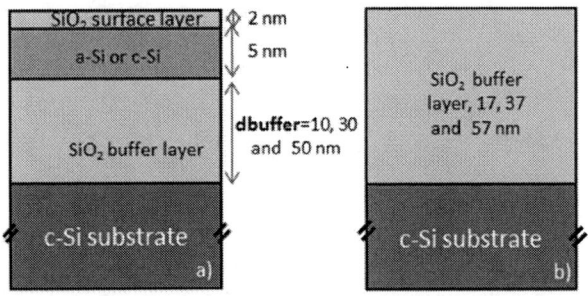

Fig. 2: Modeling of extreme coverage levels of Si : a) 100 % and b) 0 % for 1D UV-NLA simulations. In the case a), Si-NPs are replaced by a blanket Si layer. In the case b), only the SiO_2 matrix remains

III. NUMERICAL SIMULATIONS

A. One dimensional simulations (1D)

Numerical simulations were performed using LIAB simulation software described in [9]. This tool solves self-consistently the heat equation coupled to the time harmonic solution of Maxwell equation (UV laser light coupling), including temperature dependency of materials parameters and phase change (here using an enthalpy-based approach). The laser annealing system for our future experiments being the SCREEN LT-3100 platform, we used the corresponding laser pulse with 160 ns duration and a 308 nm wavelength. In a first approach, 1D numerical simulations were carried out with the simplified theoretical structures shown in Figure 2. All simulation conditions are summarized in table I.

N°	5 nm Si layer	SiO₂ buffer thickness (nm)
1	a-Si	10
2	a-Si	30
3	a-Si	50
4	c-Si	10
5	c-Si	30
6	c-Si	50
7	–	17
8	–	37
9	–	57

Table I: Framework of 1D numerical simulations performed in this work in order to delimit the UV-NLA energy densities and avoid the substrate melt. In grey simulations are performed with a full sheet Si layer and no Si layer for the others. In this case, the Si layer is replaced by SiO₂ in order to preserve the same structure thickness which is important for laser heating.

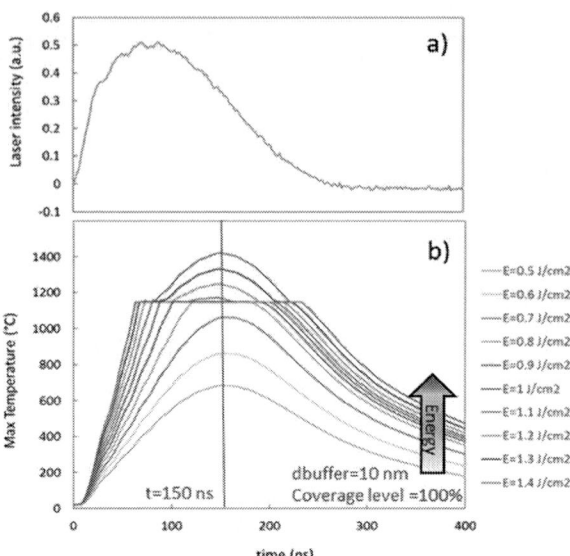

Fig. 3: a) : laser pulse profile used for 1D numerical simulations. b) : max temperature into the a-Si layer versus time and laser energy (J/cm²). The thickness of SiO₂ buffer layer dbuffer=10 nm. The maximum temperature is reached at 150 ns.

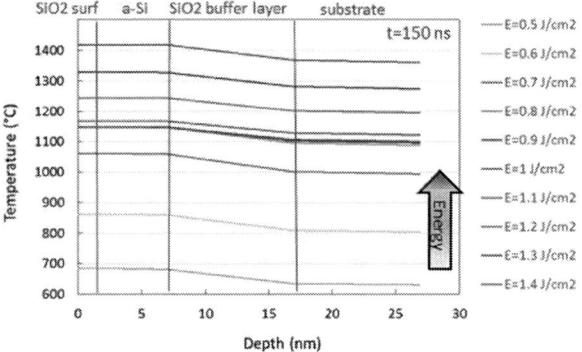

Fig. 4: 1D numerical simulations. Temperature as a function of depth in the structure a) of Fig.2 and laser energy density at t=150 ns and dbuffer=10 nm for a a-Si layer. This graph provides information on the temperature in both in the a-Si layer and substrate.

These simulations take also into account the initial phase of the Si film (crystal or amorphous) as well as temperature and phase dependent optical and physical material parameters at the given wavelength. The temperature in the top Si layer as a function of time is plotted in Figure 3. The temperature goes up as the energy density increases until reaching a plateau. This one corresponds to the beginning of the a-Si layer melt. The plateau becomes larger when the energy rises until the totality of the a-Si layer is melted. Afterwards, the temperature still increases in the Si layer. This graph allows the identification of the energy density corresponding to the Si layer melt onset and its full melt. Then, when the maximum temperature is reached in the Si layer (t=150 ns), we can also access to the spatial temperature depth profile shown in Figure 4. Thus, we can determine the temperature both in the top Si layer and the Si substrate.

Figure 5 shows the predictions for different buffer layer thicknesses: 10, 30 and 50 nm and in both cases of amorphous (full lines) and crystalline Si layer (dotted lines). Laser energy densities shown in the graph are related to the beginning of the melt in the top Si layer. Note that these energy densities decrease as the buffer oxide becomes thicker. Indeed, in that case, the heat is less well evacuated by the substrate due to increased thermal insulation from the buffer layer. The temperature shift between c-Si and a-Si based structures is attributed to the melting temperature difference: Tm(a-Si) =1147°C and Tm(c-Si) =1414°C [10]. Melting Si-NPs without melting the Si substrate already appears much easier for a-Si than for c-Si-NPs. Besides, Figure 6 shows the critical energy densities leading to the melt of the a-Si layer (full lines) or the substrate (dotted lines). The energy densities allowing the Si layer melt and keeping the substrate below 1414°C, constitute the process window (PW). At first, the choice of a thick buffer layer seems to be relevant to keep a sufficient temperature difference between the top Si layer and the substrate. However, in the case of thicker SiO₂ buffer layer, the temperature increases severely near the surface structure and can reach the melt temperature of the SiO₂ matrix (1713°C). Again, this is due to the high thermal insulation from the presence of thick buffer layer as we saw above.

The conclusion is that it is preferable to work with a buffer layer with a maximum thickness of 30 nm. PWs determined with this 1D approach are summarized in table II for a-Si and c-Si layers. We also give schematic cross sections to illustrate physical phase in the multilayer structures. The energy densities shown in red represent the limitation due to the matrix melt and not the substrate melt.

is obvious that using a thick buffer layer (>30 nm) leads to a poor heat diffusion and to a severe temperature increase at the surface.

	a-Silayer (100%)		c-Silayer (100%)		SiO2 (0%)
Buffer thickness (nm)	E(a-Si NPs melt) J/cm²	E(substrate melt) J/cm²	E(c-Si NPs melt) J/cm²	E(substrate melt) J/cm²	E(substrate melt) J/cm²
10nm (17nm for 0% Si layer)	0.80	1.50	1.10	1.40	1.30
30nm (37nm for 0% Si layer)	0.50	1.00	0.80	1.10	1.00
50nm (57nm for 0% Si layer)	0.45	0.70 Limited by SiO2 melt (1713°C)	0.76	0.85 Limited by SiO2 melt (1713°C)	1.00

Table II: Summary of 1D numerical simulations performed in this work. Schematic cross sections indicate crystalline phase of the top Si layer: solid in red and liquid in blue. Energy densities (J/cm²) defined a PW which also takes into account the SiO₂ melting point when the buffer layer is too thick.

B. Two dimensional simulations (2D)

2D simulations complete this work by using the model described in Figure 1. For various SiO₂ buffer thicknesses, Figure 7(a) shows the onset of the Si-NPs phase change during the laser annealing with the corresponding energy densities. Figure 7(b) illustrates the onset of the substrate melt. Molten areas are indicated in blue whereas solid ones are in red. The predicted PW (0.90 - 1.20 J/cm²) for this NPs coverage (33.7 %) is narrower than the one found for a 100 % coverage (0.80 - 1.50 J/cm²). The limitation due to the heat confinement at the surface for buffer layer thicknesses beyond 30 nm is observed in the case of 50 nm where the whole "SiO₂ embedded matrix" becomes liquid as the same time as the substrate.

Figure 8 gives the simulated PW thanks to simulations performed with various 5 nm diameter a-Si and c-Si-NPs coverage levels, indicated in the insert part and for a 10 nm thick SiO₂ buffer layer. As the number of Si-NPs increases, the energy necessary to melt the nanoparticles and the substrate decreases. The SiO₂ matrix is transparent to the UV-NLA but the more silicon in the matrix the more energy radiation is absorbed in the matter.

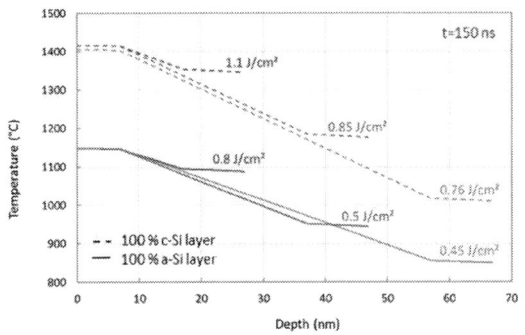

Fig. 5: 1D numerical simulations. Temperature as a function of depth at t=150 ns for dbuffer= 10 (red), 30 (blue) and 50 nm (green). The indicated energy densities correspond to ones necessary to reach Si-NPs melt.

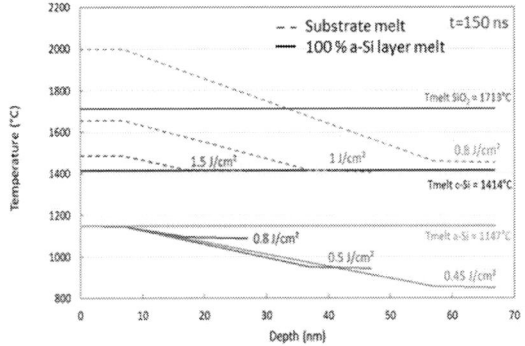

Fig. 6: 1D numerical simulations. Temperature as a function of depth at t=150 ns when dbuffer=10 (red),30 (blue) and 50 nm (green) and for an amorphous 100 % a-Si layer as an example (full lines). Dotted lines show the temperature and energy densities corresponding to the substrate melt. Here, it

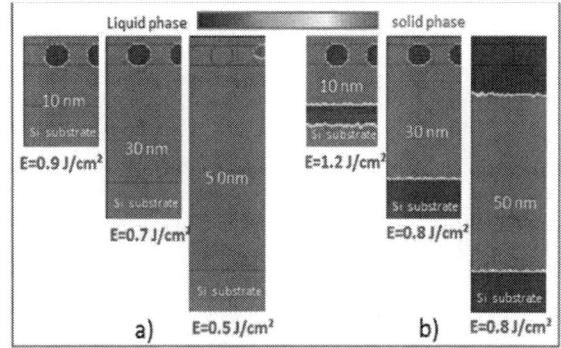

Fig. 7: 2D simulation of laser annealing for three 5 nm diameter aSi-NPs, 5 nm spaced (coverage level of 33.7 %); melting areas at the onset of a-Si NPs melt (a) and substrate melt (b) and corresponding laser energy densities. In the case of a 50 nm thick SiO₂ buffer layer, the substrate melt is accompanied by the matrix one.

978-1-7281-0941-1/19 $31.00 © 2019 IEEE

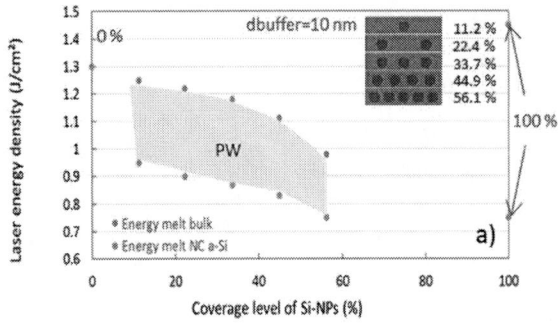

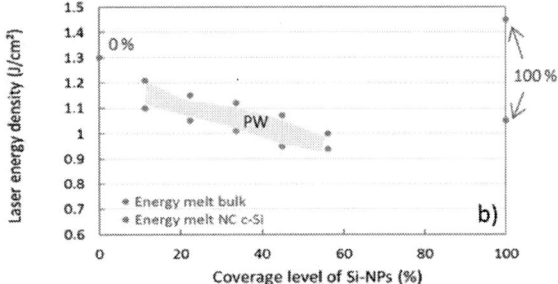

Fig. 8: Process window defined with various coverage levels of aSi-NPs (a) and cSi-NPs (b).

This absorption fosters the temperature elevation in the Si-NPs. However, we can observe that in the case of a full Si layer instead of NPs (case of 100% of Si), the energy density to melt the substrate is higher compared to the other coverage levels. We can explain this by a change in optical characteristics namely the reflectivity of the structure. To verify this point, we have plotted in Figure 9 the evolution of the reflection coefficient at 308 nm as a function of the coverage levels.

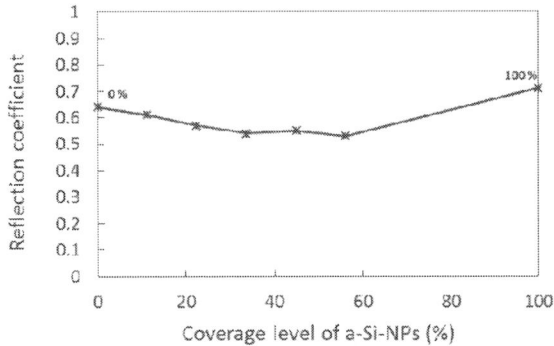

Fig.9: Reflection coefficient as a function of coverage levels of Si in the SiO₂ matrix. Extreme cases 0% and 100% also appear in the graph to compare

This coefficient is given for a laser beam with an energy density level below the melting threshold.

IV. CONCLUSION

In order to prepare experimental nanosecond laser annealing of plasmonic structures based on Si-NPs embedded in a SiO₂ matrix, we performed 1D and 2D numerical simulations taking into account electromagnetic and thermal aspects of such a process. These simulations provide guidelines on the design of the structures to be annealed and on the expected process windows. Such process windows depend at first order on the phase of the Si-NPs just before laser annealing (crystalline or amorphous), then on the oxide thickness between the Si-NPs and the Si substrate, and finally on the Si-NPs surface coverage. Amorphous Si-NPs lead to a wider process window compared to their crystalline counterparts thanks to the much lower melting point of a-Si. Si-NPs located 2 nm below the surface should preferably be separated from the Si substrate by a buffer SiO₂ layer thinner than 30 nm in order to reach Si-NPs melting point (for optimal dopant activation) before that of the Si substrate or the top oxide layer.

REFERENCES

[1] D. Rowe, J.S. Jeong, K.A Mkhoyan and U.R. Kortshagen, "Phosphorus-doped silicon nanocrystals exhibiting mid-infrared localized surface plasmon resonance", Nano Letters, vol. 13 (3), pp.1317-1322, 2013.
[2] H. Atwater and A. Polman, "Plasmonics for improved photovoltaic devices", Nature materials vol.: 9 (3), pp. 205-213, 2010.
[3] S. Ossicini et al., "Simultaneously B- and P-doped silicon nanoclusters: formation energies and electronic properties", Applied Physics Letters, vol. 87 (17), 173120, 2005.
[4] M. Peregro, C. Bonafos and M. Fanciulli, "Phosphorus doping of ultra-small silicon nanocrystals", Nanotechnology, vol. 21 (2), pp. 1-6, 2010.
[5] S. Gutsch *et al*, "Electronic properties of phosphorus doped silicon nanocrystals embedded in SiO2", Applied Physics Letters, vol. 106 (11), 113103, 2015.
[6] C. Bonafos *et al*, "Controlled fabrication of Si nanocrystal delta-layers in thin SiO2 layers by plasma immersion ion implantation for nonvolatile memories", Applied Physics Letters, vol. 103 (25), 253118, 2013.
[7] H. Rinnert, O. Jambois, and M. Vergnat, "Photoluminescence properties of size-controlled silicon nanocrystals at low temperatures", Journal of Applied Physics, vol.106 (2), 023501, 2009.
[8] K. Huet, F.Mazzamuto, T.Tabata, I. Toqué-Tresonne, and Y. Mori, "Doping of semiconductor devices by laser thermal annealing", Materials Science in Semiconductor Processing, vol. 62, pp.92-102, 2017.
[9] S.F. Lombardo *et al*, "Theoretical study of the laser annealing process in FINFET structures", Applied Surface Science vol. 467-468, pp. 666-672, 2019.
[10] A. La Magna, P. Alippi, V. Privitera, G.Fortunato, M. Camalleri, B. Svensson, "A phase-field approach to the simulation of the excimer laser annealing process in Si", Journal of Applied Physics, vol.95 (9), pp.4806-4814, 2004

Parallelized Level-Set Velocity Extension Algorithm for Nanopatterning Applications

Michael Quell
Christian Doppler Laboratory
for High Performance TCAD
Institute for Microelectronics, TU Wien
Wien, Austria
quell@iue.tuwien.ac.at

Alexander Toifl
Christian Doppler Laboratory
for High Performance TCAD
Institute for Microelectronics, TU Wien
Wien, Austria
toifl@iue.tuwien.ac.at

Andreas Hössinger
Silvaco Europe Ltd.
Cambridge, United Kingdom
andreas.hoessinger@silvaco.com

Siegfried Selberherr
Institute for Microelectronics, TU Wien
Wien, Austria
selberherr@tuwien.ac.at

Josef Weinbub
Christian Doppler Laboratory
for High Performance TCAD
Institute for Microelectronics,TU Wien
Wien, Austria
weinbub@iue.tuwien.ac.at

Abstract—We present a parallelized algorithm for accelerating the velocity extension calculations in a level-set method, which is essential for surface velocity based topography simulations, such as etching or deposition simulations for nanopatterning applications. The proposed algorithm improves the prevailing fast marching method by optimizing the heap data structure and efficiently reordering the calculations. We implemented the algorithm into Silvaco's Victory Process simulator, which is utilized for evaluating our algorithm with a three-dimensional simulation of an ion beam etching process used for spin-transfer torque magnetoresistive random access memory devices. Our results show a significant serial speed-up by a factor of at least 1.4 and a total speed-up by a factor of up to 8 using 8 threads for the velocity extension.

I. INTRODUCTION

Modern semiconductor devices are progressively employing non-planar geometries, e.g., high-aspect ratio pillars, cavities or fins. The thus necessary fabrication processes are based on complex patterning techniques on the nanoscale, essentially requiring high control over geometry parameters. For example, the devices proposed in the field of emerging memory techologies [1] particularly demand optimized nanopatterning to enable a small feature size and high density memory cells in order to replace conventional complementary metal-oxide semiconductor (CMOS)-based random access memory (RAM) [2], [3].

Process simulations employing the level-set method [4], [5], [6] enable highly robust and accurate simulations of the temporal evolution of the wafer surface during such fabrication processes of semiconductor devices. The level-set method naturally handles topological changes in three dimensions and allows for the simulation of a variety of processing techniques via a general velocity function.

A level-set topography simulation involves several steps which are summarized in Fig. 1(a). Starting with the construction of the level-set function from the initial geometry, the surface (front) velocity has to be calculated according to a process-specific physical model. Subsequently, the surface velocity is extended to the computational domain (velocity extension), which allows to move the surface according to the velocity by employing a time integration scheme for the level-set field. After the time stepping the level-set field is converted to an explicit mesh, which enables the visualization of the resulting surface.

The velocity extension step, which has to be performed in every time step, significantly contributes to the overall runtime. In this work we present a shared-memory parallelized velocity extension algorithm which considerably reduces the simulation runtime. The proposed algorithm was implemented into Silvaco's Victory Process simulator [7], which we utilize here to test our algorithm by simulating an ion beam etching (IBE) step required for the production of spin-transfer torque magnetoresistive RAM (STT-MRAM) devices [8].

II. VELOCITY EXTENSION

The level-set method implicitly represents the wafer surface Γ as the zero level-set of the function $\phi(\vec{x}, t)$ and the time evolution is determined by the level-set equation

$$\frac{\partial \phi}{\partial t} + V|\nabla \phi| = 0. \tag{1}$$

V is a velocity field which is defined on the entire computational domain, modeling the motion of the surface. However, most process-specific physical models only provide the surface velocity V_{surface} on Γ. In order to solve (1), V_{surface} has to be extended to the entire computational domain. We employ

978-1-7281-0941-1/19 $31.00 © 2019 IEEE

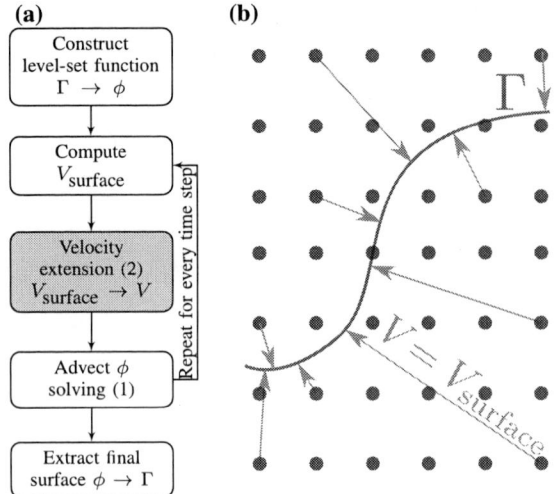

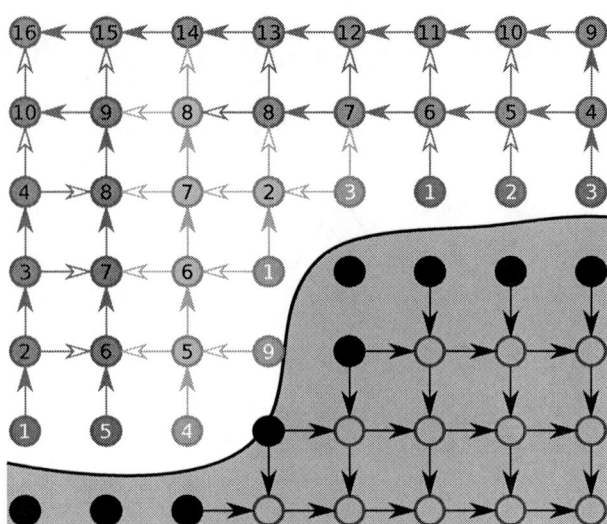

Fig. 1: **(a)** Flow diagram of the computational tasks in a level-set simulation. **(b)** Velocity extension according to (2) yields for every grid point the velocity of the closest surface point. Thus the velocity along the orange arrows is constant.

Fig. 2: Processing order of the grid points for one side of the surface by the proposed algorithm employing three threads (red, green, and blue). Grid points with a white number next to the interface are the grid points which constitute the *runs*. Filled arrows show successful updates of the neighbor, while bare arrows show that the neighboring grid point either had an unprocessed upwind neighbor or has been already processed by another thread.

the signed-distance property preserving approach, which is characterized by constant velocity values along the surface normals [9]. This is shown for selected grid points in Fig. 1(b). The associated partial differential equation

$$\nabla\phi \cdot \nabla V = 0, \text{ and } V\big|_\Gamma = V_{\text{surface}} \qquad (2)$$

is typically solved with the fast marching method (FMM) [4]. As a prerequisite for the FMM the velocity is set on the grid points in direct proximity to the surface [9]. The surface divides the set of grid points into two subsets, allowing for the subsequent independent application of the FMM for both sides.

The FMM processes the grid points in ascending order by their level-set value, enabling the usage of an upwind difference scheme to approximate the gradients in (2). Thus, only information from grid points which are closer to the surface is utilized, resulting in a one pass algorithm. In particular, the processing of a grid point consists of two tasks: (a) calculate the velocity and (b) insert all unprocessed neighbors into the minimum heap. To track the processing order of the grid points, a global minimum heap data structure sorted by the level-set values is used. Consequently, the FMM is an inherently serial algorithm.

III. PARALLELIZED VELOCITY EXTENSION

Our proposed algorithm overcomes the limitation of the FMM by executing a modified FMM for every grid point in direct proximity to the surface. We refer to the execution of the modified FMM based on one of those grid points and the corresponding minimum heap (from the modified FMM) as a *run*. The *runs* are processed one after another, which

effectively reorders the calculations and yields a speed-up by reducing the effective heap sizes.

The introduction of the non-global minimum heaps creates cases in which grid points have unprocessed upwind neighbors. These cases need special treatment: We remove grid points which have an unprocessed upwind neighbor from the current minimum heap without processing it. Nevertheless, it is ensured that all grid points are processed, because once the formerly unprocessed upwind neighbor is processed (by a different *run*) the removed grid point is visited again, allowing for the application of the upwind scheme.

In summary, some of the grid points are visited several times by our algorithm (depending on the number of upwind neighbors). However, the velocity is calculated only once, using the same upwind neighbors as in the FMM. Consequently, we ensure that calculated extended velocities are identical compared to the FMM.

The *runs* enable an efficient approach to parallelize the velocity extension algorithm, because they are dynamically assigned to different threads in parallel. Once the currently assigned *run* is completed (i.e., its minimum heap is empty) a new unprocessed *run* is assigned. The dynamic assignment of the *runs* to the threads is necessary as the associated computational load varies strongly between them. Fig. 2 shows the processing order of the grid points for an exemplary surface topology. One side of the surface is processed by three threads.

Each thread is assigned a *run* (in Fig. 2, a grid point with a white number) and creates its own minimum heap containing

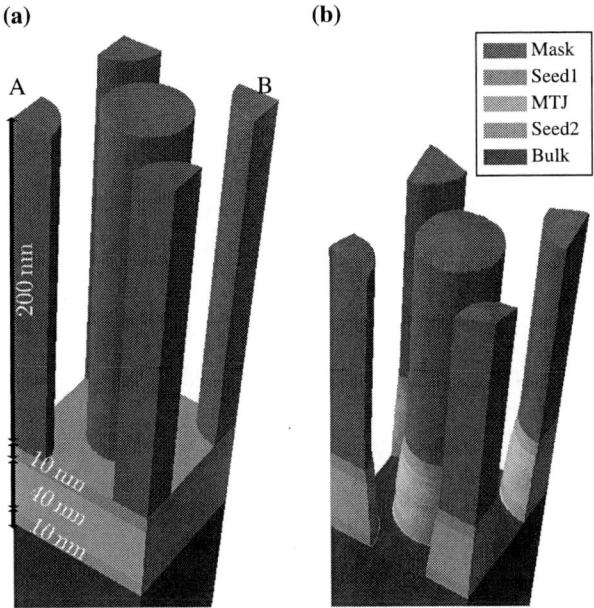

Fig. 3: Initial wafer topography consisting of the magnetic tunnel junction (MTJ) and the associated seed layers deposited on the substrate: **(a)** prior to the IBE process and **(b)** final pillar topography for the STT-MRAM device.

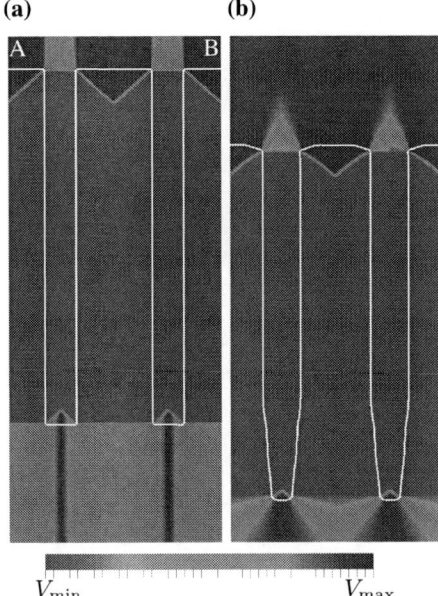

Fig. 4: Extended velocity field in the plane containing points A and B in Fig. 3 **(a)** prior to and **(b)** after the IBE process. The velocity is constant along the surface normals as imposed by (2). The white line depicts the wafer surface.

only this grid point. Then the thread processes its minimum heap analogously to the traditional FMM implementation with the previously described modification, while taking care of unprocessed upstream grid points. Once the minimum heap is empty, a new *run* is assigned. The blue thread in Fig. 2 demonstrates the varying load: The first and second *run*, (blue grid point 1 and 2) only processes a single grid point, while the third *run* (blue grid point 3) processes 14 grid points in total.

The proposed algorithm does not use explicit synchronization during the processing of the *runs*. This is a trade-off between the cost of the explicit synchronization and avoiding redundant operations, i.e., computing the velocity of a grid point more than once. It is not an issue if the velocity for a grid point is computed more than once, because the algorithm always enforces the usage of the same upwind neighbors for all threads. Thus, for any grid point the computational result is the same for every thread. The redundant operations may occur along the border of two regions processed by different threads (cf. bare arrows pointing to a differently colored grid point in Fig. 2). Nevertheless, the accuracy of the final extended velocity field does not depend on the choice between explicit and not explicit synchronization, as both threads use the same upwind neighbors.

IV. RESULTS AND DISCUSSION

We employ the proposed algorithm for simulating an IBE process which is essential for the fabrication of STT-MRAM

devices. The considered structure consists of a magnetic tunnel junction (MTJ) and the associated seed layers which have been deposited on a substrate. IBE allows to fabricate an array of MTJ pillars (shown in Fig. 3a and Fig. 3b). We apply the proposed velocity extension algorithm to the IBE step. The extended velocity fields for the initial and final surface are shown in Fig. 4(a) and Fig. 4(b) for a vertical plane slicing the computational domain diagonally (containing Points A and B in Fig. 3(a). The velocity is constant along the surface normals. Furthermore, the corners, where the velocity abruptly changes and consequently the solution is discontinuous, are well resolved.

We assessed the implementation of the algorithm (for implementation details see [11]) on a compute node of the Vienna Scientific Cluster 3 (two Intel Xeon E5-2650v2 Ivy Bridge EP processors, 64 GB main memory) [10], by comparing the runtimes for a single velocity extension step. The code uses C++11 and was compiled with gcc-7.3 with -O3 optimization.

Fig. 5 shows that the serial runtime for the velocity extension is reduced by a factor of 1.4 (1.5) for a grid resolution of $2\,\mathrm{nm}$ ($0.5\,\mathrm{nm}$). The runtime reduction is mainly caused by the splitting of the global heap into smaller heaps, resulting in less time-consuming sorting operations. The cost of visiting grid points twice is negligible, because the velocity is calculated only once.

Furthermore, we evaluated the parallelized algorithm (implemented with OpenMP 4.5), which has a parallel efficiency of approximately $60\,\%$ ($66\,\%$) for 8 threads and thus results

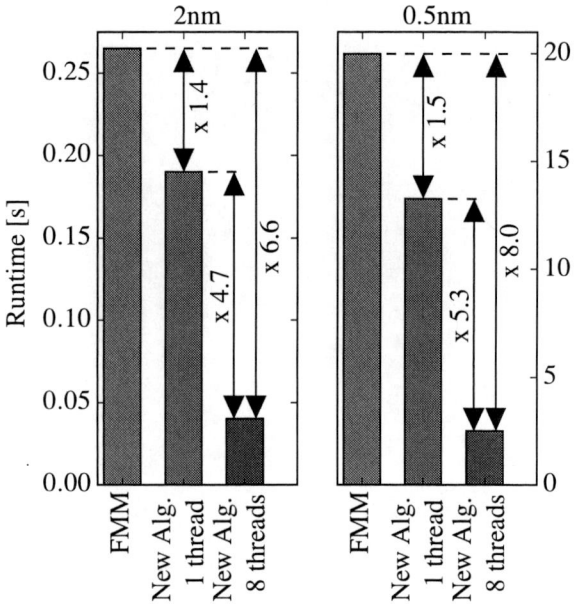

Fig. 5: Runtime of the velocity extension for a single time-step measured on a compute node of the Vienna Scientific Cluster 3. The proposed algorithm is compared to the original algorithm (FMM) for a spatial resolution of 2 nm and 0.5 nm.

in a parallel speed-up of 4.7 (5.3). Each of the 8 threads is assigned to a separate core. In our implementation, the grid points are accessed using OpenMP atomic operations, which is required to enforce a consistent view of the memory between the threads.

Avoiding explicit synchronization leads to less than 1 % (0.1 %) of the grid points processed twice (see Section III). The lower percentage of the redundantly processed grid points in the higher resolution simulation (i.e. the 0.5 nm case) is caused by a square-cube law of grid points on the borders between threads and total processed grid points. The grid points on the borders between threads scale by a power of two and the total processed grid points by a power of three, therefore, their ratio decreases with increasing size.

The serial and parallel speed-up combined yield a total runtime reduction of the velocity extension step by a factor of 6.6 (8.0) for 8 threads, which demonstrates the excellent performance of our approach. Our algorithm results in the same velocity field as the reference FMM implementation, which we confirm by the L^∞-Norm (maximum over the velocity value differences for all grid points) yielding 0.

V. SUMMARY

We have presented a parallelized velocity extension algorithm improving on the serial fast marching method, by efficiently reordering the calculations. This is achieved by replacing the global heap data structure associated with the fast marching method with many non-global minimum heaps (e.g., unprocessed upwind neighbors) The efficient velocity algorithm enables level-set topography simulation of complex three-dimensional non-planar geometries as they commonly appear in nanopatterning applications. We demonstrate the capability of the proposed algorithm with the velocity extension step in an ion beam etching process of a spin-transfer torque magnetoresistive memory device. Our implementation does not show any difference in the resulting velocity field compared to the reference fast marching method. Furthermore, we have investigated the serial and parallel speed-up, where the combined speed-up is in the range from 6.6 to 8.0 for 8 threads.

ACKNOWLEDGMENT

The financial support by the *Austrian Federal Ministry for Digital and Economic Affairs* and the *National Foundation for Research, Technology and Development* is gratefully acknowledged. The computational results presented have been achieved using the Vienna Scientific Cluster (VSC).

REFERENCES

[1] S. Bhatti, R. Sbiaa, A. Hirohata, H. Ohno, S. Fukami, and S. N. Piramanayagam, "Spintronics Based Random Access Memory: A Review," *Materials Today*, vol. 20, no. 9, pp. 530–548, 2017.

[2] D. Apalkov, B. Dieny, and J. M. Slaughter, "Magnetoresistive Random Access Memory," *Proceedings of the IEEE*, vol. 104, no. 10, pp. 1796–1830, 2016.

[3] V. D. Nguyen, P. Sabon, J. Chatterjee, L. Tille, P. V. Coelho, S. Auffret, R. Sousa, L. Prejbeanu, E. Gautier, L. Vila, and B. Dieny, "Novel Approach for Nano-Patterning Magnetic Tunnel Junctions Stacks at Narrow Pitch: A Route Towards High Density STT-MRAM Applications," in *IEEE International Electron Devices Meeting (IEDM)* , pp. 38.5.1–38.5.4, 2017.

[4] J. A. Sethian, "A Fast Marching Level Set Method for Monotonically Advancing Fronts." *Proceedings of the National Academy of Sciences*, vol. 93, pp. 1591–1595, 1996.

[5] J. A. Sethian, "Evolution, Implementation, and Application of Level Set and Fast Marching Methods for Advancing Fronts," *Journal of Computational Physics*, vol. 169, no. 2, pp. 503–555, 2001.

[6] S. Osher and R. Fedkiw, *Level Set Methods and Dynamic Implicit Surfaces*, Springer, 2011.

[7] "Silvaco Victory Process," https://www.silvaco.com/products/tcad.html, (accessed March 27, 2019).

[8] M. Gajek, J. J. Nowak, J. Z. Sun, P. L. Trouilloud, E. J. OSullivan, D. W. Abraham, M. C. Gaidis, G. Hu, S. Brown, Y. Zhu, R. P. Robertazzi, W. J. Gallagher, and D. C. Worledge, "Spin Torque Switching of 20nm Magnetic Tunnel Junctions with Perpendicular Anisotropy," *Applied Physics Letters*, vol. 100, no. 13, p. 132408, 2012.

[9] D. Adalsteinsson and J. A. Sethian, "The Fast Construction of Extension Velocities in Level Set Methods," *Journal of Computational Physics*, vol. 148, no. 1, pp. 2–22, 1999.

[10] "Vienna Scientific Cluster 3," http://vsc.ac.at/systems/vsc-3/, (accessed June 18, 2019).

[11] M. Quell, P. Manstetten, A. Hössinger, S. Selberherr, and J. Weinbub, "Parallelized Construction of Extension Velocities for the Level-Set Method," in *International Conference on Parallel Processing and Applied Mathematics (PPAM)*, Springer LNCS, 2019.

Process Simulation in the Browser: Porting ViennaTS using WebAssembly

Xaver Klemenschits, Paul Manstetten, Lado Filipovic, and Siegfried Selberherr
Institute for Microelectronics, TU Wien, Gußhausstraße 27-29/E360, 1040 Wien, Austria
Email: klemenschits@iue.tuwien.ac.at

Abstract—We introduce a client-side browser application for high performance process simulation. The codebase is taken from ViennaTS, an open-source C++ process simulation engine, and completed with JavaScript based components, such as an editor to configure the simulation parameters and a result viewer. The C++ codebase is ported to the browser by compiling to a portable standard for an abstract instruction set: WebAssembly. We demonstrate the capabilities and performance of the application by performing several configurable simulations, including the emulation of the fabrication process of a stacked nanosheet field effect transistor. The simulations conducted in the browser application are only slower by a factor of 1.6 to 3.6 compared to native, single-thread simulations. Therefore, WebAssembly presents a promising format for portable and widely accessible high performance process simulations.

I. INTRODUCTION

The introduction of three-dimensional metal oxide semiconductor field effect transistors (MOSFETs), such as FinFETs and nanowire transistors, has created the need for intricate fabrication techniques incorporating new materials and processes. As the fabrication of devices for advanced technology nodes becomes increasingly complex and costly, technology computer aided design (TCAD) is more and more important to investigate possible new fabrication techniques. Especially software development has benefited strongly from open source initiatives, creating productive global platforms for innovation. Although open-source tools for process simulation exist, their wide and active adoption and thus the creation of creative development platforms is often limited by a narrow application focus pursued by a small user group. As large numbers of users are beneficial to open source projects, deployment, accessibility, and visibility are highly important for a wider acceptance of such tools.

With the advent of WebAssembly (WA) [1], an open standard describing a binary format expressing assembly-like instructions for a virtual processor, a framework for portable high performance code was created. The WA design rationale is to establish a portable executable format which can be compiled quickly for any target system, in a fast single-pass fashion. The WA framework therefore allows client-side browser applications to be built using existing C/C++ codebases, enabling portable high performance applications to be deployed and accessed at any scale without requiring backend investments.

ViennaTS [2], an open-source, high performance process simulator written in C++, was ported to the browser, since it offers a wide array of simple as well as highly complex process models describing modern semiconductor fabrication steps. Here, we present the results of compiling ViennaTS using WA combined with a user interface to edit parameters and view simulation results in a unified application [3], which is easily deployable at scale.

II. PROCESS SIMULATION USING VIENNATS

ViennaTS [2] is an open-source, C++ based process simulator which supports two- and three-dimensional simulations of semiconductor fabrication processes. The materials and their interfaces are represented with a sparse-field level-set framework discretised on a Cartesian grid [4]. Large numbers of materials can therefore be represented accurately with minimal memory requirements. The level-set is stored in a hierarchical run length encoded (HRLE) data structure, optimised for level-set operations, such as re-distancing, velocity extension, and advection on a sparse data set. A set of advanced physical etching and deposition models is available, which include particle transport at feature-scale and intricate chemical models for the description of surface reactions. Particle transport at feature-scale is modelled using a Monte Carlo ray tracing approach via an explicit representation of the surface with partially overlapping disks [5]. Accelerated geometric models to emulate various process steps are also available, which advance the material interfaces solely by geometric considerations, offering efficient, albeit less accurate, process descriptions. Therefore, many different processing techniques with varying accuracies can be combined and tuned to match a wide field of applications.

III. WEBASSEMBLY

WebAssembly is an open standard for a *virtual instruction set architecture (ISA)*[1]. Basic components of an ISA are supported data types, an instruction set (for control flow, arithmetic/logic, and memory operations) and its encoding. The advantages of the *virtual* ISA of WA over a *native* ISA of a specific processor is that it has a portable hardware-, platform-, and language-independent design, which aims to execute in a sandboxed memory-environment with near-native speed. The interfacing/interoperation with the environment provided by the execution platform has a simple universal design and is not limited to web-platforms (i.e., the JavaScript engine of a browser). Further design goals are the ability to efficiently decode, validate, and compile the instructions of a WA program on the execution platform.

Since 2017 WA support has been widely adopted across most major browsers and JavaScript engines, and about 85% of currently installed browsers support it [6]. When compared to JavaScript, the execution performance and performance predictability of WA profits from the fact that all types are

978-1-7281-0941-1/19 $31.00 © 2019 IEEE

statically declared and programs can be compiled ahead-of-time and the execution is not interrupted by a garbage collector.

Additionally to the integration of WA support in all major browsers and JavaScript engines, there is also active development of WA-runtimes for desktop applications and efforts are ongoing to standardise the communication between WA and it's embedding platform with a WebAssembly System Interface (WASI)[7]. Due to the aforementioned universal interfacing of WA with its environment WASI, each platform implements its own interfacing strategy. The WASI aims to allow for a modular specification to provide a portable modular WA-interfacing.

The motivation for WA is due in part to the success of its predecessors *NaCl*[8] and especially *asm.js*[9]. The JavaScript library *asm.js* can even be seen as a blueprint for the WA standard, since it allowed native C/C++ to JavaScript compilation, which resulted in performant execution due to the use of a limited subset of JavaScript language features. Only the features which can be accessed by the ahead-of-time optimisation of all common JavaScript engines are part of this subset. Realising the potential of this approach, browser vendors implemented optimisations specifically for *asm.js* in their JavaScript engines, increasing the performance further.

The infrastructure for compiling C/C++ for a web-browser is currently actively developed in *emscripten* [10] and *LLVM*[11]. This framework aims to keep the necessary changes to the C/C++ codebase at a minimum by providing all implementations of commonly used interfaces to the environment, e.g., by providing a virtual file system and tailored standard libraries to map system calls to common browser engine instructions.

Currently, the main limitations, when porting an existing C/C++ codebase to WA, include a maximum memory of 4GB (WA has a 32bit address space), no wide support for C++ threads/OpenMP, and limited support for dynamic linking.

In the following, we describe the portable application built around our ViennaTS WA module. Porting the C++ codebase to be compiled to WA with *emscripten* required 570 lines to be added or modified, from a total of 28,986 lines of the entire project. Therefore, the change required to port this existing high performance C++ codebase to WA only corresponds to about 2% of the total code base.

IV. IN-BROWSER APPLICATION

Fig. 1 provides an overview of the system components typically involved on a desktop computer when using ViennaTS to simulate a process step. After preparation of the process parameters in an *Editor*, *ViennaTS* (`vts.exe`) is started by providing the parameter file (`par.txt`) which references the initial geometries (`geo.vtk`). The results (`res.vtk`) can then be visualised by loading them into a *Result Viewer*.

Analogously to Fig. 1, Fig. 2 provides an overview of the components necessary for the In-Browser Application: The user interface is executed in the *UI Thread* of the browser and contains an *Editor* and a *Result Viewer*, which are both pure JavaScript applications. The *Web Worker*, executed in a different thread than the user interface, loads *ViennaTS* (`vts.wasm`) and compiles it for the current host system in a single pass. When the simulation is started, the parameter

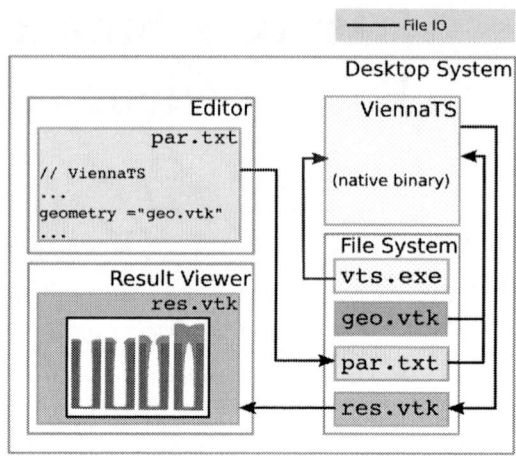

Fig. 1: Components typically involved in the preparation, execution and inspection of simulations conducted with the native desktop version of ViennaTS [2].

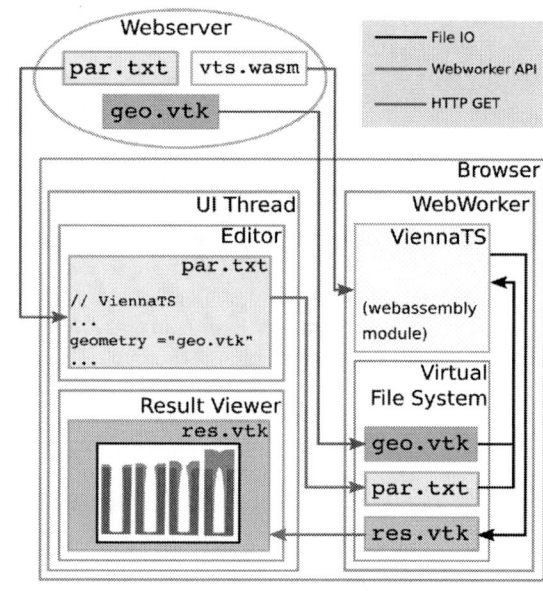

Fig. 2: Components involved in conducting a simulation with the ViennaTS in-browser application available at hpcwasm.github.io/viennats.

file (`par.txt`) is transferred to the *Web Worker* and stored in its *Virtual File System*. The initial geometries (`geo.vtk`), referenced in `par.txt`, are then fetched from the *Webserver* by the *Web Worker*. Once all required files are available, the *Web Worker* executes ViennaTS which writes the simulation results (`res.vtk`) to the *Virtual File System*. As soon as the results have been written, they are automatically transferred to the *UI Thread* and visualised in the *Result Viewer*.

The full user-interface can be seen in Fig. 3 which shows the most important parts needed to start a simulation. The "Settings" section allows the user to load and manipulate existing parameter files using an *Editor*. Once the user has populated the parameter file, the simulation can be started

978-1-7281-0941-1/19 $31.00 © 2019 IEEE

using the controls in the "Simulation" section, which also shows text-based simulator output and simulation progress. In the "Results" section, the generated output meshes are shown in the *Result Viewer*, providing basic mesh inspection and manipulation functionality as well as the ability to download the results.

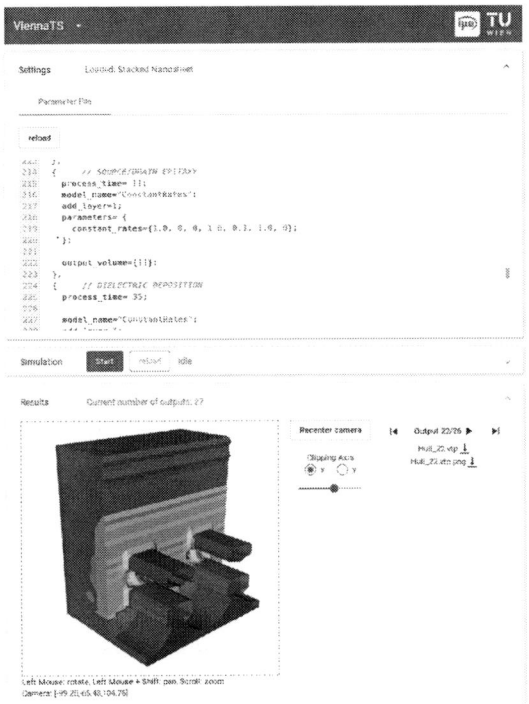

Fig. 3: Graphical User Interface of the application. The *Editor* in the "Settings" section is used to create a parameter file used as input for ViennaTS. The "Simulation" section provides additional console output when expanded and the "Results" section shows the output geometries as they become available in the *Result Viewer* and provides mesh manipulation tools.

V. BENCHMARKS & PERFORMANCE EVALUATION

In order to evaluate the performance differences between the WA and native versions of ViennaTS, numerous benchmark simulations were conducted. Some of these apply only emulations, advancing material interfaces purely by geometric parameters, while others use Monte Carlo (MC) ray tracing methods to model the particle transport inside a reactor at the feature-scale and provide chemical models to describe surface reactions. All benchmark simulations are available and can be executed online at [3].

Table I shows the different execution times obtained on a desktop computer and a smartphone for several single-threaded simulations. On the desktop, the runtime performance gap between native and browser execution varies between a factor of 1.6 to 3.6, depending on the simulation. The runtime performance gap between native execution and mobile browsers is even larger. The mobile version is slower by a factor of 10.0 to 18.6. The performance gap is caused firstly by optimisation differences, since two separate compilation stages are necessary for WA, and secondly by the more complex memory management of a browser compared to an operating system. The latter results in a larger runtime gap for simulations which are more memory intensive. On the other hand, Monte Carlo enabled simulations (labelled MC in Table I), which are rather computationally than memory intensive, show smaller runtime gaps.

Simulation	Native	Firefox	Chrome	Mobile
Geom. Hole Etch	6.97	16.97 (2.4)	20.46 (2.9)	128.62 (18.5)
Phys. Hole Etch(MC)	11.95	30.46 (2.5)	42.88 (3.6)	188.11 (15.7)
Geom. Air Gap Depo	12.27	30.10 (2.5)	36.08 (2.9)	228.56 (18.6)
Phys. Air Gap Depo (MC)	12.99	21.23 (1.6)	22.58 (1.7)	130.01 (10.0)
Geom. HAR Etch	5.26	12.77 (2.4)	15.90 (3.0)	92.88 (17.7)
Phys. HAR Etch (MC)	12.00	31.21 (2.6)	39.91 (3.3)	189.64 (15.8)
Geom. Deposition	6.54	15.57 (2.4)	19.84 (3.0)	121.51 (18.6)
Phys. Deposition(MC)	34.65	52.34 (1.5)	58.61 (1.7)	–
Stacked Nanosheet	9.61	25.19 (2.6)	27.79 (2.9)	–
Avg. Perf. Gap	–	2.3	2.8	16.4

TABLE I: Runtime (without output) comparison for the showcased examples of process emulations (Geom.) and physical simulations (Phys.) on a desktop computer (Linux; Intel i7-3770 processor) natively (*Native*) and in the browser (*Chrome 73.0, Firefox 66.0*). Additionally, *Mobile* benchmarks (*Firefox Mobile 66.0*) were conducted on a smartphone (Android; Kirin 650 processor). The relative runtime gaps to the native execution are shown in brackets. Models labelled (MC) use physical models with Monte Carlo ray tracing methods.

VI. STACKED NANOSHEET FABRICATION PROCESS

In order to highlight the full capabilities and performance of the ported simulator, the fabrication process of an advanced-node stacked nanosheet FET was emulated starting from a blank SOI wafer. Since this geometry is comparably large, the high memory requirements result in a relatively large runtime gap, even exceeding the capabilities of the tested mobile browser. Therefore, all models used to create the final structure were purely geometric and MC models were not applied. These models and their input parameters are shown in Table II.

Experimental data from [12] and [13] were used to calibrate the process steps, resulting in a geometry which matches a typical structure in use today. Fig. 4 shows the resulting geometry at the most important process steps: Starting from a blank SOI wafer, Si, SiGe, and Si layers are deposited epitaxially. These layers are then patterned to fins, using self aligned double patterning, creating the structure shown in Fig. 4a. After removing the mask, a dummy gate is created by depositing Poly-Si isotropically over the fins. As can be seen in Fig. 4b, the resulting Poly-Si is then patterned to the required gate dimensions [14]. Hf is deposited around the gate to form insulating spacers between the gate and source/drain regions. Directional etching is then applied to generate the required vertical spacers, as shown in Fig. 4c. The fins extending out from under the gate and spacer are removed to expose the bottom Si layer. This Si layer is subsequently used as a seed material for the epitaxial growth of Si, forming the source and drain by connecting the two silicon channels of each fin, which results in the structure shown in Fig. 4d. Next, an interlayer dielectric is deposited and the surface is levelled using chemical mechanical planarisation (CMP). After the removal of the dummy gate, the SiGe layer separating the two

Physical Process	ViennaTS Model	Geometric Parameters
Si Epitaxy	ConstantRates	Si=+7nm
SiGe Epitaxy	ConstantRates	SiGe=+8nm
Si Epitaxy	ConstantRates	Si=+7nm
Fin Mask	Mask	Add predefined mask
SADP Mask Growth	ConstantRates	SADPMask=+15nm
Pattern SADP Mask	DirectionalRates	SADPMask=(0,0,-20nm)
Fin Mask Removal	BooleanOp	Remove Fin Mask
Fin Patterning [a]	DirectionalRates	Si/SiGe=-30nm, Spacer=-15nm
SADP Mask Removal	BooleanOp	Remove SADPMask Material
Dummy Gate Deposition	ConstantRates	PolySi=+55nm
Dummy Gate CMP	Planarization	PolySi
Gate Mask	Mask	Add predefined mask
Dummy Gate Patterning [b]	DirectionalRates	PolySi=-90nm
Gate Mask Removal	BooleanOp	Remove Gate Mask
Gate Spacer Deposition	ConstantRates	Hf=+12nm
Gate Spacer Patterning [c]	DirectionalRates	Hf=(0,0,-35nm)
Remove Fins at S/D	DirectionalRates	Si/SiGe=-20nm
S/D Epitaxy [d]	ConstantRates	Si=+11nm
ILD Deposition	ConstantRates	ILD=+35nm
ILD CMP	Planarization	ILD
Dummy Gate Removal	ConstantRates	PolySi=-80nm
NW Release [e]	ConstantRates	SiGe=-10nm
Gate Dielectric Deposition	ConstantRates	Hf=+2
Gate Metal Deposition	ConstantRates	TiN=+4nm
Gate Electrode Deposition	ConstantRates	W=+20nm
Final CMP [f]	Planarization	All materials

TABLE II: Summary of the geometric models used to create the structure, as shown in Fig. 4. The model "ConstantRates" isotropically deposits(+) or etches(-) the respective material. "DirectionalRates" deposits/etches the material only in a specified direction. "Mask" adds a predefined mask geometry, "Planarization" flattens the topology at a certain height, and "BooleanOps" is used here to completely remove materials. Superscript letters refer to the subfigures in Fig. 4. The exact parameters used in the simulation can be found at hpcwasm.github.io/viennats/#/simulation/stackednanosheet.

Si nanowires is etched away leaving the nanowires suspended in the air, as shown in Fig. 4e. Finally, the gate dielectric (HfO$_2$), the gate metal (TiN), and the gate contact material (W) are deposited to create the final structure shown in Fig. 4f.

VII. CONCLUSION

The C++ codebase of ViennaTS was compiled to WebAssembly and executed in the browser. By doing so, it was shown that porting complex C++ codebases to a performant portable format can be achieved with few changes to the original code. WebAssembly therefore presents a promising platform for the deployment of scientific software.

Native performance cannot be matched due to the technical limitations of portable code execution. However, we could show that the average performance gap for the process simulator ViennaTS is only a factor of 2.3 for Firefox and 2.8 for Chrome compared to native desktop execution. Since all computations are carried out on the client-side in the browser, this enables portable high performance process simulations, accurately describing complex state-of-the-art fabrication techniques commonly applied in the semiconductor industry.

REFERENCES

[1] WebAssemblyCommunityGroup(W3C). Webassembly Specification. [Online]. Available: https://webassembly.github.io/spec/

[2] O. Ertl et al. ViennaTS - The Vienna Topography Simulator. [Online]. Available: https://github.com/viennats/viennats-dev

[3] P. Manstetten and X. Klemenschits. ViennaTS - Webassembly Port. [Online]. Available: https://hpcwasm.github.io/viennats

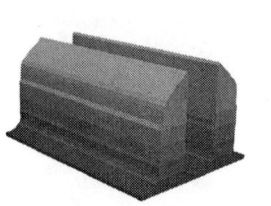

(a) Epitaxial growth and double patterning.

(b) Dummy gate patterning.

(c) Spacer formation and patterning.

(d) Source/drain epitaxy.

(e) Channel release.

(f) Final geometry after gate material deposition.

Fig. 4: Full stacked nanosheet process emulated in the browser using ViennaTS for WebAssembly. The figures show the geometry after the process steps described in each caption.

[4] R. T. Whitaker, "A Level-Set Approach to 3D Reconstruction from Range Data," *International Journal of Computer Vision*, vol. 29, no. 3, pp. 203–231, 1998.

[5] O. Ertl and S. Selberherr, "Three-Dimensional Topography Simulation using Advanced Level Set and Ray Tracing Methods," in *2008 International Conference on Simulation of Semiconductor Processes and Devices*. IEEE, 2008, pp. 325–328.

[6] A. Deveria. caniuse.com. [Online]. Available: https://caniuse.com/#search=webassembly

[7] WebAssemblyCommunityGroup(W3C). Webassembly System Interface. [Online]. Available: https://github.com/WebAssembly/WASI

[8] Google. Google Native Client. [Online]. Available: https://chromium.googlesource.com/native_client/src/native_client.git

[9] A. Zakai, Mozilla. asm.js. [Online]. Available: http://asmjs.org

[10] A. Zakai, "Emscripten: an LLVM-to-JavaScript Compiler," in *Proceedings of the ACM International Conference Companion on Object Oriented Programming Systems Languages and Applications Companion*. ACM, 2011, pp. 301–312.

[11] LLVM Developer Group. LLVM. [Online]. Available: http://llvm.org

[12] H. Mertens et al., "Gate-All-Around MOSFETs based on Vertically Stacked Horizontal Si Nanowires in a Replacement Metal Gate Process on Bulk Si Substrates," in *2016 Symposium on VLSI Technology*. IEEE, 2016, pp. 1–2.

[13] N. Loubet et al., "Stacked Nanosheet Gate-All-Around Transistor to Enable Scaling beyond FinFET," in *2017 Symposium on VLSI Technology*. IEEE, 2017, pp. T230–T231.

[14] S. Barraud et al., "Tunability of Parasitic Channel in Gate-All-Around Stacked Nanosheets," in *2018 IEEE International Electron Devices Meeting (IEDM)*. IEEE, 2018, pp. 21.3.1–21.3.4.

978-1-7281-0941-1/19 $31.00 © 2019 IEEE

A model of the interface charge and chemical noise due to surface reactions in Ion Sensitive FETs

Leandro Julian Mele*‡, Pierpaolo Palestri* and Luca Selmi†

*DPIA, University of Udine, 33100, Udine, Italy.

†DIEF, University of Modena and Reggio Emilia, 44100, Modena, Italy.

‡Corresponding author. Email: mele.leandrojulian@spes.uniud.it

Abstract—**We present a model of arbitrary chemical reactions at the interface between a solid and an electrolyte, aimed at computing the interface charge build-up and surface potential shift of ion-sensitive FETs in the presence of interfering ions. An expression for the rms value of the surface charge fluctuation and the resulting uncertainty in the ion concentration is derived as well. Application to nanoelectronic ISFET-based sensors for ions and proteins is demonstrated.**

Index Terms—**ISFET, surface binding reactions, cross-sensitivity, charge fluctuations, chemical noise.**

I. Introduction

The sensing of ions and biomolecules using CMOS technology opens numberless possibilities of low-cost portable sensors for chemical screening [1], epigenetics [2] and has drastically cut the costs of DNA sequencing [3]. Physics-based modelling of the transduction mechanism provides useful insights for the optimization of the sensor sensitivity and signal-to-noise ratio. In DC potentiometric sensors, transduction takes place at the interface between a solid material and an electrolyte and the essential physics of such interface and its first order site-binding chemical reactions can be approximately described with commercial TCAD [4]. However, sensitivity to a given ion is affected by the presence of interfering ions; since TCAD uses electrons and holes to mimic ions, it cannot describe complex electrolytes and surface reactions that involve multiple ions in a set of coupled equations [5].

In this paper, we derive expressions to account for complex interface reactions between ions and binding sites. The expression of the surface charge is coupled to the equilibrium Poisson-Boltzmann (PB) equation to obtain the corresponding potential variation. In addition, we derive a useful expression of the noise induced by chemical fluctuations (hereinafter simply chemical noise) due to the stochasticity of binding/unbinding events.

II. Model description

Consider the surface sites at the solid/liquid interface of an electrochemical Field Effect Transistor (FET) and assume they can interact with different ionic species dissolved in the electrolyte. The binding/unbinding with one charged species entails a change of the site's state and of its net charge. Each state, i, is thus characterized by a probability f_i (i.e. how

This work was supported by Italian MIUR and Flag-ERA through the project CONVERGENCE via the IUNET Consortium.

many sites over the total are in such state) and a net signed number of elementary charges z_i which may be non-integer. The simplest binding reaction is the first-order Langmuir adsorption [6] shown in Fig. 1 which can be seen as the elementary constituent of a generic set of chemical reactions.

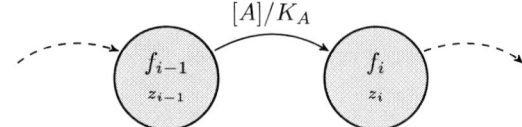

Fig. 1. Single arrow graph representing the steady state relationship between the $i-th$ state and the previous one, two arbitrary states of a site.

Here the site, from a complexed state $i-1$, binds the analyte A (whose volume concentration and dissociation constant of the reaction are $[A]$ and K_A respectively) leading to the complexed state i. In Fig. 1, nodes represent the states and the branches carry the coefficient that transforms one state probability with the previous when equilibrium is assumed. Reaction constants and ionic concentrations at the interface control the probability of the site to be in a given state and so the net surface charge expected at equilibrium.

A site with N configurations has $N-1$ reactions transforming one state into another. An extra equation is then given by the normalization of the f functions. This leads to a linear system with N equations and N unknowns:

$$\begin{cases} \vdots \\ f_i = \frac{[A]}{K_A} f_{i-1} \\ \vdots \\ \sum_{i=1}^{N} f_i = 1. \end{cases} \tag{1}$$

The surface charge density Q_S due to N_S identical sites per unit area interacting with the species, is

$$Q_S = q N_S \sum_{i=1}^{N} z_i f_i. \tag{2}$$

If different sites coexist, (1) and (2) should be solved for each site with the corresponding set of reactions, i.e. we have a matrix equation for each type of site. This general case is discussed in [7]. As for the electrostatics, if we assume equilibrium (i.e. the presence of at most one faradaic contact to the electrolyte [8]) and neglect steric effects, then the surface charge obviously influences the potential and the ionic distribution in the electrolyte according to the PB equations:

978-1-7281-0941-1/19 $31.00 © 2019 IEEE

$$\frac{d}{dx}\left[\varepsilon(x)\frac{d\psi(x)}{dx}\right] = -\left[\rho_m(x) + Q_{S,tot}\delta(x - x_S)\right] \quad (3)$$

$$\rho_m(x) = \sum_{l=1}^{N_{sp}} z_l q \left[\left[A_B^l\right]\exp\left(-\frac{z_l q}{k_b T}\psi(x)\right)\right] \quad (4)$$

where ε is the dielectric permittivity, ψ the electrostatic potential, $Q_{S,tot}$ is the sum of the surface charge densities of all different types of sites, z_l the $l-th$ ion/analyte net number of elementary charges, N_{sp} is the number of mobile species, $\left[A_B^l\right]$ the ion/analyte concentration in the bulk of the electrolyte. The interface ion concentrations depend on the voltage drop between the bulk of the electrolyte and the surface: a self-consistent solution of the coupled PB and the surface reaction equations is then necessary and has been implemented in this work.

III. VALIDATION OF THE MODEL

Our general modelling approach reproduces special cases proposed in the literature, including the site-binding model (SB) [9] at the basis of pH sensing or the surface complexation of chloride ions described by the modified SB (mSB) model [10]. In our model, this modification yields an additional state, linked by a chemical reaction concerning a deprotonation plus the chloride binding. Figure 2 illustrates the good agreement between our implementation and the experiments in [10].

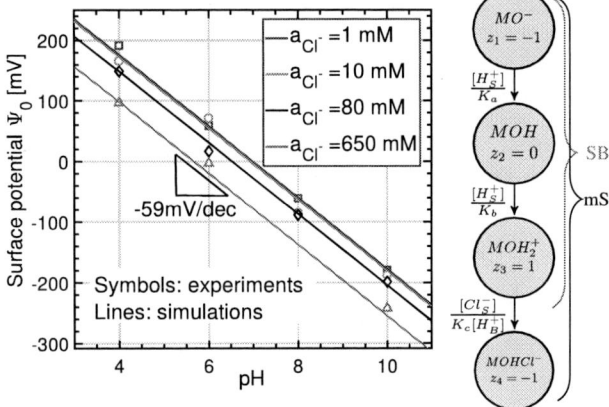

Fig. 2. Left: Comparison between our model and experiments from [10]. The parameters $K_a = 10^{-7}$ M, $K_b = 10^{-7}$ M $K_c = 3.3 \cdot 10^{-6}$ of the three chemical reactions and the number of sites $N_S = 10^{19}$ m^{-2} are taken from [10]. Right: SB and mSB reactions.

To prove that our approach can describe competing reactions, we compared our simulations with models found in the literature. The first one [11] was developed to estimate the surface charge generated at the gold/electrolyte interface of a sensor for the FimH protein. Here, the small oxidation of the gold surface leads to a pH response (SB) that electrostatically interferes with the attached mannoside ligands (Langmuir-like reactions), designed to capture FimH proteins.

Figure 3.a compares our simulations with the results obtained in [11] when using a constant double layer capacitance C_{DL} or the full PB model. We employ the same reaction parameters, number of hydroxyl groups per unit area and ionic

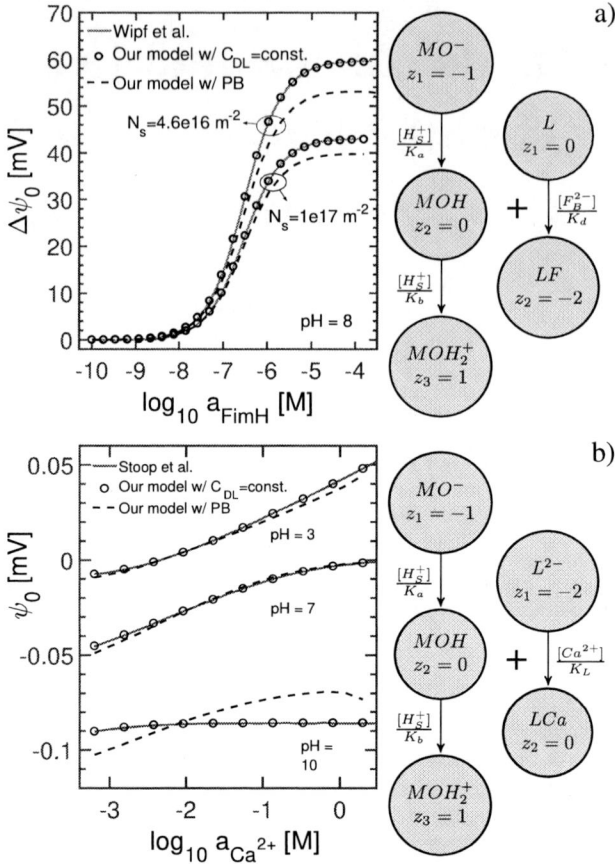

Fig. 3. Our model is applied to two cases of competing reactions using either the constant capacitance model and the full PB, and compared with results from the literature. In both cases the competing sites are SB and Langmuir-like reactions. a) Surface potential shift simulations for $C_{DL} = 0.1$ F/m^2 (symbols) and with the full PB equations (dashed lines) compared with the model in [11] (solid lines) for two densities of hydroxyl sites (SB). b) The surface potential is plotted versus CaCl$_2$ molar concentration for three different bulk pH values. Our model is compared with [12], using $C_{DL} = 0.16$ F/m^2.

strength as in the original work [11]. We find excellent agreement when embracing the same simplification as in [11] while the full PB solution predicts a lower plateau at high protein concentrations, likely because the ionic strength (especially at low concentrations such as in [11]) plays an important role in the C_{DL} which cannot be considered constant.

A second comparison is with the model in [12] for calcium Ca^{2+} sensing devices. Here the competing reactions are given by the SB model for the pH response and a Langmuir adsorption for the calcium sensing. To be consistent with the assumptions in [12], we consider the simple SB model instead of its modified version. The simulations have been performed both assuming a constant double layer capacitance, as in [12], and with the full PB system. Results are reported in Fig. 3.b. The comparison between the simplified (constant C_{DL}) and the more precise model shows that, once again, the assumption of constant double layer capacitance does not always hold true, as is here the case at high pH levels. The reason is that for large surface potentials the screening becomes more relevant and plays an important role while at low pH values the two competing reactions counterbalance their charge resulting in small surface potential and thus smaller variations in the diffusive capacitance.

IV. IMPACT OF CHEMICAL NOISE ON pH SENSING

The discrete nature of binding/unbinding events due to surface reactions produces noise on the sensor response. Here we use our general approach to estimate the fluctuation coming from the stochastic nature of the binding interface, which results in an additional random telegraph noise superimposed to the 1/f noise from the oxide/silicon interface [13]. If we assume that the read-out circuit has a bandwidth larger than the reaction rates, the integrated noise is given by the RMS deviation of the surface charge Q_S from its average value. Using f_i from (1), this is

$$\sigma_{Q_S} = q \sqrt{\frac{N_S}{WL} \left[\sum_{i=1}^{N} z_i^2 f_i - \left(\sum_{i=1}^{N} z_i f_i \right)^2 \right]}, \quad (5)$$

where W and L are the dimensions of the sensing area.

Taking for instance an ion-sensitive-FET (ISFET) as sketched in Fig. 5.d, we convert the RMS value σ_{Q_S} in a voltage fluctuation and then in an error on the pH reading as illustrated in the block diagram of Fig. 4. Firstly, we consider the surface potential variation due to a small surface charge perturbation q_p added at the electrolyte/oxide interface and compute the function G that converts charge fluctuations into surface potential fluctuations. Then, we compute the ISFET sensitivity to the measured bulk pH, H and eventually estimate the RMS value of the pH fluctuation. Figure 5 plots H,

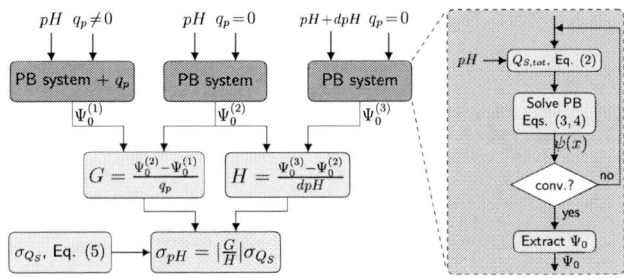
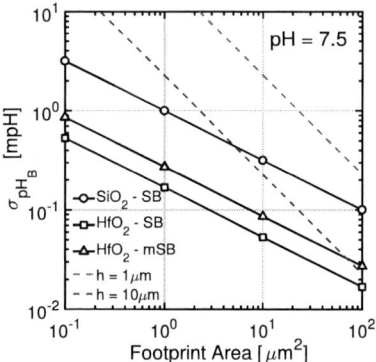

Fig. 4. Block diagram for the estimation of the chemical noise in an ISFET for pH sensing.

G and σ_{Q_S} as a function of the pH computed solving the self-consistent PB equation in a 1D electrolyte/oxide structure

(shown at the right). We see that H is larger for HfO$_2$ than for SiO$_2$ and almost perfectly Nernstian, ($H \approx 60mV/pH$) due to the larger number of sites (10^{19}m^{-2} [10] vs, $5 \cdot 10^{18}$m^{-2} [14]). At the same time, G is smaller for HfO$_2$. Therefore with HfO$_2$ the transfer function $|G/H|$ decreases but σ_{Q_S} increases. The net effect is that σ_{pH} is larger for SiO$_2$ than for HfO$_2$ (see, e.g., Fig. 6). The use of the mSB model for HfO$_2$ [10] (parameters for SiO$_2$ are not available to the best of our knowledge) results in changes of H and G and in an increase of σ_{Q_S} and σ_{pH}.

The estimation of the chemical noise can be relevant for instance for large arrays of pH sensors such as in the Ion Torrent platform for DNA sequencing [15]; in particular, the use of smaller and smaller μwells renders the binding more stochastic. Figure 6 shows the RMS value on the pH for different footprint areas (assuming sensing area = μwell area) and compares it to the signal the sensor should detect: dashed lines show the pH variation caused by a single proton inside the well, with the assumption that it diffuses towards the sensor surface. We see that σ_{pH_B} is negligibly small compared to single protonation events if the chamber is thin ($h = 1 \mu$m) but becomes comparable or larger for wells with larger footprint area devices while large areas require small well volume (thin well). The use of SiO$_2$ increases the chemical noise, and so does the account of the Cl$^-$ ions (compare the SB and mSB models). Note that Figure 6 includes only the surface chemistry as noise source.

Fig. 6. RMS error of the pH for different sensing areas (symbols) for the three cases in Fig. 5. Dashed lines: variations induced by the addition of one proton, assuming heights of the μwell of 1 and 10 μm. pH and Cl concentration are from [16].

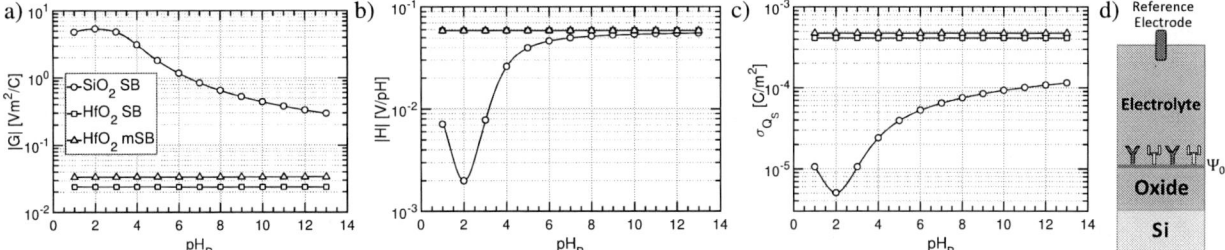

Fig. 5. The transfer functions G and H (see Fig. 4) are plotted for different bulk pHs, a) and b) respectively. c) the RMS noise computed using (5). SiO$_2$ and HfO$_2$ materials are compared using SB model and, for the latter material, the mSB model is also reported. d) sketch of the ISFET-like device simulated.

978-1-7281-0941-1/19 $31.00 © 2019 IEEE

V. CHARGE FLUCTUATION IN COMPETING REACTIONS

We apply our model to competing reactions as in [17], where the same algorithm as in Fig. 4 holds with the analyte concentration in place of the pH. Here, the bovine serum albumin (BSA) is used as functionalization to sense its antibody (aBSA) at the nano-molar range of concentrations. These specific sites coexist with those typical of mSB. For simplicity, we consider all the binding reactions to take place at the same surface, thus neglecting the size of the proteins and the formation of a Donnan potential [18]. The computation of the net charge carried by BSA and aBSA is based on the pH and buffer ionic strength of the solution, as in [17]. Uniform spatial distribution of the charge inside the protein is assumed, and only the fraction within a Debye length is considered in our model. No Boltzmann distribution is used for aBSA protein since its dimensions are much larger than the Debye length [11]. We first fit the voltage shift experiments in [17] (Fig. 7); then, we evaluate the noise (Fig. 8). We see that chemical noise sets both a lower and an upper limit to the sensitivity when saturation is reached. Furthermore, smaller areas show progressively smaller range of usability.

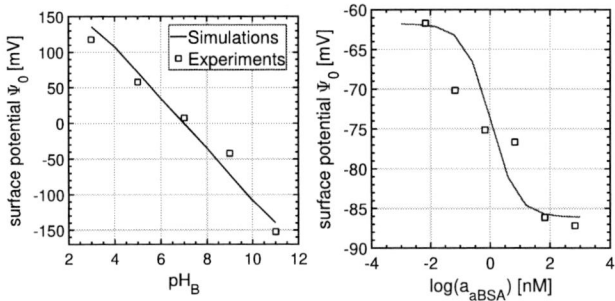

Fig. 7. Left: simulation of bare gold electrodes and data in [17] using K_a, K_b from [12] and K_c from [19]. Right: fit of the data with the addition of BSA/anti-BSA using $K_d = 1.2$ nM and $N_L = 1.7 \cdot 10^{16}$ m^{-2}.

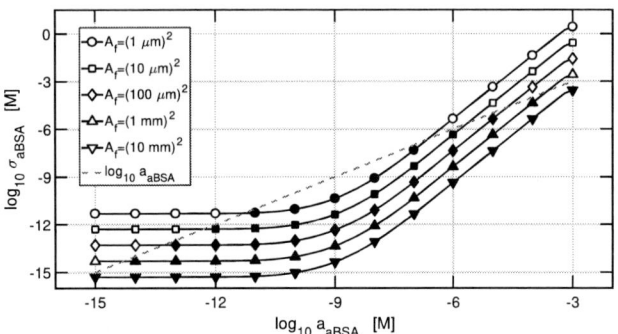

Fig. 8. Estimated RMS noise on the aBSA concentration for different sensing areas A_f. The red dashed line shows the relation $\sigma_{a_{aBSA}} = a_{aBSA}$. The filled symbols identify the useful operation range of the sensor.

VI. CONCLUSIONS

Our general model to compute the surface charge and chemical noise for arbitrary reactions at the biochemical sensor interface has been applied to ions and protein sensing. We found good agreement with existing ad-hoc models and went

further analysing the chemical noise. Results suggest that the latter may play an important role in determining the useful range of potentiometric sensors.

REFERENCES

[1] N. Moser, T. S. Lande, C. Toumazou, and P. Georgiou, "IS-FETs in CMOS and emergent trends in instrumentation: A review," *IEEE Sensors J.*, vol. 16, no. 17, pp. 6496–6514, 2016. doi: 10.1109/JSEN.2016.2585920

[2] D. Ma et al., "Adapting ISFETs for Epigenetics: An Overview," *IEEE Trans. Biomed. Circuits Syst.*, vol. 12, no. 5, pp. 1186–1201, Oct 2018. doi: 10.1109/TBCAS.2018.2838153

[3] M. Margulies et al., "Genome sequencing in microfabricated high-density picolitre reactors," *Nature*, vol. 437, no. 7057, p. 376, 2005. doi: 10.1038/nature03959

[4] A. Bandiziol, P. Palestri, F. Pittino, D. Esseni, and L. Selmi, "A TCAD-Based Methodology to Model the Site-Binding Charge at IS-FET/Electrolyte Interfaces," *IEEE Trans. Electron Devices*, vol. 62, no. 10, pp. 3379–3386, Oct 2015. doi: 10.1109/TED.2015.2464251

[5] F. Pittino, P. Palestri, P. Scarbolo, D. Esseni, and L. Selmi, "Models for the use of commercial TCAD in the analysis of silicon-based integrated biosensors," *Solid-State Electron.*, vol. 98, pp. 63–69, 2014. doi: 10.1016/j.sse.2014.04.011

[6] I. Langmuir, "The adsorption of gases on plane surfaces of glass, mica and platinum," *J. Am. Chem. Soc.*, vol. 40, no. 9, pp. 1361–1403, 1918. doi: 10.1021/ja02242a004

[7] L. J. Mele, P. Palestri, and L. Selmi, "General approach to model the surface charge and chemical noise induced by multiple surface reactions in potentiometric FET sensors," unpublished.

[8] A. J. Bard, L. R. Faulkner, J. Leddy, and C. G. Zoski, *Electrochemical methods: fundamentals and applications.* Wiley New York, 1980, vol. 2.

[9] D. E. Yates, S. Levine, and T. W. Healy, "Site-binding model of the electrical double layer at the oxide/water interface," *J. Chem. Soc. Farad. T. 1: Physical Chemistry in Condensed Phases*, vol. 70, pp. 1807–1818, 1974. doi: 10.1039/F19747001807

[10] A. Tarasov et al., "Understanding the electrolyte background for biochemical sensing with ion-sensitive field-effect transistors," *ACS nano*, vol. 6, no. 10, pp. 9291–9298, 2012. doi: 10.1021/nn303795r

[11] M. Wipf et al., "Label-Free FimH Protein Interaction Analysis Using Silicon Nanoribbon BioFETs," *ACS Sensors*, vol. 1, no. 6, pp. 781–788, 2016. doi: 10.1021/acssensors.6b00089

[12] R. L. Stoop et al., "Competing surface reactions limiting the performance of ion-sensitive field-effect transistors," *Sensor. Actuat. B-Chem.*, vol. 220, pp. 500–507, 2015. doi: 10.1016/j.snb.2015.05.096

[13] E. Accastelli et al., "Multi-Wire Tri-Gate Silicon Nanowires Reaching Milli-pH Unit Resolution in One Micron Square Footprint," *Biosensors*, vol. 6, no. 1, 2016. doi: 10.3390/bios6010009

[14] R. Van Hal, J. C. Eijkel, and P. Bergveld, "A general model to describe the electrostatic potential at electrolyte oxide interfaces," *Adv. Colloid. Interfac.*, vol. 69, no. 1-3, pp. 31–62, 1996. doi: 10.1016/S0001-8686(96)00307-7

[15] B. Merriman, I. T. R&D Team, and J. M. Rothberg, "Progress in Ion Torrent semiconductor chip based sequencing," *Electrophoresis*, vol. 33, no. 23, pp. 3397–3417, 2012. doi: 10.1002/elps.201200424

[16] J. Go and M. A. Alam, "The future scalability of pH-based genome sequencers: A theoretical perspective," *J. Appl. Phys.*, vol. 114, no. 16, p. 164311, 2013. doi: 10.1063/1.4825119

[17] A. Tarasov et al., "Gold-coated graphene field-effect transistors for quantitative analysis of protein–antibody interactions," *2D Materials*, vol. 2, no. 4, p. 044008, nov 2015. doi: 10.1088/2053-1583/2/4/044008

[18] P. Bergveld, "A critical evaluation of direct electrical protein detection methods," *Biosens. Bioelectron.*, vol. 6, no. 1, pp. 55–72, 1991. doi: 10.1016/0956-5663(91)85009-L

[19] M. Wipf et al., "Selective Sodium Sensing with Gold-Coated Silicon Nanowire Field-Effect Transistors in a Differential Setup," *ACS Nano*, vol. 7, no. 7, pp. 5978–5983, 2013. doi: 10.1021/nn401678u PMID: 23768238.

Investigation and Modelling of Single-Molecule Organic Transistors

Fabrizio Torricelli[1], Eleonora Macchia[2,3], Paolo Romele[1], Kyriaki Manoli[2], Cinzia Di Franco[4], Zsolt M. Kovacs-Vajna[1], Gerardo Palazzo[2,5,6], Gaetano Scamarcio[4,5], Luisa Torsi[2,3,6]

[1] Department of Information Engineering, University of Brescia, 25123 Brescia, Italy
[2] Dipartimento di Chimica, Università degli Studi di Bari "Aldo Moro", 70125 Bari, Italy.
[3] The Faculty of Science and Engineering, Åbo Akademi University, 20500 Turku, Finland.
[4] CNR, Istituto di Fotonica e Nanotecnologie, Sede di Bari, 70125 Bari, Italy.
[5] Dipartimento InterAteneo di Fisica "M. Merlin", Università degli Studi di Bari "Aldo Moro", 70125 Bari, Italy.
[6] CSGI (Centre for Colloid and Surface Science), 70125 Bari, Italy.

email: fabrizio.torricelli@unibs.it – luisa.torsi@uniba.it

Abstract— **Biofunctionalized organic transistors have been recently proposed as a simple wide-field single molecule technology. The further development and engineering of this disruptive technology urgently requires the understanding and modelling of the device operation. Here we show a physical-based numerical model of single molecule organic transistors. The model accurately reproduces the measurements in the whole range of protein concentrations with a unique set of parameters. The model provides quantitative information on the bioelectronic device operation. It is an important tool for further development of transistor-based single molecule.**

Keywords—modelling, single-molecule, organic transistor, biosensor

I. INTRODUCTION

Single molecule electronic detection is triggering a great deal of attention because it has the potential to revolutionize the current approach to diagnostics. Recently, a label-free single molecule detection technology based on millimeter-sized organic transistors with a gate electrode bio-functionalized with ~10^{12} bio-probes has been proposed [1,2]. Analogously to systems in nature [3,4] single molecule organic transistors (SiMoTs) provides a high interaction cross-section by means of a large number of highly packed receptors. SiMoT technology finds relevant application in several fields, as for example early diagnostic and personalized medicine, where a specific set of biomarkers has to be detected at the lowest concentration possible [5,6]. A low limit of detection, ideally single molecule, is the key for an effective and non-invasive early diagnostic. To meet the needs of these applications, the SiMoT technology has to be integrated in arrays and electronic systems. Moreover, the further development and engineering of the SiMoT technology urgently require a clear investigation and understanding of the impact of the biorecognition event on the device parameters.

In this work, we show a physical-based numerical model of SiMoTs. The electrical characteristics of SiMoTs are accurately reproduced when the ligand concentration ranges from 0 M up to 10^{-13} M with a unique set of physical and geometrical parameters. The model reveals that nano-scale single-protein interaction results in a macro-scale variation of the gate electrode work function measured as a shift of the transfer characteristics. In addition, the model shows that a single binding event affects several binding sites due to a

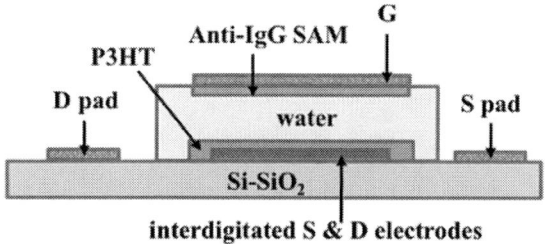

Fig. 1. Schematic side-view of a single-molecule organic transistor (SiMoT).

cooperative field-assisted mechanism and quantifies the number of binding sites affected by the protein binding.

II. SiMoT FABRICATION AND MEASUREMENTS

The SiMoT structure is shown in Fig. 1. Titanium/Gold (5nm / 50nm) source and drain electrodes are evaporated and patterned with photolithography on a Si/SiO$_2$ substrate. Before the electrodes deposition the Si O$_2$ surface is carefully cleaned in an ultrasonic bath of acetone and isopropanol for 10 minutes. The transistor channel width and length are W = 12800 µm and L = 5 µm, respectively. Poly(3-hexylthiophene-2,5-diyl) (P3HT) organic semiconductor (OSC) is dissolved in 1,2-dichlorobenzene (2.6 mg/ml), filtered with 0.2 µm PTFE filter, deposited by spin coating at 2000 r.p.m. and patterned ($A_{OSC} = 6.4 \ 10^{-3} \ cm^2$) to form the channel. To improve the polymer morphology the devices are annealed 1h on a hot plate at 80 °C. A polydimethylsiloxane well is glued and filled with 300 µl of water (HPLC grade) acting as gating medium. A Kapton foil covered by e-beam evaporated gold is used as a gate ($A_G = 0.6 \ cm^2$). The gate is positioned on the top of the water, in front of the transistor channel.

The biofunctionalized gate is obtained by covering a bare Au gate with densely-packed anti-human Immunoglobulin G (anti-IgG). According to the protocol reported in [1], the gate is cleaned in an ultrasonic isopropanol bath for 10 minutes, dried with nitrogen and cleaned with UV/ozone for 10 minutes. The chemical self-assembled monolayer (chem-SAM) is obtained by mixing a 10 mM solution of 10:1 3-MPA to 11-MUA in ethanol. The gate is immersed in the chem-SAM solution overnight in nitrogen. Subsequently, the chem-SAM formed on the gold surface of the gate electrode is

978-1-7281-0941-1/19 $31.00 © 2019 IEEE

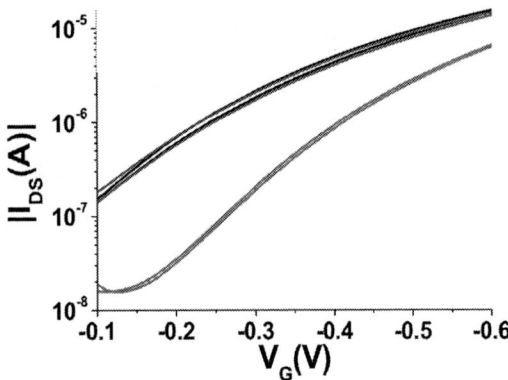

Fig. 2. Measured transfer characteristics of a single molecule transistor (SiMoT) at V_D= -0.4V before (blue and black curves) and after (red curve) biofunctionalization of the gate.

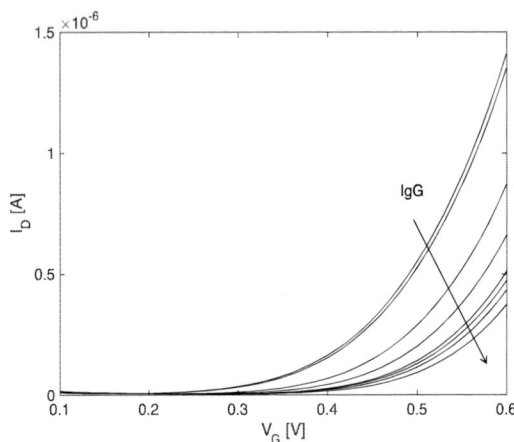

Fig. 3. Measured transfer characteristics of SiMoT at various protein concentrations. V_D= -0.4V.

activated with a solution of 200 mM EDC and 50 mM sulfo-NHS in aqueous solution for 2h. Then the gate is rinsed with PBS and incubated in a solution of phosphate buffer solution (PBS, pH 7.4, KCl 2.7 mM, NaCl 137 mM) with 100 µg/ml of anti-immunoglobulins (anti-IgG) for 2h. The gate is rinsed with PBS and incubate in a PBS solution with 1M ethanolamine in order to block the unreacted sulfo-NHS groups. Finally, the gate is immersed in a PBS solution with 100 µg/ml of BSA for 1h. The covalently bounded anti-IgG and the adsorbed BSA formed the bio-SAM. Typical transfer characteristics measured with a bare Au gate (blue and black curves) and with a biofunctionalized gate before the biosensing are shown in Fig. 2 (red curve). The comparison shows that after the gate biofunctionalization the transfer characteristic shift to more negative voltages, which result in a lower maximum drain current. The transistor with a biofunctionalized gate shows a maximum drain current larger than $6\ 10^{-6}$ A and the OFF current is about $2\ 10^{-8}$ A, resulting in an ON/OFF current ratio larger than 10^2. This is typical for P3HT electrolyte-gated transistors, thus confirming that the biofunctionalized gate does not negatively affect the transistor performance.

The SiMoT biosensing response is obtained by diluting the IgG protein in PBS at concentrations ranging from 10^{-21} M to 10^{-13} M. The gate is incubated in the analyte for 10 minutes and then measured with SiMoT, according to the SiMoT configuration showed in Fig. 1. Fig. 3 shows the measured transfer characteristics I_D-V_G of a SiMoT sensor as a function of the protein concentration (symbols). We found that the measured drain current reduces by increasing the nominal protein concentration. In addition, as a control experiment we incubated the gate biofunctionalized with anti-IgG with another protein, namely IgM. We found that after incubation of the gate with IgM the transfer characteristic is unaffected, thus proving the bioelectronic selectivity of the SiMoT.

III. SiMoT Model

To investigate the key physical and electrical device parameters, we developed a SiMoT physical model. In high-molecular-weight polymers charge transport occurs through an interconnected network of ordered regions while the amorphous fraction of the film does not participate to the transport. As charges reside in the ordered regions, the structural disorder in these regions define the electronic properties and it can be quantitatively measured by the

paracristallinity parameter g. In P3HT g = 3-7% [7] indicating the coexistence of localized and delocalized states: in paracrystalline aggregates the charge is transported by a mechanism where mobile charge is temporarily trapped in localized states, akin the multiple trapping and release [8-10].

According to this physical background, we calculate the drift-diffusion transport equation including a density of states DOS(E) = g_t(E) + g_b(E) where the localized states are described as:

$$g_t = \frac{N_t}{E_t} exp\left(\frac{E-E_{HOMO}}{E_t}\right) \quad (1)$$

Where N_t is the total density of localized states, E_t is the energy disorder, and E_{HOMO} is the highest occupied molecular orbital (HOMO) energy.

The delocalized states are describe as:

$$g_b = \frac{N_{HOMO}}{k_BT} \sqrt{\frac{E-E_{HOMO}}{k_BT}} \quad (2)$$

in the energy range $E > E_{HOMO}$. N_{HOMO} is the total density of HOMO states, k_B is the Boltzmann constant, and T is the temperature.

The hole density as a function of the Fermi energy level is calculated by solving the Fermi-Dirac integral and reads:

$$p(E_F) = \int_{-\infty}^{+\infty} DOS(E)[1 - f_d(E, E_F)]dE \quad (3)$$

where the DOS is given by the sum of Eq. (1) and Eq. (2), and the Fermi-Dirac distribution reads:

$$f_d(E, E_F) = \frac{1}{1+exp\left(\frac{E-E_F}{k_BT}\right)} \quad (4)$$

The drift-diffusion transport equation is integrated along the channel length and thickness and results [11,12]:

$$I_D = \frac{W}{L} \int_{V_S}^{V_M} \int_{V_{ch}}^{\varphi_s} \frac{\sigma_0 exp\left[\frac{q(\varphi-V_{ch})}{k_BT}\right]}{\sqrt{\frac{2q}{\varepsilon_s}\int_{V_{ch}}^{\varphi} p(\Psi,V_{ch})d\Psi}} d\varphi dV_{ch} \quad (5)$$

where ε_s the OSC permittivity, σ_0 = q μ_0 N_{HOMO} exp[E_G/(2k_BT)], E_G = E_{LUMO}-E_{HOMO} is the energy gap, μ_0 is the hole mobility in the delocalized states, E_{LUMO} is the lowest unoccupied molecular orbital energy level, q is the elementary charge, φ is the potential, φ_s is the surface potential, V_{ch} is the the Pseudo-Fermi potential. It is worth to note that in Eq. (5)

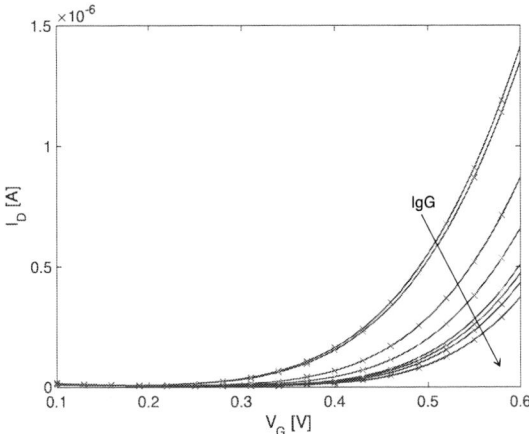

Fig. 4. SiMoT transfer characteristics as a function of IgG concentration. The protein concentration is varied from $6 \cdot 10^{-21}$ M to $6 \cdot 10^{-13}$ M. Symbols are the measurements, full lines are calculated solving Eqs. (1)-(7).

$E_F = E_{F0} + q \, (\varphi - V_{ch})$, where E_{F0} is the OSC Fermi energy level at equilibrium conditions. $V_M = \min\{V_G - V_T, V_D\}$ V_G is the gate voltage, V_T is the threshold voltage, and V_D is the drain voltage.

Eqs. (1)-(2) are numerically solved by applying the Gauss' law to the water/semiconductor interface. The electric field into the semiconductor can be calculated as:

$$F_x(\varphi_s) = \frac{C_{EDL}}{\epsilon_s} (V_{EL} - \varphi_s) \qquad (6)$$

where $C_{EDL} = \epsilon_0 \, \kappa_w / t_{EDL}$ is the electrolyte/semiconductor electric-double-layer (EDL) capacitance per unit area, κ_w is the relative permittivity of electrolyte, t_{EDL} is the thickness of the EDL at the electrolyte/semiconductor interface, $V_{EL} = (C_G A_G / C_T) (V_G - V_{FB}) + (C_{EDL} A_{osc} / C_T) \varphi_s$, $C_G = (\epsilon_0 \kappa_{SAM} / t_{SAM})$ is the gate/ electrolyte capacitance per unit area, κ_{SAM} and t_{SAM} are the relative permittivity and the thickness of the self-assembled monolayer, respectively, and $C_T = C_G A_G + C_{EDL} A_{osc}$.

The measured capacitances is equal to $C'_G = C_G A_G = 6 \, \mu F$ and $C'_{osc} = C_{osc} A_{osc} = 69 \, nF$ and therefore V_{EL} can be approximated as $V_{EL} \approx (C_G A_G / C_T) (V_G - V_{FB})$, where V_{FB} is the SiMoT threshold voltage and it accounts for the biofunctionalized gate work-function, the electrochemical potential of the electrolyte (in our case water) and the equilibrium Fermi energy of the OSC.

Finally, φ_s can be calculated from the continuity of the displacement at the electrolyte-semiconductor interface:

$$\frac{C_{osc}}{\epsilon_s} (V_G - V_{FB} - \varphi_s) = \sqrt{\frac{2q}{\epsilon_s} \int_{V_{ch}}^{\varphi} p(\Psi, V_{ch}) d\Psi} \qquad (7)$$

The integral expression (5) coupled with the electrolyte/semiconductor boundary condition (7) are numerically solved for each set $\{V_G, V_D, V_S\}$ and the drain current of SiMoT is calculated.

IV. Results and Discussion

Fig. 4 shows the comparison between the measurements and the model. Eqs. (1)-(7) accurately reproduces the drain current measured by varying the nominal ligand concentration with a single set of geometrical and physical parameters. More

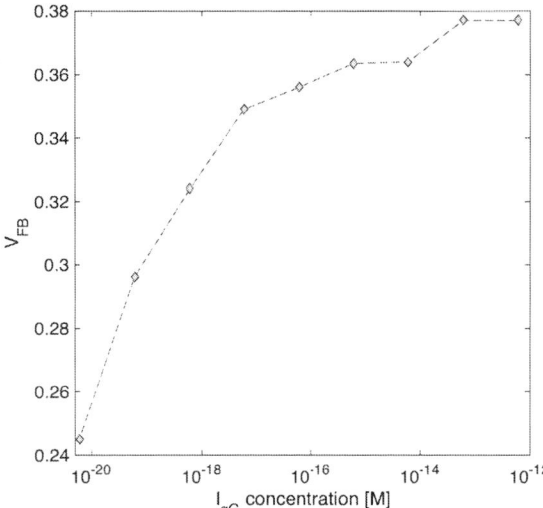

Fig. 5. SiMoT flat-band voltage extracted with the model, Eqs. (1)-(7), as a function of the nominal protein concentration IgG, which ranges from $6 \cdot 10^{-21}$ M to $6 \cdot 10^{-13}$ M.

in detail, the model parameters are obtained as follows. W, L, A_{OSC}, A_G, and T are measured. $\epsilon_s = 3 \, \epsilon_0$, $E_{LUMO} = 3.2$ eV, and $E_{HOMO} = 5.1$ eV are taken from [13]. The total density of states $N_{HOMO} = 1.28 \cdot 10^{20}$ cm^{-3} is taken from the density functional theory calculations [14]. The EDL capacitance at the water/semiconductor interface $C_{EDL} = 7.5 \, \mu F$ cm^{-2} was measured by means of electrochemical impedance spectroscopy. By fitting a single transfer characteristic measured after incubation with PBS with the drain current model Eqs (1)-(7) we obtained $\mu_0 = 0.058$ cm^2 V^{-1} s^{-1}, $N_t = 1.75 \cdot 10^{20}$ cm^{-3}, $E_t = 72 \cdot 10^{-3}$ eV, and $V_{FB} = 0.237$ V. It is worth noting that, according to [7], the extracted E_t yields a paracristallinity $g \sim 5\%$, which is fully consistent with the multiple trapping and release transport model here used.

We used the model to predict the drain current of SiMoTs as a function of the protein concentration. As shown in Fig. 4 the model (lines) accurately predicts the measurements (symbols) in the whole range of gate voltages and ligand concentrations. By varying the protein concentration, we found that only V_{FB} changes (Fig. 5) while all the other model parameters are the same. Since the biorecognition takes place at the surface of the gate electrode in contact with water, the variation of the flat-band voltage can be attribute to a variation of the gate work function that, in turn, is caused by a redistribution of charge along the bioprobes immobilized to the gate. Fig. 5 shows that V_{FB} significantly changes when the protein concentration is in the range $6 \cdot 10^{-21}$ M to $6 \cdot 10^{-18}$ M while at larger concentrations a modest response is shown and at IgG concentrations larger than $6 \cdot 10^{-13}$ M V_{FB} becomes constant. The flat-band saturation obtained when the incubated in PBS with a large number of IgGs is readily explained by considering that at large protein concentrations all the binding sites available on the gate have been affected by the various biorecognition events.

We can estimate the fraction (η) of binding sites immobilized on the gate affected by the biorecognition at a given concentration as: $N_{bioprobes} = \eta \, A_G / A_{anti-IgG}$, where $A_{anti-IgG} = 100$ nm^2 is the average area of a bioprobe obtained by means of surface plasmon resonance measurements and η is calculated by solving the following equation: $V_{FB}(c) = V_{FB0} (1 - \eta) + V_{FBmax} \eta$, where $V_{FB}(c)$, V_{FB0} and V_{FBmax} are V_{FB}

obtained with the model after incubation at the protein concentration c, in PBS and at the maximum concentration (viz. $6 \cdot 10^{-13}$ M), respectively. We found that about the 40% of the binding sites are affected by a single biorecognition event and the incubation of the biofunctionalized gate with tens of IgG proteins affects more than 60% of the binding sites.

The single-molecule sensitivity of the SiMoT could be explained by considering a cooperative effect taking place on the highly-packed SAM [1]. It is well known that the antigen/antibody binding is an exothermic reaction and the energy of the anti-IgG/IgG reaction is of the order of kJ/mol [15]. Therefore, it is reasonable to assume that part of this huge energy is transferred from the bio-SAM to the chem-SAM after a biorecognition event, and this results in a local desorption of the alkanethiols. By considering that after the ethanolamine functionalization step a hydrogen-bonding network is formed in the chem-SAM, the local defect can be propagated under the effect of the gate electric field. This results in a variation of the dipole orientation over several micrometer-square area of the gate and reflects in an appreciable variation of the gate work-functions which, in turn, is displayed by the SiMoT flat-band voltage.

V. CONCLUSION

A physical-based numerical model of single molecule organic transistors is proposed. The measured transfer characteristics are accurately described in the whole wide-range of protein concentration assessed. The physical and material bio-transistor device parameters are obtained. The model shows that the biorecognition events results in a variation of the transistor flat-band voltage, which enables the quantification of the number of bioprobes anchored to the gate involved in the biorecognition process. The proposed model is a valuable tool for the understanding, simulation and further development of the SiMoT technologies, including for example multi-modal biomarker assays and array integration.

ACKNOWLEDGMENT

The authors would like to acknowledge the financial support of the European Commission for the projects SiMBiT (Horizon 2020 ICT, contract n°824946).

REFERENCES

[1] E. Macchia et al., "Single-molecule detection with a millimitre-sized transistor," Nat. Commun., vol. 9, no. 3223, pp. 1-10, Aug. 2018. DOI: 10.1038/s41467-018-05235-z.

[2] E. Macchia et al., "Label-Free and Selective Single-Molecule Bioelectronic Sensing with a Millimeter-Wide Self-Assembled Monolayer of Anti-Immunoglobulins," Chem. Mater., Article early access, Jan 2019. DOI: 10.1021/acs.chemmater.8b04414

[3] T. Strünker, et al., "A K⁺-selective cGMP-gated ion channel controls chemosensation of sperm," Nat. Cell Biol., vol. 8, pp. 1149-1154, Jul. 2006. DOI: 10.1038/ncb1473

[4] T. Leinders-Zufall, et al., "Ultrasensitive pheromone detection by mammalian vomeronasal neurons," Nature, vol. 405, 792-796, Jun. 2000. DOI: 10.1038/35015572

[5] S. Sorgenfrei, et al., "Label-free single-molecule detection of DNA-hybridization kinetics with a carbon nanotube field-effect transistor," Nat. Nanotechnol., vol. 6, pp. 126-132, Jan. 2011. DOI: 10.1038/nnano.2010.275

[6] J. K. Rosenstein, S. G. Lemay, and K. L. Shepard "Single-molecule bioelectronics," Nanomed. Nanobiotechnol., vol.7, pp. 475-493, Dec. 2015. DOI: 10.1002/wnan.1323

[7] R. Noriega, et al., "A general relationship between disorder, aggregation and charge transport in conjugated polymers," Nat. Mater., vol. 12, pp. 1038-1044 Aug. 2013. DOI: 10.1038/nmat3722

[8] J. F. Chang, H. Sirringhaus, M. Giles, M. Heeney, and I. McCulloch, "Relative importance of polaron activation and disorder on charge transport in high-mobility conjugated polymer field-effect transistors," Phys. Rev. B, vol. 76, no. 20, pp. 205204, Nov. 2007. DOI: 10.1103/PhysRevB.76.205204

[9] F. Torricelli, L. Colalongo, L. Milani, Zs. M. Kovács-Vajna, and E. Cantatore, "Impact of energetic disorder and localization on the conductivity and mobility of organic semiconductor," IEEE 2011 Int. Conf. on Simul. of Semicon. Proc. and Devices (SISPAD), no. 12289431, Sep. 2011. DOI: 10.1109/SISPAD.2011.6035084

[10] Y. Yamashita, et al., "Transition Between Band and Hopping Transport in Polymer Field‐Effect Transistors," Adv. Mater., vol. 26, pp. 8169-8173, Oct. 2014. DOI: 10.1002/adma.201403767

[11] F. Torricelli, "Charge Transport in Organic Transistors Accounting for a Wide Distribution of Carrier Energies – Part I: Theory," IEEE Trans. Electron Devices, vol. 59, no. 5, pp. 1514-1519, May 2012. DOI: 10.1109/TED.2012.2187830

[12] F. Torricelli, K. O'Neill, G. H. Gelinck, K. Myny, J. Genoe, and E. Cantatore, "Charge Transport in Organic Transistors Accounting for a Wide Distribution of Carrier Energies – Part II: TFT Modeling," IEEE Trans. Electron Devices, vol. 59, no. 5, pp. 1520-1528, May 2012. DOI: 10.1109/TED.2012.2184764

[13] K. Hong, S. H. Kim, A. Mahajan, and C. D. Frisbie, "Aerosol Jet Printed p- and n-type Electrolyte-Gated Transistors with a Variety of Electrode Materials: Exploring Practical Routes to Printed Electronics," ACS Appl. Mater. Interfaces, vol. 6, no. 21, pp. 18704-18711, Oct. 2014. DOI: 10.1021/am504171u

[14] J. E. Northrup, "Atomic and electronic structure of polymer organic semiconductors: P3HT, PQT, and PBTTT," Phys. Rev. B, vol. 76, no. 24, pp. 245202, Dec. 2007. DOI: 10.1103/PhysRevB.76.245202

[15] M. Oda, K. Kozono, H. Mori, and T. Azuma, "Evidence of allosteric conformational changes in the antibody constant region upon antigen binding," Int. Immunol. 15, 417-426 (2003). DOI: 10.1093/intimm/dxg036

Advances in 3D CMOS image sensors optical modeling: combining realistic morphologies with FDTD

Benjamin Vianne[a], Axel Crocherie[a], Sofiane Guissi[b], Daniel Sieger[c], Stéphane Calderon[c], D. Rideau[a], Hélène Wehbe-Alause[a]

[a]STMicroelectronics, *Crolles*, France
[b]Lam Research, *Meylan*, France
[c]Coventor a Lam Research Company, *Villebon sur Yvette*, France
E-mail: benjamin.vianne@st.com, Sofiane.guissi@lamresearch.com

Abstract— **This paper describes an innovative methodology to investigate the relationship between device morphology and the optical performance of CMOS image sensors. By coupling a FDTD-based 3D Maxwell solver with silicon-accurate process modeling software, we have been able to analyze the sensitivity of image sensor quantum efficiency with respect to statistical variations in nm-scale device topology. Additionally, we studied pyramidal silicon structuration for quantum efficiency enhancement as proposed in [1].**

Keywords—*CMOS image sensor, process modeling, 3D nanophotonic simulation*

I. INTRODUCTION

Increasing the quantum efficiency (QE) of high-end CMOS Image Sensors (CIS) is very difficult, due to the impact of dark current on device noise, especially in the near-IR. Nanometer scale variability in device topology can substantially (and unexpectedly) affect both device noise and quantum performance. This impact of pixel stack morphology changes on the optical response of a CMOS sensor can be investigated using 3D FDTD nanophotonic simulation. However, complex 3D shapes built in silicon devices are difficult to model using conventional 3D design modeling tools. In addition, novel pixel structures being developed to increase QE are highly susceptible to degradation of potential gains in light sensing due to uneven surface morphology and

emphasize the need for more advanced 3D photonic modeling [1].

II. METHODOLOGY

A. Process flow modeling

In this study, we have simulated the optical response of advanced back-side illuminated sensors combining realistic structures emulated with a virtual fabrication platform [2] (Fig. 1) with 3D FDTD simulations [3]. This former 3D voxel-based technology can efficiently account for process-related local (but also statistical) variations of the device morphology (such as edge rounding, interface fluctuations, and non-conformal layer depositions) that can lead to additional parasitic light diffraction and interferences. Our simulation domain consisted of a 2x2 pixel architecture in a Bayer pattern. The process flow is integrally described within the modeling tool in a step-by-step manner, from Deep Trench Isolation (DTI) and Shallow Trench Isolation (STI) to back-side thinning. We note that accurate process modeling requires a rigorous calibration procedure. Nonetheless, as shown in Fig. 2, the simulated structure can accurately reproduce TEM cross sections, including roughness, irregularities and distortions.

B. Samples description

We executed a Delaunay meshing algorithm [4] after building the final 3D structure (Fig. 3a). This meshing

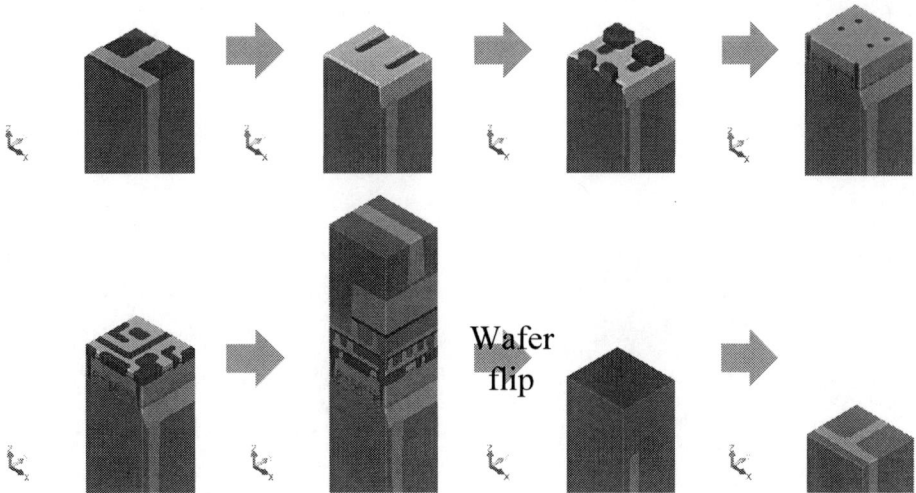

Fig. 1 3D process flow modeling of test case.

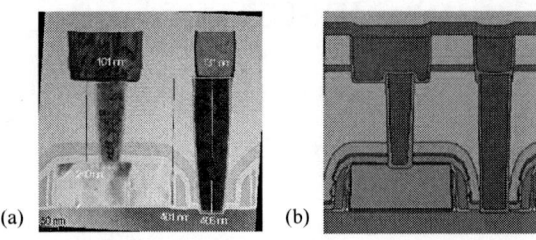

Fig. 2 (a) TEM cross section in FEOL and (b) corresponding 3D structure emulated in SEMulator3D®.

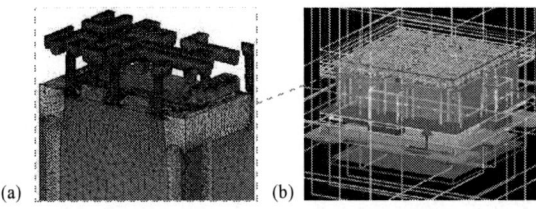

Fig. 3 (a) Magnified view of mesh structure of CIS devices and (b) its integration within the optical solver (substrate and dielectrics not shown).

algorithm consists of two main steps: a pre-processing step that extracts multi-material interfaces from the 3D voxel model, and the Delaunay mesh step that directly creates triangular surface and tetrahedral mesh elements that match the extracted surfaces. The topology of key devices e.g. DTI, STI, CMOS polysilicon gates and first level of Cu-interconnections (Metal1 or M1) are then imported into the optical design modeling tool as STL files (Fig. 3b), generated by the meshing.

Ideal pixel topology, which relies on building all devices from extrusion of GDS mask layers, is used as the reference model, as shown in Sample A (Fig. 4a). The same structure with fine grain topology, based upon the mesh import, is

displayed in Fig. 4b and shown in Sample B. The latter will be compared to the reference case in the following section. Sample C is identical to sample B except that voids are voluntarily introduced in the vicinity of DTI. Indeed, although depending upon the DTI process used, the filling of high aspect ratio trenches with oxide can induce the formation of voids in the vicinity of the trenches. This voids, which size does not exceed 40nm in width, are emulated using non-conformal deposition model (Fig. 4c). Finally, this methodology has been extended to evaluate design and process options to increase the QE. One solution consists of structuring the back-side surface of the Si substrate to increase the optical path for light in the photodiode [1,5]. To do so, a KOH-based chemistry is emulated to reproduce pyramidal patterns, as depicted in Sample D (Fig. 4d).

III. OPTICAL SIMULATION RESULTS

Unless specified, identical simulation conditions are set for the samples described in the previous section. 3D electromagnetic FDTD method is ran using a monochromatic collimated source from 400nm to 1000nm with steps of 20nm. Light is set with an incidence angle of 0° and two orthogonal polarizations are averaged incoherently to reproduce unpolarized light. Finally, Bloch boundaries conditions are applied and the maximum mesh step settings are 20nm in z and uniform at 10nm in x/y.

A. Sensitivity on QE between ideal and realistic topologies

The optical responses of an ideal pixel structure (Sample A) and a realistic pixel structure (Sample B) are shown in Fig. 5. Minor topological changes to an ideal CIS structure can generate a small percentage variation in absorption for red and NIR wavelengths. The transmission above M1 is particularly sensitive to topology changes for wavelengths greater than 600nm. This drift is well correlated with the differences of absorption in M1 layer (Fig. 6), emphasizing that the etch profile of metal interconnections and corner rounding play a non-negligible role in light absorption in the whole pixel. Discrepancies between ideal and realistic structures are clearly evidenced in cross sections of light propagation in two pixels at 920nm (Fig. 7). Therefore, the absorption in metal lines is critical and must be accurately simulated as less absorption in metal lines will lead

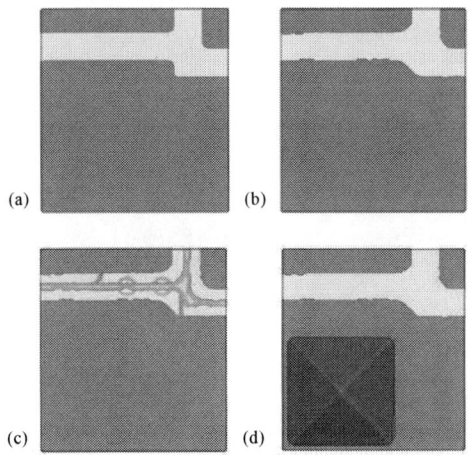

Fig. 4 Process emulation of a region of interest, viewed from the back side with (a) an ideal topology with planar Si surface (Sample A), (b) a realistic topology with planar Si surface (Sample B), (c) a realistic topology with planar Si surface and voids inside DTI (Sample C, filling oxide being transparent for clarity purpose), and (d) a realistic topology with structured Si surface (Sample D). Light blue color represents the DTI oxide and red color the Si active area. The pyramidal structure can be seen in (d).

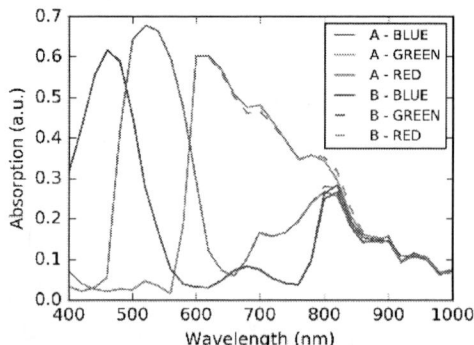

Fig. 5 FDTD simulations of light absorption in blue, green and red pixels in an ideal structure (Sample A in solid line) and in a realistic structure with expected manufacturing variability (Sample B in dashed line).

978-1-7281-0941-1/19 $31.00 © 2019 IEEE   352

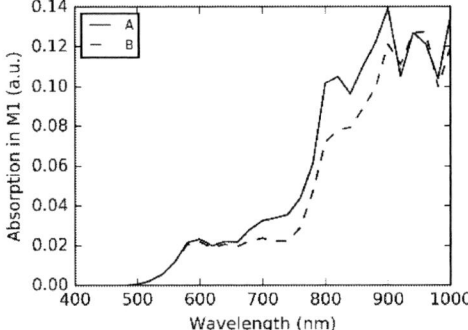

Fig. 6 Absorption of light in the first layer of metal interconnections in an ideal structure (Sample A in solid line) and in a realistic structure with expected manufacturing variability (Sample B in dashed line).

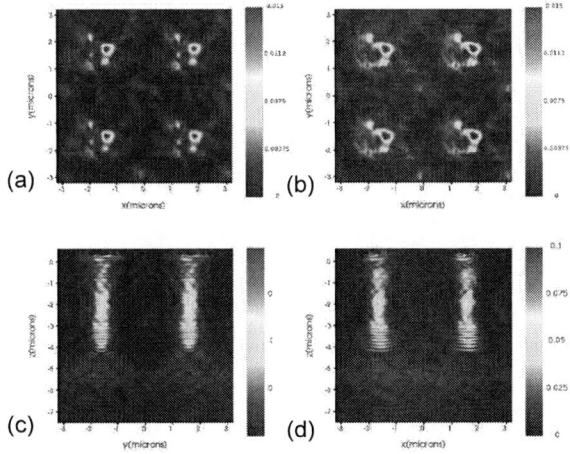

Fig.. 7 Comparison of light propagation simulated in two pixels: XY cross-sections above M1 between ideal (a) and realistic structure (b), YZ cross-sections between ideal (c) and realistic structure (d).

to more reflections back to the photodiode, i.e. more QE in the pixel.

B. Sensitivity on QE between structure with filled DTI and structures with voids inside DTI

The presence of voids in oxide filling DTI and their impact on optical performances are investigated in Sample B. In this case, incidence angle is set at 10° to force the light to go through the DTI. We discovered that 40 nm wide voids do not significantly affect the optical performance of a pixel (Fig. 8). This result highlights the minimal impact of void defects, which should in principle depend on the statistical occurrence of voids as well as their average size.

C. Sensitivity on QE between structure with planar back-side Si surface and structured back-side Si surface

QE enhancement solution with back-side silicon surface texturing is evaluated with Sample D. Process simulations are coupled with design variations of inverted pyramids pitch and size. Optical simulations show a x1.6 gain in absorption at 940nm for an optimal geometry consisting of a 3x3 array of pyramids with a square base of 400nm and a spacing of 100nm between each pyramid. This is compared to the reference case, Sample B, with a planar back-side surface (Fig. 9).

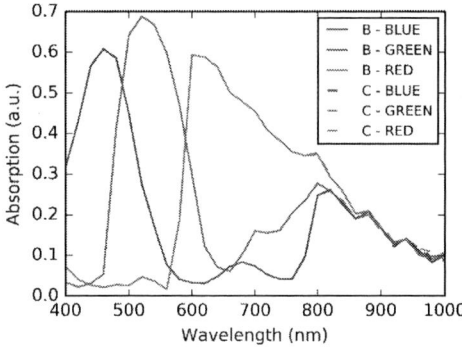

Fig. 8 FDTD simulations of light absorption in blue, green and red pixels in a realistic structure with perfectly filled DTI (Sample B in solid line) and in a realistic structure with 40nm wide voids inside DTI (Sample C in dashed line) at 10° incidence angle.

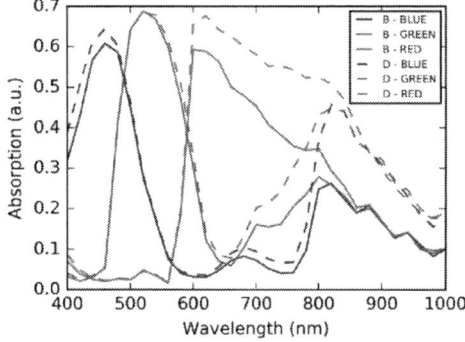

Fig. 9 FDTD simulations of light absorption in blue, green and red pixels in a realistic structure with planar back-side surface (sample B in solid line) and in the same structure with a 3x3 array of inverted pyramids (sample D in dashed line).

IV. CONCLUSIONS

Coupling a FDTD-based 3D Maxwell solver and a 3D virtual fabrication platform, using well calibrated process modeling of CIS, enables understanding of process variations effects which are not taken into account in conventional photonic modeling. This paper demonstrates the need to have an accurate morphological description of metallic interconnections since they play a major role in the absorption of light at pixel level, hence in the estimated QE. Moreover, the methodology used in this paper is particularly well-adapted for studying novel architectures using new Si structuration solutions to improve the optical performance of next generation CIS.

Full interoperability of the process modeling tool and 3D FDTD nanophotonic simulation will enable new and advanced studies. For example, the resilience of the quantum efficiency gain with respect to process variations (such as lithography misalignment, corner rounding of pyramidal patterns, as well as conformality of passivation layers deposited in etched cavities) will be characterized.

ACKNOWLEDGMENT

The authors would like to warmly thank the support teams from Coventor and Lumerical who participated in improving the interoperability between SEMulator3D® and FDTD.

REFERENCES

[1] S. Yokogawa et al., "IR sensitivity enhancement of CMOS Image Sensor with diffractive light trapping pixels," Scientific Reports, vol. 7, no. 1, p. 3832, Jun. 2017.

[2] www.coventor.com/semiconductor-solutions/semulator3d/

[3] www.lumerical.com/products/fdtd-solutions/

[4] J. Shewchuk, T. K. Dey and S.W. Cheng, Delaunay mesh generation, Chapman and Hall/CRC, 1st Edition, 2016.

[5] L. Frey, M. Marty, S. André and N. Moussy, "Enhancing Near-Infrared Photodetection Efficiency in SPAD With Silicon Surface Nanostructuration," in IEEE Journal of the Electron Devices Society, vol. 6, pp. 392-395, 2018.

Simulation of quantum dot based single-photon sources using the Schrödinger-Poisson-Drift-Diffusion-Lindblad system

Markus Kantner, Thomas Koprucki, Hans-Jürgen Wünsche and Uwe Bandelow

Weierstrass Institute for Applied Analysis and Stochastics
Mohrenstr. 39, 10117 Berlin, Germany
Email: kantner@wias-berlin.de

Abstract—The device-scale simulation of electrically driven quantum light sources based on semiconductor quantum dots requires a combination of the (semi-)classical semiconductor device equations with cavity quantum electrodynamics. We present a comprehensive quantum-classical simulation approach that self-consistently couples the (semi-)classical drift-diffusion system to a Lindblad-type quantum master equation. This allows to describe the spatially resolved carrier transport in complex, multi-dimensional device geometries along with the fully quantum-mechanical light-matter interaction in the quantum dot-cavity system. The latter gives access to important quantum optical figures of merit, in particular the second-order correlation function of the emitted radiation. In order to account for the quantum confined Stark effect in the device's internal electric field, the system is solved along with a Schrödinger–Poisson problem, that describes the envelope wave functions and energy levels of the quantum dot carriers. The approach is demonstrated by numerical simulations of a single-photon emitting diode.

Keywords—Single-photon sources, quantum-confined Stark effect, device simulation, quantum-classical coupling

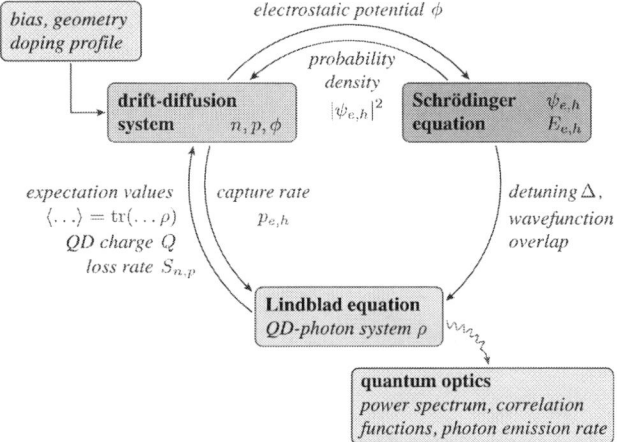

Fig. 1. Schematic illustration of the building blocks and the coupling structure of the Schrödinger–Poisson-Drift-Diffusion-Lindblad system (1)–(5). Adapted, with permission, from Ref. [3]. © SPIE 2019

I. INTRODUCTION

The currently unfolding "second quantum revolution" aims at the development of novel quantum technologies that exploit inherent quantum mechanical phenomena for communication and information processing tasks. Many applications, such as eavesdropping-secure encryption methods and optical quantum computers, rely on efficient quantum light sources that emit single photons on demand [1]. Semiconductor quantum dots (QDs) are promising optically active elements for such devices, as they provide an atom-like discrete energy spectrum and can be directly integrated into semiconductor-based photonic resonators by standard growth techniques. In the interest of compactness and scalability, electrical carrier injection is highly desirable to overcome the need for external excitation lasers. The theoretical analysis of the (semi-)classical carrier transport in quantum light emitting diodes can contribute significantly to their optimization, as the numerical simulation facilitates the understanding of counter-intuitive phenomena like rapid spreading of the injection current, which arise under the typically extreme operation conditions (cryogenic temperatures, very low current densities) [2].

On the step from basic research to real world applications, mathematical modeling and numerical simulation can assist the development and optimization of novel device designs. In many well-established simulation tools for optoelectronic devices (e.g., conventional laser diodes, LEDs etc.), the drift-diffusion model is coupled with semi-classical models for the light-matter interaction (e.g., Maxwell–Bloch equations, rate equations) to describe the optically active region. For devices operating in the quantum optical limit, however, fully quantum mechanical models are required to describe the light-matter interaction [4]. To meet this requirement, we have developed a hybrid quantum-classical model system [5], that self-consistently couples the drift-diffusion system to a Lindblad-type quantum master equation [6], which describes the microscopic QD-cavity system in second quantization (dissipative Jaynes–Cummings model). In this paper, we extend our approach by including a self-consistent Schrödinger–Poisson problem, to account for the energy shifts of the bound QD carriers in the device's internal electrostatic field via the quantum confined Stark effect.

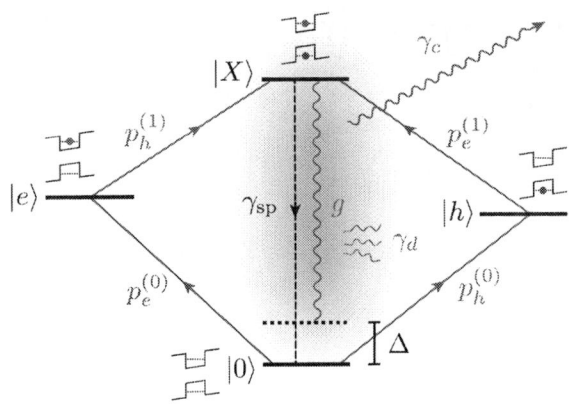

Fig. 2. QD-photon system described by a dissipative Jaynes–Cummings model with 4 electronic states: empty QD $|0\rangle$, single electron $|e\rangle = e^\dagger|0\rangle$, single hole $|h\rangle = h^\dagger|0\rangle$ and bright exciton $|X\rangle = h^\dagger e^\dagger|0\rangle$. The detuning Δ between the exciton energy and the cavity resonance is controlled by the device's internal electrostatic field. Adapted, with permission, from Ref. [3]. © SPIE 2019

II. HYBRID QUANTUM-CLASSICAL MODELING APPROACH

We describe a comprehensive modeling approach for the simulation of quantum light emitting diodes. The approach is based on the hybrid quantum-classical model system proposed in Ref. [5] and is extended by a self-consistent Schrödinger–Poisson problem modeling the envelope wave functions and energy levels of the QD carriers [3]:

$$-\nabla \cdot \varepsilon \nabla \phi = q\left(C + p - n\right) + Q, \quad (1)$$

$$\partial_t n - \frac{1}{q}\nabla \cdot \mathbf{j}_n = -R - S_n, \quad (2)$$

$$\partial_t p + \frac{1}{q}\nabla \cdot \mathbf{j}_p = -R - S_p, \quad (3)$$

$$H_\alpha^0 \psi_\alpha = E_\alpha \psi_\alpha \quad (\alpha \in \{e, h\}), \quad (4)$$

$$\partial_t \rho = -\frac{i}{\hbar}[\mathcal{H}, \rho] + \mathcal{D}\rho. \quad (5)$$

The system comprises the semiconductor device equations (1)–(3) for the transport and recombination dynamics of the quasi-free electrons and holes, a stationary one-particle Schrödinger equation (4) for each the QD bound electrons and holes, respectively, and a Lindblad-type quantum master equation (5) for the quantum statistical operator ρ. A schematic illustration of the "Schrödinger-Poisson-Drift-Diffusion-Lindblad system" (1)–(5) and the interconnection of its building blocks is shown in Fig. 1.

A. Semiconductor device equations

The electrostatic interaction between the freely moving and bound carriers of the system is described by Poisson's Eq. (1), where ϕ is the electrostatic potential, n and p are the densities of (continuum) electrons and holes, C is the doping profile, Q is the charge density of the QD carriers, q is the elementary charge and ε is the material's dielectric constant. The carrier

densities are related to the electrostatic potential ϕ and the quasi-Fermi energies $\mu_{c/v}$ by the state equations

$$n = N_c F_{1/2}\left(\frac{\mu_c + q\phi - E_c}{k_B T}\right), \quad p = N_v F_{1/2}\left(\frac{E_v - q\phi - \mu_v}{k_B T}\right),$$

where k_B is Boltzmann's constant, T is the absolute temperature, $E_{c/v}$ and $N_{c/v}$ denote the band edge energy and effective density of states of the conduction band/valence band, respectively, and $F_{1/2}$ is the Fermi–Dirac integral. The current densities $\mathbf{j}_{n/p}$ (3) are driven by the gradients of the quasi-Fermi energies

$$\mathbf{j}_n = M_n n \nabla \mu_c, \qquad \mathbf{j}_p = M_p p \nabla \mu_v,$$

where $M_{n/p}$ denotes the respective mobilities. The (net-)recombination rate R includes spontaneous emission, Shockley–Read–Hall and Auger recombination processes [7].

B. Schrödinger equation and wave function model

The energy levels and (envelope) wave functions of the QD carriers are determined by the stationary Schrödinger equation (4), where the Hamiltonian

$$H_\alpha^0 = -\frac{\hbar^2}{2}\nabla \cdot \frac{1}{m_\alpha^*}\nabla + U_\alpha \pm q\phi, \quad \alpha \in \{e, h\}, \quad (6)$$

involves a position-dependent effective mass m_α^*, the QD confinement potential U_α and the electrostatic potential ϕ given by Eq. (1). The confinement potential combines a finite potential barrier (in growth direction) and a harmonic in-plane confinement as typically assumed for lens-shaped InGaAs-QDs [8]. The Schrödinger Eqs. (4) are solved with outgoing wave conditions on a subset $\Omega_0 \subset \Omega$ of the full domain. In general, this is a non-Hermitian eigenvalue problem that yields complex eigenvalues $E_\alpha \in \mathbb{C}$ (quasi-bound states).

C. Lindblad master equation

The many-body Hamiltonian $\mathcal{H}$ in the quantum master Eq. (5) describes the one-particle energy contributions of the QD carriers, the energy of the quantized radiation field, the quantum-mechanical light-matter interaction and the Coulomb interaction between the bound carriers in second quantization:

$$\mathcal{H} = \varepsilon_e e^\dagger e + \varepsilon_h h^\dagger h + \hbar\omega_0 a^\dagger a + \hbar g\left(e^\dagger h^\dagger a + a^\dagger h e\right) - V_{e,h} e^\dagger h^\dagger h e. \quad (7)$$

Here, a and $a^\dagger$ are the bosonic annihilation and creation operators of the cavity photons and e (h) and $e^\dagger$ ($h^\dagger$) are the respective fermionic operators for the QD-bound electrons (holes). The single-particle energies $\varepsilon_{e/h} = \mathrm{Re}\left(E_{e/h}\right)$ are taken as the real values of the complex eigenvalues determined by Eq. (4). Moreover, $\hbar\omega_0$ is the resonance energy of the cavity and

$$g = d_{c,v}\sqrt{\frac{\omega_0}{\hbar\varepsilon_0 n_r^2 V_0}}\int_{\Omega_0} \mathrm{d}^3 r\, \psi_e^*(\mathbf{r})\,\psi_h(\mathbf{r})$$

is the light-matter coupling constant with the interband dipole moment $d_{c,v}$, mode volume V_0 and refractive index n_r. Finally, $V_{e,h}$ is the QD exciton binding energy.

978-1-7281-0941-1/19 $31.00 © 2019 IEEE

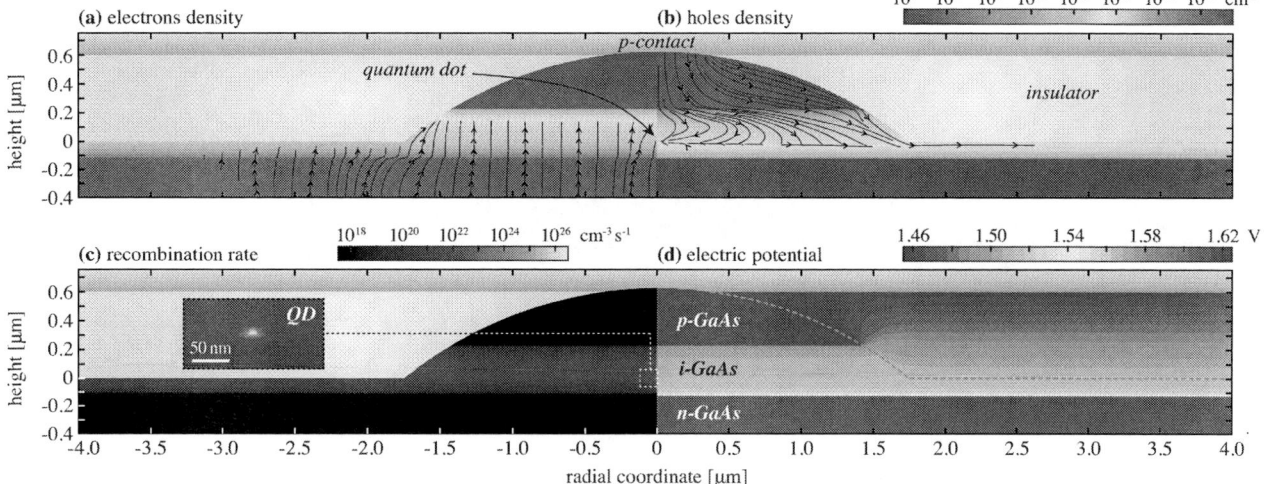

Fig. 3. Stationary injection at $I = 2.75\,\text{nA}$ ($U = 1.49\,\text{V}$) and $T = 30\,\text{K}$. The QD is located on the symmetry axis at $(r, z) = (0, 0)$. (a) Electron density n (color coded) and current density $\mathbf{j}_n$ (arrows indicate the direction of particle flux) and (b) hole density p (color coded) and current density $\mathbf{j}_p$. In the low-injection regime, the scattering of continuum carriers to the QD contributes notably to the current guiding. (c) Recombination rate and scattering losses $R + S_n + S_p$ of continuum carriers. (d) Electrostatic potential ϕ in the diode and the insulator domain. Adapted, with permission, from Ref. [3]. © SPIE 2019

The dissipation superoperator $\mathcal{D}(\rho)$ in Eq. (5) models the irreversible coupling of the quantum system to its macroscopic environment. For the sake of coupling to the semiconductor device equations, we separate the full dissipator

$$\mathcal{D}(\rho) = \mathcal{D}_e(\rho) + \mathcal{D}_h(\rho) + \mathcal{D}_0(\rho) \qquad (8)$$

into processes that change the charge of the QD ($\mathcal{D}_{e/h}(\rho)$) or leave it invariant ($\mathcal{D}_0(\rho)$). The individual terms are

$$\mathcal{D}_e(\rho) = p_e^{(0)} L_{e^\dagger(1-n_h)}\rho + p_e^{(1)} L_{e^\dagger n_h}\rho, \qquad (9a)$$

$$\mathcal{D}_h(\rho) = p_h^{(0)} L_{h^\dagger(1-n_e)}\rho + p_h^{(1)} L_{h^\dagger n_e}\rho, \qquad (9b)$$

$$\mathcal{D}_0(\rho) = \gamma_c L_a \rho + \gamma_{\text{sp}} L_{eh}\rho + \gamma_d L_{n_e n_h}\rho \qquad (9c)$$

where $L_A\rho = A\rho A^\dagger + \frac{1}{2}\left(A^\dagger A\rho + \rho A^\dagger A\right)$ is the Lindblad superoperator [6]. The capture of electrons and holes into the QD is described by $\mathcal{D}_e(\rho)$ and $\mathcal{D}_h(\rho)$, where $p_{e/h}^{(0)}$ is the capture rate for scattering into an empty QD. If the QD is already charged by a single carrier, the enhanced capture rate $p_{e/h}^{(1)}$ applies. The capture rates are driven by the continuum carrier densities n, p in the vicinity of the QD

$$p_e^{(i)} = \Gamma_e^{(i)} \int_{\Omega_0} \mathrm{d}^3 r\, |\psi_e|^2 n, \quad p_h^{(i)} = \Gamma_h^{(i)} \int_{\Omega_0} \mathrm{d}^3 r\, |\psi_h|^2 p,$$

$i \in \{0, 1\}$, with capture coefficients $\Gamma_{e/h}^{(i)}$ fitted to microscopic calculations. The charge neutral dissipative processes in $\mathcal{D}_0$ are the emission of cavity photons with loss rate γ_c, the spontaneous emission of the QD exciton to waste modes with decay rate γ_{sp} and phenomenological pure dephasing with γ_d.

D. Coupling terms in macroscopic system

Finally, due to charging of the QD and scattering of continuum carriers to bound states, the quantum system couples back to the semi-classical transport system (1)–(3). The QD charge

density Q in Poisson's Eq. (1) is given by the expectation value of the field operator $\Psi(\mathbf{r}) = \psi_e(\mathbf{r})e + \psi_h^*(\mathbf{r})h^\dagger$

$$Q(\rho) = -q\langle\Psi^\dagger\Psi\rangle \approx q|\psi_h|^2\langle h^\dagger h\rangle - q|\psi_e|^2\langle e^\dagger e\rangle, \quad (10)$$

where $\langle\ldots\rangle = \text{tr}(\rho\ldots)$, i.e., the QD carrier's spatial probability distributions (from Eq. (4)) multiplied by the occupation probability obtained from the quantum master equation (5). Following Ref. [5], the loss terms (due to capture of continuum carriers to the QD) on the right hand side of the continuity equations (2)–(3) are modeled using the charge number operator $N = e^\dagger e - h^\dagger h$ and the dissipators (9a)–(9b) as

$$S_n = +|\psi_e(\mathbf{r})|^2 \text{tr}(N\mathcal{D}_e(\rho)), \qquad (11a)$$

$$S_p = -|\psi_h(\mathbf{r})|^2 \text{tr}(N\mathcal{D}_h(\rho)). \qquad (11b)$$

The definitions (10)–(11) guarantee charge conservation [3, 5].

III. NUMERICAL METHOD AND SIMULATION RESULTS

The hybrid quantum-classical model system (1)–(5) is applied to simulate the stationary operation characteristics of the single-photon emitting diode shown in Fig. 3. The diode features a monolithically fabricated microlens structure on top, which is optimized for a high coupling efficiency to an external optical fiber [9]. The top-contact is assumed to consist of an optically transparent material (e. g., ITO), placed on an thick insulator layer with relative permittivity $\varepsilon_s = 3.0$. There is only a small semiconductor-contact interface, see Fig. 3. On the insulator domain, only Poisson's Eq. (1) is solved. The GaAs material parameters for the drift-diffusion system are taken from Ref. [10], the doping densities are assumed as $N_D^+ = 2 \times 10^{18}\,\text{cm}^{-3}$ and $N_A^- = 1 \times 10^{19}\,\text{cm}^{-3}$. Despite the low temperature $T = 30\,\text{K}$, full ionization of the dopants is assumed due to the metal-insulator transition. For the carrier mobilities $M_{n,p}$ a temperature and doping dependent low field

(a) QD occupation/ photon generation rate

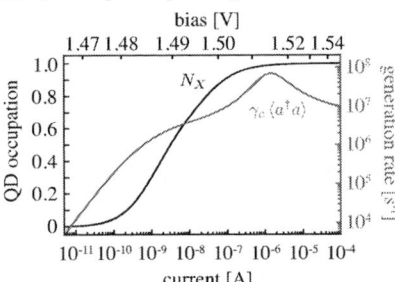

(b) spectrum

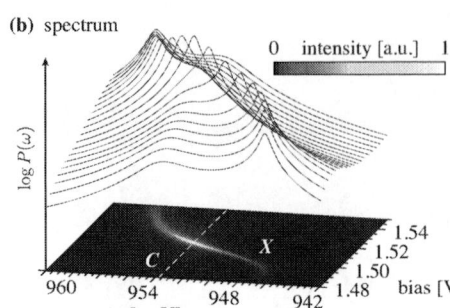

(c) second-order correlation function

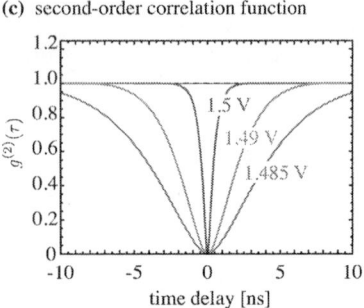

Fig. 4. (a) Exciton occupation probability N_X and photon emission rate $\gamma_c \langle a^\dagger a \rangle$ as a function of the injection current (or applied bias). (b) Power spectrum $P(\omega)$ calculated from Eq. (12) vs. applied bias. The exciton energy (X) is blue-shifted with the increasing bias due to the quantum confined Stark effect. The resonance with the optical cavity mode (C) appears at $U \approx 1.516\,\mathrm{V}$ and yields a maximum photon emission rate of about 70 MHz. (c) Time-resolved second-order correlation function $g^{(2)}(\tau)$ at different applied voltages. Adapted, with permission, from Ref. [3]. © SPIE 2019

model is used [11]. The full system (1)–(5) is solved by iterating (1)–(3)→(4)→(5)→(1)–(3)→ ... until convergence is reached (cf. Fig. 1 (a)). The drift-diffusion system (1)–(3) is discretized using a finite volume Scharfetter–Gummel method for Fermi–Dirac statistics [12, 13]. We use the temperature embedding method described in Ref. [14] to cope with the ill-conditioned discrete system at cryogenic operation temperature and low bias. Schrödinger's Eq. (4) is discretized using a second order finite difference scheme.

Figure 1 (a) shows the QD occupation and photon generation rate as a function of the applied bias. The power spectrum

$$P(\omega) = \frac{1}{2\pi} \int_{-\infty}^{\infty} d\tau\, e^{-i\omega\tau} \langle a^\dagger(\tau) a(0) \rangle \qquad (12)$$

is shown in Fig. 1 (b). At $U \approx 1.516\,\mathrm{V}$ the QD exciton is tuned into resonance with the cavity mode (via the quantum confined Stark effect), which yields a maximum single-photon generation rate of about 70 MHz, see Fig. 1 (a). At high injection currents, excitation-induced dephasing leads to a notable broadening of the emission line. Slightly below the diode's threshold voltage, the loss terms $S_{n/p}$ are the dominant terms on the right hand side of the continuity Eqs. (2)–(3), see Fig. 3 (c), such that the QD appreciably contributes to current guiding, see Fig. 3 (a, b). The second-order correlation function

$$g^{(2)}(\tau) = \frac{\langle a^\dagger(0)\, a^\dagger(\tau)\, a(\tau)\, a(0) \rangle}{\langle a^\dagger(0)\, a(0) \rangle^2} \qquad (13)$$

describes the single-photon purity of the device and is plotted in Fig. 1 (c) for different voltages. The characteristic dip at $g^{(2)}(0) \approx 0$ indicates *anti-bunching* of the emitted photons, which is a truly non-classical feature of the optical field.

IV. OUTLOOK AND CONCLUSIONS

The hybrid quantum-classical model system for the simulation of electrically driven quantum light sources introduced in Ref. [5] has been extended by a self-consistent Schrödinger–Poisson system. The extended model allows to describe important phenomena such as the quantum confined Stark effect and resonances with the cavity mode. It might be used to investigate spectral diffusion of the emission energy due to stochastic fluctuations in future works.

ACKNOWLEDGMENT

This work was funded by the German Research Foundation (DFG) under Germany's Excellence Strategy – EXC2046: Berlin Mathematics Research Center MATH+ (grant AA2-3).

REFERENCES

[1] P. Michler, ed., *Quantum Dots for Quantum Information Technologies*. Series in Nano-Optics and Nanophotonics, Cham: Springer, 2017.

[2] M. Kantner, U. Bandelow, T. Koprucki, J.-H. Schulze, A. Strittmatter, and H.-J. Wünsche, "Efficient current injection into single quantum dots through oxide-confined p-n-diodes," *IEEE Trans. Electron Devices*, vol. 63, no. 5, pp. 2036–2042, 2016.

[3] M. Kantner, "Hybrid modeling of quantum light emitting diodes: Self-consistent coupling of drift-diffusion, Schrödinger–Poisson and quantum master equations," *Proc. SPIE*, vol. 10912, p. 109120U, 2019.

[4] W. W. Chow and F. Jahnke, "On the physics of semiconductor quantum dots for applications in lasers and quantum optics," *Prog. Quantum Electron.*, vol. 37, no. 3, pp. 109–184, 2013.

[5] M. Kantner, M. Mittnenzweig, and T. Koprucki, "Hybrid quantum-classical modeling of quantum dot devices," *Phys. Rev. B*, vol. 96, no. 20, p. 205301, 2017.

[6] H.-P. Breuer and F. Petruccione, *The Theory of Open Quantum Systems*. Oxford: Oxford University Press, 2002.

[7] S. Selberherr, *Analysis and Simulation of Semiconductor Devices*. Vienna: Springer, 1984.

[8] T. R. Nielsen, P. Gartner, and F. Jahnke, "Many-body theory of carrier capture and relaxation in semiconductor quantum-dot lasers," *Phys. Rev. B*, vol. 69, p. 235314, Jun 2004.

[9] P.-I. Schneider, N. Srocka, S. Rodt, L. Zschiedrich, S. Reitzenstein, and S. Burger, "Numerical optimization of the extraction efficiency of a quantum-dot based single-photon emitter into a single-mode fiber," *Opt. Express*, vol. 26, no. 7, pp. 8479–8492, 2018.

[10] V. Palankovski and R. Quay, *Analysis and Simulation of Heterostructure Devices*. Series in Computational Microelectronics, Vienna: Springer, 2004.

[11] M. Sotoodeh, A. H. Khalid, and A. A. Rezazadeh, "Empirical low-field mobility model for III-V compounds applicable in device simulation codes," *J. Appl. Phys.*, vol. 87, no. 6, pp. 2890–2900, 2000.

[12] T. Koprucki, N. Rotundo, P. Farrell, D. H. Doan, and J. Fuhrmann, "On thermodynamic consistency of a Scharfetter–Gummel scheme based on a modified thermal voltage for drift-diffusion equations with diffusion enhancement," *Opt. Quantum. Electron.*, vol. 47, pp. 1327–1332, 2015.

[13] P. Farrell, N. Rotundo, D. H. Doan, M. Kantner, J. Fuhrmann, and T. Koprucki, "Drift-diffusion models," in *Handbook of Optoelectronic Device Modeling and Simulation: Lasers, Modulators, Photodetectors, Solar Cells, and Numerical Methods* (J. Piprek, ed.), vol. 2, ch. 50, pp. 733–771, Boca Raton: CRC Press, Taylor & Francis Group, 2017.

[14] M. Kantner and T. Koprucki, "Numerical simulation of carrier transport in semiconductor devices at cryogenic temperatures," *Opt. Quantum. Electron.*, vol. 48, no. 12, p. 543, 2016.

978-1-7281-0941-1/19 $31.00 © 2019 IEEE

A generalized multi-particle drift-diffusion simulator for optoelectronic devices

Daniele Rossi[†], Matthias Auf der Maur[†,*], Aldo Di Carlo[†]

[†]Dept. of Electronic Engineering, Università degli Studi di Roma "Tor Vergata", Via del Politecnico 1, 00133 Rome, Italy
[*]Email: auf.der.maur@ing.uniroma2.it

Abstract—**We present a generalized multi-particle drift-diffusion model capable to overcome the limitations imposed by the classic drift-diffusion model. It was designed as flexible and reusable tool that takes into account explicitly multiple carrier populations, whether charged and neutral, allowing to consider also e.g. exciton transport or ionic motion, crucial for a relevant number of device structures.**

Index Terms—**Band-to-band transition, drift-diffusion, exciton transport, ion migration, organic light emitting diodes (OLEDs), organic optoelectronic, semiconductor device modeling.**

I. INTRODUCTION

The simulation of electronic devices is nowadays a task that requires to go far beyond the simple picture of semiconductors with electrons and holes transport in corresponding bands. In the last two decades several applications have been intensively studied and developed. Indeed, devices as organic light emitting diodes (OLEDs), organic photovoltaics (OPVs) and organic field effect transistors (OFETs), exhibit complex electronic properties due to the presence of amorphous or regular assembly of polymer/small molecules in such materials. Besides electrons and holes, in devices for OPVs, such as dye-sensitized solar cells (DSSCs) and perovskite solar cells (PSCs), the ions transport plays a role in the device operation.

The traditional approach for semiconductor transport modeling in electronic devices is based on semi-classical transport equations known as drift-diffusion model (DD), or the van Roosbroeck equations [1]. In the DD model the set of equations is formulated for an electron and a hole population, assumed each in local thermal equilibrium with the host material.

Due to the complexity of mechanisms involved in new technology devices, over the years, several extensions of the DD model have been developed, but they are particularly designed for specific situations, such as for transport modeling in quantum dots (QDs) [2], or in multi quantum wells (MQWs) LEDs simulation [3].

The purpose of this work is to formulate and implement a generalization of the semi-classical drift-diffusion model extendible for multi-particle systems, flexible enough to allow the investigation of a wide range of devices for which this modeling is applicable.

II. THE MODEL

We propose a multi-particle drift-diffusion model (mp-DD) [4] that allows to overcome all limitations imposed by having only two-carrier transport model, as in traditional drift-diffusion. This is done by splitting the total particle population of the system into sub-populations, that are weakly coupled with each other compared to the corresponding relaxation time. The modeling approach is based on two ingredients:

- the model is extended to more than two carriers population, each individually assumed in a local thermal equilibrium and characterized by a local quasi-Fermi level;
- transitions between populations are formulated with strictly thermodynamic consistency, appearing as generation-recombination terms in the coupled system of equations.

The mp-DD allows to define any number of carriers, each one with its own properties, as spin, charge and density of states (DOS). Sticking to the stationary case, the system of equations reads as

$$\nabla \cdot (\varepsilon_0 \varepsilon_r \nabla \varphi - \mathbf{P_0}) = -q \sum_i z_i n_i - qC \tag{1a}$$

$$\nabla \cdot (\mu_i n_i \nabla \phi_i) = \text{sgn}(z_i) R_i \ , \ \forall i. \tag{1b}$$

The Poisson equation in Eq. (1a) is used to calculate the electrostatic potential φ by considering the overall charged carrier densities and fixed charges, respectively indicated by n_i and C. z_i and q represent the i-th carriers charge number and the elementary charge, while $\mathbf{P_0}$ indicates the fixed polarization that models e.g. piezoelectric and spontaneous polarization. The set of continuity equations in Eq. (1b) describes the transport of all carriers included in the system, both charged and neutral. On the left side of Eq. (1b), the carrier flux is written in terms of gradient of the quasi-Fermi potential ϕ_i, μ_i is the carrier mobility, while the term $sgn(z_i)$ adjusts the sign according to the definition of the quasi-Fermi potential as $-q\phi_i = E_{F,i}$. The term R_i is the total recombination-generation rate of the i-th carrier and generally depends on different carrier densities (n_i), therefore couples different carrier populations.

The implementation is done using quasi-Fermi potentials as primary variables so that we do not need to explicitly use the generalized Einstein relation, or specially treat spatial dependent material properties.

$$n_i = \int_{-\infty}^{\infty} D_i(E) \frac{1}{\left(\frac{E + \text{sgn } z_i \ q\phi_i - z_i q\varphi}{k_B T}\right) \pm 1} \, dE. \quad (2)$$

Here, $D_i(E)$ represents the DOS of the i-th carrier, while $\pm$ denotes the possibility to use of both the Fermi-Dirac and Bose-Einstein statistics. The latter is required in order to describe quasi-particles like excitons, thus z_i can be also 0. In analogy to chemistry, all recombination/generation rates are written as reactions between different species in the system.

$$\sum_j \alpha_j n_j \rightleftharpoons 0 \rightarrow \sum_j \alpha_j \phi_j = 0 \quad (3a)$$

$$r_i = \alpha_i \left[1 - e^{-\frac{q}{k_B T} \sum_j \alpha_j \phi_j} \right] \sum_{\{\theta_j\}} \gamma(\{\theta_j\})$$

$$\times \prod_j \left[\frac{1}{2}(1 - \text{sgn}(\alpha_j)) \pm f_j(\theta_j) \right]^{|\alpha_j|}. \quad (3b)$$

Equation (3a) shows the reaction and the corresponding thermodynamic equilibrium condition in the isothermal case. Here, α_j is an integer that represents the stoichiometric coefficient of the j-th particle, which determines the number of particles involved in the reaction for population j. Furthermore, its sign determines whether the particle is destructed or generated in the reaction.

Equation (3b) originates from the inter-band scattering terms appearing in the Boltzmann transport equation, using however equilibrium distribution functions. It gives the general expression for the net recombination rate of a specific recombination process in which the i-th carrier is involved, such that $R_i = \sum r_i$. $f_j(\theta_j)$ is the equilibrium distribution of carrier j, depending on a set of degree of freedoms (θ_j), like e.g. spin or crystal momentum. From Eq. (3b) any recombination model can be derived, e.g. from the simple radiative decay of excitons up to three-carriers mechanisms such as Auger recombination [4]. The model is implemented in the simulation software TiberCAD [5] using the Galerkin finite element method (FEM), and calculates solutions of the non-linear equations with Newton method.

In the following, we present two different contexts where the multi-particle drift-diffusion model can be exploited.

III. SIMULATION OF PHOSPHORESCENT OLED

The need of very wide bandgap and the short lifetime of the blue emitters limit the performance of RGB OLED displays. For this reason, the fabrication of highly efficient deep-blue OLEDs is quite challenging. Since the first attempt of OLED fabrication [6], over the years further attempts have been done. The second-generation OLEDs, based on phosphorescent emitters, represent one of the most common design strategies employed for blue-emission. This technology exploits the inclusion of organometallic complexes to harvest both singlet (S) and triplet (T) excited states by enhancing the spin-orbit coupling (SOC). It allows ideally achieving nearly

100% IQE and EQE $\simeq$19%, as demonstrated by Adachi et al. [7] in 2001.

Here we model both the electrical and optical operation of a typical phosphorescent OLED, as depicted in Fig. 1.

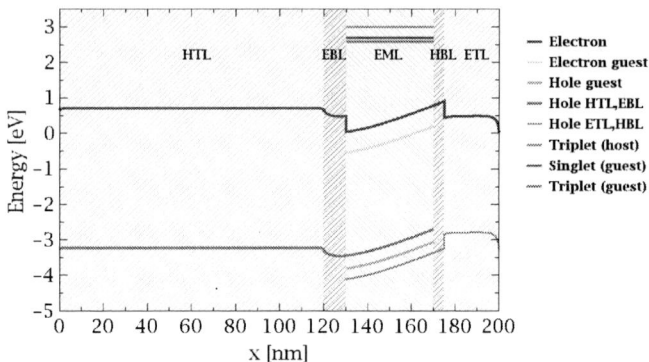

Fig. 1. Energy levels of carrier sub-populations included in the PHOLED system: respectively 12 electrons/holes and 3 excitons singlet/triplets

The emitter (EML) consists of a host-guest matrix system made by the mixture of EBL, HBL and guest materials (with a molar ratio of 70:20:10) The energy levels alignment is designed as it to foster the inter-system crossing (ISC) process contributing to the phosphorescent emission through triplets radiative decay. The structure is completed by 2 blocking/transport layers for the electrons and holes injection. In each material, we include carrier populations by explicitly defining their own properties as charge, spin and DOS. Given the organic nature of materials, we include the molecular disorder for both charged and neutral carriers by using Gaussian DOS profiles centered in the HOMO and LUMO levels. The charge transfer rate (R_{bb*}) between host and guest populations is accounted by using the following band-to-band recombination model

$$R_{bb*} = C_{bb*} n_b \left(1 - \frac{n_{b*}}{N_{b*}} \right) \left[1 - \left(\frac{E_{f,b*} - E_{f,b}}{k_B T} \right) \right]. \quad (4)$$

Here, the constant rate C_{bb*} regulates the rate of transitions depending on the carrier concentrations (n_b, n_b^*) and corresponding quasi-Fermi energy levels ($E_{f,b}, E_{f,b*}$), while N_b^* represents the overall number of available states. For this system we set $C_{EBL,guest}$=10^{-13} cm$^3 \cdot$ s^{-1}, $C_{HBL,guest}$=10^{-10} cm$^3 \cdot$ s^{-1} and $C_{EBL,guest}$=10^{-14} cm$^3 \cdot$ s^{-1}.

Concerning excitons, we consider T_1(h)=2.84 eV, S_1(g)=2.56 eV and T_1(g)=2.49 eV and we model the related radiative (rr) and non-radiative (nr) decay processes. Figure 2(a) shows schematically all main kinetic mechanisms included in the simulations involving excitons.

The emission is due to the triplets radiative decay and it is aided by ISC that allows increasing the triplet density (n_T) available. We model this process with

978-1-7281-0941-1/19 $31.00 © 2019 IEEE

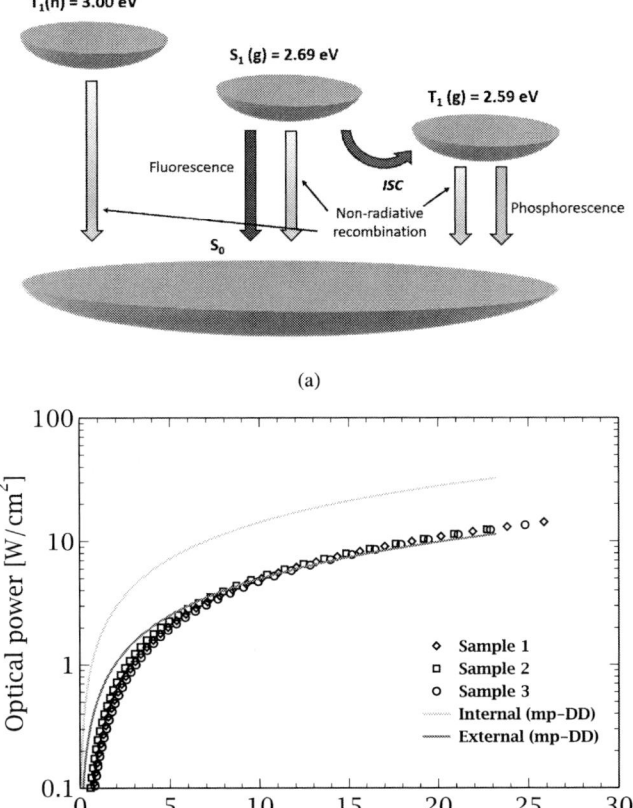

(a)

(b)

Fig. 2. a) Scheme of exciton states and decay rates accounted within the host-guest system. The exciton decay processes are modeled by setting: $\tau_{nr,T_1}(h)=10^{-6}$ s, $\tau_{rr,S_1}(g)=10^{-9}$ s, $\tau_{nr,S_1}(g)=10^{-9}$ s, $\tau_{rr,T_1}(g)=10^{-6}$ s and $\tau_{nr,T_1}(g)=10^{-4}$ s; b) internal and external optical power comparison between experimental results and mp-DD calculations.

$$R_{ISC} = C_{ISC} n_S \left(1 + \frac{n_T}{N_T} \left[1 - \left(\frac{E_{f,T} - E_{f,S}}{k_B T} \right) \right] \right), \tag{5}$$

where, n_S indicates the singlet density, $E_{f,T}, E_{f,S}$ the quasi-Fermi levels of both exciton species and C_{ISC} represents the constant rate, for which we set a value of 10^{-11} cm$^3 \cdot$ s^{-1}. The exciton generation is then included using constant rates $C_G(h) = C_G(g) = 10^{-22}$ cm$^3 \cdot$s^{-1}, and respecting the typical 1:3 ratio between singlets and triplets according to spin statistics.

We examined the device operation and we calculated the optical power generated internally at different current density bias. From comparison shown in Fig. 2(b), the internal optical power (red) is quite consistent with the curve trend of the external optical powers experimentally measured (black symbols). Furthermore, we calculated a light extraction efficiency (LEE) of around 35%, that allows to obtain a reasonable match with experimental results on the external optical power. This calculation is roughly done by assuming that LEE is constant

for both λ_{S_1}=461 nm and λ_{T_1}=479 nm.

IV. ION MIGRATION IN PSCs

The origin of JV hysteresis effect in perovskite solar cells is a topic still debated by scientific community. This anomalous effect is attributed to different phenomena, among the many, the pre-conditioning treatments of the device before scan, the presence of large defect density at transport layer interfaces and the ion migration aided by vacancies [8], [9].

Here, we exploit the mp-DD model's ability to investigate the ion migration effect on the JV curve hysteresis. To do this, we simulated a PSC structure made by FTO/TiO$_2$/MAPbI$_3$/Spiro-OMeTAD/Au, where the 500 nm perovskite (MAPbI$_3$) layer works as absorber, while 50 nm-thick TiO$_2$ and 200 nm-thick Spiro-OMeTAD materials work respectively as electron and hole transport layer. Due to its low activation energy, the iodide ion (I^-) is the mobile specie, that we include in the mp-DD system Eq. (1) as additional transport equation. Since the response of mobile ions in MAPbI$_3$ is on the order of few seconds and the typical scan rates provide V_{scan} >0.12-0.14 V$\cdot$s^{-1}, during both the reverse and forward scan ions have not enough time to diffuse through the perovskite layer. For this reason, we performed drift-diffusion simulations in fast-scan limit condition ($V_{scan} \gg 0.14$ V$\cdot$s^{-1}) to calculate the reverse and forward JV characteristics of the solar cell. This assumption allows us to model ions as fixed charged traps during the reverse and forward voltage sweep, and modeling them as mobile particles only at short-circuit current and at maximum applied voltage operation conditions. This is done to simulate the slow response than free carriers during the scan and their migration mechanism when the scan process stops.

Using this modeling concept, we investigated the PSC operation by varying the ion density (N_I) in the range of 1×10^{15}-5×10^1 cm^{-3}, and considering a diffusion coefficient of 10^{-12} cm$^2 \cdot$s^{-1}. The charge carrier generation profile is calculated from the Lambert-Beer model by assuming illumination from the cathode (FTO). We account for recombination losses, both the defect-mediated (ShockleyReadHall) and direct, by setting typical rate constant and carrier lifetimes for perovskite materials (τ_{SRH}=10^{-8} s and C_{dir}=10^{-9} cm$^{-3}\cdot$s^{-1}).

Figure 3(a) shows the comparison of the most representative results obtained. While for the bottom value of N_I used there is no effect, the increase of N_I leads to have different JV curves for the reverse and forward scan. This depends on the quantity of ions accumulate at interfaces between perovskite and transport layers. Observing the substantial drop of short-circuit current density (J_{SC}) respect to the slight variation of the open-circuit voltage (V_{OC}), is clear that the ions accumulation mainly affect the device operation at J_{SC}, where the density of free carriers is lower and can be comparable with N_I. In fact, as depicted in Fig. 3(b), the more we increase N_I and greater is the ions accumulation at interfaces, thus more effective will be the hysteresis effect. Comparing the blue and red curves we can conclude that the progressive accumulation of ions at interfaces concurs to the introduction of an effective

978-1-7281-0941-1/19 $31.00 © 2019 IEEE

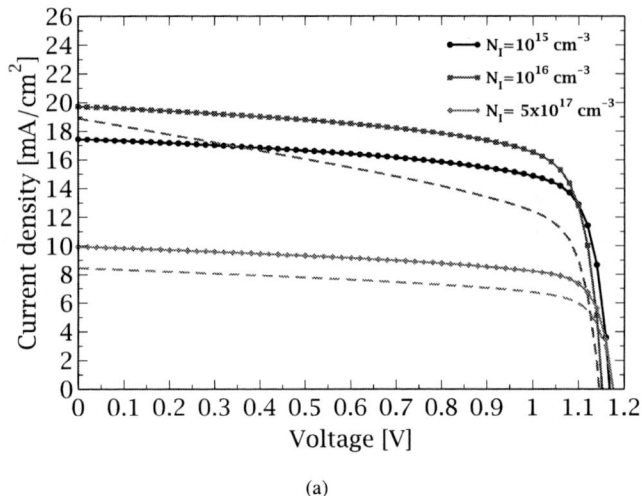

(a)

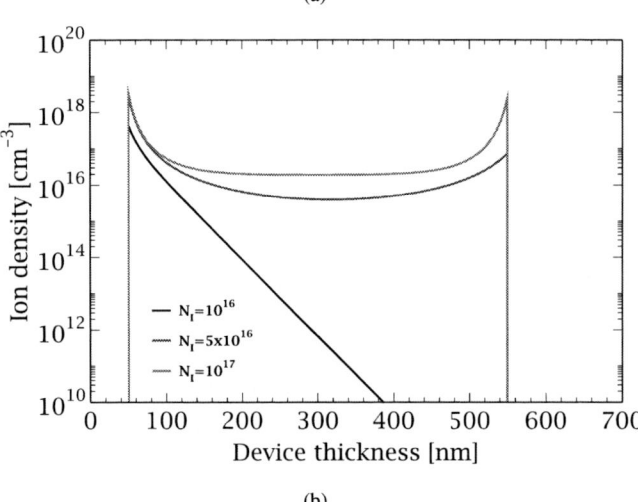

(b)

Fig. 3. a) Reverse (full lines) and forward (dashed lines) J-V characteristics and b) ion density profile at short-circuit current density for different values of ion concentration distributed within the 500 nm thick perovskite absorber.

shunt resistance, which determines a considerable drop of the solar cell performance.

V. CONCLUSIONS

We have presented an implementation of a generalized multi-particle drift-diffusion model capable to overcome the limitations typically imposed by the two-carrier transport based drift-diffusion model. The model allows considering multiple carrier sub-populations, each assumed in thermal local equilibrium and guarantees a strictly thermodynamically consistent formulation for both the generation and recombination between populations. We proved its flexibility and ability by describing systems with different application contexts.

In Sec III we have demonstrated the model's ability on the calculation of both the charge carriers and excitons transport. The modeling of the inter-system crossing process allowed us to calculate the internal optical power emitted by a typical phosphorescent OLED structure. Then, by comparing calcula-

tions with experimental measures we have estimated the light extraction efficiency.

In Sec IV, instead, we have investigated the effect of ion migration on the perovskite solar cells performance. The inclusion of iodide ions within the MAPbI$_3$ absorber allowed us to calculate the transport, and concluding on their accumulation effect on the JV characteristics.

Besides the examples shown, the model presented is suitable for a wide number of systems, for which ions, excitons or electron/hole sub-populations play a fundamental role in the device behavior.

ACKNOWLEDGMENT

This work was supported by the European Unions' Horizon 2020 Research and Innovation Programme under Grant 737089 (CHIPSCOPE) and Grant 826013 (IMPRESSIVE).

REFERENCES

[1] W van Roosbroeck. Theory of the flow of electrons and holes in germanium and other semiconductors. *Bell Labs Technical Journal*, 29(4):560–607, 1950.

[2] Markus Kantner, Markus Mittnenzweig, and Thomas Koprucki. Hybrid quantum-classical modeling of quantum dot devices. *Physical Review B*, 96(20):205301, 2017.

[3] Friedhard Römer and Bernd Witzigmann. Effect of auger recombination and leakage on the droop in ingan/gan quantum well leds. *Optics express*, 22(106):A1440–A1452, 2014.

[4] Daniele Rossi, Francesco Santoni, Matthias Auf Der Maur, and Aldo Di Carlo. A multiparticle drift-diffusion model and its application to organic and inorganic electronic device simulation. *IEEE Transactions on Electron Devices*, 66(6):2715–2722, 2019.

[5] M. Auf der Maur, G. Penazzi, G. Romano, F. Sacconi, A. Pecchia, and A. Di Carlo. The multiscale paradigm in electronic device simulation. *IEEE Transactions on Electron Devices*, 58(5):1425–1432, May 2011.

[6] Ching W Tang and Steven A VanSlyke. Organic electroluminescent diodes. *Applied physics letters*, 51(12):913–915, 1987.

[7] Chihaya Adachi, Marc A Baldo, Mark E Thompson, and Stephen R Forrest. Nearly 100% internal phosphorescence efficiency in an organic light-emitting device. *Journal of Applied Physics*, 90(10):5048–5051, 2001.

[8] Henry J Snaith, Antonio Abate, James M Ball, Giles E Eperon, Tomas Leijtens, Nakita K Noel, Samuel D Stranks, Jacob Tse-Wei Wang, Konrad Wojciechowski, and Wei Zhang. Anomalous hysteresis in perovskite solar cells. *J. Phys. Chem. Lett*, 5(9):1511–1515, 2014.

[9] Giles Richardson, Simon EJ O'Kane, Ralf G Niemann, Timo A Peltola, Jamie M Foster, Petra J Cameron, and Alison B Walker. Can slow-moving ions explain hysteresis in the current–voltage curves of perovskite solar cells? *Energy & Environmental Science*, 9(4):1476–1485, 2016.

An Efficient Method for Modeling Parasitic Light Sensitivity in Global Shutter CMOS Image Sensors

Federico Pace[1,2], Olivier Marcelot[1], Philippe Martin-Gonthier[1], Olivier Saint-Pé[2], Michel Breart de Boisanger[2], Rose-Marie Sauvage[3] and Pierre Magnan[1]

[1]ISAE-SUPAERO, Université de Toulouse, 10 Avenue Edouard Belin, 31055 Toulouse, France
[2]Airbus Defence & Space, 31 Rue des Cosmonautes, ZI du Palais, 31400 Toulouse, France
[3]Direction Générale de l'Armement, 60 Boulevard du Général Martial Valin, 75509 Paris, France
E-mail: federico.pace@isae-supaero.fr, Tel: +33 5 61 33 87 94

Abstract—Parasitic Light Sensitivity (PLS) is a key performance parameter for Global Shutter CMOS Image Sensors (GS-CIS), which quantifies the sensor sensitivity to light when the shutter is supposed closed. Its modeling and understanding would allow for an optimization in developing future sensors. This paper aims to present an efficient method for 2D modeling PLS in GS-CIS through separation of the optical problem from the carriers motion one. The optical problem is solved thanks to Finite-Differences Time-Domain (FDTD) simulations, while solution to the carriers motion problem is given through the application of the Boltzmann Transport Equation (BTE). This method is presented as a faster alternative to the coupled use of FDTD and TCAD simulations: since it is supposed that the two problem solutions are independent, the two simulations can be performed in parallel. The results show good match between the developed method and the TCAD solutions, thus showing fair agreement with experimental data, probably due to a poor knowledge of the back-end process.

Index Terms—CIS, GS, FDTD, BTE, PLS, Shutter Efficiency

I. INTRODUCTION

Global Shutter CMOS Image Sensors (GS-CIS) [1]–[4] are increasingly becoming attractive for a wide spectrum of appli-

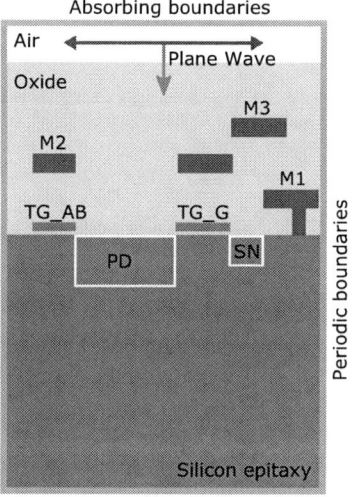

Fig. 2. Example of the structure simulated with the FDTD method. As a case of study, only M3 is used in the pixel to screen the SN. Blue line shows one of the two simulated Plane Wave polarization. Only normal light incidence has been simulated for this study, but oblique light incidence could also be considered. Structure of pixel can also be appreciated.

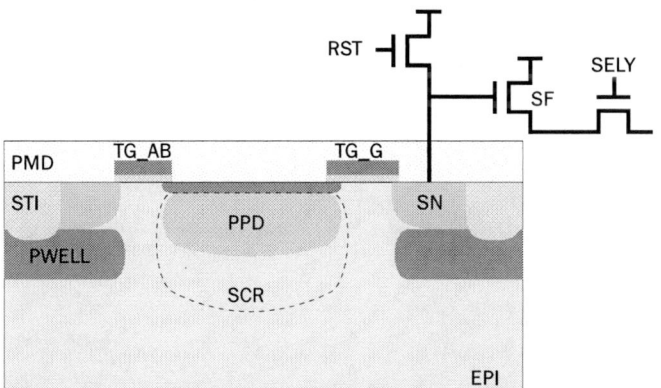

Fig. 1. Example of a GS 5T pixel cross section. Pinned PhotoDiode (PPD) is visible at the center of the figure with its pinning layer (depicted in violet just above). Storage Node (SN) is situated right to the PPD and acts also as the Floating Diffusion (FD) of the 5T GS-CIS. Transfer Gate is of Global type (TG_G) and an Anti-Blooming Transfer Gate (TG_AB) is visible at the left-hand side of the figure.

cations, ranging from automotive [5] to space applications [6], having been conceived for reducing motion artifacts.

An example of a pixel that can operate in Global Shutter mode is presented in Fig. 1. In Global Shutter mode, the incoming light gets collected by the PhotoDiode (PD) for a certain amount of time, then the stored charges are transferred from the PD to the Sense Node (SN) in a synchronous way throughout the entire matrix and stored for readout. Due to the matrix dimensions, synchronous global readout of the stored signal is not feasible, thus requiring a rolling-shutter fashion readout that is performed line by line. Given that the SN sensitivity to parasitic charges is non-negligible, there could appear some unwanted signal that is not constant throughout the entire matrix and degrade the Global Shutter performances.

In order to evaluate the system capacity to screen the SN from external perturbations, a figure of merit has been identified, defined as Parasitic Light Sensitivity (PLS), and

978-1-7281-0941-1/19 $31.00 © 2019 IEEE

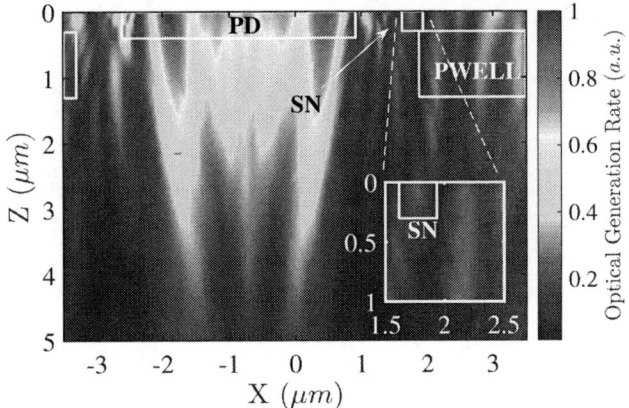

Fig. 3. Example of Optical Generation Rate map at $\lambda = 650\,nm$. The map extends in the z-direction from $0\,\mu m$ (silicon surface) to $5\,\mu m$ (end of silicon epitaxy). Zoom in the SN region is also shown. Non-negligible generation rate is found in and close to the SN zone. Screening of pixel is therefore not perfect because of light diffraction, becoming increasingly important in smaller pixels and that has to be taken into account when designing GS-CIS.

described in (1), with $\mathcal{S}_m$ being the sensitivity of node m.

$$PLS = \mathcal{S}_{SN}/\mathcal{S}_{PD} \qquad (1)$$

Since limited work has been proposed so far and only for specific cases, such as BackSide Illumination pixels or just PLS simulation results without explanation of the method [7], [8], this paper aims to propose a faster alternative 2D method to TCAD simulations for modeling PLS in GS-CIS. Results are shown for a $7\,\mu m$ pitch pixel and compared to TCAD simulations, knowing that the method can be applied to a large variety of pixel dimension and structures.

II. METHOD FOR MODELING PLS

The presented 2D method exploits Finite-Differences Time-Domain (FDTD) simulations for photo-generation of carriers and aims to treat carriers motion through Boltzmann Transport Equation (BTE). Silicon has a negligible thermal recombination at moderate illumination strengths, carriers transport is supposed independent on the carriers density thus transport problem and optical problem can be considered uncorrelated. This brings some advantages in terms of computational time in comparison with TCAD simulations:

- it allows running the optical and the carrier transport simulations in parallel, which would not be possible when exploiting TCAD simulations because the optical problem solution must be given as input to perform the carrier transport simulation;
- it allows running a lesser number of simulation when optical conditions are to be changed (for example incidence angle, light intensity, etc.), there is no need in re-performing the carrier transport simulations.

In a general case, this method can be preferred to TCAD simulation for a faster analysis of the device in various conditions.

The following will briefly describe the formulation and the solution of the two problems previously stated.

A. Photo-Generation of Carriers

In order to solve the optical problem and calculate the photo-generation rate, a commercial software exploiting the FDTD method has been exploited [9]. The FDTD method is generally used to solve Maxwell's equations for light in visible and near-visible wavelength range through use of a rectangular finite-difference mesh [10], [11].

Fig. 2 shows an example of the FDTD simulated pixel structure. In order to reproduce the behavior of a pixel inside a matrix, periodic boundary conditions of the Bloch type are applied on the right-hand side and left-hand side of the simulated structure, where it is supposed identical pixels would take place. On the other hand, absorbing boundary conditions are placed at the bottom of the silicon epitaxy and at the top of the simulates structure. It is therefore supposed that all the light reflected by the pixel or the light moving towards the substrate (below the epitaxy) would not be useful for our simulation. Plane Waves with normal incidence has been used for the presented simulation, even though other incidence angle could also be tested. Non-polarized light has been achieved through averaging of two orthogonal polarization simulations. Source incoherence has been achieved through spectral averaging.

The simulations result in an Optical Generation Rate Density (OGRD) map per each considered wavelength, describing the photo-electron generation rate density per each point at the epitaxial silicon level. An example of OGRD is shown in Fig. 3. It is possible to appreciate that non-negligible direct light is coming onto the SN because of diffraction and multiple reflections.

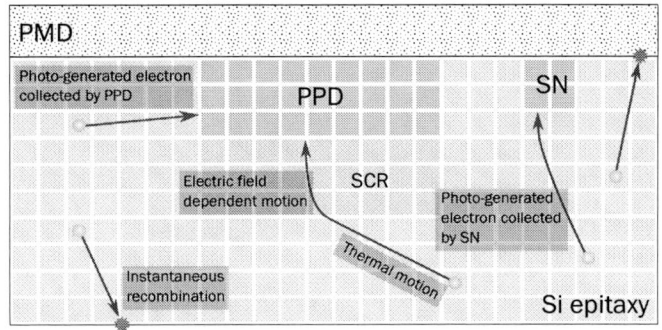

Fig. 4. Example of simulation exploiting the Boltzmann Transport Equations under the given hypothesis. A carrier is created in a voxel, its initial direction is randomly chosen. Then, electron moves according to the BTE. The electron motion is not perturbed by any scattering nor recombination effects, except when reaching a simulation border (silicon/PMD interface for example) where its recombination is instantaneous and considered lost. Electron motion also ends when reaching the PD or SN node, and the electron is taken into account for the arrival node.

978-1-7281-0941-1/19 $31.00 © 2019 IEEE

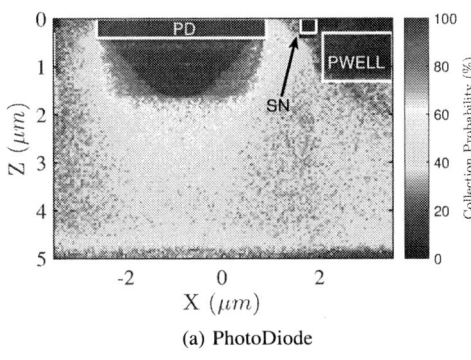

(a) PhotoDiode

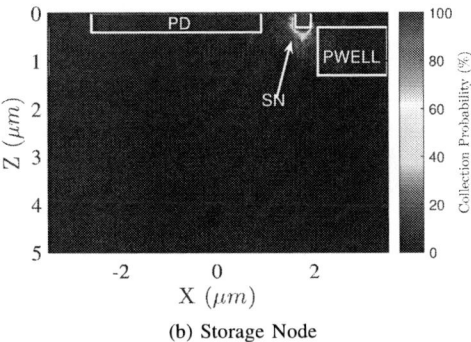

(b) Storage Node

Fig. 5. Example of Collection Probability map of (a) PD and (b) SN, simulated with the Boltzmann Transport Equations at $\lambda = 650\ nm$. In (b) it is possible to appreciate non-negligible collection probability far from the SN zone, showing the interest in our study.

B. Photo-Generated Carriers Motion

BTE is used to model the behavior of particles in semiconductors under non-equilibrium state as function of time and the local electric field [12], [13].

As for the assumptions, no scattering has been taken into account, as well as instantaneous recombination at surfaces and wavelength dependent initial carrier energy. These assumptions bring to the behavior represented in Fig. 4.

BTE are then used to create a Collection Probability (CP) map for both PD and SN. A CP map describes the probability of a carrier, created at a point in space, to be collected by either the PD or the SN. For this purpose, 100 carriers per point in space are simulated. Fig. 5 shows an example of CP map for the PD.

The OGRD and CP maps are finally used to compute the nodes sensitivity $\mathcal{S}$ through (2) per each wavelength. Since simulations are performed in 2D, a geometry correction factor has to be applied and it has been identified in the node's third dimension (w_{PD} and w_{SN} for PD and SN respectively). Finally, the desired PLS can be computed through (3).

$$\mathcal{S}_{PD} = w_{PD} \cdot \iint OGRD \cdot CP_{PD} \cdot dxdz,$$
$$\mathcal{S}_{SN} = w_{SN} \cdot \iint OGRD \cdot CP_{SN} \cdot dxdz \tag{2}$$

$$PLS = \frac{\mathcal{S}_{SN}}{\mathcal{S}_{PD}} = \frac{w_{SN} \cdot \iint OGRD \cdot CP_{SN} \cdot dxdz}{w_{PD} \iint OGRD \cdot CP_{PD} \cdot dxdz} \tag{3}$$

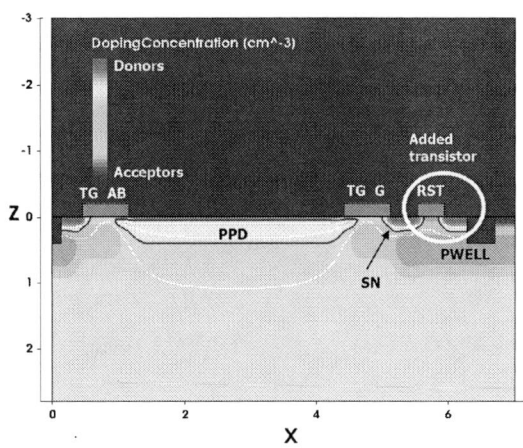

Fig. 6. Example of the pixel structure simulated exploiting TCAD, here represented thanks to doping concentration. White lines represent boundaries of Space Charge Regions (SCR) when the PPD is emptied. As it can be appreciated in the white circle, a transistor is added in order to allow the SN to be floating.

III. Coupled FDTD+TCAD simulations

As reference, TCAD simulations have been exploited. In order to start from the same optical assumptions, optical generation data from Lumerical have been imported to the TCAD simulator (for our purposes Synopsys Sentaurus has been used) through a MATLAB routine using a linear interpolation from the finite difference mesh to the finite element one [14].

The 2D simulated structure is shown in 6. It can be appreciated that an additional transistor (i.e. RST) has to be added in order to allow the potential of the SN to be floating. However, in some pixel structures the RST transistor can be placed distant in space from the SN node and connected to the latter via a metallic connection; 3D TCAD simulations become necessary to correctly simulate the floating behavior of the SN.

PD and SN sensitivities are calculated through the slope of the electron density change, in the nodes, as function of time.

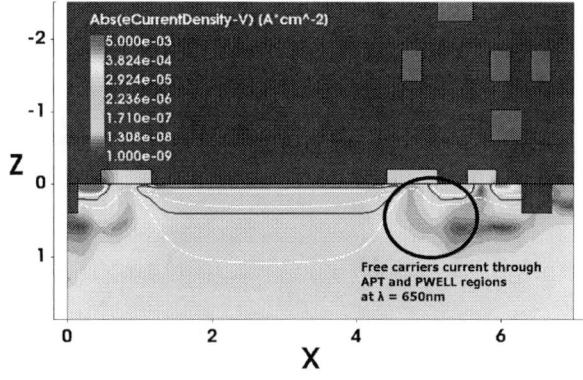

Fig. 7. Absolute electron current density while under illumination at $\lambda = 650\ nm$. It is possible to appreciate how the main electronic current flowing towards the SN (shown by the black circle) passes close to the PWELL region, where p-doping decreases.

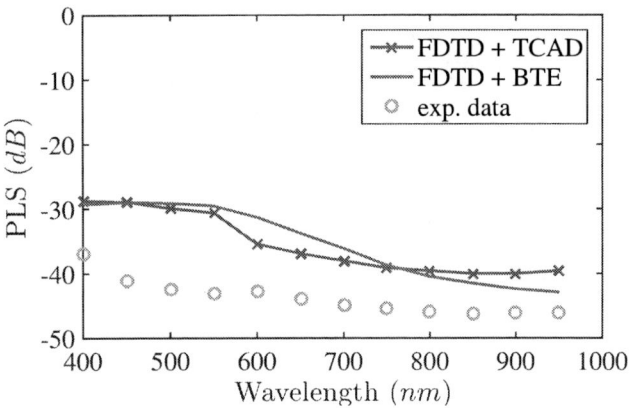

Fig. 8. PLS simulation results are shown for both coupled FDTD+TCAD simulations and FDTD+BTE method. A good agreement of the results can be appreciated between the two methods, although simulations show discrepancies with the experimental data curve. As for the experimental part, node's sensitivity is measured through variation of its output voltage as function of the impinging light irradiance. PLS is finally calculated as the ratio of the two sensitivities.

PLS is then calculated using (3).

Moreover, a further comparison of the two methods could be done. Taking a look at the absolute electronic current flow in the pixel resulting from TCAD simulations in Fig. 7, it is possible to appreciate that the main current flowing towards the SN comes from free carriers close to the end of the PWELL area, where the Anti-PunchThrough (APT) region is present and being lightly doped compared to the PWELL region. Similarly, it can be appreciated from Fig. 5b that the probability of carriers being collected by the SN is non-negligible in the same region explained for TCAD simulations.

IV. SIMULATION RESULTS AND EXPERIMENTAL DATA

Figure 8 shows the comparison between the coupled FDTD+TCAD simulation and our method exploiting FDTD simulations and BTE. Our model shows good match to the TCAD modeling, with a similar behavior in the entire tested wavelength range.

The simulations results have moreover been compared with experimental data taken from a matrix of $7\ \mu m$ pitch pixels with Global Shutter functions, on which the simulated structure have been modeled. As it can be appreciated in Fig. 8, simulations show some discrepancies with the experimental data, the reason still being under investigation. One hypothesis is that the discrepancy can be caused by poor knowledge in the Back-End fabrication process, where diverse oxide layers can strongly modify the propagation of light inside the pixel and thus modify the optical generation pattern at the silicon level. This can be especially justified from the fact that the discrepancy is stronger at shorter wavelength, where light penetration is mainly superficial and carriers diffusion has a lesser impact.

V. CONCLUSIONS

We have presented a new method for modeling PLS in GS-CIS. The method shows its interest when used to model PLS at

different optical condition (e.g. incidence angle, light intensity, ...). Compared to coupled FDTD+TCAD simulations, where a TCAD simulation has to be done per each optical condition, our model results faster, where only FDTD simulations have to be re-performed, and allows parallelization, showing the interest in our study.

Simulation results have been moreover compared to experimental data, showing some discrepancies, especially at longer wavelengths. This could be explained by a poor knowledge in the Back-End fabrication process.

Finally, our method could be further improved by creating a more realistic electric field chart for BTE calculations and with a better comprehension and modeling of the Back-End layers. 3D simulations could also improve results, taking into account the complexity of the pixel structure.

REFERENCES

[1] G. Meynants, J. Bogaerts, X. Wang, and G. Vanhorebeek, "Backside illuminated global shutter CMOS image sensors," in *Int. Image Sensor Workshop (IISW)*, 2011, pp. 305–308.

[2] A. Krymski, "A High Speed 1 MPix Sensor with Floating Storage Gate Pixel," *Proc. Int. Image Sensor Workshop (IISW)*, 2015.

[3] S. Velichko, J. J. Hynecek, R. S. Johnson, V. Lenchenkov, H. Komori, H. W. Lee, and F. Y. J. Chen, "CMOS Global Shutter Charge Storage Pixels With Improved Performance," *IEEE Transactions on Electron Devices*, vol. 63, no. 1, pp. 106–112, Jan. 2016.

[4] M. Kobayashi, Y. Onuki, K. Kawabata, H. Sekine, T. Tsuboi, T. Muto, T. Akiyama, Y. Matsuno, H. Takahashi, T. Koizumi, K. Sakurai, H. Yuzurihara, S. Inoue, and T. Ichikawa, "A 1.8e $\^-_\mathrm rms $ Temporal Noise Over 110-dB-Dynamic Range 3.4 $\mu\ textm$ Pixel Pitch Global-Shutter CMOS Image Sensor With Dual-Gain Amplifiers SS-ADC, Light Guide Structure, and Multiple-Accumulation Shutter," *IEEE Journal of Solid-State Circuits*, vol. 53, no. 1, pp. 219–228, Jan. 2018.

[5] A. Lahav, A. Birman, D. Perhest, A. Fenigstein, Y. Grauer, and E. Levi, "A global shutter sensor used in active gated imaging for automotive," in *Proc. IISW*, 2015, pp. 1–4.

[6] G. Lepage, A. Materne, and C. Renard, "A CMOS image sensor for Earth observation with high efficiency snapshot shutter," in *Proc. Int. Image Sensor Workshop (IISW)*, 2013.

[7] L. Stark, J. M. Raynor, F. Lalanne, and R. K. Henderson, "A Back-Illuminated Voltage-Domain Global Shutter Pixel With Dual In-Pixel Storage," *IEEE Transactions on Electron Devices*, vol. 65, no. 10, pp. 4394–4400, Oct. 2018.

[8] T. Yokoyama, M. Tsutsui, M. Suzuki, Y. Nishi, I. Mizuno, and A. Lahav, "Development of Low Parasitic Light Sensitivity and Low Dark Current 2.8 um Global Shutter Pixel," *IEEE Sensors*, vol. 18, no. 2, p. 349, Jan. 2018. [Online]. Available: http://www.mdpi.com/1424-8220/18/2/349

[9] Lumerical Inc. [Online]. Available: https://www.lumerical.com/

[10] K. Yee, "Numerical solution of initial boundary value problems involving maxwell's equations in isotropic media," *IEEE Transactions on Antennas and Propagation*, vol. 14, no. 3, pp. 302–307, May 1966.

[11] J. Vaillant, A. Crocherie, F. Hirigoyen, A. Cadien, and J. Pond, "Uniform illumination and rigorous electromagnetic simulations applied to CMOS image sensors," *Optics Express*, vol. 15, no. 9, pp. 5494–5503, 2007.

[12] C. Jacoboni and L. Reggiani, "The Monte Carlo method for the solution of charge transport in semiconductors with applications to covalent materials," *Rev. Mod. Phys.*, vol. 55, no. 3, pp. 645–705, Jul. 1983. [Online]. Available: https://link.aps.org/doi/10.1103/RevModPhys.55.645

[13] Y. Yamashita, M. Uchiyama, and D.-N. Yaung, "Quantum Efficiency Simulation with Boltzmann Transport Equation for Motion Modeling of Individual Particles in Photodiodes," *Proc. Int. Image Sensor Workshop (IISW)*, 2017.

[14] M. D. Kelzenberg, *Silicon microwire photovoltaics*. California Institute of Technology, 2010.

978-1-7281-0941-1/19 $31.00 © 2019 IEEE

Author index

Acosta-Alba Pablo	331	Cueto Olga	5	Hubert Guillaume	219
Adamu-Lema Fikru	49, 81, 105	Curvers Benoit	331	Huet Karim	331
Afzalian Aryan	199	Daniel Luca	287	Hurley Paul	315
Alexander Blessy	271	de Gironcoli S.	211	Hylin Carl	53
Andee Yogadissen	263	Del Linz Leonida	97	Ida Jiro	291
Ansh P.	175	Delalleau Julien	255	Iizuka Takahiro	235
Apsangi Priyanka	53	Della Marca V.	255	Ikegami Tsutomu	41
Arai Yasuo	291	Di Carlo Aldo	359	Ilatikhameneh H.	215
Arreghini Antonio	61	Di Franco Cinzia	347	Innocenti Jordan	255
Arutchelvan G.	283	Ding Jie	49, 155	Ishikawa Kiyoshi	299
Asai Hidehiro	129	Dobrzynska Jagoda	243	Jeon Seonghyeok	303
Asenov Asen	49,81,85,89,105,155,275	Donati Guerrieri S.	239	Jiang Xiangwei	149
Asenov Plamen	223	Driussi Francesco	279	Jiang Zhengping	215
Asoudegi Nima	295	Du Gang	307	Jiménez David	163
Auf der Maur M.	359	Ducry Fabian	69	Joung Saehoon	267
Avila Herrera F.	235	Dutta Tapas	81, 275	Jungemann C.	25
Ayyagari Buvna	271	Eckl Christian	247	Kamei Tatsuya	33
Babayan Artem	327	Ender Johannes	57	Kantner Markus	355
Baccarani Giorgio	243	Endo Hiroki	291	Kargar Zeinab	25
Badami Oves	49, 81, 89, 155	Esseni David	73, 279, 287, 315	Karner Markus	61
Balestra Luigi	243	Fang Tao	145	Kataria Santender	279
Bandelow Uwe	355	Filipovic Lado	323, 339	Kavanagh Karen	145
Bani-Hashemian M.	69	Fiorentini Simone	57	Kayama Yasuyuki	299
Bankapalli Y. S.	21	Floyd Liam	315	Kerdilès Sébastien	331
Barbé Jean-Charles	223	Fukuda Koichi	41, 129	Khakbaz Pedram	279, 295
Barnaby Hugh James	53	Furnémont Arnaud	61	Khomyakov Petr A.	203
Barraud Sylvain	223, 275	Gahoi Amit	279	Kikuchihara H.	235
Bazizi El Mehdi	271	Gambi Alberto	279	Kim Dae Sin	93,191,215,299
Berrada Salim	81, 89	Gao Xujiao	167	Kim Jongchol	191, 215
Bettetti Fabrizio	97	García Ruiz F. J.	153, 163	Kim Kyoung Yeon	179
Bhagdikar Sharang	13	Georgiev Vihar	49,81,85,89,105,155,275	Kim SoYoung	267
Biswas Arnab	133	Ghetti Andrea	29	Kim Youngkwon	125
Blaise Philipe	211	Ghione Giovanni	239	Kim Yunho	125
Blom Anders	203	Giacomazzi Luigi	137, 211	Kimpton Derek	65
Bonafos Caroline	331	Giannozzi Paolo	279	Klemenschits Xaver	339
Bonani Fabrizio	239	Gity Farzan	315	Klüpfel Fabian	223, 275
Bourdet Léo	195	Gnani Elena	243	Kobayashi K.	33
Breart de Boisanger M.	363	Gnudi Antonio	243	Koprucki Thomas	355
Brech Helmut	247	Godoy Medina A.	153, 163	Kopyt Pawel	187
Brown Andrew	223	Goes Wolfgang	57	Korzenietz Andreas	251
Brunetti Rossella	37	González Marín E.	153, 163	Kovacs-Vajna Z.M.	347
Cai Linlin	1, 307	González Medina J.	153	Kozicki Michael	53
Calderon Stéphane	351	Gounder Jowesh	141	Kramer Aaron	207
Carrillo-Nunez H.	81, 85, 89	Guan Yunhe	81, 85	Krysicki Mateusz	187
Caruso Enrico	97, 315	Guichard Eric	65	Kudo Takuya	141
Chen Chao-Hsuan	259	Guissi Sofiane	351	Kumar Jeevesh	175
Chen Wangyong	1, 307	Han Seung-Cheol	45	Kumashiro S.	33
Chen Ying	141	Hashizume T.	129	Kwon Uihui	93,191,215,299
Cheng Binjie	223	Hattori Junichi	41, 129	La Torre Camilla	227
Cheng Ming-Cheng	77	Hayase Shigeaki	101	Lacord Joris	219
Cherkaoui Karim	315	Hemeryck Anne	183, 211	Law Mark	117
Cho Jin	65	Hennigan Gary	167	Lee Hyeongu	113
Cho Yucheol	303	Herrero Saboya G.	183	Lee Jaehyun	81
Choi Jaehee	93	Herrmann Tom	263	Lee Jonghyun	125
Choi Sang-Jin	125	Heyns Marc	283	Lee Maeng-Eun	203
Choi Woosung	191, 215	Hille Frank	251	Lee Ming-Yi	121
Chuang Min-Hui	121, 171	Hinkle C. L.	207	Lee Seon-Young	299
Colombeau Benjamin	271	Hiroki Akira	33	Lehmann Steffen	263
Cotorogea Maria	133	Hong Sung-Min	45	Lemme Max	279
Coulié Karine	255	Hössinger Andreas	327, 335	Li Jing	195
Cristiano Fuccio	331	Huang Andy	167	Li Yiming	121,171,259
Crocherie Axel	351	Huang Jun Z.	149	Li Zunchao	85

Lin Jun	315	
Lin Samuel	271	
Linn Tobias	25	
Liu Teren	145	
Liu Xiaoyan	1, 307	
Locati Jordan	255	
Long Yuxiong	149	
Lorenz Jürgen	223, 275	
Low Kain Lu	17	
Ludwig Daniel	133	
Luisier Mathieu	69, 231	
Luo Jun-Wei	149	
Mabuchi Takuya	159	
Macchia Eleonora	347	
Magnan Pierre	363	
Mahapatra Souvik	9, 13, 109	
Manoli Kyriaki	347	
Manstetten Paul	339	
Manzanarez Hervé	5	
Marcelot Olivier	363	
Martin-Gonthier P.	363	
Martin-Samos Layla	137, 183, 211	
Martinie Sébastien	219, 223	
Mattausch Hans Jurgen	235	
Medina-Bailon Cristina	81, 89, 275	
Medina-Rull Alberto	163	
Meersha Adil	175	
Mele Leandro Julian	343	
Menzel Stephan	227	
Millar Campbell	223	
Min Jeong-Guk	299	
Mittal Sushant	271	
Mitterbauer Ferdinand	61	
Miura Hideo	141	
Miura-Mattausch M.	235, 319	
Monaghan Scott	315	
Monga Udit	93	
Moreau Stéphane	5	
Mori Takayuki	291	
Musson Lawrence	167	
Muthuseenu K.	53	
Narang Vinod	17	
Natarajan Sanjay	271	
Negoita Mihai	167	
Niedernostheide F.-J.	251	
Niel Stephan	255	
Niquet Yann-Michel	195	
Nishizawa Yutaka	299	
Ollier Nadege	137	
Orio Roberto	57	
Pace Federico	363	
Pae Sangwoo	93	
Paillard Vincent	331	
Pal Ashish	271	
Pala Marco	73	
Palazzo Gerardo	347	
Palestri Pierpaolo	97, 315, 343	
Parihar Narendra	9	
Park Byung-Gook	179	
Park Hong-Hyun	191, 215	
Park Yonghee	93	
Pasadas Francisco	163	
Patrick Erin	117	
Paulet Olivier	255	
Pearton Stephen	117	
Pederson Dylan	125	
Penazzi Gabriele	203	
Pin D.	97	
Poiroux Thierry	219	
Pourfath Mahdi	295	
Pourghaderi M. A.	191, 215	
Pourtois Geoffrey	199	
Quell Michael	327, 335	
Radu Iuliana	283	
Raja Laxminarayan	125	
Reggiani Susanna	243	
Regnier Arnaud	255	
Reiser Korbinian	247	
Ren Fan	117	
Richard Nicolas	137, 183, 211	
Rideau Denis	183, 351	
Rivero Christian	255	
Rollo Tommaso	287	
Romele Paolo	347	
Rossi Daniele	359	
Rostand Neil	219	
Royet Anne-Sophie	331	
Rozeau Olivier	219, 223	
Rudan Massimo	37	
Ryu Seung-Min	125	
Sachid Angada	271	
Sadi Toufik	49, 155	
Saint-Pé Olivier	363	
Salles Nicolas	211	
Salski Bartlomiej	187	
Samukawa Seiji	121, 259	
Sandow Christian	251	
Sano Nobuyuki	311	
Saremi Mehdi	271	
Sauvage Rose-M.	363	
Scamarcio Gaetano	347	
Schanovsky Franz	61	
Schenk Andreas	231	
Schlenvogt Garrett	53	
Schmidt Alexander	299	
Schrag Gabriele	251	
Segatto Mattia	97	
Selberherr Siegfried	327, 335, 339	
Selmi Luca	315, 343	
Seo Junbeom	113	
Sharma Ribhu	117	
Sharma Uma	109	
Shim Hyewon	93	
Shimizu Mitsuaki	129	
Shin Mincheol	113, 303	
Shrestha Niraj Man	259	
Shrivastava M.	175	
Sieger Daniel	351	
Sklenard Benoit	211	
Smidstrup Soren	203	
Soleimani Maryam	295	
Speciale Nicolò	37	
Stanojevic Zlatan	61	
Steiner Klaus	61	
Stokbro Kurt	203	
Stradi Daniele	203	
Subramani Nandha	263	
Sung Wen-Li	171	
Suwa Takeshi	101	
Suzuki Ken	141	
Sverdlov Viktor	57	
Takatsuka Hirotaka	235	
Tarng Jenn Hwan	259	
Teo Chea-Wei	17	
Thakor Karansingh	9	
Thareja Gaurav	271	
Thean Aaron V.-Y.	17	
Thirunavukkarasu V.	49, 81	
Tiwari Ravi	9	
Toifl Alexander	327, 335	
Tokumasu Takashi	159	
Toral-López A.	153, 163	
Torricelli Fabrizio	347	
Torsi Luisa	347	
Townsend Mark	53	
Tsai Zuo-Min	259	
Tsukahara Kohei	311	
Twynam John	247	
Uene Naoya	159	
Uh Jiho	125	
Valant Matjaz	137, 211	
Van de Put Maarten L.	207	
Van den bosch Geert	61	
Vandenberghe William	207	
Vej-Hansen Ulrik G.	203	
Venica Stefano	279	
Venitucci Benjamin	195	
Verreck Devin	61, 283	
Vianne Benjamin	351	
Vobecky Jan	243	
Vuttivorakulchai K.	231	
Wachutka Gerhard	251	
Wang Liping	223	
Wehbe Alause Hélène	351	
Wei Zhongming	149	
Weigel Robert	247	
Weinbub Josef	327, 335	
Wellendorff Jess	203	
Wong Hiu Yung	9, 21	
Wünsche Hans-Jürgen	355	
Xia Guangrui	145	
Yang Jiancheng	117	
Yasuhara Shigeo	159	
Yoshida Katsuhisa	311	
Yu Hongyu	145	
Zaitsu Masaru	159	
Zaka Alban	263	
Zhang Qinqiang	141	
Zhang Xing	1	
Zhao Yudi	1	
Zhao Zhixing	263	
Zurhelle Alexander	227	